PLASTICS MOLD ENGINEERING HANDBOOK

PLASTICS MOLD ENGINEERING HANDBOOK

FOURTH EDITION

Edited by

J. Harry DuBOIS

and

Wayne I. PRIBBLE

Pribble Plastics Products, Inc.
New Haven, Indiana

 SPRINGER SCIENCE+BUSINESS MEDIA, LLC

Copyright © 1987 by Springer Science+Business Media New York
Ursprünglich erschienen bei Van Nostrand Reinhold Inc. 1987
Softcover reprint of the hardcover 1st edition 1987
Library of Congress Catalog Card Number 86-24666

ISBN 978-1-4684-6580-8   ISBN 978-1-4684-6578-5 (eBook)
DOI 10.1007/978-1-4684-6578-5

16 15 14 13 12 11 10 9 8 7 6 5 4 3 2 1

Library of Congress Cataloging-in-Publication Data

Plastics mold engineering handbook.

    Includes bibliographies and index.
    1. Plastics—Molds. 2. Plastics—Molding.
I. DuBois, J. Harry (John Harry), 1903-
II. Pribble, Wayne I.
TP1150.P543 1987   666.4'12   86-24666

Contributors

Charles C. Davis, Jr.
Research Devisers
P.O. Box 84
Lawrenceville, New Jersey 08648

Robert D. DeLong
(Address unknown)

J. Harry DuBois
(deceased, March 1986)

Leon R. Egg
(Address unknown)

Paul E. Ferland
Retired
640 Summerset Court
Kernersville, North Carolina 27284

Brooks B. Heise
President
Heise Industries, Inc.
196 Commerce Street
East Berlin, Connecticut 06023

Sidney Levy, P.E.
Sidney Levy and Associates
P.O. Box 7355
LaVerne, California 91750

Robert J. Musel
Musel Enterprises, Inc.
Rural Route 6, 709 Dune Drive
Pleasantville, New Jersey 08232

Wayne I. Pribble, Editor
President
Pribble Enterprises, Inc.
P.O. Box 221
New Haven, Indiana 46774

Sumner E. Tinkham
Retired
1 Buckley Hill Road
Morristown, New Jersey 07960

Preface to Fourth Edition

About three years ago, the publishers of *Plastics Mold Engineering Handbook* suggested we review the need for a fourth edition. Much has transpired since the third edition (1978). A cursory inspection of this edition, by those familiar with the third edition, will reveal that the *basic data* in the first six chapters has had little, if any, revision. The obvious reason is that *basic principles remain the same regardless of any complexities which may develop*. Thus, it is the last half of the present edition which has seen the most changes. For example, the chapter on blow molding is completely new. Several contributors to the third edition are no longer associated with plastics and so we had to find new contributors. This edition also covers the special considerations in designing molds in the multiton range of weight. I am reminded of firms like R. O. Schulz Co., of Elmwood Park, IL, who foresaw the future when they installed fifty-ton capacity cranes in their present plant, which was built in the 1940s. The development of very large machines capable of handling mold blocks weighing 20,000 to 40,000 pounds, and the combining of this equipment with computers, has matched the capability of the molding machine builders. We hesitate to specify numbers because the size increases with every issue listing in *Modern Plastics Encyclopedia*.

In all cases, we have tried to give credit for information which various companies have felt to be proprietary, and we have respected their confidence by the process of only advising of the existence of such information. Fortunately for the plastics industry, the management of most progressive companies has realized, over the years, that sharing enhances their own knowledge as well as that of others in the same segment of the plastics industry. Some information and data is copyrighted by other sources. In all cases, proper credit is given for permission to publish in this edition. Specifically, we refer to figures 11.61, 11.62, and 11.63, supplied by The Society of Plastics Industry, Inc., and the Plastics Bottle Institute. John Malloy's letter with permission to publish says: ''it has consistently been the intention . . . that this information be widely disseminated.'' This policy of ''wide dissemination'' is a marked change in the policy of the bottle industry compared to the policy of only fifteen or twenty years ago, when it was almost impossible to obtain data about threads for bottles.

Ever since our first edition (1946), Mr. DuBois and I have been in the forefront of advocating the change from ''art'' to ''science.'' For this reason this text beginning with the first edition has been used in educational institutions for courses in plastics and tooling for plastics.

I want to thank J. Harry DuBois for his encouragement in spite of his incapacitation caused by a stroke in 1984. During the course of the work on this fourth edition, I conferred with him four times in Green Brook, New Jersey. Each time, he contributed thoughts on how to make *Plastics Mold Engineering Handbook* the most useful text ever published on this subject. Unfortunately Harry DuBois passed away in March 1986.

The fact of a fourth edition, which should be in print during the 41st anniversary of the 1st *edition* (Chicago: American Technical Society, 1946) should be evidence of the need and value of an update to carry serious students and readers well into the 1990s. Even now, I am urging my successor (whoever that might be) to begin collecting and compiling the mountain of data that would need to be sifted through for a fifth edition.

WAYNE I. PRIBBLE
New Haven, IN

Preface to Third Edition

Plastics Mold Engineering Handbook is concerned with the particular analysis, decisionmaking, and work required to effect a "good" mold design. Almost any kind of a mold will produce usable parts, but *only* a well-designed mold can meet the requirements, expectations, and desires of the modern molder who expects long mold life, economy of production, and trouble-free operation.

A well-made mold is of *no* value unless it is built to a "good" design. It is readily apparent that good design, good manufacture, and good operation must complement each other. One is no good without the other.

The data in this book, which have been accumulated over a 60-year period by a great many persons, have also been tested in actual practice. In addition, the experiences of many people have been drawn upon to insure that you receive the very best available data to guide you in your own design efforts.

It would be practically impossible to enumerate all those who have contributed in some way to this text. We are indebted to the review referees, whose practical comments, insistence on perfection, and casual needling have caused us to make a special effort to be exact in meaning, precise in detail, and critical in inclusion. We have endeavored to cover all this in a manner that should create dedication, enthusiasm, and follow-through by the engineer, designer, and moldmaker. For, if any of these three fail to do their job, final failure or intolerable delay may result.

The purpose of the chapter arrangement is to enable the reader to build step-by-step in logical sequence. A close study of the text will reveal the detailed analysis needed to effect a good design. You will also find explicit instructions on the use of the materials. The early chapters cover the generalities of plastics materials and the principles of good designing, drafting, and engineering practice. We believe detailed analysis of a particular design problem is the *best way* to educate, train, and edify the newcomer in the field. We also believe the "old timers" need to have a compilation of rules, data, and reminders to keep them aware of some *critical items* at just the right time to prevent error. Thus, an extensive checklist is presented. It will insure consideration of the potential hazards, weaknesses, and misunderstandings that face mold designers, engineers, and builders.

There are, of course, many variations of molds, whatever their general classifications. Naturally, it is not possible in a presentation on mold fundamentals to describe in detail the very complex designs that sometimes evolve. However, you can be sure that any complex design can be broken down into its simplistic fundamentals as outlined in this text.

We have tried to mention all mold-design and moldmaking methods— even those that are rarely used. Our purpose here is to stimulate interest and to encourage original study.

We wish to thank the many users of the previous editions for their helpful suggestions for changes and improvements in the text. Since many pieces of equipment that are obsolete by present standards continue to be used, we have described mold types for some of them. For instance, this text is used in parts of the world where very primitive equipment is employed. There, the people need data on molds for simple processing equipment, and to use the supply of moldmaking materials, which may be available in these localities but far removed from suppliers of standard mold parts.

We are indeed grateful for the widespread acceptance and distribution of this text since it was first published in 1946 by the American Technical Society. We appreciate the obligation this places on us to be accurate, precise, and factual. In preparing this new edition, we have carefully researched the intervening developments and have made every effort to provide serious readers with a body of knowledge that they can carry confidently with them well into the 1980s. Fundamentals do not change; the *application* of fundamentals changes. Thus, it becomes necessary to update periodically even a basic work like *Plastics Mold Engineering Handbook*.

J. HARRY DuBOIS
Morris Plains, New Jersey

WAYNE I. PRIBBLE
New Haven, Indiana

Contents

Introduction to
Plastics Processing

Revised by Wayne I. Pribble

The many plastics materials have found unlimited markets and an amazing variety of applications in divers manufacturing fields (Fig. 1.1). These materials offer many desirable characteristics such as interchangeability of parts, excellent finish, desirable electrical and mechanical properties, variety of color, light weight, thermal insulation, rapid production and low cost. Many of the good properties of the finished product are dependent on the quality of the tool-maker's work. The molds and dies used are the all-important factor in the continuous and low-cost production of quality plastics products.

MOLDING METHODS

In the production of molded products, many types of dies or molds are used to confine the plastic mass of molding material while it "sets" or hardens to the desired shape. These molds are mounted in some type of machine that will open and close the mold, apply high pressure as needed, and facilitate filling the mold by external means. The plastic material is held under pressure in the mold until it hardens sufficiently to hold its shape after ejection. In many cases these units take the form of automatic molding machines with completely self-contained pressure pumps, automatic facilities for heating the material, and timers to control the various operations and sensors to signal malfunctions.

Steam, hot water, oil, gas or electricity are used to heat the mold and maintain temperature control. The kind of heating or cooling to be used for a given job is governed by the means available and the character of the job. Large compression molding plants find considerable economy in the use of

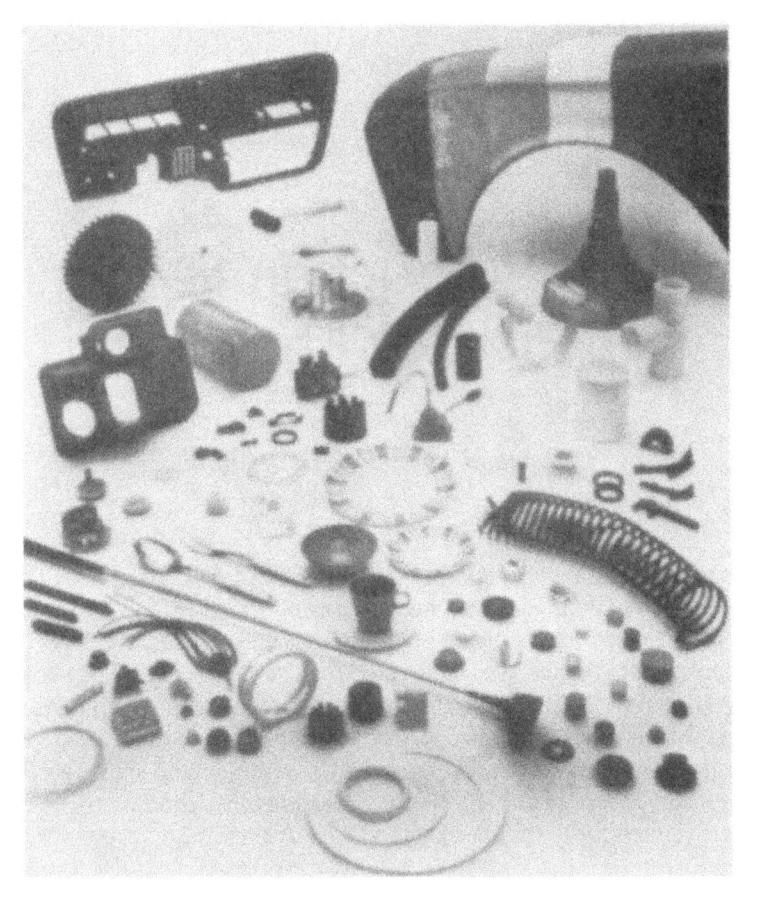

FIG. 1.1. A variety of molded and extruded plastics products. (*Courtesy Allied Chemical Corp., Morriston, NJ*)

steam. High melt temperature dies and molds are heated by electricity or gas. Injection molds are electrically or hot-oil heated or cold-water cooled.

Other molds must be cooled by circulating water or refrigerant. Special mold temperature units are available for maintaining constant temperature in the molds. Some injection molds must be heated rather than cooled when the materials melt at high temperatures.

SYSTEMS OF MOLDING

Many processing systems are in use today owing to the variety of materials available, and because manufacturers were reluctant to scrap old equipment and techniques as new methods were developed. Compression and transfer molding are used extensively for materials which harden or set under heat and pressure. Such materials are called thermosetting or hot-set plastics. The thermosetting materials may also be injection molded in injection machines. The hardening of thermosetting materials is effected through chemical change in the resin which takes place under heat and pressure. The press and mold hold the compound under great pressure while this action is in process. After the chemical change is effected, the cured or set resin will continue to hold the filler particles of the compound in compression when the mold pressure is released. The piece has then been formed to the shape of the mold.

Injection molding is used primarily for materials which are plastic when heated and solid after cooling. These materials are called thermoplastic or cold-set plastics. In molding thermoplastic products, the material is melted or made plastic by heat, mechanical work, and pressure. The plastic mass of hot material is injected into a relatively cold mold, which causes the material to harden and retain the shape of the mold.

When thermosets are injection molded, the hot screw-plasticized compound is injected into a hotter mold, which advances the hardening of the product while the material is held under pressure in the mold.

Compression Molding

The molds used for compression molding consist of a cavity and a force or plunger, as shown in Fig. 1.2. Guide pins maintain the proper relation between these members. The mold cavity forms one surface of the molded part and the molding compound is generally loaded in this member. The mold plunger forms the other surface of the piece being molded, and serves to compress the compound when the mold is closed. The molding compound is thus confined to the open space between plunger and cavity while it

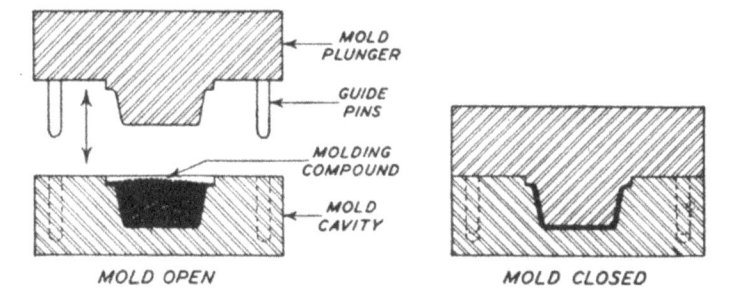

FIG. 1.2. In the compression molding process the material is placed in the mold cavity by the press operator. When mold is closed, molding compound is compressed to shape of piece and held in this form until it hardens.

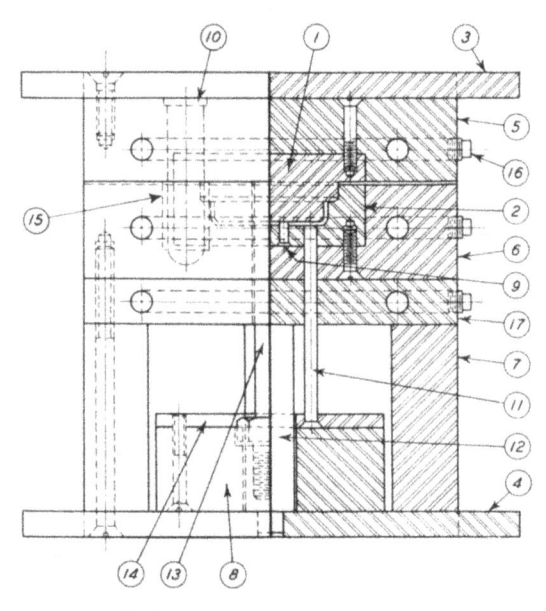

Fɪɢ. 1.3. Landed plunger compression mold, showing details of mold construction and names of mold parts.

1. Plunger or force	7. Parallels	13. Safety pins or rods
2. Cavity	8. Ejector bar	14. Pin retainer plate
3. Top plate	9. Mold pins	15. Guide pin bushing
4. Bottom plate	10. Guide pins	16. Pipe plugs
5. Top retainer shoe	11. Knockout or ejector pins	17. Steam plate
6. Bottom retainer shoe	12. Support pins	

hardens. Numerous features are incorporated in the molds to facilitate production. Provision for heating is often included in the mold design. Some presses are equipped with heating platens which transfer the heat to the molds. In some mold designs, the cavity and plunger, or the top and bottom retainer shoes (parts 2, 1, 5 and 6, respectively, in Fig. 1.3) are drilled out to permit steam or hot water to circulate for heating. Knockout or ejector pins (part 11) are often used to push the piece away from the cavity or plunger.

The only essential difference between the mold of Fig. 1.3 and a design in 1986 would be the use of socket head cap screws instead of the fillister head screws.

Transfer Molding

During the compression molding process, the molding compound is plasticized by the heat and by the pressure exerted when the plunger is closed on the cavity. This creates a considerable force and material flow which will distort and break small mold pins or intricate mold sections. The transfer process (Fig. 1.4) differs from this in that the preheated material is loaded

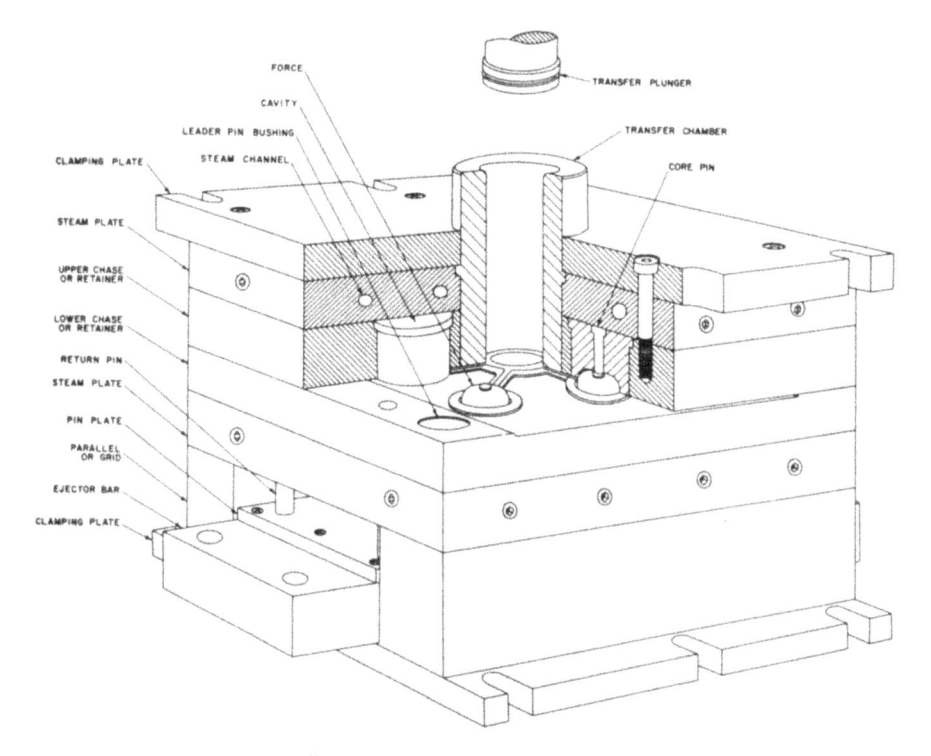

FIG. 1.4. Plunger transfer mold.

in an area external to the cavities, and is then forced into the cavities by a high-pressure ram that enters the loading area to form the molded product. The pressure is maintained until cure is accomplished and the piece hardened. In the transfer process, the material enters the mold as a fluid, with greatly reduced force being exerted on the mold members and inserts. Extremely complex parts having long or complicated inserts, small pins and inserts, or removable wedges or side cores are molded advantageously by the transfer process. This process is especially advantageous when it is necessary to mold parts which combine thick and thin sections.

Injection Molding

In the injection molding process, a considerable bulk of material is held in the heating chamber and a small quantity is pushed or injected into the closed mold, Fig. 1.5. The machines used are automatic injection molding machines, which perform all operations in sequence. For thermoplastic materials, the heating cylinder or screw plasticizing area of the injection machine is supplied with external heating bands that control the temperature up to its plastic

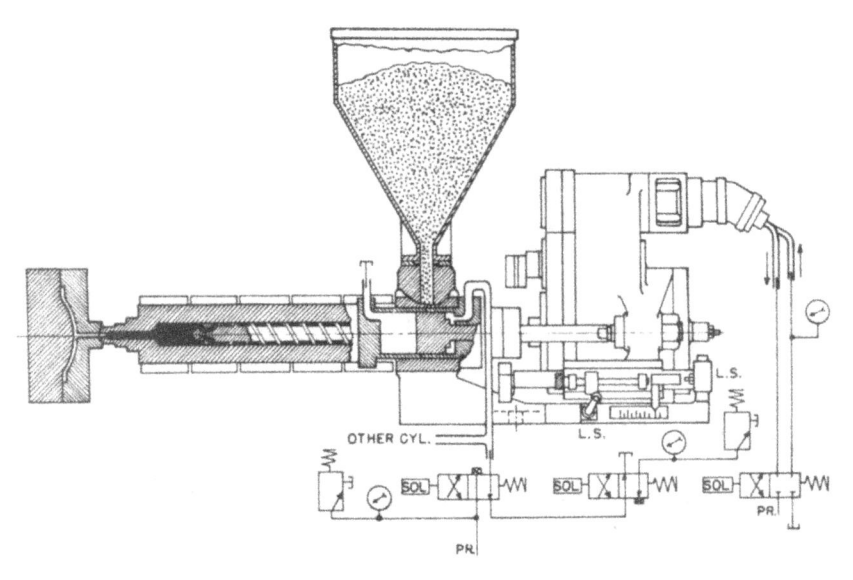

FIG. 1.5. An injection machine with reciprocative screw plasticizer. (*Courtesy Egan Machinery Co.*)

state where it can be pushed into the mold. The mold is kept at a temperature below that of the material so that it will chill or harden the compound after it is injected. Automatic transfer molding is sometimes called "injection" molding. When thermosets are being injection molded, the hot plastic compound is injected into a hotter mold for completion of the hardening or "cure."

FIG. 1.6. This 2500-ton plunger injection press was updated by the addition of a reciprocating-screw plasticizer greatly increasing its capacity and versatility. (*Courtesy Egan Machinery Co. & Engineered Plastics Machinery Co.*)

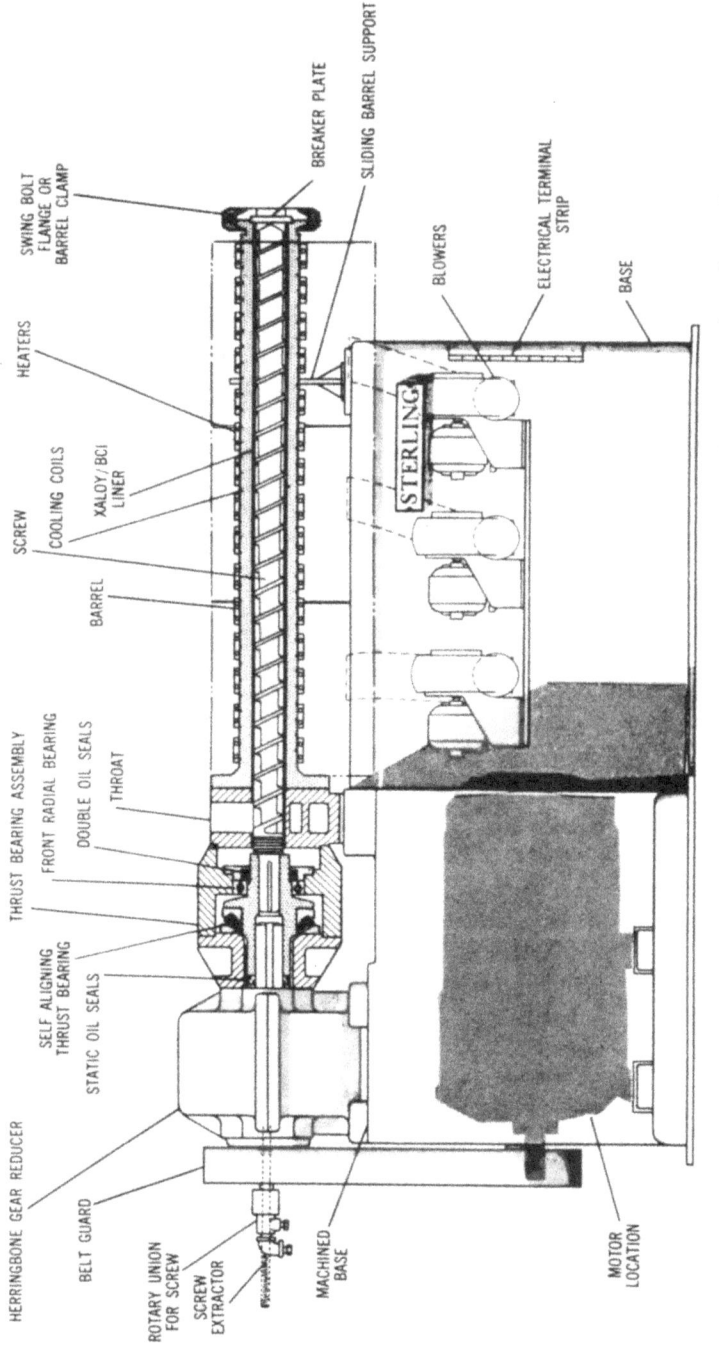

Fig. 1.7. The screw-type extrusion machine is used for making continuous strips or tubes of regular section.

Continuous Extrusion

Plastics materials are extruded in continuous strips of regular section, Fig. 1.7. This is done by a machine which operates much like a sausage stuffer. The raw material is placed in a hopper, where it is moved into and through a heating chamber by a screw feed. At the die end of the heating cylinder the material (which has been heated and compressed to a plastic mass) is forced through a die which shapes the extruded section. A moving belt carries the section away from the die, and the final dimension of the part is governed by the speed of this take-off belt. The extruded piece is stretched to a reduced section area by the take-off belt.

The extrusion dies are relatively simple and inexpensive and are quite similar to extrusion dies used for the low-melting-point metals. Figure 1.8 shows the rear or screw side of an extrusion die used to make a rectangular strip. Note the tapered entry.

Blow Molding

Bottles and other hollow articles are extrusion blow molded of thermoplastic materials. For this, a tube, called a parison, may be extruded and this hot thermoplastic tube is clamped between the faces of a blow mold. Air pressure is immediately applied in the clamped tube to expand it and fill out the mold contour (Fig. 1.9).

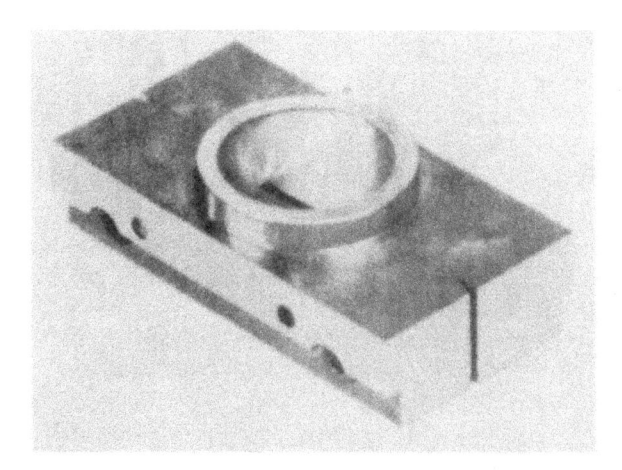

Fig 1.8. An extrusion die used for production of continuous strips of thermoplastic material.

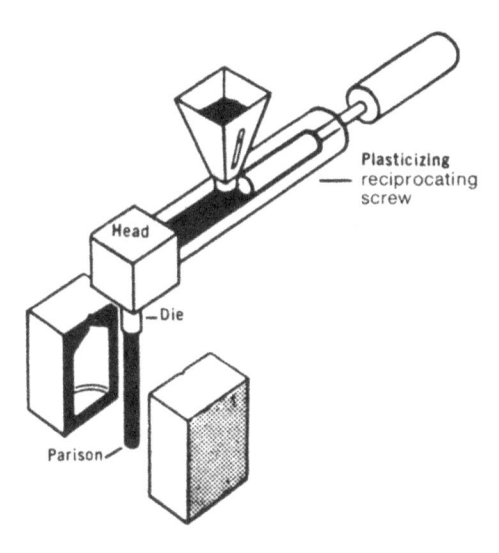

Plasticizing
reciprocating
screw

Head

Die

Parison

FIG. 1.9. Basic extrusion blowing principle for blow molding. The parison is a tube of molten plastics which is clamped between the die halves and expanded to the die shape by air pressure.

An alternate process, called *injection-blow molding* forms the parison in the first position and then moves quickly into the blow mold position for expansion by internal air pressure as shown by Fig. 1.10. In the final position, the blown object is stripped from the core pin while the mold is open.

Rotational Molding

Rotational molding machines are equipped to rotate the molds continuously in the vertical and in the horizontal axis during the molding cycle. This processing procedure facilitates the production of integral hollow parts of almost any size or shape—open or closed—rigid or flexible. In this process, a measured amount of liquid or powdered plastics material is placed in each mold cavity. With the mold halves closed, they are rotated in a heated area while rotating continuously in the two planes until the entire inner mold surface is coated with the molten plastics. (Thermosetting compounds cure during this heating cycle.) After the desired thermoplastics hardening takes place or a suitable "shell" of thermoplastics has formed in the mold surface, the mold is cooled by a water spray or air blast while rotation continues. Production molds Fig. 1.11 are fabricated by aluminum casting, machining, stamping of sheet steel or electroformed nickel. The molds are vented by a small tube containing a porous filler. Details on the simple mold designs for this process are supplied by the machine makers.

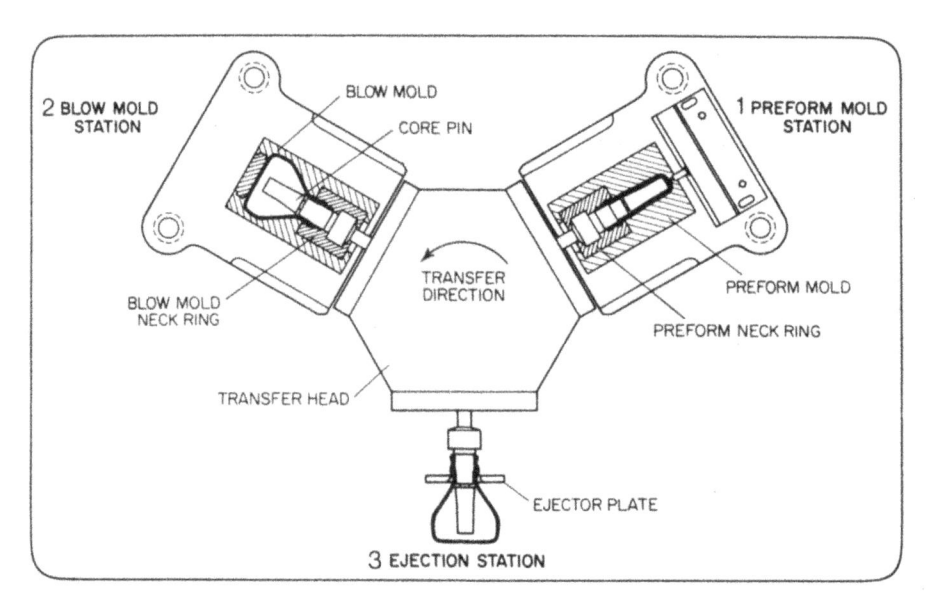

FIG. 1.10. The majority of injection blow molders commercially available are three-station types: (1) preform injection molding, (2) blow molding, and (3) ejection.

Cold Molding

Many materials are pressed or formed to shape in unheated molds. After leaving the molds, the formed pieces are hardened in baking operations. One type of cold-mold material is made from cement and asbestos mixed with a small amount of clay. After being molded to shape, this material is hardened by baking in a wet steam atmosphere that cures the cement and hardens the molded piece.

The molds used for cold molding are relatively simple, since they serve merely to form the piece. Large area ejector pads are used because the material is soft as it comes away from the pressing operation. These ejector pads raise the piece up out of the mold so it may be removed without distortion. The cold-mold materials are abrasive and cause wear, thereby necessitating frequent replacement of the mold sections.

Vacuum Forming

Thermoplastic sheet materials are converted into simple shapes by a thermoforming process often called vacuum forming. In this process, the sheet is supported over the mold and is heated to a temperature at which it can be reshaped. This mold is provided with multiple holes so that a vacuum may

FIG. 1.11. A piece being removed from a rotational mold. (*Courtesy Barberton Plastic Products, Inc., Barberton, OH*)

be drawn that will pull in the softened sheet to the mold contour where it hardens to the desired shape. Many variations of this process are used. A simple form of vacuum forming is shown in Fig. 1.12.

Expanded Polystyrene (EPS)

The expanded styrene products are often molded by the use of pre-expanded beads that are further expanded and fused in a mold. Steam is commonly used in the pre-expanders from which the molds are filled by the use of air pressure. Steam is then used for the final expansion of the pre-expanded beads in the mold. The steam enters through holes in the mold which is subsequently cooled (Fig. 1.13). Expanded styrene products are extensively used for ice buckets, novelties, packaging, etc.

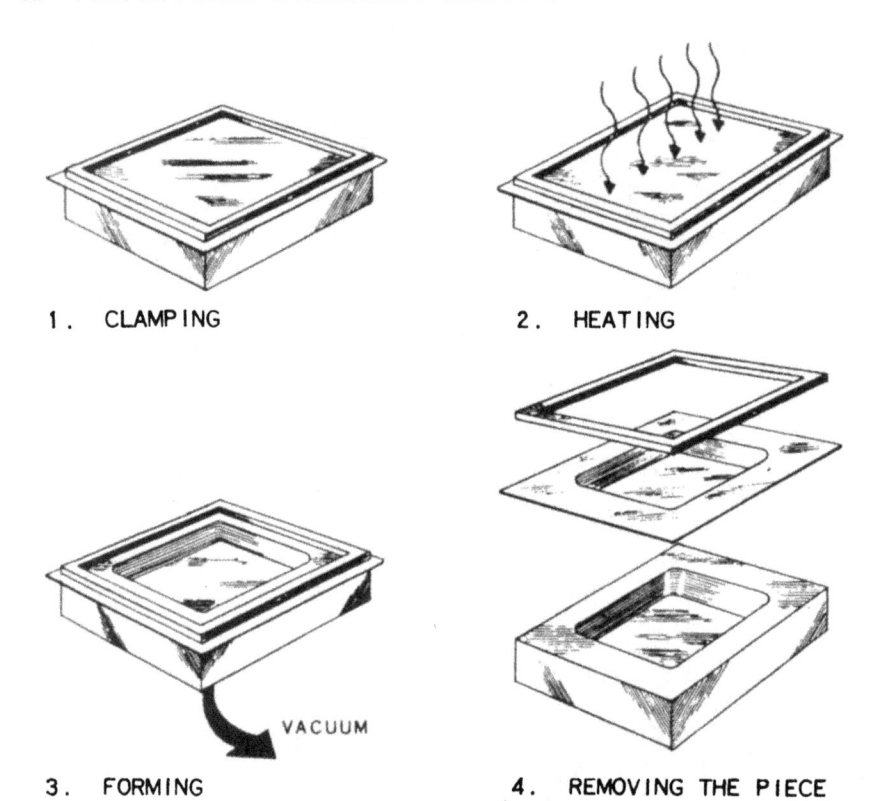

1. CLAMPING

2. HEATING

VACUUM

3. FORMING

4. REMOVING THE PIECE

FIG. 1.12. Thermoforming or vacuum forming of thermoplastics sheet.

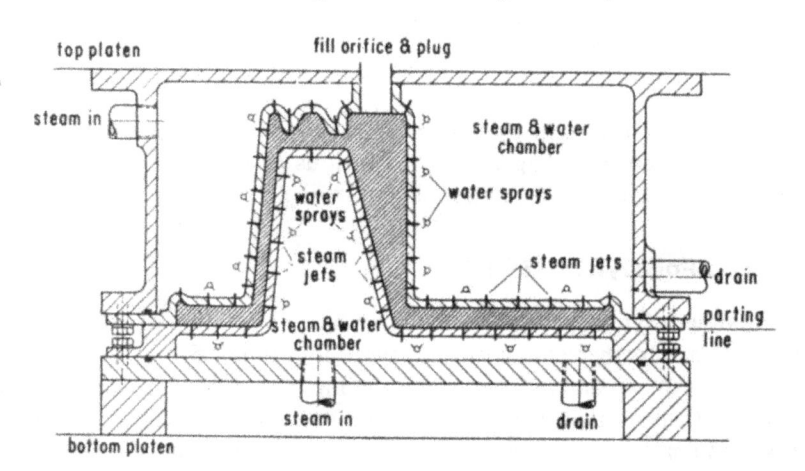

FIG. 1.13. Square steam chest type of mold for expandable polystyrene foam mold. Steam is admitted to cavity from ports equally spaced on both sides of the cavity. Mold walls carry clamping forces.

Reinforced Plastics

Only a few years ago, the term *reinforced plastics* was used to refer to a group of low-pressure molding materials fabricated by using a thermosetting resin to coat or impregnate a cloth, random fibers, or continuous filaments. Now, the term is used to designate any resin type, either thermosetting or thermoplastic, which has some type of reinforcing material incorporated with the resin. The result is a compounded raw material ready to be processed by injection, compression, transfer, or extrusion. If the raw material is thermosetting and in sheet form, it may be stamped, die cut, or hand cut to make layer type preforms for molding in a hot mold. Reinforced thermoplastic resin is generally pelletized for hopper loading into an injection machine. It should be obvious that the term reinforced plastics covers a wide range of materials and a wide range of applications. Switchgear parts, automobile components (particularly body parts), airframes, and aircraft panels are only a few of the many possible applications. The fibrous reinforcements covers the entire range of fibers, such as aramid, glass, carbon, thermoplastic (such as acrylic and nylon), and the hybrid combinations of any of the foregoing.

Note that fillers and extenders may be fibrous in nature, but the resulting material would not be termed reinforced plastics. Fillers and extenders generally refer to low-cost (as compared to the cost of the resin) materials such as clay, wood-flour, minerals, etc. Reinforcing fibers may be considerably more expensive than the base resin, and in all cases increase the difficulty of processing into marketable molded parts. The requirements of the precise application are the determining factors in the choice of the base resin. Then, the degree and purpose of reinforcement determines the choice of fiber and its method of preprocessing prior to compounding with the resin. Books on fibers and reinforcement are readily available to cover this extensive field.

Calendering

Calendering is a rolling process combined with extrusion for coating webs such as plastics, fabrics, paper and other materials with a continuous plastics coating. It is essentially a continuous extrusion process which feeds into rolls that establish the thickness and surface characteristics of the applied coating. A simple form of calendering is shown in Fig. 1.14.

Structural Foamed Products

A plastics foamed product is one having a cellular core and a solid integral skin. In the forming process, the solid skin is formed when the injected foaming mass is chilled by the cool mold surfaces. This skin then provides

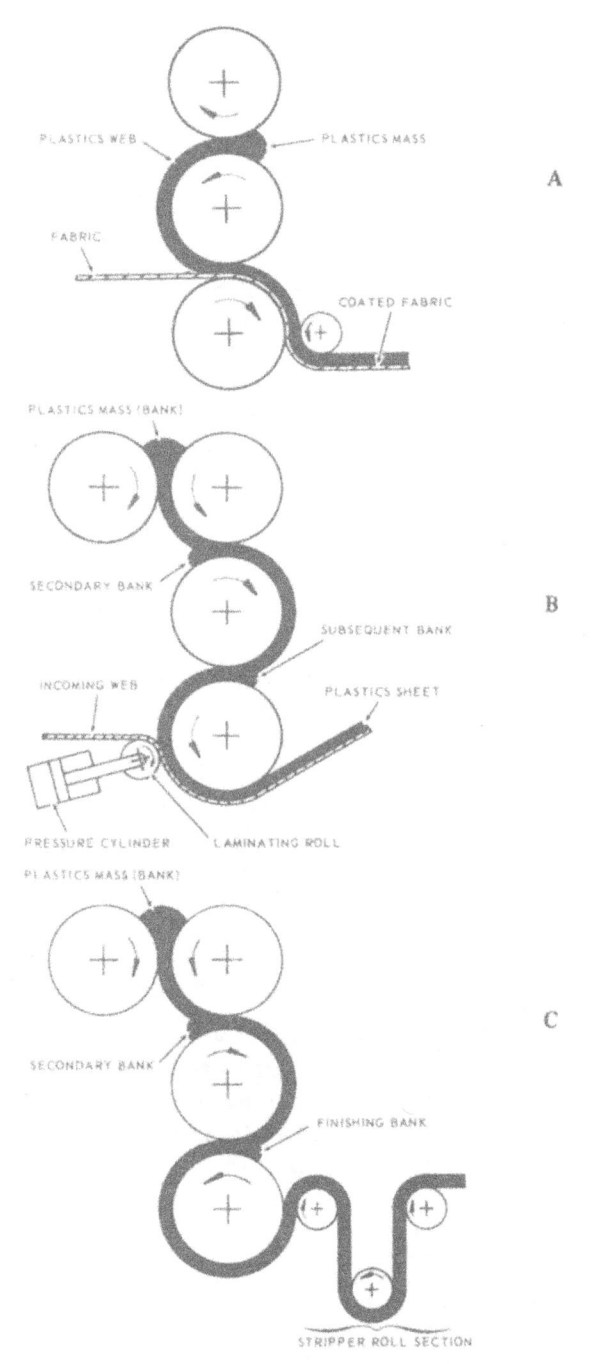

FIG. 1.14. Three calendering processes: A. Coating fabric; B. Carrier web; C. All plastic sheet.

an insulating barrier that permits the central mass to complete its expansion, forming the inner foam core. All thermoplastics may be foam molded or extruded.

In the high pressure foam molding process, the mold is completely filled with the foaming plastic mass at the conventional injection machine pressure. The expansion type of mold then opens slowly under proper controls to permit the desired expansion. In the low pressure process, conventional injection machines are used with resins containing preblended foaming agents or resins that are expanded by a "piped in" gas.

TOOLS FOR PLASTICS

The entire processing industry is dependent on the mold-maker. Good tools give continuous production, excellent finish, easy ejection of molded part without distortion, close dimensions, flawless pieces, and low cost cleaning of parts after molding. A good mold will give many years of satisfactory service. The success of any molding operation may be measured by the efficiency of the mold design and the quality of the mold construction. Molding surfaces are often chrome plated for better finish and wear.

The best of molds are broken in production from time to time, and the mold-maker must be prepared to make repairs and replace broken mold sections quickly. Molds wear, as shown in Fig. 1.15, and frequently require dimensional correction and repolishing and replating after continuous service over long periods. Many times the mold-maker is required to change the design of a part and make mold changes that are complicated and involve risk.

FIG. 1.15. Left: Illustration shows wear in recess of plastics mold caused by plunger striking sides of mold. Right: Worn plastics mold part after atomic-hydrogen arc welding has added sufficient stock to permit remachining to original dimensions.

The plastics molder of thermosets uses many tools which must be designed and built by his mold-maker. Loading fixtures are constructed to facilitate accurate and fast insertion of the right amount of material into the mold cavities. On some jobs, they are used to fix in place the metal inserts frequently incorporated in molded products. Many types of unloading fixtures are used to remove molded pieces from the molds. Unscrewing fixtures are frequently constructed to remove threaded sections from the mold. After molding, it is often desirable to place the piece on a cooling or shrink fixture in order to hold shrinkage to fixed limits. These shrink fixtures are accurately designed and built to permit contraction of the part from the mold size to the close dimension required after cooling.

Some molded parts are subjected to finishing operations after molding. These operations serve to remove "gates," "flash," or "fins." The gates are the feeders which connect the molded piece to the source of material in transfer and injection molding. "Flash" is the excess material squeezed out of the mold cavity as a compression mold closes, or as the pressure is applied to a transfer or injection mold. A "fin" is formed by the material which flows into the small gap between movable parts of the mold, Fig. 1.16. This excess material must be removed from molded parts during the finishing operation. Trimming and punching dies are frequently used for this operation. These dies punch out the fins in holes and on irregular surfaces. Other special fixtures serve to perform such operations as grinding dimensions to close limits or drilling unmoldable holes. Gauges for checking dimensions, and jigs for filing and drilling, are built by the mold-maker.

Heavy production schedules often justify considerable expense in devising

FIG. 1.16. Molded pieces as removed from the press. Note "fins" where mold members join and at parting lines. Fins are removed in the cleaning operation.

FIG. 1.17. A modern preforming machine with vibratory feed, overhead hopper, and dual stroke pressure application will compact the compounds that have difficult flow characteristics. (*Courtesy Mykroy Ceramics Co.*)

special fixtures to expedite finishing and after-molding operations performed by the molder.

Many of the molding compounds are preformed before molding, as shown in Fig. 1.17. This is done in special presses with "pill" dies. These preforms or "pills" facilitate loading the molds since each unit contains a fixed amount of material. Pills are often made special for the specific material or application, therefore the mold-maker must be prepared to design and build "pill" dies.

Thermal and ultrasonic bonding techniques are used for the thermoplastics and the tool-maker is often called on to provide fixtures for this and other postmolding operations such as decorating, adding inserts, combining parts, and machining.

Plastics Tooling

The metal working industries make many dies, jigs, and fixtures out of plastics, and these products are identified by the name *plastics tooling*. The plastics processing industries refer to their tools as molds, dies, and fixtures.

REFERENCES

Baer, Eric, *Engineering Design for Plastics*, New York: Van Nostrand Reinhold, 1964.

Beck, Ronald D., *Plastics Product Design*, 2nd ed., New York: Van Nostrand Reinhold, 1980.

Bernhardt, E. C., *Processing of Thermoplastic Material*, New York: Van Nostrand Reinhold, 1974.

DuBois, J. H., *Plastics History—USA*, Boston: Cahners Books, 1972.

DuBois and John, *Plastics*, 6th Ed., New York: Van Nostrand Reinhold.

DuBois and Levy, *Plastics Product Design Handbook*, New York: Van Nostrand Reinhold, 1977.

DuBois and Pribble, *Plastics Mold Engineering*, 1st Ed., Chicago: American Technical Society, 1946.

The Handbook of Plastic Optics, Cincinnati: U.S. Precision Lens, 1973.

Mock, John, Plastic product design—A controlled balancing act, *Plastics Eng.*, p. 17, Dec. 1983.

Modern Plastics Encyclopedia, 1983–84 ed., New York: McGraw-Hill, 1984.

Olesky and Mohr, *Handbook of Reinforced Plastics*, New York: Van Nostrand Reinhold, 1964.

O'Toole, J. L., Design guide (systematic approach to material selection and design), in *Modern Plastics Encylopedia*, 1983–84 ed., pp. 392–396, New York: McGraw-Hill, 1984.

Peach, Norman, Places for plastics in electric/electronics (E/E), *Plastics Eng.*, p. 21, Apr. 1984.

For further reading, we suggest:

Smoluk, George R., Look to lasers for a practical way to do the tough fabricating jobs, *Modern Plastics Magazine*, p. 61, May 1984.

Chapter 2 / Basic Mold Types and Features

Revised by Wayne I. Pribble

In order to introduce the reader to the designing of molds, it is necessary first to explain what a mold is expected to do. Throughout this book, we use the term "molds" to indicate any device used to provide the final shape and dimensional details of a product. A mold is only one item in a series of devices that act to shape a plastics material. The vast majority of molds are made up essentially of two halves which open and close. One half of the mold is called the *force*, *plunger* or *core*, and is known as the male half of the mold. The other half is called the *cavity*, and is known as the female half of the mold. Obviously then, the male half forms the inside of a part, whereas, the female half forms the outside contour of a part.

The raw material that is to be molded is an end product of some other organization. There are separate companies that specialize in resins, fillers, catalysts, colors, curing agents, and a long list of other components that are used by the compounder to make a molding material. The final material is truly a "compound" because it is formulated from many different components combined in accordance with a "recipe" and processed through very precise chemical and mechanical processes. The chemical reaction processes are not within the province of this text. Reference should be made to the Glossary to gain familiarity with materials. There are books available on each of the components, processes and devices. We will make no effort to tell you how to develop and operate a chemical process, or how to build and operate a press. However, the designing of a mold is closely related to the equipment that will be used for the mold to operate in the desired fashion. Regardless of the process used, some kind of a mold is required if the end product is to be formed into the specified shape.

This device, called a *mold*, can range from a simple wooden or plaster form to a very complicated and extremely large (10 or 20 ton) mechanism.

The extremely simple molds of wood and plaster are usually used by the hobbyist who wants to experiment with plastics. Some readers will question this assertion because break-away plaster is very common in the epoxy and fiberglass layup and in the molding of air ducts and intricate shapes for the aircraft industry.

Plastics Mold Engineering will not be able to cover everything. This prelude to basic mold types and features will give you an initial understanding of the molds only. We urge you to collect catalogs and house organs which describe and promote new methods or new combinations of old methods that enterprising persons have put together to make a fascinating device that probably will do a job previously considered impossible. Our point here is—The most complicated mold ever built was made up of the simple components and actions described in this text. Most inventions are simply the application of known engineering principles to a new problem, the result being a device or method to lighten man's work, reduce his effort to achieve a wanted goal, or reach his goal in shorter time.

The primary function of a mold is to shape the finished product. In order to do this it must have some means of introducing the plastics material to be formed; it must have some means of forming the inside, outside, and both sides of the product; it must have some means of maintaining the temperature desired in the process; it must have some external or internal mechanism for operating the various features of the mold; it must have some means for allowing the finished product to be removed from the sections which were used in the forming process (this is called *ejection*); after ejection, it must provide for easy removal of any and all excess material that may have been left during the previous cycle; it must be designed and constructed with adequate strength in the various sections to resist the alternate application and release of pressure.

All plastics processing requires three elements, in varying proportions, depending upon the processing characteristics of the material to be molded or processed. These three elements are time, temperature, and pressure. Time ranges from fractions of a second to hours, or days. Temperature ranges are from freezing to 1200 or 1500° F (650–815° Celsius). Pressure ranges are from negative (vacuum) to 50,000 psi.

The purpose of this particular statement is to show the impossibility of detailing or mentioning every contingency that may be encountered. In this chapter, we will show examples of the basic mold features and the different types that can be used in any combination that will perform the job required by the mold. First we will discuss press operation, second, general mold types and, finally, mold construction features. Using these three elements, an engineer should be able to convey to a designer precisely what is to be designed. A good designer will make a rough sketch or preliminary layout to determine the various wall thicknesses, runner layouts (the path the material will take into the mold), and the features that he wishes to have incorporated into the mold.

Before proceeding with a description of the specific mold types and features, it would be well to discuss the press and operations of the mold in the press in order to understand just what function the mold plays in the process of plastics fabrication. For your convenience, the different types of machines that are used in conjunction with the molds detailed in later chapters are listed below:

1. Compression molding presses (machines). A "press" is a simple unit that holds the mold and supplies opening and closing pressure. A "machine" is a self-contained press unit with the necessary pumps, timers, etc., to achieve automatic operation.
2. Transfer molding presses.
3. Injection molding machines.
4. Continuous extrusion machines.
5. Blow molding machines.
6. Vacuum forming machines.
7. Rotational molding machines.

The various presses and the operation of the mold in the press or machine will be discussed in the following pages.

INJECTION MOLDING MACHINES

The two halves of an injection mold are bolted in place to the stationary and movable die plates. Most injection machines operate in a horizontal position. The injection press makes use of two hydraulic cylinders, and the mold is closed by the action of a hydraulic cylinder. Some presses use straight hydraulic action, while others use a toggle mechanism. Figure 2.1 shows a toggle mechanism, but a straight hydraulic action serves the same purpose. The purpose of the mechanism on the left side of Fig. 2.1 is to open and close the mold and to provide the *clamping pressure* to hold it closed during the injection part of the cycle. The injection cylinder at the right side of Fig. 2.1 serves to force the compound through the "shot" or injection. This cylinder has nothing to do with the opening or closing of the press. Several methods of heating or plasticizing the material are in current use, but none of them have any direct bearing on the type or design of mold required.

The ejector plate or knockout bar (Fig. 2.2) is operated by an ejector rod, which stops the plate as the mold opens and before the end of the "open" stroke is reached. Thus, the mold cavity is pulled back after the ejector or. knockout pins are stopped so that the molded piece comes free as the pins extend out of the mold cavity.

TYPES OF MOLDS

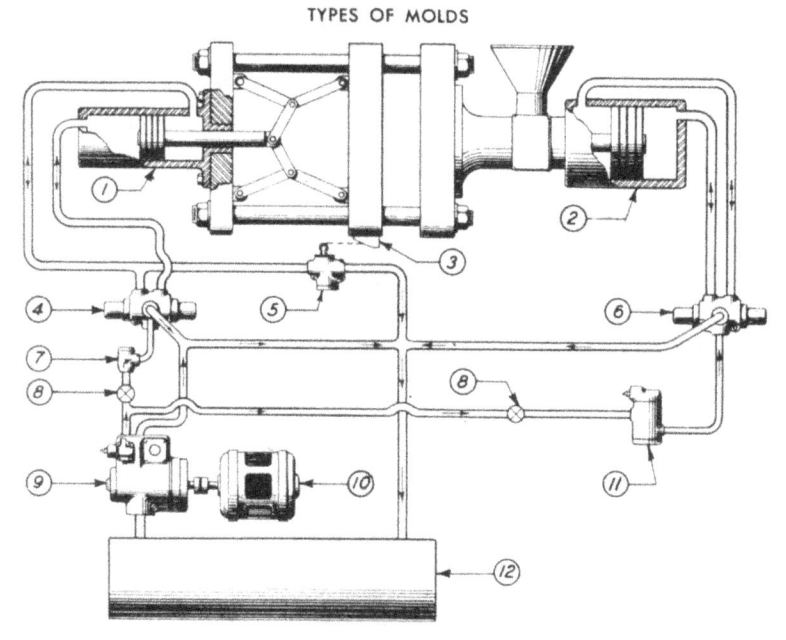

FIG. 2.1. Diagram of a simple hydraulic circuit for a plunger injection molding machine (See also Fig. 1.5), showing the following parts:

1. die cylinder;
2. injection cylinder;
3. cam;
4. four-way reversing value for die cylinder (pilot or solenoid operated);
5. cam operated valve (for automatically dropping pressure between cycles);
6. four-way reversing valve for injection cylinder (pilot or solenoid operated);
7. check valve;
8. globe valves (for set-up purposes only);
9. double pump and combination valve unit for automatic volume control and adjustable pressure regulation;
10. motor;
11. pressure reducing valve for independent adjustment of injection pressure;
12. oil reservoir (in machine base).

(*Courtesy Vickers, Inc., Detroit, MI*)

The ejector bar and pins are pulled back into molding position by springs, or they are pushed back into molding position by various devices that are described elsewhere, such as early return pins, hydraulic or air cylinders, or safety pins. This return action takes place as part of the "open time" or as part of the "closing time cycle." This is called *resetting the ejector*. Many other operations or actions may be included in the mold design. Some of these are pulling side cores, unscrewing threaded sections automatically or collapsing cores for internal undercuts.

COMPRESSION AND TRANSFER PRESSES

Conventional compression and transfer molding is done in hydraulic, air, or mechanically operated presses. The facilities or methods of mounting

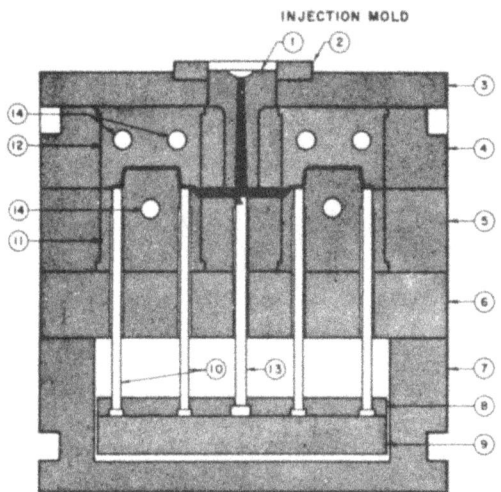

INJECTION MOLD

1 SPRUE BUSHING
2 LOCATING RING
3 TOP CLAMPING
   PLATE
4 FRONT CAVITY PLATE
5 REAR CAVITY PLATE
6 SUPPORT PLATE
7 EJECTOR HOUSING
8 EJECTOR RETAINER
   PLATE
9 EJECTOR PLATE
10 EJECTOR PINS
11 CORE INSERT (male
   section)
12 CAVITY INSERT (female
   section)
13 SPRUE PULLER PIN
14 WATER LINES

FIG. 2.2. Various components of a simple two-plate injection mold used for injection molding. (*Courtesy Dow Chemical Co., Midland, MI*)

molds and operating them are similar in all presses of this type. Since hydraulic presses are most frequently used, the following description is given of the operation of a mold in a hydraulically operated press. Two general types of presses are used: the upstroke press, and the downstroke press. The downstroke press makes use of an overhead cylinder so that the ram and top platen are moved downward to apply pressure to the mold. This type of press is widely used in hand layup or sheet molding operations for such parts as truck cabs, fender skirts, automobile fascia, etc. The advantage of this press is that the working area remains at a constant level and allows the operator to walk around the mold and on the platen. Platens 10 foot square are not uncommon in these operations. (See section on Large Molds in Chapter 6.)

The upstroke press has the cylinder positioned below, and the ram moves the platen upward toward the stationary platen. This press is considered to provide greater safety for the operator in the event that something malfunctions in the hydraulic circuit. In that event, the lower platen will drift downward and is not likely to trap anyone. Most hydraulic presses used in compression and transfer molding have a main cylinder or ram to apply the pressure. This may be either single-acting (applies pressure in one direction only and depends on gravity or auxiliary cylinders to return the main ram to its fully open position) or double-acting (one cylinder uses pressure in one direction to "close" and pressure in the other direction to "open").

In compression molding, another cylinder is frequently used to operate the ejector bars or stripper plates. This cylinder is conveniently located above the press and uses air or hydraulic pressure as its operating force. The compression press shown in Fig. 2.3 is more or less typical of presses used for the past fifty years in this phase of the plastics industry. This particular illustration is used because all the component parts are "in the

FIG. 2.3. Upstroke press used for compression molding small plastic parts. The various features shown are:

1. cross bar;	11. stress rod (2);
2. air cylinder (2-way);	12. daylight of mold;
3. U-washer;	13. bottom ejector bar;
4. push-back rod (2);	14. push-up rod;
5. air valve;	15. outlet valve;
6. adjusting nut;	16. bottom platen (movable);
7. top platen (stationary);	17. inlet valve;
8. grid;	18. hot plate;
9. top ejector bar;	19. air hose nozzle;
10. clamp bolt;	20. pull-down rod.

open" and not covered by sheet metal dressing. The air cylinder is shown with the two-way valve which controls the direction of force exerted by the cylinder.

The piston rod of the air cylinder is connected to the cross bar which, in turn, transmits the motion and force to the push back rods and then to the ejector bar and ejector pins. A top ejector bar or a bottom ejector bar may be used, or, in some cases both are used (see Fig. 2.31).

The mold is bolted in place on the press platens with clamp bolts at front and back. Although molding presses may develop from 5 tons to 1000 tons or more, the only pressure exerted on the mounting (clamp) bolts is the weight of the mold and the force required to open the mold. Sheets of laminated phenolic, glass-bonded mica, or transite asbestos, are often used as insulating spacers between the mold and the press platen. Such insulation reduces the heat transmitted to the press, and gives better thermal uniformity within the mold.

The mold shown in the right-hand press in Fig. 2.3 is a double ejector bar mold. The upper ejector bar is moved down and up by the air cylinder on top of the press. The lower ejector bar is raised by the push-up rod as the press platen moves downward in the "opening stroke." The push-up rods stop the downward motion of the ejector bar while the main part of the mold is still moving down. This relative motion causes the mold cavity to move below the level of the ejector pins which elevate the molded part out of the cavity when the press reaches the bottom end of the stroke. On the upward stroke, springs cause the ejector bar to return to its seated position where it rests in contact with the bottom clamping plate (or stop buttons) after the mold travels up to where it is free from the push-up rods. The push-up rods operate freely through clearance holes in the press platens.

The top platen of the press may be raised or lowered as required by means of adjusting nuts, thus providing for molds of varying height. Variation of several inches in mold height can be compensated by the ram travel without platen adjustment. Extremely high or low molds may require some adjustment of the platen to secure economical operation. Mold designers must never overlook the fact that the available daylight opening in an automatic or semiautomatic mold, after it is installed in the press, must be greater than the depth of the molded piece. The molded piece is ejected into the daylight opening and, of course, this opening must be large enough to permit removal of the piece from the mold. The top press platen must be level and parallel (within a few thousandths) with the lower platen to insure proper closing of the mold.

The air hose, shown in Fig. 2.4, is provided to "blow out" the mold for the removal of flash on completion of each cycle. The nozzle of the air

Fig. 2.4. Showing use of air hose for removal of flash from mold. Note that bottom knock-out bar of this semiautomatic mold is raised by push-up rod. Parts are: (1) guide pins; (2) brass nozzle on air hose; (3) push-back rod, which pushes knockout bar down as mold closes; (4) safety pin, which pushes knockout bar down as mold closes if other means fail; (5) push-up rod, which raises knockout bar.

hose must be copper, brass, or other soft material to avoid damage that may result from contact with the mold.

It will be noted that the molds shown in Fig. 2.3 are heated by steam. The cavity is connected in series with the plunger, as shown in Fig. 2.5, by means of a flexible metal hose. A reducer is often included in the steam line in order that the temperature of the mold may be reduced and regulated.

The essential difference between the compression press just described and the top ram transfer press is the replacement of the top air cylinder with a hydraulic cylinder having a piston rod operating downward and attached to a plunger or ram. The operation and general design of a transfer mold will be discussed later in this chapter.

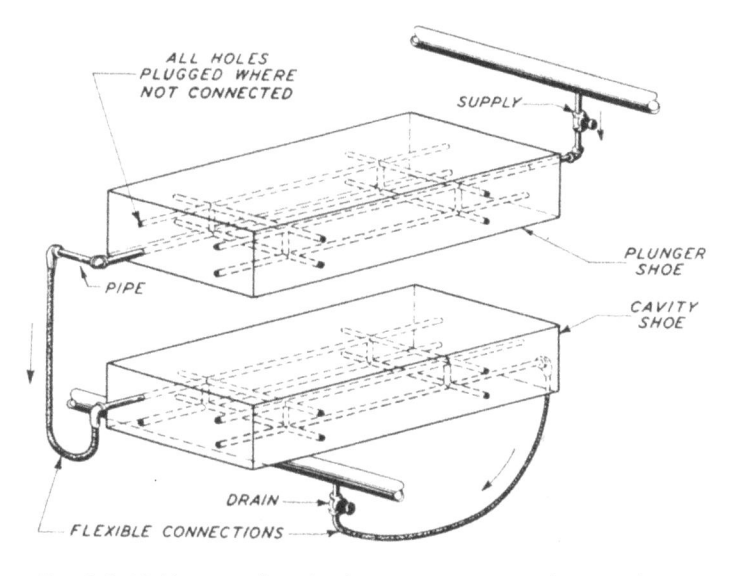

FIG. 2.5. Mold perspective, showing water, steam, or oil connections.

CONTINUOUS EXTRUSION MACHINES

A continuous extruder does not make use of a mold in the usually accepted terminology. Rather, it uses a die that provides a specific shape to the material being forced through it in a continuous stream. These details are covered in Chapter 10.

BLOW MOLDING MACHINES

Blow molding tooling is very much like compression molding because the blow mold generally closes on a hollow shape that has been deposited in between the halves of the mold. Here the similarity to compression molding stops. Air or gas pressure is introduced into the center of the heated shape. This internal pressure then causes flow of the heated material into intimate contact with the relatively cold mold sections. Here the plastics material solidifies and is subsequently ejected. Chapter 11 discusses the details of the molds and covers multiple tooling in common use.

VACUUM FORMING MACHINES

The mold used in vacuum forming (simplest type) is similar to the female half of a compression mold or a blow mold. Auxiliary equipment on the machine heats the sheet material and drapes it over the force or cavity as indicated by the precise technique chosen. A corollary to the cavity half is the draping of material over a force or male half. The techniques of vacuum forming are the subject of an entire book and no attempt will be made to cover them in this text. They are a very specialized section of the art and the designs of the molds are more in the pattern and casting field. The mold making follows procedures suggested for blow molds (Chapter 11). The trimming equipment is simple shearing or clicker dies used on standard presses made for this type of operation. Vacuum forming machine builders provide adequate mold design data for their process.

ROTATIONAL MOLDING MACHINES

The tooling for rotational molding can be compared to the split cavity techniques of blow molding; their molds are similar except that no cooling areas are required (see Fig. 1.11). The machine provides the means of holding the tooling and rotating it about two axes at the same time. This rotation causes the powdered material that was loaded into the mold, before it was clamped into the machine, to solidify and cling to the wall of the mold. The thickness of the deposit depends on the time allowed for rotation and the amount of material placed in the mold.

GENERAL MOLD TYPES

The variety of molding materials and molding methods has necessitated the development of many mold types in order that full advantage of the material possibilities might be secured. Three general types of molds are used and these may be subdivided into several classes. The three general types are compression molds, transfer molds and injection molds.

These three systems, described in Chapter 1, will be reviewed here. There is no particular significance to the order in which they are presented. Historically, compression molds were the very first types to be used in the middle 1800s. The injection molds came into being in the 1920s for the thermoplastics processing and the transfer molds came into use in the 1930s. For a history of the development of the industry, reference should be made to *Plastics History, U.S.A.**

Compression Molds

Compression molds make use of a mold cavity for receiving the compound when the mold is open, and a force or plunger for compressing the compound as the mold closes. These molds are generally used for the thermo-

*J. Harry DuBois, published by Cahners Book Co., Boston, MA, 1972.

setting and cold-molded materials only. The molds may be either hot or cold. They are used hot with the thermosetting phenolics, ureas, etc., and cold with the cold-molding compounds. Compression molds are seldom used for the thermoplastic materials because of the long period required for heating the material to the plastic stage and the necessity for chilling the mold immediately thereafter in order to harden the piece.

Transfer Molds

Transfer molds are of two distinct types. A *plunger* or *auxiliary* ram transfer mold, Fig. 2.6, is the type most often used. It has a built-in transfer pot

Fig. 2.6. This 300-ton transfer press has a top ram and a bottom transfer ram to facilitate mold design. It will be noted that the bottom ram is mounted below the lower platen and rides up and down with the clamping ram.

Fɪɢ. 2.7. This 3-plate transfer mold is used for production of thermostat base. A removable cavity is used because of metal inserts which must be loaded prior to molding. (*Courtesy Pribble Enterprises, Inc., New Haven, IN*)

or tube which is separate from the cavities and forces. This tube is open at both ends. Clamping pressure and transfer pressure are applied by two separate rams. The tube receives the molding compound, and the mold is closed tight at the parting line by the clamp pressure. Then the transfer plunger is actuated to apply pressure to the material in the tube. The transfer ram, entering at one end, drives the molding compound out the other end of the tube, through the runners and gates, and into the cavity. The amount of pressure generated inside the cavity is entirely dependent on the transfer ram pressure.

The *integral* or *three-plate* transfer mold, Fig. 2.7, uses a floating third plate to carry the closed-bottom transfer pot. The same pressure which clamps the mold is also used to generate the transfer pressure in the pot. The pot plunger is built into the mold. As the clamping ram advances, the pot plunger enters the pot and applies pressure to the molding compound. Continued application of heat and pressure causes the molding compound to flow through a restricted opening, called a sprue, into the runners, through the gates, and into the cavity. Transfer molds generally are used for molding the thermosetting compounds only.

The transfer-type mold is selected for many difficult jobs since it can be used for work that cannot be molded readily with conventional compression molds. Among the many desirable features obtainable by the use of transfer molds are the following.

1. Intricate section and difficult side cores may be molded.
2. Delicate and complicated inserts can be used.
3. Small and deep holes may be molded.
4. Parts molded in this manner will have more uniform density.
5. Closer tolerances may be held. This is especially true of those dimensions that are dependent on the opening and closing of the mold.
6. Parting lines will require less cleaning or buffing since there is almost no fin. This is extremely important when the rag-filled materials are used.
7. The heavy fabric or cloth or glass-filled thermosetting materials may be molded more easily and with less flash problems by using a transfer mold. However, the loss in impact strength may be a factor to be considered before choosing a transfer mold over a compression mold. See also Chapter 7 which describes special injection presses for these materials.

The cost of transfer molds is slightly above that of conventional molds since they have the pot as an added part. Piece prices also increase because of the extra material required for cull, sprues, runners, and gates. The cost

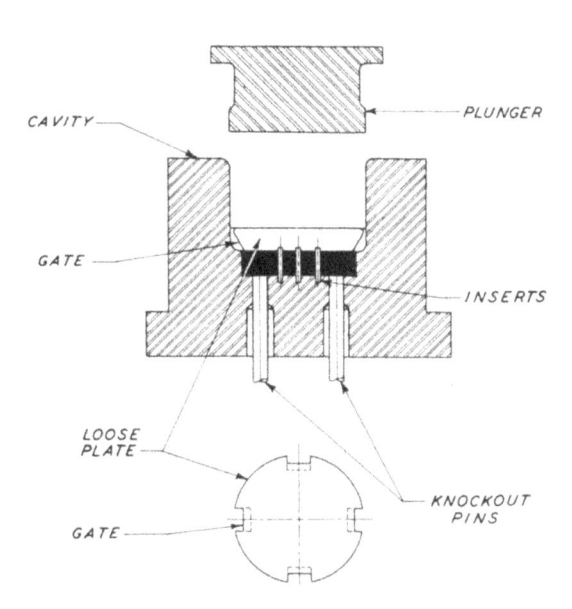

Fig. 2.8. Loose-plate or hand-transfer mold for supporting vertical inserts at both ends.

of gate removal usually will be under the normal finishing cost of compression molded parts. In most cases which involve molding problems such as those itemized above, the lower final cost of the part, after all finishing, will more than justify the additional mold expense.

A variation of the full size transfer mold is the hand-transfer illustrated in Fig. 2.8. These molds usually have a loose plate and are relatively small in size. They are used where inserts must be held at one or both ends and where quantities are in the hundreds. The primary purpose is to hold inserts and transfer the material into a closed cavity and to keep tool costs to a minimum.

In this type of mold, the gates are conveniently located around the edges of the loose plate, and the well above the plate is used as the pot for the compound. This simple mold is often less expensive than the conventional type of transfer mold and gives excellent results.

Injection Molds

The injection mold is essentially a closed mold (the mold is closed without any material in it). After the application of pressure to close the mold and hold it tightly clamped against injection pressure; the molten plastics material is forced into the closed cavity by a source of pressure other than that which caused the mold to close. The melting of the plastics material in the injective machine cylinder is called *plasticizing*. Figure 2.9 shows a molded part as it comes from the injection mold. The runner clearly shows as the cross-bar in front of the operators left arm. One gate is indicated by his left thumb. The molten material passes through the runner and gates (2) on its way into the cavity. The point at which the molten plastics material passes from the runner into the cavity is called the *gate*. You will note that we refer to "flowing into the cavity." This cavity means the space between the male section and the female section into which the molten plastics will eventually form into the desired shape and detail. The point at which the core and cavity separate or move apart when the mold is opened is called the *parting line*. Chapters 7 and 8 detail the different manners in which the material can be introduced into the cavity through a gate or gates in various locations. Each location has its advantage and disadvantages. The proper choice of gating is one of the essential fundamentals of mold engineering that must be mastered by the mold designer.

Injection molds are used for molding either thermosetting or thermoplastic materials. In the case of thermosetting material the mold is run hot, (Chapter 7), i.e., hotter than the plastics material that goes into the mold. In the case of thermoplastic material the mold is run cold, (Chapter 8), i.e., colder than the plastics material that goes into the mold. Ranges of temperature would be meaningless here because of the very wide range, meaning several hundred

FIG. 2.9. The rear end panel injection molded at Millington Plastics Company, Upper Sandusky, OH, from polyester resin weighs slightly over three pounds.

degrees between the different types of the same classification of materials. The designer should read every data sheet and manufacturing technique book that he can obtain from the manufacturer of the raw material that will be molded in the mold under design.

OTHER TOOLS FOR PLASTICS

Another plastics tool is the extrusion die. Technically speaking, this cannot be classified as a mold since it is completely different in function. The extrusion die serves as an orifice, constricting and shaping the molten material as it is forced through under pressure. In the fiberglass processing trade, a pull-trusion die is used. This means glass fibers are threaded through the die after being wetted with a resin mixture. A mechanical device then pulls the material through the die. A typical extrusion die is shown in Fig. 2.10 and its design is detailed in Chapter 10.

Blow molds are another specialized tool used in processing plastics. The

FIG. 2.10. An extrusion die used for production of continuous strips. (*Courtesy General Electric Co., Pittsfield, MA*)

original blow molds were used to blow glass bottles and shapes. Blow molding may be compared to inflating a balloon inside a box (mold), the difference being that thermoplastic material is used instead of rubber and it hardens when it contacts the cold mold surface. The inflating occurs inside a mold, usually split lengthwise, and the inflated object retains the size and shape of the inside of the mold. The blow mold is a variation of the split-wedge or split-cavity mold. The major difference is that in this case the entire mold is split, and temperature control is provided within each half of the mold. The clamping of the split halves is accomplished by action of the blow molding machine. The design and construction are quite different from those of conventional molds. This subject is covered in Chapter 11 as a separate and special problem.

The expanded-plastics, reaction injection and foam molds are treated separately in Chapter 13. Structural foam is an element of the expanded plastics and requires unique mold designs, as well as specially adapted injection machines. Here, again, the serious designer must pay close attention to the trade literature available for the many specific materials that can be foam molded. Essentially, the molds used for foam molding confine the charge while it is expanded by heat, or gas pressure liberated in the heating process, filling the space between the cavity and core to form the desired molded part.

SPECIFIC TYPES OF MOLDS

The second manner in which molds are described is by the manner in which they are to be operated. Molds can be classified as a hand molds, semiautomatic molds or automatic molds.

Hand Mold (*H.M.*)

The hand mold is described here as an historical type mold and not because many are in use at this time. The original molds from the 1850s up to 1920 were of this type because parts were quite small and quantities were limited. When hand molds are used, it is almost invariably with thermosetting materials.

As the name signifies, the hand mold is usually manipulated by hand. Hand molds are removed from the press for loading and unloading. An arbor press is frequently used to open the mold and/or to remove the molded piece after the mold has been opened. Heat is transmitted by conduction from heating platens fastened to the press. A typical method of opening a hand mold by means of die lifts is shown in Fig. 2.11. When the bottom press platen is down, it will be noted that it is level with the work table so the press operator can slide the mold in or out of the press easily. A portion of the work table forms a hot plate on which mold sections are placed to keep them hot while the operator is working on any one section.

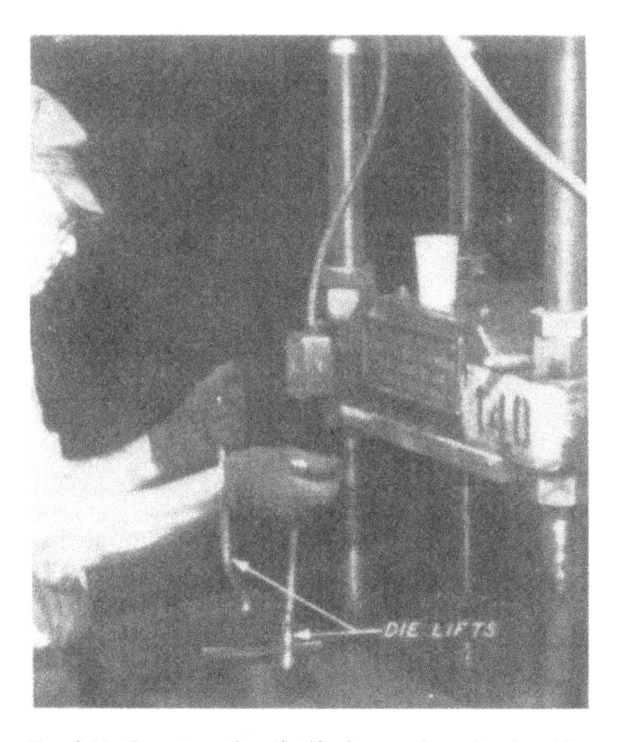

FIG. 2.11. Operator using die lifts for opening a hand mold.

Hand molds are used for small runs or experimental jobs that require minimum mold cost. They are also used advantageously for complex shapes which incorporate a number of pull pins and loose wedges. A hand mold that would weight more than 75 lb should not be designed. Obviously, such a mold would be impracticable because of the excessive physical effort required to use it. Hand molds are slow in operation and require greater labor expense than do other types of molds, and these factors serve to produce a higher molding cost. Moreover, these molds are more easily damaged by misalignment and other errors which may result from manual opening and closing of the mold. Many molders build and keep on hand standard mold frames for hand molds so that semiautomatic mold-operation (as shown in Fig. 2.34 of this chapter) may be secured for low runs.

Semiautomatic Mold (S.A.)

A semiautomatic mold is shown in Fig. 2.4. It is fastened in the press for the duration of the run. The molded piece is automatically released as the press opens, thus permitting ready removal by the press operator. Semiautomatic molds are used for production jobs where the saving in labor and molding time justifies the added cost of the mold frame and knockout mechanism.

In using a semiautomatic mold, the press operator is needed for only a brief portion of each molding cycle. The duties of the operator include excess flash removal, insert and compound loading, removal of the piece from the mold, and opening and closing of the press.

Automatic Mold (Auto.)

Automatic molds, an example of which is shown in Fig. 2.12, are similar to semiautomatic molds. They have additional mechanical features (often external) which serve to perform all operations automatically in sequence when used in an automatic press, as shown in Fig. 2.13. The compound is measured into the mold from a hopper by mechanical devices that are set in motion by a master timer. The timer operates the valves or linkage required to close the press and open it again when the molding cycle is completed. Ejector pins remove the piece from the mold cavity or core so that it may be picked up or blown into a receiving pile by external devices.

Automatic molds can only be used when automatic presses are available. They are seldom used when inserts are required to be molded in the piece. These molds are generally more expensive than semi-automatic molds but their operating cost is considerably lower. Automatic molds eliminate human error, but they may be difficult to keep in adjustment for certain

Fig. 2.12. Fully automatic, four-cavity self-degating injection mold. Pieces are stripped from core pins as mold opens, ejected out of the cavity, degated and are ready for assembly. (*Courtesy Stokes Trenton, Inc., Trenton, NJ*)

jobs. They are best adapted to jobs which can be kept in almost continuous and long run production.

MOLD CONSTRUCTION NOMENCLATURE

Molds are classified in accordance with special construction features as well as by type. Designs of this kind are given names descriptive of their unique features, and a mold of this category may fall into any of the broad classifications. It is important for the mold designer to become familiar with these features in order that he may select the best combination for each job and production requirement.

Flash Molds (*Fl.*)

A flash mold is defined as a mold type wherein the parting line is at right angles to the direction of the force that has been applied to close the mold

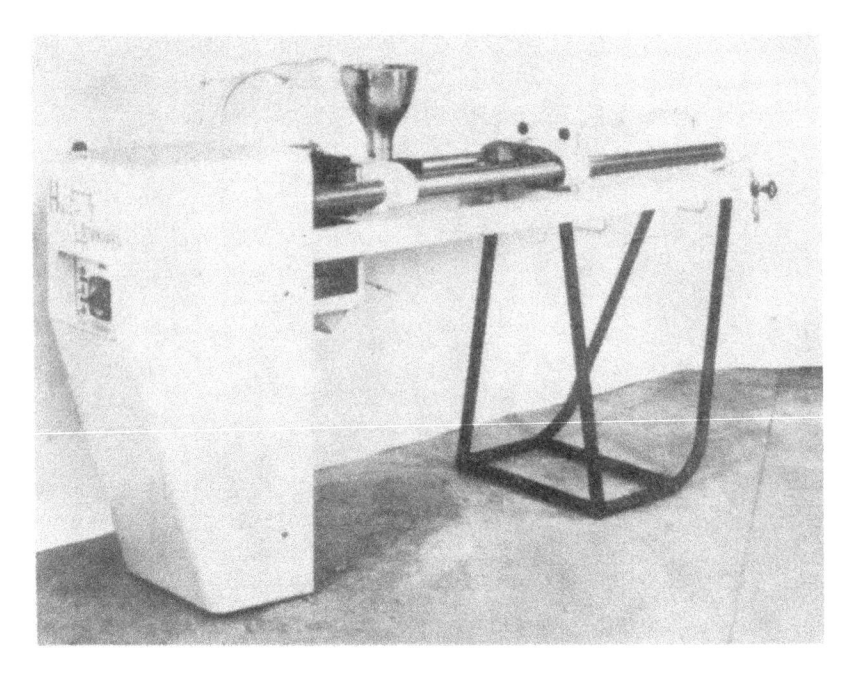

FIG. 2.13. This in-line fully automatic injection press for small thermosetting parts was first built about 1957. Although no longer a production model in the U.S.A., several units are still in service. New units have recently been made in the European market. (*Courtesy Hull Corporation, Hatboro, PA*)

and keep it closed during the curing or set time. *Curing* means the polymerization or hardening time required for a thermosetting material to harden sufficiently to be ejected from the mold. We suggest that the reader investigate in other publications the phenomenon of polymerization and the temperature control of thermosets in the manufacturing process. The design of heating channels and cooling channels is very important in the design of a mold and is discussed at the appropriate place. It becomes doubly important when maximum production is desired over long periods of time— doubly important because when cycles are counted in seconds, the saving of even one second in the cycle time can be an important percentage. Obviously, as the cycle time approaches zero, one second becomes a very high percentage of increased production.

Injection molds, transfer molds, and compression molds make use of the flash-type parting line. You will soon note the essential difference between flash mold in compression molding and a flash type parting line in injection and transfer molding. The difference is: the molding material is placed in the mold before applying pressure to close a compression mold. Whereas, *molding material* is introduced into the mold after pressure closes the injection and transfer molds. The material introduced into the cavity is injected by

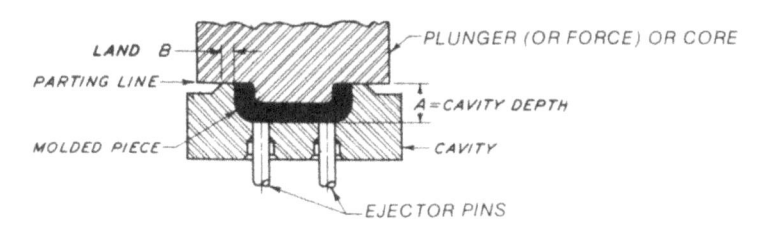

FIG. 2.14. Cross section of a simple flash mold.

pressure from some source other than that which closes the mold and keeps it closed.

A flash mold for compression molding is designed in a manner that permits excess material to escape easily as the pressure is applied. A cross section of a simple flash mold is shown in Fig. 2.14. The depth of the mold cavity A is near the depth of the finished molded part. The mold cavity is loaded with an excess amount of compound that will be squeezed out, passing over the land area B. The land is narrow, usually about ⅛ in. in width. The resistance to the flow of material provided by this constricted section is the only force which opposes the flow from the mold. This does not permit the compound to be compressed to the density desired for many classes of work.

The flash is always horizontal, extending out over the land. If the mold is closed too slowly, a heavy flash will result. This is costly to remove, and presents an unsightly appearance. If the flash mold is closed too fast, the density will be low and the parts will not have full strength.

Flash molds cannot be used for compression molding of impact materials since the loading space available is insufficient for these more bulky materials. Another disadvantage is the loss of material that results from overloading the charge to insure fair density. The flash mold is entirely dependent on the guide pins for alignment of the plunger and cavity, and, if uniform wall thickness is important, flash molds may not be satisfactory for the job. Pills or preforms are generally used for loading the flash type molds because of the small amount of space available for loading, and so that easier handling of the molding compound may be gained.

Advantages of flash-type compression molds are: lower mold cost and easy accessibility to cavities. They are especially satisfactory for small parts where high density is not particularly important. Thus a disadvantage is low density in molded parts.

Landed Plunger Mold (L.P.)

The landed plunger mold (sometimes called semipositive or landed positive mold) is similar to a flash type mold except that it has an added cavity or loading well. Figure 2.15 shows a cross section of this type of compression

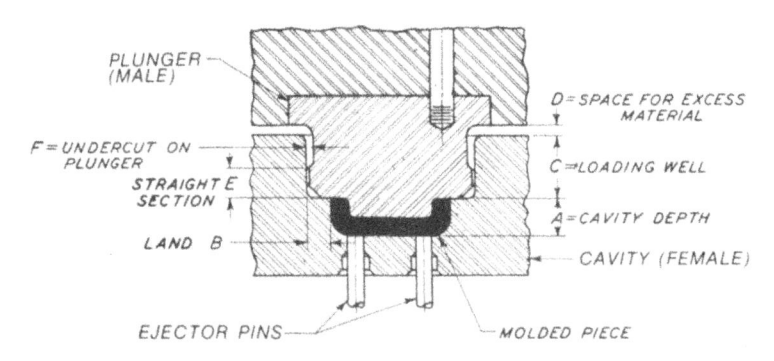

FIG. 2.15. Cross section of a landed plunger mold.

mold. The depth of the cavity A is near the size of the finished part. The added loading well is shown at C. The land B is usually 3/16 in. wide, and the space D provides for the excess of material as it is squeezed from the cavity and surrounds the plunger.

A landed plunger mold will produce parts of higher density than is obtainable with a flash type mold. Its ample loading space permits the use of the more bulky materials. Pieces containing small pins may be molded with powder in a landed plunger mold because of its greater loading capacity. For such pieces, landed plunger molds are more suitable than the flash type as the latter may necessitate the use of preforms which may break small pins and fine mold sections.

The landed plunger mold also permits control of the amount of overflow without leaving a heavy flash. This is accomplished by removing portions of the straight section of the plunger, at E, to nearly the same depth as the undercut F. Excess material will flow through the gap or overflow slot thus provided. The amount of material escaping can be regulated by the number and size of the slots.

Since the plunger does not rub along the side wall of the mold cavity proper, as it does in other mold types, there is no scoring or scratching of the cavity wall to mar the surface of the molded piece as it is ejected from the mold. This mold type saves compound when compared with the flash type, and an overload can cause no damage. The landed plunger mold is generally very satisfactory and offers loading and operating convenience.

The principal disadvantages of the landed plunger mold are its higher cost when compared with the flash mold and its unsuitability for the high-impact materials. The cost disparity is created by the extra time required in fitting the plunger to the cavity and the additional expense of machining the greater cavity depth. Landed plunger molding is not recommended for the high-impact materials because the compound which lies on the land area may absorb the pressure and fail to give a satisfactory pinch off at the parting line. Any excess of compound will produce a heavy flash, which will be found difficult and costly to remove.

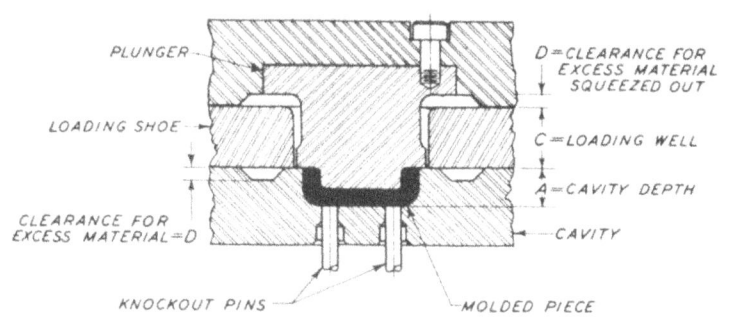

FIG. 2.16. Cross section of a loading shoe mold.

Loading Shoe Mold (*L.S.*)

The loading shoe mold is not a basic type, but a variation of the flash type mold. It probably evolved through the necessity of providing the greater loading space demanded in using the high-impact materials. This type

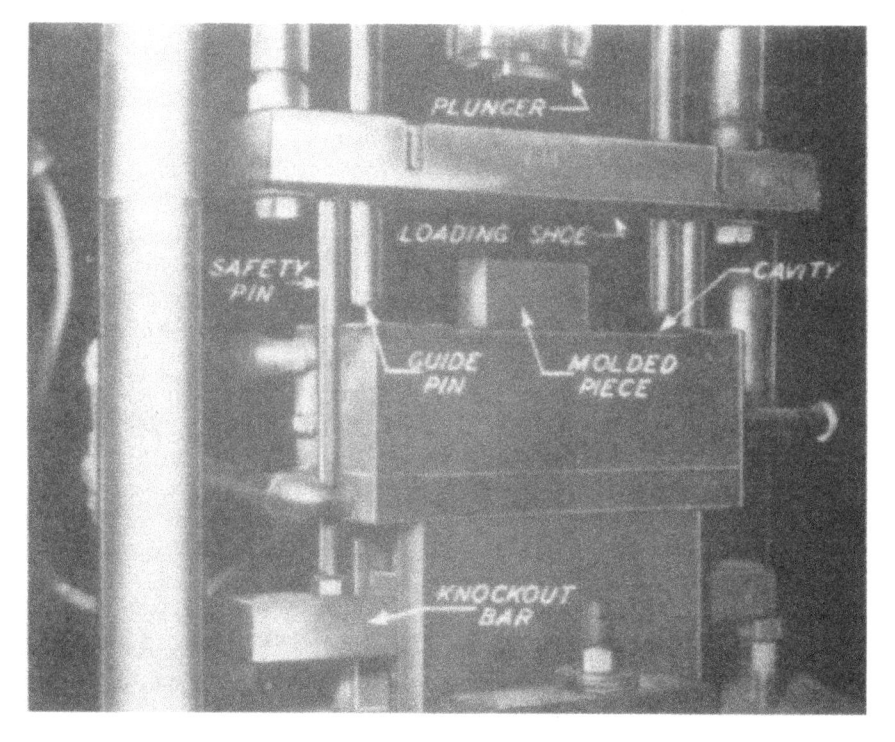

FIG. 2.17. Loading shoe mold with molded piece partially ejected. (*Courtesy General Electric Co., Pittsfield, MA*)

of mold, Fig. 2.16, consists of a cavity, a plunger, and a loading shoe. The loading shoe is merely a floating plate suspended midway between the plunger and cavity when the mold is open. Either preforms or powder may be loaded in this type of mold. Figure 2.17 shows a loading shoe with molded piece partially ejected.

The loading shoe mold offers many advantages for certain types of compression molding. The cavity is more accessible than is that of the landed plunger mold, and inserts may be loaded easily in it. The height of the cavity well is lower in this classification, but the mold will never-theless cost about the same as a landed plunger mold because of the added shoe. High-impact materials may be molded in this type of mold, therefore it is frequently used for work which specifies these materials, although the positive mold next described may provide even greater advantage.

Multiple-cavity loading shoe molds are not recommended because temperature differentials may cause binding of the loading shoe. Flash may cause the loading shoe to stick to the top plunger when the mold is open, thus creating danger of accident should the operator reach in before the loading shoe arrives at its normal position.

Stripper Plate Mold

This type of mold construction is similar in some respects to that used for the loading shoe mold and the removable plate mold. It is functionally operated in the same manner as a loading shoe mold.

The fundamental differences between a stripper plate mold and a loading shoe mold is that the stripper plate fits the core at the inside of the molded part (or somewhere between the inside and outside, but never larger than the outside dimensions of the molded part). Conversely, the loading shoe is nearly always larger than the molded part dimensions and it performs no function in ejecting the part from the mold. The primary purpose of a stripper plate is to eject the part from the mold without distorting it or without the presence of objectionable ejector pin marks. The stripper plate is frequently used for parts with thin wall sections (meaning .10 to .40 in. approx.) when the part cannot be ejected by means of ejector pins. These pins may pierce the thin wall if the part does not readily release from the mold. This trouble is frequently encountered in injection molds used for the thermoplastic materials, which are often quite soft and easily pierced or distorted when removed from the mold. Stripper plate molds are not often required for the thermosetting materials because the finished piece is hard and consequently ejector pins will serve satisfactorily for ejection of parts. When ejector pin marks would be inimical to appearance, decorative or functional parts will require stripper plate mold construction.

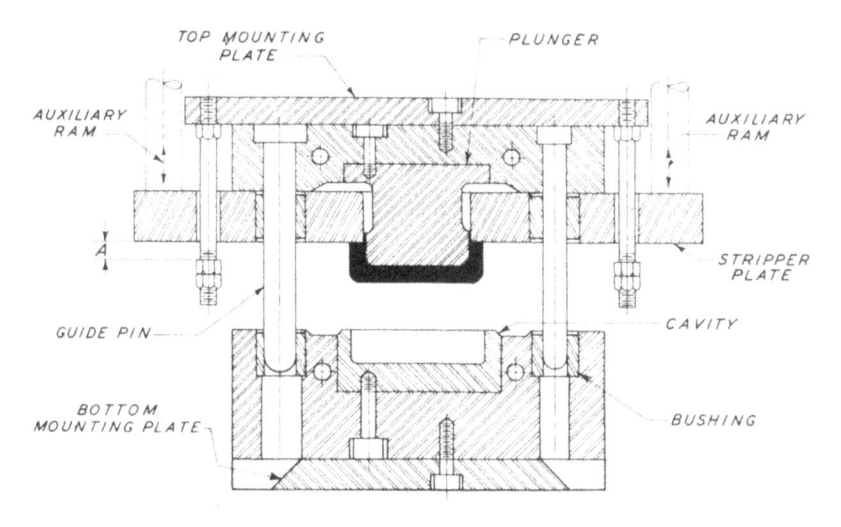

Fig. 2.18. Diagram showing construction used for a stripper plate mold. The stripper plate is pushed down by auxiliary press equipment, causing molded parts to be "stripped" from plunger.

Molding by this method should be confined to units which contain only a small number of cavities, as temperature differentials may cause binding of the plates. Conversely, large numbers of cavities would require special attention to minimize differential expansion problems.

The action of the stripper plate in a compression mold is shown in Fig. 2.18. As the press opens, the stripper plate will be moved down. This motion pushes or strips the molded pieces off the mold plunger. The area of free action of the stripper plate is limited, as indicated at A. This control prevents the plate from being pushed too far down. The stripper plate may be operated by an auxiliary ram or it may be operated by the opening and closing of the press.

Figure 2.19 shows a stripper plate injection mold in which the stripper plate is operated by the opening and closing of the press. (B) shows how the stripper plate fits around the mold parts; (A) shows the mechanism used for moving the stripper plate. Note that the molded part would be in the right end at (A), and in the left side at (B).

Uniform ejection of the molded piece is at all times important. Much of the dimensional accuracy of the piece may depend on uniformity of the ejecting force. Proper ejection of the piece from the mold always presents a problem, and it has been said wisely, "One piece can always be made in a mold, but getting the part out of the mold in one piece is another problem."

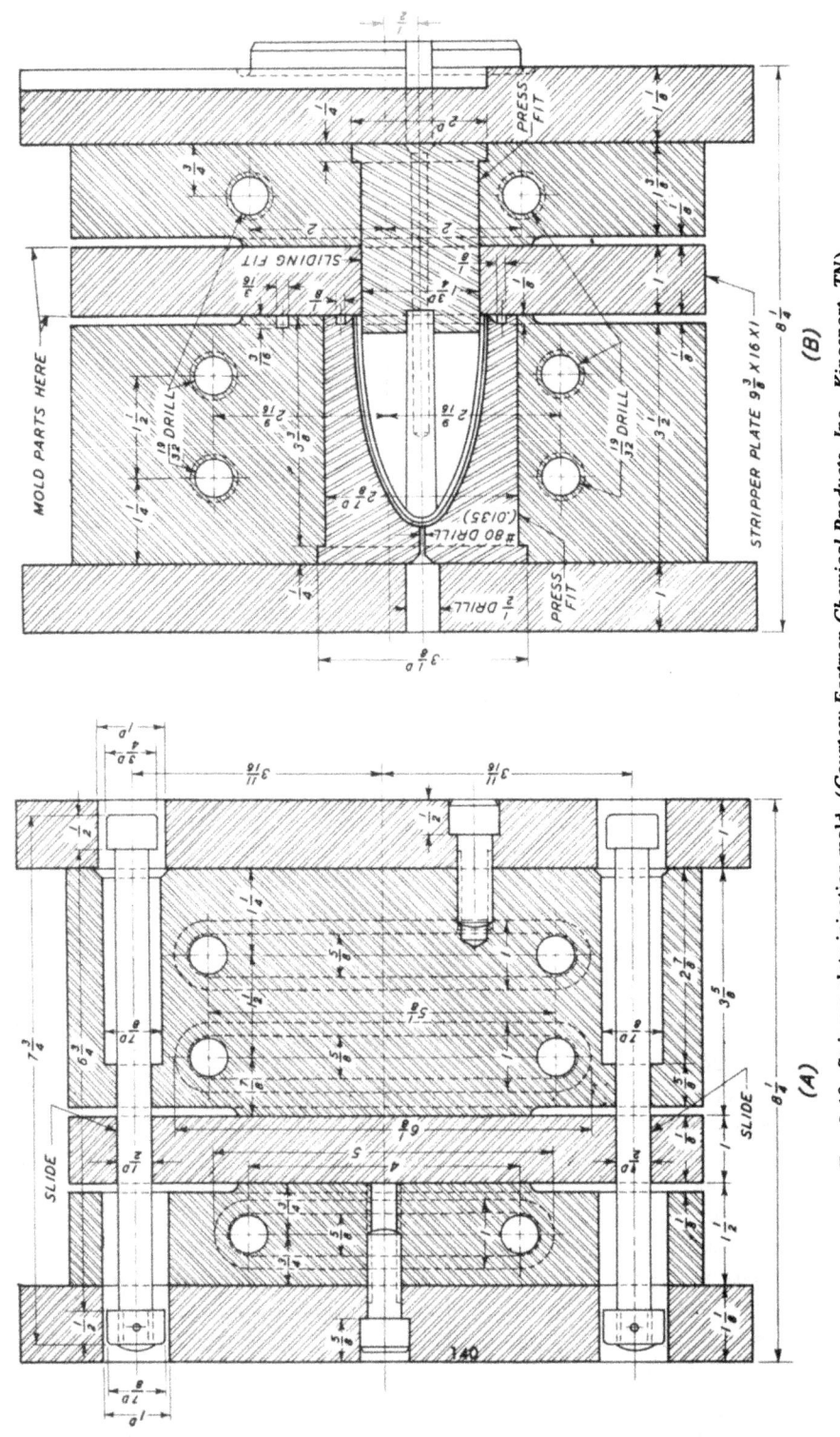

FIG. 2.19. Stripper plate injection mold. (*Courtesy Eastman Chemical Products, Inc., Kingsport, TN*)

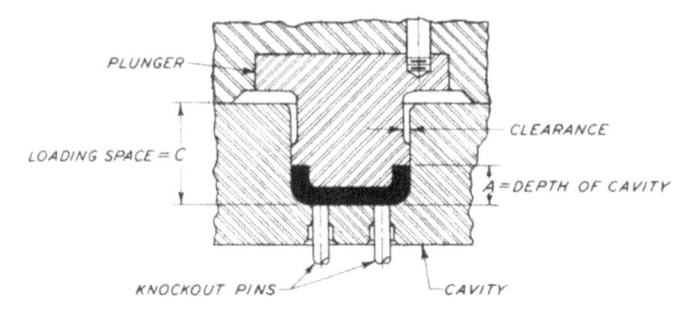

FIG. 2.20. Diagram showing construction used for a positive mold. Ejector pins could extend through plunger instead of through cavity.

Positive Mold (*Pos.*)

The positive mold, shown in Fig. 2.20 operates like a piston in a cylinder—the cavity being the cylinder and the plunger the piston. It is used chiefly for the cloth-filled materials and for parts having a long draw, such as radio cabinets. In using a positive mold, the compound must be weighed carefully since there is little means of escape for any excess compound. The plunger fits the cavity closely, with approximately 0.003 in. per side clearance. The positive mold is preferable to any type of landed mold for molding the cloth-filled materials. In attempting to mold cloth-filled materials in a landed mold, it will be seen that the small pieces of cloth will not pinch off in the land area and so will absorb pressure and prevent closing of the mold. This will cause a heavy horizontal flash line to be produced. The positive mold flash line is relatively light and has the advantage of being vertical and more easily removed. Since there is no land in a positive mold, all the pressure is used for compressing the compound. The flash produced in a mold using cloth-filled materials is difficult to remove because the fabric must be cut away—it cannot be broken by tumbling.

The positive mold should not be used generally for simple designs or for parts made from free-flowing material because other types of molds will be found better adapted to these jobs. An important disadvantage in the use of the positive mold results from the plunger rubbing the side wall of the cavity. This scores the cavity wall, and may mar the molded piece as it is ejected. Considerable difficulty is experienced in operating multiple-cavity positive molds since there is no means of equalizing the pressure differentials caused by unbalanced loads of compound. Other types of molds equalize the unbalanced loads by the overflow which they are able to accommodate at the flash line.

It is also essential in the design of a positive mold to allow wall thickness in the bottom and sides of the cavity. It is possible, for example, in a four-cavity mold, to have the material unequally loaded in the individual cavities and thus all of the press pressure falls on the cavity with the most material in it. Guard against this error by overdesign in the mold.

Semipositive Mold (S.P.)

The semipositive mold for compression molding, as shown in Fig. 2.21, is a combination of landed plunger and positive molds. The cavity well B (right-hand sketch) may be large enough for any desired compound. The positive portion of the mold, at A, usually is dimensioned 1/16 or 3/32 in. This mold design permits the load of compound to vary within reasonable limits and still produce good molded parts. The excess material is permitted to escape from the cavity up until the positive portion of the plunger enters the positive portion of the cavity. During the travel of the plunger through A, the mold is positive, and all the pressure is now applied to the compound, which is trapped in the cavity.

The semipositive mold serves a wide variety of applications, and is especially desirable for parts requiring a long draw, for parts which have a heavy section at the bottom of a deep piece or parts which must combine heavy and thin sections. This mold tends to produce a better surface finish on the medium-impact materials because of their higher density, and also because the greater pressure appears to force the resin to the surface. Another advantage of the semipositive mold is that it produces a vertical flash line which may be removed easily by finishing on a belt sander. The landed plunger mold produces a horizonal flash line that can be removed only by spindling or filing, and these operations are far more costly than sanding.

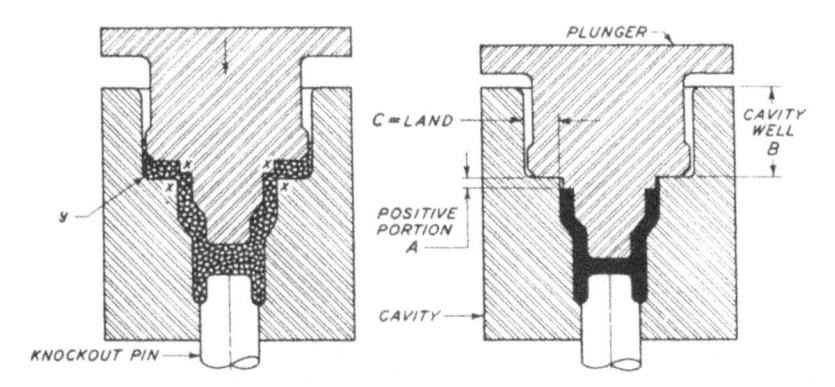

Fig. 2.21. Semipositive mold as it appears in partly closed position (left), just before it becomes positive. Compound trapped in area y escapes upward through overflow grooves in plunger. Mold in closed position (right). When corners at x pass, mold is positive, and practically all of entrapped compound is compressed.

Molds used for the melamine and urea compounds should be semipositive or positive if transfer molds are not required. These compounds require both heat and pressure to bring them to the plastic state, and good molded parts cannot be obtained without causing the compound to flow under continuous pressure, or without keeping it sealed in the cavity during the entire curing cycle.

The several types of molds described in the foregoing pages are the fundamental types which, with various modifications, are used for all mold construction. Many variations have been devised to meet special problems. These molds were developed to meet the compression molding problems which require the mold to compress the material and at the same time prevent its escape as the mold closes. Transfer and injection molds are closed tightly before the compound is introduced into the cavity, therefore they may be disregarded in discussion of these special problems. As stated previously, the transfer and injection molds usually adopt designs similar to the flash type mold since no extra loading space is required and the simple butt type of parting line is satisfactory.

SPECIAL MOLD CLASSIFICATIONS

A variety of mold types has been developed for special classes of work. These special types were devised to reduce costs or improve operating conditions, or to facilitate the molding of complex shapes which may not be molded easily in the more simple molds.

Subcavity Gang Mold

As the name implies, this mold consists of a group, or "gang," of cavities located below a common loading well, and it is used in compression molding. A hundred or more cavities may be contained in each gang of a mold designed for small pieces. Such molds are frequently built with from three to six gangs, each gang containing fifty to one hundred cavities. The cavities are located at the bottom of a loading space, as shown at *A* in Fig. 2.22. This loading space is large enough to hold the compound required by all cavities. As the mold closes, the compound is plasticized by the heat and pressure so that it flows into the cavities, leaving a thin layer of flash connecting all the pieces, as shown in Fig. 2.23. This flash enables the press operator to pick up the load from each gang as a unit instead of picking up the individual pieces. The flash is then removed by tumbling or some other finishing process.

The subcavity gang mold is especially advantageous for small pieces since it facilitates loading the many cavities and picking up the finished

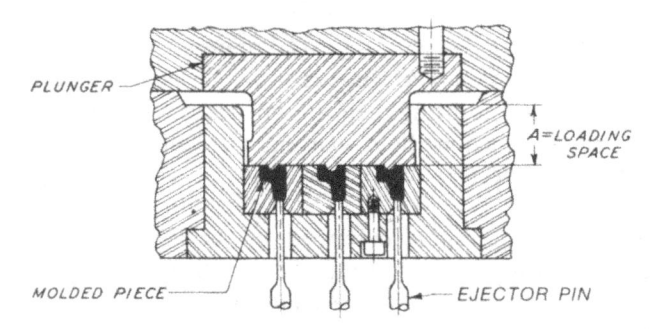

FIG. 2.22. Diagram showing construction of a subcavity gang mold containing three cavities.

pieces. A single preform may serve as a charge for each gang, or for the entire mold charge. Materials that are difficult to preform are easily loaded in bulk in this mold. Mold cost is low inasmuch as machining operations for provision of separate loading spaces above each cavity are saved. Many simple designs permit the use of a single plunger for each gang and this effects considerable saving as it eliminates the costly work required to locate the plunger for each cavity and saves the cost of machining individual plungers.

Some disadvantages are found in the subcavity gang mold since the density of parts made in this type of mold may not be as high as that of parts made in other types of molds. Large area loading wells may produce light parts because of the long flow required to bring the compound to the end or corner cavities, although this advantage may be overcome by using a preform of the approximate size and shape of the loading space. The area of the land should not be much over one half the area of the piece for many medium-sized pieces. A good general rule to follow, is to keep the cavities as close together as tool-strength calculations permit, and to keep the size of the loading well to a maximum dimension of 2½ to 3 in. by 4 to 5 in. This will provide an area of 10 to 15 sq in., which experience has shown to be generally satisfactory.

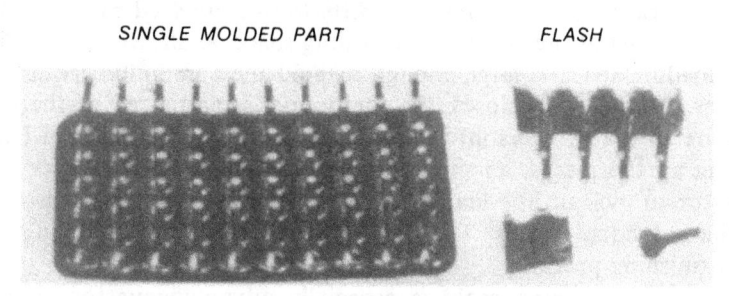

FIG. 2.23. Left is one molding from a 60 cavity subcavity gang mold. These pieces will tumble free from connecting flash. This construction avoids the necessity for loading each cavity individually.

It must be remembered that it is impossible to squeeze all the material from between two pieces of steel. No matter how great the unit pressure may be, there is still a limit to the thinness at which flash may be held at the pinch-off point. This limit has been found to be 0.003 to 0.005 in. for the free-flowing wood-flour phenolics. Therefore, if the land area of a sub-cavity gang mold is increased to much over one half the total surface area of the molded part, a heavy flash will result and, of course, this means increase in molding and finishing costs as well as a greater number of rejections. An alternative, in the event of excessive land area, is to use a larger press and greater pressure. This, however, provides greater unit pressure on the land area and causes breakdown or deforming of the land. When breakdown occurs, it will be necessary to make new cavities because finishing costs would be exorbitant otherwise, or because the molded piece itself, as a result, will not meet the required dimensional tolerances.

Split-Wedge or Split-Cavity Mold

Split-wedge or split-cavity molds are used in all types of work to facilitate the molding of pieces that have undercuts or projections which prevent removal by a straight draw. A split-cavity mold is shown in Fig. 2.24. This

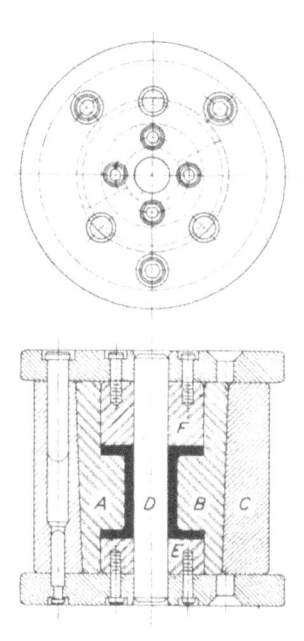

FIG. 2.24. Cross section of a split-cavity mold for producing a molded spool. A and B form the two halves of cavity; chase, C, holds the halves together. Other parts are: D, core pin; E and F, bottom plate and force plug, respectively.

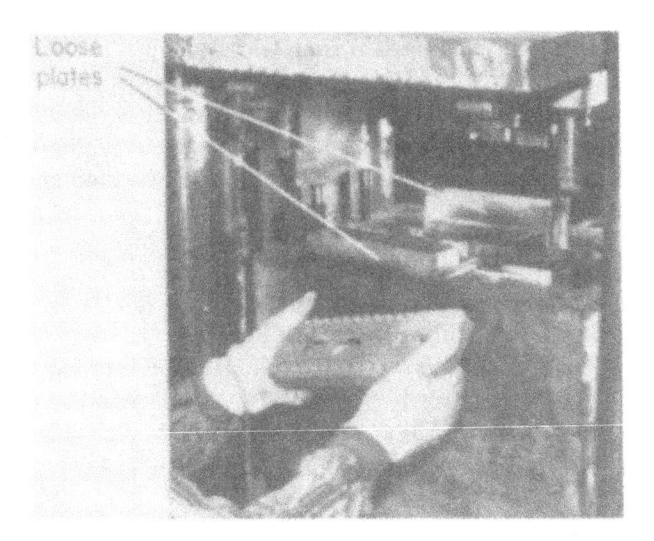

FIG. 2.25. The loose plates shown above are ejected with the molded piece and then reinserted in the mold. These loose plates make the depressed numerals on the side of the part. A tapered section locks them tightly against the central section. (*Courtesy Geo. K. Olson Studio, Florham Park, NJ*)

mold is used for the production of a spool. The entire cavity sections, *A* and *B*, are extracted from the mold as a unit and then separated for removal of the molded part. It will be noted that these sections are made with a taper to fit the chase or housing, *C*. Tapered sections are favored over straight sections because the force of the molding pressure wedges the two halves tightly together and reduces the thickness of fin that can form at the parting lines. As mold wear occurs, compensation may be made in many parts by the removal of a few thousandths of an inch from the bottom of the sections. This allows the taper to seat more deeply in the housing. Mold sections and other loose parts are generally called wedges. Figure 2.25 shows a split wedge or loose wedge mold used for producing a switch base of diallyl phthalate compound with lettering on the edge that prohibits vertical or "straight draw" of the part for ejection. The front and rear cavity wedges come up with the molded piece and are subsequently reinserted in the mold after removal from the piece.

Figure 2.26 shows the type of wedge used in molding side protrusions and side holes. The side boss *A* and the ring *B* would prevent this part from being molded without the use of split wedges. Note that these wedges have two locating pins, *C*, each of different size. These serve three purposes:

1. They reduce the possibility of damage to the edges of the wedge sections since these two sections must be put together with a straight push over the locating or guide pins.

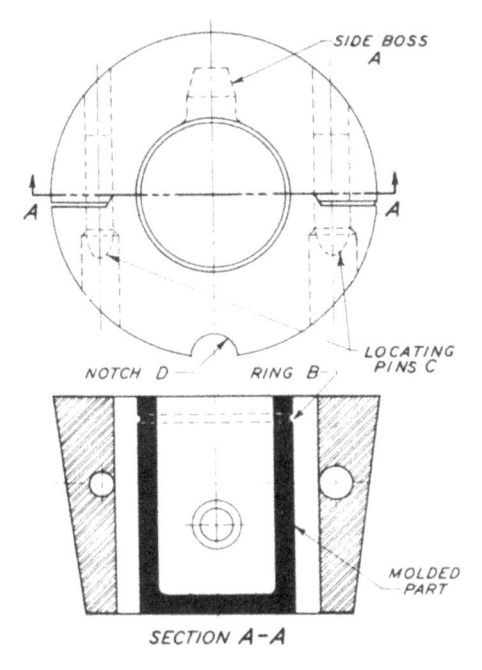

SECTION $A-A$

FIG. 2.26. Split-wedge mold section used to mold part which contains side protrusion.

2. They make it easier for the operator to keep the wedges together while removing or replacing them in the mold.
3. They prevent the operator from trying to insert the two halves in the mold with one half upside down.

The notch D registers with a projection in the wedge retaining shoe accurately to locate the side boss A with respect to other sections of the mold.

The molded sides holes, as shown in Fig. 2.27, are generally molded with a removable wedge to produce the hole, as this will obviate the necessity for using split cavity construction. This type of mold is easy and inexpensive to operate because its lighter weight minimizes operator fatigue and there are fewer parts to be handled.

Figure 2.27 at (A) and (B), shows molds of identical construction except that at (B) the wedge is not fully removable when the mold opens. At (B), the wedge must be raised high enough to permit the molded piece to be pulled free. As the press opens, the plunger withdraws from the cavity; then, as the press continues to open, the pin attached to the knockout bar rises and pushes the wedge, with molded piece attached, up out of the cavity. When the press is fully open, the wedge is up, free from the cavity, where the piece may be removed readily by sliding it to the right.

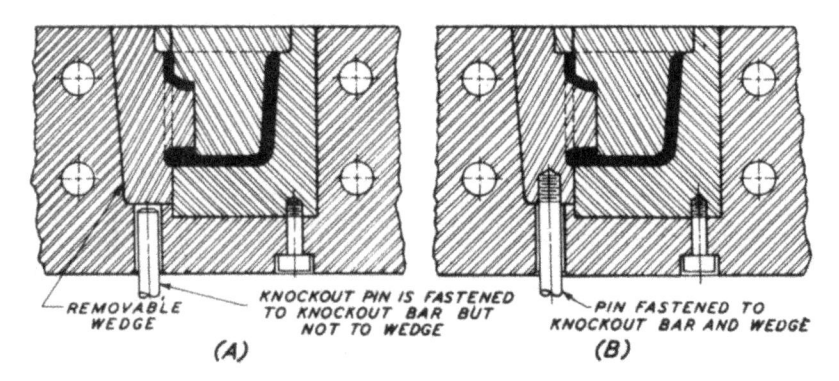

FIG. 2.27. Wedge-type mold used for producing side holes in molded pieces. A removable wedge is shown at (A); a fixed wedge, at (B). The knockout pin raises wedge out of cavity for removal of part.

For most applications, the construction shown at (*A*) is slightly better than that of (*B*), as it is less difficult for the operator to make certain that all the flash has been blown out of the mold.

Removable Plate Mold

The removable plate mold construction is used most frequently for the production of parts have molded threads. This construction, Fig. 2.28, makes use of a plate which is taken out of the mold where it may be worked on for release of the molded pieces by unscrewing them or by using special ejector fixtures. To facilitate production, two plates are used in most cases, as the parts may be removed from one plate during the curing period of the other. These extra plates are used extensively when several inserts must be threaded into the plates. The use of the extra plate will, in most instances, give a fifty per cent increase in production.

Molds which use this construction must not be too large as excessive weight will make the plate unwieldy and overheavy to handle. Twenty pounds is a desirable maximum weight for the removable plate mold, although fifteen pounds is considered better. When heavier plates are required, they should be designed to slide from the mold onto a track so the operator will not have to lift them. Hinge designs may be added, enabling the operator to slide the plate out of the mold and swing it to the vertical position where the molded pieces will be readily accessible.

The extra plate is kept on a hot plate while out of the mold so that it will, as nearly as possible, retain mold temperature. This is done so as not to lengthen the molding cycle when the plate is put back in the press.

FIG. 2.28. Removable plate or tray-type mold construction is used for compression, transfer, and injection molds to facilitate loading of inserts. In this transfer mold, the Station (X) withdraws mold core pins and ejects pieces. Inserts are then loaded at this station while the duplicate tray is in the press for the molding cycle. (*Courtesy Shaw Plastics Co., Berkeley Heights, NJ*)

Removable Plunger Mold

A variation of the removable plate mold is the removable plunger mold. This mold type makes use of a sliding-plate construction to permit removal of the mold plunger during the molding cycle. This design is useful for jobs which require many inserts to be loaded on the plunger. It may be used advantageously to permit the removal of deep parts from the plunger. This is especially helpful when the press can not be opened up to the full depth required for removal of the part.

Swing Mold

The swing mold is a variation of the removable plate mold, one half of the plate being held in the mold while the other half (which contains a duplicate set of cavities) extends out of the mold, thus permitting easy removal of the parts. This plate revolves on a central axis. Its principal advantage over the removable plate mold is that it does not have to be lifted by the operator. Instead, it is raised by action of the press, and little effort is required to rotate it to the opposite position. The swing mold is generally selected when the extra plate would be too heavy to be handled easily.

Stack Mold

When two transfer or injection molds are operated in tandem in the same molding machine, they are called *stack molds, two-level molds* or *tandem molds*.

Spring Box Mold

The spring box mold is a variation of the preceding mold types. Its basic construction may be landed plunger, loading shoe, positive, flash, etc. This construction is frequently used to solve difficult molding problems involving inserts.

These problems usually have to do with holding the insert in its proper location. As mentioned previously, the plastic material flow may exert a great force in the mold. This force may be sufficient to dislodge or carry with it any fragile projections which lie in its path. An obvious solution of this problem in compression molding (where the greatest trouble is experienced) is to keep the inserts out of the path of the flow. When the flow of compound has practically ceased, the inserts may be moved into the still plastic compound, where they will be held securely on completion of the compound cure.

Figure 2.29 at (*A*), shows a spring-box mold in its normally open position; at (*B*), the same mold fully closed. In the open position, the insert and the molding compound have been loaded in place by the conventional procedure and the mold is ready for the pressure to be applied. A spacing fork is inserted between the mold cavity and the parallels, as shown. This fork will prevent compression of the die springs and will hold the insert below the bottom of the cavity by the distance *A*. As the plunger (not shown) enters the cavity, a complete molded shape will be formed except that the insert will not be in its proper place. After the compound has reached its plastic state, and before curing begins, the pressure is momentarily released by the operator so that the spacing fork may be pulled out, as shown in Fig. 2.30. Immediately after, the pressure is applied again. This time the die springs will be compressed and the cavity assembly forced down so the insert will remain in its proper position (see *B*, Fig. 2.29) throughout the cycle. After the part has been cured, release of the pressure will permit the die springs to open the gap between the cavity assembly and the parallels so the fork may be placed in position for the next cycle.

A spring box mold may also be used for other purposes. Some parts may require spring box construction to provide greater density where the bottom of a part is long and thin-walled. The spring box action in this case

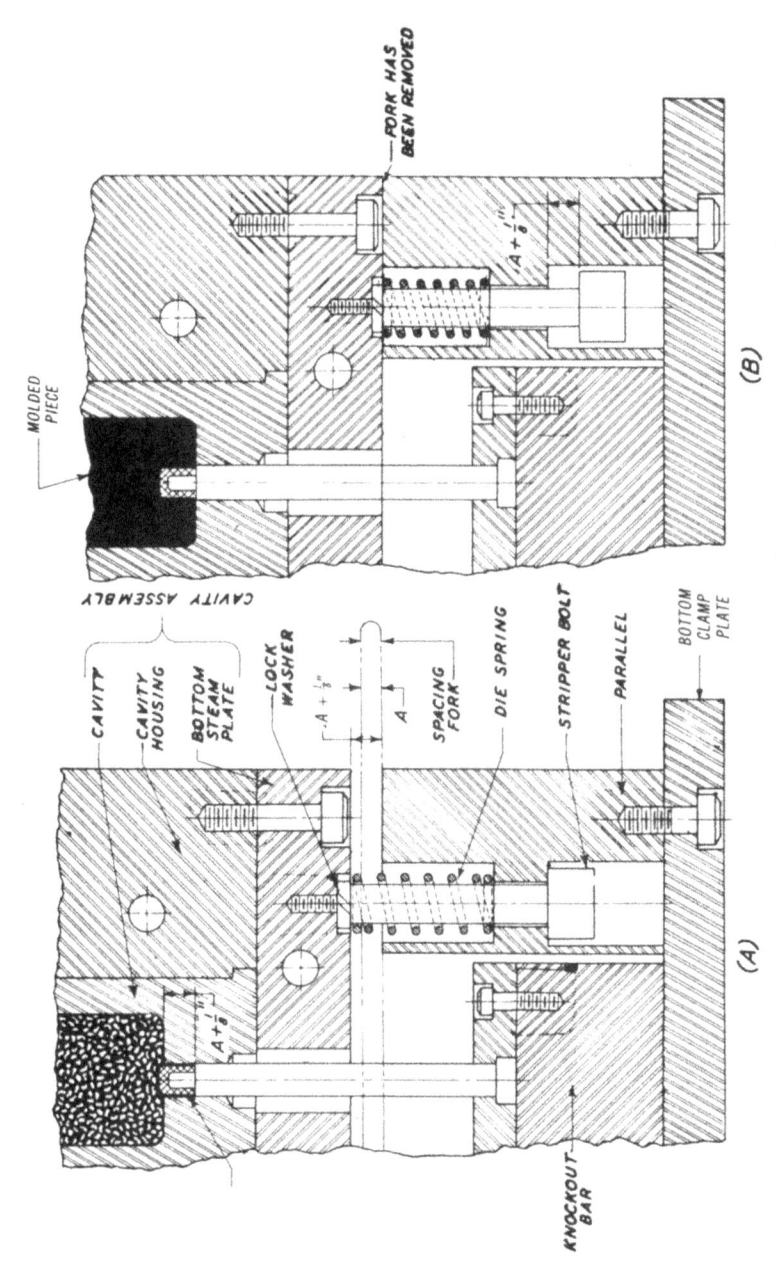

FIG. 2.29. Diagram showing construction of a spring-box mold. (A) Mold open, without pressure applied. (B) Mold closed, with pressure applied.

FIG. 2.30. Pulling out spacing fork used in spring box mold. (*Courtesy General Electric Co., Pittsfield, MA*)

provides the extra pressure needed to insure full density after the normal flow takes place.

Double-Ejector Molds

It is generally possible to design a mold so that the molded part will stay on the plunger or in the cavity. In some cases it is desirable to provide a double-ejector arrangment in order that the piece may be ejected from the cavity or plunger. The design of the piece may not permit the use of pickups, and, therefore, the piece may stick to either part of the mold.

Double-ejector designs are also desirable when inserts are to be molded in the top and bottom of a piece and the length of the plunger will interfere with the loading of inserts in the top. This is illustrated in Fig. 2.31. The top ejector pins extend down to the bottom of the plunger when the mold is open so that inserts may be loaded readily on the pins. In like manner, the bottom ejector pins extend up out of the cavity when the mold is open to permit easy loading of the inserts.

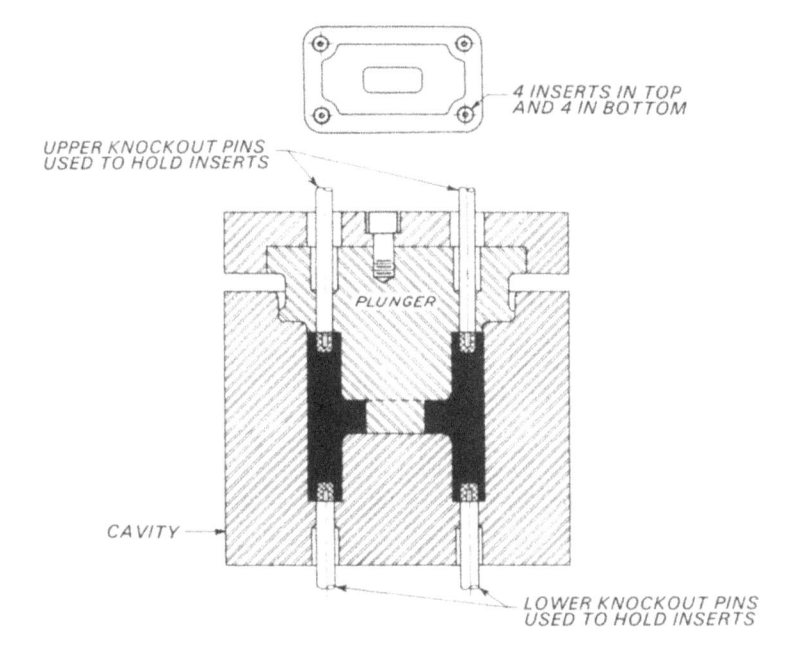

FIG. 2.31. Diagram showing double-knockout mold construction used to hold top and bottom inserts. It would be very difficult to reach behind long plunger to load inserts, hence ejector pins, which can be lowered for loading, are used.

In some cases the top pin is used as a hold-down device to prevent compound from entering the threads of a through insert, or for keeping the face of the insert clean. A diagram illustrating this construction is shown in Fig. 2.32, and the mold is shown in Fig. 2.33.

Standard Frame Molds

Many molders and mold-makers utilize standard mold frames for reasons of economy. A savings in cost may be achieved through use of these standard frames since they can be produced in production lots. This permits the skilled mold-maker to devote his time and equipment to the cavity and force details. Some molders with large tool shops design a line of standard frames for their general requirements and make them up in production runs at regular intervals. Trade literature is the best source for determining the availability of suitable standard frames for compression, transfer, and injection molds. A wide variety of sizes, shapes, and materials is available either as standard or near-standard stock items.

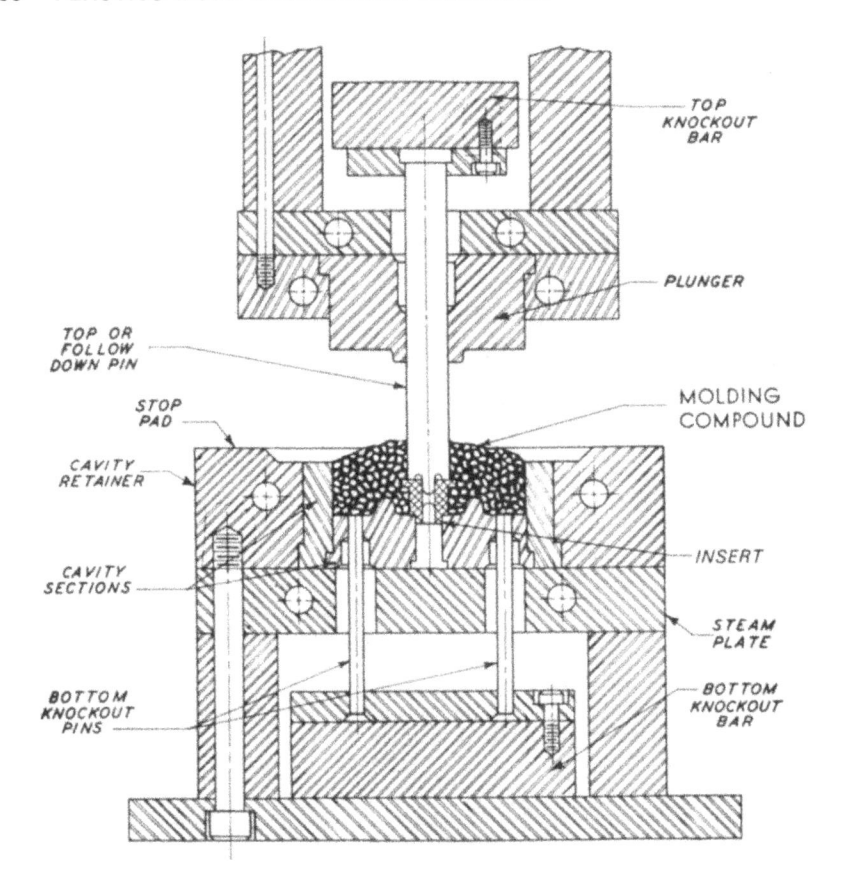

FIG. 2.32. Diagram showing use of upper and lower ejector pins. Top pin is not used as an ejector, but serves to keep face of insert free of flash.

One type of standard frame mold is more correctly called a standard frame for hand mold. This is actually a conversion unit which serves to permit the use of a hand mold as a semiautomatic mold. These units are usually limited to small multiple-cavity or medium-sized single-cavity molds. Ejector pins are made a uniform length so that the frames will fit any of the several molds which may be inserted.

These systems offer substantial economy for small compression molds. The initial mold cost is comparatively small as there are fewer parts to make. The common standard conversion frames are adaptable to various molds. These molds can be operated economically since they operate as semi-automatic molds and the mold investment is limited to the hand mold cost.

FIG. 2.33. Two-cavity, double-ejector mold. Lower ejector pin is used to hold insert and eject molded part. Top ejector pin is then dropped to cover top of insert before compound is loaded in cavity. This system keeps inserts clean. Parts are: 1. Upper ejector bar; 2. plunger; 3. top ejector pin; 4. cavity; 5. compound; 6. bottom ejector bar; 7. inserts; 8. molded parts. Note that inserts are free of flash.

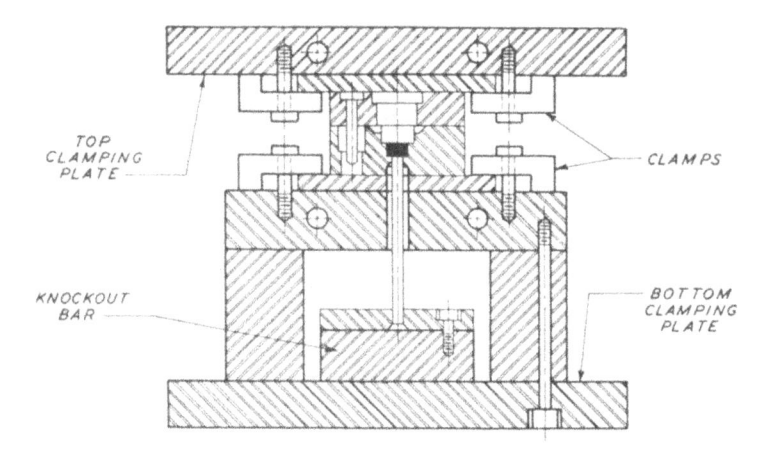

FIG. 2.34. Diagram showing design of a standard frame for hand molds. A variety of hand molds may be constructed for use in this frame.

Mold storage space is reduced by the use of these hand molds, which are mounted in standard frames. The principal disadvantage is found in having to employ a tool-maker to mount the mold in the frame. This usually requires one to two hours' time.

Other molders make use of mold plate assemblies which are mounted with grids in the press. These grids are frequently cast iron spacer blocks which are made in several sizes to permit the use of knockout pins of varying length. The grid type of construction obviates employment of highly skilled tool-makers in making the set-up, although this work still requires skilled or semi-skilled set-up men. Molds which are set up in this manner are usually multiple-cavity molds for small pieces, or single-cavity molds for larger parts.

Low-pressure Molds

The low-pressure or reinforced-plastics molding process makes use of very simple molds since the pressures exerted are low. These molds are not often heated and rubber bags are frequently used for one of the mold members. The forms (or moids) used for this work may be built up from wood, plaster or metal. These mold sections are often produced as metal spinnings and they are frequently cast from iron by conventional foundry practice. Molds for this work may also be constructed by the electro-

Fig. 2.35. Custom designed molds. Left and center: Standard frame unit with unit mold insert for plunger transfer machine. Right: Automatic compression mold halves for Stokes automatic machines. Comb and loading board are shown in the rear. Right foreground: Semi-automatic injection mold with split wedges operated by finger cams. (*Courtesy Tech Art Plastics Co. Morristown, NJ*)

plating process. The rubber bags are sometimes formed in these mold sections, and, in some cases, they are tailored from sheet stock. This type of mold design is relatively simple and therefore will not be described in detail in this book.

Individual Preference

Each molder adopts some general practice which is predicated on the type of work he does, the equipment and labor available, and the experience he has had in application of the various methods. In all cases the mold design selected must produce quality molded parts at minimum cost. Many features of mold design are flexible, therefore procedures may be determined by the materials and molding methods available. For some jobs there is only one right way to design the mold and the tool-maker must weigh the relative advantages of every possible design before he decides which is best for the job at hand. A variety of contemporary molds is shown in Fig. 2.35.

REFERENCES

Glanville, A. B., and Denton, E. N., *Injection Mold Design Fundamentals*, 2nd Ed., New York: Industrial Press, Inc., 1965.
Modern Plastics Encyclopedia, 1986–87 Ed., New York: McGraw-Hill, 1986.
Sors, Laszlo, *Plastics Mold Engineering*, New York: Pergamon Press, 1967.
Thayer, Gordon, *Plastics Molds*, Cleveland: Huebner Publications, 1944.

Chapter 3 / Tool-Making Processes, Equipment and Methods

Revised by Paul E. Ferland

The mold-making industry is comprised largely of shops that have developed special services. Custom mold makers are specialists since they provide services most frequently required by their customers in a particular area or market. Metal-working machines are available in great variety and new processes and equipment are developed every year. Few shops can afford the investment in each type of equipment that is most efficient in building a specific mold. It would be impossible to have all the human skills of mold making represented in one shop.

Specialization by processes and related skills keeps the cost of modern mold making at a reasonable level. For example, there are shops that build only standardized or custom mold bases and components for sale to the mold-making industry (see Chapter 8). The custom shop does not need to invest in the large equipment such as the large millers or rotary surface grinders needed to build precision mold bases. With over a thousand different sizes of standard bases for selection from a catalog, and fast deliveries from area warehouses, the local mold shop can concentrate on preparing cavity inserts and the final assembly steps in building a complete mold.

Custom mold shops are generally equipped for fabrication of molds by the basic and popular process known as *machining* or *metal removal*. However, most shops frequently depend upon other specialists in the trade for services such as hobbing, EDM machining, electroforming or casting of cavity inserts that cannot be machined economically. Hobbing is a metal displacement process and electroforming is a metal deposition process. Local mold suppliers are cooperative in explaining the use and application of their equipment and skills to the mold designer. They can also explain, or help the designer to obtain information about other special processes as they affect mold design, mold efficiency, and cost.

The principal ways in which metal can be processed into components of a finished mold are:

1. Metal-cutting which is the removal of metal from a piece of stock by mechanical application of force on a tool with one or more cutting edges.
2. Metal-displacement processes where mechanical forces are used, but no cutting tool is involved.
3. Metal-deposition processes where a pattern or master shape is utilized and the metal is deposited over or around the master to form a metal reverse shape.
4. Casting, pressure casts and Shaw Process.
5. Chemical erosion or photoengraving.
6. Electrical discharge machining (EDM).
7. Miscellaneous processes such as bench polishing and hand engraving.

FIG. 3.1. Power hacksaw.

CUT-OFF EQUIPMENT

A power hacksaw (Fig. 3.1) or band cut-off saw is most commonly used for cutting bar stock to the desired rough size. A mold maker may or may not invest in a power cut-off saw depending upon the availability of steel stock in a local metal supply warehouse. The supplier will certainly have one (Fig. 3.2). These machines are motor-driven and apply a metal-cutting blade or band to the bar stock. Some of the newer machines feature an automatic feed between successive cuts of the same length, with a pre-set number of cuts.

An abrasive cut-off machine is found in some shops. In this machine a narrow motor-driven abrasive wheel moves into the stationary bar, making a very smooth cut with little waste. This is a most efficient way to cut small pre-hardened steel bars, certain alloys, or hardened standard ejector pins to approximate length before final machining and fitting in a mold assembly (Fig. 3.3).

FIG. 3.2. Band cut-off saw. (*Courtesy Ethyl-Marland Mold, Pittsfield, MA*)

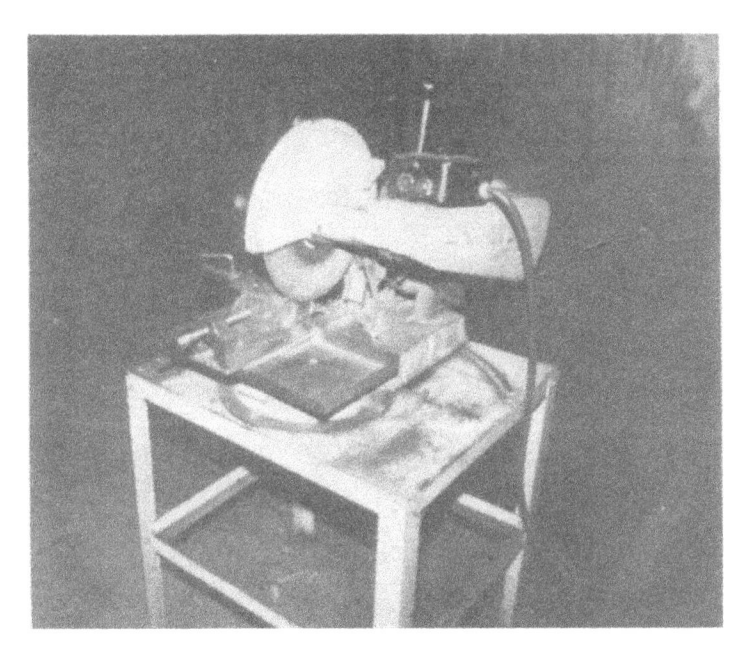

FIG. 3.3. This abrasive cut-off machine makes use of an abrasive wheel, making a smooth cut in pre-hardened stock.

Oxyacetylene torches are used to burn or cut low-carbon steel plate bars to rough size for special mold frame components. The torch is more often used in repair of molds by welding or brazing, and should never be used to cut alloy or tool steel, as their structure is changed by the application of heat.

METAL CUTTING PROCESSES (MACHINING)

In these processes, a cutting tool with one or more cutting edges is mechanically applied to the metal work piece at a controlled rate and with a steady force. Chips of metal are removed until a desired dimension is obtained. Metal is also removed with an electrical spark as in EDM. Equipment for metal cutting is considered essential in the majority of custom mold-making shops. The examples which follow show or describe representative machine tools most frequently used for mold making.

Shaper and Planer

The shaper, planer and milling machine are used to square up rectangular blocks of steel to the desired size. A shaper has a moving cutting tool that passes back and forth over the work, taking a cut on each forward stroke.

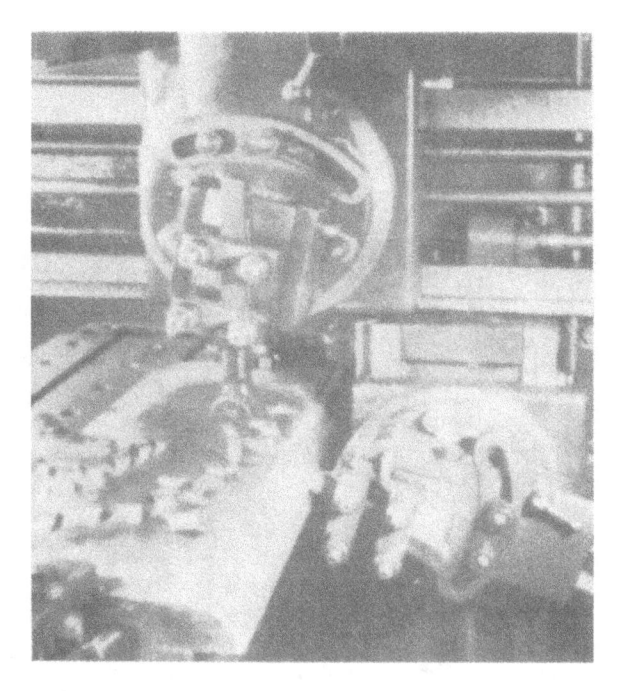

FIG. 3.4. Large chips can be removed in this powerful hydraulic planer here seen taking a bevel cut from one edge of a steel plate.

Shapers are built in a variety of sizes from small high-speed units to large machines that take 36-inch blocks.

A planer does the same job as a shaper but uses fixed cutting tools. The work is placed on a moving table that passes under the cutting tools, as shown in Fig. 3.4. This is a powerful machine that takes large cuts from one or more surfaces at every stroke. In most shops, a shaper is used for the finishing of blocks and plates requiring a work stroke for $\frac{1}{2}$ to about 20 in. A planer is commonly used where the work stroke varies from 1 or 2 feet up to several feet. A planer may be used for finishing several plates of the same size and setup, just as the shaper.

Generally, a shaper is used, in preference to a planer, for work within its capacity. The shaper operates more rapidly than the planer, and is more efficient for the jobs that it can handle. In recent years the shaper and planer have been replaced by milling machines with carbide tooling which are much more productive.

Lathes

The lathe is the most common piece of tool room equipment. A standard tool room lathe is shown in Fig. 3.5. Sizes range from the small bench models

FIG. 3.5. Standard tool room lathe. (*Courtesy Ethyl-Marland Mold, Pittsfield, MA*)

FIG. 3.6. A large mold cavity is shown in the lathe and being ground to the correct outside diameter.

to large engine lathes. Lathes are used for cutting round shapes and internal and external threads, and for boring, grinding, polishing, etc. Figure 3.6 shows a large mold cavity section set up for grinding the outside diameter. The small bench lathe is a high-speed tool used largely for producing the small round pins used in mold making (Fig. 3.7). Some tool shops have high-speed polishing lathes, but the average shop uses a high-speed bench lathe for this work.

Copying or tracing lathes can be used for cutting complex profiles in a cavity or on a core. They have tracing heads which will follow the profile of a template or pattern fixed firmly to the lathe bed. By electric or hydraulic means, the tracing stylus controls the motion of the lathe tool which then duplicates the geometry of the pattern on the work piece. Figure 3.8 shows a standard lathe with a hydraulic-operated attachment for tracing. Another type of tracer lathe is shown in Fig. 3.9. Equipment of this type can save many hours of calculation in design or fabrication in the shop. It greatly reduces the possibility of error or variation from part to part on a number of duplicate mold parts.

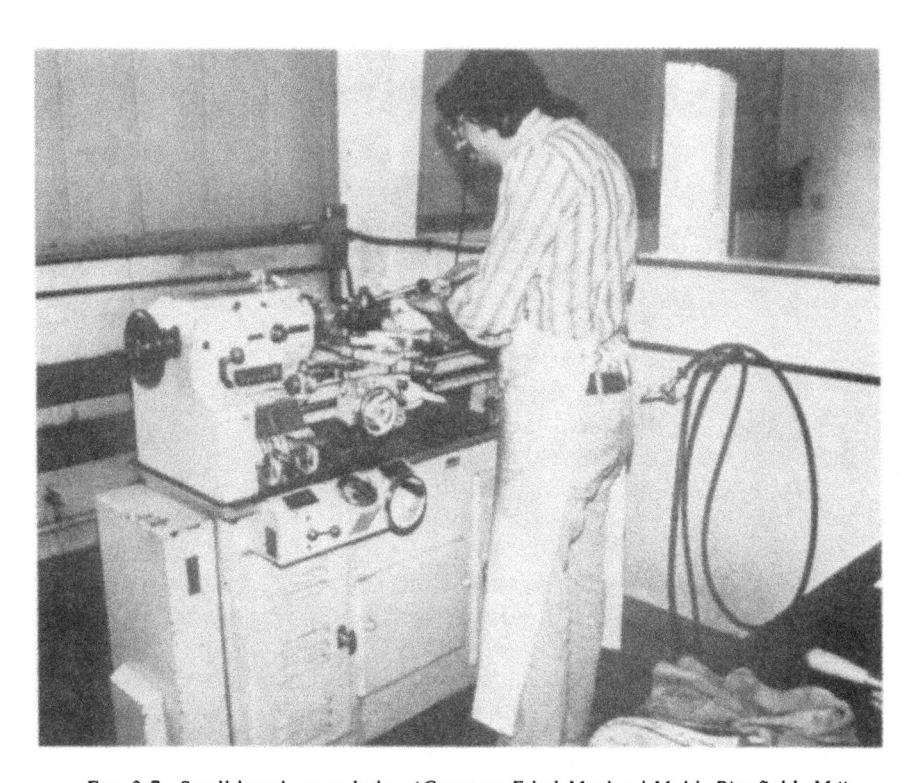

Fig. 3.7. Small bench press lathe. (*Courtesy Ethyl-Marland Mold, Pittsfield, MA*)

Fɪɢ. 3.8. Lathe equipment with automatic tracing. Note the computer-control box mounted to right of operator. (*Courtesy Ethyl-Marland Mold, Pittsfield, MA*)

Some shops now have tape operated lathes which have increased the productivity of the mold maker even further. A tape operated lathe utilizes a program punched into a paper tape. Once a work piece is loaded in the chucking device, the tape automatically machines the piece in accord with the punched program. Obviously, a machine running automatically can produce a finished piece much more quickly than an operator who has to continually adjust and set the machine. Some molds may require 100 or more cavities and forces of a cylindrical nature, and this dictates a tape controlled machine for cost savings, as well as shortened delivery time.

Drilling Machines

Drilling is considered by many to be the fastest and most economical metal-cutting or machining process. A small drill press which will take ½-in. diameter drills is a popular and inexpensive machine for drilling, reaming, lapping, tapping, spot-facing, or counterboring small holes. Holes as large as 3 in. in diameter are drilled on larger machines. The radial drill press is

FIG. 3.9. Tracer lathe machining mold component. (*Courtesy Ethyl-Marland Mold, Pittsfield, MA*)

a very useful heavy-duty machine which has a drill head mounted on a movable arm that travels to the position desired for drilling operations. Because of this maneuverability, the work may be clamped on a stationary table while holes are drilled at various points, as shown in Fig. 3.10. Mold plates often require many parallel holes for the circulation of heating or cooling media.

For deep hole drilling, a *gun drill* is normally used to assure a straight and true hole. A straight hole becomes essential when a drilled hole may be only $3/16$ to $3/8$ in. away from a mold pocket 20 in. or more from the edge of the plate where the drilled hole started. A gun drill (sometimes called spade drill) has cutting edges at the tip, and relieved spiral flutes. A small hole for cooling and flushing exists through each lip of the drill. Special gun drilling machines have several drilling heads, adjustable for spacing, and operate very much in the same manner as a horizontal boring mill.

Jig Borer

A jig borer is a combination of drill press and vertical mill, and is capable of very precise work. The movable work table has adjusting feeds which permit the locating and spacing of holes with extreme accuracy. Dial indicators

Fig. 3.10. Radial drill press with movable arm.

may be used for measuring distances to 0.0001 in. Holes may be located and drilled accurately and reamed or bored to size in this machine. The advantage gained by the use of a jig borer derives from its ease in locating any number of holes with extreme accuracy. Odd sizes of holes may be bored to the correct size easily in this machine as shown in Fig. 3.11. Jig borers are especially helpful for layout work which demands precision.

Grinding Machines

When steel is hardened some distortion invariably occurs. This necessitates the machining of critical dimensions after the hardening. Grinding is the most effective means known for machining hardened steel. Grinders are frequently used to put a smooth finish on blocks of steel before layout work is begun.

Rotary or Vertical Surface Grinder. The rotary grinder, Fig. 3.12, is used for the rough finish-grinding of large flat surfaces. The motor and grinding wheel are mounted in a vertical position and the face of the wheel rotates

FIG. 3.11. Mold maker using a jig borer for the precision location of a hole in a mold section.

FIG. 3.12. In the rotary surface grinder, the work is placed on magnetic chuck so that it may be rotated under the horizontal grinding wheel as the chuck moves into grinding position. This grinder is used for rough grinding and fast removal of stock.

in a horizontal plane. The work to be ground is placed on a round magnetic chuck, which also rotates in a horizontal plane, but opposite in direction to the rotation of the grinding wheel. The thickness of the work can be controlled closely by the micrometer wheel feed provided. Several pieces, to be ground to the same size, may be placed on the magnetic chuck and ground with one setting of the wheel. This grinder is usually operated with a spray of cooling solution running over the work.

Surface Grinder. The surface grinder is used for grinding soft or hardened steel pieces to the desired finish. Grinding is an inexpensive means of finishing to close dimensions and assuring that opposite faces will be parallel. Diamonds may be used to shape the cutting wheel so that angles or radii may be ground. These machines, Figs. 3.13 and 3.14, use a magnetic chuck to hold the work as it passes back and forth under the grinding wheel. Micrometer feeds are provided so that the depth of the cut may be controlled closely. These machines may be either "wet or dry" grinders.

Cylindrical and Internal Grinders. The universal cylindrical grinder (Fig. 3.15) is used for grinding cylindrical shapes which may be rotated on centers. Tapers may be ground by the use of attachments. The universal grinder will

Fig. 3.13. Medium size surface grinder. This grinder is more precise than the one shown in Fig. 3.12. (*Courtesy Ethyl-Marland Mold, Pittsfield, MA*)

Fig. 3.14. Large surface grinder.

grind the inside or outside of round shapes. An internal grinder may be used on work that cannot be rotated on centers. Internal grinders are used for grinding bores, internal radii and other round shapes requiring close accuracy.

A face plate or a chuck forms a base to which the work is fastened in an internal grinder. A jig grinder, such as the one shown in Fig. 3.16, is similar to a jig borer and it has a grinding head instead of a boring head. The jig grinder is used to grind holes to precise size and to space them accurately, similar to the way it is done in a jig borer.

Milling Machines

The milling machine is the most widely used machine in mold making. Milling is a metal-cutting process used to machine all shapes. All milling machines have a work table that will move in three coordinate directions relative to a horizontal or vertical spindle. Milling machines, made in a variety of sizes

FIG. 3.15. The universal cylindrical grinder used to grind outside diameters.

and, with various accessories or attachments, are the most versatile machine tools in mold making.

Vertical milling machines are often called *die-sinking* machines, because the rotating spindle supporting the cutting tool, or end mill, will move along its vertical axis and lower the cutter into the work piece. In horizontal milling machines, the spindle axis is parallel to the plane of the work table. A universal milling machine is a horizontal type with an additional swivel movement of the table in the horizontal plane.

Universal and Horizontal Milling Machines. These perform some of the operations common to the surface grinder, shaper or planer. They use milling cutters somewhat like a circular saw with a wide face. One or more cutters are mounted on an arbor which is mounted into, and driven by, the horizontal spindle. The arbor is supported against excessive deflection by a heavy over-arm and outboard support as shown in Fig. 3.17.

A different set up with a shell cutter mounted directly in the spindle for squaring the sides of the steel block is shown in Fig. 3.18. Other stub arbor tools permit the use of end mills, and in these instances the arbor and over-

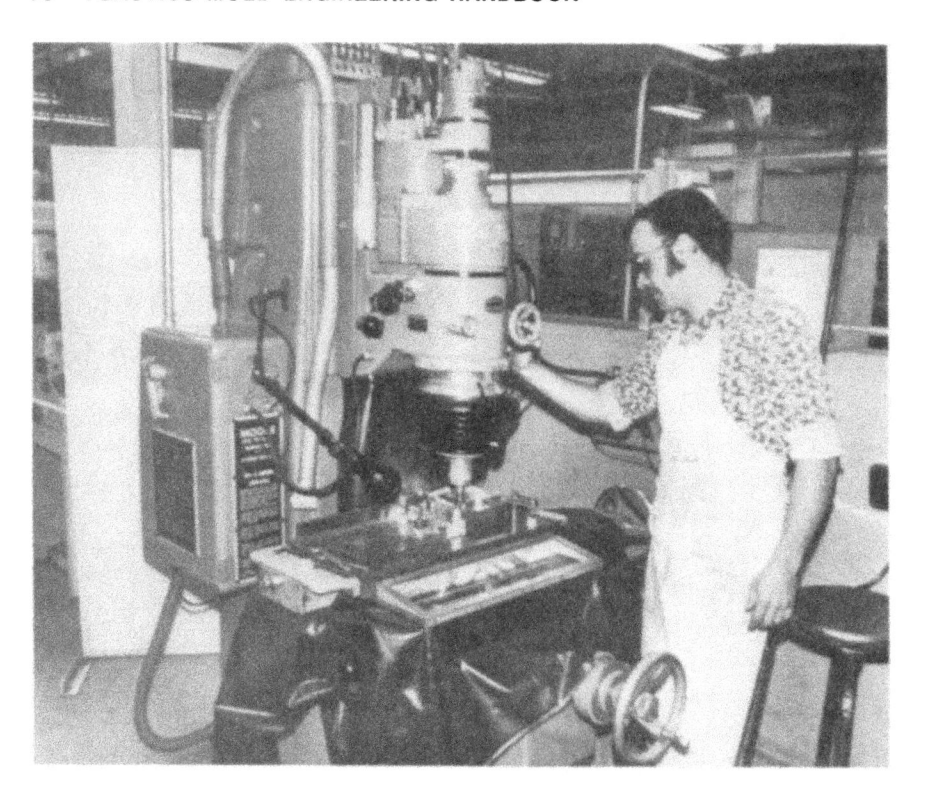

FIG. 3.16. Jig grinder. This machine is used to grind holes to close tolerances on size or location or both. (*Courtesy Ethyl-Marland Mold, Pittsfield, MA*)

arm support are not used. Another version of a horizontal milling machine, called a *jig mill* is shown in Fig. 3.19, finishing pockets in a mold retainer plate. Note that the horizontal spindle can traverse in a vertical plane, while the table moves only in the horizontal plane.

Vertical Milling Machines. These are used for most of the final metal-cutting operations in rectangular cavity surfaces, and on mating cores or forces. These machines are fluted cutters called *end mills*, with cutting edges on the diameter and end (see Fig. 3.20). The movable work table has feed screws for motion in the vertical and two horizontal directions. Dials on the table feed screws are calibrated for each .001 in. of table movement. Normal spindle movement is in the vertical direction and can be controlled with accuracy equal to the table movements. Some vertical mills have the spindle driving head mounted on dual swivel joints, so that the cutter axis may be set at a compound angle relative to the work table. Attachments include hydraulic tracers for profile milling or duplicating, rotary tables, sine bars, and dividing

FIG. 3.17. A flat surface is being milled in this plain milling machine.

heads. Other attachments are available for vertical milling machine versatility. With these accessories, the mold maker can meet the challenge of the most complex designs. Conversely, the capabilities of milling machines and related accessories, coupled with the ingenuity of experienced mold makers, present a challenge to the designer. The practical limits of the milling process are the physical size of the largest machine tools, and the strength of the smallest end mills. Mold cavities for television cabinets have been cut in single forged blocks of steel weighing 10 tons (40 cu ft in volume). End mills are available commercially as small as 1/64 in. diameter.

Rotary Head Milling. This machine, shown in Fig. 3.21, provides control of spindle movement so that the cutter may move along straight, angular, or radial paths in addition to vertical motion along its axis. This feature, in addition to the usual horizontal table movements, can generate contours that would require a rotary table on a regular vertical mill with a fixed spindle. This can save much time in setups on multiple-cavity molds.

Duplicating Milling Machines. Duplicating machines are modified vertical milling machines that reproduce the contours of a master pattern in a cavity or core insert. Irregular or complex shapes, which may be impossible to

FIG. 3.18. Squaring block with a stub-arbor, shell cutter in a plain horizontal milling machine. (*Courtesy Tooling Specialties, Inc., Denver, CO*)

dimension on a drawing, can be generated from suitable model or pattern. Figure 3.22 shows a duplicator setup machining a cavity for a blow mold. The tracing head on the right operates a servo-control valve, controlling hydraulic circuits to cylinders which power the three coordinate movements of the work table and cutter spindle. Both the master pattern and the work piece are fastened securely to the movable work table. The cutter (usually a ball nose end mill) is mounted in the power spindle and centered over the work. A tracing stylus of proper shape and size is mounted in the tracer spindle and centered over the pattern. The pattern must be the same size as the finished work piece. Duplicating ratios are 1:1, but reduction for wall

FIG. 3.19. Cutting rectangular pockets in a mold plate using a jig-mill. (*Courtesy D M E Corp., Detroit, MI*)

FIG. 3.20. Vertical mill cutting detail in a mold core. (*Courtesy Tooling Specialties, Inc., Denver, CO.*)

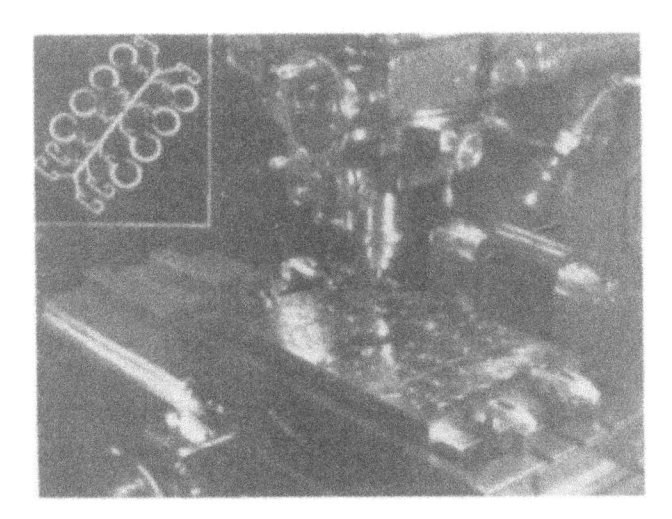

F<small>IG</small>. 3.21. Milling an 18-cavity mold on a rotary head milling machine. The use of this machine eliminates considerable layout work. (*Courtesy Kearney & Trecker Products Corp., Milwaukee, WI*)

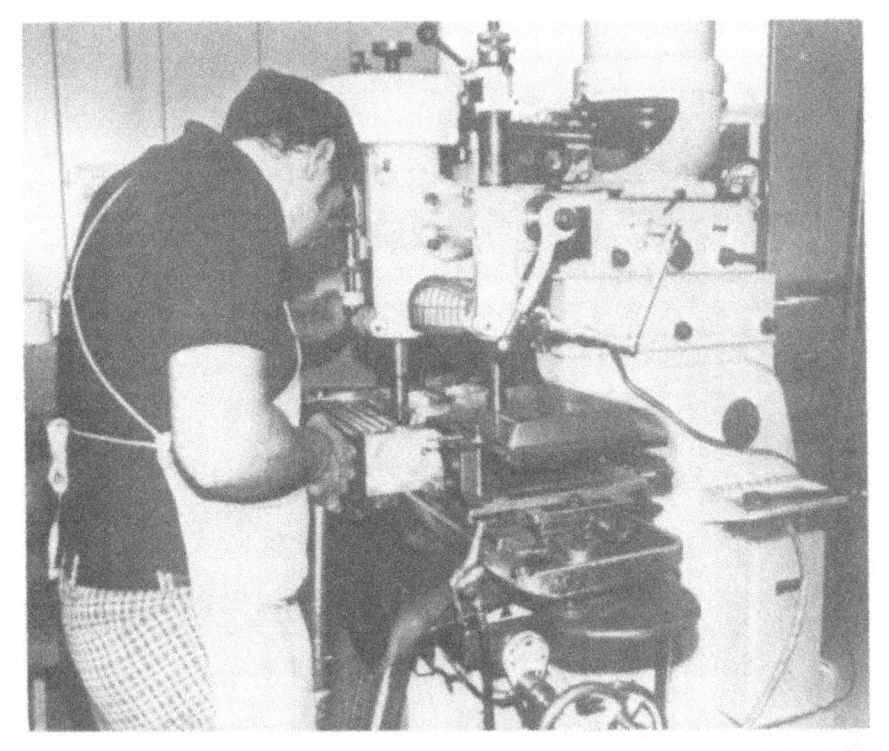

F<small>IG</small>. 3.22. Hand-operated duplicating miller. (*Courtesy Ethyl-Marland Mold, Pittsfield, MA*)

thickness or approximate enlargements for shrinkage are possible through change of stylus size relative to cutter size.

Patterns are generally metal, either cast or machined, if several duplicates are to be produced. However, cast epoxy resin, or hard plaster and fabricated maple, or mahogany patterns may be used if stylus pressure is low and only one duplicate is needed.

Tracer control may be manual or automatic. Motion of the stylus over the pattern is translated into identical cutter movements, so that the rotating end mill moves in the work piece and duplicates the path followed by the tracer. Figure 3.22 shows a manually operated duplicator, and Fig. 3.23 shows a large duplicator that has an electrically operated tracer control system. This appears to be a horizontal type of duplicator, but is really a vertical type rotated 90 degrees on its "back" for better support of large, heavy work pieces. Figures 3.24 and 3.25 illustrate Bridgeport millers modified for mold work.

Some duplicators can be fitted with special controls or attachments which reverse the table motion relative to the normal tracer control, and a "mirror-image" or right-hand contour may be machined from a left-hand pattern.

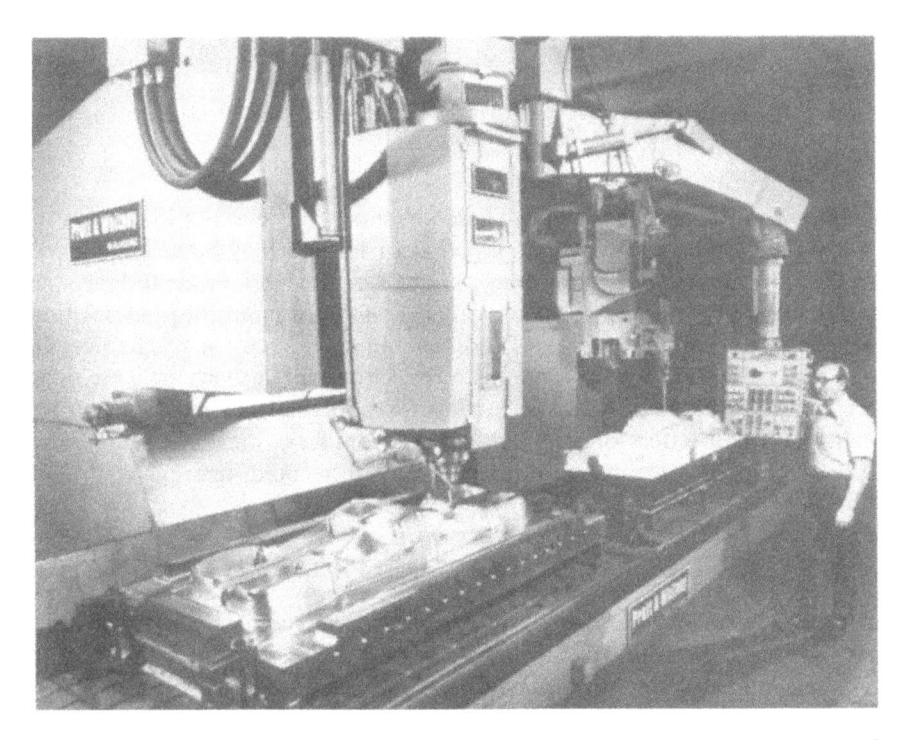

FIG. 3.23. Large duplicating mill. Work piece at left, pattern at right. (*Courtesy Pratt & Whitney Machine Tool Division, Boston, MA*)

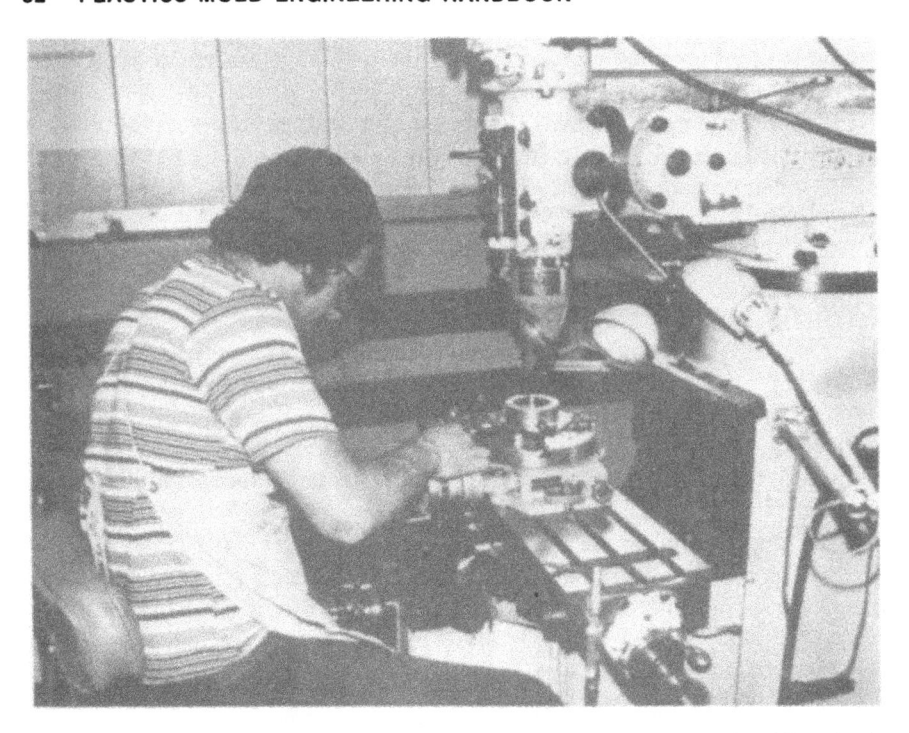

FIG. 3.24. Bridgeport vertical miller with rotary table and special angle milling head. (*Courtesy Ethyl-Marland Mold, Pittsfield, MA*)

Pantograph Milling Machines. These are similar in function to duplicating machines. However, the ratio is larger than 1:1 and may be as high as 20:1 so the pattern must be appropriately larger than the work piece. Independent tables with three coordinate movements are used for mounting and positioning the pattern and the work. These machines are used for mechanical engraving, and when set for large ratio reduction will cut very delicate detail from a large pattern. Realistic models for hobbyists are machined in this manner. Small letters or numbers are cut from large master types. Figures 3.26 and 3.27 show pantographs that are used in mold making.

METAL-DISPLACEMENT PROCESSES

These processes are more commonly called *hobbing* and *cavaforming*. Since each method, regardless of the name, involves the displacement of the metal by some means other than machining, and the use of master patterns to determine the final dimensions of the work piece, we shall consider these processes as similar. They are most frequently used in making cavities or

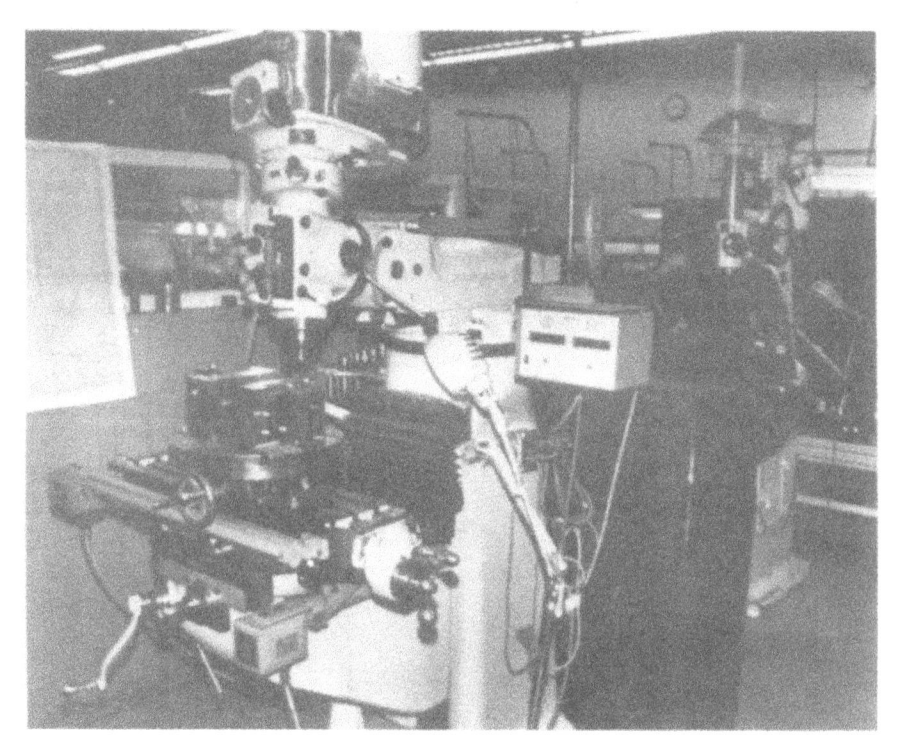

FIG. 3.25. Bridgeport vertical miller equipped with read-outs (*Courtesy Ethyl-Marland Mold, Pittsfield, MA*)

cores and are often purchased by a custom mold maker from a specialist supplier.

Hobbing

Cold hobbing is a mechanical process whereby a hardened steel master shape is forced under high pressure into a prepared soft steel blank, thus securing an impression of the *master*, or *hob*, as it is sometime called. The hobbing process is especially useful when multiple-cavity molds are to be constructed. In some cases it is economical to hob a single-cavity mold, for example, when the shape would be hard to machine as a cavity, or where depressed letters may be required. "Depressed letters" in this case is to be interpreted as cut into the hob. When the hob is pressed into the hobbing blank, the depressed letters in the hob will result in *raised letters* in the final hobbing. Obviously, the resulting depressed letters in the *molded part* will then be ready for a paint fill. Figure 3.28 shows a group of master hobs that were used for sinking mold cavities. The hobs were cut from a special steel

FIG. 3.26. Pantograph mill. The pattern at left is ten times the size of the work at right. (*Courtesy Ethyl-Marland Mold, Pittsfield, MA*)

in the conventional manner and hardened and polished. The cavity block is a prepared block of S.A.E. 3110 steel or the equivalent, and the impression is made cold. The press must exert very high pressure. Some hobbing presses develop pressures as high as 3000 tons. Many mold makers send their hobbing to outside specialists who have the large presses required for this work.

The "Cavaform"* process may be used to advantage for deep, small diameter cavities having draft and other internal configuration instead of straight round holes. The pencil barrel cavity is a typical application. A highly accurate hardened and polished male master is made. Annular mold inserts are then gun drilled to the desired depth, hardened and polished to a 4–8 microinch finish. The mold insert is then placed over the male master and reduced to its configuration by a swaging-extrusion process. Fifteen hundred cavities have been made over a single mandrel by this process. The machinery required for this process is large and expensive and such work is done on a job basis by the owners of the "Cavaform" trade name.

*Massie Tool and Mold, Inc., St. Petersburg, FL.

Fig. 3.27. Larger dimensional pantograph.

Fig. 3.28. Group of master hobs used for hobbing mold cavities. (*Courtesy General Electric Co., Pittsfield, MA*)

METAL EROSION PROCESSES

Electrical Erosion (Electrical Discharge Machining–EDM)

Again, in this metal removal process, a master pattern is required; however, it is used as an electrode and must be electrically conductive. Hardness is not a requirement so copper alloys are generally used to make the master. Cast zinc and machined graphite are also used in some instances. Figure 3.29 shows the principle of spark erosion applied to mold making. The gap between the master and the work is quite uniform and small. As the master descends, small intense sparks are generated wherever the gap is reduced. Erosion occurs on both master and work, but is a negative polarity, and the master erodes only one fourth to one tenth as rapidly as the work piece at positive polarity. The work may be hardened before the process begins so that distortion due to heat treatment is eliminated. The dielectric fluid must circulate at all times to remove the minute particles that are formed between the master and the work piece. Electrical erosion is slow compared to mechanical cutting of soft steels, but for certain conditions, such as narrow deep slots, it has great advantages. Figure 3.30 shows typical electrical erosion equipment with a control console. Note that the construction resembles that of a

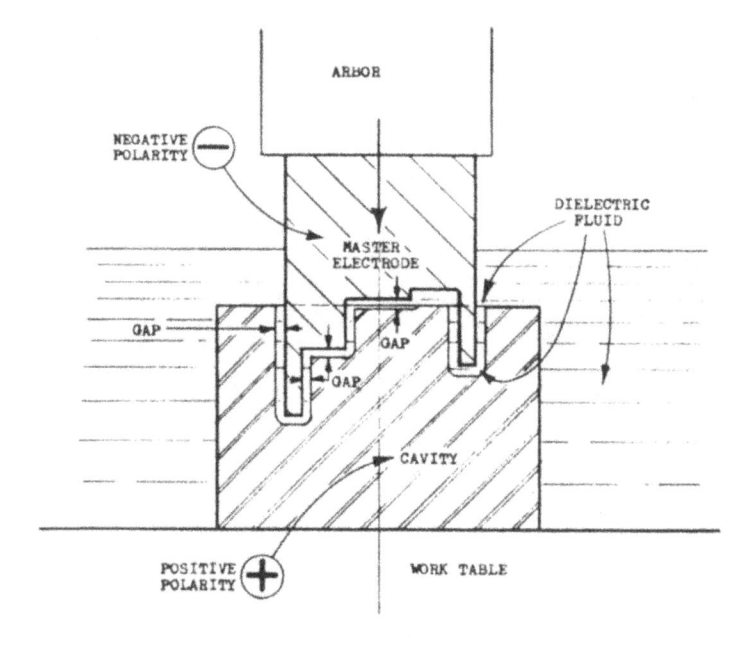

FIG. 3.29. Principle of electrical erosion. The gap or clearance generated is very uniform.

Fig. 3.30. Electrical erosion machine with power supply at right and dielectric fluid pump at rear. (*Courtesy ELOX Corp., Troy, MI*)

vertical milling machine, with the same horizontal table movements. In planning to utilize this process, the mold maker roughs out stock by machining wherever possible and then sends the work piece to the specialist in electrical erosion.

Chemical Erosion or Photoengraving

The art and skills of photoengraving have been applied to mold making. Specialists can reproduce textures such as fabrics and fine leather on molded parts by etching the reverse detail into a hard or soft metal cavity. In the photographic process, the chemically resistant coating is produced in a film applied to the molding contours and surfaces. The film is then exposed to a projected image of the desired texture. Portions of the film are chemically resistant; others are washed away to expose the metal for erosion. Detail applied in a mold by chemical erosion or displacement is of quite uniform depth and will appear as raised markings or decorations on the molded part.

CASTING PROCESSES

These processes include casting and its variations. As can be seen in any commercial foundry, gravity casting in foundry sand molds rarely provides adequate density and strength necessary for sound cavities and cores with a high polish on the molding surface. Improved casting processes use ceramic-like materials rather than sand, or means other than gravity to deposit metals in a desired shape on a suitably prepared master. A master pattern is necessary in all of these processes, and in most processes the patterns can be of any stable material such as hard plaster, carved or cast plastic, or any solid metal. The specialists who supply these services will furnish design data and information relating to their particular process. The mold designer must allow adequate draft and add a shrink factor for the particular process selected.

The Shaw Process* was developed in England and a number of specialists are licensed to provide this service in the United States and other countries. A Shaw mold is a semi-ceramic mold cast about a suitable pattern of any stable material. The mold is made in two or more casts which gel on the pattern. These rather fragile reverse casts are separated from the master and fired at a high temperature, just as pottery or porcelain is baked. The resultant structure has a controlled porosity and a very good reproduction of the surface detail of the master, in reverse. A tall vertical riser or sprue develops adequate pressure in the assembled ceramic mold as molten metal is poured in and fills the molds. This pressure and the uniform fine porosity of the entire mold produce excellent castings which are dense and sound. Any accepted tool steel alloy for mold making can be cast to shape by this process. A Shaw mold can make only one reproduction or cast, thus a separate mold is made for each insert ordered. In the right application, it can be very economical compared to machining a cavity or core. Precision castings are also produced by other specialists, using centrifugal casting machines and a one-piece semi-ceramic mold or investment. This method may be called *investment casting, centrifugal casting*, or the *lost wax process*.

The following description is similar for all of them. High temperature plaster or semi-ceramic investment material is cast in one piece around a pattern that will melt and be absorbed by the mold when it is oven-dried or fired. This porous investment, or mold, then contains hollow cavities without parting lines into which molten alloys will flow through a suitable sprue or runner system. This mold is rotated in a horizontal plane at a fixed

*See Chapter 4.

distance from the center of rotation, while molten metal is poured at the axis of the rotation. Centrifugal forces and mold porosity again provide dense sound castings with smooth surfaces and excellent dimensional accuracy, particularly in the thinner sections. The total weight of metal in a single cast is limited, thus this procedure is more suited to multiples of small cores and cavity inserts, where detail is fine and difficult to machine. Large heavy cavity castings are not produced by these processes due to machine limitations.

Hot Hobbing or Pressure Casting

This is a method used to make the best beryllium-copper cavity and core inserts. Molten alloy is poured over a hardened steel master, called a *hob*, which is placed in a heavy ring or yoke as shown in Fig. 3.31. A force plug is then inserted above the molten metal and the assembly is placed between the platens of a hydraulic press. Pressure is applied to the casting during solidification and shrinkage of the casting. Dense, sound castings result, which can be heat-treated to obtain very good physical properties. Beryllium

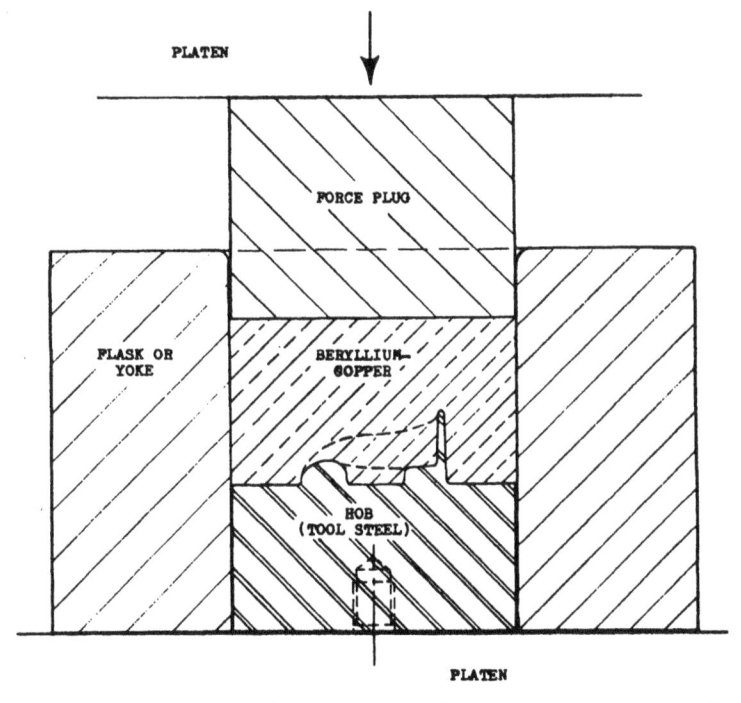

FIG. 3.31. Pressure casting principle. Molten metal exerts equal pressure on all surfaces of delicate hob section.

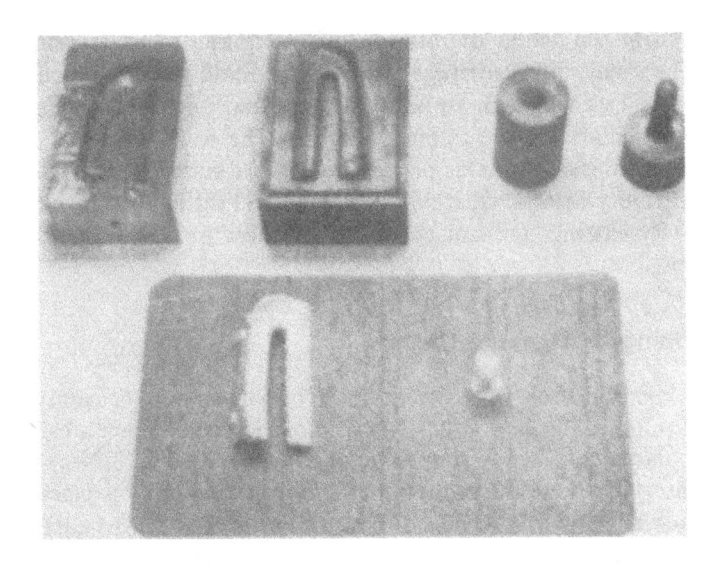

FIG. 3.32. Hobs and beryllium-copper pressure castings. Note flutes and ribs which are formed in the cavity sections. (*Courtesy Tooling Specialties, Inc., Denver, CO*)

pressure casting specialists will recommend the proper alloy steel for various hobs or masters. Configuration of the hob detail and the number of castings to be produced determine the selection of the proper heat-resisting steel. Some, with highly alloy content are expensive and difficult to machine. Cost of the master is offset by the advantages obtained by the process which can provide cavity detail impossible by machining or cold hobbing. Figure 3.32 shows examples of master hobs and castings produced from them.

Molded Cavities

"Custom molded cavities" is a 1982 development of the 3M Center, St. Paul, MN, and data concerning this process should be in the designer's file for ready reference where the process is applicable. The principal limitation is physical size of the final tool, which cannot exceed 3 in. × 3 in. × 3 in. 3M custom cavities are molded from a STELLITE® alloy using a single, rigid model (male or female) supplied by the customer. Any number of identical cavity inserts can be made from a single model (as opposed to the Shaw Process, which is destructive to the model). The application in a mold is in the same manner as machined steel, hobbed steel, or beryllium cavities.

The indicated use is for cavities with *decorative detail* of any degree of complexity. Cavities with *functional detail*, such as connector strips, are

STELLITE® is registered trademark of the Cabot Corp.

easily molded. The third use is multiple cavities of identical shapes. The 3M literature suggests using a female model if at all possible. The quality of a molded cavity is directly related to the quality of the model. The model can be of any type of *metal*, inasmuch as the molding process is nondestructive to the model. Heat or pressure is not applied in the process. Molded cavities shrink from the model size, thus, 3M should be contacted for shrink specifications.

The author believes the process has several advantages, namely:

1. reduced cavity costs,
2. reduced tooling time,
3. repeatable detail, and
4. quick replacement in the event of maintenance.

METAL-DEPOSITION PROCESS

Electroforming*

Electroforming is an electroplating process which differs from the casting techniques of metal deposition. Casting processes use heat energy in molten metal, controlled physical forces, and are rapid production methods for selected metal alloys. Electroforming deposits pure hard nickel ions on a conductive or conductive coated master. Deposition or plating on the master contour occurs at the rate of .001 to .003 in./hr. until a thickness of ⅛ in. is obtained. Copper is then plated on the back of the nickel shell, which is still on the master, until a suitable cavity wall thickness is obtained. The copper strengthens the hard nickel, which is relatively brittle, and provides good heat transfer for injection molding of thermoplastics, with fast production cycles. Electrical energy is used to deposit the metallic ions on the master from a chemical solution. This is a relatively slow process, current values are low, and little or no heat is generated. The advantages are that such materials as wax, polyethylene, other thermoplastics, and rubber can be used for a master when given a conductive coating. Radical or severe undercuts can be produced in a cavity. With a mirror finish on a soft metal master such as brass, a reverse cavity can be produced with a perfect polish on the molding surface.

It is logical to assume that flame spraying of metals will be used to make mold cavities in the near future. Molten droplets or discrete particles of metal are deposited on a master until a desired shell thickness or build-up is obtained. Equipment available uses welding gases or electrical energy to melt metal in the form of wire or powder. The deposit has a porous structure

*See Chapter 4.

which would be detrimental in high-pressure mold cavities unless impregnated with a liquid metal or resin. This could be an entirely satisfactory approach to cavity making for short-run or experimental molds, and has been used with some success in low-pressure processes such as vacuum forming dies and blow molds for limited production. Flame spraying is often called *metallizing*, and was originally developed to repair worn bearings and shafts in motors, engines and other equipment. It deserves mention with other metal-deposition processes, because it has exciting possibilities in mold making. The designer might think of this process as "progressive casting" or "mechanical plating" in considering possible applications.

MISCELLANEOUS PROCESSES

Mold-Making Procedure

The sequence of operations in the making of various mold members should be understood by the mold designer so that he may more accurately visualize the progress of a mold through the shop. The operations listed here serve to illustrate good shop practice. Actual practices and techniques are determined by the equipment available in any shop.

Mold Bases*

The construction of a standard mold base was explained earlier. The mold base manufacturer may machine in special pockets and water lines, or the custom mold maker can perform these operations after the standard frame is delivered. Delivery commitments determine the choice. In either case, the disassembled plain plate members, ground square and parallel on all surfaces, are taken to the bench for layout work. Screw holes, water or steam lines, and other holes are laid out for drilling. Pockets may be laid out for rough sawing to shape, mill, bore or lathe-turn to finished size. Figures 3.33, 3.34, 3.35 and 3.36 show the large type and variety of machine tools used in the construction of mold bases.

Cavity and Core Inserts

These are usually produced from annealed tool steel alloy bars or forgings of suitable stock size. After rough cutting to size with a cut off saw, round members are turned to approximate size on the lathe; rectangular parts are rough sized on the millers. Round or circular internal openings are

*See Chapters 5, 7 and 8.

FIG. 3.33. Custom mold base showing complete machining of holes and pockets.

cut or bored with drill press, jigborer, vertical mill, or lathe as required. Vertical milling machines or duplicators are used for machining rectangular cavity or core details. External surfaces are often finished on a grinder prior to layout or close tolerance machining. External machining procedures are the same for inserts produced by metal-displacement or metal-

FIG. 3.34. Boring mill installation used in machining mold bases. (*Courtesy Ethyl-Marland Mold, Pittsfield, MA*)

FIG. 3.35. Jig mill boring mold base. (*Courtesy Ethyl-Marland Mold, Pittsfield, MA*)

deposition processes. Hobs and master patterns are machined in the same manner as a core insert for a mold. Letters and numbers are generally stamped or engraved in the mold members after other operations are completed, and prior to polishing.

Measurement and Layout

All of the conventional hand tools are used by mold makers. The vernier height gauge (Fig. 3.37) vernier calipers, micrometers, clamps, indicators, V-blocks, parallels, surface plates, angle plates, sine bars, and gauge blocks are invaluable for layout and dimension operations, such as checking work in process. (See also measurement of surface finish - Page 105).

Hand tools necessary for bench finishing are diemaker's files and riffle files, chisels, scrapers, and engraving tools (Fig. 3.38). Abrasive materials are coated cloth and paper, graded stones, lapping compounds, and diamond paste. The entire finishing and polishing technique is one of metal cutting by hand, working out machine marks and imperfections in molding surfaces as shown in Fig. 3.39, until the surface quality appears as in Fig. 3-40.

FIG. 3.36. Jig boring mold base.

Benching and Polishing

Benching and polishing operations are often isolated, gaining freedom from vibration and noise as shown in Fig. 3.41.

Final polish to the highest luster is only done after heat-treatment and hardening of a mold insert. Liquid honing or vapor blasting removes scale and discoloration due to hardening, and greatly reduces the tedious hand labor of polishing. Liquid honing equipment is similar to sand blasting. A mixture of water, lubricant, and fine abrasive powder is propelled by compressed air onto the surfaces to be cleaned. The entire operation is done in an enclosed cabinet.

Polishing specialists are available to do this critical finishing. Optical polishing as required in plastic lens molds can be done economically only in shops with the proper measuring equipment and long experience in the art.

FIG. 3.37. A finished mold section is carefully checked with a vernier height gauge before assembly.

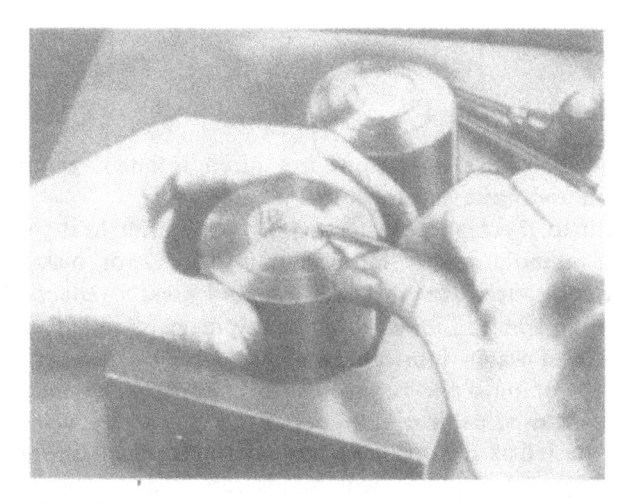

FIG. 3.38. Hand tools are frequently used for the engraving of molds.

FIG. 3.39. Most of the mold polishing is done by hand with the aid of pine sticks and polishing compound.

FIG. 3.40. This mold core shows the mirrorlike surfaces which are produced by careful polishing and finishing. For fine finish, careful work is essential.

FIG. 3.41. Benching and polishing installation. (*Courtesy Ethyl-Marland Mold, Pittsfield, MA*)

STEPS IN MOLD FINISHING

1. Machine finish (mill, grinder or lathe).
2. Use file or riffler to remove machining marks.
3. Use coarse emery or carborundum dust to remove file marks.
4. Use progressively finer abrasive powder to remove the marks left by previous coarser grade.
5. Finish with 600 grit, dry abrasive powder and hand or power felt wheel. Three important points to be observed are:
   a. Use separate soft pine sticks to apply the abrasive powders, and polish. Never mix different grits on the same stick.
   b. As soon as maximum results are achieved with the present grit, move to the next one.
   c. Be very careful that chips or shavings do not adhere to the polishing stick or felt wheel. A tiny, tiny speck can undo hours of hard work by causing a scratch in an otherwise polished surface.

Expert Polishing

Polishing specialists who have become *expert* because of handed-down techniques will tell you the "secret" of rapid polishing and mirror finishes is: *knowing when you have done all you can do with a particular grit and grade of polishing material.* Specifically, this means knowing when to go on to the next step in the process as outlined above. It will be useless to attempt high polish until after all "scratches" have been reduced to invisibility. There exists a mistaken belief that chrome plating will make up for an otherwise poor polishing job.

Welding

Helium arc welding methods are useful in both mold making and mold repairing. Rods of analysis similar to the cavity or core insert are used to add metal when an excess has been removed or when changes must be made. This equipment is invaluable for repairing worn or broken molded parts.

Hand Grinder

The hand grinder may be driven by air or electricity and is operated manually as shown in Fig. 3.42. These units operate at speeds ranging from 10,000 to 75,000 rpm. This is a very useful tool for cutting special contours and for finish grinding to the close fit required in mold assembly operations. Fitted with a polishing wheel, the hand grinder may be used in finishing and polishing operations.

Flexible Shaft Grinder

This grinder is a hand tool which has a flexible shaft thay may be operated at any angle or position desired and that can be fitted with various finishing

FIG. 3.42. The hand grinder is used for finishing operations and for cutting special contours.

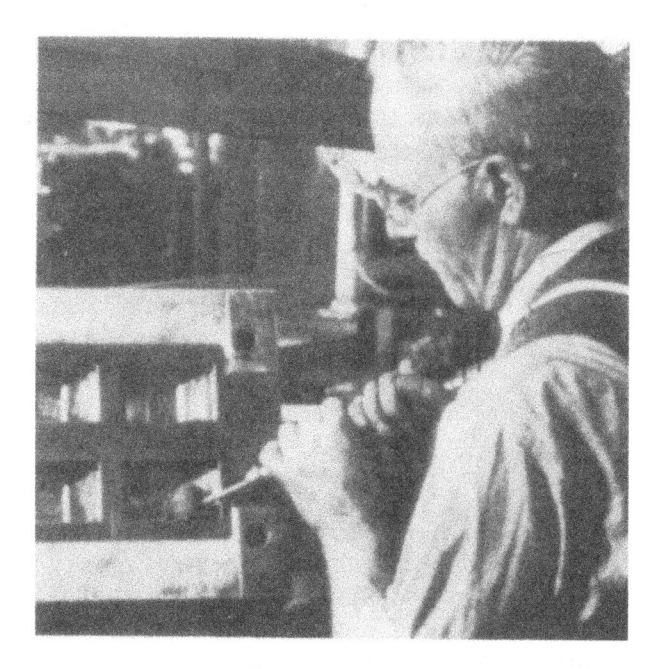

FIG. 3.43. Flexible shaft grinder fitted with a polishing wheel.

tools. It is also used as a hand grinder, but it has more power and less speed than that machine. This unit, fitted with a polishing wheel, is shown in Fig. 3.43.

Another version of the flexible shaft hand machine has a right-angle head with an adjustable stroke oscillating motion. It can be used to polish in slots or other confined spaces, where a rotary polishing motion is not suitable.

Chrome Plating of Mold Components

Most experienced mold engineers recommend plating of all molding surfaces of a production mold as soon as first samples are approved and before any production runs are made. Hard chrome plating (Fig. 3.44) is recommended for all but a few molds that need gold plating to resist the corrosive action of some plastics. Proper chrome plating is achieved by a few expert platers who are experienced in the art of plating and with the problems faced by the plastics engineers.

A qualified hard chromium plater must have a knowledge of the requirements of the mold in order to prepare properly for the plating operation. With most jobs, this involves a detailed case history, with an analysis from

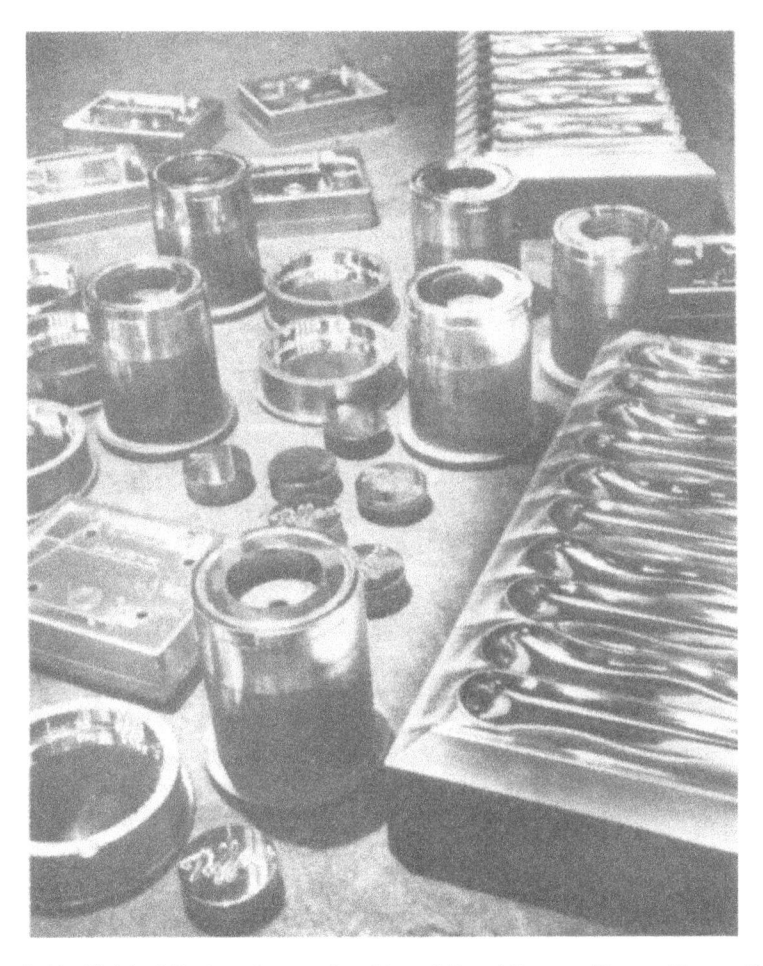

FIG. 3.44. Finished display of assorted molds and dies. (*Courtesy Nutmeg Chrome Corp., West Hartford, CT*)

the molding engineer's viewpoint of the functioning and performance requirements. He needs to know the type of metal to be plated, the need for stress relief before and after plating, and the final polish specification. With this information, the plater can issue proper job instructions for fixture selection, anode placement (Fig. 3.45), preplating treatment and the plating and polishing techniques (Fig. 3.46) that will be essential.

A visit to the plater by a knowledgeable mold engineer is desirable at the start of any plating job to make sure that all problems are anticipated. A phone call to the plater after he has the work will, in many cases, eliminate personal visits.

FIG. 3.45. Electrodes being adjusted for the chrome plating of the knife handle cavities. (*Courtesy Nutmeg Chrome Corp., West Hartford, CT*)

Heat Treating Equipment

Most plastics molds use hardened cavities, plungers and pins. Other parts of the mold are also hardened. Mold steels are generally annealed before work is begun, and they are often annealed or normalized during the mold making process. Most small mold parts are made from forgings. Both of these materials must be annealed so that they will machine easily. Mold parts are heat-treated after machining or hobbing to obtain strength, wearing qualities and distortion resistance. The equipment most frequently used (Figs. 3.47 and 3.48) consists of an annealing furnace, a tempering furnace, a carburizing furnace, a large burner and suitable quenching baths. Oil, gas and electricity are the heating media most frequently used in the heat treating of steel. The furnaces are merely fire-brick lined ovens equipped with a heating unit. Liquid baths of lead or salt serve special needs; the lead bath to draw and temper steel parts, and the salt baths to minimize

FIG. 3.46. Polishing a cavity after plating. (*Courtesy Nutmeg Chrome Corp., West Hartford, CT*)

FIG. 3.47. Vacuum furnace used for hardening and annealing. (*Courtesy Steel Treaters, Inc., Troy, NY*)

FIG. 3.48. Atmosphere gas carburizing furnace also used for hardening. (*Courtesy Steel Treaters, Inc., Troy, NY*)

FIG. 3.49. Liquid salt baths used for heat-treating tool steels. (*Courtesy Steel Treaters, Inc., Troy, NY*)

surface oxidation. Quenching baths may consist of oil, water, salt, caustic soda solution, or air.

Final Assembly

Equipment for final assembly will include heavy handling devices such as steel-topped tables, a portable die lifter, and an overhead hoist above the table. Hand tools will include C-Clamps, pry-bars, feeler gauges, and other conventional equipment that the experienced mold maker finds useful. Any errors in design or construction of components will be discovered by logical checking in final assembly. Each finished component should be checked for fit with each mating part, i.e., cores in cavities, guide pins in bushings, ejector pins in ejector holes, mechanical actions for side core pulls and ejector plate movement must be checked for ease of movement. Actions should work freely by hand on the assembly bench.

Both halves of the assembled mold must be properly aligned so that the parting line registry and mitre is complete and correct with very light pressure. In the case of molds with irregular rise and fall of the parting line, assembly can take many hours to achieve the proper mitre or contact between mold halves. Adjustments are made as required, by techniques that determine points of contact or excessive interference. Stoning, hand grinding and polishing procedures are used to eliminate interference.

Often, a last minute inspection of the completely assembled mold in the shop can save hours of press time in getting a new mold running properly. Certain details like gates, vents and polish are sometime hard to define completely on the drawings. The mold designer should cooperate with the shop and take a few minutes, if possible, to check these small but important details. Figure 3.50 illustrates a mold assembly area.

MEASUREMENT OF SURFACE FINISH, MOLDS AND PARTS

In the past, surface finish and its measurement have often been the subject of scientific rather than practical interest. With the growing demand for close tolerance parts, a necessity in these days of high productivity and interchangeability, a practical understanding at shop level of surface finish of plastics molds and its measurement has become essential.*

Surface finish should be specified by the mold designer, the degree of finish being carefully related to the use to which the component will be put. Unfortunately, the surface finish specification is sometimes arrived

*This section supplied by M. O. Nicolls, DeBeers Industrial Diamond Div. It was prepared in cooperation with Rank Precision Industries, Leicester, U.K. and Hommelwerke GmbH, Mannheim, F.R.G.

FIG. 3.50. Mold assembly area.

at by a process of trial and error. It is important to bear in mind that the best finish attainable by a skilled operator, within the limits of his machine, is not necessarily the best finish for a component.

In the past, visual appearance rather than mechanical design requirements has frequently determined the surface finish values; indeed, a surface finish governed by visual appearance could very well be over-specified, leading to unnecessarily high production costs (Fig. 3.51). While surface

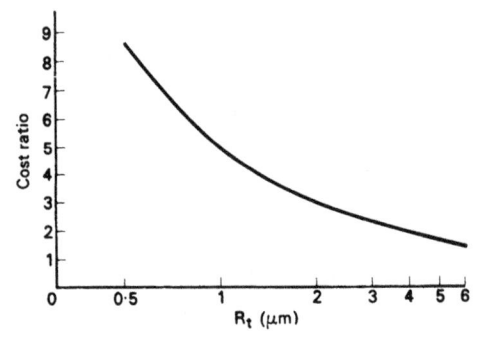

FIG. 3.51. This graph indicates the proportional increase in cost when an improved surface finish is sought.

finish can be a complex and academic subject, the basic principles are nonetheless simple enough for a practical working knowledge to be gained quite easily.

In view of the growing importance of achieving, measuring and interpreting high quality surface finishes, this basic study of surface measurement technology has been compiled.

While the grinding process has been used to illustrate various points throughout this booklet, much of what is said can be applied to other machining processes.

Terminology

It is essential that any study of surface measurement be preceded by a clear statement defining the terms in general use.

Surface Texture. This describes the overall condition of the surface, and can be broken down as follows:

Primary texture or *roughness* is that part of surface texture best defined as the marks left by the action of the production process used, such as a grinding wheel or lathe tool (Fig. 3.52). The primary texture can be measured by various constants (Fig. 3.53) that collectively help to build up an accurate picture of it. Individually, these constants are of very limited value in assessing primary texture.

Average arithmetic roughness (R_a, CLA or AA) is also known as Centre Line Average (British) and Arithmetic Average (American); for historic reasons CLA and AA are generally quoted in microinch values, while R_a tends to be quoted in microns. R_a is a mean value of the roughness (see Fig. 3.53).

Smoothing depth—R_p)—means distance between the highest point and the mean line. R_p generally results from the condition of the production tool such as a lathe tool or grinding wheel (Fig. 3.53).

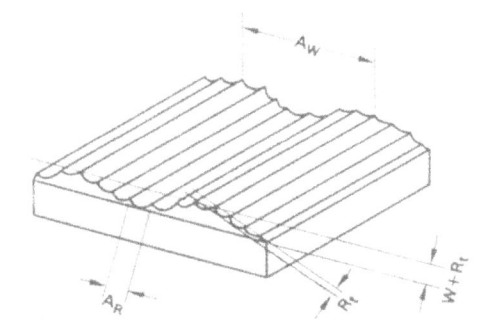

FIG. 3.52. Various components of a surface are: R_t, roughness (primary texture); A_w, waviness, (secondary texture); A_r, roughness spacing, and $W + R_t$, waviness plus roughness.

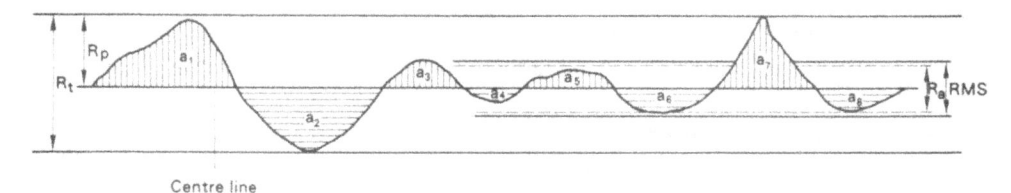

Centre line

Meter Cut-off (mm)	Traverse Length (mm)	
	Minimum	Maximum
0.075	0.40	2.00
0.25	1.25	5.00
0.75	1.40	8.00
2.50	5.00	15.00
7.50	16.00	40.00

Table I

FIG. 3.53. The various parameters R_a, R_p, and R_t and RMS are illustrated. It may seem that the centerline is that line that divides the areas such that: $A_1 + A_3 + \cdots\cdots A_7 = A_2 + A_4 + A_6 \cdots\cdots A_8$.

Maximum roughness—R_t—is the highest point to lowest point within the tracing stroke. An example of the cause of R_t and its magnitude would be the grit and its size used in a grinding wheel (Fig. 3.53).

Root-Mean-Square (RMS) is an average geometric roughness and was an American standard. In 1955, it became obsolete, but naturally enough will still be encountered occasionally. It is sufficient to say that its numerical value is some 11 percent higher than that of R_a (CLA, AA) (see Fig. 3.53).

The foregoing standards, with the exception of RMS, are in common use. Other terms will be encountered in the study of surface measurement such as R_{tm}, R_z, tp, R.

It is worth enlarging on the parameters R_{tm} and R_z, since increasing reference is being made to them by manufacturers of surface measuring devices and Standards Institutions. Both R_{tm} and R_z are parameters that give a measurement of the average peak-to-valley height, the former being intended for measurement by a machine whereas the latter lends itself to graphical determination and cannot yet be reproduced by a machine. In some cases R_{tm} and R_z can be used as alternative or supplementary parameters to R_t. (For determination of these parameters see Fig. 3.54).

In order to simplify the illustration of surface measurement principles, reference will be confined in the remainder of this article to R_a, R_p, and R_t, as it is felt that these parameters will be sufficient to develop the basic themes.

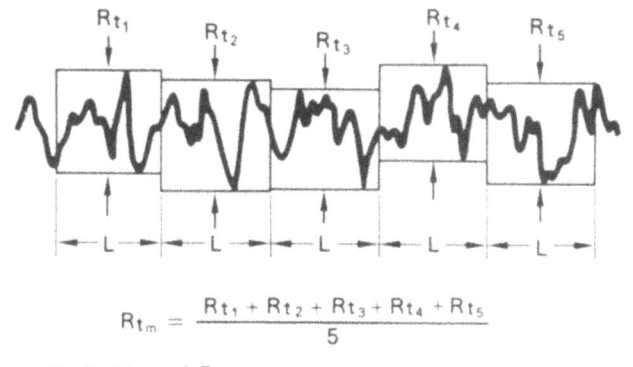

$$Rt_m = \frac{Rt_1 + Rt_2 + Rt_3 + Rt_4 + Rt_5}{5}$$

(A) *Definition of* R_{tm}

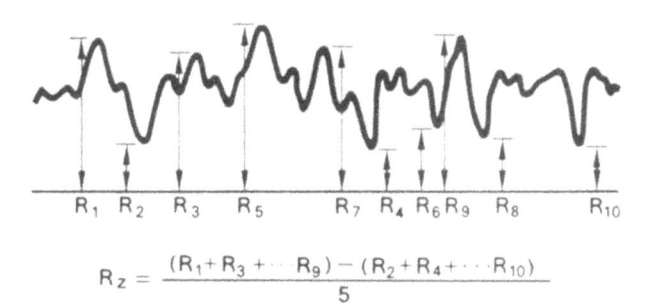

$$R_z = \frac{(R_1 + R_3 + \cdots R_9) - (R_2 + R_4 + \cdots R_{10})}{5}$$

Definition of R_z

FIG. 3.54. Definitions.

Secondary Texture.   Secondary texture is that part of the surface texture that underlies the roughness. All types of machine vibrations, for instance spindle deflection and imbalance, can be the cause; it is generally described as waviness or more simply W (Fig. 3.52).

Lay.   The production process used will form patterns on the surface. The predominant pattern direction is known as the lay.

Cut-off (Sampling Length).   Cut-off is a facility that is built in to most surface measuring devices. Its function is to suppress waviness (secondary texture) to whatever degree is required within the limitations of the cut-off unit. Clearly this facility is of great importance as it allows the effects of the production process to be studied separately from the effects of machine inefficiencies. Cut-off is a filtering operation that is performed by a frequency-dependent electronic filter. The cut-off values according to the British standard are as follows: 0.003 in., 0.010 in., 0.030 in., 0.100 in., 1.0 in.; according to the metric standards they are as follows: 0.025 mm, 0.075 mm, 0.25 mm, 0.75 mm, 2.5 mm, 7.5 mm. There is generally a choice of three or more cut-off values on a device posessing this feature.

Tracing Stroke (Traverse). This can be quite simply described as the distance covered by the stylus during the recording cycle. Many surface measuring devices are fitted with variable stroke lengths.

Standardization

We have chosen to discuss two points in some depth, because they are constantly giving rise to doubt and argument.
1. Should we standardize on a particular cut-off or should it be varied? If it is to be varied, how do we select which values to use?
2. Should we standardize on a particular traverse length or should it be varied? If it is to be varied, how do we select which one to use?

Meter Cut-off. One should not randomly apply any particular meter cut-off value; in order to establish and apply a meaningful cut-off one should analyse a surface both visually and by initial exploratory measurement with the equipment. If measurements are required of the ground surface as it actually is, in other words inclusive of primary and secondary texture, standardization on a particular meter cut-off value could well ultimately result in the oversuppression of the roughness values, which would in turn give lower values of R_a, R_p, and R_t than is actually the case. When both primary and secondary texture values are required, cut-off must be applied when R_a, R_p, and R_t have reached a point beyond which any further movement across the surface being measured by the stylus would not increase their values.

For example, in measuring a surface that has little or no waviness (secondary texture) these values will be reached relatively quickly and a short cut-off value can be applied. If study by initial sampling reveals considerable waviness, a greater cut-off value may need to be applied since it may be desirable to include secondary texture in addition to primary values.

Traverse Length. If one wishes to stick to a rule then the ISO standard should be used in accordance with the relationships quoted in Table 3.1.

If, however, one wishes to use a physical method of selection then observation of the ground surface generally reveals the predominant lay and the frequency with which it repeats itself. It is pointless to allow the stylus to traverse two or more areas having identical lays. In doing this, the machine

TABLE 3.1.

| | Traverse Length (mm) | |
Meter Cut-off (mm)	Minimum	Maximum
0.075	0.40	2.00
0.25	1.25	5.00
0.75	1.40	8.00
2.50	5.00	15.00
7.50	16.00	40.00

is merely recording the same values twice. The length of stroke should therefore be sufficient to traverse a particular feature. It should be borne in mind however, that since the surface appearance varies, it is desirable to take at least three separate traverses to allow for accurate analysis.

In the case of surface grinding with the use of wheels up to $1\frac{1}{4}$ in. in width, we have found that in the majority of cases a stroke length of about $\frac{1}{4}$ inch is sufficient. The use of a traverse length less than this may lead to inconclusive results for reasons previously highlighted. Likewise use of a traverse length greater than $\frac{1}{4}$ in. can be superfluous since the stylus may well cross similar features two or more times.

Microinch and Micrometer

English speaking countries generally record surface measurement values in terms of the microinch (μ-inch). Continental countries, being totally metric, with the exception of Holland, tend to use the micrometer (μm); the Dutch use the microinch. One micrometer is equivalent to 39.37 microinches. The micrometer is equally commonly referred to as the "micron" which is, by definition, one thousandth of a millimeter.

A point of interest is that the Dutch established a parameter known as ru, which was an attempt to integrate the metric and English systems. While this parameter is still in use in Holland, it is not used elsewhere. Unlike R_a, ru is not an average parameter, it is a unit of measurement: 1 ru $\cong$ 1 microinch. This system yields whole numbers for ease of specification and verbal description.

Micro Pickup Types

There are basically two different types of pickup available with most surface measuring devices. The most common is that in which the pickup arm is supported on the surface being measured by a sliding skid (Fig. 3.55). The

FIG. 3.55. Illustrates the micro pickup supported by a sliding skid.

second type is more complex and is one in which a datum plane is used (Fig. 3.56); this datum plane is aligned parallel to the feed slide of the gearbox. A greater degree of accuracy in recording the values of waviness and roughness will be attained using the second type, yet the difference between

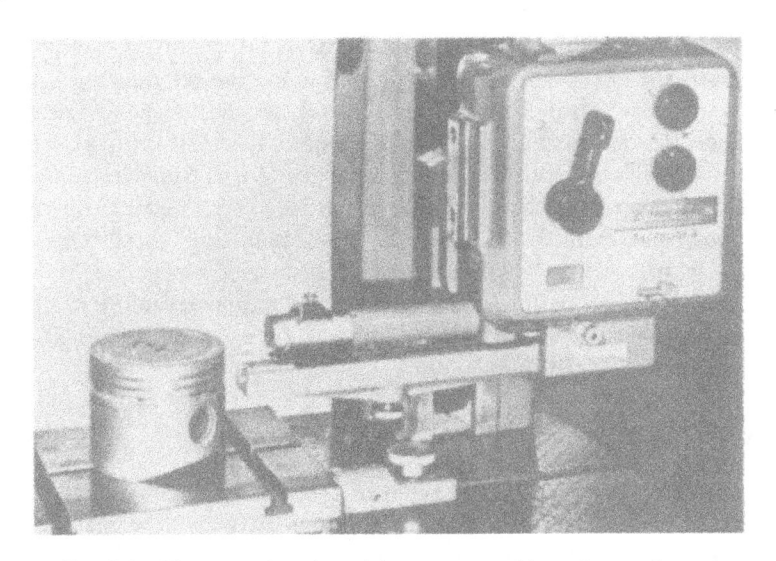

FIG. 3.56. Illustrates the micro pick-up supported by a datum plane.

the two is only marginal, and for measuring normal surfaces the sliding skid type is perfectly adequate.

SUGGESTIONS FOR FURTHER READING

Apprenticeship Standards for Tool and Die Makers, Washington, D.C: National Tool, Die and Precision Machining Associations.

Kropacek, Richard F., *Why Superfinish*, Chicago: Acme Scientific Co., 1975.

Leonard, LaVerne, Unleasing CAD/CAM potential; Interfacing product design, moldmaking, *Plastics Machinery & Equipment*, July, 1984.

Molded injection cavities cut tooling costs dramatically (editors report), *Plastics Technology*, vol. 29, Jan. 1983.

Pallante, Martin, Mold texturing adds product value at low cost, *Plastics Engineering*, p. 45, Feb. 1984.

Schurmann, Peter, Guidelines for successful hobbing, *Plastics Machinery & Equipment*, Oct. 1983.

Wilder, R. B., CAD/CAM; What it will mean to you, *Plastics Technology*, p. 55, Apr. 1982.

Chapter 4 / Materials for Mold Making

Revised by Charles C. Davis, Jr. *

Every mold designer must have a basic understanding of the various types of materials used in mold making (Fig. 4.1). The alert tool designer will keep informed on the new materials and methods introduced from time to time and note their adaptability to special requirements. In reading trade literature and in talking to vendors, he will require a basic understanding of the fundamentals of the making and heat treatment of steel, and of the terminology peculiar to the craft. Many materials other than steel are being used for molds, and any one of the several kinds may be considered and evaluated with respect to its suitability for the type of work to be done. Advances in the art of molding and mold building may at any time in the future permit the use of materials or methods not now feasible. This chapter has been prepared with the intention of presenting the factors of greatest importance in the selection of mold materials and in the design of plastics molds. **

The mold designer has many factors to consider. He must know processing conditions and production requirements. The mold contemplated must be correct, to produce parts economically, and tough to withstand the hard use to which it will be subjected (see Table 4.2). The cost of the material which goes into the mold is the least important consideration, but the hours of labor used in the construction of mold parts represent an important investment that will be lost if the design is poor or the materials unsuitable. In view of these and other considerations, the final mold design usually represents many compromises in achieving a happy balance between styl-

---

*We are indebted to William Young for the data supplied in the 3rd Edition for the material in the section entitled STEEL. Major revision by Charles C. Davis, Jr. for this edition.

**Special data on materials for blow molds are given in Chapter 3. Data on standard mold frames and mold components are given in Chapter 8.

FIG. 4.1. A six-cavity hot runner, full automatic injection mold for a crystal clear cup dispenser. (*Courtesy Stokes-Trenton, Inc., Trenton, NJ*)

ing, engineering and cost factors. The greatest care and consideration must be given to the selection of the materials used in mold building.

STEEL

Steels Used in Mold Construction

In reference to mold building, steel may be defined as a metal which has properties that make it useful as a material from which to form the main body of tools. *Tool steel* distinguishes by method and care in manufacturing, a quality of steel suitable for use in the making of tools, as opposed to structural steel, etc. Hardenable alloys of iron made with high level quality control may also be suitable for the making of molds.

There are four general classes of steel used by the mold maker. These are:

1. Low carbon steel (less than 0.20% carbon). This steel does not contain enough carbon to cause it to harden to any appreciable extent when heated and quenched.
2. Medium carbon steel (0.20 to 0.60% carbon).
3. High carbon steel (0.70 to 1.30% carbon).
4. Alloy steels. Steels of this classification contain various elements besides carbon, each element serving to contribute a definite property to the material.

**TABLE 4.1   Spectrum of Materials Used in Building Molds—
Arranged in Order of Surface Hardness.**

Material Class	Surface Hardness	Core Hardness
Carbides	68-75 Rc	68-75 Rc
Steel, nitriding	>68 Rc	>38 (varies)
Steel, carburizing	60-65 Rc	20-42 Rc
Steel, water hardening	67 Rc	40-55 Rc
Steel, oil hardening	62 Rc	40-60 Rc
Steel, air hardening	60 Rc	60 Rc
Nickel-cobalt alloy	45-52 Rc	45-52 Rc
Steel, prehardened	44 Rc	44 Rc
Beryllium copper	28-42 Rc	28-42 Rc
Steel, prehardened	28-32 Rc	28-32 Rc
Aluminum bronze	188 BHN	188 BHN
Steel, low alloy & carbon	180 BHN	180 BHN
Kirksite (zinc alloy)	80-105 BN ①	80-105 BN ①
Aluminum alloy	60-95 BN ①	60-95 BN ①
Brass	50 BN ①	50 BN ①
Sprayed metal	<50 BN ①	<50 BN ①
Epoxy, metal filled	85 RM	85 RM
Epoxy, not filled	80-110 RM	80-110 RM
Silicone rubber	15-65 Shore "A"	15-65 Shore "A"

Legend: Rc = Rockwell hardness = "C" scale; RM = "M" scale; BHN
= Brinell number 3000 kg; BN ① = Brinell number 500 kg load. (*Courtesy of Stokes-Trenton, Inc., Trenton, NJ*)

Steel is made from iron. In the making of steel, elements such as carbon, manganese, chromium, nickel, tungsten, vanadium, molybdenum, and cobalt are added, singly or in combination, in order to impart a desired quality (or qualities) to the metal.

Steel is a material which may be machined in a relatively soft state and then hardened by heat treatment to achieve improved wear resistance, toughness, strength, dimensional stability, and improved resistance to distortion under stress. The pressure in a mold may build up to a very high value under certain conditions, and this pressure may crush or distort the hardest and toughest steel mold section. Unhardened molds will not withstand high volume production in high pressure molding processes because they lack compressive strength.

Most tool steels are generally made in the electric arc furnace. Some steels are further refined by the Vacuum Arc Remelting or the Electroslag Remelting process. During the Vacuum Arc Remelting and the Electroslag Remelting process, an ingot which has been made previously by the

TABLE 4.2.   Spectrum of Materials Used in Building Molds—Arranged in Order of Surface Hardness as Commonly Used.

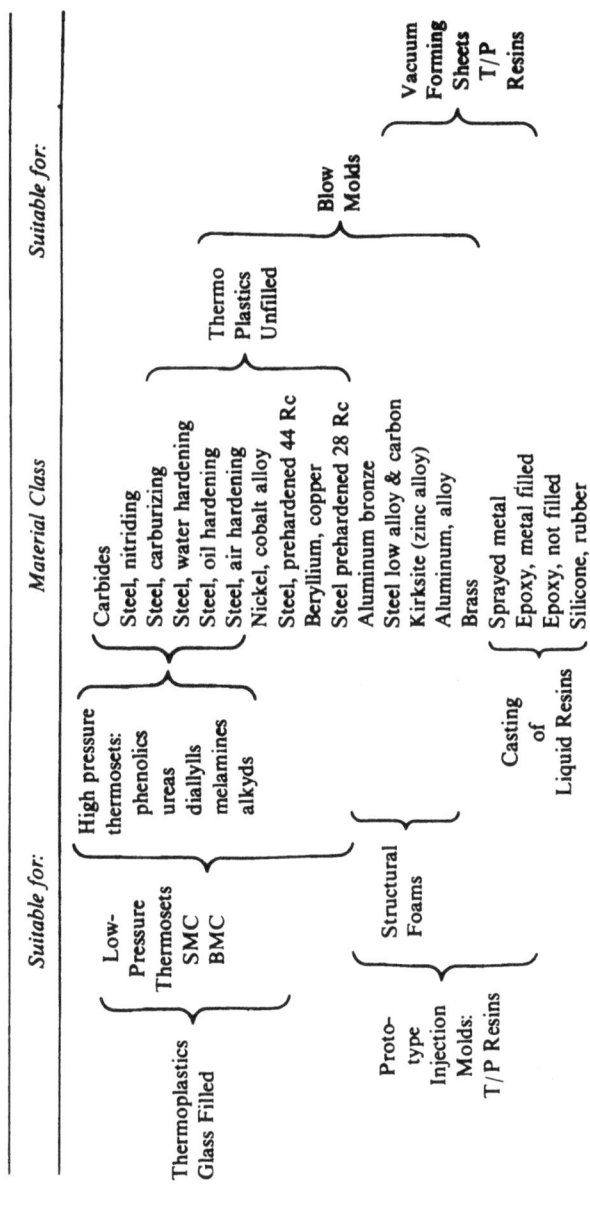

(Courtesy of Stokes-Trenton, Inc., Trenton, NJ)

electric arc melting process, is remelted by using it as an electrode. Figure 4-2 illustrates the Electroslag Remelting Process. The (ingot) electrode, which is the cathode, is slowly melted away in small droplets. These droplets float through a molten slag bath which has a purifying action and protects the liquid metal from oxidation. It solidifies quickly in the water-cooled mold, which also tends to form it in a more uniform structure. In this process, no special atmosphere or vacuum is used to protect the metal. The protection is rendered entirely by the layer of liquid slag.

In the Vacuum Arc Remelting process, no liquid slag is used to protect the metal, because the process is contained in a vacuum. It is more costly because the equipment is more sophisticated.

In preparing the charge for the electric furnace, pure iron or carefully selected scrap is used, together with alloy materials in the percentages stipulated. Inasmuch as the chemical composition of every constituent used in steels is known, various elements can be brought into association in the proper ratio to achieve a desired result. "Cold melt" electric furnace steel is made from a charge of cold materials, the name serving to differentiate it from production steels produced by charging the electric furnaces with molten steel from an open-hearth furnace.

When the steel has been melted and refined it is poured into iron molds to form ingots. Tool steel ingots may range in size from 6 in. up to 70 in.

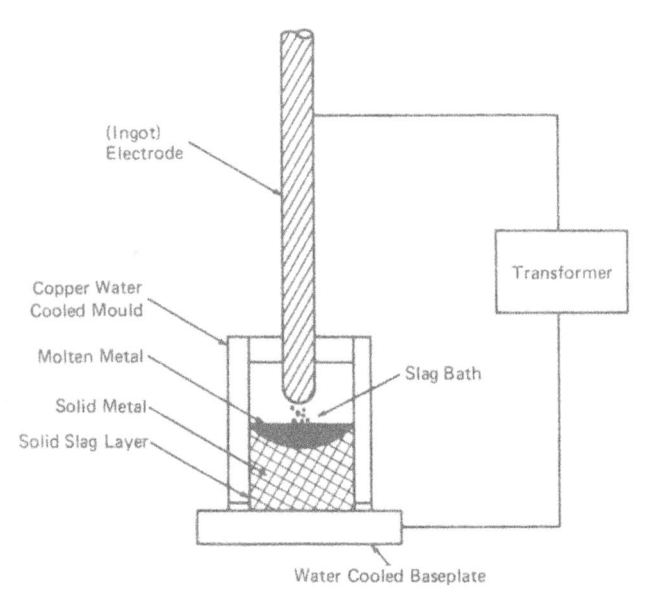

FIG. 4.2. The electroslag remelting process. (*Courtesy Crucible Specialty Metals Div., Syracuse, NY*)

in diameter. Most steel makers subject the liquid metal from the electric furnace to a degassing process which removes most of the hydrogen and oxygen dissolved in the metal. This is usually done by means of a vacuum or by introducing Argon into the liquid metal. The removal of most of the dissolved gases will result in a cleaner steel which is very desirable in plastics mold steels. There are several methods of degassing. The most popular ones are shown in Fig. 4.3.

On the left (A), the ladle of liquid steel is put into a sealed vessel from which the air has been pumped out to create a vacuum. This is called *ladle degassing*.

On the right (B) of the sketch, the ladle is sitting on top of the vacuum vessel. The liquid metal is poured into the ingot mold which has been placed into the vacuum vessel. This is called *vacuum ingot degassing*.

When the charge of steel is poured into the mold, precautions must be taken to insure uniform density in the ingot. The metal in contact with the mold solidifies first, the center of the ingot being the last to become solid. During solidification, or "freezing," the steel shrinks, and as the outer surface freezes and shrinks, metal is drawn from the center of the ingot. Unless precautions are taken to provide extra metal to fill in the portion thus reduced, the center of the ingot may be left hollow or, at least, will be of low density. The defect resulting is called *pipe*. Ingots containing pipe should never be used for making tool steel.

Ingots are converted into billets in the next stage of the manufacture of tool steel. This is done by a process known as *cogging*. In the cogging

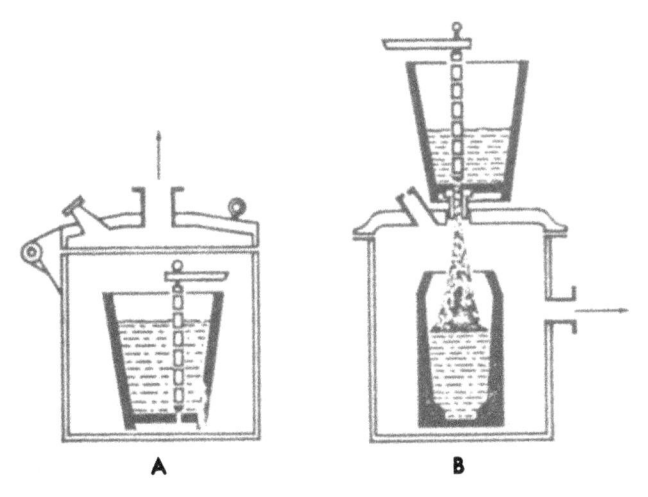

A   **B**

Fig. 4.3. Degassing methods. A Ladle degassing process. B. Vacuum ingot degassing process.

process, the ingot is heated to the proper temperature and then worked to the desired size and shape by a process of hammering, pressing, or rolling. Some types of steel, such as high-speed steel, should always be hammered or pressed in the forming of billets. These billets may be round or square. The cogging operation is used to form the billet to the shape desired and to reduce the cross-sectional area of the ingot and increase its length.

Surface defects may develop during cogging, such as seams, laps, or cracks. These must be removed before further work is done on the billets. Removal of the defects is accomplished by one of three methods, namely: chipping, grinding, or rough turning. Billets are sometimes pickled in acid to remove scale and to "point up" defects. The chipping process makes use of air-hammers and gouge chisels to remove seams. Grinding is usually done with swing-frame grinders. High-speed steel billets usually require grinding, and they are frequently ground all over. Deep seams are often removed in a milling machine operation which cuts away the defective area. Some types are milled all over by the use of a slab milling cutter which cuts away the entire surface. Billets which require removal of the entire surface are generally cogged into round shapes so they may be turned in a lathe.

Billets prepared in the manner just described are then rolled or hammered to a specified size. In the rolling of tool steel, the cross section of the billet must be reduced slowly, and this is accomplished in repeated passes through the rolls, a small reduction occurring with each pass. Care is exercised to do the rolling while the temperature of the steel is between maximum and minimum temperature limits. After being rolled, the bars must be annealed to remove the stresses and to make them soft enough for machining.

Annealing is done by heating the steel to the correct temperature and then allowing it to cool slowly. Scale will form on the surface if this is done in air, and to prevent this, the bars are annealed in an atmosphere controlled furnace or are pipe-annealed. In the pipe-annealing process, the bars are placed in large pipes and surrounded with a material that resists oxidation. (Cast iron chips are generally used for this purpose). The pipes are then sealed and the entire charge is heated to the annealing temperature. On reaching this temperature, the charge is allowed to cool as slowly as required for the type of steel being produced. This usually takes 18 to 24 hours. Today much tool steel is being annealed in atmosphere controlled furnaces. The atmosphere in the furnace is adjusted to prevent oxidation or scaling of the metal.

In the various heating cycles required for cogging, rolling and annealing, some carbon will be oxidized by air in contact with the surface of the

TABLE 4.3. Minimum Allowances for Machining and Maximum Decarburization Limits.

Rounds, Hexagons and Octagons

Minimum Allowance Per Side for Machining Prior to Heat Treatment, Inch

Ordered Size, In.	Hot Rolled	Forged	Rounds Rough Turned	Cold Drawn
Up to ½, incl.	.016	—	—	.016
Over ½ to 1, incl.	.031	—	—	.031
Over 1 to 2, incl.	.048	.072	—	.048
Over 2 to 3, incl.	.063	.094	.020	.063
Over 3 to 4, incl.	.088	.120	.024	.088
Over 4 to 5, incl.	.112	.145	.032	—
Over 5 to 6, incl.	.150	.170	.040	—
Over 6 to 8, incl.	.200	.200	.048	—

(Courtesy of Crucible Specialty Metals Div., Syracuse, NY)

metal. This forms a skin which may not retain its full percentage of carbon and consequently produces a condition known as surface decarburization, or "bark." Surface skin must be removed, and steel producers universally suggest the removal of a small amount of surface metal before using hot-rolled bar stock for tool-making purposes. The allowances given in Table 4.3 show the minimum amount which must be cut from hot-rolled bar stock for complete removal of all decarburized surface.

Some steel is prepared as *cold-drawn bars* for certain uses. In the cold-drawing operation, the annealed bars are drawn through a die in a stretching operation which reduces their cross section and provides an accurate outside diameter. In the preparation of drill rod, these bars are ground in a centerless grinder to remove all decarburized surface and provide close control of the outside diameter.

STEEL FORGINGS

Many of the larger molds are constructed from special forgings.* In making these forgings, ingots or billets are pressed or hammered to form the desired shape. Large ring-shaped pieces are frequently made up as special forgings, the cavity area being formed roughly to shape in the forging operation to eliminate the necessity for machining away a solid center area.

---

*Many steel companies now carry a stock of steel forgings in cylindrical and rectangular shapes in various sizes and annealed or prehardened.

Some forge shops carry a stock of tool steel billets which they are prepared to forge to the mold builder's specification.

MACHINABILITY

Some grades of steel will undergo certain machining operations with greater ease than others. The machinability of steels may vary with the annealing process. It is possible to anneal specially a piece of steel to give it better machinability for a given process. For obtaining the maximum amount of machinability, steel stock from which mold plungers are cut on a duplicator may require a different annealing process than a block of steel that is to be fabricated principally by boring. The machinability of a given stock will vary with the type of machining to be done, and will, to some extent, be determined by the type of annealing to which the stock has been subjected. Mold steels are usually annealed to provide the best average machinability. There is no reliable yardstick with which to measure machinability since it is influenced by a particular machinist's experience. However, as a rule, the more alloying elements (such as nickel and manganese) a steel contains, the more difficult it is to machine.

Hardness of the steel is another decisive factor. Other things being equal, cutting difficulties increase with hardness. Hardness greater than 350 Brinell usually requires special tooling and rigid machine set-ups. Machining of such high-hardness material should be limited to finishing only.

Sulfides in steel improve machinability because chips break shorter, allowing steel as hard as Rockwell "C" 45 to be machined successfully. There is, however, the possibility that sulfide particles may impair both mold polish and weldability.

HEAT TREATMENT

Heat treating a mold always introduces risks of distortion and cracking. Permanent linear movements in steel during heat treatment are to be expected. It is impossible to predict accurately the extent or direction of movement, since chemical composition, mass, geometry, design and heat treating techniques all affect the final dimensions of a mold.

Molds made from low carbon steel (0.20% or less) are usually carburized. This will increase the carbon on the outside so the mold surface will be hard enough after hardening. Few molds should be built with this elementary form of hardenable steel because the interior is soft. Molds made from the higher carbon and alloy steels have sufficient carbon to give the desired hardness, and tend to harden all the way through.

Carburizing temperatures and hardening temperatures, which vary with the type of steel being used, are given in the technical data furnished by the steel producers.

Care must be taken to protect the mold surface against oxidation. This is done by packing the mold into spent cast iron chips or pitch coke, by heating in a controlled atmosphere furnace, or by heating in a vacuum furnace.

After heating the mold to the hardening temperature, it is either quenched in a liquid such as oil, or it is allowed to cool in air, depending upon the analysis of the steel. High alloy steels harden sufficiently when cooled in air from the hardening temperature.

TEMPERING

After quenching, the mold is reheated to a designated temperature ranging from 350 to 1150°F to obtain the desired hardness and to relieve the stresses created during the hardening operation. Mold hardness may vary from 30 to 65 Rockwell C depending upon requirements.

It should be noted that the tempering process, while relieving the major stresses caused by quenching, induces minor stresses inherent to the process resulting from crystalline structure changes. Best results are obtained by a second tempering, or what is known as double tempering.

ANNEALING

Annealing, or reducing the hardness of steel, is accomplished by heating the metal to a temperature just above the critical point and then permitting it to cool very slowly.

STRESS RELIEVING

Distortion can be caused by residual stresses. Steels that have been subjected to severe grinding, hogging and cutting operations become highly stressed. Stresses induced by these operations must be relieved or distortion may occur during heat treatment.

Molds from which considerable metal must be removed should be rough machined to within ⅛ to ¼ in. of the final size and then stress relieved. The mold should be heated to 1250 to 1300°F, and held at this temperature for as many hours as the greatest inch-thickness dimension. For instance, a 10-in. thick mold should be held at the stress relieving temperature for 10 hr. After the mold has cooled to room temperature, it can be finish machined, allowing for expected material size changes that will occur during subsequent heat treatment.

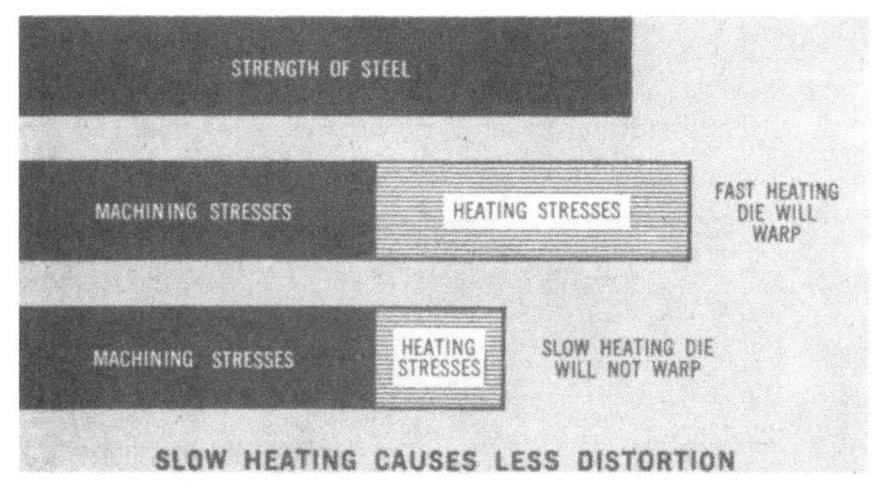

FIG. 4.4. Stresses from rapid heating after machining may cause warping of mold.

Another source of distortion is too rapid heating during heat treatment. Rate of heating should be slow enough so that all portions of the mold are at practically the same temperature.

During rapid heating, thin sections expand faster than thick sections. This causes stress in the junctions which, if greater than the yield strength of the steel, will cause the mold to distort. Distortion can also occur during rapid heating when the thin sections reach the critical temperature first and start contracting while the thick sections are still expanding.

Figure 4.4 illustrates the advantage of slow heating. During slow heating, the combined stresses are below the yield strength of the mold and no distortion occurs. During fast heating, the combined stresses are greater than the yield strength and the mold distorts. The mold cracks if combined stresses exceed the breaking strength of the steel.

Distortion problems can be prevented by using a prehardened steel. Steels such as the AISI P20 type are usually supplied with 290-330 Brinell hardness. They are easily machined and polished to a fine finish suitable for most injection molding applications. It is advisable to stress relieve molds made from prehardened steel if the mold is complicated or has sharp radii and corners.

Stress relieving should be done at about 900°F in the case of AISI P20 type steel, and the mold should be held at that temperature for as many hours as the greatest inch-thickness. The mold can then be air-cooled. The oxide film that forms during stress relieving is removed by final stoning and polishing.

HARDNESS PENETRATION

When tool steel is quenched, the outer surface, which is in contact with the quenching solution, cools very quickly. The inner core cools relatively slowly. Steels that must be quenched rapidly to give them proper hardness, will have an outer shell that is hard and an inner core that is relatively soft owing to delayed action of the quenching. This outer shell may be 3/32- to ⅛-in. deep in a water-hardening tool steel. The use of certain alloys, such as silicon, molybdenum, chromium and manganese may increase the hardness penetration. A deep-hardening steel is one that hardens to a considerable depth when cooled slowly; a shallow-hardening steel requires rapid quenching. The oil-hardening steels show greater hardness penetration than the water-hardening steels, while the air-hardening steels show the greatest degree of hardness penetration. Molds subject to large deflection should not have a high degree of penetration.

MOLD STEEL REQUIREMENTS

Plastics molds require very carefully produced grades of steel of varied types for successful mold building. Certain qualities are essential in steels that are to be used for molds. These are:

Cleanliness. A good mold steel must be clean; it should not contain nonmetallic inclusions which will cause pitting during polishing.

Soundess. The steel must be dense and free from voids and porosity.

Structure Uniformity. It must be uniform in structure and relatively free from alloy segregations. It also must be uniform in chemical analysis.

Machinability. The machinability of a grade of steel depends upon the hardness of the steel, its composition, and upon its microstructure. Softness is not necessarily assurance of machinability, and extreme softness is just as undesirable as extreme hardness. Steels which machine easily and uniformly are needed for economical mold construction.

Hobability. Hobbing steels must be very soft when annealed, and be clean and ductile as well. Mild steel may be hobbed easily, but the specially developed hobbing steels are preferable in that they offer the maximum of cleanness and uniformity. Ingot iron and the low alloy steels are easiest to hob. The higher alloy content steels (including carbon steels) offer some difficulty but give the best results in service.

Hardenability. Good mold steels must acquire the desired hardness in the heat-treating process, uniformity, a strong surface, and a tough, strong core. Ingot iron and low carbon steels will harden in water, but they may distort and become useless. Alloy steels harden in oil or air and show a minimum amount of dimensional change during hardening.

Strength and Toughness. Molds require a hard surface and a very tough core—the larger the mold, the greater the core strength needed for resisting distortion or cracking.

Heat-treating Safety. An important characteristic of a good mold steel is its ability to be hardened satisfactorily in a wide range of sections by a variety of methods while producing uniform results.

Finish. All mold steels must be able to take a mirror-like finish easily, although a dull surface is often used as the desirable final finish.

Wear Resistance. Wear resistance is a fundamental requirement of a good general-purpose mold steel. Some of the plastics cause little tool wear, others, such as the glass and asbestos-filled compounds, require the maximum amount of wear resistance.

SELECTING THE STEEL*

As the plastics industry developed and presented new materials and molding methods, larger and larger moldings were developed and the steel makers cooperated by building the larger facilities required to forge the huge blocks. They also provided stronger and tougher alloys containing a minimum of impurities available in all the larger sizes. The steel makers also met the demand for mold materials that would withstand the effects of corrosive and abrasive plastics in cooperation with the electroplating industry.

Plate Steel

Plate steel is a low carbon steel such as SAE 1020 produced by the openhearth or other inexpensive processes, wherein cleanliness is a less important factor than volume. This material is used almost exclusively for the frames of molds. Plate steel can be carburized and hardened or casehardened. It is sometimes used to make cavities and plungers, but this application is not recommended because of the low core strength of this steel, and also because structural faults, such as pipe, seams, pits, and other defects, are common to it. Plate steel should not be used for cavities or plungers on any but the cheapest of molds.

There are several qualities of plate steel available and if any pressure is to be concentrated on the plate, the better grades should be selected. Some mold builders use the cheaper grades of boiler plate for clamping plates, parallels, etc., and the better grades (something like SAE 4140) for the back-up plates, steam plates, or other members on which stresses may be concentrated. This practice requires that a large inventory of stock be carried, therefore it will be found wiser to use the better grades of plate throughout.

*See also *ASM Metals Handbook*, 8 Ed., Vol. 1, page 768 for additional data.

The saving obtained by using the cheaper grades makes a negligible difference in the total mold cost, as it amounts to only a few cents on the pound.

Mold plates are usually made from plate steel, while knockout bars and pin plates are made from machinery steel bars or cold-rolled steel. It is possible, when stock sizes can be obtained, to use cold-rolled steel, unless these parts require finishing, and in that case, machinery steel would be indicated.

Machinery steel is of the same general class as the SAE 1020 plate steel. The difference is that machinery steel is hot-rolled into flat or square bars and round rods. For many applications these bars can be used without any special finish except a surface grinding on both sides to produce flatness. With the modern cutting torch, it is usually a question of which of these two materials is more readily available, regardless of specification.

Tool Steel

Tool steel was the first material widely used in mold making. Some shortcomings and disadvantages led to the development of special alloy steels for making plastics molds. Three general types of tool steel are available: water-hardening, oil-hardening and air-hardening.

Tool steel affords only fair machinability, and it is not suitable for hobbing. After being hardened, a block of tool steel has nearly the same hardness all the way through, and may lack toughness. As a result, the mold may tend to break rather than distort when excess pressure is applied. The initial cost of tool steel is high. It is frequently used for injection molds because it does not distort or deform under high pressure as easily as other steels. This advantage is gained as a result of its being hardened all the way through. Deep hardening tool steel will give maximum life when properly applied.

Water-hardening tool steel may be used when maximum hardness is desired. It distorts considerably when hardened, therefore ample allowance for correction after hardening must be made. Water-hardening tool steels are not to be recommended generally for plastics molds.

Oil-hardening tool steel and, to a greater extent, air-hardening tool steel may be used when distortion must be held to a minimum and are recommended.

STANDARD MOLD COMPONENTS

In many situations the mold designer can save hours of decision-making in respect to material selection, if he will use standard mold frames where they are suitable, and standard mold parts, which are inevitably indicated for use, in some degree, in any mold he may build. (See Chapter 8.)

The volume of purchasing and mass fabrication facilities afforded by the principal suppliers of standard mold parts assures the mold builder of economy and reliability he can not otherwise achieve, except in unusual circumstances. He can expect that the quality of the materials, and the heat treatment, if any, employed for these products are better controlled than it is possible to do in a shop with less demand and supervision.

He must, however, be aware of the differences and intended use of the various grades of steel available in standard frames and plates. Besides the SAE 1020 plate steel mentioned above, generally supplied in an analysis up to SAE 1040, as the lowest grade of steel, most suppliers offer better choices for more severe service, of SAE 4130 to 4150, prehardened to 26 to 29 Rockwell C, and even P-20, prehardened to 28 to 32 Rockwell C.

Besides mold bases, and individual plates ground and sized to close tolerances, there are leader pins, bushings, ejector pins, sprue bushings, premachined cores and tool steel cavity blocks.

Other items which have been added to the ever-growing list, as demand increases, are such more complicated, heat-treated and assembled devices as; early returns (for the knockout system), latch-lock mechanisms to determine the order of opening of the various levels of a mold, accelerated knockout systems, hot runner probes, valve-gated hot runner sprue bushings, and even a selection of sizes of collapsing cores. The availability and quality of manufacture of these latter items allow the designer to specify them and tailor them to his specific needs, and serve to free him from having to design the mechanisms themselves. As the result of previous wide application of these devices, he can be assured that they are largely foolproof when installed by experienced personnel.

ALLOY STEELS

The alloy steels differ from tool steels in several ways. Their carbon content has been reduced and various elements have been added to give the steel the benefit their unique properties provide. Table 4.4 lists some of the ele-

TABLE 4.4. Properties of Elements of Alloy Steels.

Element	Property or Agency
Silicon	Hardness.
Carbon	Hardening agent.
Manganese	Deoxidizes during manufacturing process and adds strength.
Nickel	Toughness and strength.
Chromium	Hardness. Adds to abrasion resistance in high carbon compositions.
Vanadium	Purifier—also adds fatigue resistance.
Molybdenum	Widens heat-treating range and adds heat resistance.
Tungsten	Hardness and heat resistance.

ments used in making the various types of alloy steels and gives the properties peculiar to each.

STAINLESS STEEL

While there are many alloys classed as stainless steels, only a few need be considered for use in high-pressure molds. Because of the necessity for ability to harden, the AISI Types in the 400 Series alone are suitable. Of these, type 420 is the most commonly used. It contains 12 to 14% chromium (see Table 4.5) and can be satisfactorily heat-treated to develop a through hardness of 48 to 52 Rockwell C. It is used where corrosion and rusting are

TABLE 4.5. Stainless Mold Steels.
Identifying Elements, in Percent

Type	C	Mn	Si	Ni	Cr
T410	.15	1.00 max.	1.00 max.		12.00
T414	.05	1.00	1.00	2.00	12.00
T420	.25	1.00	1.00		13.00
T440	.60/1.20	1.00	1.00		17.50

problems. Type 420 may also be used in injection molds for corrosive materials such as the vinyls, particularly in moist or humid conditions. Gold or chrome plate is also used for further protection. Type 414 is a prehardened stainless steel having a hardness of 30-35 Rc. It does not require further heat treatment after the mold is finished. It is suitable for injection molds where pressures are relatively low. Like its Type 420 counterpart, it is recommended for molding corrosive materials where rusting is a problem. It does not need to be chrome plated. It should be noted that the thermal conductivity of stainless steel is lower than that of non-stainless steels. This causes no problem for parts of heavier cross sections, where molding cycles are slow; this property actually can be used to advantage in separating hot from cold zones, for example, in hot-runner molds. Also, because of the lower thermal conductivity, the molten plastic entering the mold doesn't receive as sudden a chill, and lower melt temperatures are used to accomplish cavity filling.

MARAGING STEELS

These newer types of steel have recently been used for some small plastics molds. The maraging steels are machined in the annealed condition. The

TABLE 4.6. Maraging Steels.

Element	C	Ni	CO	Mo	CR	Ti	AL	Cb
350 Type, 18% Nickel Maraging Steel	0.02	17.50	12.00	4.80		1.50	.10	.25
300 Type, 18% Nickel Maraging Steel	0.03	18.50	9.00	4.80		.60		
250 Type, 18% Nickel Maraging Steel	0.03	18.25	7.75	4.80		.40	.10	
200 Type, 18% Nickel Maraging Steel	0.03	18.25	7.50	4.25		.20	.10	
220 Type, 12% Chromium Maraging Steel	0.02	10.00			12.00	.35	1.30	

finished mold is then reheated to about 900° F, held for about 5 hr, and cooled in air. This treatment, called *aging*, results in an average hardness of 50 Rc, depending on the type of maraging steel (Table 4.6) which has been used. These steels shrink during aging and allowance must be made for this by the mold maker.

The mold maker and mold user should consult with the steel supplier in order to obtain all necessary information about this material.

STEEL FOR MACHINED MOLDS

Commercial low carbon machinery steels are nearly ideal when machinability is the one important consideration. These steels are usually high in phosphorus and sulfur content, however, and frequently they carry nonmetallic inclusions which mar the finish of both the mold and the molded part.

Steels which are to be used for cavities and plungers and that will be fabricated by machining are usually of the AISI-P20 or SAE 3312 (P6) type. AISI-P20 steel is used most extensively for large molds that have surfaces of considerable area exposed to the molding pressures. The P20 steel has become quite popular because of its versatility. It can be used for injection, compression and transfer molds.

For injection molds (Fig. 4.5), P20 steel is usually used in the prehardened condition, Rc 30–35, where it is readily machinable. No heat treatment is necessary. It can be chrome-plated when corrosion could be a problem. It can be nitrided when abrasion is encountered. Parting lines, or selected areas of the mold can be flame hardened if peening can cause damage such as that experienced with SMC. If a higher hardness is required, the mold can be deep hardened to Rc 45–50.

FIG. 4.5. A typical injection mold of the type often made from prehardened P-20 Steel. (*Courtesy Stokes-Trenton Inc., Trenton, NJ*)

For compression and transfer molds, the P20 mold is carburized to a depth of .030–.065 in., depending on design and requirements. The surface of the mold can attain a hardness as high as 60 Rc and the core about 45–50 Rc.

The SAE 3312 (P6) type must be carburized, resulting in a good combination of surface hardness and core toughness.

HOBS AND HOBBING

Mold cavities are often made by machining or hobbing (hubbing). In the hobbing process, a piece of steel called a *hob* (or *hub*) is used to form the actual cavity section. The hob is carefully machined to the desired shape, and after it has been hardened and polished, is used to "sink" the actual mold cavities. In this operation, the hob is pressed cold into the block of steel which is to contain the cavity, thus forming a duplicate of its own shape and dimension. Hobbing requires the application of a great amount of force. This process is generally used for forming only the mold section for the outside contour of the part. The remainder of the cavity block is finished in subsequent operations which form the "lands" and the loading space; and finish the block to the exterior size required for fitting the mold frame. Selection of the steel for the hobs and hobbing should be carefully considered. A list of the most commonly used hob steels and some of their properties is shown in Table 4.7.

TABLE 4.7.   Steels Recommended for Master Hobs.

AISI Type	Machinability Rating	Wear Resistance	Compressive Strength	Toughness	Resistance to Softening and Heat Checking	Dimensional Stability in Heat Treating
O-1	Good	Good	Good	Fair	Poor	Good
A-2	Fair	Good	Very good	Good	Poor	Very good
D-2	Poor	Very good	Very good	Poor	Fair	Very good
S-1	Good	Fair	Good	Very good	Fair	Fair
M-2	Fair	Good	Very good	Fair	Good	Good
L-6	Fair	Fair	Good	Very good	Poor	Good

HOB DESIGN

While the actual design of hobs is usually left to the individual hobber, the designer should have a general idea of the problems involved to avoid production problems and, to enable the tool maker to make the finished hobbed mold section in the best possible way. A group of small master hobs is shown in Fig. 4.6. It will be noted that very few of these hobs are made for anything except a straight push. Most hobbings require two or more "pushes" in order to sink the hob to the full depth. Generally the hob is pushed ¼ to ½ in., after which it is removed and the hobbing is furnace-annealed. This is followed by another push of ¼ to ½ in. and subsequent annealing, the process being repeated as many times as necessary to complete the job.

Where a particular dimension on a hob is very important, it is the usual practice to make one push on all blanks and then to check the hob and make the necessary adjustments in size before completing the hobbing. These adjustments usually consist of correcting dimensions which have been enlarged by the "upsetting" force on the hob.

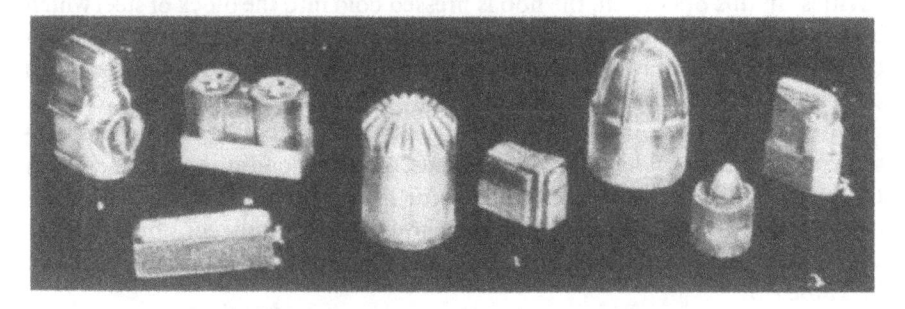

FIG. 4.6. Group of small master hobs used in the making of molds.

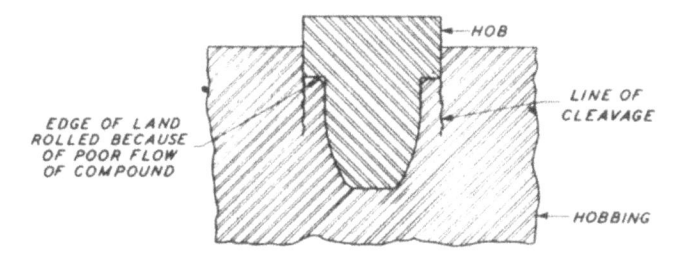

Fig. 4.7. Lines of cleavage resulting from hobbing of land area.

As a general thing, any landed plunger or similar mold should be hobbed without the land. The reason for this becomes apparent in a study of Fig. 4.7, which shows how lines of cleavage may result from the flat surface area being pushed down. These lines of cleavage may cause cracking during hardening or in service. There are, however, cases where this type of hob is used successfully. The effect of rolling the edge of the land may be overcome in either of the following ways: (1) A groove may be cut in the hob (see Fig. 4.8), thus permitting the steel to flow into the groove and later be machined to a sharp corner. (Allow a generous radius on outside corner to facilitate flow.) Or (2) an allowance of 1/16 to 1/8 in. on the depth of the well may be made for machining after hobbing. The only purpose in hobbing the land is to reduce the time required to finish-machine the hobbings.

Draft is provided on the hobs so that they may be removed from the hobbing more easily. The draft should be approximately ½° per side, although 1° per side would be much better. It is easier to hob cavities when adequate draft allowance has been made. The reverse is true when cavities are machine cut, since it is less difficult to machine a straight surface. The nearly ideal design for a hob is shown in Fig. 4.9.

The finished mold section will be approximately the same shape as the base of the hob. The size of blank used will have some bearing on the quality of the finished hobbing. Do not use a large blank where the hob has steep sides

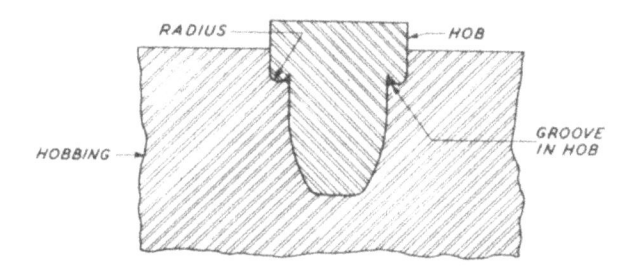

Fig. 4.8. Hob construction to allow for rolling of land.

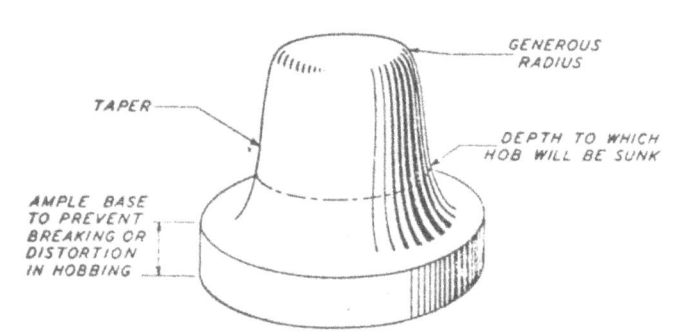

FIG. 4.9. General design of hob.

or a knurl, because the metal may flow away from the hob, with the result that the cavity will be larger than the hob and the exterior detail may be lost or rendered ineffective.

The hobbing blank must be relieved or contain a hollow section into which the displaced metal may flow during the push. Usually the relief is on the side or bottom. The side relief is usually not as effective as the bottom relief, and it may leave a "ghost ring" that cannot be removed and will show on the molded parts. As a general thing, it is desirable to remove about two-thirds of the metal which is to be displaced by the hob. It is to be remembered that hobbing is a flowing of the metal as well as a displacement. This is illustrated in Fig. 4.10, (*A*), which shows that a tearing action took place at *x* because of hindrance to flow. By elevating the retainer ring on parallels, as shown at (*B*), the displaced metal was permitted to flow evenly around the entire area of the cavity bottom.

The following suggestions offer solution of some of the problems which occur in hobbing. They are inserted by permission of The Carpenter Technology Corp. of Reading, Pennsylvania.

"In the cold-hobbing of cavities, certain methods can contribute to im-

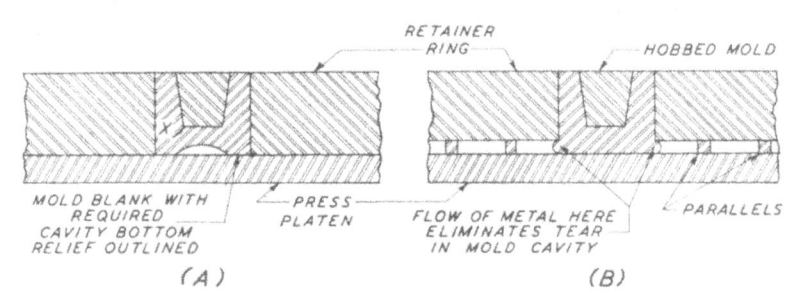

FIG. 4.10. Construction used to minimize checking during hobbing by allowance of space for material flow. (*Courtesy Carpenter Technology Corp., Reading, PA*)

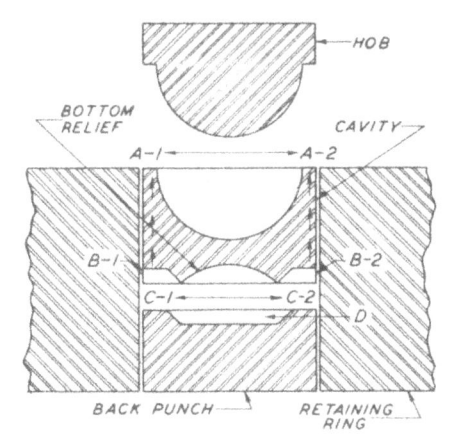

Fig. 4.11. Relief used to obtain proper flow during hobbing. (*Courtesy Carpenter Technology Corp., Reading, PA*)

proved results . . . and save considerable time. Here, for example, are two [methods] which may be found helpful. Fig. 4.11 shows what might be called a setup for cold-hobbing cavities, such as for plastic door knobs. As a rule, for a cavity such as this, the mold blank would be given the usual cupped out bottom relief. But the mold builder found that after hobbing the cavity, the dimension between points *A-1* and *A-2* opened up, becoming larger than the hob itself. So a back punch was designed to be used in connection with the retaining ring. The effect of this back punch while cold-hobbing is to push the metal up around the areas marked *B-1* and *B-2*, while the bottom relief is flattened on the back punch at *D*. Thus the reverse action of the flow tends to push the metal up along the side of the cavity, following the direction of the arrows. This causes the metal of the cavity to fit snugly on the hob at points *A-1* and *A-2*. There is no definite rule for the design of the back punch, except that the dimension between *C-1* and *C-2* should be 25 to 30% smaller than the dimension *A-1* to *A-2*.

"Figure 4.12 shows a method of cold-hobbing raised squares in the bottom of the cavity. Heretofore, it has been next to impossible in usual hobbing procedure to get these squares to come up to the proper height. This meant that some machining had to be done at the bottom of the cavity. The hobbing procedure here described overcame this difficulty. Fig. 4.12, at (*A*), shows the first step. The hob is seen to have square holes, and the special back punch slightly larger male projections. Note that the projections *C* and *D* on the special back punch have an angle of approximately 60° per side. This assembly is put under pressure in the retaining ring. The special back punch then forces the metal in the cavity to flow in the general direction of the arrows, thus helping properly to fill out the squares *A* and *B* shown in the

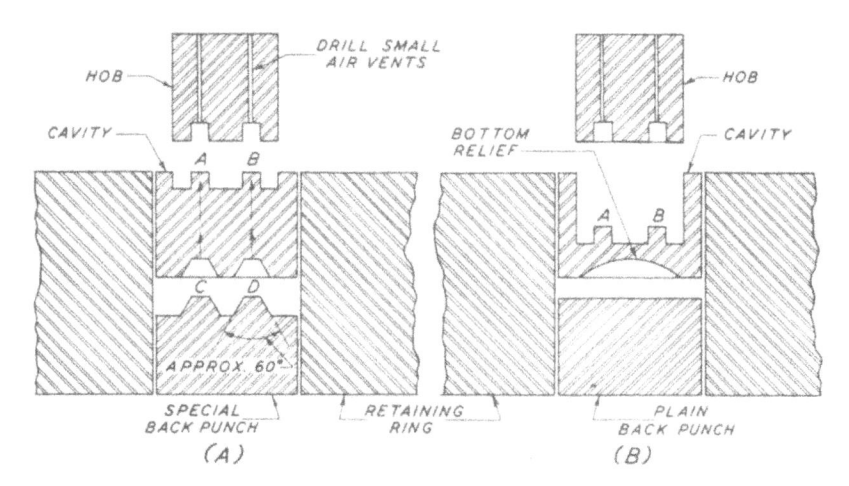

FIG. 4.12. Method used in cold-hobbing raised squares in bottom of a mold cavity. (*Courtesy Carpenter Technology Corp., Reading, PA*)

cavity. The second operation is illustrated in sketch (*B*). The partially hobbed cavity has now been cupped out with the usual bottom relief. The special back punch used for the first operation has been replaced by a plain back punch. The hob is then sunk to the desired depth. Always strain-relieve the mold by heat treatment after cold-hobbing."

The hobbing blank must be highly polished before hobbing, as this will impart a better finish to the cavity section. The lubricant generally used for hobbing is copper sulfate (blue vitriol). The hob is inserted in this solution for about five seconds to pick up the required coating of copper. Following the hobbing, the same polishing procedure used for the finishing is repeated for the cavity and plunger sections.

Some hobbers have special techniques to enable them to handle specific types of work that would otherwise be extremely difficult. One example is the production of mold cavities having proportions such that the depth of sink is many times the diameter, or smallest dimension across, at the parting line, such as cavities for pen barrels. The expert hobber proceeds a little differently from the steps described earlier, because a hob of these proportions would break, on first push, if forced into a solid blank.

Some experts carefully prepare a hole in the blank, to accept the majority of the length of the hob, before applying any pressure. Others make a series of hobs in progressive lengths and, starting with the shortest, alternately sink and anneal the blank until the shortest hob is down, and repeat this successively with each of the other hobs, in order of length. Extreme care must be taken that the hobs are finished to exactly the same contours, except for length.

Another example, long thought to be impossible, is the production of cavities with actual undercuts on the side walls, such as for decorated closures. The ability to release the hob after sinking in such a situation is the result of much study and experimentation. It depends on the expansion of the blank as it recovers from radial precompression applied prior to hobbing and made possible by ingenious design of the retainer ring.

HOBBING STEELS

Using AISI type designations, one is able to select the best potential hobbing steel alloy compatible with the expected difficulty of hobbing. Type P1 is the easiest of all types to hob, being almost pure iron. Other alloys range upward to type P6, which has sufficient alloying elements to yield a core hardness of 20 Rockwell C after receiving proper carburizing heat treatment, and which is difficult to hob. The alloys bearing P6 and higher type designation are more suitable for machining than for hobbing.

The steels most frequently used for injection molds are listed in Table 4.8. With so many different types of steel available, some mold makers and

TABLE 4.8. Steels for Injection Molds.

Steel AISI	Application
P20	Suitable for all types and sizes of machine-cut molds. Usually used in the prehardened condition Rockwell "C" 32 to 35. This should be carburized and hardened for low viscosity or glass filled plastics and for usage in excess of 100,000 pcs per cavity.
H13	Used for large and small molds when toughness and strength is required. Good dimensional stability during hardening. Hardens up to Rockwell "C" 52 but is tougher at "C" 48.
A2	For small and medium size molds when higher hardness is required as for molding abrasive materials.
D2	For small molds when abrasion becomes a problem. Also for molds operating at temperatures up to 750° F.
Type 420 Stainless	For small and large molds for molding corrosive resins, such as PVC and Delrin. Also used when rusting is problem because of "sweating" of mold surface. Heat treat precautions: after hardening, double temper at 750° F for highest toughness and best corrosion resistance. Hardness will be Rockwell "C" 48 to 52.
SAE 4140	Usually used for holders and shoes. Can be used for molds where a high finish is not necessary. Usually used in the prehardened condition Rockwell "C" 28 to 32.
M2 High Speed Steel	Use if operating temperatures are above 1000° F, but not higher than 1150° F, and the mold hardness must be higher than 60 Rc.

TABLE 4.9.   Steels for Compression and Transfer Molds.

Steel AISI	Application
P-20	Must be carburized 0.030 to 0.065% deep, depending on size. Surface hardness Rockwell "C" 60. Core hardness Rockwell "C" 45–50.
H13	Good for large compression molds if nitrided after heat treating to Rockwell "C" 48.
A2	For small molds. Good for automatic compression molds.
SAE 4145	Often used for holders and chases. Can be carburized to Rockwell "C" 60.
S1	For small and medium sized molds. Tougher than A-2 but not as stable, dimensionally in hardening. It may be oil-hardened. Use at Rockwell "C" 53–56.
S7	For medium and large molds. Has good combination of toughness and stability. Sections thinner that 2 inches will air harden—otherwise oil harden to Rockwell "C" 53 to 56.

Fig. 4.13. A compression mold composed of a cavity with a deep load space made from carburized P-20 steel, and with plungers made from through hardening S-1 steel. Chrome plated. (*Courtesy Stokes-Trenton Inc., Trenton, NJ*)

molders find it difficult to select the proper steel. It is estimated that about 90% of all high-pressure plastics molds are made from only six types of steel. It would be helpful if the industry could standardize on fewer steels. Steels could be produced more economically in larger quantities; inventories of popular types could be increased because of fewer categories, increasing their availability to the mold maker. The heat treater's task would also be easier, resulting in better heat treatment and longer mold life.

The steels most frequently used for compression and transfer molds are listed in Table 4.9.

NONFERROUS MATERIALS FOR MOLDS

Rolled or forged steels with deep hardening properties are predominant for use in compression molds and, together with surface hardenable steels, and semi-hard steels, are specified for the majority of injection molds.

As injection moldings get larger, and lower production, in most cases, is required for these larger moldings, alloys other than steel can often be safely specified.

The principal nonferrous castable alloys used in molds, including prototype or experimental molds are: alloys of copper and beryllium, alloys of aluminum, and zinc alloys.

Beryllium Copper

This series of hardenable copper alloys, with a basic composition of 2.5% beryllium and 96.5 percent copper, has been used for over 40 years for injection molds, where casting is an advantage in making molds for shape and decoration as compared to engraving or hobbing in steel.

After the general introduction of blow molding, copper casting was considered the most durable way of tooling blow molds, and was specified for many of the molds imported with the early machines from Europe (see Chapter 11 for expanded data on materials for blow molds.)

The casting technique was greatly improved in the 1940s by the "pressure casting" process, wherein a hydraulic press was used to compact the molten alloy just after it was poured over a hardened master (similar to a hob for steel hobbing) with sufficient force, e.g., 100 tons, not only to push out all the air, but to assure a dense, void-free homogeneous casting.

This development, still popular for castings up to several hundred pounds, was supplemented (but not replaced) in the late 1960s by the process in which a ceramic pattern is used in place of the heat treated steel hob; i.e., "Ceramic Casting."

Later, improved by the application of a vacuum during the casting operation, because the ceramic is unable to withstand the compacting pressure used in the earlier technique, very reliable castings, virtually void free, and with excellent reproduction of the surfaces cast against, are available for much larger molds. When the conversion of furniture to plastics in the late 1960s occurred, beryllium casters were able to mold up to 2000-pound castings.

Many alloys of copper with beryllium are available. Table 4.10 lists the five alloys most commonly used for molds.

Selection depends on the desired degree of fluidity and the mold-making process to be used. Certain alloys are for making cores and mandrels rather than molds. The table shows the alloys, their composition, principal properties and advantages.

In general the alloys with 1.7% or more beryllium—20C, 245C, 275C and 165C—have better fluidity and therefore reproduce details better. They have less tendency to form dross, thus making possible very good metallurgical structure. And the foundry can cast them at lower temperatures (1825 vs 2300°F). This is important when using a ceramic mold-making process depending on available melting equipment. At lower temperature, there is less metal-mold reaction. Also, simpler foundry equipment is feasible.

Within the class of materials having 1.7% or more beryllium, higher beryllium contents give higher fidelity of reproduction. But the higher the beryllium content, the higher the cost. The 20C and 245C alloys (see Table 4.6) are most common. But the choice can depend on pattern quality. While with an excellent pattern 20C will be satisfactory, alloy 245C can compensate somewhat for less pattern precision.

The most important mold manufacturing reasons for using the higher beryllium-content material are (1) to gain the higher fluidity and (2) to be able to pour at a lower temperature and thereby minimize erosion on the hob or metal pattern. In addition, higher beryllium content provides a longer period of fluidity for conforming to the shape and detail of the hob.

Alloys with less than 1.7% beryllium content are generally used only for mold cores and mandrels, where high fidelity of reproduction and very high strength are not necessary. In such cases, two benefits are available: lower mold material cost and higher thermal conductivity for faster heat extraction during molding.

Pressure Casting. Pressure casting consists essentially of pouring the mold alloy around a steel master hob or pattern and immediately applying

TABLE 4.10. Beryllium Copper Alloys Properties and Applications.

Composition	Berylco Copper Alloy Ingot	Density, lb/cu in.	Thermal Cond. BTU/sq ft/in./ hr/° F(68° F)	Rockwell Hardness[a]	Pouring Range, ° F	Characteristics and Applications
Be 0.45–0.75 Co 2.35–2.70	10C	0.311	1400–1600	B 40; B 96	2050–2250	High thermal and electrical conductivity, good strength and good high temperature properties. Ideal for high temperature, lower pressures, and maximum cooling or heating properties.
Be 1.65–1.75 Co 0.20–0.30	165C	0.298	750–900	B 59; C 38	1900–2050	High strength and hardness, with good resistance to corrosion. Recommended for salt water immersed applications.
Be 1.90–2.15 Co 0.35–0.65 Si 0.20–0.35	20C	0.292	650–800	B 63 C 43	1850–2050	High strength, hardness and excellent fluidity. Ideal for investment, sand and ceramic castings. Good thermal properties (2–3 times steel).
Be 2.25–2.45 Co 0.35–0.65 Si 0.20–0.35	245C	0.292	600–750	B 75; C 45	1850–2000	High strength, hardness and wear resistance with good thermal conductivity. Applicable to pressure casting, sand and ceramic castings, very high fluidity.
Be 2.50–2.75 Co 0.35–0.65 Si 0.20–0.35	275C	0.292	600–750	B 85; C 46	1850–1975	Very high castability, or fluidity, hardness and wear resistance with good thermal conductivity. For pressure cast molds, sand and ceramic casting.

[a] Values for solution heat treated before semicolon; solution heat treated and aged after.
(*Courtesy Kawecki Berylco Industries Inc. New York, NY*)

pressure while the metal is molten. The metal flows to conform to the shape and surface finish of the hob very precisely.

When making a hob, the mold maker must include fillets wherever possible and incorporate a draft angle of at least 1½ degrees. He must follow machining with polishing. Of course, he must allow for the shrinkage of BeCu from pouring temperature to room temperature. Shrinkage is predictable and consistent.

Air hardening tool steels are generally best for hobs. These metals machine readily, do not shrink excessively when hardened, and provide the necessary hardness and strength. Besides machining a hob, the mold maker can sometimes cast one via a ceramic casting process.

Control of melting and pouring temperatures is important. Overheating is particularly harmful. It is also best to minimize turbulence during pouring.

To actually pressure cast the mold, start with a hob that is thoroughly cleaned and covered with a parting agent. Heat the hob and the shim plates and top pusher plate from approximately 800 to 1150°F.

When heated, lower the chase or mold casing around the hob assembly and move the two onto a hydraulic press.

Pour the beryllium copper at the recommended temperature into the chase so that it first contacts the chase walls and then flows onto the hob. A deflector will insure this pattern of flow. When the beryllium copper has covered the hob, place the pusher plate over the molten metal and apply press force immediately. Because the molten beryllium copper will begin to solidify when it contacts the chase and hob, time and pressure are both very important. Move as quickly as safety allows.

After pressurizing, the mold maker raises the chase and loosens the side or shim and pusher plates by rapping them. It is important to remove the hob from the mold before too much cooling occurs. If cooling proceeds too far, it will be necessary to heat the hob and mold to about 1000°F to facilitate separating them.

Ceramic Casting. Ceramic casting follows any one of a number of patented proprietary procedures. The following sequence is typical:

A pattern of the part to be produced is made with proper shrinkage allowances. The mold maker casts a special, quick setting liquid rubber around the pattern. The elastic quality allows stripping from the pattern even though the pattern may have a deep grained surface. It also provides faithful reproduction.

Now a ceramic is cast into the rubber mold, producing a replica of the designed part. The special ceramics used preserve the surface texture. A proper ceramic mixture will combine good surface reproduction with relatively high permeability. This last property allows adequate outgassing during casting of the BeCu around the ceramic, resulting in a sound, dense casting.

After pouring the ceramic slurry into the rubber mold, the mold maker allows it to solidify and then fires it in an oven. He then lutes gates and risers to the ceramic. Melting and pouring of the beryllium copper around the ceramic pattern follows the steps covered under hot hobbing.

After solidification of the metal, the caster breaks away the ceramic. He then carefully controls the cooling of the metal to achieve good dimensional tolerances in the casting. See Fig. 4.14.

Wrought Alloys. Where the cavity or core shape is not complex and a casting is not indicated, high strength wrought alloys of beryllium copper are available in rod, tube, bar and plate form. These may be ordered from stock in the annealed or precipitation hardened state. Brush Alloy #25 is one of the stronger and most available wrought alloys.

Heat Treatment. The mold maker can heat treat a beryllium copper mold to the hardness that the user desires. The hardenability of beryllium copper

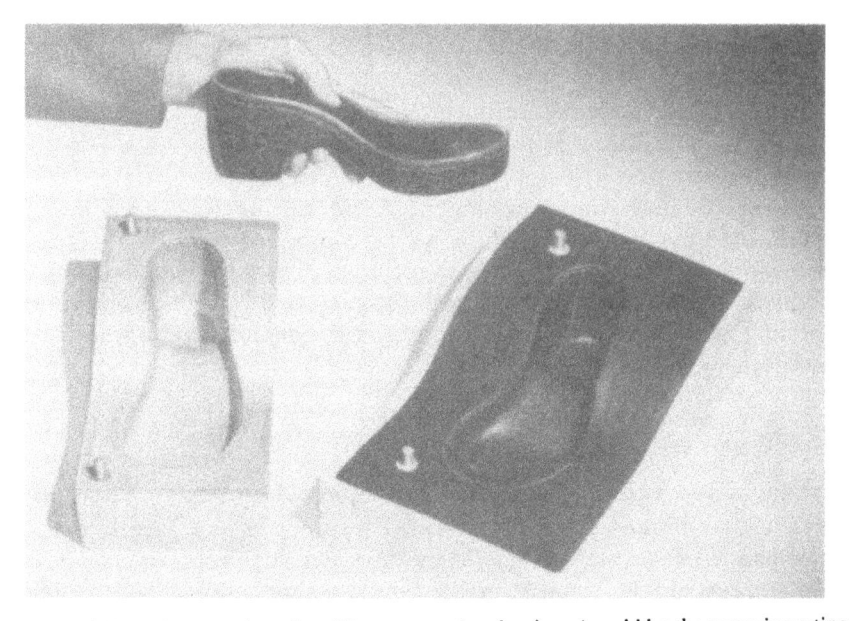

FIG. 4.14 Steps in preparing a beryllium copper (or aluminum) mold by the ceramic casting process. Original pattern or model (above). Rubber negative (right), which is then used as a mold for the ceramic intermediate (not shown). Finished cavity (left) is cast from the ceramic. (*Courtesy Unicast Development Corp., Pleasantville, NY*)

extends from B80 as high as C42 depending on the particular alloy used and the time and temperature of the hardening process.

In general, heating to 1450 to 1475° F and quenching in water (solution heat treating) produces a soft condition. Aging at 550 to 750° F produces hardness by precipitation of a gamma metal phase. Precipitation is faster at the higher end of the range. Maximum properties are attainable by first, solution heat treating and then precipitation hardening.

Machining. After casting and heat treatment, a machining stage begins. This removes gates and risers, produces smooth shut-off surfaces between mold halves and produces flat surfaces on the mold halves for mounting on clamping plates.

High-speed tools are adequate for machining BeCu. Fast cooling is necessary since overheating from tool contact can readily precipitation harden the material. Water-soluble cutting fluids will provide the highest rate of cooling during machining. But for some heavy operations, such as tapping and deep drilling, lubrication of the tool-work piece contact point is so important that use of a mineral lard oil with a 3–7% lard oil content or a sulfurized mineral lard oil with 7% lard oil and a maximum sulfur content of 3% is recommended.

Allowable machining speeds and feeds and depths of cut will depend on the hardness level to which the material has been heated. Depth of cut should be fairly shallow, yet deep enough that cutting takes place below the layer of oxide scale that forms on the surface during the heat treatment. A minimum depth should be 1/64 in.

Finishing. Manual vapor blasting or liquid honing with glass beads in water will provide a matte finish on beryllium copper molds. (Ethylene plastics require this type of finish to facilitate release of parts from molds.) This type of finishing operation will also remove all traces of any ceramic adhering to the mold surface. Polishing is optional depending on plastics and end use requirements.

Aluminum

Aluminum alloys for high-pressure injection molds are only used as proto-types or on a limited production basis. However, they are largely specified for expanded foam, structural foams (low-pressure processes), rotational, blow and cast molds. Since the greatest usage and best physical properties are required in molds for structural foams, discussion of these alloys is included in the chapter on Mold Design for Structural Foams.

Aluminum bronze alloys have been around a long time, and while seldom used, because of soft surfaces for cavity and core components, they have a considerable reputation for wear plates, slides, moveable bushings and/or plungers in contact with steel where galling may become a problem. Available in plate, bars, rounds and hollow bars (not custom castings). They exhibit good bearing properties if loading is kept within engineering limits. Typical analysis: Cu 85.5%, Al 10.5%, Fe 3.5%. Physical properties: Tensile 105,000 psi, yield 50,000 psi, elongation 14%, hardness 183–192 Brinell.

Zinc Alloys

Several alloys of zinc have been used successfully for casting molds, especially for pre-production prototypes such as for injection molding. They have the advantage of being easier to handle; pouring temperature is 850° F and yet they develop surface hardness and compressive strength equal to or better than aluminum.

In most cases, it is well to protect the cast mold by using a steel frame, which affords another advantage in that a smaller quantity of the alloy can be used, than if the castings had to be capable of withstanding clamping and injection forces without such support. Typical analysis: Zn 92%, Cu 3.5%, Al 4.0%, Mg .04%. Physical properties: Tensile 38,000 psi, elongation 3 to 5%, melting point 717° F, hardness 80 to 105 Brinell.

ELECTROFORMED CAVITIES*

A method for making cavities for plastic molds by an electroplating process is very effective for certain classes of work. The process is known as electroforming. In making cavities by this process, a plating mandrel or master is made, which master has the shape of the finished molded piece, but is larger than this molded piece by the amount of molding shrinkage that is expected. The dimensions of the plating master are exactly what are wanted on the inside of the cavity. If the master is of metal, it is coated with a parting agent, a wire is fastened to it, and it is put into a plating bath, which is almost invariably a nickel plating bath. If the master is of plastic material, which is commonly used, the plastic material is first sprayed with a reduced silver coating (similar to that used for mirrors) and the wire fastened to it, and in turn it is put into a plating bath. Both the silver spray and the parting agent used on metal masters are so thin that no detail or dimension of the original master is lost.

Once in the plating bath, the plating on of the nickel proceeds at a rate

*The Encyclopedia of Plastics Equipment, edited by H. R. Simonds, New York, Van Nostrand Reinhold, 1964.

dependent upon the shape of the master. For simple regular shapes, such as pen barrels, the plating can go very fast, and a wall thickness of 3/32 in. and over can be built up in a day or so. On plating masters that have a number of recesses, such as those for gears, the plating rate will generally be slower and sometimes as little as .010 in. is put on each day.

There are two methods used in the making of electroformed cavities. In one method a relatively thin layer of hard nickel is put on, generally less than 1/16 in., and the balance of the build-up is a softer nickel. The hard nickel runs around 500 Brinell and the softer nickel about 150 Brinell. Another method of electroforming, developed in England, builds up approximately ⅜ in. of nickel having a hardness of 450 Brinell. To build up the main mass of the cavity, copper of 220 Brinell is plated over the nickel. The copper is used because it is somewhat harder than the soft nickel generally used, it builds at a much faster rate, and it builds much more evenly so that the many machinings during the plating build-up which are usually necessary to remove the "trees" and excess "build-up" on high points are not required.

At some point during the electroforming process the master is pulled from the cavity which has been formed. In some instances, such as formation of cavities for pen barrels, the masters are pulled when the electroformed shells are about 1/16 in. thick. These masters are invariably of metal and they are started over in the cycle while the first shells formed are returned to the plating baths for continuation of the build-up. In this way a number of cavities can be made from one master in a reasonable time. On such things as gears with relatively delicate teeth, it is seldom practical to reuse the master, so the master is left in until the cavities are ready for machining. In such instances a master is required for every cavity desired.

The plating masters are made in numerous ways. A very common method is to machine them of metal. In the case of pen barrels, heat-treated stainless steel is the most commonly used material at this writing. For gears, probably the most common material is brass. Many masters are made of plastic materials. A common method of making multiple masters is to machine them first, and from this make a cavity into which can be cast or molded various plastic materials. Many shapes can be molded with almost no loss of dimension. Epoxy resins can also be used to cast into such a cavity, reproducing the original master with considerable accuracy. When original masters are made out of wood, leather, or other substances which cannot be put into the plating bath, they can be reproduced by making a cast over the original master and casting back to reproduce the original master in the desired plastic material. If the proper materials and techniques are used, no discernible loss of detail will result.

The applications of electroformed cavities are numerous, but it is difficult to make a general rule as to when they are indicated. A slight change in detail of a piece will indicate that a cavity should be made by electroforming rather than machining, hobbing or casting. A frequent reason for electroforming is the presence of delicate detail in a cavity. Since the electroformed cavity exactly reproduces the finish on the master no polishing or other work is necessary. Electroformed cavities are frequently used because of the accuracy they can achieve. A brass master is simple to make and easy to check for dimensions. Electroforming cavities will exactly reproduce the size and finish on the master. When models are available, very often an electroformed cavity is cheaper because the model can be used for the electroforming and no metal form is required such as for casting or hobbing.

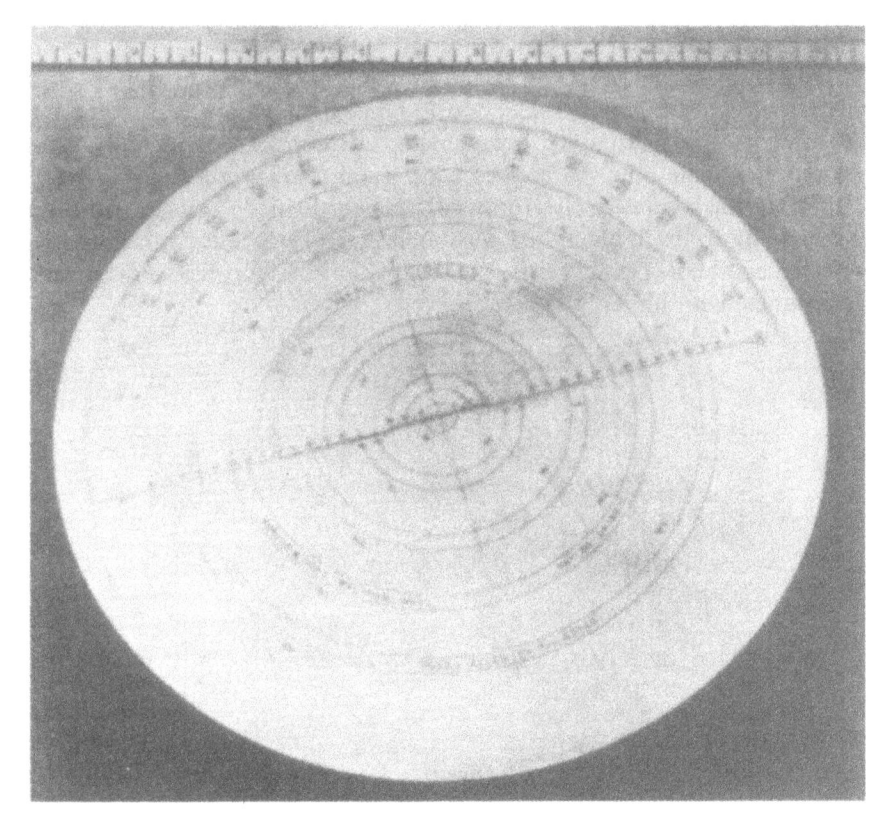

Fig. 4.15. Electroformed insert of large mold for clear plastics drafting instrument. All calibrations, numbers and letters are raised on the mold surface. This mold insert would be extremely difficult to make by any other process. (*Courtesy Electromold Corp., Trenton, NJ*)

Electroforming is an economical way to make cavities with raised details, Fig. 4.15. For example, in the making of molds for speedometer or clock dials, a piece of brass can be polished and then engraved with the numbers, and the electroforming performed over this. The result is electroformed cavities which have the numbers raised from a highly polished surface, and no further work is necessary. Cavities with delicate and undercut detail on side walls can be made.

Other methods and materials are available, usually with the result of less accuracy, to make spur gear cavities, and even helical gears, but when the helix angle is as much as 45° or more, or a worm, thread, or spiral having more than several pitches in its length has to be tooled, there is probably only one way possible—electroforming. Several examples are shown in Fig. 4.16. These parts are unscrewed, and the cavities may not be split, because if they were, the flights would be distorted or ruptured.

The use of electroformed cavities is limited by the fact that they are not as hard as fully hardened tool steel, and by inability to plate into deep recesses. This last limitation varies tremendously among the various electroformers according to the bath used. Some baths will deposit considerable metal into recesses while other baths are very limited in this respect.

In all cases, before an electroforming master is made, the electroformer who is going to use it should be consulted. Each electroformer has his own preferred method. There is little conformity in the industry as does exist, for instance, with hobbing.

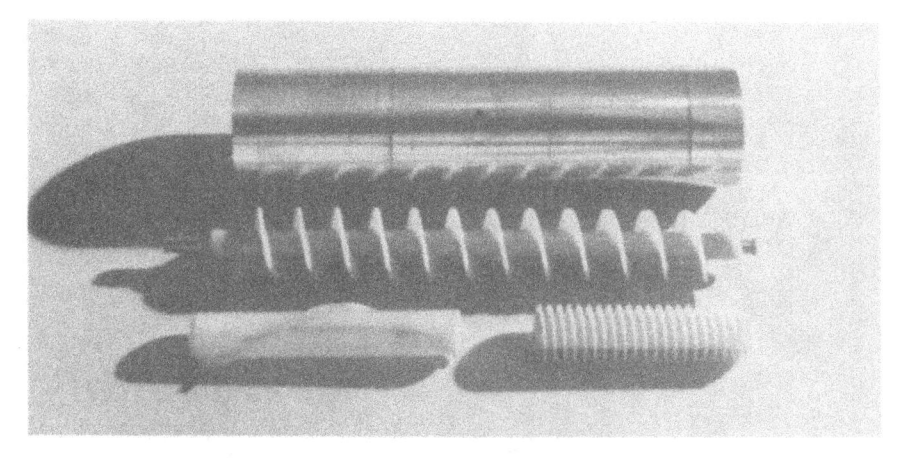

FIG. 4.16. A group of spiral moldings, front and middle, with Electroformed cavity at rear. Note the gate orifice in the center of the cavity where it minimizes length of flow. (*Courtesy Electromold Corp., Trenton, NJ*)

TABLE 4.11. Shrinkage factors for nonferrous castings and Electroforms (not including plastic molding shrinkage).

Alloy	Casting Skg.	Solution Heat Treated	Age Harden Skg.
Beryllium copper (Pressure cast)	.004 in./in.	Yes	.006–.010 ″/in.
Beryllium copper (Ceramic cast)	.008	Yes	.008–.010
Aluminum	.011	Some alloys	Before mach'g.
Kirksite	.000–.002	No	—
Electroforms	.000	No	—

TOOL STEEL CASTINGS*

Small, relatively intricate mold components in multiples are economically available in a variety of tool steels on a custom cast basis, and are therefore suitable for both thermoset and thermoplastic materials. The suppliers use the ceramic casting process to reproduce the original master patterns, similar in most respects to that used for the production of beryllium copper casts, as described on page 142. In preparing the master, the mold marker must allow for the shrinkage of the ceramic cast, the tool steel, which has a different shrinkage than that of beryllium copper, and of course the molding shrinkage of the plastic material to be used.

Because of these factors, it is important to make a test cavity (or core) and adjust the critical dimensions by altering the master, before making multiple components. The most popular alloy compatible with this process is AISI A2, with H-13 as a second choice. These are air hardening steels, and therefore have little or no size changes in quenching.

POWDERED METAL COMPONENTS

Small cavities are also available, which are so far only suitable for thermoplastic materials, custom ''molded'' using powdered metal technology.

One type is offered by the 3M Company, and while some details of the manufacture are kept proprietary, it is disclosed that the product is a sintered composite of two alloys. One alloy familiar over the years only in the form

---

*Cast Masters Inc., Bowling Green OH.

of castings is Stellite®, which is incorporated in a matrix of a copper alloy, including a dispersion of carbides. The whole combination, in the form of very fine powder, is compacted around or in the "replication" of a master, which the mold maker prepares and furnishes. Only one master is necessary, regardless of the number of cavities needed.

The limitations are: (1) size—generally not more than 9 square inches at the parting line of the cavity itself; (2) proportions—not more than a 4:1 ratio of depth to minimum cross-section dimension; (3) surface finish—20 to 25 microns as furnished, with the possibility of improving this to 4 microns by polishing; (4) apparent hardness—Rockwell C-41, which is actually deceptive on the conservative side, since the composite includes Stellite® and carbide particles, both of which by themselves have a hardness of Rockwell C-58 or more; and (5) tolerances—limited by the shrinkage allowance inherent in the process, plus polishing allowances where required. A test cavity is again recommended. But no heat treating is necessary, so no size change due to this factor need be allowed for.

Approximate analysis of the composite:

Cobalt	35%
Chromium	21
Tungsten	9
Carbon	2
Copper alloy	30
Unspecified	3

Physical properties:

Tensile strength	110,000 to 140,000 psi
Compression strength	220,000 to 280,000 psi
Modulus of elasticity	21,000,000
Apparent hardness	Rockwell C-41

Another type of P/M mold components, similar in structure, but composed of almost completely different materials, is known by the trade name of Ferro-Tic®. Originally Ferro-Tic was only available in the form of standard rounds and rectangular blanks, suitable for incorporating as inserts in high wear areas of cavities and cores, such as at or near gate locations, etc. The material is now offered as custom cavities or cores, as a reproduction of a master that the mold maker must submit. This, too, is a sintered composite of two alloys: titanium carbide particles in a matrix of alloy steel. The product is machinable in the annealed state as delivered, which includes

Stellite® is a registered Trademark of the Cabot Corporation. The 3M process is known variously as "Custom Cavities," "Replication Tooling," and "Tartan-Tool®".

Ferro-Tic® is a registered trademark of Alloy Technology International, Inc. West Nyack, NY.

tapping. After fitting to the other components, the items can be heat-treated in very much the same way as would be used for the steel alloy involved in the composite. The dimensions normally increase slightly on heat-treating and tempering.

Analysis: TiC   45%
             High carbon/High chrome steel   55%

Physical properties:
    Tensile   200,000 psi
    Compression   480,000 psi
    Elastic modulus   41,000,000
    Hardness, Rockwell
        Annealed   C-46–49
        Hardened (double tempered)   C-66–69

Because of its structure and high hardness, this material, when properly supported, is useful for small compression and transfer molds where extreme abrasive conditions are prevalent.

PLASTICS MOLD CAVITIES

A variety of materials are used for "molds" made of plastics materials. One of the most easily and widely used is the room temperature vulcanizable RTV silicones are depicted in Fig. 4.17. Wood, metal, wax or other easily made patterns may be used to cast the mold halves, using the simplest foundry techniques. This compound reproduces fine details very accurately and, because of its elasticity, undercuts will be no problem.

Casting resins* such as epoxy glass, urethane, polyester, etc., are then cast in the silicone cavities thus produced to make the molded parts. Some molds made of thermosetting resins filled with metal powders have been used for prototype injection or transfer molds when fairly low molding pressures can be used. Dental laboratories have excellent facilities for accurate small cast molds.

SURFACE TREATMENT

Upon completion of the hardening and polishing, and except in certain circumstances, the sampling or "try-out" of a mold, further development or protection of the mold surface is often required, and whether required or not, should always be considered by the mold builder, as worthwhile in many cases, especially when high production requirements are to be met.

*The Plastics Factory, NY.

Fɪɢ. 4.17. Casting RTV silicone rubber on a wood model to produce a mold for the casting of epoxy prototypes suitable for market test.

Some of the available surface treatments are given in Table 4.12. In almost every instance, a procedure lower in the table can be followed by one previously mentioned, if applicable to the same metal.

Tungsten Disulfide*

Tungsten disulfide coating may be applied to any metal, including nitrided and plated surfaces by a process involving the following steps: (1) Complete cleaning of the metal surface; (2) simultaneous protection of the cleaned surface to prevent even momentary oxidation; (3) impingement of microscopic particles of tungsten disulfide, which will only bond to the metal, not to itself and is therefore self-limiting in thickness at .00002 in.; (4) cleaning off loose particles. The effect on release of plastics from the mold surface is permanent, as for the graphite process described below.

It is successful, however, in applications where the binder used in the graphite process breaks down and/or the graphite migrates into the molded product. The principal purpose of the application, however, was intended for

*Diversified Drilube, Inc., Tulsa, OK.

TABLE 4.12.   Mold Surface Treatments.

Process	Material	Applied to	Purpose
Coating by impingement—molecularly bound	Tungsten disulfide[1]	Any metal	Reduce friction & metal to metal wear with dry film—non-migrating.
Coating by impingement—organically bound	Graphite[2]	Any metal	Reduce sticking of plastics to mold surface—can migrate.
Electrolytic plating	Hard chrome	Steel, nickel, copper alloys	Protect polish, reduce wear and corrosion (except for chlorine or fluorine plastics).
	Gold	Nickel, brass	Corrosion only
	Nickel	Steel and copper alloys	Resist corrosion except sulphur bearing compounds, improves bond under chrome, build up and repair worn or undersized molds.[5]
Electroless plating	Nickel	Steel	Protect non-molding surfaces from rusting.
	Phosphor nickel[4]	Steel and copper	Resist wear and corrosion.
Blasting	Fine mesh abrasives (liquid honing)	Any metal	Remove scale after heat treating. Provide a fine matte finish. Improves release of some thermoplastics.
	Sand	Any metal	Provide a coarse matte finish.
	Steel shot	Steel	Peen, to improve crack and fatigue resistance.
	Glass beads	Any metal	Peen, to remove fine scratches & tool marks.
Photo etch	Chemicals	Steel and aluminum	Decorate or "texture" mold surfaces
Nitriding	Nitrogen gas or ammonia	Certain steel alloys	Improves corrosion resistance, reduces wear and galling. Alternate to chrome & nickel plating.
Liquid nitriding	Patented bath[3]	All ferrous alloys	Improves lubricity and minimizes galling.
Anodizing	Electrolytic oxidizing	Aluminum	Hardens surface, improves wear and corrosion resistance

[1]Drilube®   [4]Kanigen®
[2]Microseal®   [5]Nicoform®
[3]Tuf-Triding® or Lubricasing®

reducing sliding and rotational friction between metals where no lubricants are tolerated.

Graphite Embedment on Mold Surfaces

This process makes use of impinged lubricative coating* developed for metal fabricators in the aerospace industry. Productivity increases of 10 to 20%

*Microseal Corp. Div. Great Lakes Chemical Co., Mountain View, CA.

are reported from thermoplastic molds treated by this process. Mold surfaces to be treated by this process are treated with a binder and then exposed to a high pressure (120–130 psi) spray of ultrafine graphite particles. This pressure spray impinges the graphite onto the tool surface to a depth of 0.0002 to 0.0004 in. A surface coat of .00008 in. builds up on the surface of the mold component. Lubricative plating may be applied to chrome-plated surfaces. However, to restore size and polish, surfaces must be buffed.

Chrome Plating

Many molders have found that chromium plated molds are a great asset and specify chrome plating on all mold cavities and plungers. Chrome plating is also used in mold repair work for building up worn sections. This requires a cleaning tank, an etching tank, a plating tank and a final cleaning tank. An electroplating generator and facilities for building up the anodes, as shown in Fig. 4.18, are also needed. Chrome plating equipment is very useful in the hands of an experienced workman. Plating specialists do this work for the small tool shops. See also Chapter 3. Only a few platers* are equipped to chrome plate nitrided surfaces.**

Electroless Plating

As used on plastics mold components, electroless plating is nickel plating. Electroless simply means that no electric potential is applied to the bath. The result is a much more even deposit—no undesirable extra build-up on sharp corners, and nearly perfect penetration into recesses. Furthermore, if one of several patented baths is used, the deposit may be hardened after plating by baking at 750° F.

The softer nickel deposits and even electroforms which average Rockwell C-50 can be protected against scratching and wear by hard chrome plating, if desired. Chrome plating bonds better to nickel than directly to steel. (Nickel bonds better to steel than chrome does.)

Besides the obvious situation where electroless nickel is used for molding surface protection, it is extremely valuable and unique in use for protecting the surfaces of the mold frame itself, including the drilled water lines, from corrosion resulting from acid conditions in the water and condensation on other surfaces from humid atmospheres combined with refrigerated water. Accordingly, the rear surfaces of cavities and cores, including "O" Ring grooves and cooling holes, are improved by electroless nickel plating.

*Nutmeg Chrome Corp., W. Hartford, CT.
**Armoloy Corp.—Various Locations.

FIG. 4.18. Quality chrome plating results from proper placement of the anodes. Depicted here is the finished cavity and the same cavity is shown with anodes in position for plating. (*Courtesy The Nutmeg Chrome Corp., West Hartford, CT*)

Nitriding

Nitriding improves the hardness, and therefore the wear resistance, of the surface of the mold. It is especially effective in prolonging the useful life of a mold exposed to abrasive action of glass and asbestos reinforcement in any plastic, but especially with glass-filled thermosetting resins. Where chrome plating will yield a hardness of 65 Rockwell C, and a protective coating of usually .0003 or at most .001 in. thick, nitriding of the most suitable alloys of steel will yield a hardness of 70 Rockwell C, and while the usual depth of penetration is .003 to .005 (unlike chrome it is not added to the surface); by prolonged processing time, penetration of .016 to .020 in. can be usefully obtained. The disadvantage compared to chrome is that it cannot be removed and reapplied. However it takes unusual vigilance to detect when the chrome plating has been worn away, in order to avoid subsequent accelerated erosion of the steel beneath.

Nitriding is done by heating the steel to about 975°C, and exposing it either in an atmosphere-controlled furnace to nitrogen gas, or in a molten salt bath to ammonia salts. In either case, nitrogen is absorbed into the surface of the steel, and if suitable elements are present, combines with them to form the very hard nitrides.

Steel containing aluminum in small percentages, such as 3% is particularly suitable. Otherwise, steel containing carbon of at least .4%, together with chromium, vanadium or molybdenum will respond to the treatment. Molybdenum is especially beneficial in the alloy since it reduces the characteristic brittleness of nitriding.

The normal depth of penetration of .003 to .005 in. is obtained in 12 hours, but it takes 72 hours to get penetration of .015 in.

Of the steels listed in Table 4.8 (for injection molds) the following can be nitrided: P-20, H-13, and 420 stainless; of the steels listed in Table 4.9 (for compression molds) H-13 and S-7 are nitridable.

An excellent steel, besides the through hardening Nitralloy Series, which yields optimum nitriding, is P-21. It contains aluminum and if it has been solution heat-treated as is normally done for use for molds it will age harden as it is being nitrided, to a core hardness of 38 while the surface is 70 Rockwell C.

THERMAL BARRIERS FOR MOLDS

As molds go up in temperature, a point is reached where the loss of heat going from the mold to the press platen cannot be tolerated. At ordinary mold temperatures, this problem is often minimized by multiple channels in the clamping plates, giving the effect of minimum heat transfer areas. Transite* asbestos sheet is commonly used in the intermediate temperature zones where dimensional control across the parting line is not critical. For highly accurate molding with absolutely flat and parallel press platens, glass-bonded mica,** a machinable ceramic is used because of its low thermal conductivity and its absolute dimensional stability. Glass-bonded mica can be lapped to an optical flat and will hold it indefinitely. For Situations where thermal barriers must be of minimal thickness, Nomex*** sheeting is proportionally effective. For localized areas where high physical properties are needed, alloys of titanium may give some relief.

POINTERS

- In case of doubt, use a type of steel which is better than the one you might select but are not sure it will be satisfactory.

*Transite, Johns Manville, Greenwood Plaza, Denver, CO.
**Mykroy, Mykroy Ceramics Company, Ledgewood, NJ.
***Nomex—E. I. DuPont de Nemours and Co. Inc., Wilmington, DE.

- Don't specify a hardness in excess of that recommended for the steel used for the particular application. In general, more molds are lost by cracking than are worn out by use.
- Always double temper after hardening any steel, and if the hardness is too high after the first tempering, double temper at a lower temperature.
- Don't expect plating to cover surface defects or to improve polish; plating exaggerates pits, scratches and blemishes.
- Don't try to cover up cracks by welding. If the crack is not too extensive, cut it all away, and build up the weld from sound structure.
- Do not use nickel plating in contact with rubbers containing sulfur, nor chrome plating in contact with chloride or fluoride plastics.
- Go over the sharp corners of cores and cavities after chrome plating and check for excessive build-up which may interfere on fitting and on sliding surfaces. Excessive compressive loads can result at the parting lines when clamped. Failure to do this may result in chipping of the chrome.
- Watch for "white layer" embrittlement from EDM operations on molds. There are ways to avoid this problem: (1) Slow down final EDM operation at the end, using low amperage. (2) Inspect for "white layer" and polish away. (It is seldom over .0001 or .0002 in. thick.)
- Check with your chrome plater to be sure he takes precautions against hydrogen embrittlement. Bake chrome plated parts 375° F for an hour, before putting in service or applying stress.
- Take advantage of the maraging and precipitation hardening steels; while being nitrided, these age harden to improve interior structure and hardness.
- Remember that nitriding is theoretically an irreversible process while through-hardened and pack-hardened steels can be annealed; chrome and nickel plating can be stripped.
- Do not subject a steel to a surface treatment that involves a temperature higher than that at which it has been tempered.
- Often, molds are "tried-out" before they are plated. Do not try out molds for corrosive or highly abrasive materials, unless they have been plated. It is better to repolish lightly a mold that has had to be stripped of its plating in order to make corrections, than to have to remove a substantial amount of metal because of corrosion.
- Where metal slides on metal, select materials and heat treatments for the two components so as to obtain surfaces having hardnesses separated by at least six or eight points on the Rockwell C scale.

REFERENCES

Alcoa Aluminium Handbook, Pittsburgh, PA: Aluminium Co. of America.

Bengtsson, Kjell and Worbye, John, Choosing mold steel for efficient heat transfer, *Plastics Machinery & Equipment*, Aug. 1984.

Hoffman, M., What you should know about mold steels, *Plastics Tech.*, p. 67, Apr. 1982.

Properties and selection of metals, in *The Metals Handbook*, Vol. 1, Metals Park, Cleveland, OH: American Society for Metals.

Heat treating, cleaning and finishing, *The Metals Handbook*, Vol. 2, Metals Park, Cleveland, OH: American Society for Metals, Cleveland OH.

Revere Copper & Brass Publication, New York: Revere Copper and Brass.

Shimel, John F., Prototyping: How and why, *Plastics Design Forum*, p. 75, Jan./Feb. 1984.

Stahlschlussel (The Key to Steel), Metals Park, Cleveland, OH: American Society for Metals.

Stainless Tool Steels for Molds, Uddeholm Steel Corp., 1984.

Tool Steel, Simplified, Philadelphia, PA: Chilton Publishing.

Worbye, John, Polishing Mold Steel, *Plastics Machinery & Equipment*, Feb. 1984.

Chapter 5 / Design Drafting and Engineering Practice

Revised by Wayne I. Pribble

PRINCIPLES AND RULES OF DESIGN

Mold designers and tool draftsmen follow many general rules which experience has shown are both practical and desirable. Some of these rules have been established as standards for the preparation of mold drawings; the men who follow these rules avoid many of the troublesome and unsatisfactory mold designs which result from neglect of fundamentals. This chapter was prepared to detail the principles and rules of design which, over many years, have been found to give the best results. Understanding of these rules and intelligent application of them will help the draftsman to produce drawings that will convey his design to the toolmaker in such manner that he may interpret it readily with no possibility of misunderstanding. It must be understood that the rules given here are general in their application and are to be interpreted with regard for the special conditions and existing practices of the shop where the tools will be designed, built and used. The mold designer must familiarize himself with his own shop practice and learn what limitations will modify the application of these rules.

Drawings

Drawings are the permanent record of a design from which many copies or prints may be made.

A fundamental requirement of a drawing is that it shall give the necessary information *accurately*, *legibly* and *neatly*. The tool-maker's first measure of a draftsman's ability is based on the neatness and legibility of the print which is furnished him. His final evaluation is made on the basis of the accuracy of the drawing. A drawing may carry as many as 500 dimensions and, of this number, only one may be inaccurate, but the damage done by that one wrong dimension can far outweigh the good of 499 that are correct. It is impossible to emphasize too strongly the necessity for accuracy in all dimensions and for clear presentation as well.

There is a common and very correct tendency among draftsmen who have worked for a period of time in one place to leave some items to "shop practice." This may include such things as clearances, tap drill sizes, tapped holes, etc. The consulting designer and the designer of molds which may be built in any one of several tool shops cannot do this to any large extent because plastics practice is not standardized. "Shop practice" varies widely among molders, and the molding shops where the molds are designed and the men who build the molds may be several hundred miles apart.

At several points in this text, the use of standard mold bases is detailed. Wherever possible, we recommend the use of these highly specialized components, and once again, we recommend that the designer keep a complete file of catalogs for these standard mold bases and standard mold components. (See Chapters 3 and 8.) Duplication of these catalogs in this text would serve no useful purpose. However, we do show designs based upon standard mold bases. Bear in mind that cost is an important factor in today's economy where the wages of the mold-maker represent a significant part of the overall cost of a mold. The mold bases are built as complete units by tool shops that have the varied equipment needed to fabricate these units. This equipment includes large grinders, tape controlled mills, jig borers, and similar equipment, all of which is referenced in Chapter 3. Figures 8.92, 8.93A, 8.93B and 8.94 cover the utilization of standard mold drawings to simplify and shorten mold designing time, as applied to injection mold design.

Tracings

A tracing is a form of drawing used for making prints. *Prints* are the copies of the drawings (or tracings) used in the shop as a guide in the construction of the mold or product. Most draftsmen make the drawing directly on tracing paper or tracing cloth. Others make the drawing complete and then prepare a tracing from the original. A tracing is never used for manufacturing and should not be used for reference purposes. The cost of retracing is high and the draftsman must see that the tracings receive proper care and treatment. Tracings are easily damaged by careless handling, therefore the following rules should be observed:

1. Do not use sharp instruments for erasing.
2. Do not permit tracings to become wet or soiled.
3. Do not fold or crease a tracing.
4. Do not give a tracing to anyone in the shop; instead, have a print made for the person needing it.

Plots

A *plot* is the output of a computer program. The size and type plotter used is chosen by the availability in a particular design section using CAD (Computer Aided Design). The probable availability will be an A/B plotter, or a C/D plotter. Plotters for E size are usually part of the sophisticated systems in use by larger corporations or universities. In some instances, an A/B/C/ D plotter will be available. The plotter designation A, B, C, D, or E has reference to the dimensions of the standard sized paper available for drawing, tracing or plotting. The size and scale of a plot is selected by the designer at plot time. The plotter software is configured, in advance, to operate with a particular plotter. The usual choices available are: 1) plot what is seen on the monitor screen; 2) plot to size; or 3) plot to scale. Item 1) is usually known as a *screen dump*, which may also be dumped to a dot matrix printer equipped for graphics. This plot is commonly used in a reference manual as an index of drawings available. Item 2) allows for each screen pixel becoming equivalent to some quantity on the plotter. This command must be used in the event a drawing was not drawn to a particular scale when originally generated. Item 3) allows a drawing originally made to a particular scale to be plotted to any other convenient scale.

Obviously, we cannot be totally specific about plots because of variations associated with your particular system or plotter. Our purpose here is to alert the potential new user to terms used with CAD. Inasmuch as ''overlooking the obvious'' is a common failure, we reiterate here, ''study the manual'' which comes with each of the many CAD systems on the market. Most software vendors will supply a manual at some nominal charge, then allow credit toward the purchase of their software package. A manual will (or should) include information on plotter configurations and operation. Of course, we assume that you would also ask, ''Who else is using this system.''

Here it should be noted that the jagged line effect seen on a low-resolution monitor screen will not be duplicated on a plotter. A high-resolution monitor increases the cost of a system and provides a prettier picture, but it has no effect on the quality of the plot. Angle lines and circles are examples of screen representations showing the jagged line effect, because it is impossible to show the true character on a screen made up of rectangular pixel shapes. However, a plotter is designed to plot a straight line from A to B, or to plot a circle in very small increments so a true circle results. In any case, we recommend the selection of a plotter based on the usual size of your output drawings.

Plotting is the final step for which a design is created. Thus *paper*, *vellum* or *transparent film* are choices for the final plot. The choice of plotting medium will also determine the type of pen needed for the plotting. If a one-time use in the tool shop is all that is required, paper will be satisfactory. Plots to vellum are usually used when it is desired to have a permanent plot from which prints can be made as needed. A plot to a transparent vellum or film should be made because the cost of making a print from a transparency is low compared to the time and cost required to make multiple copies of an individual plot. The choice of transparent film is dictated when accuracy of the plot is essential, such as in automotive panels, where scale measurements are taken directly from a plot or drawing. This scaling practice, while common in automotive applications, should only be done when it is clear that the designer intended for the final plot to be used in that manner. In other words, a print should never—repeat—never be scaled because handling and humidity conditions can distort paper images.

Title

The title selected for a tracing must describe the work in an accurate manner and it must not be confusing. A typical title form is shown in Fig. 5.1. The essential information for a title must include:

1. Size of mold (number of cavities).
2. Type of mold.
3. What the mold will produce.
4. A serial drawing number.
5. The names of people who worked on the drawings and the dates on which the work was done. Names must be written as signatures. Standard abbreviations may be used for the months, but the months should not be designated by number.

FIG. 5.1 Typical title form.

Projection

All drafting should follow the rules of Orthographic Projection. In the United States, most of the work is done in third-angle projection. All tracings should state clearly the projection angle used.

Sections

An adequate number of cross sections and views should be drawn to present the details clearly. It is desirable to show at least three views or sections of all except the simplest of round parts. Good drafting practice, however, will not make excessive use of views. Dotted lines should be used only when they serve to clarify the drawing and in that way reduce the chance of error. It is preferable practice to show a section rather than to use dotted lines. Cross hatching is often considered a waste of time, therefore it may be omitted except where it is considered necessary to clarify the drawing. Plastic materials are generally shown solid black in cross section, as illustrated in Fig. 5.2.

Views and dimensions should be given completely and accurately so that the tool-maker will have full information and not be required to make additional calculations or to ask questions. Tabulation of dimensions should be avoided in the actual preparation of drawings, as the possibility of error or misreading is greatly increased thereby. The complete part and all sections should be drawn before starting to dimension. This practice will avoid much erasing and give cleaner and neater prints.

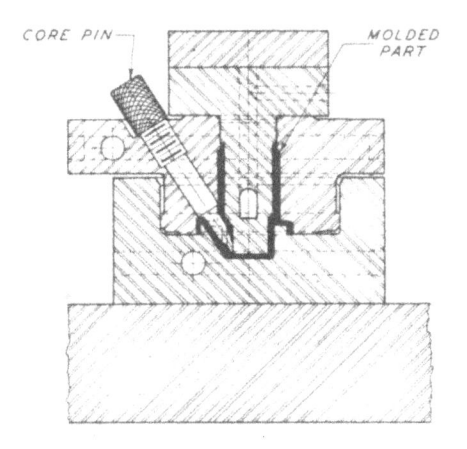

FIG. 5.2. Molded parts are usually shown solid black in cross-sectional drawings. (*Courtesy Eastman Chemical Products, Inc. Kingsport, TN*)

General Rules of Drafting Practice

The following suggestions are given for the purpose of presenting those fundamentals considered essential to good practice. Observance of these rules will serve to avoid errors commonly made in the design of molds.

1. Do not try to second guess the product designer concerning his actual needs in the final molded part. Refer all questions concerning insufficient detail or information on the product drawing to the responsible design engineer. *Always* secure authorization in writing for any changes that you believe will improve the product, reduce the cost of tooling, reduce manufacturing cost, or prevent an actual error. The best procedure is: to mark up three identical drawings showing *everything* that you have *used* or *assumed* in designing the mold—this includes suspected errors, unclear information, ejector pin locations, gate locations, drafts, tolerances and requested or suggested changes, then send two marked copies to the purchasing agent who handled the buying of the mold, and ask him to return one marked print with the design engineer's approval or comments. *Always* retain file copies of these negotiations, including the final approved print to which the mold is designed. Correcting or rebuilding a mold built to unauthorized deviations can be *very* expensive, time consuming, and frustrating to the customer.

2. Check the product drawing very carefully before mold design is begun. Redesign the product completely when necessary to make sure that the piece can be molded consistently and satisfactorily with the production methods and materials available.

3. In cases where the estimator specifies the mold design that was used as the basis for his quotation, make sure that this design is followed unless approval is given for deviation.

4. Long slender cores and mold sections should be designed as mold inserts when they cannot be eliminated by a change in the product design.

5. While positive draft is the usual practice, do not overlook the use of zero draft or negative draft when their employment may be helpful.

6. Be sure that connections for temperature control media, and the thermostat locations do not interfere with clamps, clamp bolts, strain rods, ejector rods, or other parts of the machine or press for which the mold is being designed. Make a note on your assembly drawing specifying the machine or machines for which the mold is designed.

7. Be sure to allow ample clearance between drilled holes for the temperature control media and the adjacent holes for ejector pins, screws, guide pins, bushings, etc. One-fourth inch is the minimum with which moldmakers like to work (carry a special note if it *has to be* less than ¼ in.). For holes in the 12- to 20-in. range, use ⅜ in. clearance. Use proportionally greater clearance for longer or deeper holes where steam, oil, or water is

the media (electric heaters can be close to adjacent holes with no problem of leakage of media). Look up other references in this text for information on heating and cooling channels.

8. Remember that most thermoplastic materials require large degrees of cooling. However, many of the engineering thermoplastics require heating the mold. All the thermosets require heating the mold.

Thermosetting materials are not quite so critical in relation to temperature control of the mold. However, urea and melamine materials require uniform surface temperatures for best results in molding. Give special attention to channeling in all molds where maximum production is required. It is not difficult to calculate temperature needs and the heat transfer needed in a mold. Time spent in a calculation will pay dividends. As a general rule, we empirically state that it is almost impossible to *over channel* a mold. In any case, over channel is to be preferred to *under channel*. The use of rapid conductive metals, such as beryllium copper, should also be considered. Channels in long slender core pins, is called *coring*. Consider also the use of *heat pipes** which will either heat or cool from a remote source. Air jet cooling is also frequently used, where other media is difficult or impossible to use.

9. Make use of standard lengths of screws, dowel pins, and guide pins whenever possible. Small deviations from these standards cost money.

10. Specify the type or kind of steel for all hardened mold parts. Call for the trade name or type of steel to be stamped on the back of the mold section. This practice will give the heat treater essential information if it becomes necessary to anneal or rework the piece.

11. Call attention to any unusual features or important dimensions by means of notes, so that the tool-maker's attention will be focused on these requirements. Tangent radii, negative draft, or special, sharp corners, and special hardening or tempering to be done, must be plainly indicated and detailed as necessary.

12. Do not deviate from standard design practice unless at least one other experienced designer has agreed that the changes will improve the operation of the mold.

13. Study the methods used in the tool shop where the mold is to be built so that the mold can be dimensioned in the manner best suited to the equipment available.

14. The designer should, when possible, indicate the method of setup for machining by the manner in which dimensions are placed on the drawings.

15. Give the important dimensions in three-place decimals. Show the tolerances only where required by close tolerances on the product draw-

*Heatbank, Hughes Thermal Products Div., Torrance, CA (and others).

ing. Be sure all close tolerances are actually needed, and that those specified can be met. Give the tool-maker *no more* than 50% of the tolerance allowed by the product drawing.

16. Where involved calculations are required to determine the centers of radii, hole location, contours, etc., preserve your figures and record them in such a manner that you can recalculate the dimensions easily a few weeks later when changes or checking may be required.

17. When checking dimensions, do a thorough job; assume that all dimensions are wrong until you personally prove that the calculations are correct.

18. If an error is discovered in a dimension, find out, if possible, what faulty reasoning produced the error.

19. Expect to make mistakes, and check every detail to find them; avoid making the same mistake twice. A mistake on a drawing is only a potential loss, but it becomes a real loss if it goes into the toolroom undiscovered, thus causing faulty construction.

20. Check the daylight opening in the press to be sure the molded part can be removed from the mold. *Warning:* some daylight figures given by press manufacturers include maximum stroke. Others use maximum daylight plus stroke. Be sure you known which is meant.

21. Check the platen, strain rod, and ejector layout to be sure the mold can be installed and operated in the press for which it is designed.

Shop Standards

Each design section should compile all of the data which define its shop practice and any other standards that are followed consistently. These standards will include such items as:

1. Molding press data showing capacity, mold-size limitations, daylight opening, auxiliary rams, ejector operating mechanism, clamping bolts, pressures available, and the location of holes in platens.
2. Material stock lists showing steel sizes in stock or readily available.
3. Drill sizes and tapped hole specifications.
4. Standard insert design and sizes.
5. Technical data on plastic materials showing shrinkage, bulk factor, density, draft angles, etc.
6. Spring charts showing sizes and capacities of springs commonly used in mold construction.
7. Mathematical tables and formulas.
8. Factual data on shrinkage (transverse, longitudinally, diametrally and in direction of draw) and flash build-up as actually obtained from previous molds which used the same molding material and molding methods.

Computer Aided Design (CAD)

The CAD program chosen should use a minimum number of operating commands which can be used from on-screen prompting as opposed to constant reference to a manual. For the experienced mold designer, a CAD program using the same general technique of drafting as is followed using pencil, T-square, and triangle will be most quickly learned. Look for and select a menu driven program which is also compatible with a digitizer tablet and cursor control which allows maximum speed of selection and operation. Keyboard entries should be restricted to specific dimensions or specific text. Most CAD programs in current use allow for automatic dimensioning, thus even keyboard entry of dimensions should be rarely needed. Every keyboard entry is a potential for error, as is proven by the number of retypings needed just to obtain correct copy for this book text.

In those cases where the designer is fortunate enough to have a CAD system available, it may become part of his duties to predraw many of the items listed under shop standards (above). We have also encouraged the preservation of calculations, checking dimensions, developing standards, etc. In the following text are three check lists. We particularly direct your attention to the *designer check list* covering moment-by-moment decisions required of the designer. As you become familiar with CAD systems, it will soon become evident that many of the cautions and choices will actually become selections from a *data base* which is part of developing your own CAD system. The CAD system allows drawing once, checking once, then using over and over as component parts of a total design using a *copy* or *exchange* command. Currently, much of the data for mold bases and component parts, such as plates, guide pins, bushings, hot nozzles, etc., are part of data bases available from the vendors of these items.

When you arrive at item 4 of the designer check list, and you are performing the design with a CAD system, you will note that with a CAD system, the choice of *paper size* is made at plot time. (Refer to *plots* earlier in this chapter.) With a CAD system, *page size* (do not confuse with paper size) is chosen at design time. A rule of thumb is to use the smallest page size that will conveniently show the overall of the design to be drawn. It is a designers choice whether he shows all necessary views on one page, or whether he draws a separate page for each view. In the latter case, the pages can be either assembled to one page, then plotted on the desired paper size, or each of the pages can be plotted on the same paper by specifying size and location at plot time.

It should be evident that a 12 to 19 in. *monitor* screen allows only a certain amount of observation at any one time. Using the *zoom* feature found in every CAD program, small detail can be enlarged for easy drawing or viewing. Hopefully, a color monitor will be available, otherwise the 200 to 2000 layer capability of the CAD program will be almost useless. Color monitors also assist in visualizing depth and shape just as colored pencils were frequently used to color-code a complicated part drawing to effect a 3-D image for the mind.

Well designed and documented CAD software will follow the sequence of items as called out in the designer check list. However, many of the required items will have been predrawn, either for the current design or available for copying from the *source disk* (remember—draw once and use it over and over?) For example, assume a 64 cavity mold design for a T-shaped part. Orientation for gating is four groups of 16 cavities each with the gate at the bottom of the T. By predrawing the T-shape and *filing* or *saving* it as a pictorial item, the CAD system allows copying the predrawn item, orienting in any direction, *mirror imaging*, *scaling* to any size, and placing the image in an exact position on the page as many times as desired. Compare the speed of CAD positioning with hand-drawing 64 duplicates, one at a time. Most designers drawing by hand content themselves with indicating one each of 4 different orientations—and that may be adequate. One tendency when using CAD is to overdraw by adding more detail than is necessary for any given item. For example, let us assume ten screw locations are shown in a plan view. Only one screw need be shown in a front or end elevation, but CAD easily plants a picture of a screw length. Thus, the unwary designer might place five details of a screw where one would suffice. Always ask if more detail will serve any useful purpose.

Hopefully, the choice of CAD software and the system would have been made on the basis of ease of operation, simplicity of command structure and speed of execution. Some software uses 150 or more commands, whereas other software may use only 25 or 30 commands as on-screen selectable choices. Most of those CAD programs using a large number of commands will be run with a *digitizer tablet* with cursor control. *Regeneration* to the monitor screen is a function of hardware, but it should be quite rapid to avoid operator waiting time. Finally, we recommend a CAD program which uses precision to 6 decimal places. Most NC (Numerical Control) equipment requires accuracy to four decimal places, or else it will reject the computation or entry.

The use of CAD is projected to grow at an ever-increasing rate. Currently, any particular electronic technology is valid for about three years (although useable much longer) after which newer developments will create obsolescence of the old technology. Thus, we encourage selection of a CAD program which will be periodically *upgraded* by the supplier at little or no cost to the user.

We forsee greater use of microcomputers replacing the drafting board of the 1980s. Several companies now offer the detailed analysis of any particular part as a service to end user, mold designer, mold maker or molder.

Several colleges and universities have already installed quite sophisticated CAD systems (the million dollar type) and offer courses of instruction. They also offer connect-time to local users who only occasionally need such service as finite element analysis to determine the adequacy of such items as strength, impact resistance, and flexibility. Material flow analysis, thermodynamics (heat exchange) in the mold, or 3-D modeling for aesthetic appeal are other services available through these on-line or walk-in services. We grant the desirability of having all these "goodies" at your fingertips, but the cost of an infrequently used feature is seldom justifiable to management.

ENGINEERING AND DESIGN PROCEDURES

Most engineers and designers follow some kind of orderly routine in the engineering of a mold. We recommend this practice and offer the following check lists to assist you in the procedure. Obviously, there must be a high degree of communication between the engineer and the designer who reduces to drawings the complete design. The engineer is responsible for the ultimate functioning of the mold and the related tooling. For this reason, three check lists are supplied. The first list covers the *preliminary* decisions usually made by the responsible engineer. The second list covers the moment-to-moment decisions to be made by the mold designer, and the third list covers the final answers and follow-up usually performed by the engineer. We suggest here that the mold designer who aspires to become an engineer must familiarize himself with the reasoning that goes into each of these decisions.

ENGINEERING CHECK LIST (preliminary to design)   To Do   Done

 1. Review all correspondence, quotations, orders and other data that may have any bearing on the part application or mold design.   _____   _____
 2. Advise customer of any changes needed to bring the part into conformance with quotation.   _____   _____
 3. Establish molding material by supplier and number.   _____   _____
 4. Establish specific type of mold design.   _____   _____
 5. Establish heating or cooling system.   _____   _____
 6. Establish equipment to be used and overall mold size limitations for this equipment?   _____   _____
 7. Establish number of cavities and approximate layout or arrangement.   _____   _____
 8. Re-verify material chosen to be sure it is satisfactory and useable in the application.   _____   _____
 9. Establish workable tolerances.   _____   _____
10. Establish parting lines and location of split line for wedges.   _____   _____
11. Establish best ejection method. Locate ejector pins, pads, blades, sleeves, plates, collapsing cores, two-stage ejection, double ejection, etc.   _____   _____

ENGINEERING CHECK LIST (*Continued*)	*To Do*	*Done*

12. If transfer or injection, establish gating areas and specify type of gate. ___ ___
13. Establish mold venting points. ___ ___
14. Establish mold finish required by customer, by material chosen, or by method of molding. ___ ___
15. Establish draft angles to be applied. How much and where? (don't forget negative draft is useful). ___ ___
16. Engineer review items 1 through 15. Secure customer's approval where needed. ___ ___
17. Engineer discuss with designer. ___ ___
18. Establish shrinkage factor (transverse–parallel). If, thermoset, is post-baking a requirement? Have you allowed extra shrinkage? ___ ___

DESIGNER CHECK LIST	*To Do*	*Done*

1. Review the preliminary engineering check list (with the engineer, if possible). Do you understand everything? If not—*ASK*. ___ ___
2. Review the catalog data on Standard Mold Bases and Components to select the most economical group of components for the proposed mold design. Can you use a *complete mold base?* Will you have to build up from *standard plates?* Can you use *standard components?* Will your supplier "start from scratch" with raw steel? (See also Chapter 8.) ___ ___
3. Answer the following questions:
   A. Where can "pickups" be placed if they are needed? (a "pickup" is a deliberate undercut.) ___ ___
   B. Are inserts to be molded-in or assembled after molding? In any case, get a copy of the insert drawing. Insist that inserts not be made until after the mold is designed (when the inserts are to be molded in.) ___ ___
   C. Are side inserts necessary and, if so, how are they to be supported? ___ ___
   D. Are wedges or side cores required? removable or captive? ___ ___
   E. Where will wedge split line (parting line) be located? How operate wedge or side cores? ___ ___
   F. What type of insert pins are to be used and how will inserts be held on the pins? ___ ___
   G. Do mold pins spot holes? Do they butt in center? Do they enter the matching section of the mold? ___ ___
   H. Where will mold-maker want radii for ease of machining? Will customer permit it? ___ ___
   I. Where will mold-maker want sharp corners for ease of machining or reducing cost? Will customer permit it? ___ ___
   J. Will the cavity be hobbed, machined, cast or electroplated? ___ ___
   K. Can or should the cavity (or core) be made in one piece? Where are inserted sections needed? ___ ___
   L. Where are the high wear areas in the mold? Should they be inserted or backed up with hard plates and wear pads? Is lubrication needed or provided for? ___ ___

DESIGNER CHECK LIST (Continued)

	To Do	Done
4. Now you are ready to select drawing paper size. Is size selected **large** enough to show all needed views without crowding?		
5. If using a standard mold base (or plates) draw in the complete mold base outline including location of guide pins, screws, return pins, etc.		
6. If not done in Item 5, do so now—layout horizontal and vertical center lines. Be sure to allow ample space for all views and details.		
7. Layout cavity arrangement prescribed (circles—square—rectangular). Will spacing allow temperature control media channels?		
8. Establish ejector system to be ready for item 10.		
9. Layout one molded part of each configuration in plan view and fill in force and cavity outlines.		
10. Add ejector pins (or system established in item 8).		
11. If injection or transfer mold, establish sprue, runner and gate sizes and the material route from nozzle or pot to cavity (do not overlook hot runner, runnerless, hot manifold, hot tip, etc.).		
12. If compression mold, establish land areas, loading well depth and cavity wall thickness.		
13. If transfer mold, establish pot and plunger size (transfer chamber) or sprue (if needed), runner size and path, and gate size.		
14. If automatic compression mold, establish land areas, loading board specifications and part removal board specifications.		
15. Locate the center line and size of the temperature control media channels.		
16. Draw in guide pins, return pins, screws, and stop pins. Use ample number of screws with calculated holding power to resist stresses.		
17. Add retainer plates, clamp plates, width and length of ejector bar, parallels and stop pins.		
18. Project the top layout (plan view) to front and side views to develop retainer plate thickness, length of screws, etc. *NOTE:* One thick plate is better than two thin ones. If a long running mold, use hard or prehardened plates to back-up core pins, forces, cavities and slides (see 3(L)).		
19. Add support pillars or parallels. Use adequate support to prevent plate sagging under pressure.		
20. Draw in sprues, runners, gates and ejector pins.		
21. Dimension all views and parts as required.		
22. Add part numbers, material list and general assembly notes including mold number, part number, operating press data, etc.		
23. Detail cavities, forces, core pins, wedges, slides, side cores and all other parts requiring detail drawings. (Shop practice will determine this.)		
24. Call for steel trade number or identification on a non-working surface of all hardened parts (in event of later modifications requiring heat treating or annealing).		
25. Specify tolerances where needed to assure compliance with product requirements.		
26. Carefully check all drawings for dimensional errors or reversal of detail.		

DESIGNER CHECK LIST (Continued)   *To Do*   *Done*

27. Review the engineering check list (preliminary). Have you done everything as specified? (Vents are most often overlooked.)   ——— ———
28. Be sure your supervisor, checker, or tool engineer knows *all* the little assumptions, changes, additions or deletions you have made. If necessary, mark another product drawing to show all these things and deliver to cognizant engineer.   ——— ———
29. Check all drawings just once more to be *sure:*
    - (a) that mold can be assembled.   ——— ———
    - (b) That, when assembled, all parts are in proper orientation with each other—right hand to right hand and left hand to left hand—and   ——— ———
    - (c) that mold will *fit* and *operate* in the equipment specified or intended.   ——— ———
30. File your notes. File all information obtained from others. File all unusual calculations for future reference (or for checker).   ——— ———
31. Deliver final prints to mold-maker.   ——— ———

ENGINEERING CHECK LIST (final)   *To Do*   *Done*

1. Review the *preliminary check list* (engineers).   ——— ———
2. Review the *designer's check list* and the tool drawings. Has he done everything according to your instructions (or given good reason for not doing so)?   ——— ———
   Has he *told you* (and supplied marked product drawing) about his assumptions, changes, deletions or additions?   ——— ———
3. Get final approval of *customer* on final configuration.   ——— ———
5. Review projected areas or mold, material volume capacity of equipment and molding material limitations.   ——— ———
6. Issue special instructions for mold set-up and operation if design warrants.   ——— ———
7. Review entire project with production, tool department, quality control, set-up, and management personnel.   ——— ———
8. Plan for fixtures or special installation of mold well ahead of scheduled mold delivery.   ——— ———
9. Follow mold progress at regular intervals of mold construction.   ——— ———
10. When problem arises, take necessary action, and repeat as needed.   ——— ———
11. Open the mold and visually inspect same before mounting in press.   ——— ———
12. Use thin paper and ask for a paper impression of the mold under pressure, and before attempting sampling or production.   ——— ———
13. Review upon completion of sampling and advise interested parties of adequacy or inadequacy of design and engineering. Note specifically any areas of trouble.   ——— ———
14. Finally, *ask* what can we learn from this project that will be useful for future mold designs?   ——— ———

The three check lists were prepared in collaboration with Sumner E. Tinkham, Morristown, NJ.

DIMENSIONING MOLD DRAWINGS

It is always desirable to follow the rules of good practice when dimensioning mold drawings. The following suggestions will minimize error and facilitate tool construction:

1. Locate points by coordinates rather than by degrees of an angle and radii. Most toolroom machinery is designed to move in two directions, one at right angles to the other.

2. Start dimensions from a center line and show dimensions for intermediate steps. This will help you in checking, and it will also help the tool-maker.

3. It should be understood that dimensions crossing a center line indicate symmetrical spacing from the center line unless otherwise specified.

4. Where the relationship or alignment of mating parts is involved, be sure to allow for this when dimensioning the mold parts. Call special attention to these mating parts by inserting reference notes.

5. Do not repeat dimensions except where absolutely necessary; in the event of a dimensional change, one of the dimensions may be overlooked. When repetition is necessary, insert a reference as indicated in Fig. 5.3 for the .315 dimension.

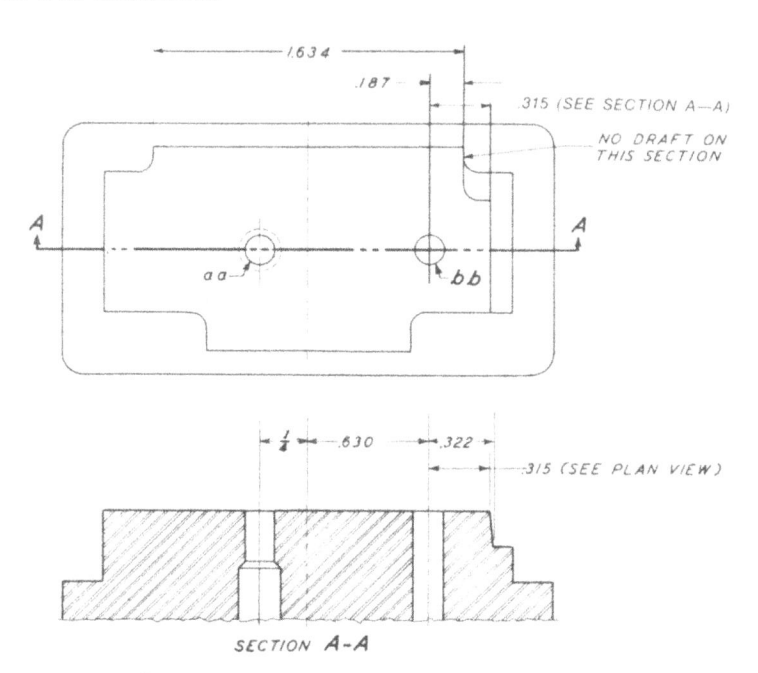

Fig. 5.3. Proper notes for repeated mold dimensions.

6. Enter all essential dimensions on every drawing. It is bad practice to permit scaling of prints, since the printing and drying process may introduce considerable distortion.

7. Keep all related dimensions together so the tool-maker will not need to "hunt" for the dimensions he needs.

8. Try to keep the dimensions between the views as much as possible.

9. Dimensions given for length of thread, depth of tapped hole, etc., are generally understood to indicate the minimum length of the full thread. Tool-makers will make the required allowance for the thread ending.

10. Use three-place decimals for all ordinary cavity and plunger dimensions, four-place decimals are used only where the extra accuracy is essential, or where it is necessary to make the component dimensions "add up." Use fractions or two-place decimals wherever ordinary scale dimensions will suffice.

11. Numerals and figures on tracings must be heavy enough to print well. Most designers use 3H pencils for layout work and 2H pencils for dimensioning. Allow space so that it will not be necessary to crowd dimensions. Decimal points must be distinct so they will print well.

12. Dimensions should be given at points from which it will be easiest for the tool-maker to work and measure.

13. Avoid repetition of figures and the use of the same combination of figures whenever possible. Typical examples to be avoided are 1.342 and 1.324, or 1.732 and 1.723. Errors often result from such combinations and a change in the last figure will eliminate the problem. The numerals zero, six, nine, and one often cause trouble when read upside down, therefore the last figure in a group containing any of these numerals should be changed. For example, .061 may be read as .190 when inverted, whereas .062 obviously would be upside down if the print were turned.

14. Where holes are designated for assembly purposes, as shown in Fig. 5.3, use a double letter such as *aa*, *bb*, *cc* etc.; this will avoid confusing the designation with a letter drill size.

15. The *plus* tolerance is *always* placed above the *minus* tolerance when they are added to the drawing. For example:

$$.125^{+.002}_{-.000}, \ .212^{+.000}_{-.002}, \ .314 \pm .002$$

This is done because it is common practice to mention the plus tolerance first when speaking.

Tolerances and Allowances

Allowances are the intentional differences in dimensions on two parts which fit together. *Tolerances* are the allowances made for unintentional variations in the dimensions. *Limits* are the maximum and the minimum dimensions which define the tolerance. *Interference* is a negative allowance which is used to assure tight shrink or press fit. A *basic dimension* is the measurement from which all variations are allowed. A *nominal dimension* is the dimension halfway between the maximum and minimum limits. Tolerances on mold dimensions are required because of the unintentional variations that occur in machining, hardening, polishing, plating, etc.

Toolroom Tolerances. Most toolrooms have established standards of allowable variation to which they work. Typical standard toolroom tolerances are given in Table 5.1.

These are reasonable working tolerances for a toolroom and the designer should take them into consideration when designing molds.

Tolerance Allocation. There are two systems of showing permissible variation from a basic gauge dimension. One is the *bilateral* system, and the other is the *unilateral* system.

In order to understand the unilateral and bilateral systems of tolerance allocation, it is necessary to conceive the idea of tolerances on tolerances. Since a tolerance is a measurable extent of magnitude, it, like any other dimension, can be accurate only within specified limits. There is no such thing as an exact dimension.

Either one of these two systems of tolerance allocation may be used in setting up tolerances for each of the four sets of gauges which may be used. It is necessary to allow some tolerance in the transference of measurement from the precision gauge block to each succeeding gauge block, as shown in Fig. 5.4 until the working gauge is made. Although precision gauge tolerances are measurable to the millionth part of an inch, it should be understood that the most accurate of measuring devices is accurate only within specified limits.

The *bilateral* system allows a variation *both* above *and* below a basic gauge dimension. A 1.500 ± .005 dimension would require a maximum

TABLE 5.1 Typical Toolroom Tolerances.

Dimension	*Tolerance**
Fractional dimensions 1/4 in. or less	±.008
Fractional dimensions over 1/4 in.	±1/64
Three-place decimals	±.005
Four-place decimals	±.001

*Unless otherwise specified.

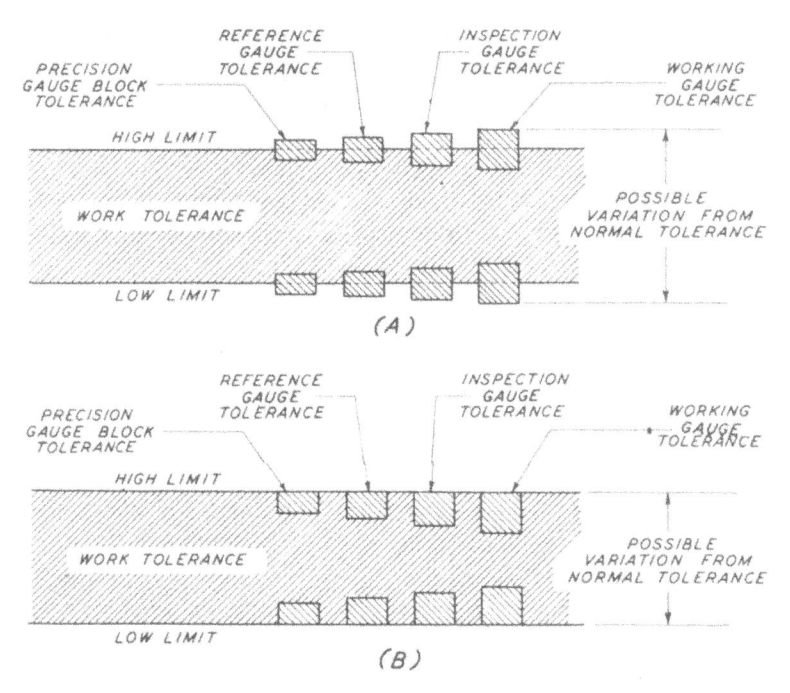

Fɪɢ. 5.4. (A) Tolerances as allocated by the bilateral system. (B) Tolerances as allocated by the unilateral system. (*Courtesy Sheffield Corporation, Dayton, OH*)

gauge of 1.505 and a minimum gauge of 1.495. In the bilateral system, one could say that the .005 tolerance must be accurate within plus or minus .0005 (one-half of one thousandth). This means that the maximum gauge would be dimensioned 1.505 ± .0005 and the minimum gauge would be dimensioned 1.495 ± .0005. Therefore, if it happened that the gauges were made to the extremes, a part could be made anywhere between 1.4945 and 1.5055 and still pass the gauge inspection. Thus the actual variation could be .011 instead of .010.

The *unilateral* system allows a variation in one direction only, that is, *either* above *or* below a basic gauge dimension, depending on whether the gauge is for a maximum dimension or for a minimum dimension. If the gauge is for a maximum dimension, the tolerance will be allowed only on the minus side. If the gauge is for a minimum dimension, the tolerance will be allowed only on the plus side. Thus, a 1.500 ± .005 dimension would require a maximum gauge of 1.505 and a minimum gauge of 1.495. Using the unilateral system of allocating tolerances to these gauges, one could say that the .005 tolerance must be accurate within .001. This means that the maximum gauge will be dimensioned $1.505^{+.000}_{-.001}$ and the minimum gauge would be dimensioned $1.495^{+.001}_{-.000}$. Therefore, the total possible variation of any part could be only .010 and still pass the gauge inspection if both gauges were made to the extremes.

In the bilateral system, the high and low limits would *bisect* the gauge tolerance zones, and the tolerance for each gauge would be allocated as plus or minus from its respective basic dimension, as shown in Fig. 5.4.

In the unilateral system, the high and low limits would *encompass* all the gauge tolerances, so that the tolerance on the gauges would be allocated as minus from the high limit and as plus from the low limit.

The bilateral system has been in use in this country for a long time and probably is adhered to at present for most general commercial work. Ordinance engineers contend that the unilateral system is more scientific than the bilateral and more effective in precision work.

Calculating Mold Dimensions and Tolerances

Mold dimensions are computed from the following general rules, which compensate for the variables. Nevertheless, remember that good judgement is always better than following a rule.

Shrinkage allowance is an "add-on" factor. Every molding material has a shrinkage factor specified by the manufacturer. (*Warning*—Some recently developed materials actually "grow" when taken from the mold, therefore shrinkage factor is a negative value.) The factors furnished by the manufacturer may be a narrow range such as .003 to .004 in./in. for mica-filled phenolic. They may also be a wide range, such as .005 to .040 in./in. for Nylon. In any case, the designer *always adds* shrinkage (except in the aforementioned *warning*). Space does not permit disposal of the argument that shrinkage is added to some parts of the mold and subtracted from other parts of the mold. *If you add shrinkage to any part of the mold, add it to all parts.*

The designer should seize every opportunity to obtain and record specific shrinkage data from his own shop. This is done by checking molded parts and mold at room temperature. Subtract the smaller dimension from the larger dimension, then divide the result by the dimension of the molded part. The results of the division is the shrinkage allowance in inches per inch, and should compare with the value given by the manufacturer. Your own history on specific materials and mixes used under your shop conditions will be far more reliable and reproducible than the manufacturer's data. At this point, let us mention the phenomena of different rates of shrinkage in the same part. Shrinkage *parallel* to flow may differ from shrinkage *transverse* to flow. Shrinkage in thin sections may differ from shrinkage in thick sections. By "differ," we mean that the *rate of shrinkage* in percent or in thousands of inches per inch or whatever other method of shrinkage rate specification is used is different. In thermosetting molding, shrinkage rate will be one value when compression molding and another value when transfer or injection molding an identical material. Thermoplastic materials attain different rates of shrinkage depending upon (1) cylinder temperatures at time of injection, (2) nozzle temperatures, and (3) temperature of the mold. Some molders select an *average shrink rate,* apply this to the mold, then set up manufacturing conditions to obtain the allowed shrinkage rate.

The dimensions which locate *holes and bosses* in the plan view of a mold should use the *nominal* dimension plus the shrinkage factor. For example, a dimension of $1.250\,{}^{+.010}_{-.000}$ for the location of a hole, and using a shrinkage factor of .008 in. per in. would be specified as 1.265 [1.255 (nominal) times 1.008 equals 1.265] or [1.255 + (.008 × 1.255) = 1.265]

Projections, *pins*, or other male parts of the mold are calculated by subtracting ¼ of the *total allowable* tolerance from the *maximum dimension* permissible. Then add material shrinkage. For example, .500 ± .010 would become .505 plus shrinkage. Tolerance on mold is given in minus direction.

Cavities, depressions, grooves, and other female parts of the mold are calculated by *adding* ¼ of the *total allowable* tolerance to the *minimum dimension* permissible. Then add material shrinkage. For example, a .500 ± .010 wide groove would become .495 in. plus shrinkage. Tolerance on the mold is given in plus direction.

Dimensional tolerances, as given on a tool drawing, should amount to no more than ½ the desired tolerance for the molded part because the mold variation is only *one* of the factors influencing the final dimension of the molded part. Other factors affecting the final part dimensions are:

1. variable material shrinkage from batch to batch
2. heat
3. pressure
4. cure or chill time

The previous rules are used because a hole may be made larger and a boss may be made smaller to achieve the desired results after trial of the mold indicates the actual shrinkage and the accuracy of the tool work.

Drafts and Taper. Wherever possible, draft should be allowed within the tolerance given by the part drawing. However, this is not always possible. Then the designer must be very careful in using draft. The customer should determine just where a dimension is to be taken and in which direction draft should be allowed.

In case of doubt, dimension the mold so *metal* can be removed to make the correction at a later date. Don't forget, it is always cheaper to ask questions than it is to guess wrong.

MOLD STAMPING

Product designers generally show the depth of letters and numerals when it is essential that proper depth be allowed for painting, good appearance, etc. Many times, the depth of such lettering is left to the mold designer and, in such case, he should submit his specifications and recommendations to the product designer for approval, thus making certain that the height of raised letters or lines will not complicate the assembly of the device. Ordinary stamped letters that are to remain unpainted will be plainly visible when they are only .005 in. high. If the characters are large, however, a height of .010 in. may be used.

Characters are often specified by reference to some standard type specimen book such as *The Book of American Types,* published by American Type Founders, of Elizabeth, New Jersey. The height of the letter and its weight and depth should be specified, as shown in Fig. 5.5. Lettering that is to be painted will be raised in the mold. The elevation of such letters should be at least one half the weight of the line. All characters must be stamped in the mold left hand. A typical designation would read: "Stamp in ⅛-in. L.H. characters .005-in. deep." Full information must be given for special characters.

Pad Length. Lettering is often placed on a removable pad in order to facilitate stamping and to permit a change in lettering when required. The length of the pad for stamped characters (letters or figures) can be calculated by using the following formula:

Pad length = number of characters × height of characters
+ the height of one character

For example, a pad for the word "Lower" in ⅛-in. letters would be calculated at 5 (number of characters) × ⅛ in. (height) + ⅛ in. (height) = ¾ in. minimum length of pad).

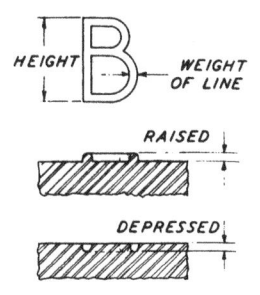

Fig. 5.5. Height of letter, weight of line, and the amount that the letters are to be raised or depressed should be specified on tool drawings.

TOOL STRENGTH

Molds are designed to give maximum life and low maintenance expense. Small, fragile sections should be designed for easy removal and low-cost replacement. Designing for adequate tool strength is always a problem, and no definite formulas can be given. Several fundamental considerations will serve as a guide in the solution of many problems. The strength required must be adequate to resist the compressive, bending, or shearing stresses set up by the highly compressed molding compound as it moves into position and hardens. Some of these stresses may be calculated when necessary, but the mechanical construction of the mold is such that most designers do not calculate the actual total stress loads on all mold sections. As a matter of fact, flow conditions during molding may introduce highly localized stresses which cannot be calculated or predicted.

There are four parts of the mold which must be considered to make sure that the mold design is satisfactory. There are:

1. The cross section of the ejector bar must be adequate in strength to resist the bending stresses.

2. The thickness of back-up plates behind the cavities, forces and other mold sections must be adequate.

3. The wall section of mold cavities, loading pots or transfer chambers must be sufficient to resist the spreading force resulting from the mold pressure.

4. The thickness of the bottom area of mold cavities must be sufficient to resist distortion and breakage.

The strength of the ejector bar increases in direct proportion to the width of the bar and as the square of the thickness. This means that the bar should be kept at the minimum width required for the ejector pins, since a small increase in thickness is much more effective than a considerable increase in width. A desirable average minimum width for the ejector bar is 2 in.

For example, if an ejector bar 1 in. thick and 4 in. wide were increased in width to 8 in., its strength would be doubled. If this same bar (1 in. thick and 4 in. wide) were increased to 1.414 it would then have twice the strength of the original bar. The mathematical proof of this statement may be taken from the formula for beam stresses. In this formula the stress is shown to vary as the section modulus. The section modulus of any given bar is determined by the formula $BD^2/6$ in which B equals the width of the bar and D equals the thickness. This readily shows that if B is doubled, the strength of the bar is doubled, but if D is doubled, the strength of the bar is quadrupled.

Table 5.2 shows values which have been found to be satisfactory for the wall thickness of mold sections. An approximate general formula fre-

TABLE 5.2. Recommended Minimum Wall Thickness for Mold Cavities and Retainer Plates.

Inside Dia. of Sets (In.)	Wall Thickness for Round Sets (In.)	Wall Thickness of Retainer Plates (In.)	Wall Thickness for Rectangular Sets (In.)	Wall Thickness of Retainer Plates (In.)
Up to 1¼	5/16	1¼	7/16	1¼
1¼–2	3/8	1⅜	½	1⅜
2–3	7/16	1½	9/16	1½
3–4	½	1⅝	⅝	1⅝
4–5	9/16	1¾	11/16	1⅞
5–7	⅝	1⅞	13/16	2⅛
7–9	¾	2	1	2½
9–12	⅞	2¼	1⅜	2¾
over 12	1	2½	1¾	3

Note: Add ⅛ in. on wall thickness when depth exceeds twice the basic wall thickness indicated in table.

quently used for the basic wall thickness of mold sections under 2-in. diameter is to use 60 per cent of the depth of the cavity, but never less than 5/16 in., nor more than ¾ in. When the section is greater than 2-in. diameter, an additional 1/16 in. should be added for each additional inch of diameter. In cases where the depth is greater than twice the basic wall thickness, add an additional ⅛ in. to the wall thickness. For example, a cavity with a 5-in. diameter loading space of 1-in. depth would require a 13/16-in. wall thickness. This is calculated as follows:

$$(1 \times .60) + 3/16 = 13/16 \text{ in.}$$

For practical purposes, the outside diameter would be dimensioned at 6¾ in. to facilitate its construction from 7-in. diameter stock by taking a ⅛-in. cut to clean up the stock (see Fig. 5.6). Designers generally calculate the size of stock required and then select the nearest larger size, giving as the final dimension the clean-up size of the stock.

Molds which have split cavities require special consideration from the standpoint of having adequate strength in the retainer plates to resist the cavity "opening up" under molding pressure. The thickness of the cavity halves is the same as that for a solid cavity, but the retainers should have 1½ to 2 times the normal wall thickness, using the higher multiple for deep cavities. Such cavities are always shrink-fitted to the retainer.

Figure 5.7 shows a four-cavity mold for the switch part which is seen at the right corner of the cavity retainer. Note that each cavity has been made in two sections and that the eight cavity sections have been shrink-

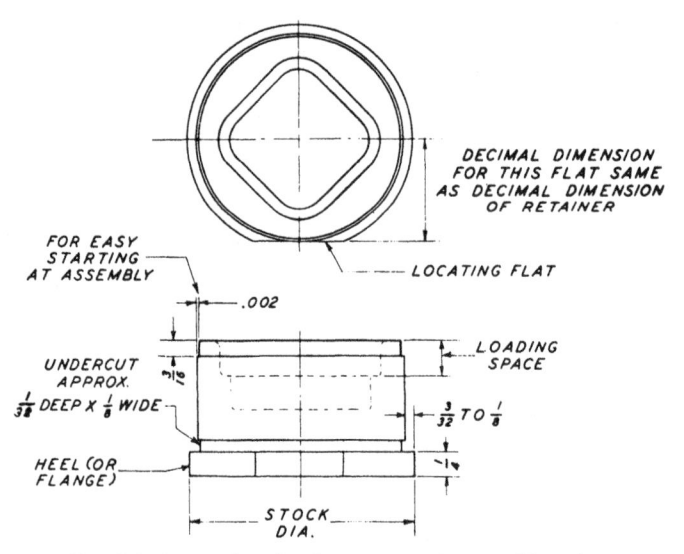

DECIMAL DIMENSION
FOR THIS FLAT SAME
AS DECIMAL DIMENSION
OF RETAINER

FOR EASY
STARTING
AT ASSEMBLY

LOCATING FLAT

.002

UNDERCUT
APPROX.
$\frac{1}{32}$ DEEP X $\frac{1}{8}$ WIDE

LOADING
SPACE

$\frac{3}{32}$ TO $\frac{1}{8}$

HEEL (OR
FLANGE)

STOCK
DIA.

FIG. 5.6. Proportions for flange (or heel) on mold sections.

fitted to the retainer. The high and thin barriers required on this part necessitated the making of the split cavities. (It would be difficult to machine such deep barriers in a solid block of steel.) The mold is shown in open position with the ejector pins raised. Six ejector pins are used, and four movable pins hold the inserts. This makes a total of ten pins for a piece approximately 2½ by 3½ in.

FIG. 5.7. Each cavity of this four-cavity semiautomatic landed plunger mold is made in two pieces and all are shrink-fitted to the retainer. Molded part is shown at right.

Noteworthy also is the fact that the split sections are located crosswise in the mold. If the split were lengthwise, the sides of the retainer would be stressed by the molding pressure and the force would tend to bow the sidewall, since the stress would be exerted over a greater area in the lengthwise direction.

SHRINK FIT ALLOWANCES

Where shrink fits are used, as in the assembly of loading pots or split sections, the standard allowance for interference of metals is as indicated in Table 5.3. The total interference should be subtracted from the size of the mold section to arrive at the size of the hole in the retainer plate. Simple press fits are specified by decimal dimensioning the mold sets the same size as the hole in the retainer plates.

The size of flanges or heels, as shown in Fig. 5.6 should be stock dimensions whenever possible, and the width should be 3/32 to ⅛ in. The flange thickness should be ¼ in. unless extreme stresses are anticipated.

Approximately 1/64 in. clearance should be allowed in the retainer plates to clear these stock dimensions, except at the locating flat, shown in Fig. 5.6.

MOLD PINS

Many kinds of pins are used in molds to form holes in the molded piece, to locate inserts, and to eject the finished piece. The insert pins and pins entering the molded piece are commonly referred to as mold pins to differentiate them from ejector or knockout pins. All pins should be made to the maximum allowable diameter and to the minimum length. Mold pins

TABLE 5.3. Interference Allowance * for Shrink Fit.

Hole Size Up to (In.)	Interference Allowance (Minus) (In.)	Hole Size Up to (In.)	Interference Allowance (Minus) (In.)
1	.0010	12	.0058
2	.0015	14	.0065
3	.0020	16	.0070
4	.0028	18	.0075
6	.0035	20	.0080
8	.0045	22	.0088
10	.0053	24	.0093

*Subtract allowance from mold section to obtain hole size in retainer plate.

which form part of the surface of the molded piece should be chrome-plated when the rest of the mold is plated. Ejector pins should be fitted loosely in the ejector pin plate. This allows the pins to align themselves with the holes in the mold sections. The ejector bar does not expand as much as the heated retainers, and this differential introduces some mis-alignment which must be compensated.

Pins with an integral cam to produce sidewall undercuts are called jiggler pins. Molds should be designed to make use of ejector pins whenever possible. The number and placement of the pins is entirely dependent on the size and shape of the molded piece. The basic function of knockout pins is to remove the molded part from the cavity or core with no distortion occurring, and it is better to have too many rather than too few pins to accomplish the desired result (see Fig. 5.8).

Figure 5.8 shows a single-cavity transfer mold and two molded parts. The part at the left shows the gates and runners still attached. Any combination of inserts may be needed for this part. Six ejector pin marks may be seen at the outside edge. An ejector pin is also located at each insert position and under each runner. This makes a total of 26 movable pins needed for this one cavity.

Ejector pins should be dimensioned so they will come .005 in. above the mold surface unless otherwise indicated on the product drawing. If stampings such as the cavity number or trade mark are desired on the ejector pin, the letters should be .005 in. deep and the pin should project .010 in. above the mold surface to make sure that the lettering does not project above the surface of the piece.

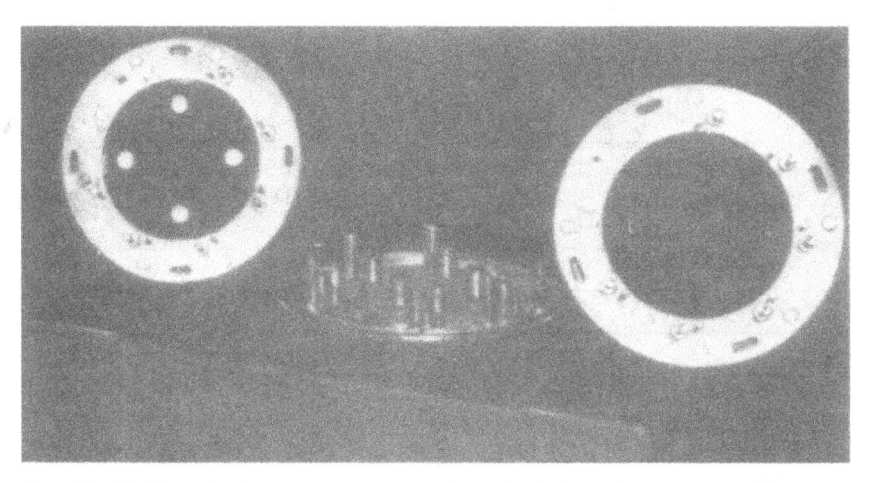

FIG. 5.8. Uniform ejection of the parts molded in this single-cavity transfer mold is assured by adequate number of well-located ejector pins. Molded part at left still has runners attached. Note ejector pin marks near outer edge.

Ejector pins must be as short as possible but must raise the piece ⅜ in. above the top of the cavity for production convenience. If wedges are raised by the ejector pins, they should be raised ¼ in. above the retainer so that they may be picked up easily.

Some designers specify that ejector pins 3/16 in. and larger be given flat gas reliefs. Pins up to ½-in. diameter have flats whose depth is equal to ⅛ the radius of the pin. For pins from ½- to ¾-in. diameter, four flats equal to 1/16 of the pin radius are used. Shown in Fig. 5.9 at (*A*) are the large gas vents often used where the ejector pin is inserted at the bottom of a deep cavity, or where it is necessary to fill out a thin-walled section and provide considerable gas relief. Such gas vents are generally flats from .003 to .005 in. deep that permit trapped air to flow through and escape. The thin gap provided will allow only a small amount of the molding compound to pass, since the plastic material will "set up" quickly in this thin opening and thereby block the flow. The addition of two grooves near the working end of the pin, as shown in Fig. 5.10 (*B*) will help keep ejector pin holes free of flash. These grooves will carry the flash forward out of the hole on each stroke. A blast of air dislodges the flash from the pin.

Two methods, as shown in Fig. 5.10, are used to form the heads on ejector pins. The head shown at (*A*) is formed by heating the rod to a red heat and then peening or swaging the head. This method is used where the

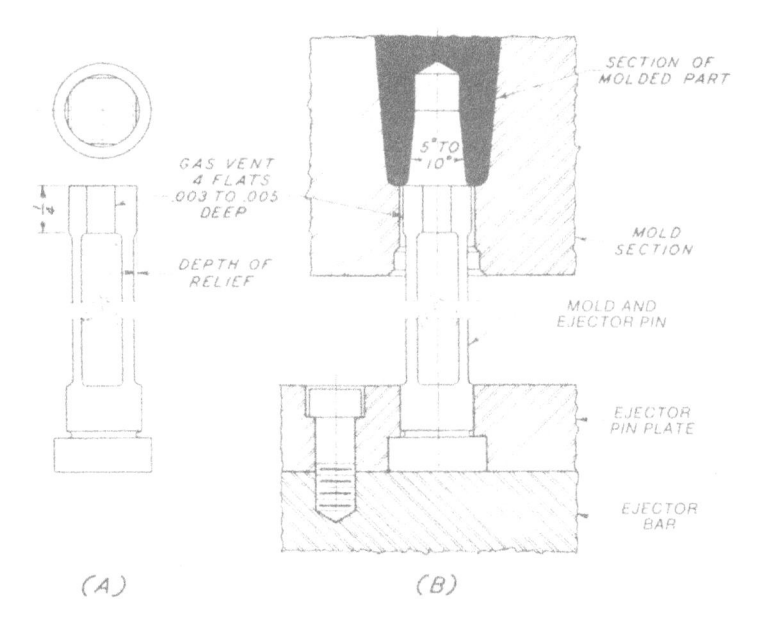

FIG. 5.9. (A) Standard gas vents and pin relief. (B) Gas vents and relief applied to mold pin. Note that 5° to 10° angle for pin strength to increase depth of hole.

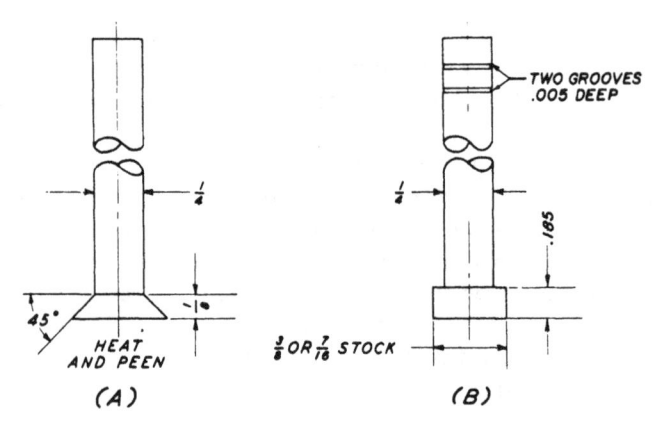

Fɪɢ. 5.10. Conventional methods of forming heads on ejector pins. (A) Peened head; (B) turned head.

length of the pin need not be controlled accurately. When the length must be held accurately, a turned head is formed by turning down the length of the pin (or butt welding, as some suppliers do) as in Fig. 5.10 (*B*). There are a number of suppliers of standard pins with turned heads. Follow their standards which have been well established, and are now widely used.

In designing the riveted-head type of ejector pin, consideration must be given to the fact that the principal stress is tensile. When the pin is pulled back, flash causes the end of the pin to bind, and considerable force may be needed to pull these pins back into place after they have been lifted to eject the molded part. The head must be strong enough for this stress. A good general rule to follow is to have the height of the riveted head equal to one-half of the pin diameter in sizes up to ¼-diameter. Pins larger than ½ in. should have turned heads. Pins which are less than 2 in. in length will generally have turned heads, as shown in Fig. 5.11 at (*A*). The center for turning or grinding is a great help to the tool-maker, who will make good use of it. Such centers should be added to the details of turned pins whenever possible. The ⅜ × ¼-in. boss shown at (*B*) is sometimes used where large pins are to be fully ground on the outside diameter after hardening. This small boss is removed to form a flat face on the end of the pin.

Pins which are to be prevented from turning make use of a flat that is tangent to the body of the pin, as shown at (*C*) in Fig. 5.11. This is done for the convenience of the mold-maker, as by this means he is permitted to mill or grind until the cutter touches the body of the pin.

Pins larger than ½-in. diameter usually are made with an undercut or groove, as shown at (*B*), Fig. 5.11. This allows the grinding wheel to travel to the heel and insure a straight pin. An alternative procedure is

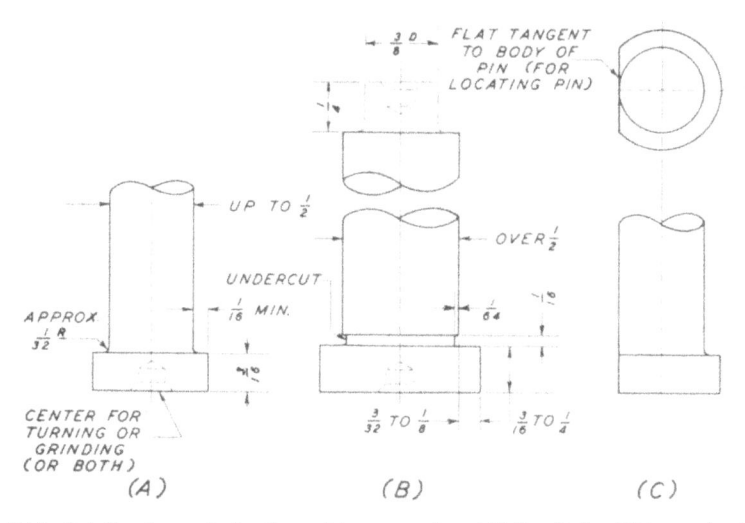

FIG. 5.11. Details of turned pins for mold construction. (A) Small pin; (B) large pin; (C) pin requiring angular location.

to leave a radius, as shown at (*A*), since it is practically impossible to grind a sharp corner.

Pins which form holes in the molded part should not project above the molding surface more than the amount indicated in Table 5.4, since the

TABLE 5.4. Ratio of Length to Diameter for Mold Pins.

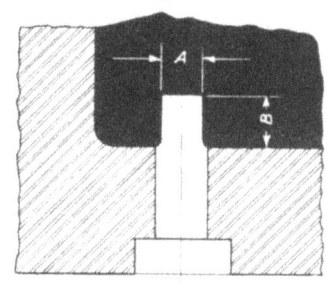

Pin Size A (in.)	*Length* B (in.)	
	Compression Molds	*Transfer or Injection Molds*
Up to ⅛₆	1 times dia.	2 times dia.
⅛₆ to ¾₆	1½ times dia.	3 times dia.
¾₆ to ⅜	2 times dia.	4 times dia.

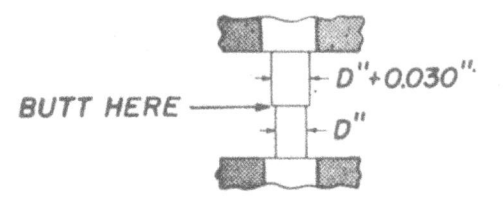

BUTT HERE

$D'' + 0.030''$

D''

FIG. 5.12. Illustrating use of butt pins for long holes.

lengths indicated have been found to be the maximum that will stand up under average conditions. Certain special conditions and molding care may permit use of longer mold pins. Holes which are formed by solid sections of the mold and which cannot be replaced in the event of breakage should be only one-half the height shown in the table. Where molded holes of greater length than those shown in the table are required, they may be formed by two pins which butt at the center, as shown in Fig. 5.12. Butt pins should be backed up by hardened steel plates or have oversize heads to prevent them from sinking in a soft plate, as sinking causes a heavy flash to form between the pins and thus increases finishing costs.

It will be noted from Fig. 5.12 that the diameter of one pin is greater than that of the other. This variance serves to compensate for slight misalignment of the mold cavity and pins. Longer holes may be molded by the use of *entering pins* which "enter" the opposite half of the mold, as shown in Fig. 5.13. Entering pins should always have turned heads, since the flash sometimes causes them to stick badly. The taper ream in the clearance holes, as shown in Fig. 5.13 allows the flash that enters around diameter *a* to move up freely when the mold is "blown out," or when it is pushed up by the entering pin. If this flash is not given an easy exit, it will build up a solid plug in the hole and, in a short time, cause the pin to stick, bend or buckle.

Where dimension *c* (Fig. 5.13) must be held to a tolerance closer than .010 to .015, the pin should be solid, rather than movable, as shown at *y*. The bearing surface *b* should not be more than 1½ to 2 times the diameter *a*. For pins ⅜ in. or larger in diameter, *b* dimension may be approximately 1 to 1½ times the diameter of *a*. The bearing surface *d* for ejector pins may be calculated the same as for *b*.

In compression molds, the length of the entering pins must be sufficient to permit entry to the force before it enters the loading space of the cavity. The flow of compound which starts when the force enters the cavity could deflect these pins so they could not enter. All mold pins should be tapered at least ¼° if possible on the molding surface. Shrinkage allowance must be added.

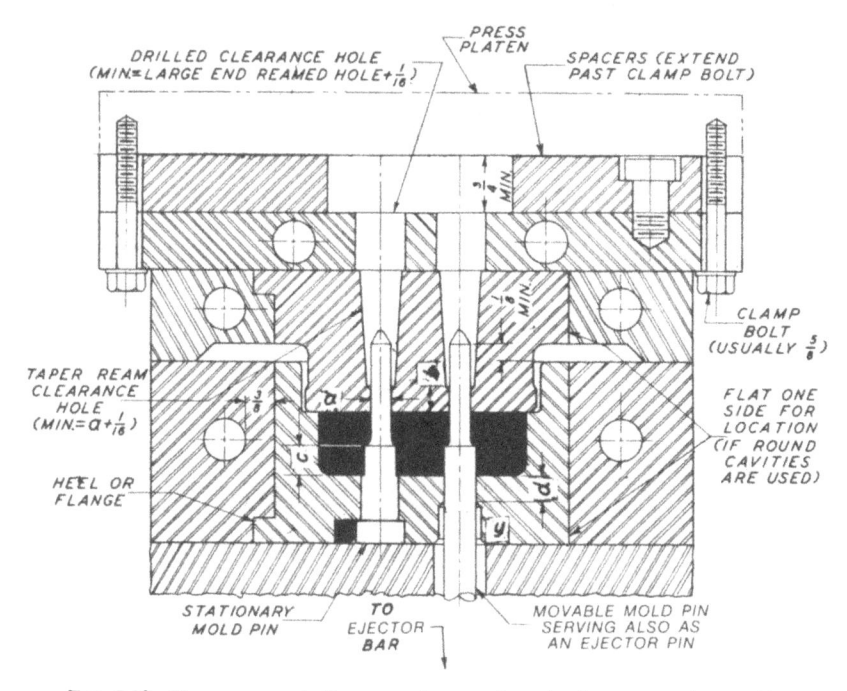

Fɪɢ. 5.13. Clearances and allowances for entering pins in compression mold.

Figure 5.14 shows good types of construction for long holes. Two pins may butt in the center, as indicated at *a*, or one pin may butt against the plunger, as at *b*, or against the cavity, as at *c*, above. Note that the minimum size of each pin is usually the nominal size plus shrinkage, plus .002 or .003 for a slightly oversize hole. This small amount of oversize may be cut down if the final molded piece shows too much clearance, whereas the pin cannot be made larger if it is made too small.

When close tolerances are specified on hole sizes, it is good practice to use ¾ of the plus tolerance. For example, the pin for a hole specified at 0.375 ± .010 in wood flour phenolic compound would be calculated as follows: 0.375 + .003 in. (shrinkage) + .008 in. (tolerance), resulting in a hole of 0.386-in. diameter. The reverse situation would be encountered for a boss having this same tolerance (0.375 ± .010), and the hole for the boss would be calculated: 0.375 + .003 (shrinkage) − .008 in. (tolerance), or 0.370 in.—the dimension used in the mold. Good designers always make mold pins to the maximum and mold holes to the minimum to insure that a sufficient amount of metal remains in case slight changes must be made to meet required tolerances.

Small mold pins that are less than 3/32-in. in diameter should be designed

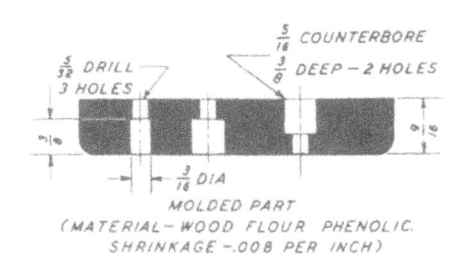

MOLDED PART
(MATERIAL - WOOD FLOUR PHENOLIC.
SHRINKAGE -.008 PER INCH)

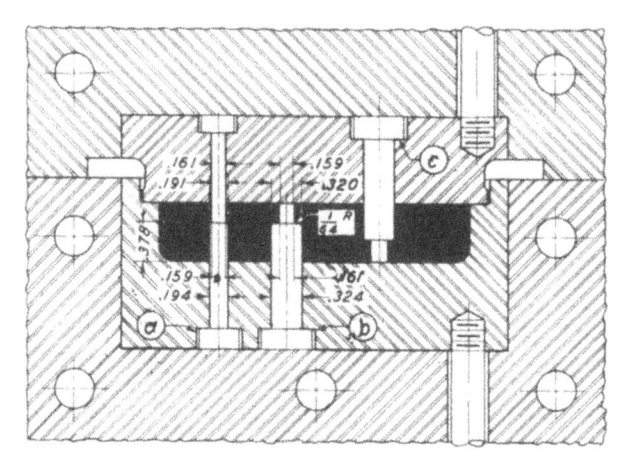

FIG. 5.14. Mold pin construction for deep holes with shrinkage allowance added.

as shown in Fig. 5.15, since it is quite difficult to drill a hole straight in these small sizes. It is also difficult to hold the diameter closely and obtain a good fit with the pin. This construction permits the tool-maker to lap or polish the short bearing area after hardening in order to fit the pin.

Small ejector pins (⅛ in. diameter or less) are often designed to be made

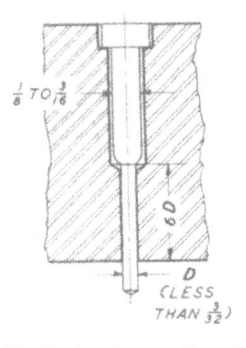

FIG. 5.15. Design for small mold pins.

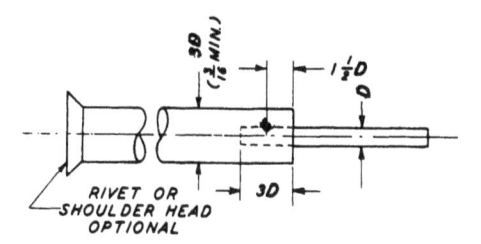

Fig. 5.16. Method of cross-pinning a two-piece ejector pin. Center line of cross pin is at edge of pin D. Minimum diameter of cross pin is 1/32 in.; maximum diameter is 1/2 D.

from two pieces, as shown in Fig. 5.16 to gain the required strength. Sufficient clearance must be allowed in the retainer plates and back-up plates to allow the large pin diameter to enter without interference. This construction is most often used when flat blade ejector pins are required. It is seldom used for round ejector pins and should not be used if any other choices are available.

OTHER MOLD PARTS

Safety Pins or Push-backs

Safety pins are extensively used where small ejector pins are required. Their name derives from their function, which is to protect the ejector pins. Figure 5.17 illustrates the action of the safety pins. As the ejector bar moves, the safety pins travel along with it, and, as the mold closes, these heavy pins push the ejector bar back to its proper position and thus the force is not transmitted through the ejector pins.

The diameter and number of these pins vary with the size of the mold, but ⅜ in. is usually the minimum size. For average molds using up to 50 or 100 ton pressure, a ⅝-in. pin will be satisfactory. Two pins may be used on narrow ejector bars, but three or four are needed where the bar is wide. A nonsymmetrical arrangement should be used so the mold cannot be assembled incorrectly. In compression and transfer molds, the clearance hole through the various plates should be 1/16 in. larger than the pin. In injection molds, the pins are usually slip fitted so they will support the weight of the ejector assembly in a horizontal position.

Male Molded Threads

The basic major diameter of the mold section is determined by subtracting ¾ of the allowable tolerance from the basic major diameter of the molded thread, then adding the proper material shrinkage. For the tolerance on the basic mold dimension, use ⅛ of the allowable tolerance (plus).

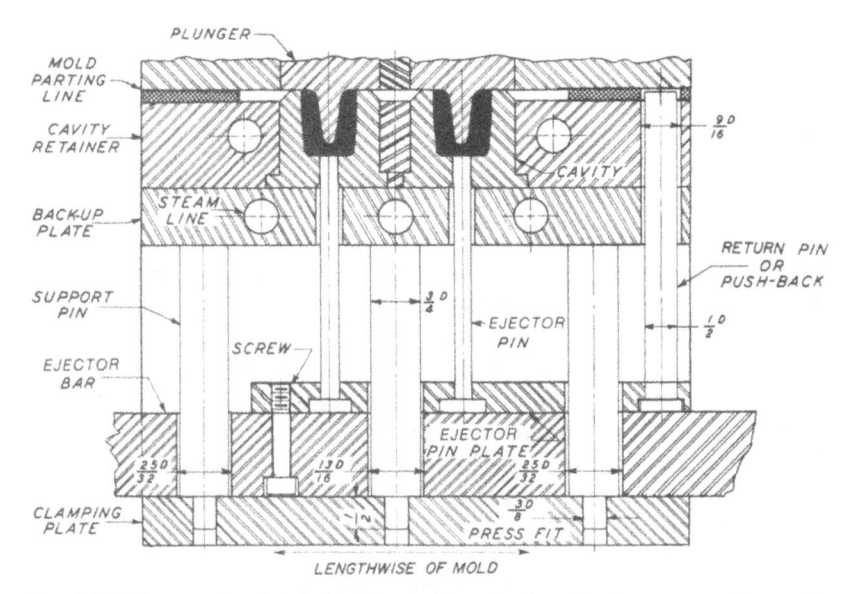

F_{IG}. 5.17. Diagram showing design of support and safety pins in a compression mold.

The basic pitch diameter of the mold (for ⅜-in. thread engagement, or less) is determined from the basic pitch diameter of the molded thread minus ¾ of the allowable tolerance, adding the necessary material shrinkage. The tolerance on the basic mold dimension is found by using ⅛ of the tolerance (plus).

When the thread engagement is greater than ⅜ in., subtract all of the allowable tolerance to obtain the basic pitch diameter of the mold and use ⅛ of the allowable tolerance (plus) for tool error. If more than ⅝-in. length of engagement is required, compensate for the shrinkage in the lead also.

For the basic minor mold diameter, subtract all of the pitch diameter tolerance, then add the material shrinkage. One-eighth of the total tolerance (plus) is allowed for tool error.

A typical example is shown in Fig. 5.18 for molding a 1 in.-8NC-1 screw which has the following basic dimensions and tolerances:

$$\text{Major diameter, } 0.9966 \,{}^{+.0000}_{-.0222}$$

$$\text{Pitch} \quad \text{diameter, } 0.9154 \,{}^{+.0000}_{-.0111}$$

$$\text{Minor diameter, } 0.8432 \text{ maximum}$$

Female Molded Threads

To determine the basic major diameter of the mold, add all of the pitch diameter tolerance plus the material shrinkage to the basic major diameter

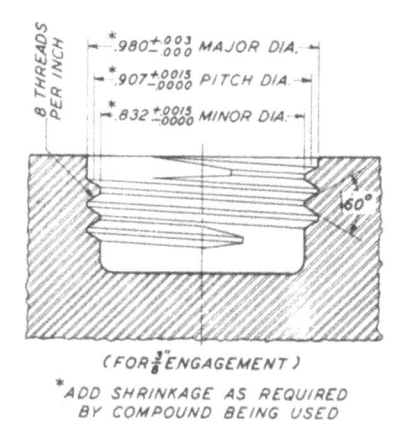

(FOR ⅜" ENGAGEMENT)

*ADD SHRINKAGE AS REQUIRED
BY COMPOUND BEING USED

FIG. 5.18. Female threaded mold section for 1 in. —8NC-1 screw produces male thread on molded piece.

of the molded thread. For the tolerance on the mold, use ⅛ of the pitch tolerance (minus).

To determine the basic pitch diameter of the mold (for ⅜-in. thread engagement, or less), use the basic pitch diameter of the molded thread plus ¾ of the allowable tolerance plus the material shrinkage. For the tolerance on the basic mold dimension, use ⅛ of the allowable tolerance (minus). Over ⅜-in. engagement, make use of all the allowable tolerance and ⅛ of the allowable tolerance (minus) for tool error. If more than ⅝-in. engagement is required, compensate for the shrinkage in the lead also.

For the basic minor mold diameter, use the basic minor diameter of the molded thread plus ¾ of the allowable tolerance plus material shrinkage. Mold tolerances are taken as ⅛ of the tolerance (minus).

A typical example is shown in Fig. 5.19 for a male threaded mold section for a 1 in.-8NC-1 nut having the following dimensions:

Major diameter, 1.0000 Minimum

Pitch diameter, 0.9188 $^{+.0111}_{-.0000}$

Minor diameter, 0.8647 $^{+.0148}_{-.0000}$

Mold Springs

Springs are extensively used in mold construction and typical applications are for the spring boxing of molds and return of ejector bars. The side springs shown in Fig. 5.20 at (A) and (B), may be used on either top or bottom ejector bars.

When the ejector bar travel is small, and where space permits, internal springs may be used, as shown in Fig. 5.21. In such applications, the ejector

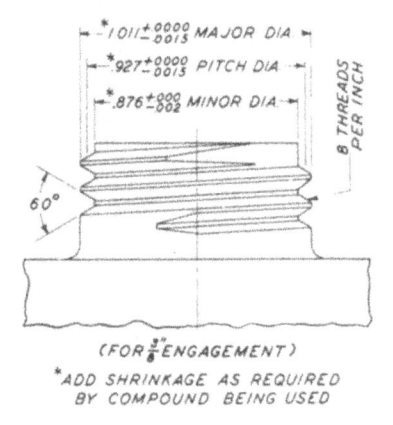

FIG. 5.19. Male threaded mold section for 1 in. —8NC-1 nut produces female thread on molded piece.

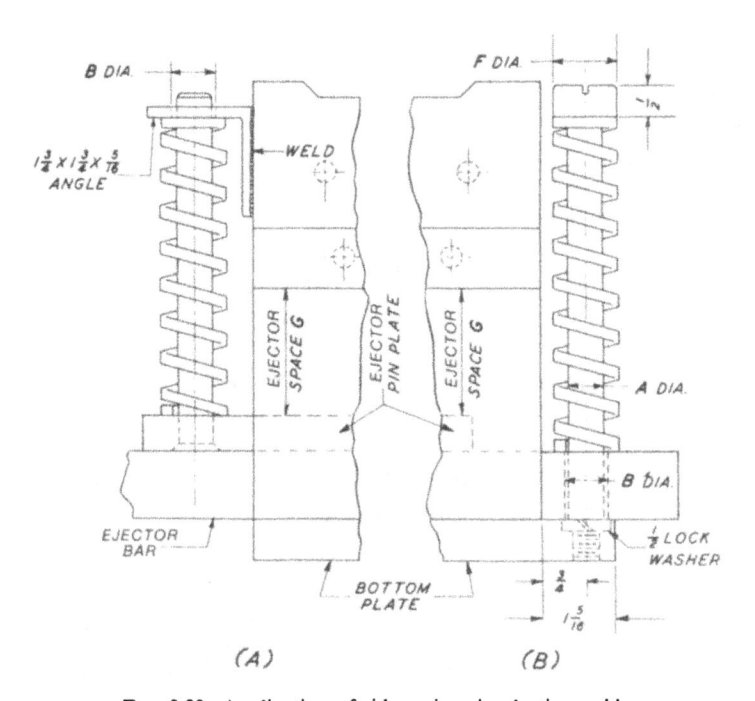

FIG. 5.20. Application of side springs in plastics molds.

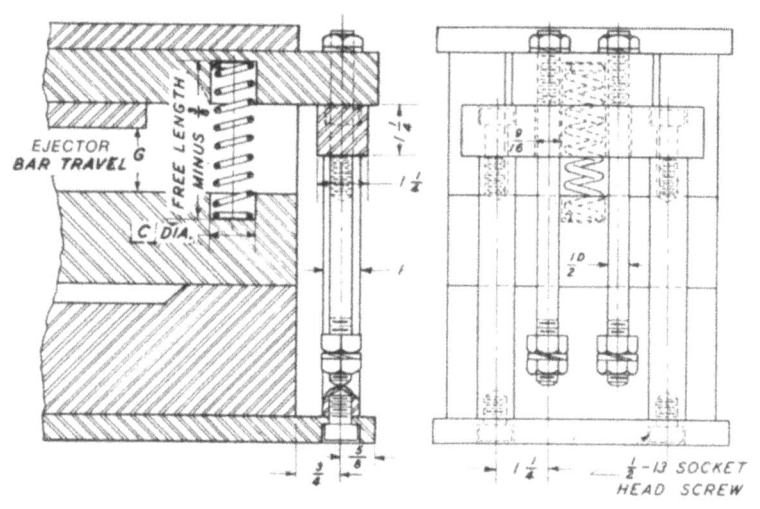

FIG. 5.21. Top ejector mold using internal springs and external pulldown rods.

bar travel G should be limited to 1½ or 2 in. This arrangement is also used where no auxiliary cylinders are available to operate the ejector bars.

The number of springs needed for any given mold is dependent upon the size of the mold and the size of the press being used. The minimum number that can be used in spring boxing is two on each side of the mold. Generally, not more than a total of six are used except in unusual applications.

For ejector operating springs (Fig. 5.21), two or three springs usually will be sufficient.

Mold Screws

Socket-head screws are best for mold work because they are easily disassembled. They also act as dowels in mold assembly, since usually they require only 1/64-in. clearance. Small screws are used in the small molds and larger screws in the larger molds. The thickness of plate has definite relationship to the size of the screws. The thickness of the head on a socket-head screw is the same as the body diameter. Thus a 5/16-in. screw requires a 5/16-in. minimum depth of counterbore. The right and wrong way to use these screws is shown in Fig. 5.22, which also indicates the proper clearances. The depth to which a screw should enter (dimension F) is the same as the outside diameter of the screw. At this depth the screw would break at about the same time the threads would strip. A rule commonly applied in determining the correct length of screw is: length of screw shall be the same

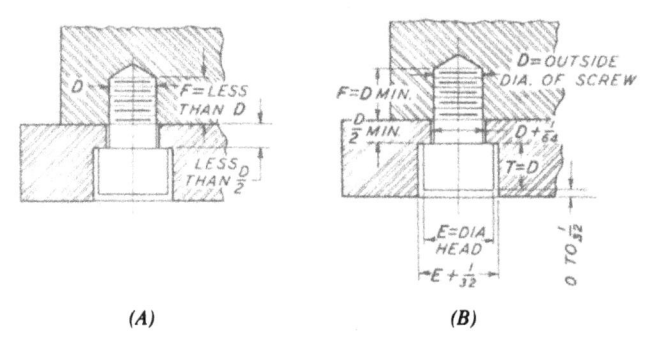

(A)   (B)

FIG. 5.22. (A) Improper application of socket-head cap screws; (B) proper application.

as the thickness of plate into which the screw head is recessed. Where greater holding power is required, more screws must be used. (See Table A.3 in Appendix.)

Parallels

The parallels should be as close together as possible under the cavity, allowing 1/16 to 3/16 in. clearance on each side of the ejector bar. The height of parallels is calculated to allow the ejector pins to push the molded piece ⅜ in. or more above the cavity. Most designers calculate this height and then add ¼ in. for the additional clearance that may be needed in the press set up.

The maximum width of center parallels should be 1/3 the total width of ejector bar. Center parallels should be provided where possible; if they cannot be used, the additional support may be provided by *support pins*. These center support pins are usually made from ¾- to 2-in. round stock, and they pass through the ejector bar. Support pins are best located midway between center of the cavities, as shown in Fig. 5.17. Clearance allowance for one pin is 1/32-in. for the ejector bar and pin plate; two or more pins use 1/32-in. clearance for the end pins and 1/16 in. for the center pins. These ends holes with the small clearance serve as guide pins for the ejector bar and minimize the horizontal thrust on the ejector pins. The 1/16-in. clearance for the central pins gives the tool-maker a little additional tolerance in his mold construction. Support pins are not often hardened when used in molds which operate vertically. Support pins and guide pins are often combined in molds which operate horizontally. It is recommended that hardened guide pins and bushing be used in the ejector assembly of injection molds, particularly if small ejector pins are involved.

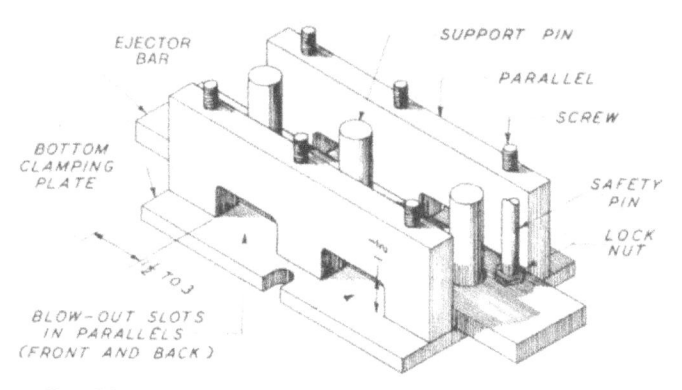

FIG. 5.23. Bottom ejector mold with blow-out slots in parallels.

Blow-out Slots

In using bottom ejector molds in compression molding, it is often important
to control the variation in the height of the ejector pins. Much of this variation
results from flash accumulating under the ejector bar, therefore it is desirable to have the operator blow out flash after each cycle before the
ejector bar is dropped. Slots are cut in the parallels, as shown in Fig. 5.23
to facilitate the blowing operation, and they are often cut in both front and
back parallels.

TEMPERATURE CONTROL MEDIA AND METHODS

We have previously stated that temperature is an essential ingredient in
every molding operation or at some point in the process. Generally, this
text deals only with controlling mold temperatures. Uniformity of heating
or cooling is the objective to be gained, and the problem justifies considerable
thought and planning. The bibliography lists texts dealing with heat transfer
and the methods used in calculating it. Heat transfer is the name of the game,
whether it be transferring heat out of the material and into the mold surface,
or transferring heat out of the mold surface and into the material. In either
case, the faster the heat can be transferred, the faster the cycle and the
faster the rate of production. Production costs are based on rate per hour,
thus it becomes obvious that whatever increases hourly production decreases
cost.

It has been emphasized previously that heat calculations are not necessarily complicated. However, this is true only if approximations are satisfactory. Chapter 14 gives specific information for the use of more sophisticated computer-based programs, for instances where exact information is

needed for the very difficult molding where flow controlled by heating or cooling is critical to success in manufacturing. The following *simplified* system may be used to calculate heating and cooling loads. You will note that the simplification advises that 20% be added for a safety margin. This means that the approximations are only accurate within 20%, which will probably be satisfactory in 80% of the cases. The authors subscribe to the 80/20 rule which, in essence, states that 80% of the benefit can be obtained for 20% of the cost (or effort). Certainly this is true in heat calculation unless there is a computer program available to perform the complicated calculations. The quantity of heat or cooling required is determined by the larger of the following two factors;

1. Quantity of heat or cold required to bring the equipment to operating temperature in a given time period.
2. Quantity of heat or cold required to maintain the desired temperature during operation.

When calculations are completed, add 20% for an error factor and as a margin of safety.

To calculate initial heat requirements:

$$\text{KW hr} = \frac{\text{Total Btu/hr}}{3412} \times 1.20 \text{ safety factor}$$

$$\text{Total Btu/hr} = \frac{A + B + C + D}{\text{Heat-up time (hr)}}$$

$$A = \text{weight (lb) of mold or platen} \times \text{specific heat}$$
$$\times \text{ temperature rise, } °F$$

$$B = \text{weight (lb) of system piping} \times \text{specific heat}$$
$$\times \text{ temperature rise, } °F$$

$$C = \text{weight (lb) of liquid} \times \text{specific heat}$$
$$\times \text{ temperature rise, } °F$$

$$D = \frac{\begin{array}{c}\text{exposed surface} \\ \text{area (sq ft)}\end{array} \times \begin{array}{c}\text{heat loss/sq ft} \\ \text{at final temp.}\end{array} \times \begin{array}{c}\text{time allowed} \\ \text{for heat up (hr)}\end{array}}{2}$$

To calculate heat required for operation:

Total Btu/hr = Btu absorbed by material added per hour

+ Btu/hr lost through radiation from molds and piping

To calculate operating cooling load:

Total Btu/hr = weight of material being processed (lb/hr)

$\qquad$ × specific heat × temp. change °F

$\qquad$ − Btu/hr losses through radiation from molds and piping

To calculate flow requirements (GPM):

$$\text{Lb/hr of circulated fluid} = \frac{\text{operating heating or cooling load in Btu/hr}}{\substack{\text{specific heat of} \\ \text{circulated fluid}} \times \substack{\text{allowable temp. difference} \\ \text{in °F across mold, roll etc.}}}$$

$$\text{Flow rate in gpm} = \frac{\text{Lb/hr of circulated fluid}}{\text{weight in lb/gallon} \times 60}$$

(Courtesy Super-Trol Mfg. Corp., Cleveland, OH)

In the following text, we refer to "channels." This should be interpreted to mean steam channels, hot oil channels, hot water channels, cold water channels, holes for electric heaters, holes for heat pipes, air channels, refrigerant channels and any or all means of controlling temperature in a mold, whether by directed flow with baffles, or non-directed flow.

The mold sets should be surrounded as well as possible by channels (see Fig. 5.17). Back-up plates are channeled so temperature control can be effected from above and below the molded parts. Various sizes of channels are used depending upon shop practice and the availability of specific types of temperature control media in a given molding shop.

For heating purposes, 9/16 in. channels are adequate for the average mold. As the mold gets larger, the channels should increase in size or number. In molds made as castings, the temperature control chambers are frequently cast in place. A 9/16 in. drill can be used up to 30 or 40 in. of depth. However, channels longer than 15 to 20 in. should be drilled from opposite sides. Heating channels are generally spaced about 2 to 3 in. apart for ordinary work and molds for presses up to 200 tons. Openings are ordinarily tapped with a pipe thread of the size needed to fit the conduits used in your molding room. Quick disconnects are suitable for water or refrigerants. Special quick disconnects are available for steam or hot oil, but should be used cautiously because of the extreme danger of rupture or accidental disconnect while under pressure. Bad burns can result.

When constructing molds which are to have set-in cavities, as shown in Fig. 5.17, the channels should be no closer than ¼ in. for holes less than 12 in. long. Longer holes require larger clearances between the channel and adjacent holes, slots and side walls.

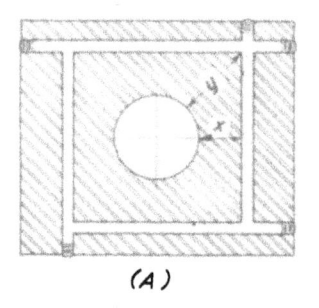 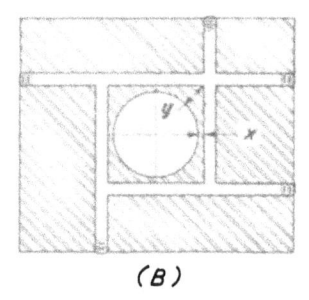

(A)   *(B)*

Fig. 5.24. (A) Proper steam channeling: x (greater than ½ y) represents shortest distance from channel to cavity; y represents greatest distance from channel to cavity. (B) Improper steam channeling: x (less than ½ y) represents shortest distance from channel to cavity; y represents greatest distance from channel to cavity. (Superheated water is sometimes used and the principles of steam heating are applicable to it.) (*Courtesy Allied Chemical Co., Toledo, OH*)

Special consideration must be given to the spacing of heating channels in molds used for urea or melamine materials. Figure 5.24 shows good proportions to use for all molds, including those for urea or melamine.

When the ejector pins enter the molded piece, channels should be drilled in the ejector bar so the temperature of the pins will be approximately the same as the rest of the mold. If the pin is large enough to permit some of the media to be channeled through it, by all means make that provision. This applies particularly to injection molds where even very small core pins must be channeled to permit fast heat transfer out of the molding material. Long or large mold plungers must be channeled or cored as shown by Fig. 5.25 to get uniform temperature with the female section of the mold.

We emphasize that molding of urea or melamine materials requires uniform heating which means direct channels in the male section of the mold, as well as, direct channels in the female section of the mold. As a general rule for all molds, it is wise to use direct channels whenever the male section is longer than twice its diameter. Reference to vendor's catalogs, will show several standard baffle types as available "off the shelf."

Most molders want the temperature control media connections on the back side of the mold, that is, away from the operator. Avoid hot pipes or wires on the operator side if there is any other way of making the necessary connections. All other channel openings should be properly plugged to prevent leakage of the media.

The assembly shown in Fig. 5.26 gives the general characteristics of a mold half using directed flow of media. The term directed flow is applied to mold designs in which there is only one path for the media to follow in passing from the inlet to the outlet. Tracing the media flow in Fig. 5.26 in a "zig-zag" fashion from inlet to outlet will show where the lines must

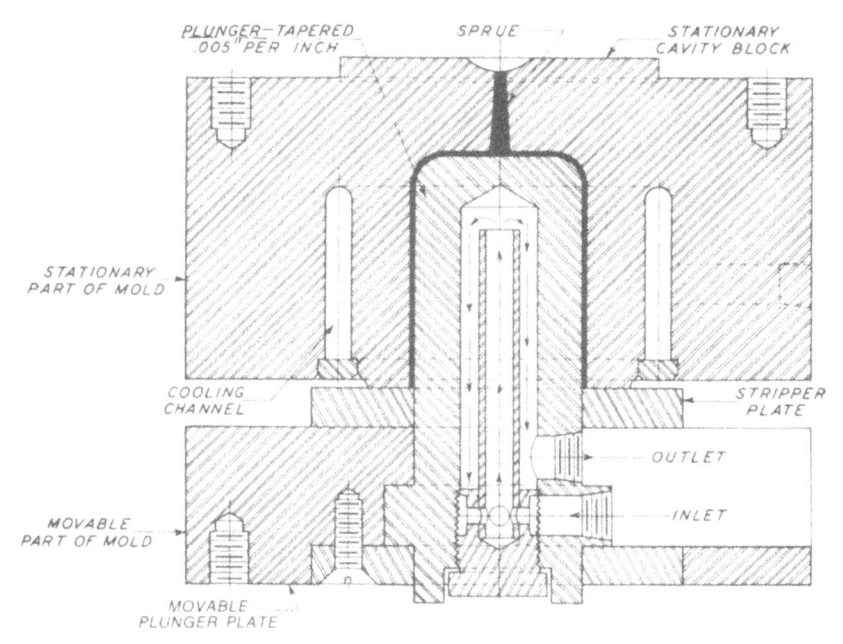

Fɪɢ. 5.25. Long mold plunger has been cored out for circulation of heating or cooling medium. (*Eastman Chemical Products, Inc., Kingsport, TN*)

be baffled to provide a one-way channel. We definitely recommend directed flow in all molds requiring cooling or using hot oil or hot water, for heating.

Compression, transfer or injection molds for thermosets, that use steam as a heating medium, make good use of undirected flow. In this case, all channels in the same plate are interconnected to allow free access of high-pressure steam. The only requirement is that the inlet be at the highest point in the mold, and the outlet at the lowest point in the mold. A *warning* is in order here. When steam channels enter the male section and the condensate must return to a higher level for discharge, be sure to use *directed flow*. A general rule, provide a channeling so that temperature variation from section to section will not exceed 20° F.

Guide Pins and Guide Bushings

Guide pins and bushings are used on all except the very simplest and cheapest of hand molds. At least two pins will be used and as many as four may be required. Where only two guide pins are used, one should always be ⅛ in. larger than the other so the mold cannot be assembled incorrectly. Three-and four-pin assemblies may use the same size pins if an unsymmetrical

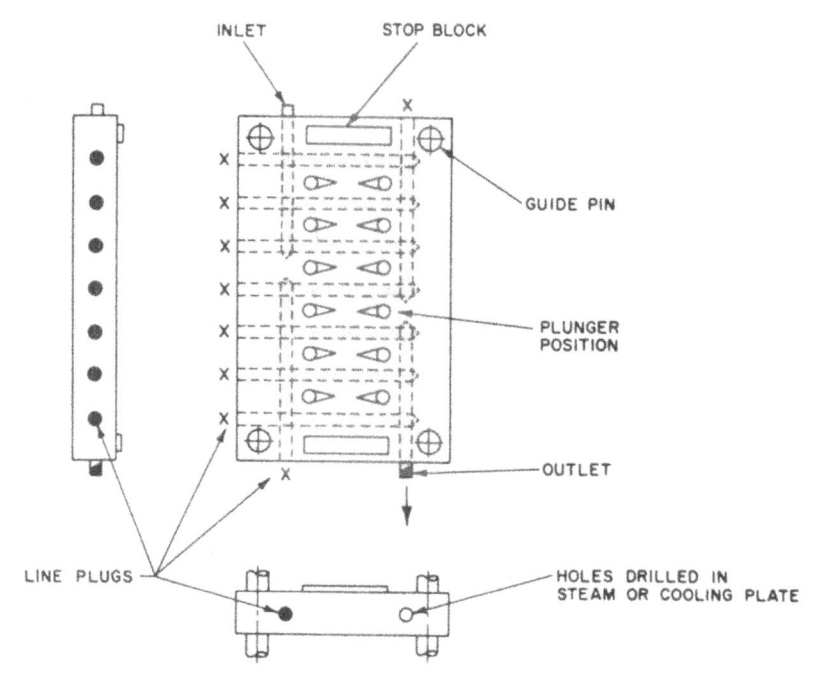

Fig. 5.26. Assembly of plungers, steam or water plate, guide pins and stop blocks. Dotted lines indicate channels for directed flow of media.

spacing is used. Guide pins are located as far apart as possible so that the effect of the clearance between the pin and bushing will be minimized.

The diameter of guide pins varies from ½ in. for small molds used in 5- to 10-ton presses up to 3-in. diameter for 1500-ton presses. A good practical size is ¾ to 1½-in. diameter for the guide pins of average-sized molds. When considerable side thrust is expected as a result of nonsymmetrical flow conditions, the larger size pins should be selected.

The length of guide pins should be such that the straight portion of the pin will enter the bushing to a depth equal to the pin diameter before the plunger enters the loading space. Guide bushings should be at least as long as the diameter of the pin and they must always be used when the plates are not hardened. When hardened plates are used, guide bushings are not absolutely necessary but they are preferred, especially if the mold will be a long running mold.

Guide pins and bushings must be press-fitted to the mold plates. Typical tolerances and construction details for these parts are shown in Fig. 5.27 at (*A*) and (*B*). The section dimensioned 1.248-in. diameter by 3/16 in. long, at (*B*), is helpful to the tool-maker because it permits him to start the bushing in the plate a short distance before the press-fitted portion comes

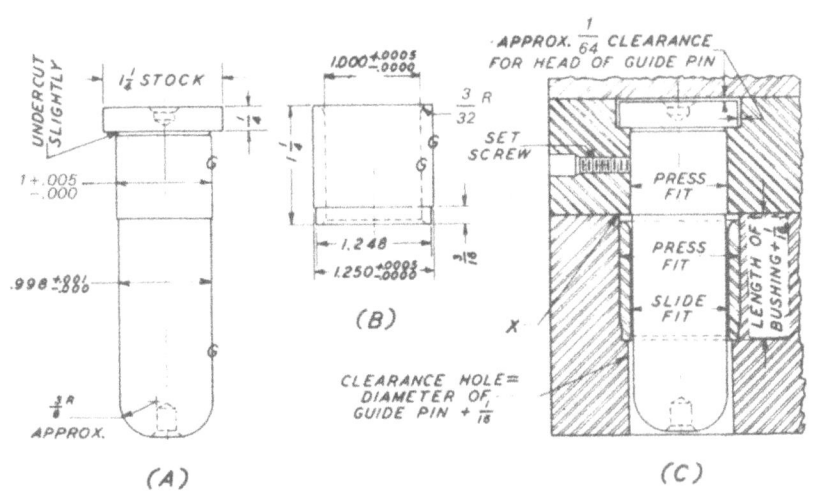

Fɪɢ. 5.27. Typical guide pins and guide bushings. (A) Guide pin; (B) guide bushing; (C) assembly.

into position. This also facilitates entry perpendicular to the plate. The edge of the hole in the retainer plate, shown at x in (C), is usually peened to make sure that the bushing does not pull out. In cases where the length of the bushing is the same as the plate thickness, allowance should be made to permit peening each end of the hole or use a shoulder bushing. Set screws are often used when the length is very short. A set screw, as shown at (C), may be used when the guide pin is not backed up by a plate. The set screw should rest against a flat on the side of the pin. There are several suppliers of standard guide pins and bushings. It is recommended that their standards be followed wherever possible.

Guide pins are seldom lubricated and frequently are used in a rusted and galled condition that causes them to stick. Thus considerable damage may result from pulling them out of the plates. These factors necessitate careful fastening of guide bushings and guide pins. (Also improved shop practice on lubrication).

HOBBED CAVITIES AND PLUNGERS

Several mold design considerations are governed by the methods used for the making of the cavities and plungers. If the job comes within any of certain classes, it is probable that hobbing will be economical for the job. These classes are:

1. When several identical sections are required.
2. When raised characters or designs are required in the mold.

3. When machining would be difficult because of irregular outlines, depth, or detail requirements.
4. When the product outline is symmetrical above and below the parting line and the same hob can be used for both parts to insure a good match at the parting line.

OPPOSED CAVITIES AND BALANCED MOLDS

When the shape of the cavity or core is such that the molding pressures will create side pressure (bending pressures) the cavity arrangement is usually "back-to-back," i.e., so that distorting pressures in adjacent cavities offset each other. This is indicated in Fig. 5.28 which also shows the hob used in sinking the cavities. In some cases this might be referred to as "balancing a mold." However, the balanced mold is most often a term used in describing a particular arrangement of cavities in an injection mold. The balanced cavity layout, when prescribed by the engineers preliminary check list, is understood to mean that each and every cavity is *exactly* the same distance (transportation distance) from the sprue. Meaning, that the path which the melted material must follow from *sprue* to *gate* is the same length. Further, runner sizes (diameters or cross section) are gradually in *exactly* the same fashion in each of the paths. The usual

Fɪɢ. 5.28. Opposed cavities equalize side-pressure. The master hob shown below is used for sinking cavities.

graduation being smaller cross section as the runner branches out. The larger cross section being the nearest to the sprue. The purpose of "balancing" is to achieve uniform and rapid fill of all cavities, at essentially the same instant

Another aspect of "balancing" is when it is not practical to place each cavity exactly the same distance from the sprue, or where parts of different size and configuration are to be molded in a family mold. In these cases, balancing is achieved by runner size, cavity placement (largest part nearest the sprue) and gate size variation. To the best of our knowledge, there are no exact mathematical formulas which will predict the desired results. In our experience, we used a weighted average calculation proportioned on part weight vs. total shot weight, runner length required, and molding characteristics of material being molded. The best way to start is to use the runner size and gate size that you would normally use on the largest part if molding it alone. Then, proportion the other runners and gates accordingly (and smaller).

Chapters 7 and 8 also present information on runner layout and gate sizes.

SURFACE FINISHES AND TEXTURED MOLDS

The Society of the Plastics Industry and the Society of Plastics Engineers have created master blocks with surface finishes clearly defined, specifiable by number, and capable of duplication by any knowledgeable moldmaker. Commercial houses have also created textured designs that can be specified by number (presumably to be applied by them). Specialized businesses provide mold finishing services and they are prepared to provide any desired finish from vapor-blast (a dull finish) to optically flat (flat and true within a few angstrom units). The surface finish required on your design will depend on (1) the end users needs or wants, (2) finish needed for release or ejection from the mold, or (3) special effects such as clarity, light reflection, light refraction, or secondary finishing (painting, vacuum depostion, plating, etc.)

Hand finishing a mold can become very time consuming. Be fair to the moldmaker by making the finish requirements known at the time you ask for his bid.

REFERENCES

CAD Users Manual, Langshorne PA: Robo Systems, 1984.

Crate, James H., More myths of plastics part design, *Plastics Design Forum*, Nov./Dec. 1984.

DuBois, J. H. and W. I. Pribble, *Plastics Mold Engineering*, 1st Ed., Chicago: American Technical Society, 1984.

Fine, Arthur and McGonical, Charles, Design of complex connector mold, *Plastics Machinery & Equipment*, Mar. 1984.

Heat Pipes, Torrance, CA: Hughes Thermal Products, 1975.

Leonard, LaVerne, Window profiles raise designers' sights, *Plastics Design Forum*, p. 62, July/Aug. 1984.

Mafilios, Emanuel P., Designing molds to cut thermoset scrap, *Plastics Engineering*, p. 35, Oct. 1984.

Mock, John A., Mold design, manufacture and control—An integrated concept, *Plastics Engineering*, Jan. 1984.

Nelson, J. D., Shrinkage patterns for molded phenolics, *Plastics Engineering*, July 1975.

Pixley, David and Richards, Peter, Thermoset or thermoplastic for electrical/electronic E/E, *Plastics Design Forum*, p. 28, Apr. 1981.

Pribble, Wayne I., Galley of goofs (phenolic part design), *Plastics Design Forum*, Nov./Dec. 1984.

Sors, Laszlo, *Plastics Mould Engineering*, Oxford: Pergamon Press, 1967.

Sors, Laszlo, General Electric launches potent design info system, *Plastics World*, p. 32, Aug. 1984.

Sors, Laszlo, Designing for producibility—A roundtable forum, *Plastics Design Forum*, p. 23, Jan./Feb. 1984.

Suggested for Further Reading

Krouse, John K., Automation revolutionizes mechanical design, *High Technology*, Mar. 1984.

Levy, Sidney, What CAD/CAM programs may not do for the designer (yet). *Plastics Design Forum*, p. 62, Nov./Dec. 1984.

Levy, Sidney, Complete CAD/CAM moldmaking software, *Modern Plastics*, July 1984.

Chapter 6 / Compression Molds

Revised by Wayne I. Pribble

A large percentage of the molds used for thermosetting plastics are the compression type. It is the older molding method, and later developments retain much of this prior art. In developing this text on mold design, it is desirable to start with the compression molds and follow with the other types. The full detailing of the design of a single-cavity hand mold that is converted to a semiautomatic 12-cavity mold will be used to introduce the reader to the fundamental calculations and basic design procedure. This detailed presentation of a fundamental mold type is followed by discussion of the unusual details and considerations that arise in the design of other compression mold types.

Hand molds are being eliminated for many applications because of the high molding labor cost. Hand molds continue to offer better answers for prototypes and complex insert assemblies in many applications. Hand molds are used for compression, transfer and injection molding. Shown in Fig. 6.1 is a standard mold set that may be used for the conversion of hand molds to semiautomatic. Standard mold bases and units should be considered for all new hand or single cavity molds.

COMPRESSION MOLDING

The reader should review the data on compression molds and compression molding in Chapters 1 and 2 at this time so that the forms of the various types of compression molds and the operation of compression molding presses will be fresh in mind. The basic molding problem calls for a mold that will compress the compound to the desired shape, and hold it under compression and heat while the chemical action which hardens it takes place. This must be done in the simplest and least costly manner, the mold being designed so

207

FIG. 6.1. Standard mold frame for single cavity compression molds. Many hand molds are being converted to semiautomatic by the use of these units. (*Courtesy Master Unit Die Products, Inc., Greenville, MI*)

that the compound and inserts may be introduced easily and the part ejected without distortion. Since the mold is idle while it is being loaded and unloaded, the efficiency of these operations, the quality of the piece, and the cost of the finishing operations will be a true measure of the quality of the mold.

The actual molding process involves the problem of forcing a bulky material into a given shape and space by the use of pressures ranging from 2,500 psi upward, accompanied by the application of heat (250–380° F) for the purpose of plasticizing the compound and causing it to flow and fill out to the mold contour. This action may produce highly localized stresses in various parts of the mold, and thereby cause serious mold breakage if the mold or parts are not properly designed.

The raw materials may be charged into the mold by at least three different methods. These are listed in the order of their preference and general use:

1. Preforms, or "pills."
2. Volumetric loading by loading board or measuring cup.
3. Weighed charge of powder or preforms.

The preforming does not change the material itself, but serves to provide a loading unit of predetermined weight. Preforms are easily handled and waste is negligible. The use of preforms also reduces the loading space required in the mold.

The production requirements of the user of the molded parts will determine the minimum number of cavities to be used. The fact that 100 cavities are needed to maintain the rate of delivery does not mean that all must be in one mold. Molds for bottle caps have contained as many as 150 cavities; molds for buttons may contain 500 cavities in a single mold. These large molds are exceptional, as the average mold will be found to contain from 5 to 15 cavities for medium-sized parts. The use of a low number of cavities has many advantages, since the smaller molds remain open a shorter period for loading, etc. Moreover, on the smaller molds, pressures are more uniformly distributed and "outage" is minimized, as repair work on a small mold means production loss from a lesser number of cavities. For example, if a 36-cavity mold were removed from the press for repair of a broken pin, all production would be stopped. If, instead, three 12-cavity molds were used and one had to be repaired, only one-third the total production would be lost.

DESIGN OF HAND MOLD

For example, assume that a mold for a lever is desired. See Fig. 6.2. Specifications call for this part to be molded from one of two materials: black wood flour filled phenolic or gray urea-formaldehyde compound. The user wants a sample mold built quickly for test, for sample to be followed by construction of a production mold capable of producing 7500 pieces a week. The quickest and least costly mold that can be built will be a single-cavity hand mold.

The first step in the design of such a mold will be the tabulation of information for the mold data card, and this involves the determination of the bulk factor and the shrinkage.

	Bulk Factor	Shrinkage
Phenolic wood flour	3	.006–.009
Urea cellulose	3	.008

Volume calculations may be made either by laying out the piece in sections and calculating the volume of each section or by computation from the weight of a model.

Weight calculations are obtained from the volume, using formula (1), which follows. (This formula is also used to calculate the volume when the weight is known.)

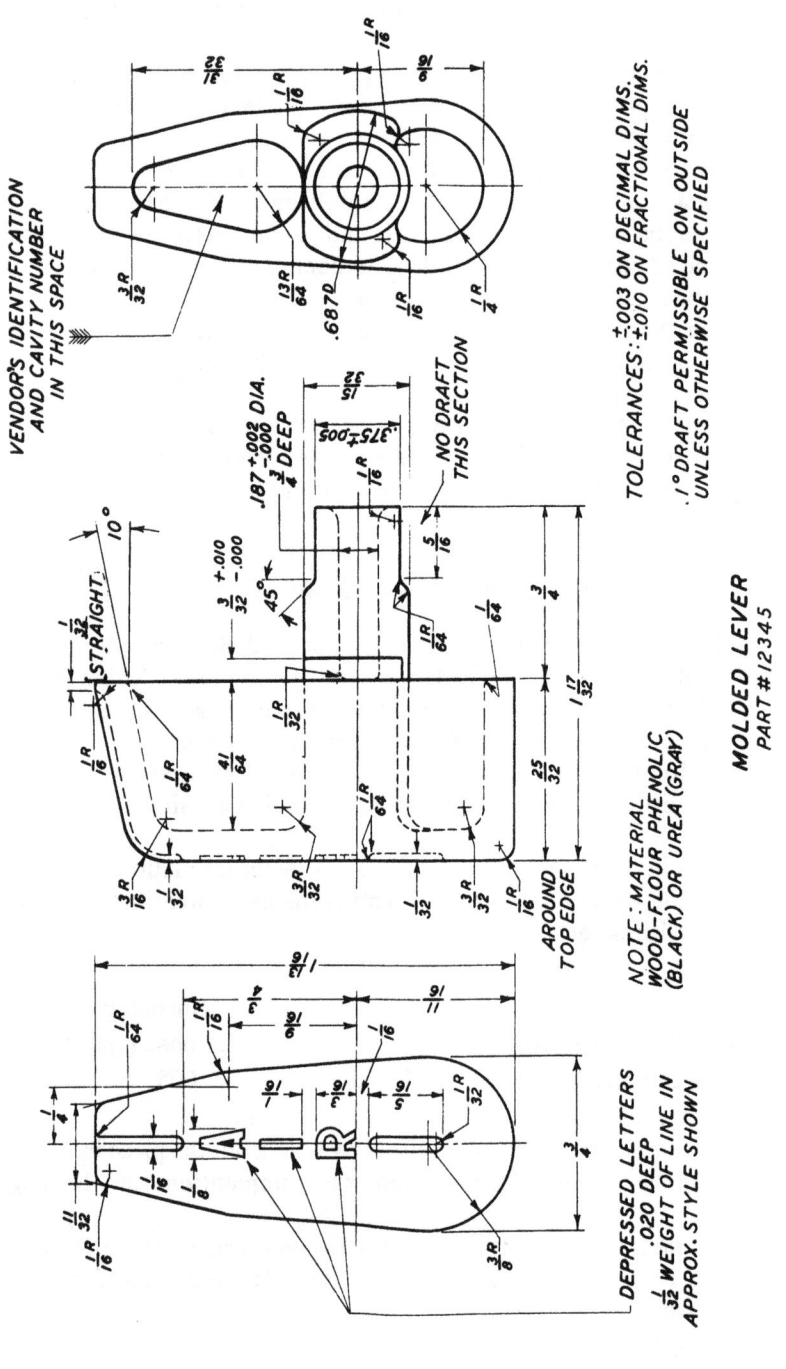

MOLDED LEVER
PART #12345

FIG. 6.2. Engineering requirements call for a single-cavity sample mold; production requirements call for a 12-cavity semiautomatic mold.

$$V \times W_U = W_T \tag{1}$$

$$V = \text{total volume of part}$$

$$W_U = \text{unit weight of material}$$

$$W_T = \text{total weight of part}$$

The volume of the lever has been found to be 0.70 cu in. We must add to this a flash factor of 10% to include the material required for the flash. Flash is the material that will be squeezed out around the plunger and through the overflow slots as the mold closes. This allowance must be made in compression molding to prevent precure at the parting line and to enable that material which lies on the land to escape outward and provide a good pinch-off. The overflow also takes care of any variation in the load by permitting the excess compound to escape.

With the addition of the 10% for flash, the total volume of compound would be:

$$.70 + 10\% = 0.77 \text{ cu in.}$$

By application of formula (1) we obtain for the gross weight of the lever:

$$V = .77 \text{ cu in.}$$

$$W_U = .76 \text{ oz/cu in. for phenolic or } .85 \text{ oz/cu in. for urea}$$

$$.77 \times .76 = .58 \text{ oz gross weight for phenolic}$$

$$.77 \times .85 = .66 \text{ oz gross weight for urea}$$

Arithmetical calculations will convert these weights to 3.6 lb per hundred pieces for phenolic and 4.1 lb per hundred pieces for urea compound.

Formula (2) is a general formula used for calculating the total volume of loose powder or preforms. Thus,

$$\frac{W \times BF}{W_1 \times 100} = V \tag{2}$$

$$W = \text{Gross weight of molded part per 100 pieces}$$

$$BF = \text{Bulk factor of compound}$$

$$W_1 = \text{Weight per cu in. of compound}$$

$$W_1 = .054 \text{ lb/cu in. for urea or } .048 \text{ lb/cu in. for phenolic}$$

$$V = \text{Total volume of compound required}$$

$$(W \text{ and } W_1 \text{ must } both \text{ be expressed in lb or oz})$$

By the application of formula (2) the following calculation is made:

$$V\,\text{urea} = \frac{4.1 \times 3}{.054 \times 100} = 2.27 \text{ cu in.}$$

$$V\,\text{phenolic} = \frac{3.6 \times 3}{.048 \times 100} = 2.27 \text{ cu in.}$$

Determining *depth of cavity well* or loading space may be calculated by formula (3).

$$\frac{V - V_1}{A} = D \tag{3}$$

$V =$ Total volume of material required (cu in.)

$V_1 =$ Volume of actual cavity space (cu in.)

$A =$ Horizontal area of loading space plus area of all lands (sq in.)

$D =$ Depth of loading space in in. from top of cavity to pinch-off land

In this formula, V_1 must take into consideration the volume of any mold pins, inserts, or projections that will subtract from the gross volume of the cavity space.

Factor A will be the horizontal area of the part plus the area of all lands. The lands (when used) generally will be at least $\frac{1}{8}$ in. wide. For this mold, a $\frac{1}{8}$-in. land will be sufficient, and by reference to Fig. 6.3C, it will be noted that the land is bounded by the 0.824-, 2.048-, 0.987- and 0.494-in. radius dimensions. In cases where the lands are irregularly shaped, the land area can be computed by tracing or drawing its outline on cross-section paper and counting the number of spaces included in the outline. A planimeter may be used for measuring this area.

By the application of formula (3):

$$D = \frac{2.27 - .83}{1.60} = 0.90 \text{ in.}$$

This calculation therefore indicates that the depth of the loading space for a powder load should be 0.90 in., which is approximately $\frac{7}{8}$ in. A powder load would be indicated for this part because of the long slender core pin that must be held straight. If this pin should bend, the lever would be out of alignment and consequently would be rejected. (If preforms had been indicated for this job, the same calculations would have been made, using the proper preform bulk factor.)

A $\frac{1}{8}$ in. land and a $\frac{7}{8}$-in. loading space have been selected and the actual design of the mold may now be begun. Whereas positive or semipositive molds are usually indicated for urea molding, this mold will be built as a landed plunger mold, as several conditions point to the desirability of using a mold of this type. Using a positive mold, a very weak section would be left where the 0.687 in. diameter comes almost tangent to the outside of the part (Fig. 6.2). Top ejectors are necessary, and two ejector pins will be required on the edge of the part. The extra width provided by the land is needed for the ejector pins. The irregular outline of this part would make it difficult to fit the cavity and the plunger of a positive mold closely enough to permit cleaning of the parts by tumbling.

The completed hand mold assembly, with material list and title block, is shown in Figs. 6.3A and 6.3B. The plunger, part 7 (Fig. 6.3A), is shown in detail in Fig. 6.3C. A thorough study of this assembly will give the reader a good understanding of the hand mold design. Important points to note are:

1. Plates overhang in opposite directions to facilitate disassembly. The overhang should be at least ½ in. and, generally, no more than 1 in.

2. The cavity will be hobbed, since the shape is best produced by the hobbing process and because production plans for a multiple-cavity mold will make additional use of the hob.

3. Note pin plate, part 12 (Fig. 6.3B), and pins, parts 10 and 11. The function of this assembly is to plug the ejector pin holes during the molding cycle. After the mold is removed from the press, this plate and pin assembly is removed from the mold. The plate and pin assembly shown in Fig. 6.4 is then substituted and, since these pins are ⅞ in. longer than parts 10 and 11 (Fig. 6.3B), the pin plate assembly may be used to push the part from the plunger, serving in the way an ejector bar does when used in a semiautomatic mold.

4. Socket-head screws are usually specified for mold work because they are easily assembled and disassembled. The hand mold uses four 5/16-18 screws to fasten the top and bottom plates in place. This is adequate, as the force required to open the mold is considerably less than that required to close it.

5. The minimum length of a guide pin should be such that the straight portion enters the bushing or bearing surface to at least ½ the diameter of the pin before any part of the plunger enters the cavity well. For hand molds it is well to increase this length since the initial alignment of cavity and plunger is entirely dependent on the guide pins. To find the length of the guide pins, part 6 (Fig. 6.3A), the calculation would be: Over-all length of plunger approximately 2½ in. plus 5/16 in. radius plus 5/16 in. entering equals 3⅛ in. minimum length of guide pin. Since the maximum length would be 3⅜ in.

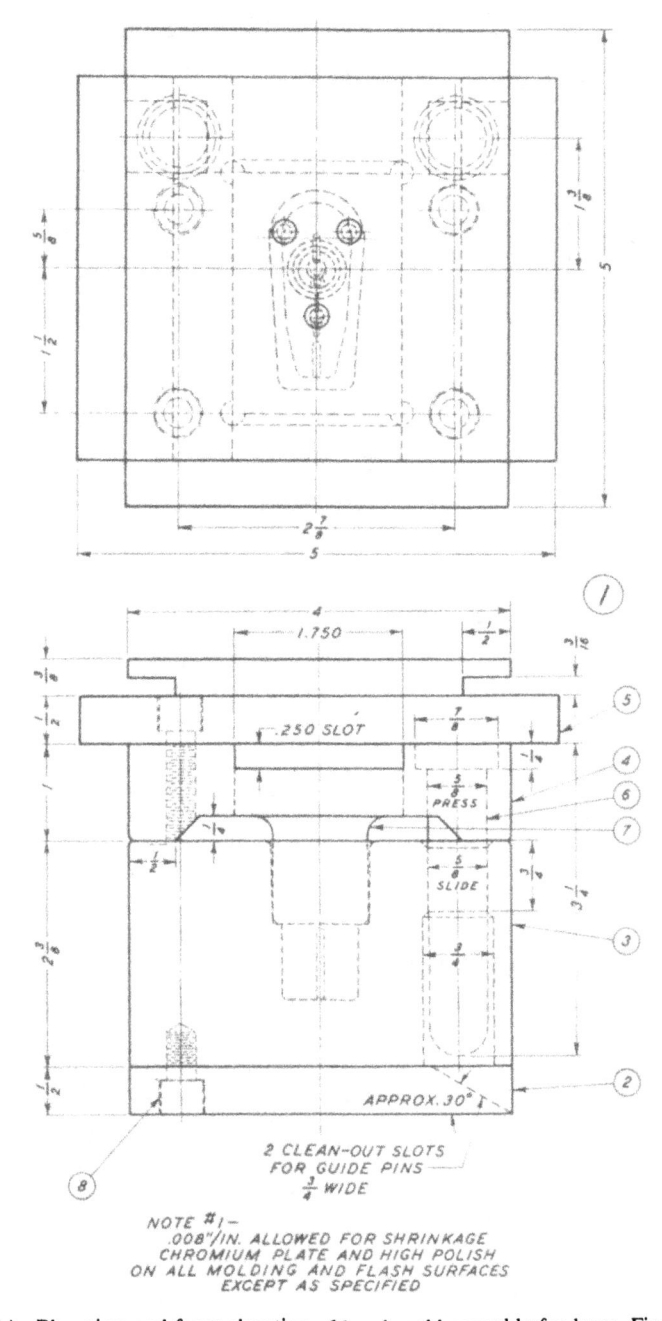

Fɪɢ. 6.3A. Plan view and front elevation of hand mold assembly for lever, Fig. 6.2.

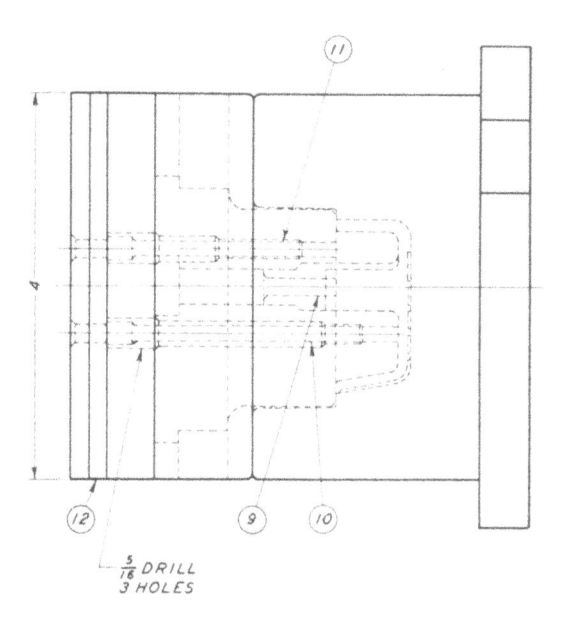

$\frac{5}{16}$ DRILL
3 HOLES

1	ST. C. R.			12	PIN PLATE		4	4	$\frac{3}{8}$
2	DR. ROD	R'WELL C-56	HARDEN	11	PIN	$\frac{1}{4}$	$5\frac{7}{8}$	(1 PC)	
1	DR. ROD	R'WELL C-56	HARDEN	10	PIN	$\frac{1}{4}$	$3\frac{5}{8}$		
1	DR. ROD	R'WELL C-56	HARDEN	9	INSERT	$\frac{5}{8}$	$2\frac{1}{8}$		
8	STEEL		$\frac{5}{16}-18$	8	SCREW (SOC. HD.)	$\frac{1}{2}$			
1	S.A.E. 3312 OR EQUAL	R'WELL C-58	CARBURIZE & HARDEN	7	PLUNGER		2	$2\frac{3}{4}$	$2\frac{3}{4}$
2	OIL HDN. TOOL STEEL		HARDEN	6	GUIDE PIN	$\frac{7}{8}$	$3\frac{3}{4}$		
1	ST. C. R.			5	TOP PLATE		5	4	$\frac{1}{2}$
1	ST. C. R.			4	PLUNGER PLATE		4	4	1
1	S.A.E. 3110 OR EQUAL	R'WELL C-58	CARBURIZE & HARDEN	3	CAVITY		TO SUIT		
1	ST. C. R.			2	BOTTOM PLATE		5	4	$\frac{1}{2}$
X				1	ASSEMBLY	D.	L.	W.	T.
						MAT'L REQ'D			

1-CAVITY HAND MOLD
(ASSEMBLY)

G. No.

GROUP DESCR.

FIRST MADE FOR LEVER   DWG. #12345

BEGUN BY   TRACED BY

FINISHED BY   INSPECTED

FIG. 6.3B. Side elevation with material list and title block for hand mold.

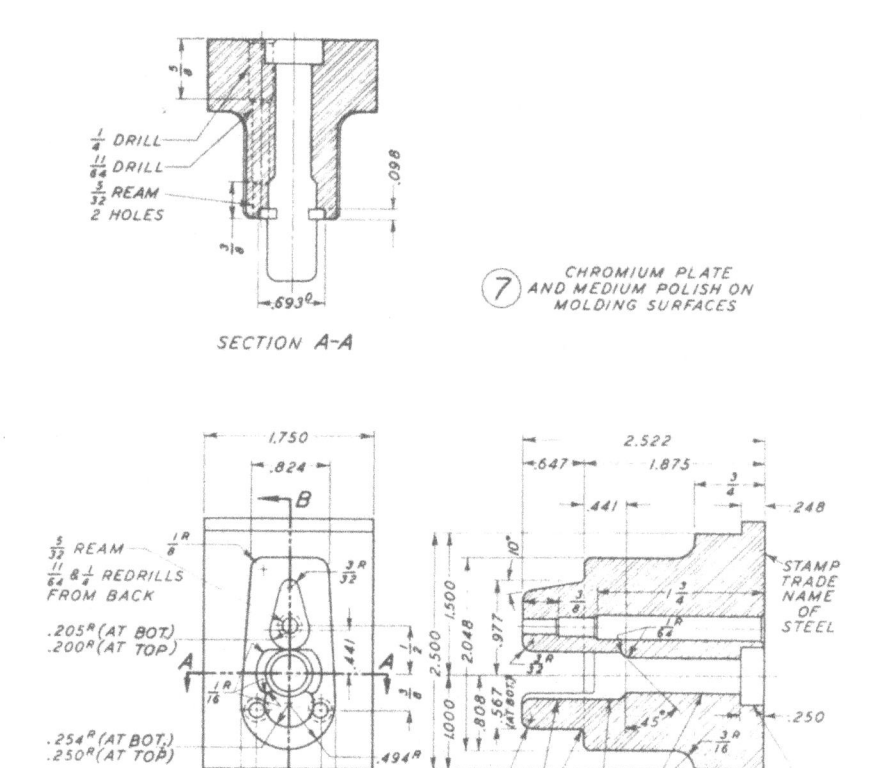

FIG. 6.3C. Detail of plunger for hand mold.

(the sum of the thickness of the cavity block and plunger plate—2⅜ in. + 1 in.), a pin between the maximum and minimum, or 3¼ in., would be used.

6. The two clean-out slots for the guide pins are quite important, as they permit flash or dirt which falls into the guide-pin holes to escape and also prevent the hole from filling to a point where the guide pin may be broken or damaged. These slots should be of ample size and be cut to an angle. Molds never should be designed without provision being made for dirt to escape from the guide-pin holes.

7. The ¼ in. clearance which is cut into part 4, Fig. 6.3A, is also very important. This clearance makes it possible for the flash to escape once it has passed the plunger. Less clearance might cause the flash to become almost as dense as the molded part, thus making its removal from these relatively

rough surfaces difficult. In cases where the cavities are set together closely in multiple-cavity molds, this clearance may be increased to ⅜ or ½ in. to provide ample room for escape of the flash.

8. The pin plate, part 12 (Fig. 6.3B), is designed with the 3/16 × ½-in. cutout on each side to permit die lifts or screwdrivers to be inserted for the purpose of prying the plate loose from the mold proper.

9. Reference to the plunger detail, part 7 (Fig. 6.3C), shows that the ejector pin holes are to have a two-step re-drill. The reason for this is that a ¼-in. re-drill to the full depth required would leave a wall of only 1/64 in. at the edge of the plunger. A reduction of this diameter to 11/64 in. would leave a wall of 5/64 in.

10. The freehand detail sketches, Figs. 6.5A–6.5E, illustrate a method of presenting the details of the frame and the parts not ordinarily detailed on the regular mold drawing. The parts generally detailed on the regular drawing are the plunger and cavity and the mold pins and ejector pins. Of course, where the assembly drawing is completely detailed, as in Figs. 6.3A, 6.3B and 6.3C, sketches will not be necessary. In Figs. 6.5A–6.5E detail sketches are presented for the purpose of showing how matching parts are specified and dimensioned.

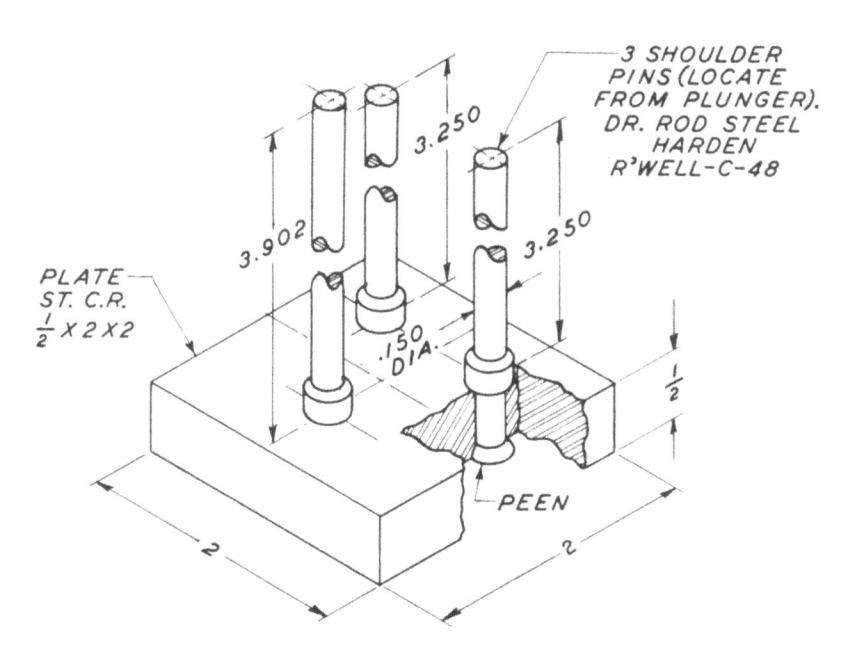

FIG. 6.4. Ejector plate and pin assembly used to extract parts from mold shown in Figs. 6.3A, 6.3B and 6.3C.

11. A study of the cavity, Fig. 6.5A, shows that control dimensions only are given, namely, the 0.785-in. depth and the shrinkage and draft allowances for the hob. Other dimensions are taken from the part drawing by the tool-maker. The $\frac{5}{8}$ (L) ream indicates that the guide-pin hole should be a sliding fit for a standard $\frac{5}{8}$-in. plug gauge. This hole diameter could be specified as 0.625 plus .0005 minus .0000 and have the same meaning. The note, "Follow Dimension Chart," refers to the chart shown in Fig. 6.6. For practical purposes, the dimensions which affect the size of the molded part directly are considered to be expressed in four-place decimals, and tool-makers will work to the closer tolerance. For those dimensions which do not affect the size of the molded part, the tool-maker may vary the size, but he must maintain the clearances so that heights, etc., will still "add up," providing proper pinch-off, overflow, and so on. For example, it is possible that, through error, the actual depth of the loading space in the cavity would be made 0.860 in. instead of 0.875 in. as specified. Since this is not a serious discrepancy, the tool-maker would proceed to reduce the length of the plunger and mold pins a corresponding amount in order that the mold might still offer a proper pinch-off. For this reason it is very important, when making replacement mold parts, to check the old piece accurately for the actual dimensions. It is not safe to assume that the piece will be exactly as specified on the drawing, as the tool-maker may deviate from the mold drawings for many reasons and still produce a mold that will make parts to specification.

12. A comparison of the dimensions of the cavity well, Fig. 6.5A, and the plunger, Fig. 6.3C, shows that 0.001 in. per side has been allowed for clearance between cavity and plunger. The allowance may be from 0.001 to 0.002 in. This small amount of clearance makes the landed plunger mold especially useful in molding parts for which uniform wall thickness is essential.

13. The $\frac{5}{8}$ (S) ream on the plunger plate, Fig. 6.5B, indicates a pressfit with the guide pin. Note that the reaming is done in the plunger plate with the cavity block, Fig. 6.5A, after assembly of the plunger, Fig. 6.3C, and the plunger plate, Fig. 6.5B. Holes for the guide pins, dowels, screws, ejector pins, etc., are nearly always finished first in parts that are to be hardened. Following the hardening, the holes are transferred to, or spotted in, the soft plates. Thus the dimensional changes which take place when sections are hardened do not destroy the alignment of the holes as they do when the holes are made while both parts are soft.

Blind dowel holes in hardened parts are reamed by driving a soft, slightly oversize plug into a hole in the hardened part, then reaming a small hole into the plug, as shown in Fig. 6.7.

14. The dimensions specified for the mold insert, Fig. 6.5C, have close tolerances, but these are essential, as the hole formed by the pin is the most critical of all dimensions on the part. The sizes shown are obtained by using

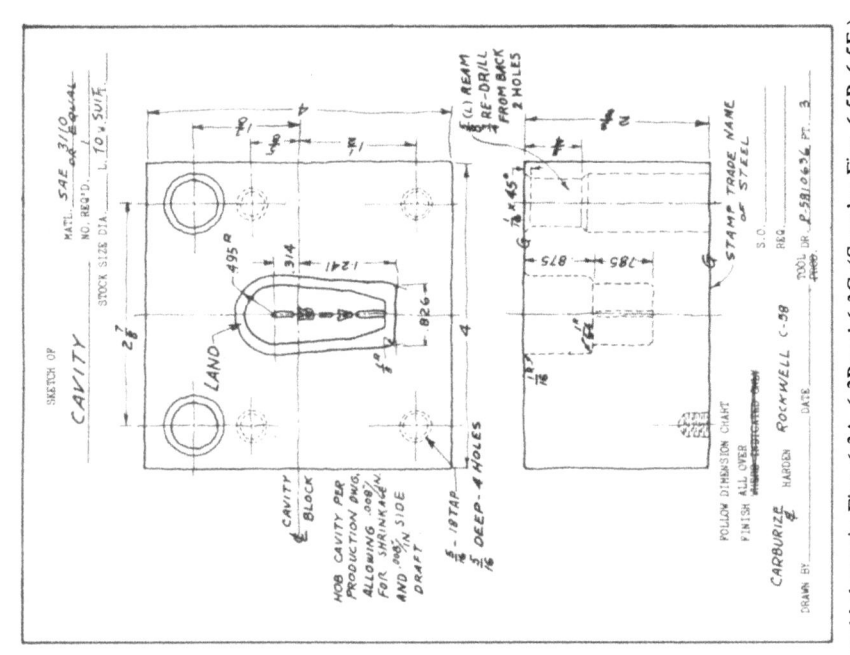

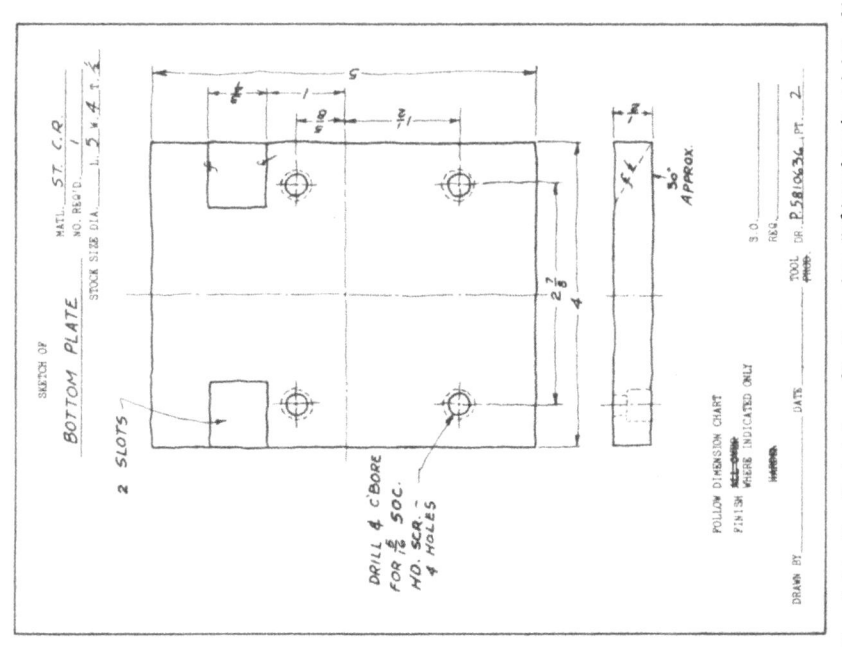

FIG. 6.5A. Detail sketches of bottom plate (left) and cavity (right) of hand mold shown in Figs. 6.3A., 6.3B and 6.3C. (See also Figs. 6.5B–6.5E.)

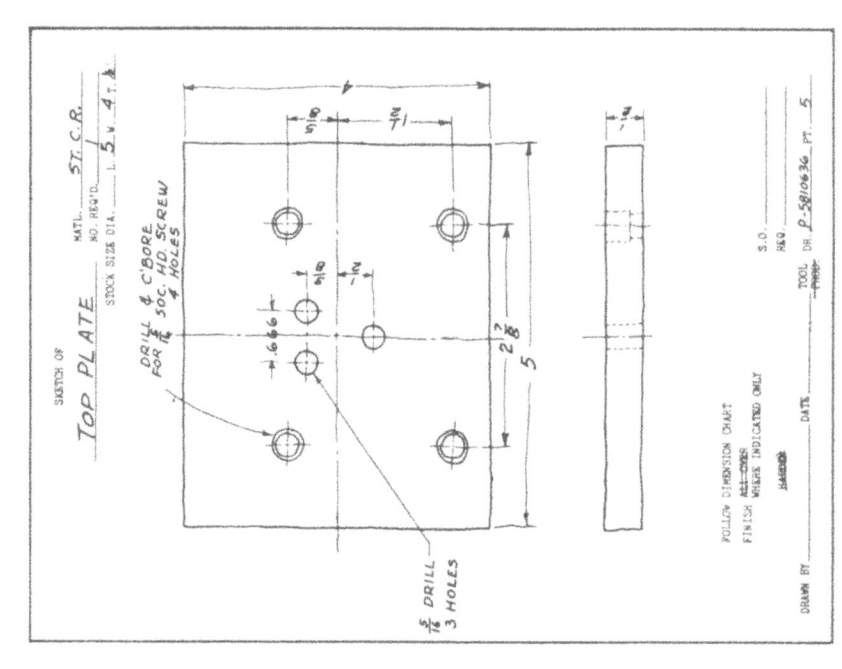

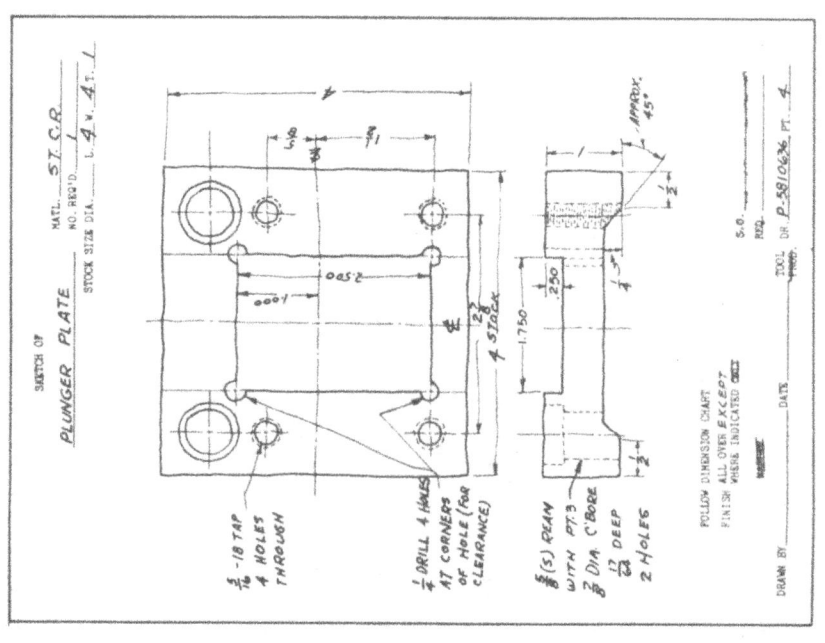

FIG. 6.5B. Detail sketches of plunger plate (left) and top plate (right).

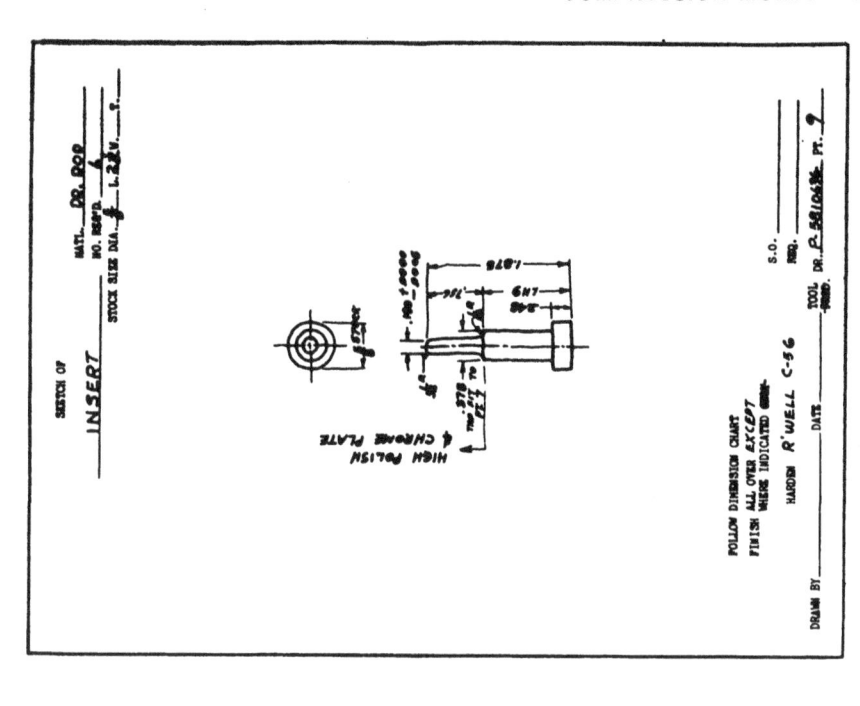

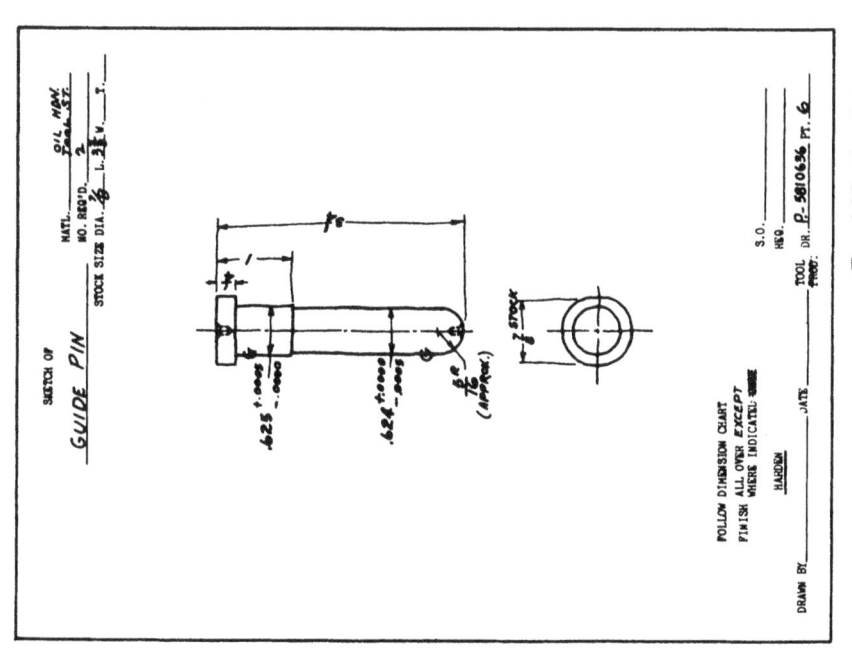

FIG. 6.5C. Detail sketches of guide pin (left) and insert (right).

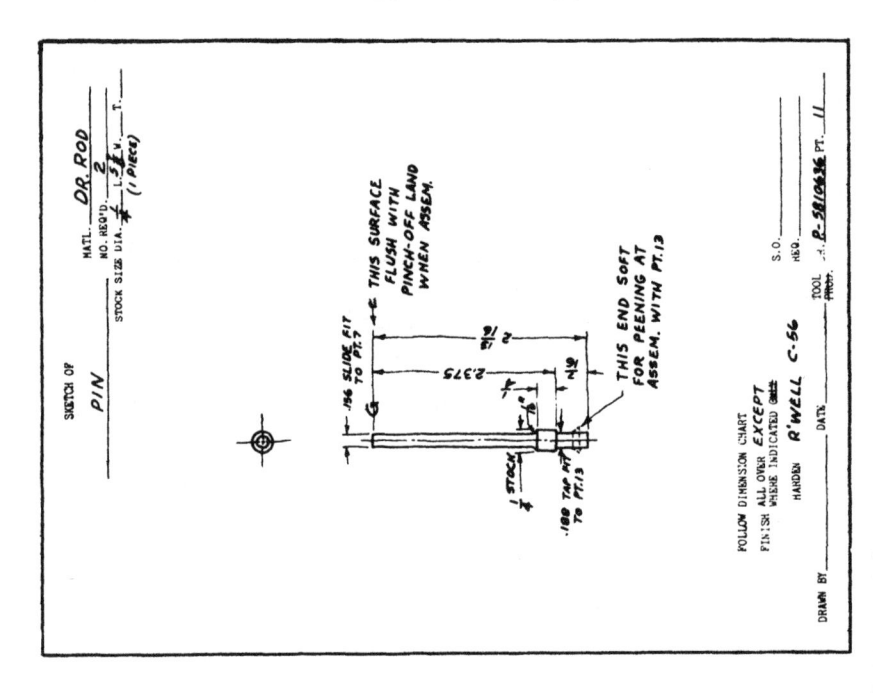

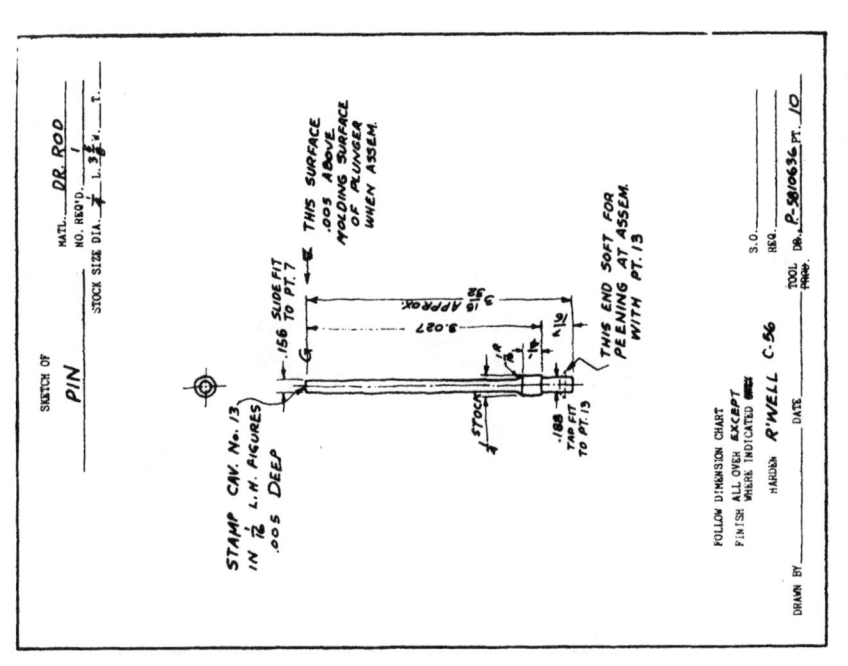

FIG. 6.5D. Detail sketches of pins.

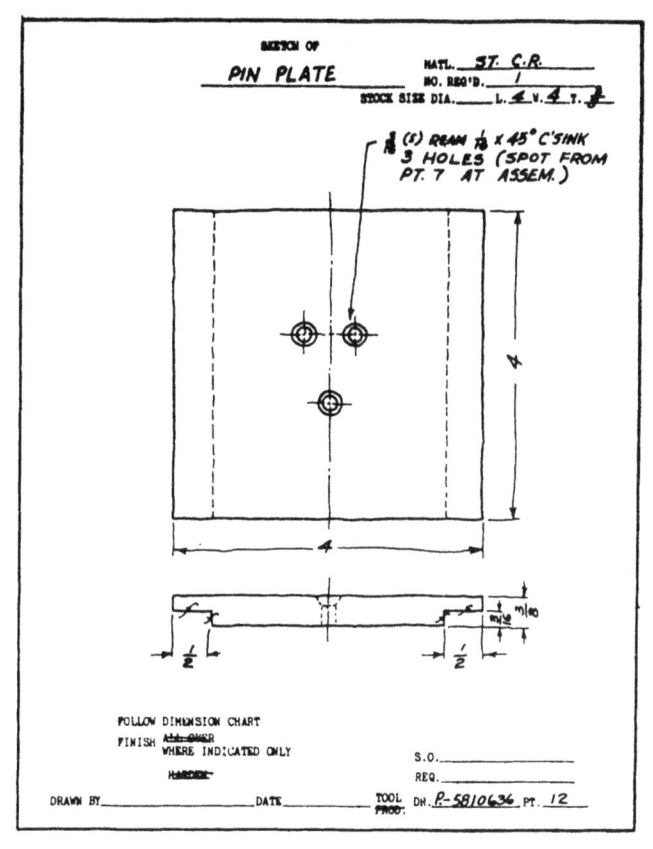

Fig. 6.5E. Detail sketch of pin plate.

the maximum dimension plus shrinkage and applying the tolerance to the minus side.

15. Cavity number 13 is stamped on the pin, Fig. 6.5D (left) as the final production mold calls for twelve cavities, and the extra cavity brings the total to thirteen.

DESIGN OF 12-CAVITY SEMIAUTOMATIC MOLD

Completion of the tests of the single-cavity mold will permit design of the production mold for the lever. Production requirements indicate the desirability of a 10- to 15-cavity mold. The available presses are 42-ton and 64-ton machines. The area of the land for this part is 1.6 sq in. Recommended molding pressures for the materials that will be used are 2.5 tons

<u>DIMENSION CHART</u>

It is to be understood that when machined dimensions are given in common fractions on <u>Tool Drawings</u> or <u>Sketches</u>, work should be done within the following limits.

$\frac{1}{4}$ or less, $\frac{1}{128}$ plus or minus

Over $\frac{1}{4}$, $\frac{1}{64}$ plus or minus

When variations less than these are necessary, or greater than these are permissible, they will be indicated accordingly.

Unless otherwise specified, show <u>machined dimensions</u> given in decimals as follows:

3-place decimals, .005 plus or minus
Example: .563, .375

4-place decimals, .001 plus or minus
Example: .5630, .7500

<u>Dowel and guide-pin holes</u> will be considered reamed to plus .0000, minus .0005, and should be indicated by the letter S (small). A slide fit will be indicated by the letter L (large), meaning plus .0005, minus .0000.

Standard shrinkage allowance for plastics molds should be specified on the drawing.

Press-fit for guide pins and bushings should be .001 to .002.

FIG. 6.6. Typical toolroom dimension chart showing tool-maker's tolerances.

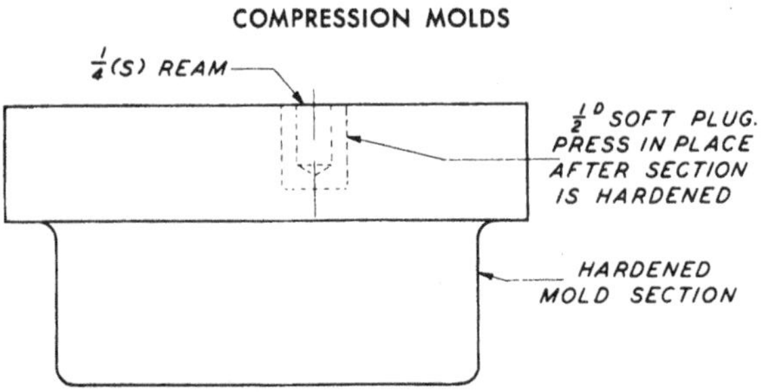

COMPRESSION MOLDS

FIG. 6.7. Typical manner of reaming blind dowel holes in hardened mold sections.

per sq in. for phenolic compound and 3 tons per sq in. for urea. By multiplying these factors the minimum pressure is obtained.

$$1.6 \text{ (land area)} \times 3 \text{ (tons/sq in.)} = 4.8 \text{ tons/cavity}$$

If 12 cavities are selected, the total pressure required would be $12 \times 4.8 = 57.6$ tons, which is well within the capacity of the 64-ton press. The long shank on the lever will necessitate a little extra pressure to make sure that it will "fill out." Experience in operating a single-cavity mold indicates the possibility of difficulty at this point and suggests that a movable pin would offer a distinct advantage. The movable pin would provide a free area around the pin through which trapped air and gas could escape. All of these considerations indicate that a 12-cavity landed plunger, semiautomatic, spring-box, top ejector mold is the proper selection. See Figs. 6.8A–6.8D.

The approximate size and arrangement of the cavities in the mold should be decided upon as the first step in the design. To determine the approximate sizes of the individual cavities, refer to the Table of Recommended Minimum Wall Thickness (Table 5.2). Here it is found that the recommended wall thickness should be 7/16 in. when the rectangular cavity well is approximately 1 in. wide. For this part, consider the width rather than the length of the cavity well, since the cavity is narrow in relation to length, and there is a full radius at one end. This will strengthen the cavity considerably. Where a number of cavities are nested together, the wall thickness of the cavity well can be reduced somewhat. This is possible inasmuch as the wall of one cavity will support that of the adjacent cavity. These considerations permit selection of ¼-in. wall thickness, and the cavity size will be 1½ × 2⅝ in. approximately.

There are six possible arrangements of a 12-cavity mold, as shown in Fig. 6.9. Usually it is wise for the designer to make tentative layouts or, at least, some scale drawings to determine the best arrangement. Factors determining the arrangement of cavities are: (1) Operator's visibility and convenience. Can the operator see into the cavity to be sure that all is as it should be? In case long, narrow cavities are used, it is logical to have the long way of the cavity front to back in the mold. This reasoning would rule out sketches (C), (D) and (E) of Fig. 6.9. (2) A second consideration arises from the necessity for holding the width of the ejector bar to a minimum. This is necessary to minimize the mold plate distortion which might result from pressure concentrated in the center area. It is accomplished by holding the parallels under the cavities or plungers as closely as possible. The arrangement shown at (A) in Fig. 6.9 will permit use of a 5-in. ejector bar, whereas the arrangement pictured at (B) would require a 7½ in. bar. The layout (A), therefore, indicates the most desirable arrangement and the one chosen for this mold. This arrangement will permit a steam line to run lengthwise in the center of the mold, thus providing the uniform heating required to mold the urea compound. An alternative arrangement would be that pictured at (C), as this permits the use of a 4½-in. ejector bar and steam* lines can be drilled front to

---

*Wherever "steam" is mentioned it includes any suitable media such as hot oil, superheated water or electric cartridges.

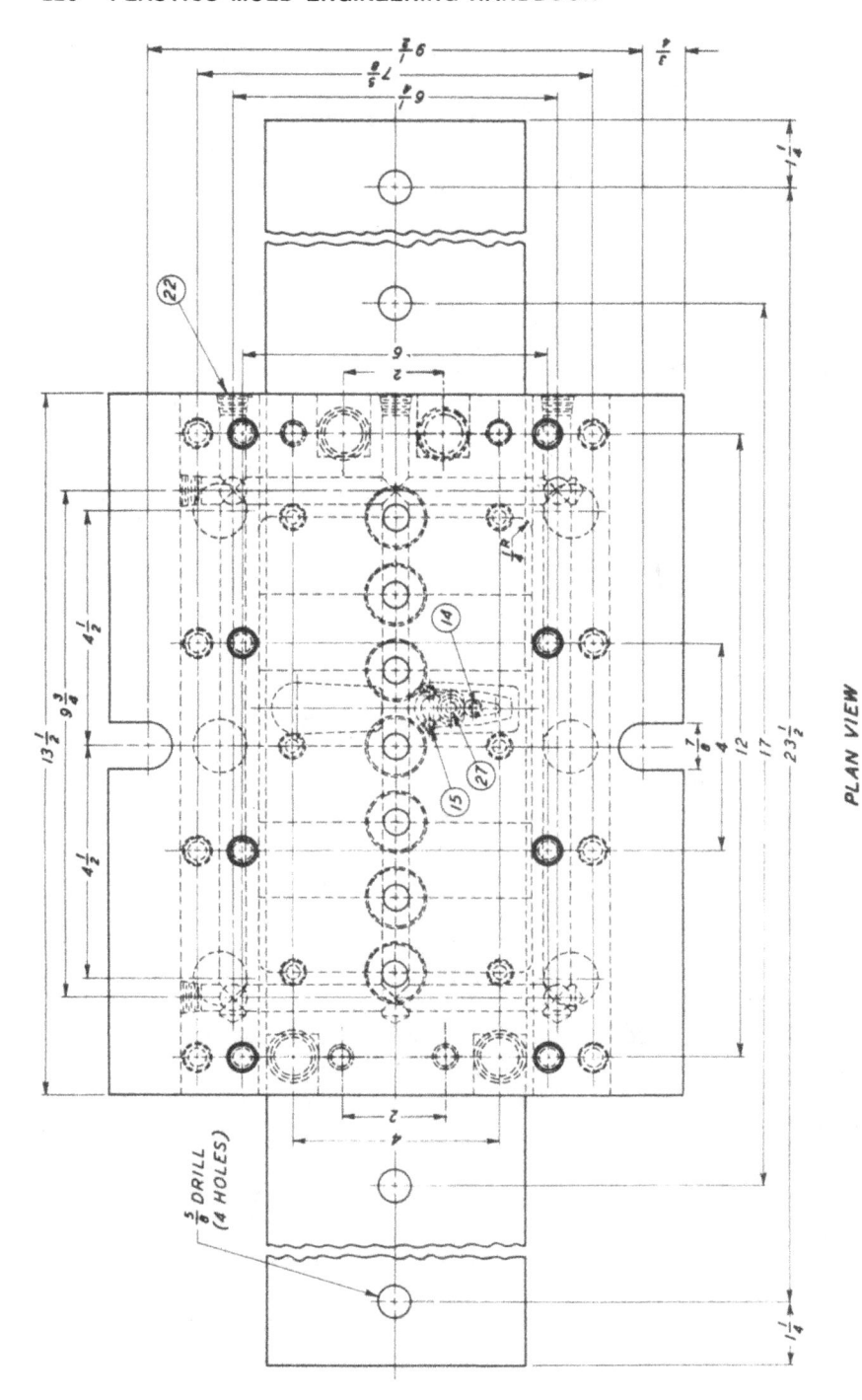

PLAN VIEW

FIG. 6.8A. Plan view of 12-cavity landed plunger semiautomatic mold for lever shown in Fig. 6.2.

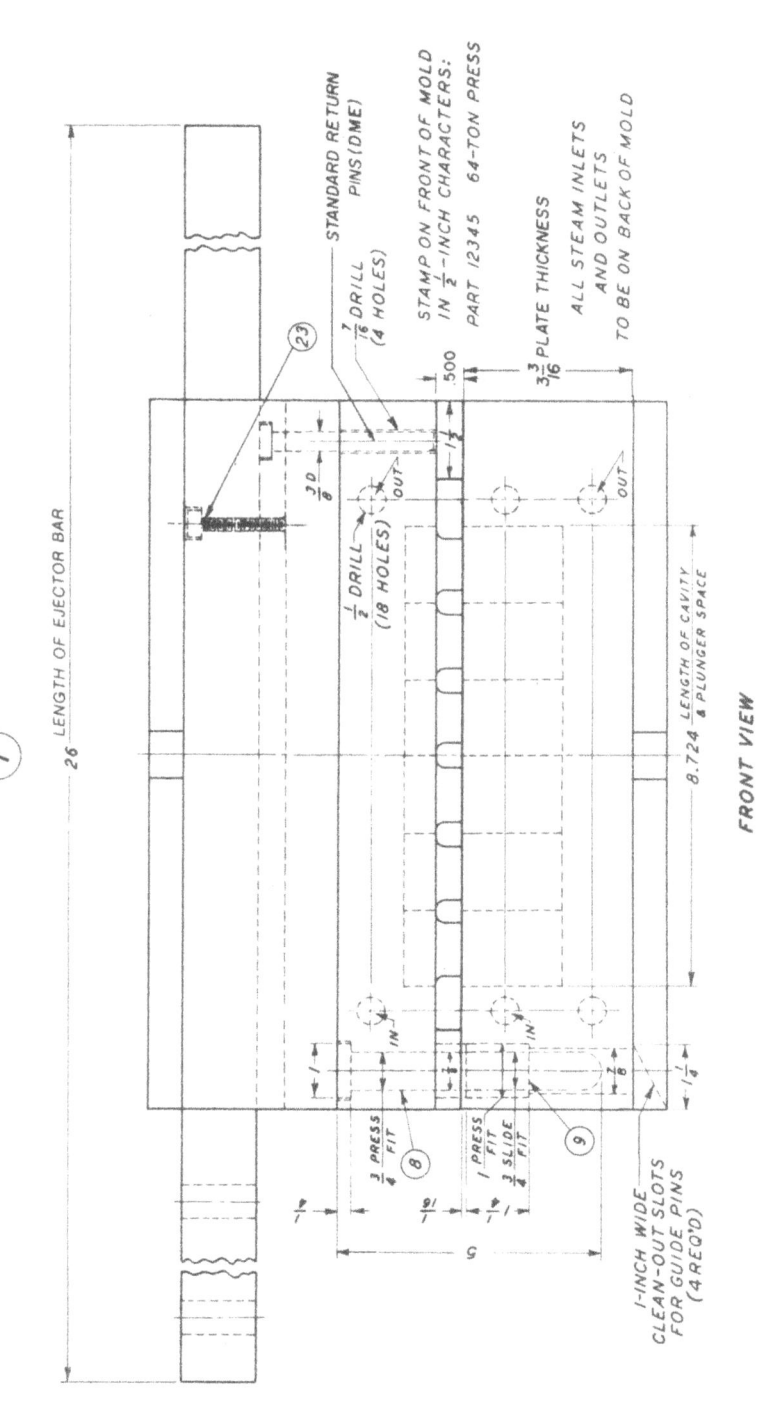

Fɪɢ. 6.8B. Front view of 12-cavity mold. (See also Figs. 6.8A and 6.8C.)

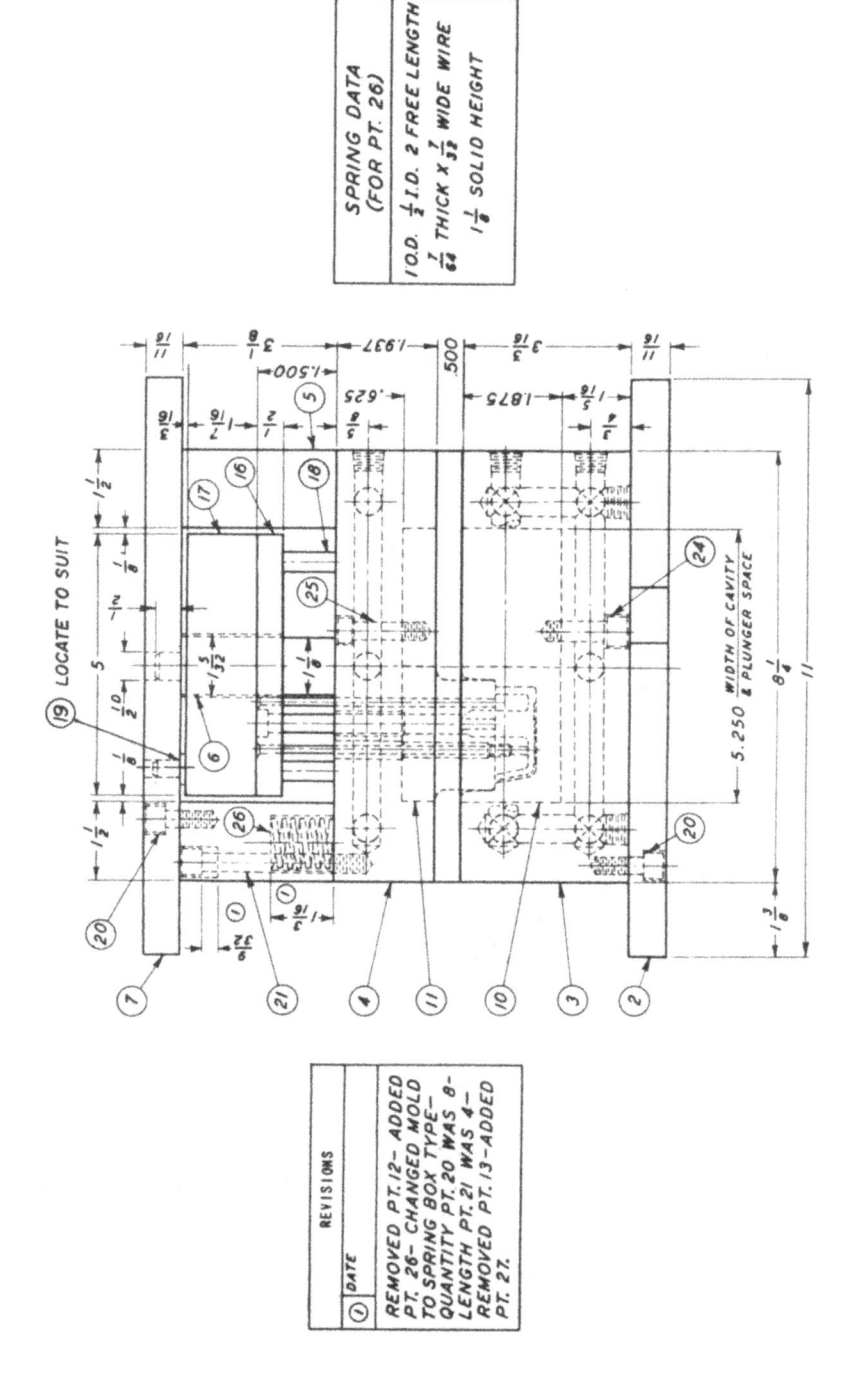

END VIEW

FIG. 6.8C. End view of 12-cavity mold (see Figs. 6.8A and 6.8B). Revision block at left shows changes that have been made from original design to provide for "spring-boxing" mold.

	No.	Material			Part No.	Description		D	L	W	T
	2	FLAT GROUND TOOL ST.		HND	28	STOP PAD		8 3/8	1 1/2	5/8	
①	12		PT. 5	728	27	EJECTOR PIN					
①	6				26	SPRING (SEE NOTE)					
	24	STEEL		5/16 - 18	25	SCREW SOC. HD.		1 1/2			
	24	STEEL		3/8 - 16	24	SCREW SOC. HD.		1 1/4			
	6	STEEL		1/4 - 20	23	SCREW SOC. HD.		1 1/2			
	14	STEEL		3/8 STD.	22	PIPE PLUG SOC. HD.					
①	8	STEEL		3/8 - 16	21	SCREW SOC. HD.		3/4			
①	16	STEEL		3/8 - 16	20	SCREW SOC. HD.		1			
	8	STEEL PURCHASE STD.			19	STOP PIN					
	4	PURCHASE STD.			18	RETURN PIN					
	1	MACH. ST.			17	EJECTOR BAR		26	5	1 1/2	
	1	ST. C. R.			16	EJECTOR PIN PLATE		13 5/8	5	1/2	
	24		PT. 4	728	15	EJECTOR PIN					
	12		PT. 3	728	14	EJECTOR PIN					
①	~~12~~		~~PT. 1~~	728	13	INSERT BODY					
①	~~12~~		~~PT. 6~~	728	12	INSERT					
	12		PT. 2	728	11	PLUNGER					
	12		PT. 1	728	10	CAVITY					
	4	OIL HDN. TOOL ST.		HARDEN	9	BUSHING		1 1/8	4 1/2		(1 PC.)
	4	OIL HDN. TOOL ST.		HARDEN	8	GUIDE PIN		1	5 5/8		
	1	MACH. ST.	S.A.E. 1020		7	TOP PLATE		13 1/2	11	3/4	
	7	ST. C. R.			6	SUPPORT PIN		1 1/8	3 3/8		
	2	MACH. ST.			5	PARALLEL		13 1/2	3 1/4	1 1/2	
	1	MACH. ST.	S.A.E. 1020		4	PLUNGER PLATE		13 5/8	8 3/8	2	
	1	MACH. ST.	S.A.E. 1020		3	CAVITY BLOCK		13 5/8	8 3/8	3 1/4	
	1	MACH. ST.	S.A.E. 1020		2	BOTTOM PLATE		13 1/2	11	3/4	
	X				1	ASSEMBLY		D.	L.	W.	T.
								MAT'L REQ'D			

G. NO. / GROUP DESCR.

12-CAVITY SEMIAUTOMATIC MOLD
(ASSEMBLY)

FIRST MADE FOR LEVER   DWG. 12345

BEGUN BY	TRACED BY	
FINISHED BY	INSPECTED	727

FIG. 6.8D. Material list and title block for 12-cavity mold. Part numbers refer to Figs. 6.8A, 6.8B and 6.8C.

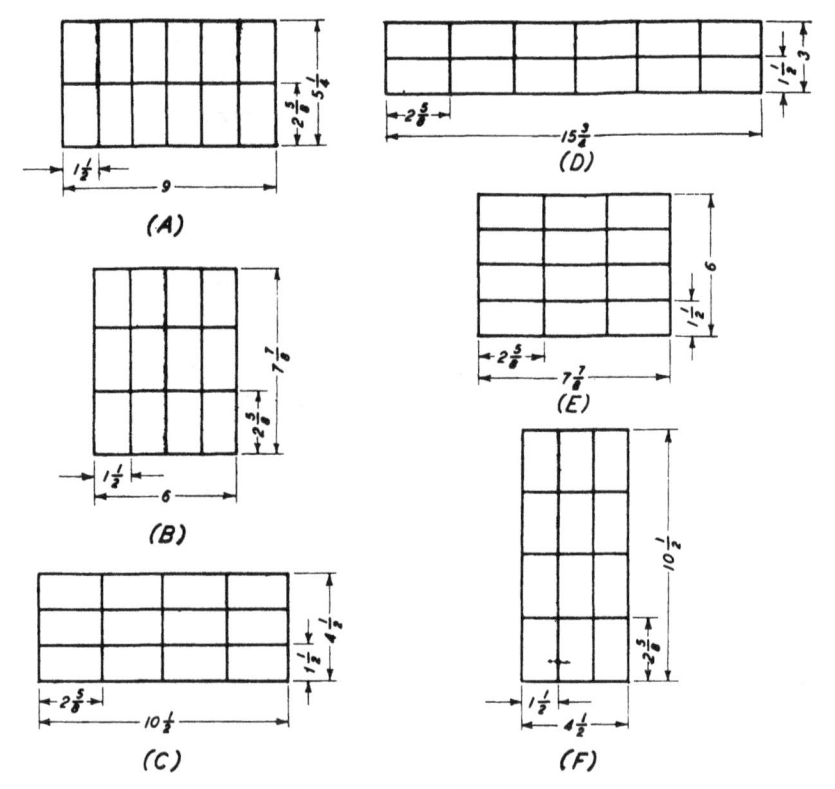

FIG. 6.9. Six possible arrangements of twelve rectangular cavities.

back under the cavity junction points. This arrangement sacrifices visibility however.

In starting to lay out this mold (Figs. 6.8A–6.8D), the horizontal and vertical center lines of the plan view are drawn first. The sizes 1-29/64 × 2⅝ work out conveniently for the outside dimensions of cavity and plunger and are "blocked in," giving a total of 5.250 in. width and 8.724 in. length of cavity and plunger space. The slight reduction from 1½ to 1-29/64 was made to permit the use of an 8¼ × 13½ frame size. The outline of the molded part showing all coring out together with land, ejector pins and mold pins is then drawn in place. This shows the tool-maker the exact position of the cavities and plungers when the mold is fully assembled. Be sure to show the position of the molded part in both front and back rows, as it is often desirable to reverse the part position in the rows to get better layout or balance side pressures. The outline of the land is sufficient for the back row. Next, allow a minimum of ¼ in. between the edge of the cavity recess and the edge of the nearest steam lines. (Most tool-makers would prefer an allowance of ⅜ in.) In this mold the length of the cavity space is 8.724, so the center distance across end steam lines can be 9¾ in., which allows a clearance slightly over ¼ in. The lengthwise steam lines can be on 6¼-in. centers. The center steam line may now be drawn in place. Note that the steam lines are drilled from the back and right-hand end of the mold. Steam connections are usually at the back of a mold, where they are out of the way. The lengthwise steam lines may be drilled from either end or from both ends, although most designers choose the right end as standard practice. A tapered pipe tap is shown where each steam channel terminates, and the ½-in. drilled hole requires a ⅜-in. standard taper pipe tap.

An allowance of ⅜ in. for clearance beyond the steam line is made to locate the edge of the guide-pin bushing. (One-fourth inch is the minimum allowance.) A guide-pin bushing of 1 in. diameter will be used, and it is found that these add up to provide a 12-in. center distance between guide pins. Again, ¼ in. is allowed between the outer edge of the guide-pin bushing and the edge of the cavity plate, giving an over-all length of 13½ in. for the mold frame.

The width layout chosen had a 6¼-in. center distance for the steam lines and, by making an allowance of ¼ in. for clearance, 7⅝ in. is obtained as the center distance for the fastening screws. The screws selected are ⅜—16. A minimum of ⅛ in. should be allowed between the outside diameter of the screw and the edge of the plate into which the screw is threaded. Preferred practice indicates that one-half the diameter of the screw should be allowed. This ⅛-in. minimum allowance gives the total cavity and plunger plate widths as 8¼ in.

The 64 ton press selected for this mold has one set of clamp bolt holes at 9½ in. spacing in the platens, and this indicates the proper location for the centers of the clamp slots in the top and bottom plates. The plate should extend a minimum of ¾ in. past the center of these slots, which gives a width of 11 in. for the top and bottom plates.

The ejector bar may be drawn in next and its width specified as 5 in. The width was determined by the essential spacing of the ejector pins, the designer having first made sure that none of the pins would come closer than ¼ in. to the edge of the bar. A 5 in. bar with ⅛ in. clearance on either side and parallels of 1½ in. width are just right for the 8¼ in. width of the plunger plate. The length of the ejector bar depends on the press to be used which, in this case, requires that a 26 in. length of ejector bar be used. The press allows 17-in. or 23½-in. center distance for the operating rods, therefore ⅝-in. drilled holes are located on these centers. This will permit a ½ in. bolt to pass freely through the bar so that the push-back rods may be fastened to the ejector bar.

The next step is the location of the guide pins and safety pins with respect to the horizontal center line. In the location of the guide pins it is necessary to make sure that all steam lines are cleared by at least ¼ in. For a mold of this size at least four guide pins should be used, and they should be located as far apart as possible to minimize the effect of the clearance between guide pin and bushing. Where only three guide pins are used, it is customary practice to locate two at the left side and one at the right, behind the center line. Inasmuch as most press operators are right-handed, this keeps the guide pin out of the way as far as possible.

The return pins (Fig. 6.8B) push the ejector bar into its proper location as the mold closes, thus insuring that no strain shall be applied to the ejector pins if the external mechanism for moving the ejector bar should fail. The safety pins also serve to hold the ejector bar in place while the mold is being moved to and from storage. These pins are held in place by counter-bored holes in the ejector pin plate. Figure 6.10 shows a typical design for shouldered pin retention.

Support pins, part 6 (Fig. 6.8C), are placed at the corners of the cavities to provide ample support. These support pins are usually about one fourth to one fifth the width of the ejector bar. The support pin is made with a turned section (usually ⅜ to ½ in. diameter) which is press-fitted to the top or bottom plate.

Location of the screws is the next item of importance. The ⅜ in. screws were selected when consideration was given to the width of the mold and

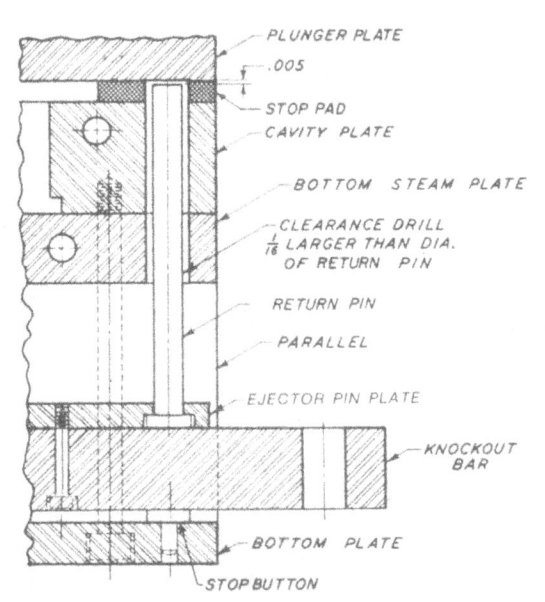

FIG. 6.10. Typical design for return pin having a shouldered head.

to the steam-line clearances. The screws at either end are usually on the same center line as the guide pins and safety pins. the right to left spacing is often 3 to 5 in. For this mold, four screws on 4 in. centers on each side of the mold can be used. The screws on 7⅝ in. centers front to back are used to fasten the bottom plate and cavity block together and also to fasten the parallels and plunger plate together. Another set of screws is used to fasten the top plate to the parallels. This set of screws may be located at any point so long as the heads of the screws in the parallels are cleared.

Three springs on each side complete the plan view. These springs are located in the center of the parallels and may be located on any satisfactory center lengthwise. The springs selected are 1 in. outside diameter and 2 in. long. Reference to the end view will show that they are to have an initial compression of 13/16 in.

The end view can now be started in the same manner as the plan view, that is, by laying out the molded part in its proper position. Since this is to be a top-ejector mold, the bulk of the mold will be in the top portion. The parting line is drawn first. The height of the cavity from the pinch-off land to the bottom is calculated as 1.250. This allows approximately the same amount of steel under the molding surface as is used in the cavity wall. Uniform wall sections will help to avoid the distortion which can occur in hardening if uneven wall sections are used.

The sampling of the single-cavity mold for this part has shown that the ⅞ in. depth of loading space will not be needed, but that ⅝ in. depth will be sufficient, and it is desirable to hold this space to the minimum. Using the ⅝-in. loading space, the height of the cavity becomes 1⅞ in. Since the plungers are set closely, a flash clearance of ½ in. will be allowed and a total depth of 2.375 from the top of the stop pads to the bottom of the cavity recess will be used. By using 3¼ in. SAE 1020 machine steel or boiler plate, the thickness of steel will be 1-5/16 in. below the cavities, which amount will allow ample room for the ½-in. steam line to be drilled at ¾ in. from the bottom of the plate. The 3-3/16-in. finished dimension allows for grinding to thickness without additional machining. It should be noted here that the width of all plates, bars, cavities, etc., has been determined in the plan view, and these lines are to be drawn in lightly in the end view until the thickness of the various members has been determined, at which time these widths may be drawn in solidly.

Eleven-sixteenths in. clamping plates allow use of ⅜-in. socket-head screws for assembly purposes; 13/16 in. plates would be used for ½-in. socket-head screws.

The depth of the plunger recess is 0.625. This dimension may vary between ⅜ in. and ¾ in., with ½ in. being the best maximum. Some designers do not use set-in plungers, preferring to have the plunger mounted on top of the steam plate. There are arguments for both methods, but the matter is largely one of personal preference and shop practice.

Plungers set in a recess—

1. Require close fit, but can be ground as sets with the cavity.
2. Will probably heat a little faster but no more uniformly.
3. Eliminate any spaces between the plungers that are likely to fill with flash. It takes time to remove flash and the operation increases molding costs.

Plungers set on top of the steam plate are—

1. Easier to locate and dowel in place.
2. Easier to assemble, since snug fit between plungers is not required for alignment.

Since the depth of the plunger recess is 0.625, a 2 in. plate ground to 1.937 in. may be used for the plunger plate. This allows 1-5/16 in. space for the 1/2-in. steam line, which is drilled 5/8 in. below the top surface of the plate. The reader may wonder why the 5/8 in. dimension was selected for the plunger plate when 3/4 in. was chosen for the cavity block. The intention is to keep the steam lines near the center of the available space, as this affords the tool-maker some leeway in case his drill runs slightly out of line. The 5/8 in. dimension is close to the center of the 1-5/16 in. thickness. In the cavity block, four vertical steam lines are drilled at each of the intersections of the horizontal lines. These vertical lines permit free passage of steam and condensate from the top to the bottom level of the steam lines. The point at which steam lines enter the mold must be plugged or must serve as the connecting point, and the 1/2-in. steam holes require 3/8-in. standard taper pipe taps. The recommended length of engagement to produce a leakproof joint for a 3/8 in. pipe is 1/2 in. This indicates that the horizontal steam line must be at least 3/4 in. from the bottom of the plate as, otherwise, the pipe plug in the vertical hole would partially close the horizontal hole. These considerations led to selection of 3/4 in. spacing for the cavity block and 5/8 in. spacing for the plunger plate.

Two considerations govern the height of the parallels. These are the thickness of the ejector bar and the amount of free movement (travel) required to clear the part from the mold. In this case a 1-7/16-in. ejector bar is selected. The bar can be ground from 1 1/2 in. stock (1-7/16 in. is the minimum thickness which could be used and 1-11/16 in. is in stricter conformance with good practice). The number of ejector pins is also a factor in the thickness determination, since four or five large pins would not require as strong a bar as twenty or thirty small ones, the resistance of the larger number being greater. Travel of the ejector bar should be sufficiently ample to eject the part at least ¼ in. above the mold surface. The longest projection on the plunger is approximately ⅝ in., so a 1-in. travel will be sufficient. This inch of travel plus ½ in. of ejector pin plate plus 1-7/16 in. thickness of ejector bar plus 3/16 stop buttons add up to 3⅛-in. height of parallel. The edges can be finished from 3¼-in. stock and the other surfaces remain unfinished.

The 11/16-in. top clamping plate is the next to be drawn in place. It is good practice to show considerable detail in this end view, as it will usually indicate more clearly the exact relation of the parts. Note that all screws (except one), ejector pins, springs, support pins and steam lines are outlined clearly in this view.

The clearance holes for ejector pins should be at least 1/16 in. larger in diameter than the pin, and they should be enlarged in each succeeding plate to allow for any misalignment in assembly.

The front view merely serves to show the length of the ejector pin plate, details of guide pin and bushing, clean-out slots for the guide pins, safety pin, location of cross steam lines, flash clearance, and the general outline of the front of the mold.

It is very important to keep all views as simple as possible. Avoid trying to put too much detail into any one view. Make extra views when necessary or whenever a view will be useful for the purpose of clearer presentation of details.

The question will arise: When shall decimal dimensions be used and when will fractional dimensions be satisfactory? A general rule will not satisfy this question. The best solution will be found in consideration of (1) whether or not the particular dimension is vital to the fit of mold parts, and (2) whether it will affect the assembly. Reference to the assembly, Fig. 6.8C, will show that the only dimensions given in decimals are the length, width and depth of the cavity and plunger space, the thickness of the plunger plate, and the distance from the face of the knockout bar to the top of the plunger plate. The fact that these dimensions are given in decimals does not mean that they are to be adhered to rigidly, but indicates that the tool-maker must compensate for any deviation he introduces by varying another dimension on some detailed part. Obviously it would be unnecessary and confusing to dimension top or bottom plates or designate the exact location of screws, etc., in decimals, as variation in these dimensions will not make any difference in assembly of the mold. *Decimal dimensions must not be used unless they are absolutely essential.*

The last work on the assembly drawing is the title block and stock list. It is important to enter in the list enough information to enable the tool-maker to order all necessary material for the mold, including screws, dowels, springs, etc. The mold should be designed to use as many stock or standard items as possible. Sizes given in the stock list must be rough sizes and they should include necessary machining allowances. All parts which are identical should carry the same part number and specify the quantity used.

Other important notes to be placed on the drawing are: amount of shrinkage allowance, chrome plate information, and all other data that will be of use to the tool-maker or the person using the mold. Shrinkage allowance notes should be inserted as changes may be made later, and without this information, recalculation will be necessary.

Undoubtedly, the question will arise as to why the author did not use standard plates or standard components available from the catalog items. The reason is: the author wished to detail the thinking behind each of the selections and to design as compact a mold as practical. Once the beginner understands the concept of mold design and the theoretical material behind it, it becomes an easy matter to adjust his thinking to include standard mold bases, standard mold plates and standard components—meaning by "standard" items that are available from a catalog as a ready-made item, usually "out of stock."

Stock Allowances

Minimum allowances for finishing of mold parts should be made as shown in Table 6.1. Although this table shows ⅛ in. allowance on the diameter of mold sections, this does not hold true where the cavities and plungers have shoulders or flanges. Whenever possible, the outside diameter of all mold parts having a shoulder should be a standard stock size.

TABLE 6.1. Minimum Allowances for Finishing Mold Parts.

Parts	Diameter (in.)	Length (in.)	Width (in.)	Thickness (in.)
Steam plates		⅛	⅛	1/16
Cavity and plunger plates		⅛	⅛	Up to 2 in., 1/16 Over 2 in., ⅛
Clamping plates		None	None	1/16
Safety, ejector, support and mold pins	None	⅛		
Guide pins and bushings	⅛	⅛		
Ejector bars		None	None	1/16
Parallels		None	1/16	None
Mold sections (plungers, cavities, etc.)	⅛	⅛	⅛	⅛

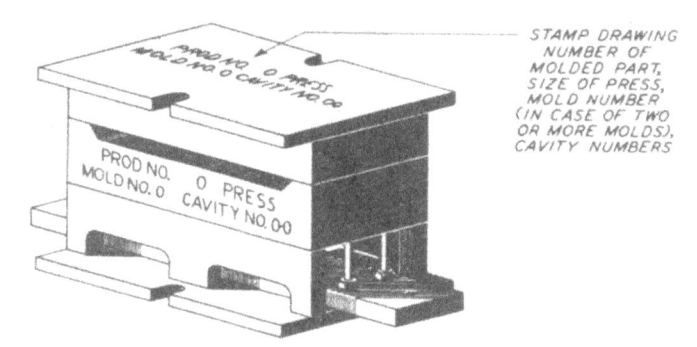

FIG. 6.11. Molds are identified easily when adequate information is stamped on the frame.

Mold Identification

Fig. 6.11 shows the identification data which should be stamped on all molds. Proper stamping saves much questioning by the men in the press room and mold storage room in that it provides ready identification of molds. The size of the stamping should be approximately ½ in. in order that it may be read easily in dark corners, where molds are so commonly stored. Mold parts that are to be heat-treated should be stamped on some non-functioning surface with the name or type of steel used.

Drawing Revisions

It will be noted that the 12-cavity mold, Figs. 6.8A–6.8D, was not a spring box mold originally. (Refer to pages 245–247 for discussion of "spring-box" molds.) Revisions in design are inserted in a panel which usually is placed in a corner of the drawing (see Fig. 6.8C). Here any additions, removals, or changes are listed, and each is explained in relation to the original tool drawing. Every dimension or part that was changed is identified by a small circle (3/16 in. diameter is a good size) containing the proper "change number." All persons who were furnished the original prints should receive copies of the revised drawings. Properly identified changes are very important because they make it unnecessary for persons receiving the revised drawings to check all dimensions found on the originals.

When changes are made in mold cavities or plungers, the customary procedure is to revise the tool drawing so that sizes will be correct for future production needs. Sketches are made as shown in Fig. 6.12 to show the sections in which changes occur. All purchase orders for molds should show the drawing number and the latest change date to avoid loss resulting from construction of a new mold to an old design.

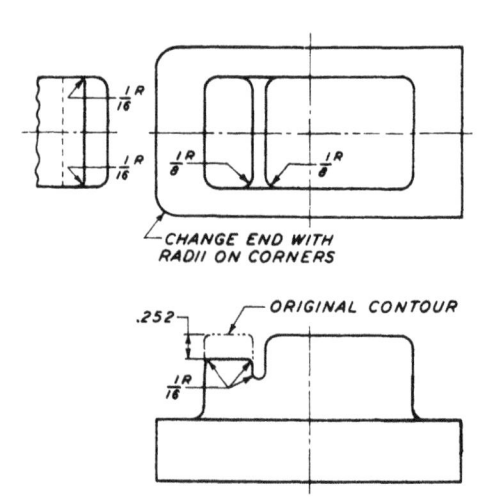

FIG. 6.12. Changes to be made in existing mold sections are indicated by showing both old and new contours.

Detailing Mold Parts

Some designers prefer to make the detail drawings first. It is generally desirable, however, to complete the mold assembly drawings first so that the plunger and cavity sections are fully developed before they are detailed. Figures 6.13A–6.13D show the cavity, plunger, ejector pin, and mold pin details for the lever. It should be studied with Fig. 6.2 and Figs. 6.8A–6.8D. In practice it is well to do the larger pieces first and then fit the smaller parts into blanks left on the sheet. For smaller or very highly detailed parts it is generally considered good practice to draw them to a larger scale (usually double size). In making this drawing, the plan view and all sections are drawn to scale before starting to add the dimensions.

Cavity Calculations

(See Figs. 6.13A–6.13D.) The outside dimensions of the cavity were determined previously for the assembly drawing and the centers of the radii laid out at 1-27/32 from the outside end of the cavity. This dimension was set up on the drawing as 1.842 and it becomes the starting point for all other dimensions on the cavity and plunger detail. This is because it is the center of the radii and the only point common to the details of the cavity and plunger. The width of land selected for this cavity is 0.113, and it is small. For practical purposes, a land width of 0.125 or more would have been better. The effect of a narrow land is to create a high unit stress at this point, and the top edge of the cavity may be crushed as a result. On

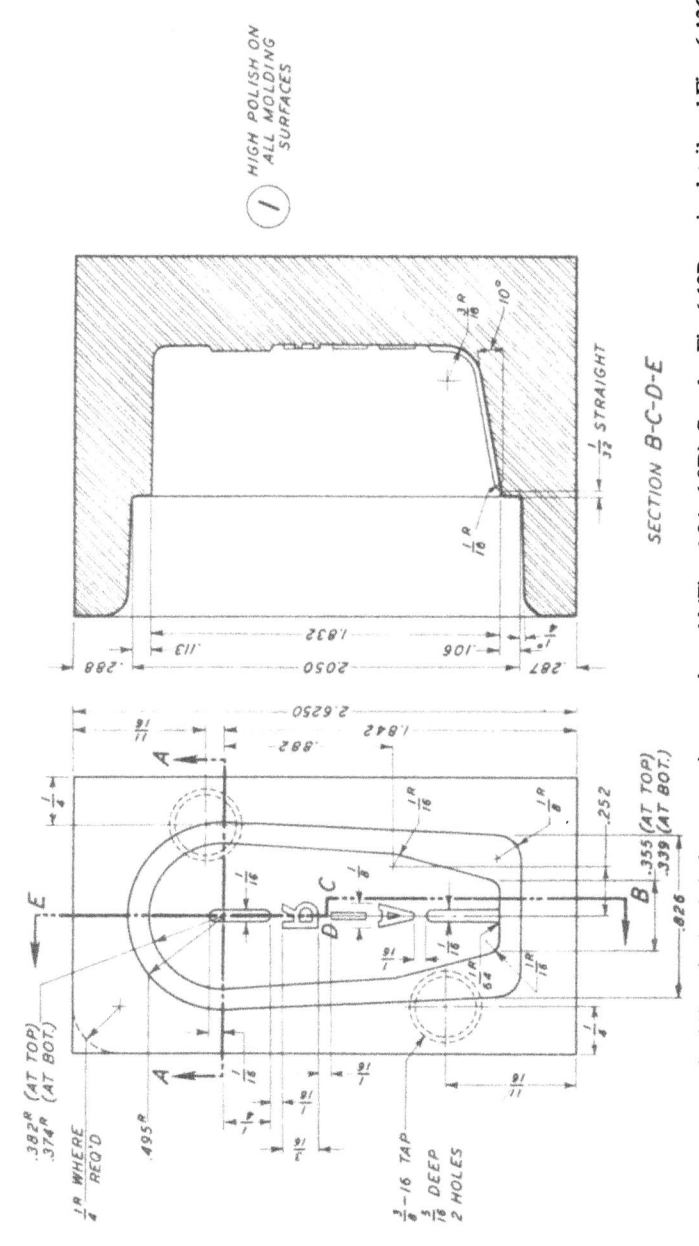

FIG. 6.13A. Cavity detail of 12-cavity landed plunger semiautomatic mold (Figs. 6.8A–6.8D). See also Fig. 6.13B, cavity detail, and Figs. 6.13C and 6.13D, mold pin and plunger details.

239

FOR MOLD ASSEMBLY SEE 727

.008"/IN. ALLOWED FOR SHRINKAGE

							# PCS	D.	L.	W.	T.
12	R.D.S.	R'WELL C-58	HARDEN	5	K.O. PIN				$\frac{1}{2}$	$4\frac{3}{4}$	
24	DR. ROD	R'WELL C-52	HARDEN	4	K.O. PIN						
12	DR. ROD	R'WELL C-52	HARDEN	3	K.O. PIN		6	$\frac{5}{32}$	36		
12	S.A.E.-3312 OR EQUAL	R'WELL C-58	CARBURIZE & HARDEN	2	PLUNGER		12	$2\frac{3}{4}$	$2\frac{1}{4}$	$1\frac{5}{8}$	
12	S.A.E.-3110 OR EQUAL	R'WELL C-58	CARBURIZE & HARDEN	1	CAVITY		12	TOSUM			

MAT'L REQ'D

12-CAVITY SEMIAUTOMATIC MOLD
(DETAILS)

FIRST MADE FOR LEVER   DWG. 12345   728

TRACED BY
INSPECTED
BEGUN BY
FINISHED BY
GROUP DESCR.   G. NO

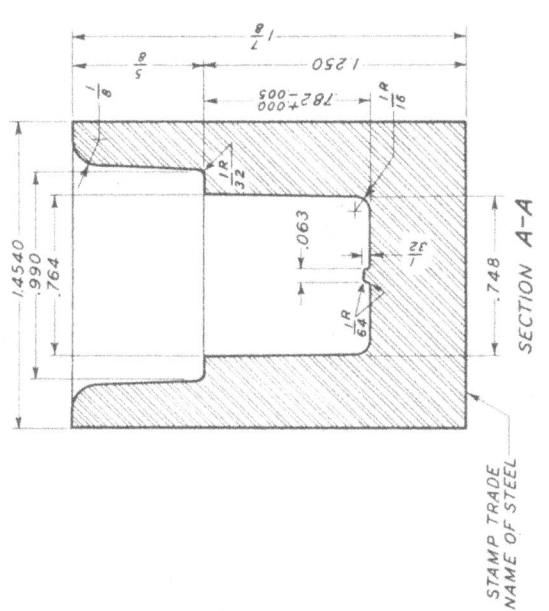

SECTION A-A

STAMP TRADE NAME OF STEEL

FIG. 6.13B. Cavity detail and title block for detailed parts.

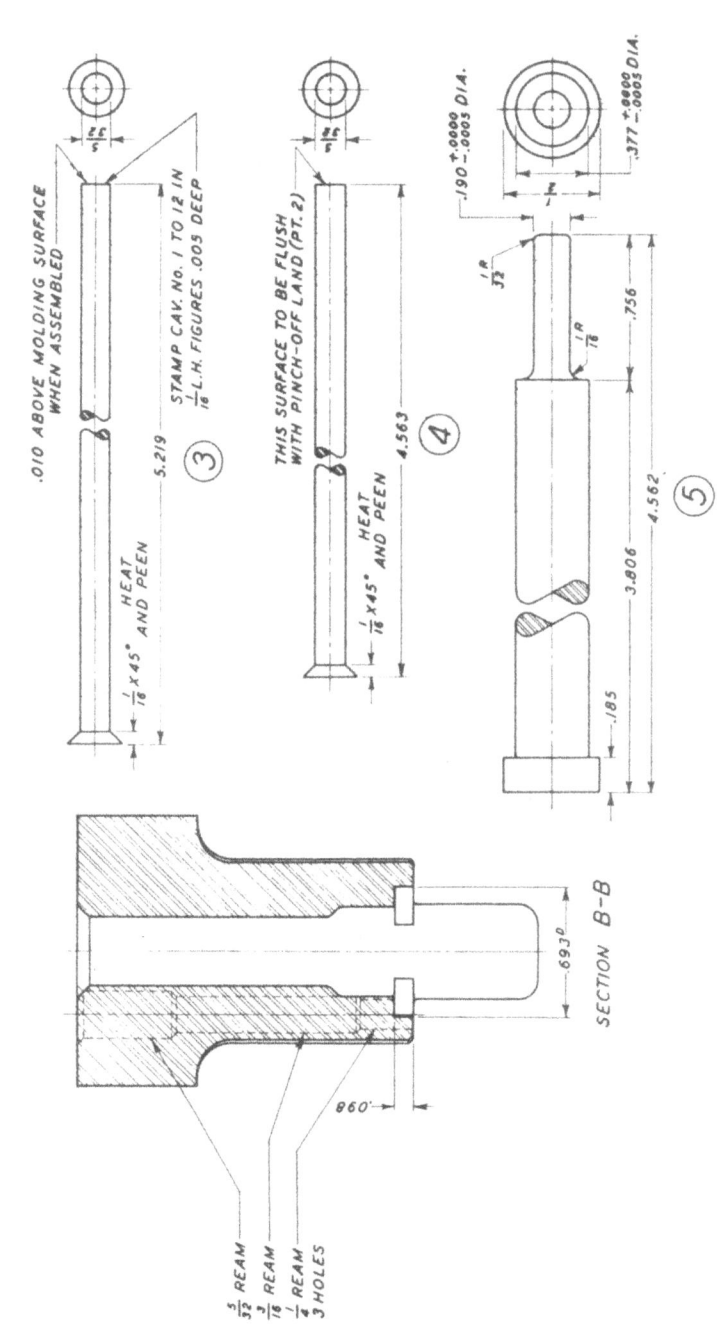

Fig. 6.13C. Mold pin detail and plunger section. See Fig. 6.13D for plunger detail.

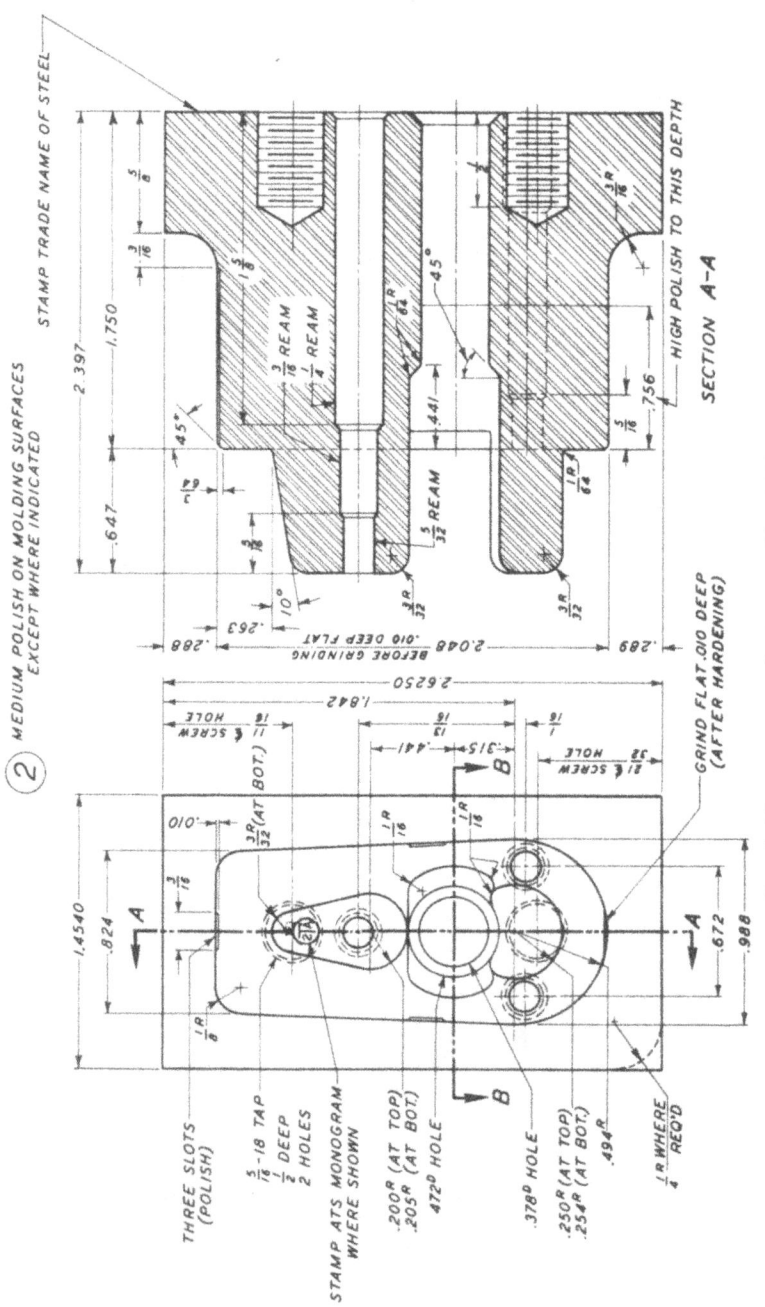

FIG. 6.13D. Plunger detail. (See also Fig. 6.13C.)

the other hand, a land that is overwide will absorb too much pressure and thus produce a heavy flash. The maximum land for molds of average size should be about 3/16 in., although lands up to ½ in. have operated well.

A comparison of the part drawing, Fig. 6.2, and the cavity details, Fig. 6.13A and 6.13B, will show the manner in which draft and shrinkages are allowed. For most parts, where dimensions are given in fractions, as are the outside contour dimensions of this lever, the draft is divided. A point halfway between the top and bottom of the piece should be the exact size specified on the production drawing.

Shrinkage Calculations

Calculations for the cavity dimensions are based on a shrinkage allowance of 0.008 in./in.

Basic width = 0.750 + 0.006 (shrinkage) = 0.756 in.
Bottom width = 0.756 − 0.008 (½ total draft) = 0.748 in.
Top width = 0.748 + 0.016 (total draft) = 0.764 in.

$$\text{Radius at top} = \frac{0.764}{2} = 0.382 \text{ in.}$$

$$\text{Radius at bottom} = \frac{0.748}{2} = 0.374 \text{ in. radius}$$

11/32 in. width = 0.344 + 0.003 (shrinkage) = 0.347 in.
11/32 in. width at bottom = 0.347 − .008 (½ draft) = 0.339 in.
11/32 in. width at top = 0.339 + 0.016 (total draft) = 0.355 in.
Basic depth of cavity = 0.781 (base dimension) + 0.006 (shrinkage)
= 0.787
*Flash thickness correction = 0.787 − 0.005 (flash) = 0.782

A tolerance of + 0.000 − 0.005 is given on the 0.782 depth of the cavity. This 0.005 is ½ the minus tolerance allowed on the molded part. The 0.782 dimension might be expressed as 0.777, omitting the tolerance. This would allow the tool-maker to use his normal tolerance.

Flash Thickness Allowances

Allowances for flash thickness (build-up) in compression molds, using thermosetting molding materials are:

For *rag-filled, high-impact compounds* (Military CFI-10 or over), allow 0.010 in.

---

*The flash thickness adds to the total thickness of the part and this thickness must be subtracted from the basic cavity depth in order that the finished piece may have the desired thickness.

For **cotton-flock compounds** in large molds, allow 0.007 in.

For *wood-flour compound* in small molds, allow 0.003 in.

When positive molds are used (no lands), the depth of cavity should be the minimum dimension given on the production drawing plus standard shrinkage for the material being used.

For all other molds (except as previously noted) and for all other compounds, allow 0.005 in.

Since the cavity thickness from the back to the pinch-off land when added to the 1.750 plunger dimension must equal the sum of the 0.625 recess in the plunger plate, Fig. 6.8C (part 4), plus the .500 stop pads, plus the 1.875 depth of recess in the cavity plate (part 3), it is given as a decimal dimension. The depth of the cavity well and the over-all height is given as a fraction, as these dimensions are not critical.

Noteworthy also is the fact that the screw holes in both cavity and plunger are located in positions where they will clear the steam lines in the cavity and plunger plates.

Plunger Dimension Calculations

The same general procedure used for dimensioning the cavity is followed in dimensioning the plunger except that only ½ the draft is allowed. In this case, it is desired that the molded part stay on the plunger since top ejectors will be used. (Ejector marks would be very undesirable on the cavity side of this part.) The use of only half as much draft on the plunger as in the cavity plus a medium-dull finish on the plunger should be sufficient to make this part hold without the aid of pickup marks, which can be supplemented if needed. The finish on the plunger should be only smooth enough to prevent the flash from sticking.

The clearance between plunger and cavity well on landed plunger molds is usually from 0.001 to 0.002 in. per side. Since the proper amount of overflow slot width may have to be determined by trial, the slots should be made small and later enlarged as required for satisfactory mold operation.

For the 3/32 plus 0.010 minus 0.000 dimension (Fig. 6.2), use 0.094 plus 0.001 (shrinkage) plus 0.003 = 0.098 where the 0.003 represents approximately ¼ of the tolerance allowed. The 0.375 plus or minus 0.005 in. diameter shank (Fig. 6.2) uses a mold dimension of 0.378, as this is the nominal dimension plus shrinkage. Note that a 3/64 bevel at 45° is specified on the edge of the plunger to clear the 1/32 radius in the cavity.

Pin Details

A close study of details of the other parts will show the calculations which follow the instructions given in Chapter 5.

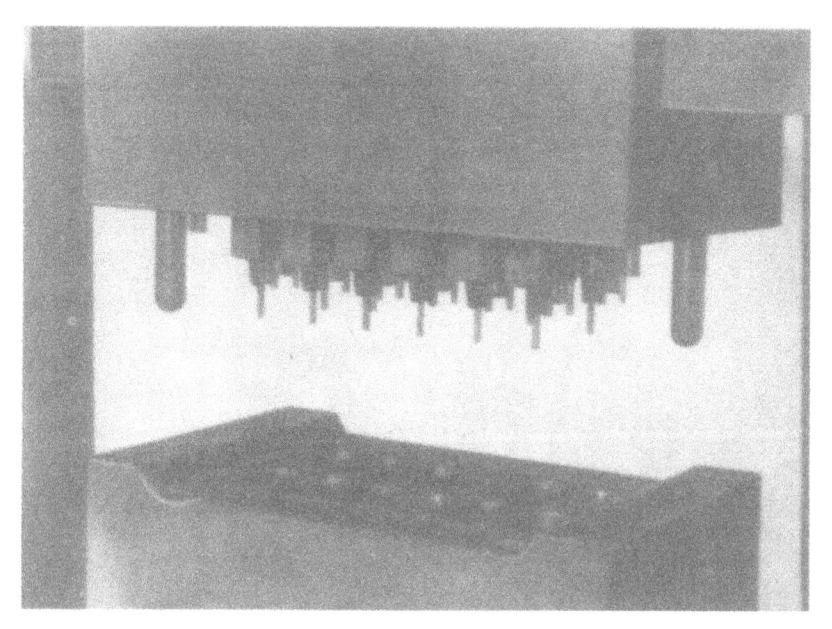

Fig. 6.14. Twelve-cavity landed plunger semiautomatic mold used for molding lever (Fig. 6.2).

The completed mold is shown in Fig. 6.14. One cavity has been taken out for repair and a block inserted to support the cavities on either side while repairs are made.

SPRING-BOX MOLDS

One type of design used for a spring-box mold was illustrated in Fig. 2.29. Alternative assembly methods, as shown by Figs. 6.15 and 6.16, are also used.

In a design problem under consideration it will be assumed that it is desirable to spring-box an insert which is 11/16 in. over-all in length. As a general rule 2/3 of the insert length is used for spring-boxing, ½ in. being selected as the nearest whole fraction in this case. One-eighth in. extra is allowed to cover the space required for insertion of the spacing fork. This totals ⅝ in. for the spacing between the top of the parallels and the bottom steam plate. By reference to the Spring Data Chart (Table A.4A, Appendix). Any one of three springs may be used. For average jobs, such as that under consideration, the medium size may be selected. Light springs are used where small pins are to be spring boxed or inserted in small molds. By using spring W.S.-19 and following down the "free length" column to the ½-in. dimensions, the length of the spring is found to be 3⅛

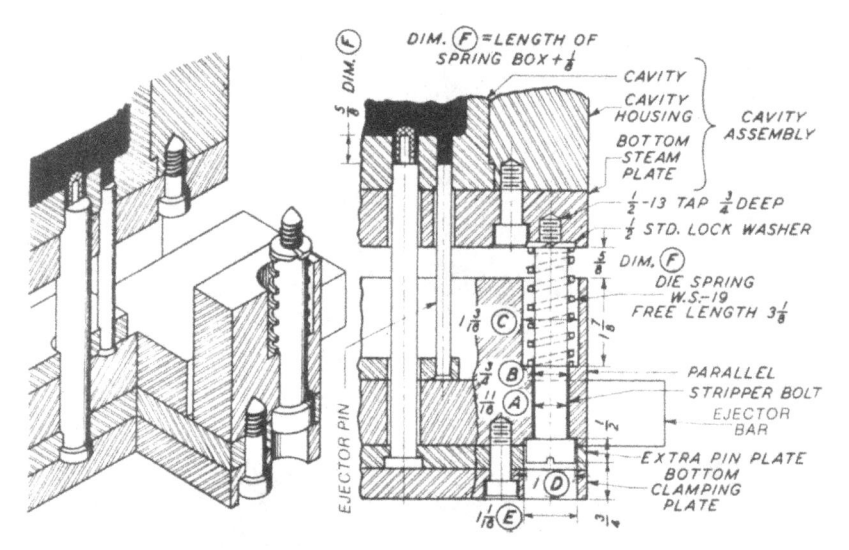

FIG. 6.15. Alternate assembly for spring-box mold. (Dimensions taken from the Spring Data Chart, Table A.4A, Appendix.)

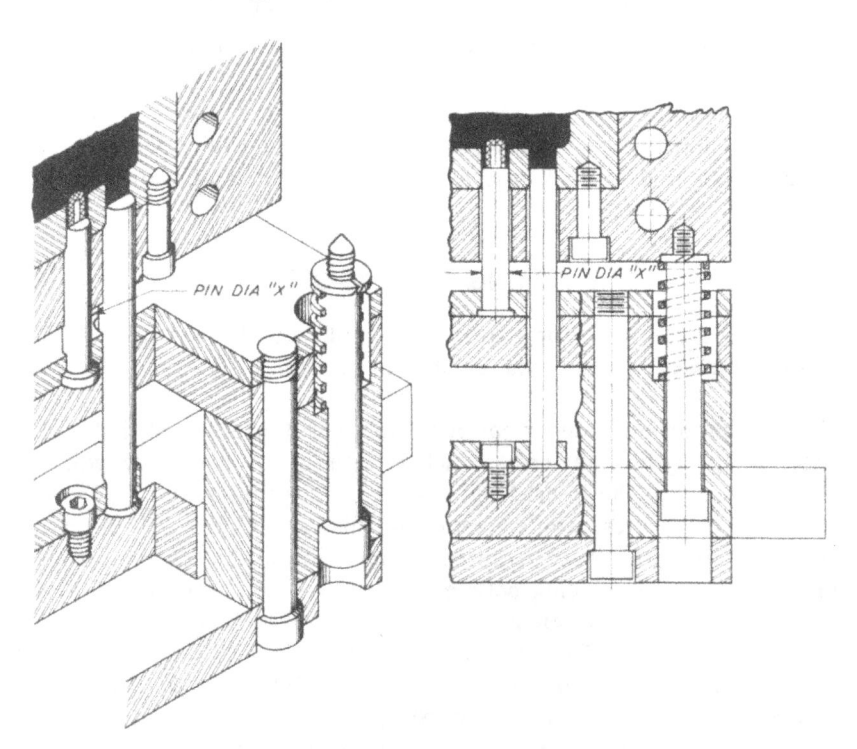

FIG. 6.16. Alternate assembly for spring-box mold.

in. Under "Depth of Counterbore (H)," the correct size for the ½-in. dimension is found to be 1⅞ in. The other dimensions given in the Spring Data Chart under W.S.-19 are:

> Dimension A, diameter of stripper bolt, 11/16 in.
> Dimension B, clearance for stripper bolt, ¾ in.
> Diameter C, counterbore for spring, 1-3/16 in.
> Diameter D, diameter of stripper bolt head, 1 in.
> Diameter E, counterbore for head, 1-1/16 in.

When the insert pin is not needed as an ejector pin, this assembly may be designed as shown in Fig. 6.16. In this case, the pin diameter should be at least 1/64 in. larger than the maximum diameter of the insert being spring-boxed.

LOADING SHOE AND STRIPPER PLATE MOLDS

Stripper plate molds are used more extensively for injection molding than for compression molding. The loading shoe mold is a type most extensively used in compression molding to provide additional loading space for the bulky materials. The fundamental principles and uses for both types of molds were discussed in Chapter 3 and it would be well to review this matter before proceeding with these design details.

The general design details of these molds follow the standards described for the landed plunger molds discussed earlier in this chapter. The same procedures and calculations are indicated for the frame, heating media line clearances, ejector bar and pins, mold and safety pins, shrinkage, etc., of loading shoe and stripper plate molds.

A captive loading shoe may be used as shown in Fig. 6.17 (A). The cutaway section at the limit screw (B) shows how the screw is assembled. The loading shoe or stripper plate may hang midway between the cavities and plungers, as shown in Fig. 6.18.

The captive loading shoe mold shown in Fig. 6.17 may be used when there is no necessity for reaching into the cavities to load inserts. It incorporates a desirable safety feature. The shoe travel is limited by the screws to ⅜ in. or ½ in. above the cavities. This small gap is sufficient to permit the operator to blow out the flash that forms between the loading shoe and the cavities. The flash that forms between the sides of the plunger and the loading shoe causes binding and usually is heavy enough to cause the shoe to stick to the plunger as the mold opens. When the captive loading shoe system is used, there is no danger of the plate dropping unexpectedly and trapping the operator. Safety latches may also be used to eliminate this risk.

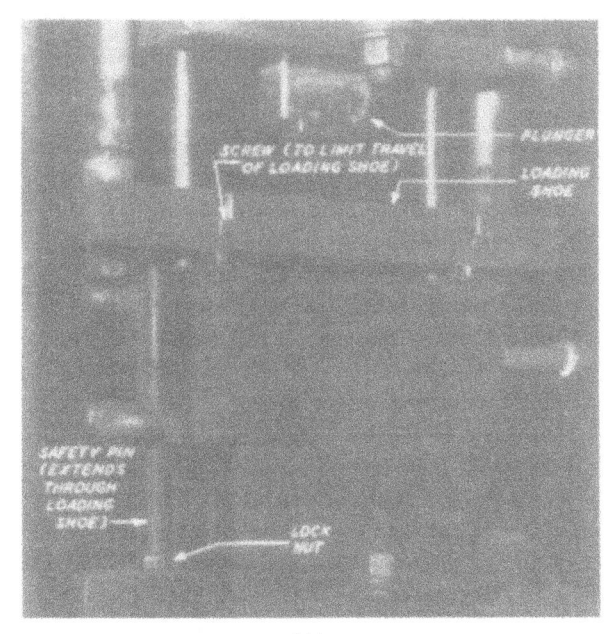

(A)

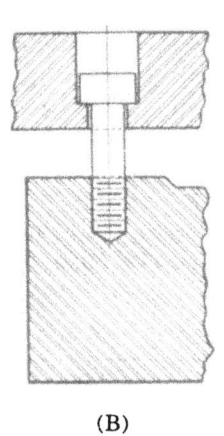

(B)

FIG. 6.17. (A) Single-cavity captive loading shoe mold. (B) Section showing assembly of limit screw.

The safety pins used for captive loading shoes must extend through the shoe and butt against the plunger plate. This permits free movement and full travel of the ejector bar.

The stripper plate mold shown in Fig. 6.18 is a 12-cavity mold for the production of the small tube shown in front of the mold. It is molded from

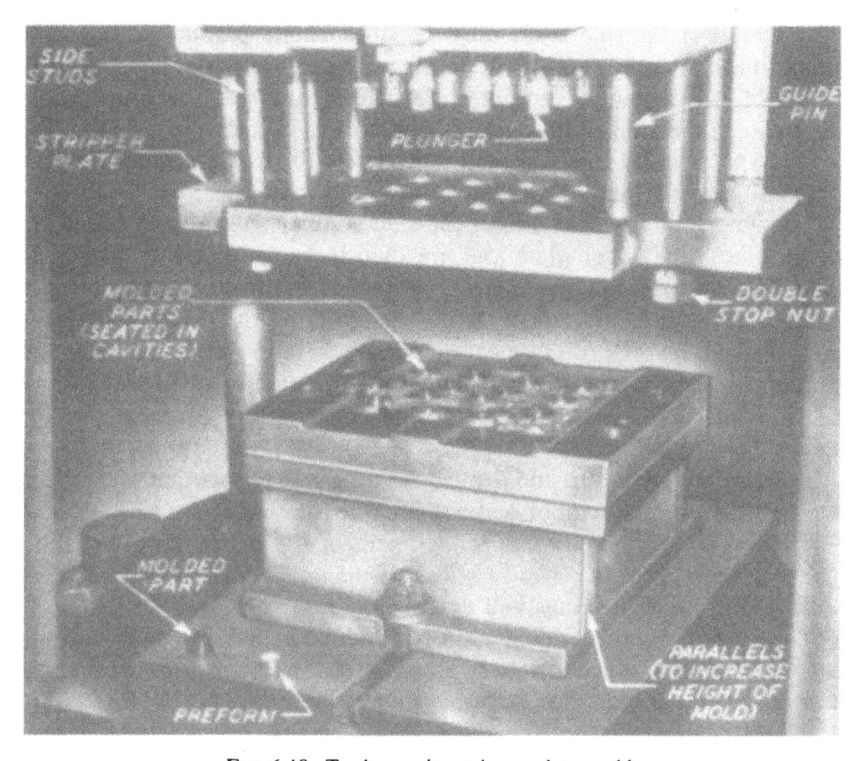

FIG. 6.18. Twelve-cavity stripper plate mold.

a cotton flock phenolic compound. The tube has only a thin wall and a small head, as shown in Fig. 6.19, and this size does not permit the use of knockout pins.

It will be noted that the molded parts have been stripped from the plunger, and, as shown, are partly seated in the cavities, where they may be picked up easily.

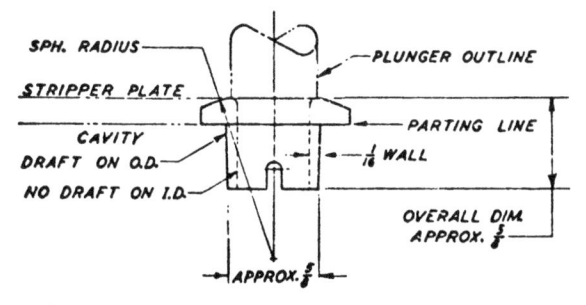

FIG. 6.19. Part as molded in the 12-cavity mold shown in Fig. 6. 18.

The length of the side studs must be sufficient to permit the top of the stripper plate to come at least 2½ in. below the end of the plunger. A larger number of cavities might require more space at this point in order to provide greater accessibility in loading. Pill loading boards may also be used for this type of mold to eliminate the necessity for reaching into the mold. Pill loading boards permit quick loading of all cavities.

In designing this type of mold, the side studs are usually threaded into the top plate or plunger plate and are then locked in place by a lock nut situated above or below the plate. The lock nut must not come above the clamping plate and interfere with the clamping of the mold in the press. The side studs pass through clearance holes (1/16 in. larger than the stud diameter) in the shoe or stripper plate and terminate in a double stop nut. The stop nuts and lock nuts may be fastened tightly by means of lock washers.

The guide pins for this mold must be long enough so that the stripper plate or loading shoe cannot leave them when the mold is open. Thus the guide pins also serve to maintain alignment of the plates. Guide bushings must be used in all soft plates or shoes. If the entire shoe or plate has been hardened the guide bushings will not be needed.

It will be noted that parallels have been added below the cavity although there is no ejector mechanism. In this case the parallels serve to increase the height of the mold. This increase is necessary because the length of the side studs is greater than the thickness of the cavity plate and bottom steam plate. The end of the side studs must come at least ⅛ in. above the bottom of the mold when the mold is closed.

The principal differences between loading shoe and stripper plate molds are:

1. The inside opening of a loading shoe mold is larger than the molded part. The stripper plate mold is smaller than the maximum dimensions of the molded part.

2. The loading shoe has nothing to do with ejection of the part, whereas the stripper plate serves to remove the part from the plunger.

3. The loading shoe mold has a conventional ejector mechanism, and the parts are usually designed to stay in the cavity. The stripper plate mold makes use of minimum draft or pickups in order that the pieces shall remain on the plunger or core.

4. The thickness of the loading shoe must be sufficient to allow for the bulk factor of the compound, and it may be a very heavy plate requiring safety pins for safety. The stripper plate need be only thick enough to give sufficient tool strength to resist bending.

It should be recognized, however, that there are cases where a combination loading shoe and stripper plate may be used.

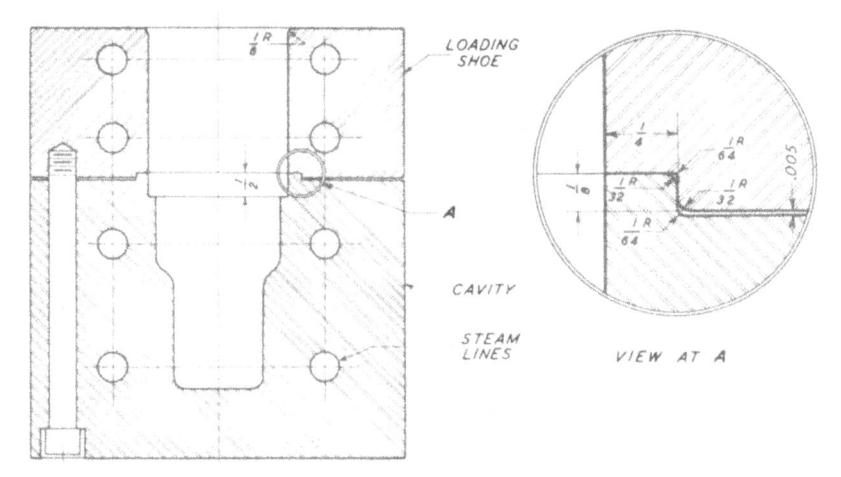

F<small>IG</small>. 6.20. Two-piece cavity construction, showing loading shoe attachment used when shoe is to remain stationary during operation of mold.

The floating loading shoe permits the use of shorter ejector pins and offers reduced travel of the ejector bar, since the parts will be ejected into the space between the cavities and the loading shoe.

In some cases the loading shoe is attached permanently to the cavity and therefore cannot move. Use of high-impact materials, which require considerable loading space, may produce a condition in which the combined depth of the cavity is greater than may be machined by conventional methods. The shoe may be fastened to the cavity plate with screws to form a two-piece cavity assembly, as shown in Fig. 6.20. The 0.005 in. space between the loading shoe and cavity insures a tight seal at the ¼ in. land. To hold these two sections together, use the same number and diameter of screws as are used in the top and bottom plates. All loading shoes should be carburized and hardened as would any other mold section coming in direct contact with the molding compound.

POSITIVE MOLDS

Positive molds which operate in a manner similar to the action of a piston in a cylinder necessitate correct clearances between the cavity and the plunger. If the clearance is too large, the compound will escape and unfilled parts may result. These unfilled parts are sometimes called "NFO" or "short shots." Let us digress a moment and explain that a mold engineer should never forget that deliberate short shots is a good engineering tool when the flow pattern in a part cannot be determined by simple observation of a complete part. The "short shot technique" consists of starting with approximately 25% of what would be the total part weight on the first shot. Then, on each subsequent shot, increase the material weight slightly. Continue this process until a full part is achieved. The purpose being to get visual evidence of the exact flow pattern of material as it moves around in the mold. It also points out exactly where the last fill takes place, and this is the point at which cavities need to be vented. By "venting" we mean a deliberate opening in the mold into which the air trapped in the cavity can be exhausted ahead of the incoming material. Venting is more thoroughly discussed in transfer and injection molds (Chapters 7 and 8). Paradoxically, no amount of heat, or extra pressure will make up for lack of venting.

Consideration must be given to "clearances" between cavity and plunger or core. When the clearance is too little, material will not escape and air gas may be trapped in the part causing blisters. When blisters appear on compression molded parts it is an almost sure sign of trapped air or gas or undercure. The first remedy to try is "breathing" or "bumping" the mold. *Breathing* consists of releasing the closing pressure of the press and allowing the mold to come apart at the parting line. The space allowed varies from just cracking the molds to as much as ½ in. of opening. Time of opening varies from 1 to 2 seconds to as much as 10 or 15 sec. Multiple operations of breathing are called *bumping*.

A clearance that is too little may also cause one side of the male section to rub the cavity wall, thereby scoring or roughening the surface. A scored wall will mar the surface of the molded part as the part is ejected. Most users will object to this marring.

Practical experience has shown that 0.003 in per side clearance between the cavity and plunger will give good molding results.

The problem of designing a mold to produce the piece shown in Fig. 6.21, will be considered a typical example of positive mold design. This

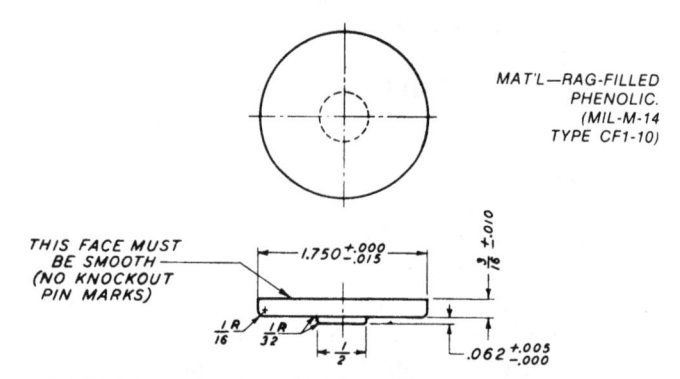

FIG. 6.21. Molded spacer to be molded in positive mold, Figs. 6.22, 6.23 and 6.24.

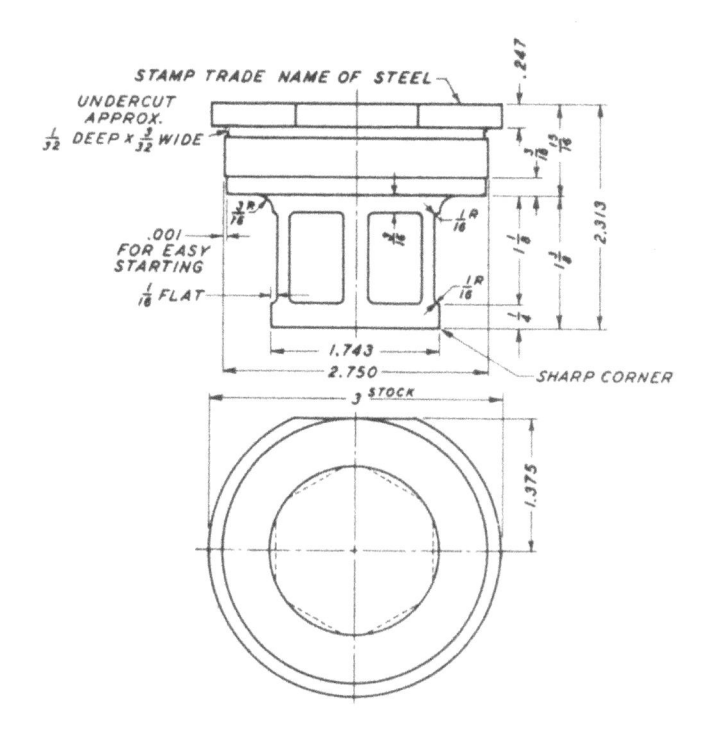

FIG. 6.22. Detail of plunger for positive mold shown in Fig. 6.24.

part is to be made from a rag-filled phenolic, and a positive mold is indicated because it will yield a dense part and hold the flash line to the minimum. The cavity for this simple round shape may be fabricated easily by being turned in a lathe. If several cavities were to be made, they could be made by hobbing.

The plunger and cavity details are shown in Figs. 6.22 and 6.23, and the assembly in Fig. 6.24.

The rag-filled phenolic compound may have a bulk factor ranging from 5:1 to 8:1 and a shrinkage of 0.004 in./in. Assuming that the material used will have a bulk factor of 7:1, and using the formulas given at the beginning of this chapter, the total depth of the cavity is calculated as 1-5/16; this means that the plunger will enter 1⅛ in. The necessary flash clearance is ¼ in. between plunger and cavity. A 15/16-in. plunger plate (1 in. stock) will allow room for a steam line around the plunger. (See Fig. 6.24.)

The 1/16-in. deep flash shown in Fig. 6.22 is quite essential. It could be in the form of radial relief, but it should not be a full annular undercut as this would permit the formation of a solid ring of flash which would have to be broken for removal. When the relief is interrupted by unrelieved portions, the flash becomes a series of heavy and thin sections

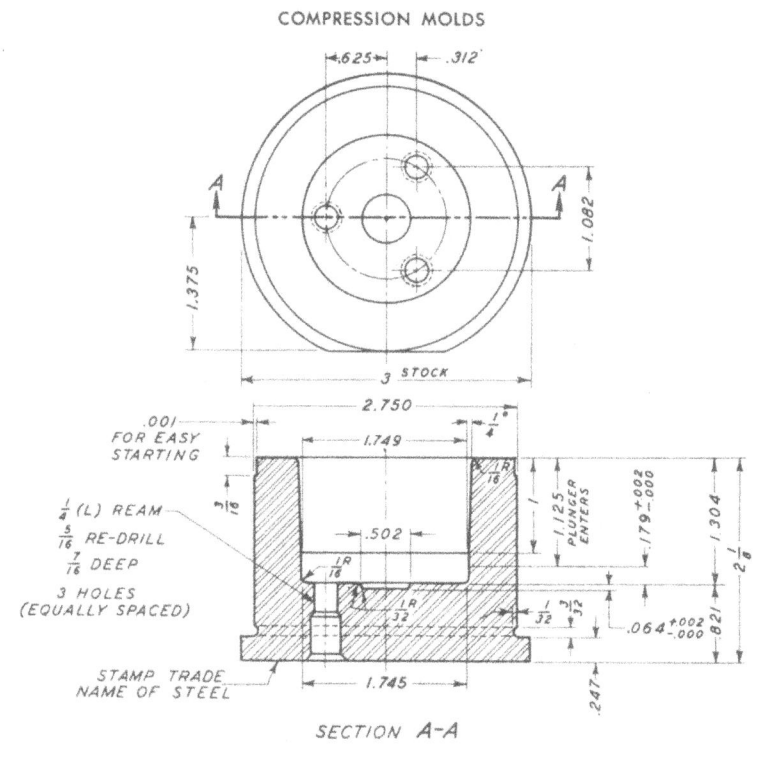

FIG. 6.23. Detail of cavity for positive mold shown in Fig. 6.24.

which may be blown easily from the plunger with a blast of air. The unrelieved portion should be one-fourth to one-third of the total side area of the plunger.

The cavity, as shown in Fig. 6.23, is to be mounted in a 2⅜-in. cavity plate (2½ in. stock), as this will leave approximately 13/16 in. between the mold surface and the bottom of the cavity, which amount will be adequate when the cavity is backed up properly by heat plate and support pins. It has been determined that the plunger will enter 1.125. The depth of the piece is 0.187, and the added shrinkage 0.001. From this total 0.009 of the thickness tolerance is subtracted, leaving 0.179. This is added to the 1.125 to arrive at the cavity depth of 1.304. By starting the mold in this manner, it can be corrected after trial indicates the amount necessary. This calculation provides that with a minimum load of compound, and with the mold entirely closed and resting on the stop blocks, a part will be produced gauging 0.178 (near the minimum tolerance) after shrinkage. If trial of the mold indicates that this is too close to the minimum dimensional limit, it can be corrected easily by grinding off a few thousandths from

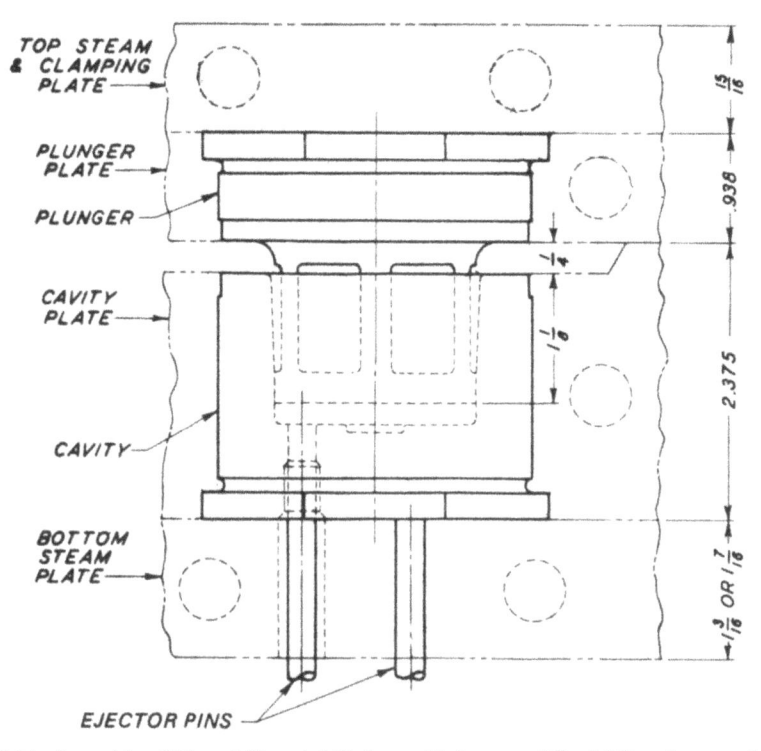

FIG. 6.24. Assembly of Figs. 6.22 and 6.23 for molded spacer (Fig. 6.21), using a positive mold.

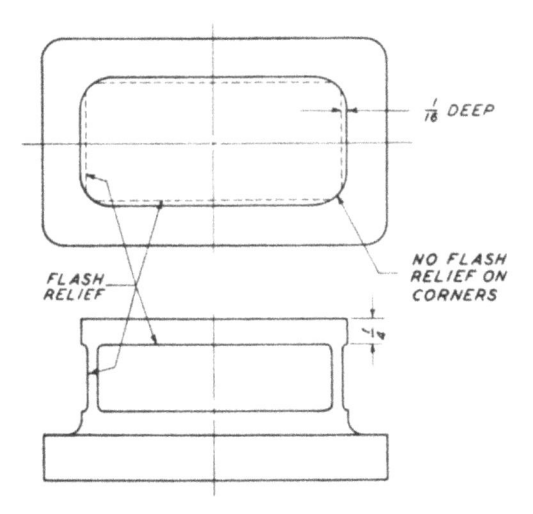

FIG. 6.25. Flash relief on a rectangular plunger for a positive mold.

the face of the plunger. Using the above calculations, a build-up as great as 0.018 may be obtained without exceeding the upper tolerance specified for this spacer. It is necessary to take advantage of this tolerance range since the load must be kept on the plus side to insure adequate density of the part. To inform the tool-maker that 0.179 is the minimum dimension, a tolerance of plus 0.002 and minus 0.000 is indicated.

The dimension which governs the height of the cavity is the 0.821 dimension, since 0.821 plus 0.179 plus 2.313 (from plunger) must equal 2.375 (cavity plate) plus 0.938 (plunger plate). Checks of this type must be made on all calculations to make sure that errors are eliminated.

The calculations for the diameter of the cavity are made as follows:

1.750 in. $\pm$ 0.007 in. (shrinkage) = 1.757 in.
1.757 in. $-$ 0.015 in. (tolerance) = 1.742 in. (minimum dimension).
1.742 in. $+$ 0.003 in. (safety factor) = 1.745 in. diameter at bottom.
1.745 in. $+$ 0.004 in. (taper in 5/16 in.) = 1.749 in. diameter at top.

The ejector pins may be located on any convenient radius, approximately halfway between the 0.502 in. diameter bore and the 1/16 in. radius at the outside. Actually they are located nearer the outside because that is where the greatest drag will be. All other dimensions and details should be self-explanatory, as they follow previously discussed methods.

Flash relief for the rectangular plungers used in positive molds is most often done in accordance with the method shown in Fig. 6.25.

SEMIPOSITIVE MOLDS

The semipositive mold is used to obtain certain operating conveniences and to eliminate the use of a spring box in securing full part density. This mold action serves to trap and compress the charge.

A valve chamber for a milking machine is shown in Fig. 6.26. The mold section used to produce the threads is also shown, and the mold construction is illustrated in Fig. 6.27. In this case, the movable thread ring is attached to the ejector bar, which serves to push the molded piece off the plunger after it is withdrawn from the cavity. The piece is easily unscrewed after it is pushed free of the plunger. It is always good practice to withdraw any cores from a molded part before attempting to unscrew them from the mold. This is particularly advisable when an external thread is used in combination with a cored hole. An instance is recorded in which a 12 in. wrench failed to move a 1 in. diameter part that had an outside thread and a ⅝-in. diameter plug inside. After the mold was changed to withdraw the plug before unscrewing, the part was easily removed with a 2 in. lever.

It will be noted also that the plunger in Fig. 6.27 has been drilled for heating. When long plunger sections are used, heating is necessary for mini-

FIG. 6.26. Mold section (left) used for molding internal threads on valve chamber (right).

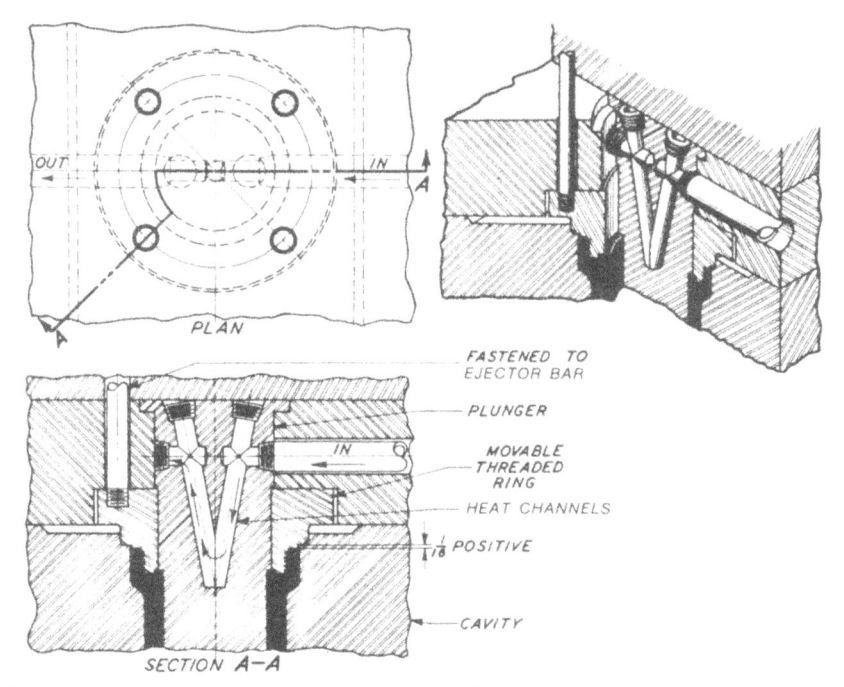

PLAN

OUT

IN

A

A

SECTION A–A

FASTENED TO
EJECTOR BAR

PLUNGER

IN

MOVABLE
THREADED
RING

HEAT CHANNELS

$\frac{1}{16}$ POSITIVE

CAVITY

FIG. 6.27. Schematic diagram of a semipositive mold which uses molded thread ring shown in Fig. 6.26.

mum cure time and best quality production. When the urea and melamine materials are molded, uniform heating is even more important, and plungers as small as ½ in. or ⅝ in. diameter may have to be cored out to provide direct heat.

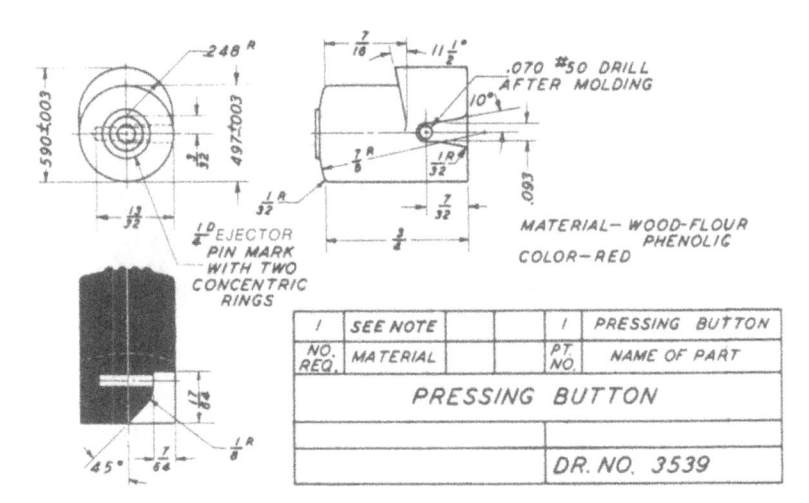

FIG. 6.28. Molded pressing button to be made in a 7-cavity mold (Fig. 6.29).

SUBCAVITY GANG MOLDS

The uses of the subcavity mold and its general construction were outlined in Chapter 2. The small pressing button shown in Fig. 6.28 is a typical example of parts produced in this type of mold. This mold can be mounted in the standard frame illustrated by Fig. 2.34, and this limits the actual mold construction to the cavity, plunger and pins, since the frame is a stock item for mold of ths size. The design for this subcavity, landed plunger gang mold is shown in Fig. 6.29A, 6.29B and 6.29C. Previously outlined principles were used to calculate the amount of loading space, shrinkage allowance and draft required. The student should prepare all necessary calculations for design of this mold.

WALL BRACKET MOLD

A common mold design problem is the thin wall, deep draw part similar to the wall bracket shown in Fig. 6.30. Figures 6.31A through Fig. 6.31J show a design for a semiautomatic, semipositive mold having a nominal depth of 1.125 for the loading well. However, the parting line is not a uniform diameter or size because of the ⅞ in. high tab for hanging (on the back side). The land width allowance is .188 in. and the entry of the positive section is .062 in. In molds of this type, the cavity well is not actually required as a loading space because the compounds used ordinarily have a low bulk factor, and because the amount of material used is small in comparison to the total volume of the cavity space. Thus, the loading well

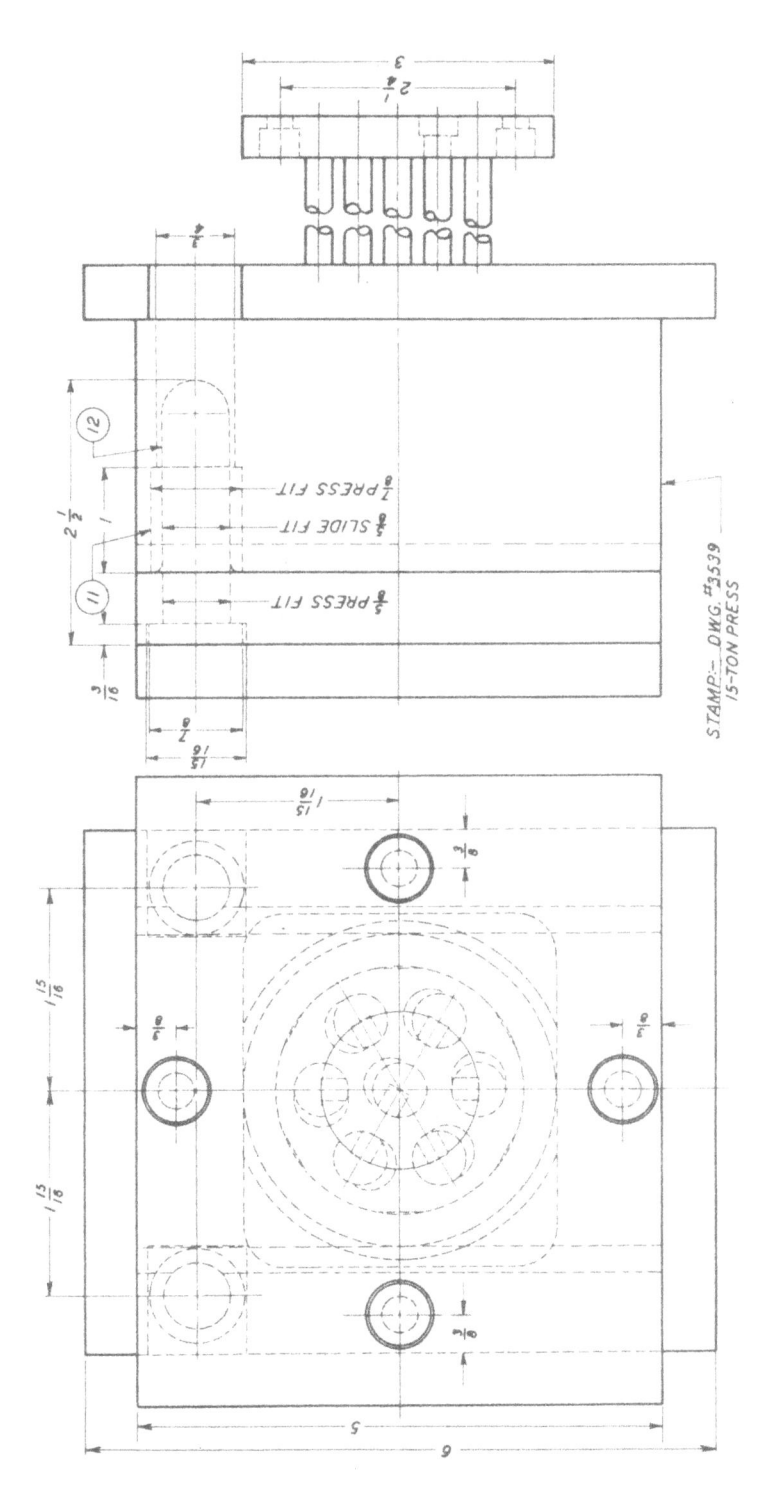

FIG. 6.29A. Plan view (left) and end view (right) of 7-cavity gang mold for part in Fig. 6.28. (See also Fig. 6.29B.)

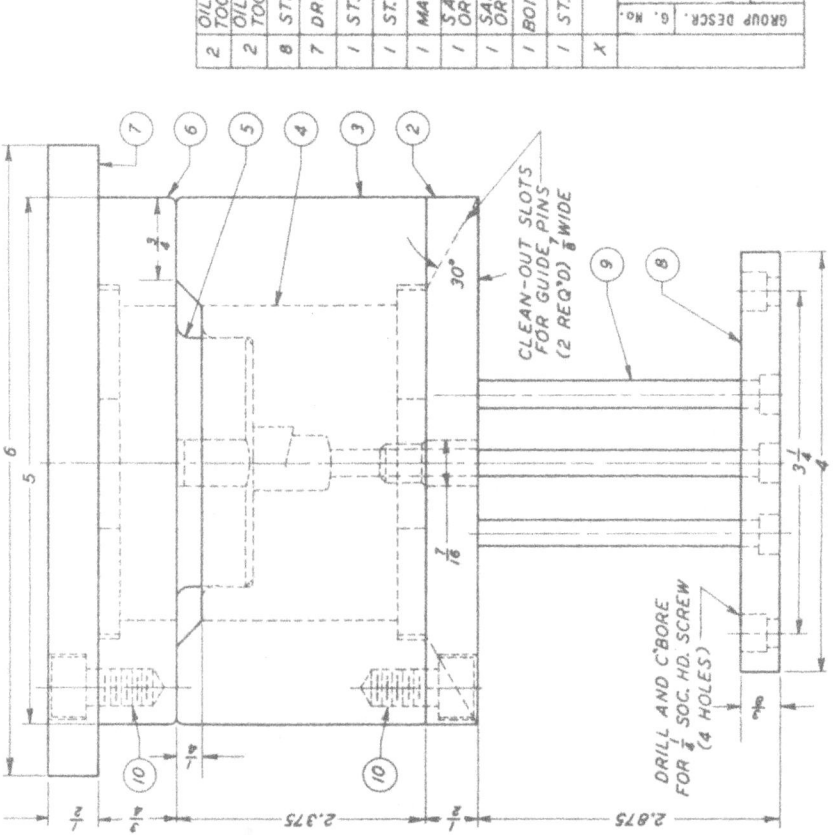

2	OIL HDN. TOOL ST.		12	GUIDE PIN	HARDEN	$\frac{7}{8}$	$2\frac{5}{8}$	
2	OIL HDN. TOOL ST.		11	BUSHING	HARDEN	1	$1\frac{1}{8}$	
8	STEEL		10	SCREW (SOC.HD)	$\frac{5}{16}$ -18	$\frac{3}{4}$		
7	DR. ROD		9	K.O. PIN	RWELL C-52 HARDEN	$\frac{3}{8}$	$4\frac{3}{8}$	
1	ST.C.R.		8	K.O. PIN PLATE		4	3	$\frac{3}{8}$
1	ST. C.R.		7	TOP PLATE		6	5	$\frac{1}{2}$
1	MACH. ST.		6	PLUNGER PLATE		5	5	$\frac{3}{2}$
1	SAE 3312 OR EQUAL	RWELL C-58 & HARDEN	5	PLUNGER		$3\frac{1}{4}$	2	
1	SAE 3312 OR EQUAL	RWELL C-58 & HARDEN	4	CAVITY		$3\frac{1}{2}$	$2\frac{1}{4}$	
1	BOILER PL		3	CAVITY BLOCK		5	5	$2\frac{1}{2}$
1	ST. C.R.		2	BOTTOM PLATE		6	5	$\frac{1}{2}$
X			1	ASSEMBLY		D.	L.	W.	T.

7-CAVITY SUBCAVITY GANG MOLD

ASSEMBLY AND DETAILS

FIRST MADE FOR   PRESSING BUTTON   #3539

BEGUN BY

FINISHED BY

TRACED BY

INSPECTED

GROUP DESCR.   G. NO.

CLEAN-OUT SLOTS FOR GUIDE PINS (2 REQ'D) $\frac{7}{8}$ WIDE

DRILL AND C'BORE FOR $\frac{1}{4}$ SOC. HD. SCREW (4 HOLES)

FIG. 6.29B. Front view and material list for 7-cavity gang mold. (See also Fig. 6.29A.)

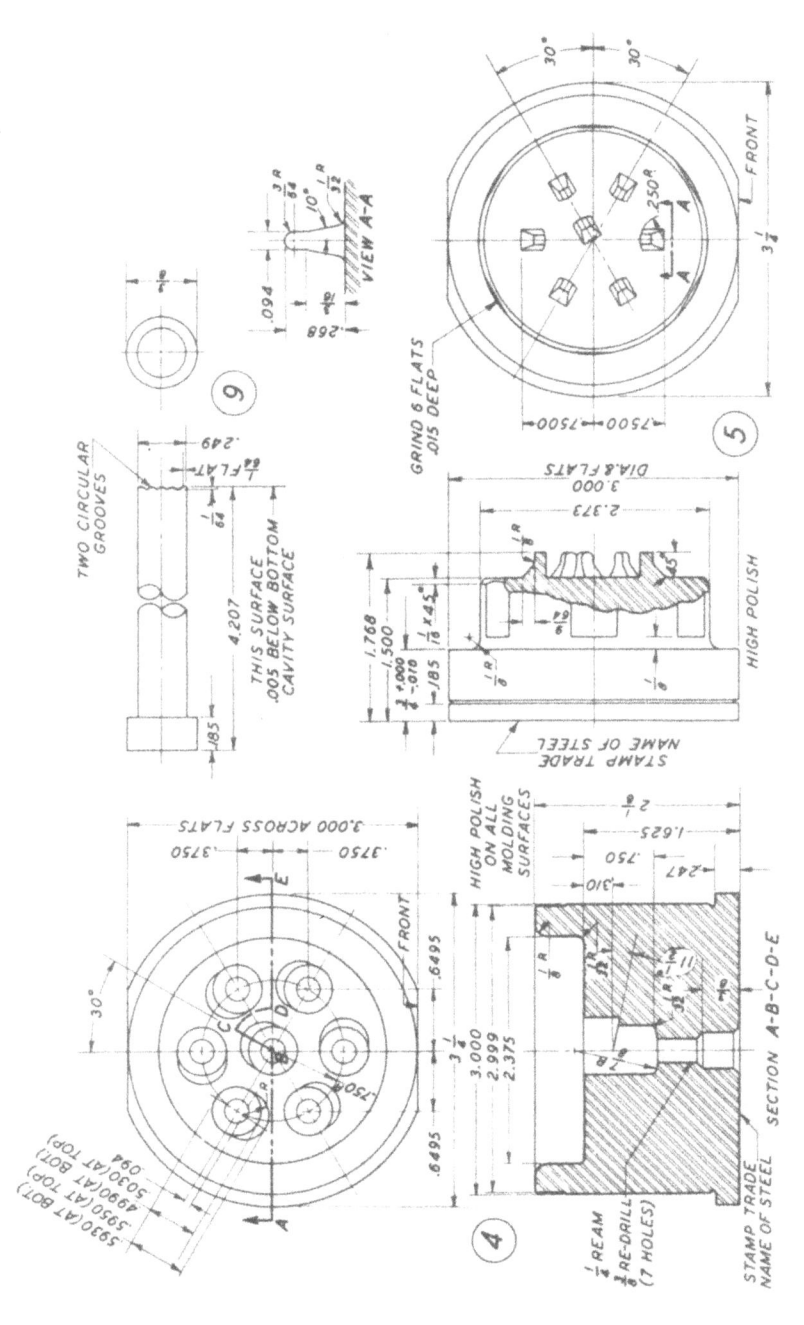

FIG. 6.29C. Cavity, plunger and knockoiut pin detail. Part numbers are identified in material list, Fig. 6.29B.

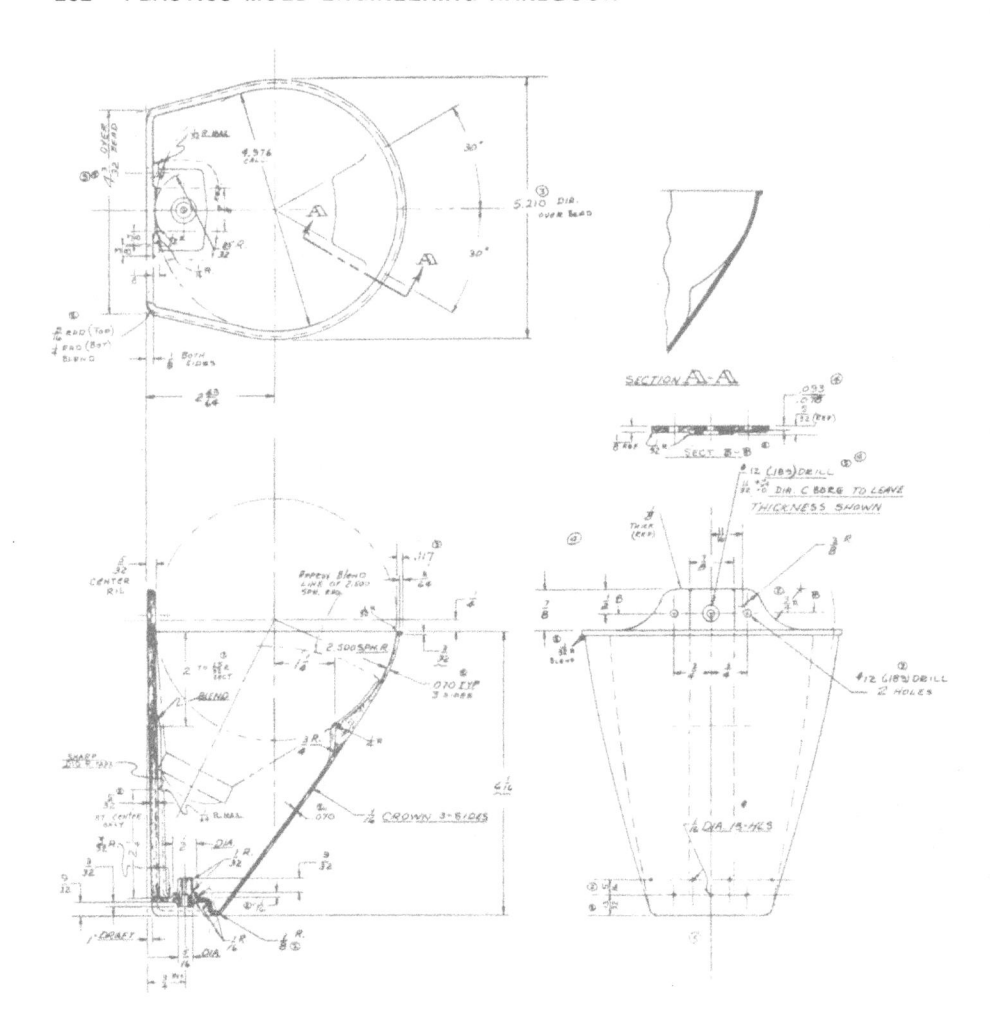

FIG. 6.30. Product drawing of a urea formaldehyde wall bracket.

really serves as a guide for the plunger or force, helping to insure uniform wall thickness by preventing sidewise displacement of the plunger. This displacement being caused by the resistance of the molding material to flowing as pressure is applied. The .062 in. positive entry is specified on this mold because urea molding material is used. Uniform back pressure is essential to the successful molding of urea or melamine materials. Obviously, uniform back pressure cannot hurt anything if used on phenolics or other thermosets. So, it is best to prepare for the worst condition likely to be encountered. This preparation is just in case the user changes his

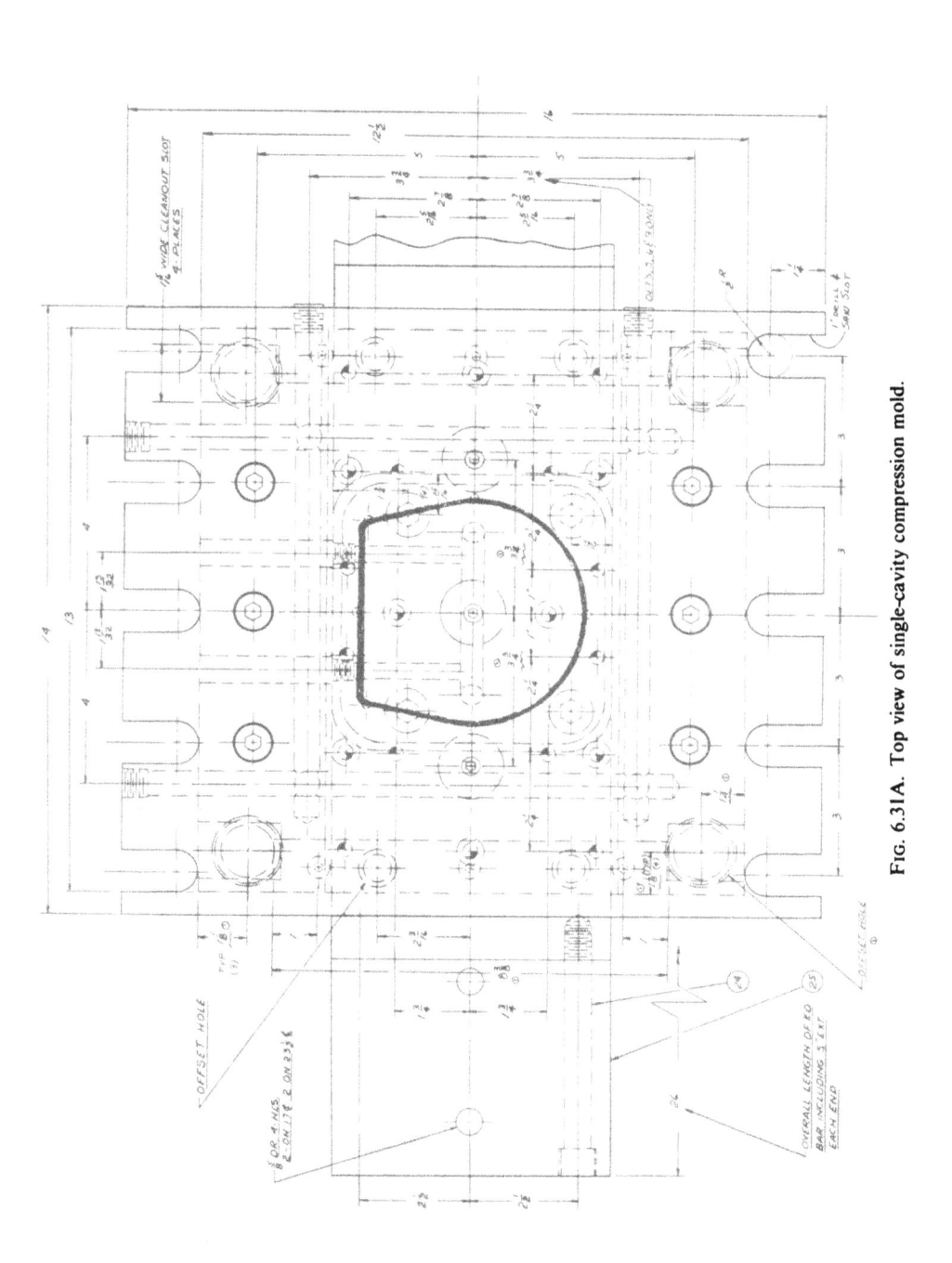

FIG. 6.31A. Top view of single-cavity compression mold.

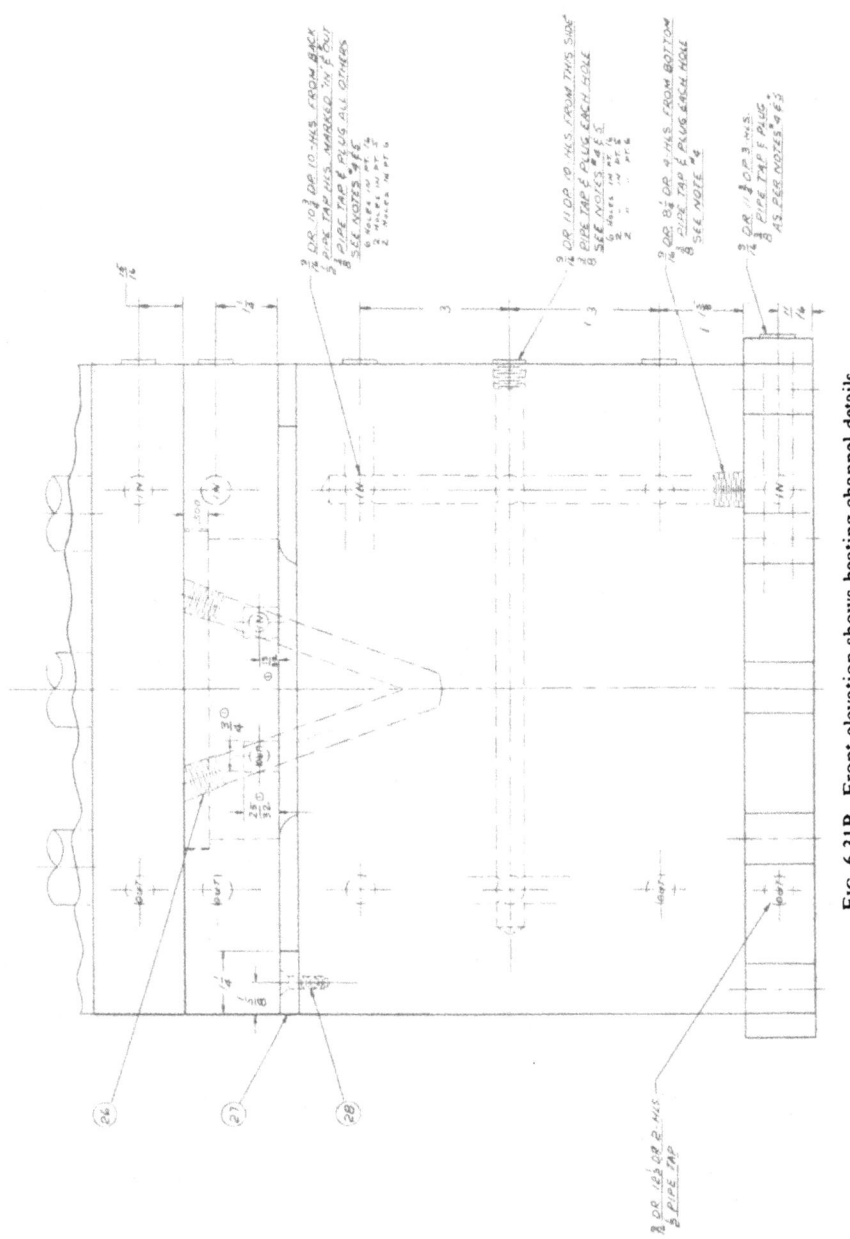

FIG. 6.31B. Front elevation shows heating channel details.

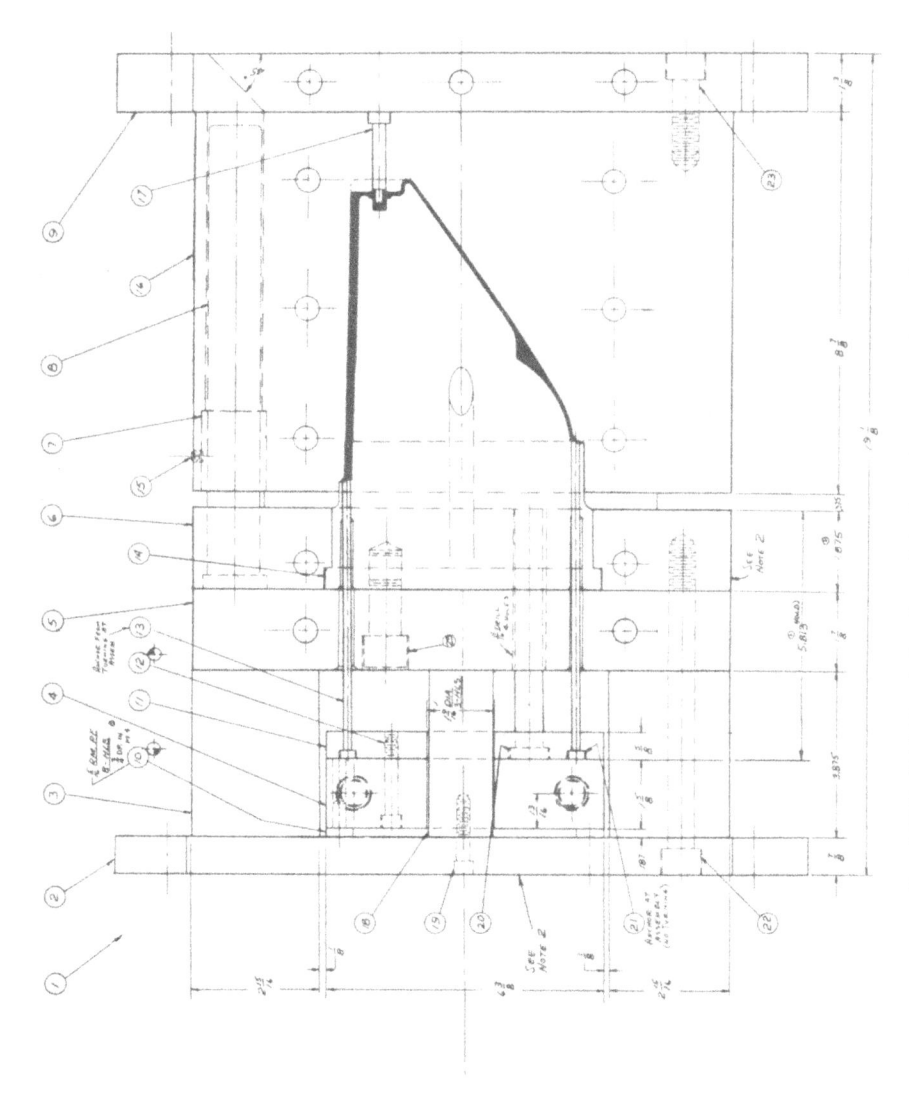

FIG. 6.31C. Right hand elevation of single-cavity compression mold.

NOTES-

1- MOLD TO OPERATE IN P.P.P. PRESS #4

2- .008 IN/IN. SHRINKAGE ALLOWED

3- STAMP ON TOP & FRONT OF MOLD
IN LARGE CHARACTERS:
    MOLD #517 CAVITY #1
    HOUSING #D-1717-11

4- STEAM LINES MUST BE THOROUGHLY
CLEANED BEFORE PIPE PLUGS ARE
ASSEMBLED

5- PIPE PLUGS MUST PROJECT $\frac{1}{16}$" FOR
FUTURE TIGHTENING

6- HIGH POLISH ALL MOLDING & FLASH SURFACES

7- MOLDING PC. MUST STAY ON FORCE
HIGH DRAW POLISH CAVITY PT. #19 & CORE
PIN PT #17 FOR GOOD RELEASE.
MEDIUM DRAW POLISH FORCE PT. #14
TO HOLD 'PC. ON FORCE

8- K.O. PINS FIT .001 LOOSE AT MOLDING SURFACE
.005 LOOSE IN PIN PLATE & .002 LOOSE ON
DEPTH OF HEAD.

FIG. 6.31D. Explanatory notes.

mind (after the mold is made) concerning the material to be used. The positive entry also "turns the flash up" in a vertical direction, and maintains a uniform flash thickness which can be removed easily and inexpensively by one pass with a file on the molded part.

We are not going to take you through all the calculations and decision-making involved in the design of this particular mold. If you have learned your lessons up to this point, you should have no difficulty in developing a similar mold design if you pay attention to the following special points in connection with Fig. 6.31A to Fig. 6.31J.

1. The heating channels (which could be steam, hot oil, or hot water heat) are used generously and their spacing is reasonably uniform. Always try for uniform heating, particularly when urea or melamine is to be molded. Reference to the right elevation (Fig. 6.31C) will show that the heating channels completely surround the part.
2. The plunger (or force) is channeled for directed media flow. The support plate and the retainer plate are also channeled so that the heat introduced to the force will stay there and not be conducted into the surrounding steel plates.
3. Four guide pins 1-¼ in. in diameter are used. These are a little larger than normal for a mold of this size (normally 1 in. would suffice).

①	4	STEEL		3/4-10	29	SOC HD CAP SCREW		2		
	6	STL.		1/4-20	28	FLAT HD. CAP SCR		1/2		
①	2	FLAT GR. TOOL ST.			27	STOP PAD		8 1/2	1/4	.375
	23	BRASS ONLY		3/8 STD.	26	PIPE PLUG				
	2	M. S.			25	K.O. EXTENSION		5 1/8	6 1/2	1 3/4
	4	STL.		5/8-11	24	SOC. HD. CAP SCR.		5		
①	6	STL.		5/8-11	23	SOC. HD. CAP SCR.		1 1/2		
	6	STL.		5/8-11	22	SOC. HD. CAP SCR.		7		
①	5	D.M.E. CAT EX. 13	M=10	D-1723-11	21	EJECTOR PIN	3/16			
①	4	D.M.E. CAT. EX. 37	M=6	CUT TO 5.812	20	RETURN PIN	5/8			
①	3	STL.		5/16-18	19	SOC HD. CAP SCR.		1		
	3	C.R.S.			18	SUPPORT PILLAR	1 1/2	4		
	1	DR. ROD		E-5272-22	17	CORE PIN	1/2	2 1/4		
	1	CRUCIBLE CSM 2		E-5272-22	16	CAVITY		13 5/8	12 5/8	9
	4	STL.		1/4-20	15	FLAT POINT SOC. SET SCR		1/4		
	1	AIR KOOL 8R EQUIV.		D-1723-11	14	FORCE		9 3/4	6 1/2	6 1/2
	1	D.M.E. CAT. #EX. 13	M=10	CUT TO 6.505	13	EJECTOR PIN	3/16			
	12	STL.		5/16-18	12	SOC. HD. CAP SCR.		1 3/4		
①	1	M. S.			11	PIN RETAINER		13 1/8	6 1/2	3/4
	8	D.M.E. CAT. #7100			10	STOP PIN				
	1	M. S.			9	BOTTOM PLATE		16 1/8	14 1/8	1 1/2
	4	D.M.E. CAT. #5316			8	GUIDE PIN				
	4	D.M.E. CAT. #5505			7	BUSHING				
	1	M. S.			6	FORCE RETAINER		14 1/8	12 5/8	2
	1	M. S.			5	BACKUP PLATE		14 1/8	12 5/8	2
	1	M. S.			4	K.O. BAR		16 1/8	6 1/2	1 3/4
	2	M. S.			3	PARALLEL		13 1/8	4	3
	1	M. S.			2	TOP PLATE		16 1/8	14 1/8	1
	X				1	ASSEMBLY				
	QTY	MAT'L.		DETAIL	PT.	DESCRIPTION	D	L	W	T

MATERIAL LIST

STK. SIZES ALLOW FOR FINISH UNLESS OTHERWISE NOTED

TITLE— (1) CAVITY SEMI-AUTOMATIC COMPRESSION MOLD

FIRST MADE FOR— WALL BRACKET D-1717-11

R. STEVENS	R. STEVENS	WZP	DWG.
5-3-60	5-7-60	June 1-60	NO.

FIG. 6.31E. Materials list for single-cavity compression mold. Part numbers 1 through 29 refer to Figures 6.31A through 6.31J.

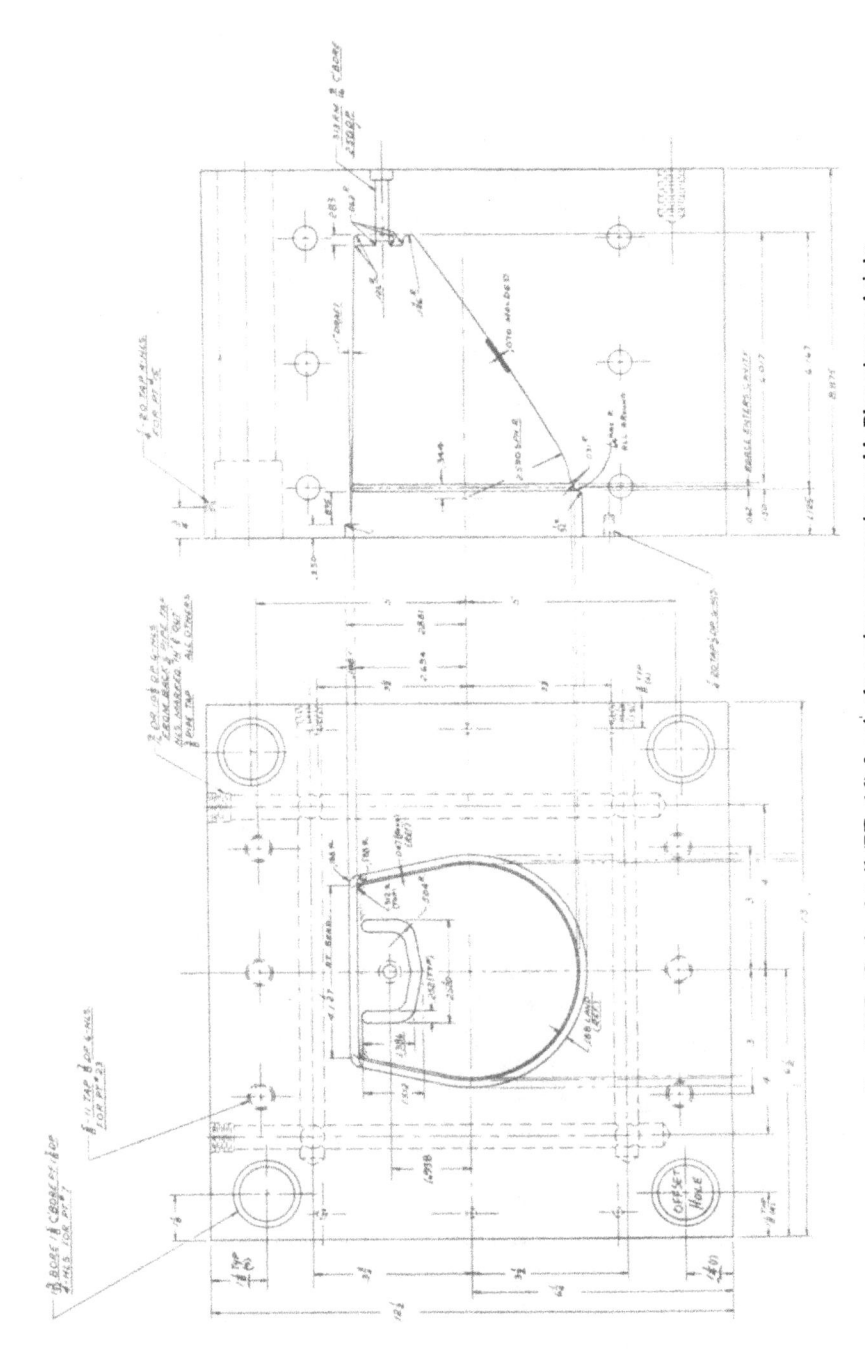

FIG. 6.31F. Cavity detail (PT. 16) for single cavity compression mold. Plan view and right hand evelation.

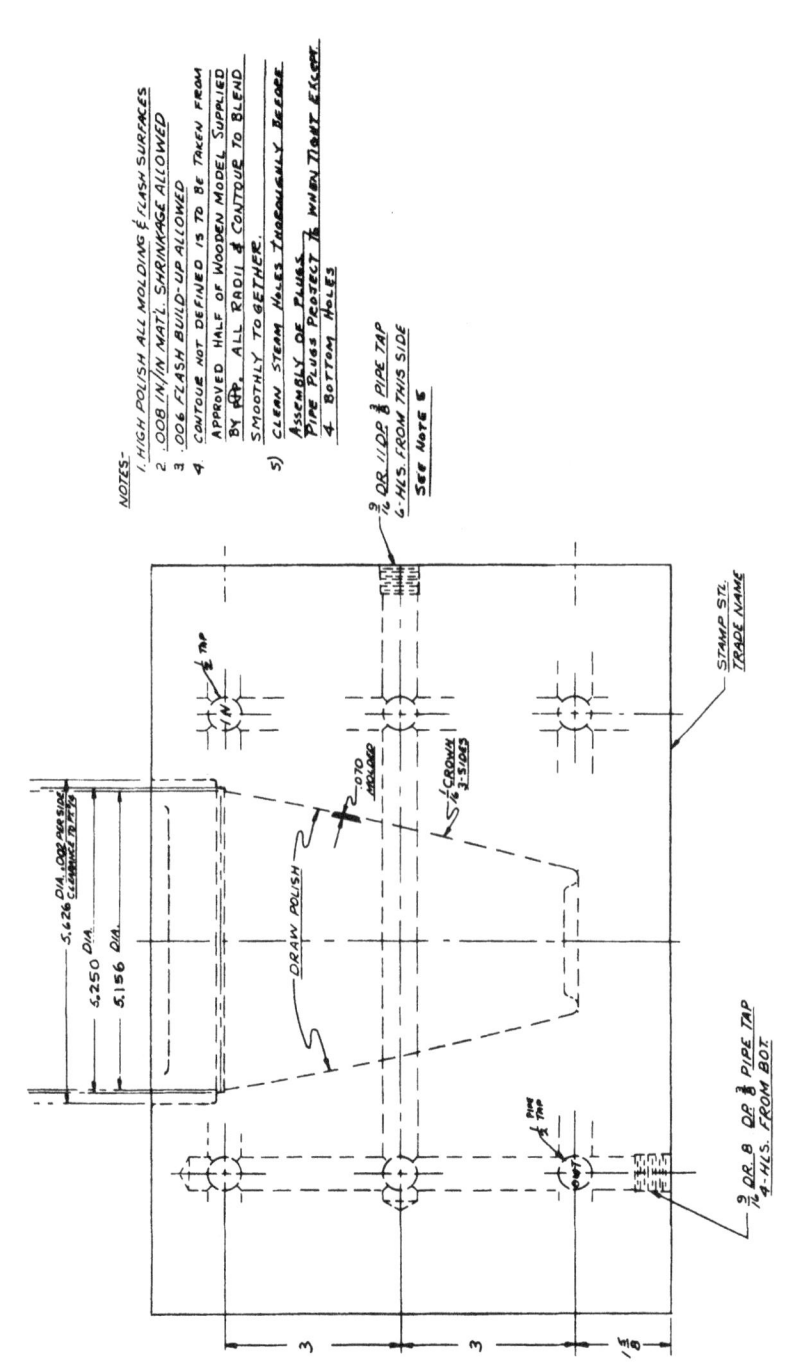

FIG. 6.31G. Front elevation of cavity (refer Fig. 631F).

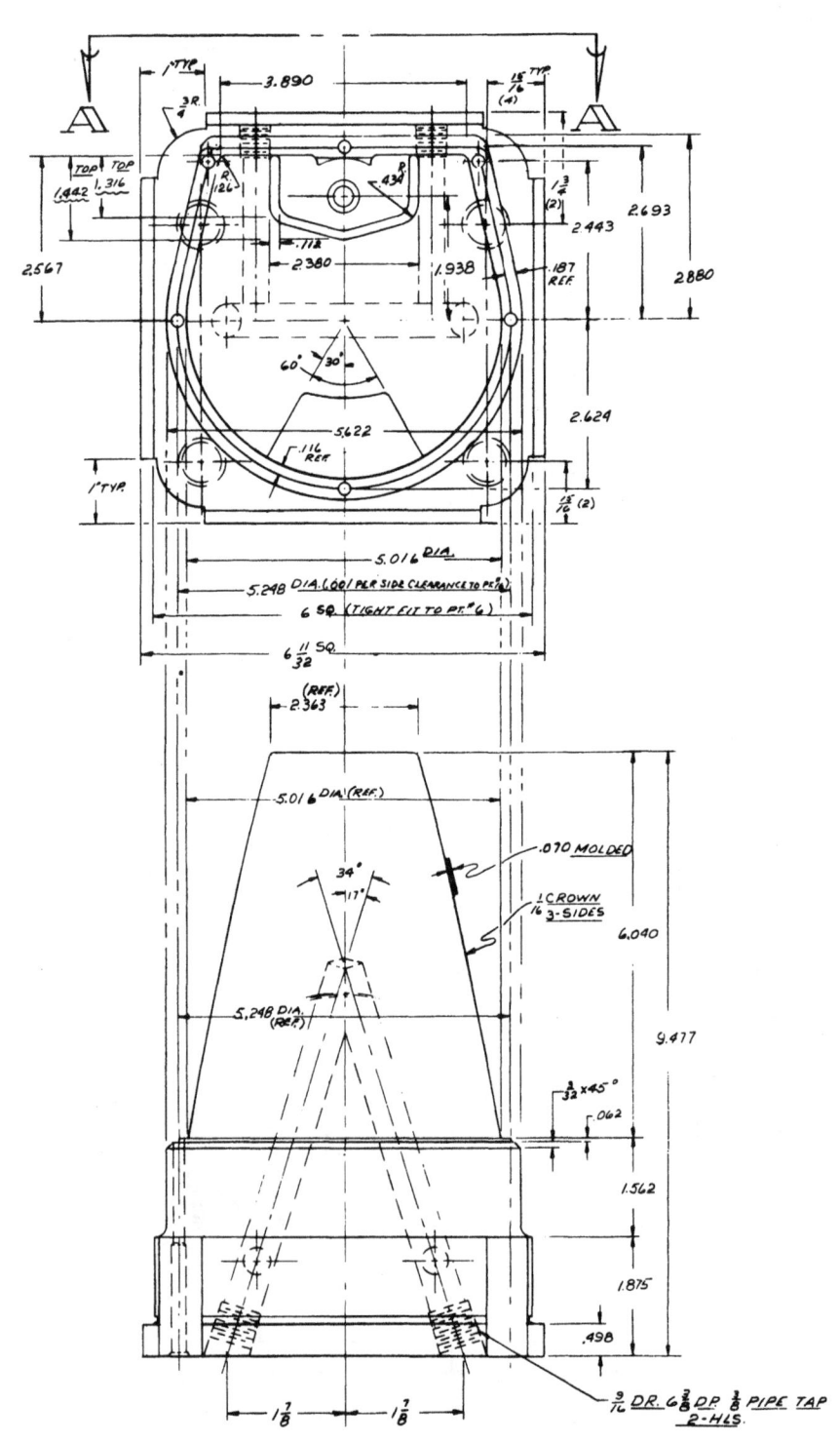

FIG. 6.31H. Two views of force (plunger) part no. 14. Front elevation and plan view.

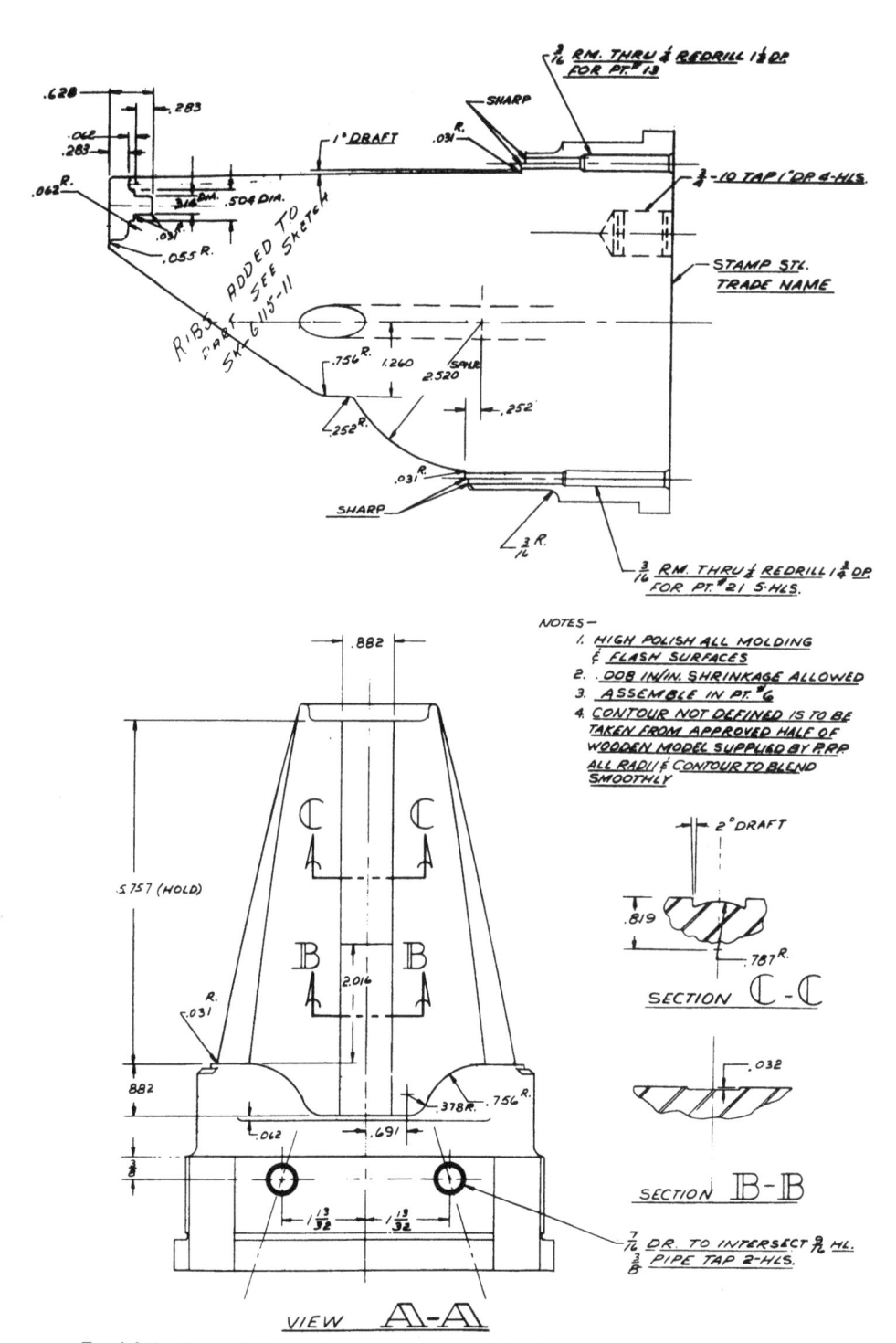

FIG 6.31I. Two additional views of force (part 14). Upper view is cross section on center line (front to back Fig. 6.31H).

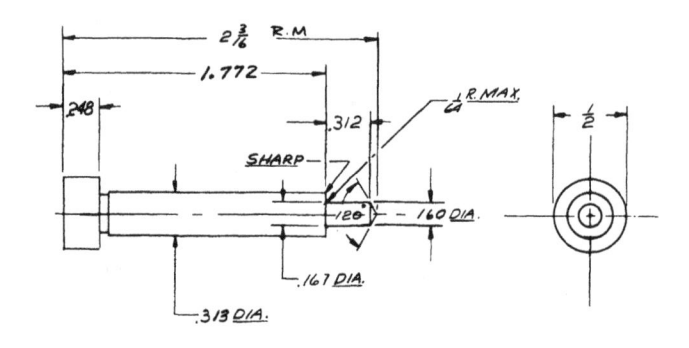

1. HIGH POLISH ALL MOLDING
   & FLASH SURFACES
2. .008 IN/IN. MAT'L. SHRINKAGE
   ALLOWED
3. ASSEMBLE IN PT. #16

FIG. 6.31J. Core pin detail. Assemble in Fig. 6.31F.

However, note that the part contour is not symmetrical and that side thrust will be generated. Oversize guide pins will help overcome this problem in molding. Also note, guide pins project beyond the force for two reasons: (1) to provide guidance for the force before it contacts molding material in the molding operation; (2) to prevent some careless person from lowering the force half of the mold onto rough surfaces that would damage the polished force. The rule here, is to always consider the handling of the separate parts of the mold. One pin is offset so handling personnel cannot even start to put the two halves together in reverse.

4. Four ⅝ in. return pins (safety pins) are used.
5. A greater number of large size screws are used in the frame. Always proportion screw diameter to the size of the mold.
6. Note support pillars to prevent plate distortion under pressure.
7. When this mold was originally built in 1960, it was considered in the "medium to large" class relative to size as compared with the "average." By today's standards this is a small mold. Later in this same chapter, we give data on large mold design and show examples of parts.

MOLD ASSEMBLY

A series of illustrations has been prepared to show the various steps in assembly of a mold. Study of the illustrations, Fig. 6.32 (A–E), will be

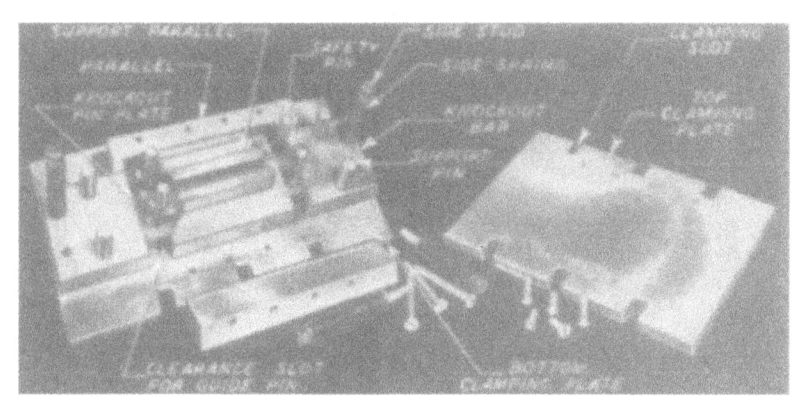

FIG. 6.32A. Subassembly of Fig. 6.32E, showing ejector bar, parallels and other parts of the ejector arrangement.

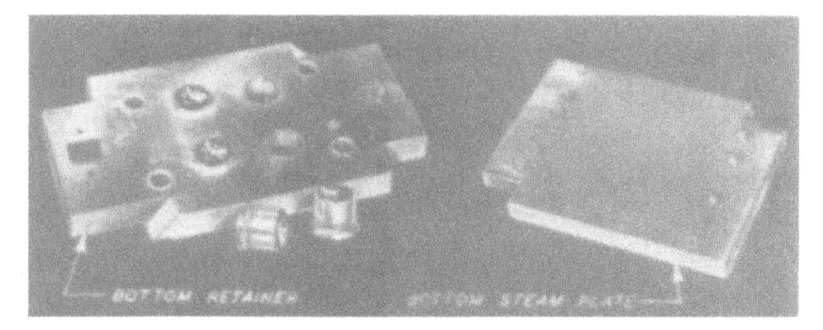

FIG. 6.32B. Bottom retainer with two cavity sections in place.

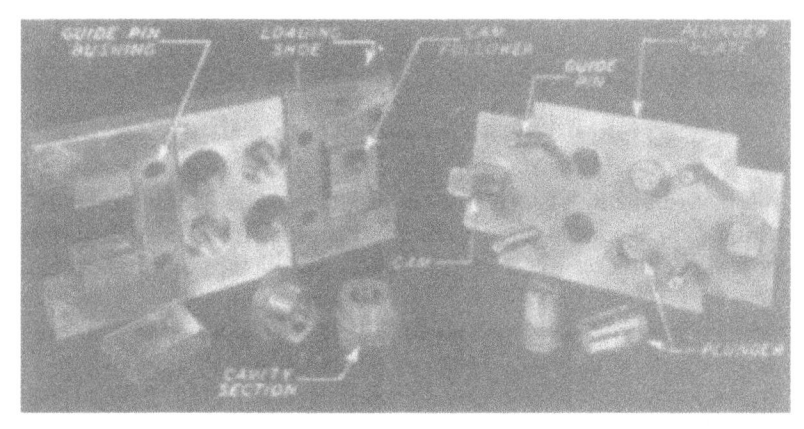

FIG. 6.32C. Loading shoe with two cavity sections in place, and plunger plate showing cams.

FIG. 6.32D. Top and bottom sections of mold ready to be assembled as shown in Fig. 6.32E.

FIG. 6.32E. Completed assembly of 4-cavity mold.

helpful in visualizing the various mold parts. The mold illustrated was constructed for the molding of vacuum tube bases from a low-loss phenolic compound. Cams are used to pull the side pins before the piece is ejected.

AUTOMATIC COMPRESSION MOLDS

Automatic molds, an example of which is shown in Fig. 6.33, are similar to semiautomatic molds. They have features which contribute towards maintaining a completely automatic compression molding operation.

FIG. 6.33. Three-cavity automatic mold. Comb mechanism removed from press.

A shuttle type feed board shown in Fig. 6.34 volumetrically measures the granular or nodular material from a hopper into the cavities of the open mold. Most compounds can be heated with an automatic infrared hopper, which will decrease between 15 and 25%, the cure time of the part. Bump breathe or time breathe can be used if it is necessary. The mold is closed and held there for the full cure time, which is set by an adjustable timer.

When the cure is completed the mold will open, and, if it is arranged for bottom ejection, top hold-down pins will assure keeping the molding in the lower half of the mold. It is important in automatic molding that all parts remain in either the bottom or the top half of the mold, not both. All motions are sequence controlled, so after the mold is fully opened the ejector pins raise the parts from the cavities, a slotted plate or comb moves in under the parts, and the knockout pins are withdrawn leaving the parts resting on the comb (Fig. 6.35). The shuttle feed board pushes the comb back as it moves in over the mold to feed compound for the next molding until the comb is clear of the mold. The comb withdraws further under its own power wiping the parts off the slotted plate and discharging them down a chute into a tote box. (See Fig. 6.36 for a typical 75-ton automatic compression molding press.)

An air blast, just after ejection and just prior to loading, removes any pieces of flash or granular material which might remain in the cavity loading chamber or on the forces. Positive removal of flash is absolutely essential to trouble-free operation of automatic molds.

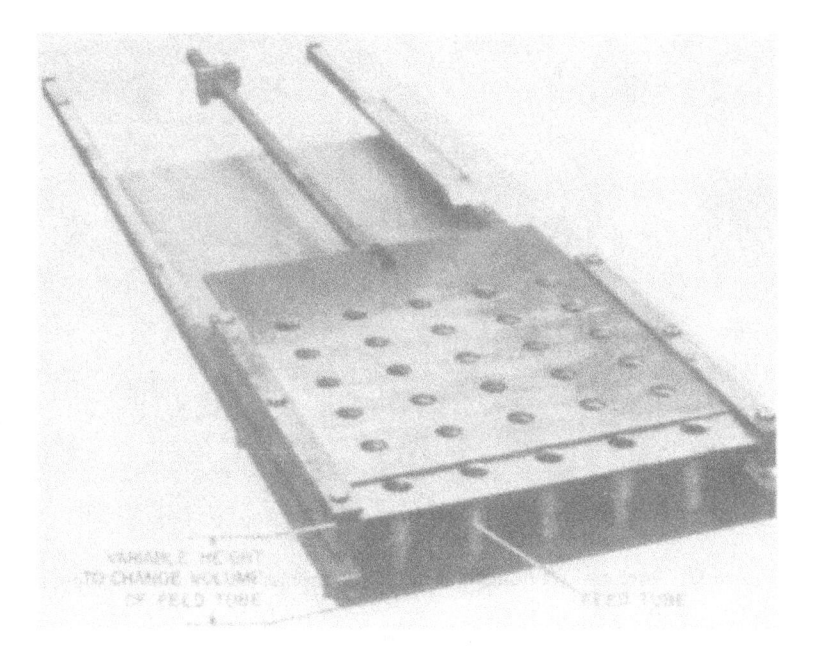

FIG. 6.34. Typical feed board from automatic press for volumetric loading of cavities.

FIG. 6.35. Cores withdrawn, molded parts ready to be removed from between mold halves.

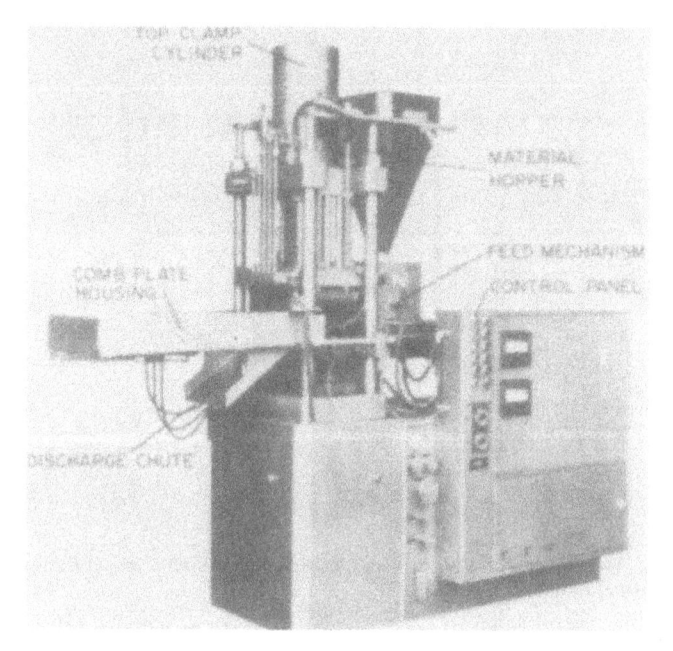

FIG. 6.36. A typical 75-ton compression molding press.
(*Courtesy Wabash Metal Products, Wabash IN*)

Positive type molds can be successfully used in an automatic operation, but should be limited to single cavity molds. It is difficult to maintain a uniform part density with positive multi-cavity molds, unless extremely accurate preforms are used for loading of materials. Flash type molds can be used when automatic molding thin wall closures or bottle caps. Thin wall closures or caps are usually large enough to accept all the granular material required to mold the part without the use of any additional loading chamber. The density obtained with flash type molds is satisfactory for this type of part. For all other applications, semi-positive or landed plunger with loading well is recommended.

Figure 6.37 shows a drawing section of a typical 2-cavity automatic compression mold using bottom ejection with top hold-down. The top hold-down pins should be positioned flush with the surface of the parts. Bottom knockout pins should be buried in the pieces approximately .010 to .020 in. when ejected from the cavity (with the customer's permission, of course). This will keep the parts located on the knockout pins until the comb is moved into position. Sufficient knockout pin stroke must be provided to raise the molded parts at least ¼ in. above the top of the loading chamber, so as to permit the comb to move in under the ejected parts. Ejector pins for the flash should also be considered. Locate ejector pins in line so a minimun number of slots must be cut in the comb plate.

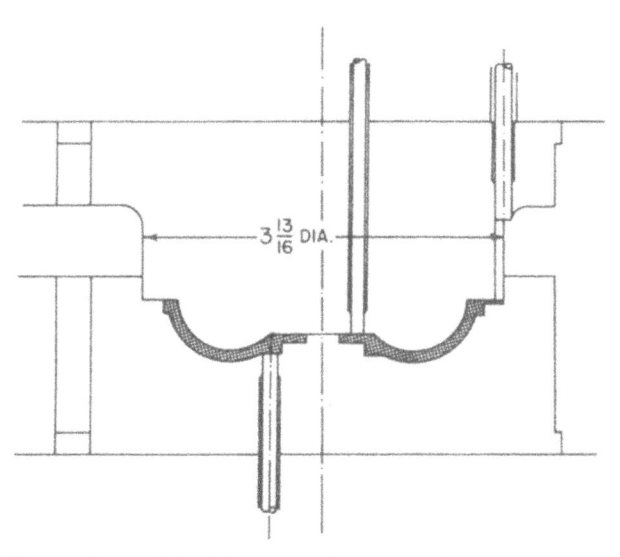

FIG. 6.37. Two-cavity mold for fully automatic compression press. (*Courtesy Stokes Div., Pennwalt Corp., Philadelphia, PA*)

Sharp corners (or no more than .010 radius) should be provided both at the top and bottom of the loading chamber. it is recommended that .002 in. clearance be maintained between the force and loading chamber walls. A ⅛-in. wide land area all around the part is usually sufficient, although it must be great enough to keep the pressure on the land below to 18,000 psi, should the mold be closed without material in the cavity.

Clearance of ¾ in. should be provided between the two mold halves when mold is closed to make sure there is no restriction to normal flash overflow.

Electric cartridge heaters are most commonly used, although steam may also be used. With the proper molding temperature and proper pressure on the press, the mold will close until there is a normal flash thickness of .002–.004 in. This is thin enough to be removed by tumbling, yet strong enough to be ejected from the mold with the part.

By all means, request the press manufacturer to supply a set of design prints demonstrating the design of molds as applied to his equipment. These drawings should supply mold heights, maximum length and width, minimum spacing between feed tubes, size and location of thermocouples, method of clamping to press, and ejector dimensions.

We take space here to describe and illustrate special machines available in some molding shops. When designing molds for these machines seek the help of the manufacturer. At least, get his approval of the design before building the mold, particularly, if you are unfamiliar with the machine.

If your shop has one of the older models, i.e., without the latest improvements, we suggest you encourage management to consider the advantages that can be achieved by updating them. The combination of automatic molding and latest equipment is considered to be the lowest cost molding process, where it is applicable to the product being made.

The updated press in Fig. 6.38 is a 10-station rotary compression press. It is now marketed complete with an extrusion plasticizer that eliminates the need for preforms because it handles powders and high bulk materials on a basis similar to an injection machine adapted to thermosetting materials. The best features of screw plasticizing and compression molding are combined in one machine.

Another machinery manufacture has developed a multiple extrusion plasticizer for use with his fully automatic press. The working end of this unit features three extruders in a side-by-side arrangement as shown in

FIG. 6.38. A 10-station automatic compression molding machine is shown with a preplasticizing extruder. The new extruder readily handles powder, nodular, and high bulk materials without preforming. (*Courtesy Alliance Mold Co., Inc., Rochester, NY*)

Fig. 6.39A. Preforms on nozzles prior to being cut off. (*Courtesy Stokes Div., Pennwalt Corp., Philadelphia, PA*)

Fig. 6.39B. Preforms loaded into mold. (*Courtesy Stokes Div., Pennwalt Corp., Philadelphia, PA*)

Fig. 6.39. The screw extruders have operated to make a preformed charge in a preheated condition (temperatures around 250° F) and ready to be cut-off and drop into the feeder mechanism. This heated and plasticized charge is then dropped into the mold cavity as shown in Fig. 6.39B. Obviously, mold design is limited to multiples of three cavities. Any system such as that illustrated will offer a reduction in curing times as compared to the previous cold powder molding techniques. Because of introducing material already in the plasticized state, mold wear is reduced and cycle reduction as great as 75% can be realized.

SPECIAL DESIGN FEATURES

Center Guide Pins

It is sometimes desirable to eliminate the guide pins from the retainers by placing them inside the mold cavity. In such cases, the guide pins act much like entering pins except that they perform no function in forming the molded part. This system is limited to product designs having a relatively large hole molded through, as do housings. The single-cavity mold pictured in Fig. 6.40 makes good use of this system by locating guide pins in the center of the mold cavity. Such guide pins serve two purposes: (1) The amount of flash in the molded piece opening is reduced by their use; (2) Greater uniformity is obtained in the mold wall section.

FIG. 6.40. Two large guide pins are located in the center of this radio cabinet mold to insure maintenance of the relation between mold cavity and mold plungers. (*Courtesy Tech Art Plastics Co., Morristown, NJ*)

Fig. 6.41. A cam is used to operate the side-pull pin in this compression mold. (Left) Side-pull pin in place for molding. (Right) Pin retracted for ejection of part.

Cam Side Pulls

It is frequently desirable to use an automatic device to operate the pins which form side holes, as these must be withdrawn before the part can be ejected. An arrangement for obtaining this action is shown in Fig. 6.41. The mold is shown partly closed in the left-hand view, ready for the loading of the compound. The side pin is in place and the ejector bar is down. At the right, the mold is shown fully opened. The cam-operated block attached to the pull pin has been moved to the right far enough to allow the ejector pins to eject the part, and these pins are shown just above the top of the cavity. The cam is about $1\frac{1}{2}$ in. square, which size provides adequate strength to resist the bending action. (Another cam of this type is shown after Fig. 6.32C and Fig. 6.42.)

LARGE MOLDS[1]

An increase in the use of sheet molding compounds (SMC) and bulk molding compounds (BMC), composed of resin and fiberglass mat, has resulted in molds ranging 8 to 10 ft square and 4 to 8 ft in overall height. It is evident that the mold size will continue to increase. It is now theoretically possible

[1]Figures 6.43A through 6.43E and the data on large molds were prepared with the help of Messrs. Sorenson and Nachtrab of Modern Plastics Tool Division of LOF Engineered Products, Inc., Toledo, OH.

Fig. 6.42. Six-cavity automatic compression mold with heating channels between the cavities. This necessitates mounting each cavity separately in a well in the mold retainer. Noteworthy also are the cams that serve to pull the side cores and permit fully automatic mold operation. (*Courtesy Stokes Trenton, Inc., NJ*)

to make a molded part as large as a house if someone will make the mold and the press in which to operate it. These large molds, now being made, operate in presses developing tonnages in the range of 1,000 to 4,000 tons. The press platens are so large that technicians must climb into the press to work on molds the size of those shown in Figs. 6.43A and 6.43B. Fig. 6.43A shows the ejector half of the mold, still on the duplicator mill and being finished to final shape, after which it will be filed and polished like the mold cavity shown in Fig. 6.43B. The large object behind the workman in Fig. 6.43B is the plaster cast used on the duplicator mill, from which the ejector half was duplicated. The molded part which will be made from this mold is a hood and fender panel for a truck.

A continuing study of Figs. 6.43C, 6.43D and 6.43E will reveal some of the pertinent design data points which you will then readily observe in Figs. 6.43A and 6.43B. Fig. 6.43C shows a cross section of a typical compression mold for sheet molding compound. Please realize that this schematic drawing illustrates a mold that may be 4 to 8 ft wide and 6 to 10 ft in length. The overall height may be 4 to 8 ft also.

Special design considerations for these large molds are as follows:

1. Baffle steam or oil channels to direct the heat as uniformly as possible around the molded part and to maintain uniform expansion of the mold. A good thermal grid pattern is essential to uniform heat necessary to make good moldings.

2. Use thermocouple locations in both halves of the mold to permit continuous monitoring of the mold temperature. Several points may be necessary to tell the whole story of temperature.

3. The wear plates mounted on the heel area are of dissimilar metals. This particular mold uses steel for one wear plate and aluminium bronze for the other. The heel areas are to offset the side thrust exerted during mold closing on the irregular contour of the part. It should be apparent when the mold is closed on SMC or BMC considerable force must be exerted to make the material flow and move into ribs and pockets and over shear edges. This force tends to split the mold cavity or the force, or to move them sidewise with respect to each other. The resolution of forces is nowhere more evident than in a mold which has been split because of a weakness in design. Either splitting or side movement is highly undesirable. Either is prevented by the heel area engagement in the last 2 in. of press travel.

The wear plate size is usually 1 by 2 in. in cross section and as long as space permits. The heels are installed within 2 in. of the cavity and the two wear plates equal 2 in. in thickness. An additional 3 to 4 in. of metal should back up the wear plates and provide an area to mount the stop pads. These allowances total 6 to 8 in. beyond the product size. In general terms, we can say that a molded part 24 by 60 in. would require a minimum mold dimension of 40 by 72 in.

4. Removable stop pads are located around the periphery of the mold and between the force and cavity blocks. Stop pads are located in whatever flat area may be available, but they should be far enough away from the molding material space to prevent the molding flash (material overflow) from flowing under the stop pad. Such misfortune would cause the mold to stand open (fail to close) and a reject part would most likely result. The total area of stop pads needed is calculated from the formula:

$$\frac{\text{Press pressure in tons}}{10} = \text{stop pad area (sq in.)}$$

This allows 10 tons/sq in. pressure on the stop pads. They should be heat treated (Rockwell C-30 to C-55) and SAE-6150 steel is recommended.

5. Support pillars between the back of the ejector male and the rear clamp plate must be used wherever possible to prevent distortion of the ejector half under molding pressure. The recommended total area of all support pillars is based on 1 sq in. of support pillar for every 28 sq in. of molded part

area. As in all other molds, compression or otherwise, placement of the support pillar is according to the designer's experience and the limitations of the ejector layout used for the particular mold. Use cylindrical pillars or rectangular parallels placed in whatever uniform spacing can be achieved in the available area after ejector locations are selected and known.

6. Guided ejector systems are essential. Four guide pins are located somewhere near each of the four corners of the ejector plate. These pins may be 5 or 6 ft away from each other. The usual diameter is 2 to 3 in., and the anchoring head is in the rear clamp plate. Use bronze bushings in the ejector plate and use plated guide pins that fit .005 in. loose. Use bronze bushings that are .005 in. larger than the guide pin and anchor the bushing in the ejector plate. Grease fittings are included so that guide pins and bushings can be kept lubricated during the molding operations. You will naturally use hardened guide pins to go with the bronze bushings.

7. Heating channels must be located in the ejector plate and the bottom (rear) clamp plate to assure even expansion of those parts with the ejector

FIG. 6.43A. Ejector half of mold for truck hood and fender panel. It is on the duplicator mill being roughed to shape. This mold will be used with sheet molding compound. (*Courtesy Bethlehem Steel Corp., Bethlehem, PA* and *Modern Tools Div., Toledo, OH*)

FIG. 6.43B. Final polishing on cavity side of mold. The object behind the toolmaker is a duplicator model. (*Courtesy Bethlehem Steel Corp., Bethlehem, PA* and *Modern Tools Div., Toledo, OH*)

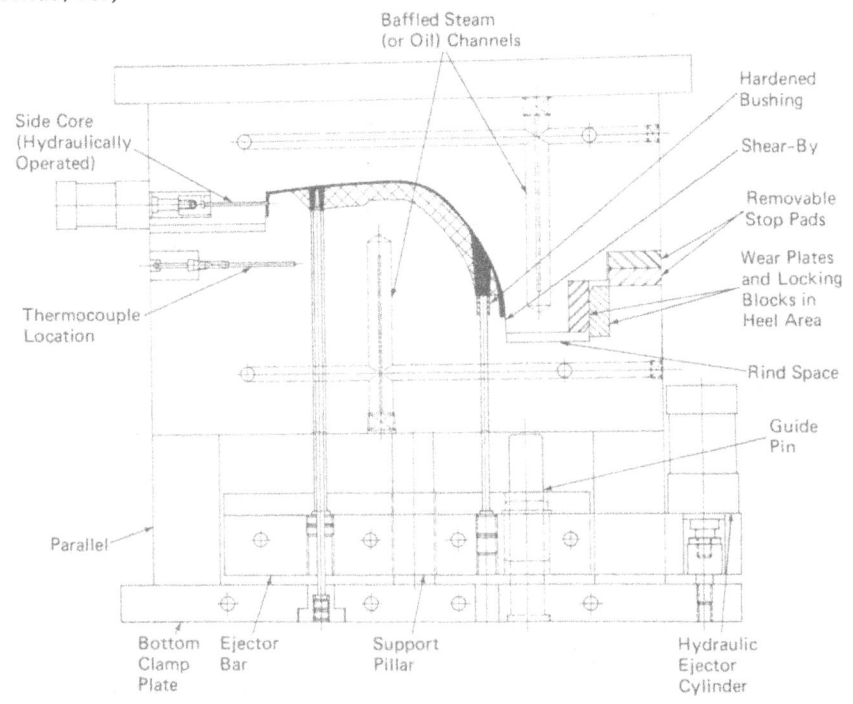

FIG. 6.43C. Schematic cross section of a large mold for sheet molding compound products.

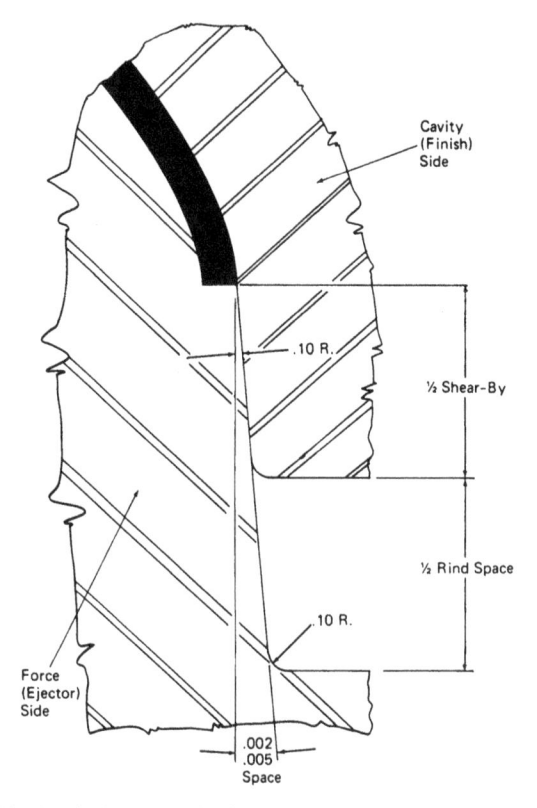

Cavity
(Finish)
Side

.10 R.

½ Shear-By

½ Rind Space

.10 R.

Force
(Ejector)
Side

.002
.005
Space

Fig. 6.43D. This sketch shows the details of the shear-by and rind space of Fig. 6.43C.

male (force). Differential expansion would bind the ejector pins and cause the ejector assembly to fail in operation.

8. Figure 6.43C shows a hydraulic cylinder in position to operate the ejector assembly independently of the motion of the press platen. (position shown is schematic) In practice, either two or four cylinders are used and these would be located at opposite ends of the ejector plate as it extends beyond the basic mold blocks. For molds of the size made up to this time, four cylinders, operating at the same pressure as the press circuits, and of $3\frac{1}{4}$ in. bore size have been found adequate. The minimum ejector stroke is that required to clear the draft and the bosses from the male half. The maximum stroke needed is determined by the method used for removal of the molded part (hand, stripping forks, removal trays, etc.)

9. Side cores, when needed, are usually actuated by a hydraulic cylinder as shown schematically in Fig. 6.43C.

10. Figure 6.43C shows the ejector pin hole fitted with a hardened bushing (similar to a drill bushing) to provide a wearing surface harder than the mold steel surrounding the bushing. A press-fit is used to hold the bushing in place. The ejector pin at the right side and the *sleeve ejector* around the

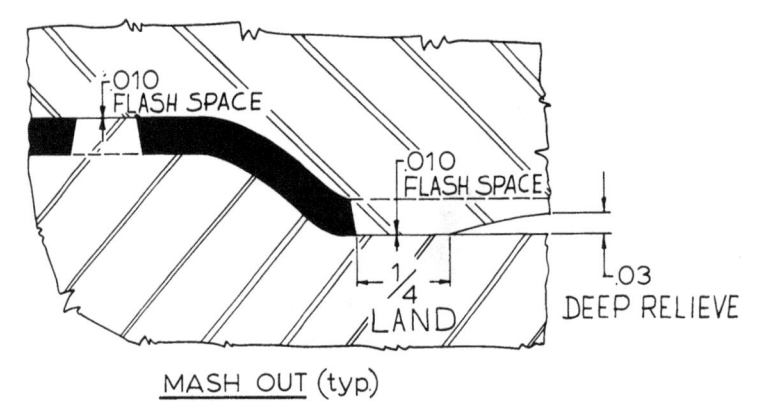

FIG. 6.43E. This illustrates the details of a mash-out area where holes are to appear in the molded part. (Wear points and mash-out should be replaceable areas.)

core pin at the left side are mounted through the ejector retainer plate and held in place by double set screws from the rear of the mold. Such construction allows replacement of ejector pin, core pin, or sleeve ejector without the disassembly of the entire ejector half of the mold. It seems scarcely necessary to point out the expensive logistics of handling a 10 to 20 ton mold just to change a broken or worn pin, or make similar repairs. The key is advance planning by the mold designer, the mold builder, and the mold user. All ejector pin holes and sleeve ejector holes should be gun drilled to keep the hole square with the base to achieve long service life.

11. Figure 6.43D gives the details of the shear-by and rind space. Note that the contour of the molded part is determined by the right hand member of the mold (*cavity* side) coming over the outside of the curve. This member will determine the dimension to which the mold is made (to make part to tolerance). Thus the .002 to .005 in. clearance space will be obtained by removing metal from the section marked "force side." Wherever possible, extend the rind space to the outside of the mold.

12. Minimum thickness of metal behind the cavity contour, or at the thinnest point on the ejector half, is 3 in. In case of doubt, make it 5 in., minimum. Proper portioning of sections is one of the engineering judgments demanded of a mold designer.

Figure 6.43E shows the technique used for mash-out areas where holes are to appear in the molded part. When molding SMC or BMC, some allowance must be made for the fact that the glass fiber content of the material is no more compressive than the steel. Therefore a depth-for-depth match would create extremely high unit pressures (limited only by press capacity) on these areas and damage to the mold would be a direct result. The .010 in. flash space is a deliberate opening between the two halves of the mold (when closed empty). Molding material coming between the halves at this point will be completely devoid of resin. Thus a dry, thin flash is left on the molded part and this flash is pushed out easily with a hand tool. As in all other high production molds (long running and trouble free) use removable replaceable and easily removable inserts at all critical wear points such as fragile pins or mash-out areas.

The material commonly used in the force and cavity of long running molds where highest quality is desired is P-20.

For medium production molds and those not requiring a high surface finish use SAE-4140 in a prehardened state.

For short runs and where finish is not critical, use SAE-1045 in its annealed state.

The data given in Tables 6.2 and 6.3 provide additional facts concerning design standards of molds and their materials as applied to molds for SMC, BMC, Preform or mat molding.

For reader concept of large parts, we have included Fig. 6.44 showing one of the large presses now being used for these large moldings. Bear in mind that we are talking about compression molds and compression molding in this chapter. Chapters 5, 8, 11 and 12 give information about large molds being used in those particular processes. Figure 6.45 shows large molded parts that could have been molded on the press shown in Fig. 6.44.

TABLE 6.2.

Material	Suitability of Use and Comments
Cast Plastics and plastics faced molds	Suitable for short runs and prototypes (see chapter II-4).
Meehanite cast molds (semi steel)	Extensively used in early RP/C molding, but rarely used to-day. The material is not durable and develops too much porosity due to casting. Requires continual patching in service.
Kirksite—cast	Has greater expansion than steel when used in combination, and also "grows" in continual service at molding temperatures. Not 100% suitable.
1040 and 1045 steels (50 Carbon type)	Good machineable mold material. Suitable for parts where post-molded surface finish is not critical. Suitable for parts where appearance is a prime requirement. Has some porosity.
P-20 steel or equivalent	Best mold material available. Almost no troublesome porosity, and has good polishability. Better grade of steel and is preferred for extremely long runs. Cost is approximately 2.5 times that of 1045.

Courtesy SPI Handbook of Technology and Engineering of Reinforced Plastics/Composites, Second Edition, New York: Van Nostrand Reinhold Co., 1973.

TABLE 6.3. Comparison of Major Structural and Operational Elements for Standard Vs. SMC Molds for Matched-Die Molding of Reinforced Plastics.

Type or Method of Molding	Matched Metal Dies for Preform or Mat Molding	Matched Metal Dies for Molding SMC
Mold Component:		
Surface finish	High polish satisfactory; can be chrome-plated if desired; non-chromed surface hides "laking"	Chrome plating preferred over 325–1200 grit finish, buffed and polished
Shear-edge	Flame hardened to resist pinching and dulling due to glass	Flame-hardened or chrome-plated to reduce abrasive wear
Guide pins	Required 0.001 in. clearance on diameter	Extra-strong and accurate guide pins required to resist sidewise thrust due to off-center charge or asymmetrical mold; must protect shear edge
Ejector pins	Not necessary for most matched-die molding	Generally required for SMC; air-blast ejection preferred; cellophane preferred to cover ejector head during molding
Telescoping at shear edge	Travel should be 0.040–0.050 in.	SMC requires 0.025–0.8 in. telescope for developing proper back pressure and best mold fill-out
Clearance at pinch-off	0.002–0.005 in.	0.004–0.008 in.
Landing or molding to stops	Needed to properly define part thickness	Not necessary; part thickness determined by weight of charge
Optimum part thickness	0.090–0.125 in. optimum = 0.100 in.	0.125 in.
Molding temperature	235–275° F	1 sec, per 0.001 in. at 275°–280° F Range = 260–290; 340° F for thin parts
Molding pressure	200–500 psi	500 for flat to 1000 for deep draw; slow close required for last ¼ in. travel.

(*Courtesy SPI Handbook of Technology and Engineering of Reinforced Plastics/Composites, Second Edition, New York: Van Nostrand Reinhold Co., 1973.*)

FIG. 6.44. Removing molded seat from mold cavity.

FIG. 6.45. These large automobile parts are compression molded of fiberglass filled polyester. Truck cabs are molded in one piece by this process.

SIDE-RAM MOLDS

Many molders make use of side-ram presses (Fig. 6.46) for difficult molding problems which require the withdrawal of cores from two or more directions, as shown in Fig. 6.47. The construction of molds for these presses is similar to that of standard compression molds except that the cavity is made in two sections and the bottom cores are mounted on a movable section which is situated in the open between the cavity halves, as shown in Fig. 6.47 when the mold is open. These molds may be any of the common types, i.e., landed plunger, positive, semipositive, etc.

Critical considerations in the design of these molds are provision of sufficient side-locking pressure, good guides and interlocking between sections. The interlocks should make the mold sections an immovable unit when they are locked in place, especially the cavity cores and the cavity halves. Provision is made for moving the bottom cores into the open space, where the mold parts can be removed easily.

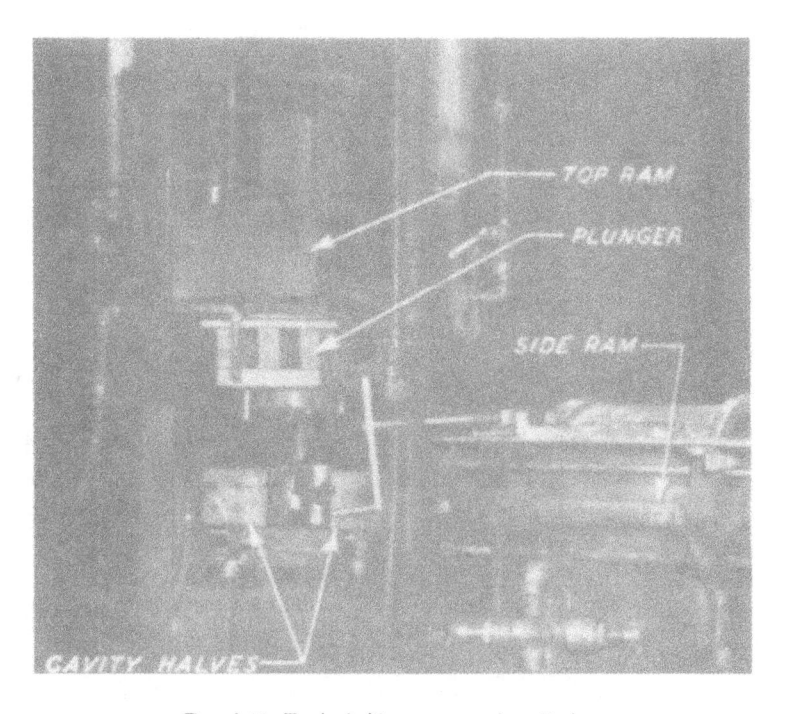

FIG. 6.46. Typical side ram press installation.

FIG. 6.47. Five-cavity split mold cavity halves installed in a side-ram press.

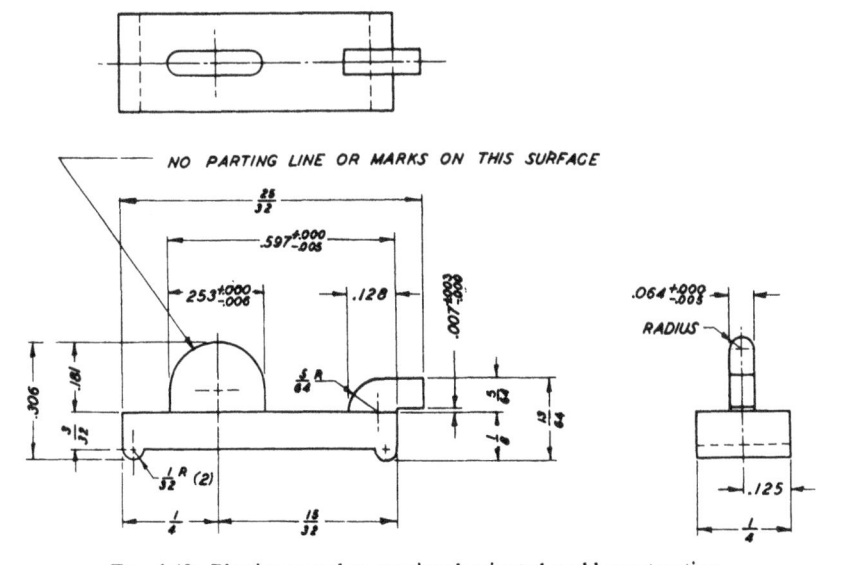

FIG. 6.48. Plastics part that requires laminated mold construction.

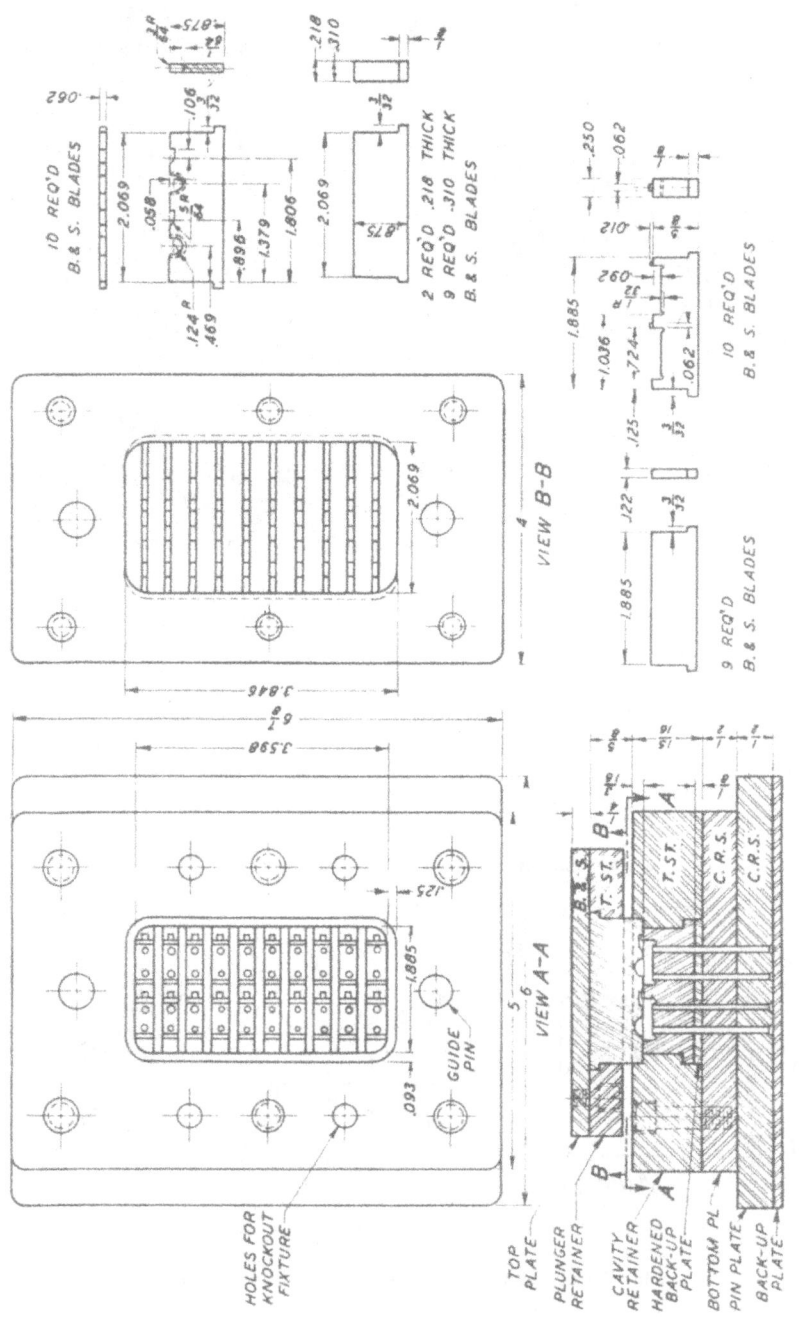

FIG. 6.49. Twenty-cavity hand mold designed for laminated construction. The part shown in Fig. 6.48 is made in this mold.

LAMINATED MOLD CONSTRUCTION

Molds that require extremely close tolerances or difficult-to-machine sections have been assembled successfully from a series of pieces. The part shown in Fig. 6.48 would require intricate mold construction if constructed by conventional processes, the radius on the 0.064-in. section being particularly difficult. The mold construction used is shown by Fig. 6.49. All parts of this mold were machined and hardened. The mold parts were then ground to final tolerances of plus/minus .0005 in., thus providing a minimum of variation in the total "add-up" dimension. The finished mold is shown in Fig. 6.50 after many thousands of pieces had been produced successfully. See also Fig. 6.51 for exploded view of similar mold.

This mold design and construction is indicated for many molds that require closely held tolerances because the final grinding-to-size after hardening permits delicate adjustment of the mold dimensions. An additional gain results from the easy accessibility of these mold sections for machining. Broken, damaged, or changed sections are easily replaced. Many designs that are difficult to machine or hob in deep cavities may be constructed easily

FIG. 6.50. Twenty-cavity hand mold shown in Fig. 6.49 after severe use.

by this "laminated" mold construction. A large semiautomatic compression mold is shown in Fig. 6.52. Holes are jig-bored by means of carboloy tools after hardening (finish cut only). The key in the top and the milled step in the steam plate hold the mold in place sideways. The tie rods pull everything together the long way of the mold.

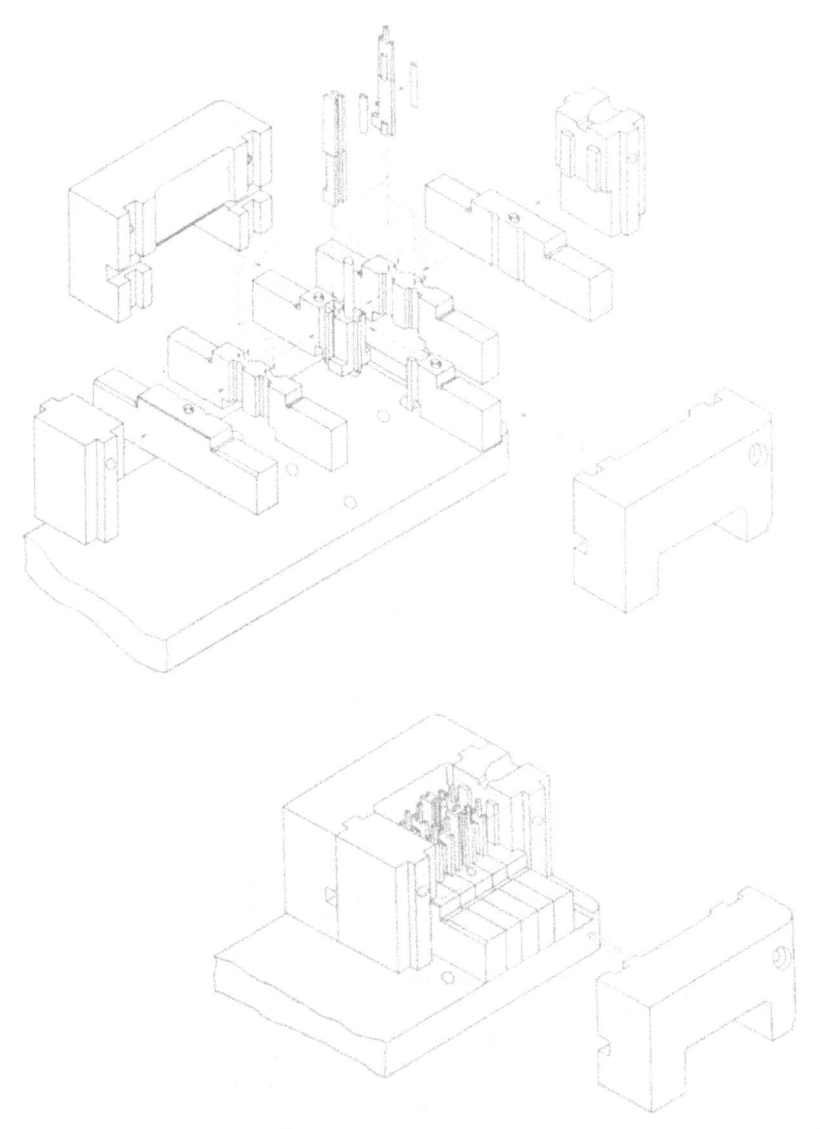

FIG. 6.51. Laminated construction of molds is an economical procedure for complex molds with a very high degree of accuracy. (*Courtesy, Dai-/chi Seiko Co., Ltd., Kyoto, Japan.*)

Thirty-two cavities have been constructed for this part, and the length of the piece is 6.500 plus/minus 0.005 in. No known type of mold construction other than this would permit the holding of such tolerances at reasonable cost. Cooling fixtures and uniformly blended materials contribute to this important method for producing parts having close dimensional tolerances.

OTHER COMPRESSION MOLD CONSIDERATIONS

1. Molds which are made with cavities or plungers in a section require especially good backing, and hardened back-up plates are recommended.

2. Loose pins used in molds may stick, therefore provision should be made for the addition of an ejector pin behind the loose pin.

3. When ejector pins butt against the opposite half of the mold, use hardened strips under the pins to prevent sinking.

4. A loose wedge must have sufficient engagement so that material cannot flow under the wedge if it should rise as the mold closes.

5. Unless considerable care is used in the location of wedge parting lines, excessive finishing operations will result. Good and bad parting lines are shown in Fig. 6.53. A parting line located near the lugs is bad because it is

FIG. 6.52. Two-cavity semiautomatic compression mold constructed in laminated sections and held together by tie rods. (*Courtesy Honeywell, Minneapolis, MN*)

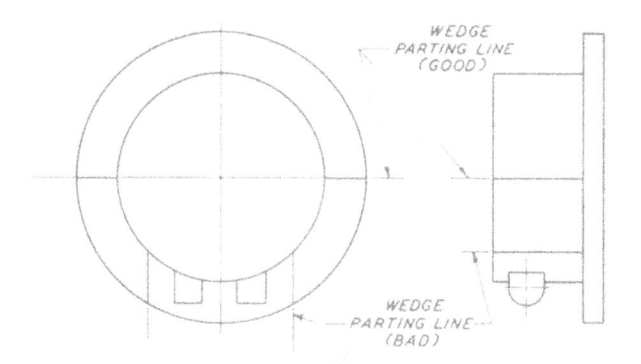

FIG. 6.53. Good and bad wedge parting lines are shown above.

difficult to buff close to the lugs. Finishing operations will be simplified by proper location of the parting line.

6. A solid part will shrink more than a part having thin walls, therefore additional shrinkage allowance must be made.

7. Removable plate molds must be designed with stop blocks between the top and bottom so the mold cannot be fully closed when the plate is out.

8. All feather-edges must be eliminated in mold designs. Molded threads must be designed without feather-edges if low cost operation is desired. Feather-edges will require frequent mold replacements because of breakage.

9. Do not pioneer a radical mold design until you have discussed it with other mold designers and with the molding plant foreman. Any designer will do well to heed the adage "Be not the first by whom the new is tried, nor yet the last to lay the old aside."

REFERENCES

Butler, J., *Compression and Transfer Mould Engineering*, New York: Pergamon Press, 1967.

Bebb, R. H. *Plastics Mold Design*, London: Iliffe, 1962.

DuBois, J. H. and Pribble, W. I., *Plastics Mold Engineering*, Chicago: American Technical Society, 1946, New York: Van Nostrand Reinhold, 1978.

Fleischmann, J. J. and Peltz, J. H., Compression molding, *Modern Plastics Encyclopedia*, 1983–84 Ed., p. 200.

Halliday, W. M., *Molds for Plastics*, London: The Engineering Universities Press, 1948.

Modern Plastics Encyclopedia, 1984–85 Ed., New York: McGraw-Hill Publishing Co.

Nachtrab, William C., Molds for sheet molding compound, *Newark Section SPE: Preprint for Symposium, Plastics/Molds/Die—Design/Construction*, 1977.

Oleesky, S. S. and Mohr, J. G., *Handbook of Reinforced Plastics*, New York: Van Nostrand Reinhold, 1964.

Sachs, C. C., and Snyder, E. H., *Plastics Mold Design*. New York: Murray Hill Books, 1947.

Sandy, A. H., *Moulds and Presses for Plastics Molding*, London: Lockwood, 1947.

Mold makers and designers—The future of the plastics industry, Retec Symposium Preprint, Grand Rapids, MI: Western Michigan Section SPE, Nov. 1983.

Chapter 7 / Transfer and Injection Molds for Thermosets

Revised by S. E. Tinkham and Wayne I. Pribble

TRANSFER MOLDING

The term "transfer molding" was chosen by the originator of the process (L. E. Shaw), "to identify the method and apparatus for molding thermosetting materials whereby the material is subjected to heat and pressure and then forced into a closed mold cavity by this same pressure and held there under additional heat and pressure until cure is complete." In the original presentations it was stated that the object was to introduce the plastics material into the mold in a highly plastic state in order to secure free flow. An injection mold for thermosets is an automatic transfer mold.

At the time of the invention of transfer molding, thermoplastics were rare and little known. The thermoplastics were molded initially in compression molds and in integral transfer molds: a mold heat and chill process was used in a slow cycle. With the introduction of fully automatic molding machines, the "transfer molding" of thermoplastic materials was called *injection molding*. Much later, as a result of the invention of the reciprocating screw plasticizer-injector and the development of thermosetting compounds for this process, the contemporary injection machines were modified to handle the thermosets. This was intially called *automatic transfer molding* but the trade has also called this type of thermosetting molding *injection*. In each case the material is plasticized external to the mold and the plastic mass is then pushed into the mold where thermoplastics harden by cooling and the thermosets harden by the continued application of heat and pressure in the mold.

The cavities, plungers, and ejecting mechanism are generally similar to those of any flash-type mold, with added provision for material flow from the plasticizing cylinder, or transfer tube, through the sprue, runners and gates. Many transfer molds will follow this construction and a series of general rules for mold design may be applied for many jobs using this process. Special data are given on the design of injection molds for thermosets.

The operation of a primitive type of transfer mold is explained in Fig. 7.1. Reference to sketch (*A*) will show that the charge is held in the transfer chamber, also called a pot, and the cavity and plunger are closed. As the mold continues to close, the compound will be compressed into a slug. The

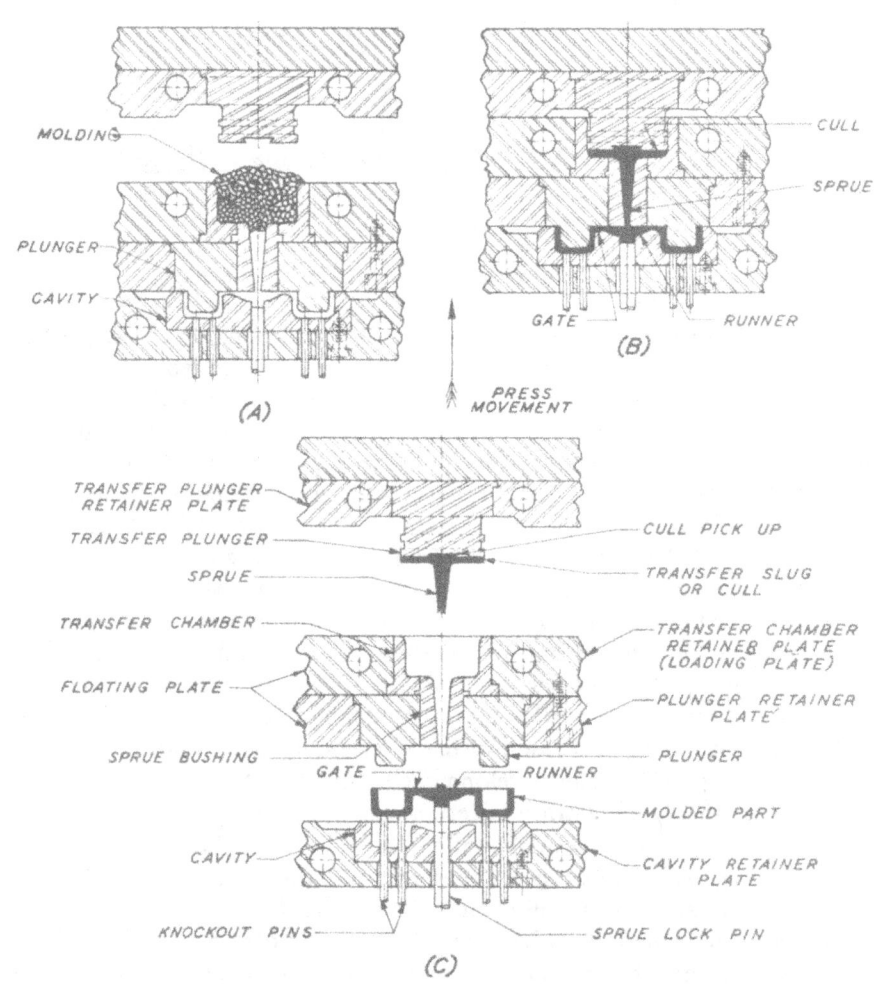

FIG. 7.1. (A) Transfer mold ready for pressure to be applied to compound to chamber. (B) Mold fully closed. The compound has been plasticized by heat and transferred from the chamber through the sprue, runner and gates to the cavity. (C) Mold fully open and ready for operator to remove molded parts and cull.

slug will absorb the entire pressure and the press travel will be stopped until the heat has plasticized the compound to the point of flow. This portion of the cycle is usually between 15 and 45 seconds in length unless special pre-heating methods are employed. If the heat-plasticizing period is too short, the compound may be forced into the cavity before it is sufficiently plastic, thus causing damage to mold pins and inserts. If the period is too long, the compound may "set up" too much before flow begins and produce poor flow conditions or failure-to-knit lines, especially where the material flows around pins and inserts. These knit lines result in poor mechanical and poor dielectric strength.

A typical pot-type transfer mold is shown in Fig. 7.2. The operator is seen removing the molded piece, which still contains the core pins. The pins are removed from the molded piece on the bench in this case and inserted in position again in preparation for the next charge. The operator will knock out the *pot slug*, or *cull*, with an air hose, load the new charge in the transfer chamber and open the valve to start the next pressing cycle. Preforms or powder may be loaded in the chamber, and the charge is often preheated to secure a fast operating cycle. This preheating is very effective, as the compound may be heat-plasticized to the flow point before being placed in the mold so that flow will commence as soon as the pressure is applied. Thorough preheating, as secured from high-frequency units or from steam heaters

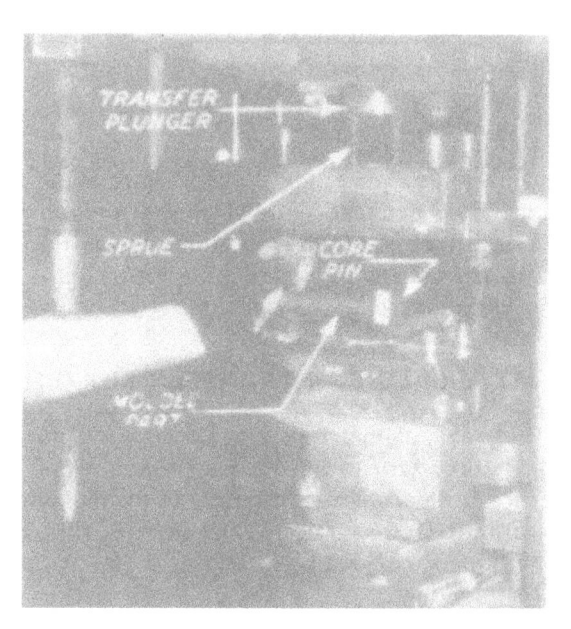

FIG. 7.2. Single-cavity transfer mold in fully open position with molded part and core pins being removed.

or infrared lamps, may reduce closing time from 30 to 90 sec to 5 to 15 sec. It is important in using pot transfer molds to make certain that the heating area of the chamber is adequate in order that the compound may flow readily around fragile inserts and mold pins without distorting or damaging them. The use of proper preheating equipment eliminates a large-area heating chamber, so that transfer molds similar to the "pressure-type" die casting dies may be used to achieve the desired results. Molds of this type have also been called plunger transfer molds and are described later in this chapter.

Transfer Mold Cavity Considerations

From the foregoing material, and by reference to the various types of molds described in Chapter 2, it will be noted that the transfer mold combines features of the flash-type mold, the loading shoe mold and the injection mold. Noteworthy also is the loose plate mold (Fig. 2.8). The initial mold design example is for an integral transfer type mold. This differs from a plunger transfer mold only in the pot and gating.

The initial considerations in the design of transfer molds are the same as those described in Chapter 6 for compression molds. The product design must be studied carefully to make sure that the design selected is moldable. The following questions must be answered before starting the layout:

1. Where can parting lines be located most advantageously?
2. Where will the gate be located, and what kind of gate shall be used?
3. What material will be used?

Parting-Line Location

The parting-line location is determined by the shape of the piece, and, whenever possible, it should be made a straight line at the top. This permits the use of a simple flash-type cutoff and a bottom ejector arrangement. In other cases the parting line must come at the center of the part, making use of a half cavity in each mold section.

MATERIALS FOR TRANSFER MOLDING

Because of the variety of materials and their widely different processing requirements it is necessary to have a better than average understanding of their properties and peculiarities essential to the mold design. Important basic factors are: bulk factor, mold release, ejection, shrinkage, outgassing, rigidity, venting, tapers, preheat needs, curing cycles, gating, and the pressure required for molding.

Specific mold designs must be developed for the materials that have been selected for the product; design compromises are essential if several materials may be used.

Gate Location

Gate location is governed by many considerations. Several important points must be considered in determining the location for gates. Well-designed gates must permit proper flow of the material after it enters the mold and easy removal of the gate after molding. Gates must be located where they can be removed easily and buffed when necessary. Best results are ontained by gating into or near heavy sections of the part. The maximum flow of material should be limited to 8 in. when possible, and this is especially true when the part has more than one gate. When an overlong flow distance is used with two or more gates, the material may not knit properly where the two streams of compound come together. It is also desirable to locate the gates away from points that will be used in the functioning of the molded part, since irregularities resulting from gate removal might necessitate additional finishing expense.

It is often convenient to gate into a hole in the molded part. When this is done, a heavy flash is left in the hole to serve as a disk gate. The hole will be drilled through in the finishing operation. This gating method is also used when it is desirable to produce a piece that will show no gate mark after finishing. Various types of gates for transfer and injection molds are described later in this Chapter.

Venting of Transfer Molds

All transfer molds must be vented in order that trapped air may escape. The size and location of the vents provided will depend upon the size and design of the piece and also upon the location of pins and inserts. Each part is to be studied with respect to these separate considerations. In general, the vent is a small space or orifice which affords free passage of air or gas but which is too small to release any appreciable amount of the molding compound.

The fin formed by the small amount of compound that does pass through may be removed easily in the tumbling or finishing operations. In certain cases it has been found necessary to provide a large opening and permit a considerable amount of compound to pass out in order to avoid flow marks and trapping of gas in a critical area. This is especially necessary in molds designed for the melamine compounds when dielectric strength is of prime importance. By permitting a quantity of material to flow past the critical point and knit at a point that is not critical, the probability of electrical failure from puncture at the knit line is minimized. Actual tests may be necessary to decide whether this type of mold needs venting and where the vents are to be placed.

Vents should be from 0.002 to 0.005 in. deep and from ⅛ to ¼ in. in width. It is normal practice to start with the smaller sizes and enlarge as required. The location of vents is a most important consideration, and by study, practice, test, and observation the designer can learn to determine from visual inspection of the part drawing the points at which venting will be needed.

Vents are often placed at the following points:

1. At the far corners.
2. Near inserts or thin-walled sections where a knit line will be formed.
3. At side-pull pins which form holes in bosses.
4. At the point where the cavity fills last.
5. In insert holding pins.
6. Around ejector pins.

The venting shown in Fig. 7.3 was applied to the mold sections shown in Fig. 2.26. The parting line vents are essential, since the cavities fill last at this point. The side pin in this case is used to secure an insert, and the vent is provided by the small hole perforating the pin. Since the insert covers the hole, it cannot fill up with compound. If no insert were used in this part, it would be necessary to machine a flat on the body of the pin next to the threads. The pin would then be removed after each shot so that the small amount of compound in the vent could be removed. In is usually necessary

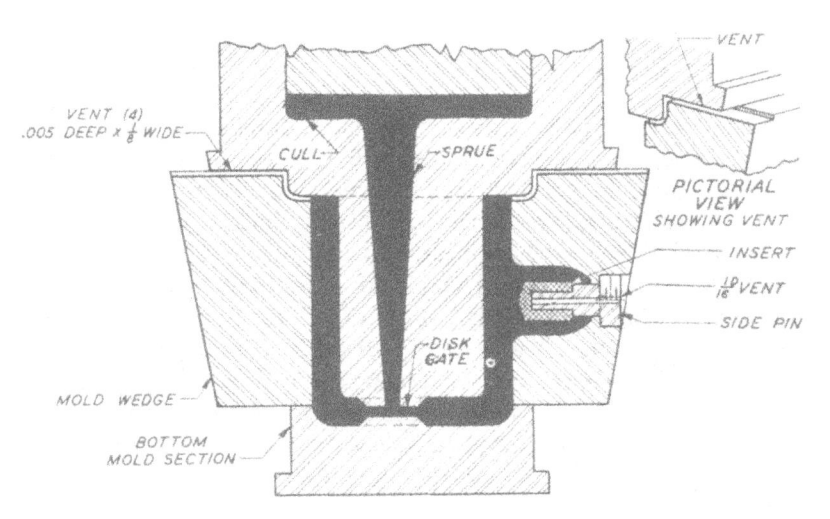

FIG. 7.3. Typical method of venting. Notice disk gate where hole will be drilled in finishing. Note that pot and force are made as one piece in this transfer mold.

to remove pins of this type after each shot for the reason that the opening fills and thus becomes ineffective in succeeding shots.

Guide Pins, Safety Pins, and Ejector Pins

The standards given for proper proportion of guide pin dimensions for compression molding also apply in transfer molding. As in other molds, the guide pin must be sufficiently long to enter the guide bushings before the transfer plunger enters the chamber. The guide pins when anchored in the cavity plate must be long enough to enter the guide bushing before the horizontal core pins enter their tapered seat.

It is also common practice to put guide pins in the transfer plunger retainer plate. These guide pins provide the alignment between the cavity sections and between the transfer chamber and plunger. Pins of this type must have sufficient length to project far enough through the loading plate to enter the guide bushings in the lower cavity plate before these two plates come together.

Guide pins must always enter their guide bushings before any of the other mold parts are engaged.

Safety pins are used on transfer molds in the same manner as on compression molds, and as described in Chapter 6. When used in conjunction with side springs, they give positive assurance that the ejector bar and pins will seat properly. When the press is not equipped with side springs below the platen, it will be practically necessary to have them mounted directly on the mold.

Ejector pins may be relatively short usually, since the flash type of parting line is used. The pins must have enough length to permit use of a stripping fork when necessary. A good minimum amount of ejector pin travel will lift the molded part about ½ in. above the flash line. This is especially true when some of the ejector pins also serve as insert holding pins, as this combination will usually necessitate the use of a stripping fork for removal of the part.

Interlocks Between Mold Sections

Generally, integral transfer molds should offer some means of insuring section alignment, especially in the case of split molds, where parting lines must match. In the case of split wedge molds, these extra guides, usually called interlocks, may be a simple dowel-pin alignment, or it may be a more intricate tongue-and-groove arrangement designed to prevent wedges from becoming misaligned or improperly assembled as depicted in Fig. 7.4.

The dowels may not be used in the operation of the mold, but they are a

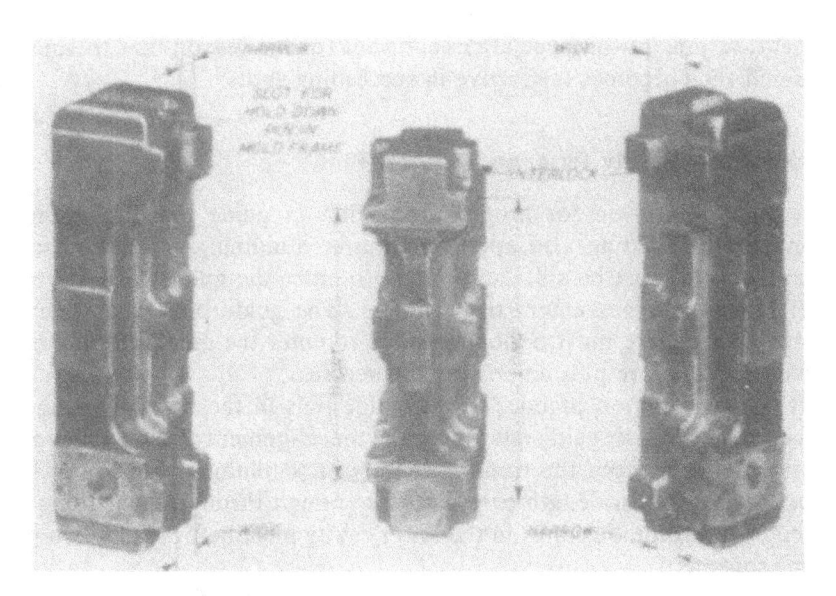

FIG. 7.4. Interlocks insure alignment of wedge sections. Note arrangement which prevents wrong assembly.

necessity from the mold-maker's standpoint, as they enable him to maintain the proper alignment while machining the cavity sections.

It is best practice when using tapered interlocks to make them so that all locking surfaces can be ground on the surface grinder. When they are not ground, it may be necessary to provide so much clearance that the interlocks become ineffective as a result of the distortion that occurs during hardening.

Insert Considerations

Through inserts which are to be molded in transfer molds make use of shoulders, as shown in Fig. 7.5. A total tolerance of 0.005 in. is the maximum allowance on the shoulder length. When the molds are made so that the distance between the plunger and the cavity in the closed position is 0.002 in. under the minimum insert length, the faces will remain free of compound. Since most inserts of this type are made of brass, they will deform easily by this amount without being damaged.

When the compound is molded into a metal shell by the transfer process, it is recommended that all cavities be fed from a single chamber, since this will equalize the pressure in all cavities. If separate chambers were to be used for the individual cavities, the one having the heaviest load of compound would receive more than its quota of pressure and the chamber might burst.

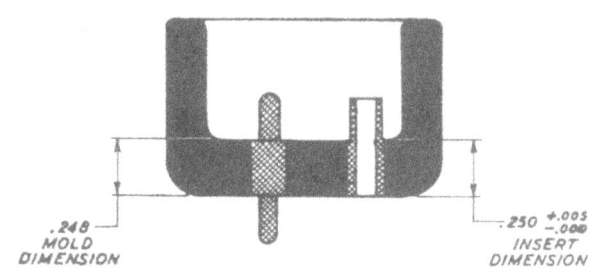

FIG. 7.5. Shoulder inserts in a transfer mold.

PLUNGER TRANSFER MOLDS

A plunger transfer mold is one that utilizes molding press equipment containing a main clamping cylinder to open and close the mold, and having an independently controlled, double-acting auxiliary cylinder to operate a transfer plunger for the purpose of transmitting the molding material into the cavity and plunger areas of the closed mold. The mold is held in position by the main clamping cylinder and the material flows from the loading chamber through the runners and gating system into the cavity area.

Three typical general press designs are used for semiautomatic plunger transfer molds.

Top Plunger

The auxiliary hydraulic cylinder assembly is mounted on the top or "head" of the press with a double-acting piston which has sufficient stroke to enter a loading chamber with the necessary volume to permit adequate pressure and material capacity to form the shape and density of the mold cavity and plunger geometry Fig. 7.6. This is the most widely used type of semiautomatic transfer molding equipment, possibly because it was the most economical conversion for the older compression type of equipment. This also offers better ejection of parts and ease of loading inserts, which are commonly used on the interiors of molded plastics product designs. If necessary, top ejection may be designed into the mold for specific applications, and inserts may be positioned in the top half of the mold as well as in the bottom half.

Bottom Plunger

The bottom plunger is basically the same as the top plunger, except that the plunger cylinder assembly is mounted below the lower clamping platen, or in some designs it is installed as a separate independent assembly above the lower clamping press platen.

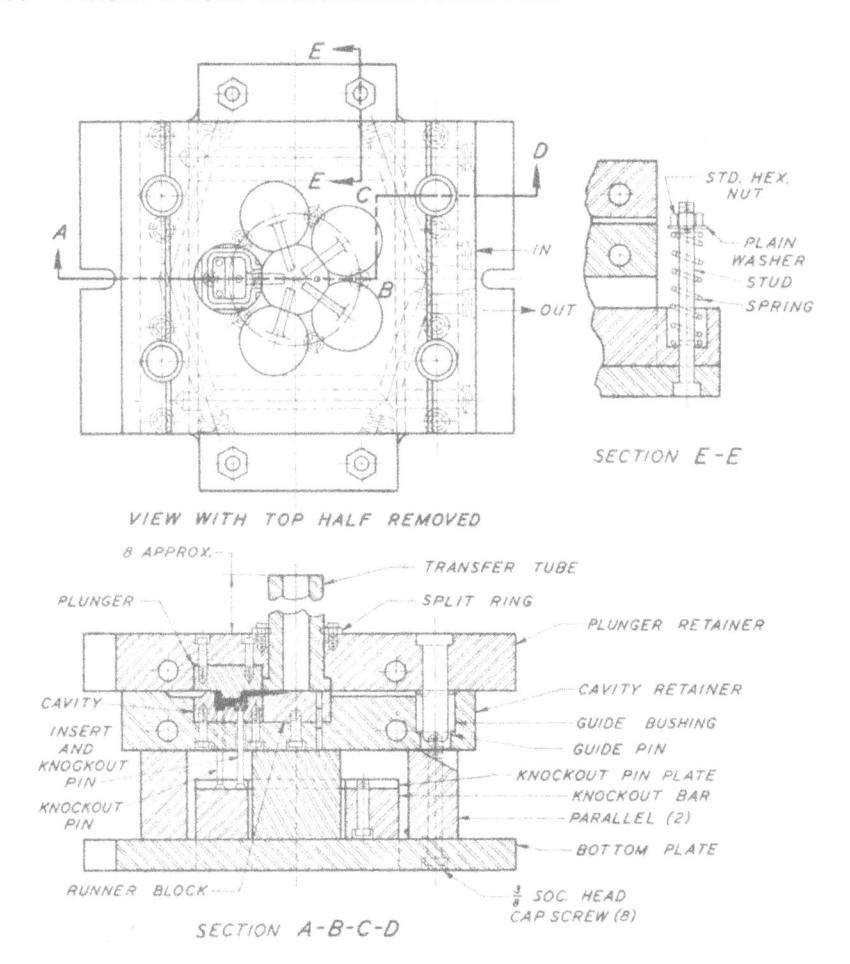

FIG. 7.6. Schematic diagram of a plunger-type transfer mold.

This design permits good cycling, minimum plunger transfer stroke, and top ejection of the parts from the mold.

Horizontal Plunger

The horizontal plunger assembly is mounted parallel to the press platen, and is employed for special molding product designs. This type of design is not widely used since it requires special equipment and the normal advantages are not too great except for very unique designs. One typical design with a distinct advantage is shown in Figs. 7.7 and 7.8 permitting tandem operations of two molds.

FIG 7.7. Conversion unit designed to convert conventional compression presses to transfer presses, using auxiliary side-rams to effect the material transfer. Numbered parts are: (1) Thrust lock; (2) upper plunger; (3) upper cavity; (4) side transfer ram; (5) lower plunger; (6) lower cavity; (7) main ram for clamping only; (8) transfer chamber; (9) side transfer ram. (*Courtesy Plastics, Inc., St. Paul, MN*)

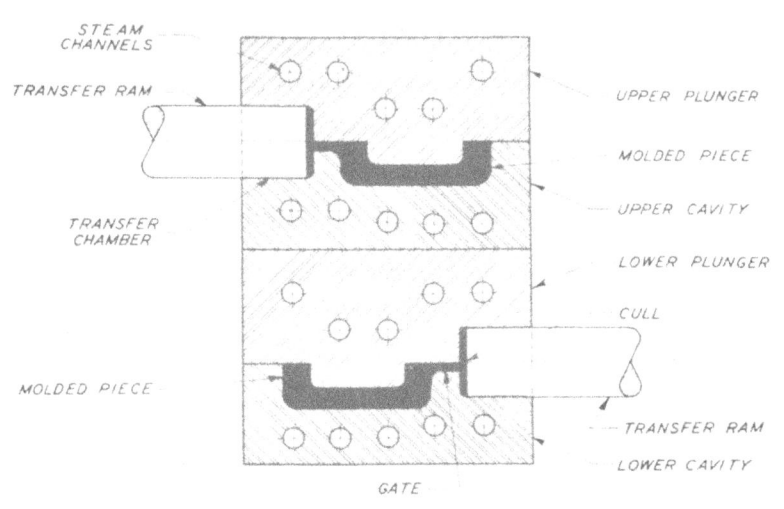

FIG. 7.8. Tandem transfer molds are opposing side rams to transfer the compound. Main press ram supplies vertical clamping pressure. (*Courtesy Plastics, Inc., St. Paul, MN*)

Other versions of multiple molding have been accomplished with a singular clamping ram press, utilizing twin auxiliary, vertical, double-acting plungers for efficient and economical operations on specialized designs.

The two common ways to convert compression equipment are:

1. Rebore the press head and permanently assemble a double-acting transfer cylinder on the top of the press.
2. Assemble an auxiliary cylinder assembly complete with parallels and mounting plates, and fasten it to the bottom or top press platen. This will reduce the existing "daylight" of the press, and the molds are mounted above or below the auxiliary cylinder unit.

Transfer-plunger equipment presses are generally designed with a clamping pressure ratio of *4 or 5* to *1* for the transfer pressure. The self-contained hydraulic systems permit complete flexibility of ratios in clamping or holding pressure and the transfer pressure and speed of transfer.

The unit pressure applied to transfer the material from the loading chamber into the cavities must be kept below the unit clamping pressure to prevent the opening of the mold since the molding compound is in fluid condition during the transfer period.

Transfer-molding design fundamentals must be followed for an efficiently operating mold. The basic fundamental are product design, material, and equipment. The transfer mold considerations should be reviewed again at this time.

PRODUCT DESIGN CONSIDERATIONS

1. Complete understanding of the product design must be achieved. Areas in question must be resolved in a manner favorable to the mold, material, and production essentials.

2. Review of tapers, tool strength, ejection, gating, and parting lines will disclose essential product changes.

3. This study must include the end-use components or mating parts that are used in the final assembly.

4. Review dimensional and tolerance requirements to see that they are appropriate for the material and desired mold.

5. Functional parts do not need the careful planning essential to the appearance of decorative parts.

6. If inserts are to be molded into the part, they must be checked to insure proper fitting tolerances, and, if possible, designed to seal molding material from flowing into threads or areas that will necessitate secondary flash removal operations.

7. If design permits, position of gating must be considered and designed

for the most economical removal after molding. A gate location and size that permits breaking off after molding or breaking at the time of ejection in the mold is the least costly. Gating into a hole for removal by drilling is a good alternate choice. If the gate position is such that it requires machine removal and buffing, it may add 30 to 50% of the actual cost of molding to the price of the product.

8. Finally, from the product and end-use functional requirements the designer must have a sound understanding of the quality standards expected of the molded product. This study will help design and deliver a mold that will produce satisfactory parts with minimum tool cost and minimum molding expense.

Consultation with material manufacturers or molders will be helpful in an effort to define these requirements more accurately for specific materials, since many values effect the pressure required, i.e. flow of material, speed of transfer, preform shape, preheating, length of runner systems, gating, and design of parts.

There are other thermosetting materials which may be transfer molded, and in specific cases the designer should check with materials manufacturers for essential information. If, for any reason, after the mold is constructed it is found necessary to change the molding material, experimental molding trials will indicate the tooling design changes essential to obtain an efficient production design. It is sometimes necessary to build experimental shape molds to determine shrinkage factors when new materials are to be molded. Shrinkage values for a given material are different for each molding method.

Figure 7.9 shows the comparable values of various size clamping pressures with appropriate plunger transfer pressures and the associated transfer-plunger diameters.

These calculations were developed on 100% hydraulic ratings, and in all cases should be reduced to compensate for pressure reductions or other area factors based on historical studies for a given piece of equipment. A 10% reduction factor is often used.

You will notice that in many cases, the calculations do not permit the use of maximum transfer pressure since it exceeds the predetermined maximum transfer pressure allowed by the available clamping pressure.

In many cases compensation can be provided by changing the transfer plunger diameter, or, in an extreme case, changing the transfer cylinder and plunger.

Figure 7.9 shows the actual projected molding area (including runners and cull areas) that is available under these specific capacities and these data are fundamental to the mold design.

Formulations of various materials in specific categories, preheating techniques, moisture content, speed of transfer, gating, venting, and runner

Press size (Tons)		Clamping pressure (lbs)	Transfer pressure (lbs)	Plunger diameter	Maximum transfer pressure (psi)	Projected area at maximum pressure (sq in.)	Projected Molding Areas (Sq in.) at Various Transfer Pressures									
Clamp	Transfer						3000 psi	4000 psi	5000 psi	6000 psi	7000 psi	8000 psi	9000 psi	10,000 psi	11,000 psi	12,000 psi
50	10	100,000	20,000	1½	11,300	8.84	33.3	25	20	16.6	14.4	12.5	11.1	10	9.1	8.3
75	15	150,000	30,000	1¾	12,500	12	50	37.5	30	25	21	19	17	15	13.6	12.5
100	25	200,000	50,000	2¼	12,594	15.9	66.7	50	40	33.3	28.6	25	22.2	20	18.2	16.7
150	32	300,000	64,000	2¾	10,774	27.8	100	75	60	50	42.9	37.5	33.3	30	27.3	25
200	50	400,000	100,000	3¼	12,044	33.2	133	100	80	66.6	57	50	44.4	40	36.4	33.3
300	75	600,000	150,000	4	11,925	50.3	200	150	120	100	85.7	75	66.7	60	54.5	50
300	75	600,000	150,000	4½	9,434	63.6	"	"	"	"	"	"	"	"	"	"
300	75	600,000	150,000	5	7,645	78.5	"	"	"	"	"	"	"	"	"	"
500	125	1,000,000	250,000	5½	10,521	95	333.3	250	200	166	144	125	111	100	91	83

FIG. 7.9. Transfer press capacity fundamentals.

design will have a direct effect on actual pressures required and the above data provide adequate guide lines for mold designing in spite of these variables.

Generally speaking, some of the physical properties of a given material processed by transfer molding are slightly lower than the properties that may be obtained from compression molding. It is desirable on critical molding designs requiring the ultimate in product properties, to confer with the various materials manufacturers for available laboratory test results that will affect the mold or product design.

PLUNGER TRANSFER MOLD DESIGN

The initial step in any mold design is to review the product drawing to establish the parting line, ejection location, gating location, and venting areas.

The parting line is designated at the perimeter which permits the best location for mold construction and operation in accordance with press movement, ejection (withdrawal from the mold) and ease of finishing after molding.

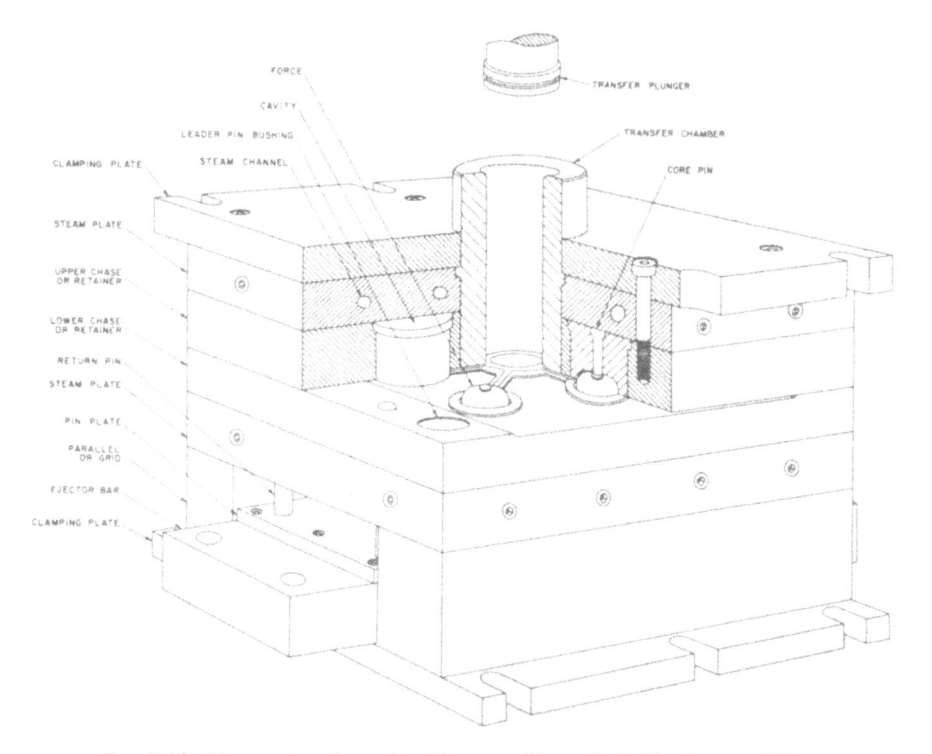

FIG. 7.10. Plunger transfer mold. (*Courtesy Wayne F. Robb, Denver, CO*)

Ejection locations release the molded part from the mold after the curing cycle has been completed, and must be located dimensionally to cover a uniform area of ejection travel, with size of ejection being sturdy enough to withstand the numerous press movements in production. A second ejection function is to permit the venting of air and gas at these positions through the medium of machined areas of the ejector pins or mold sections that will permit the escape of air and gas.

Venting is to be provided in areas where gas and air will concentrate. Such areas may be vented by the use of ejector pins, with clearance areas between the pin and mold section. The parting line area should be vented opposite the gating area, and also, on larger parts, intermittently around the top surface of the cavity at the flash line.

Venting areas in the mold must be located so as to insure that the air or gas will easily escape into the atmosphere to produce repetitively good moldings. Any obstruction to this area will appear as an unfilled or poor surface on the molded part.

Many times after a sample molding run, it is necessary to provide additional venting; however, it is good designing practice to predict the areas that may be troublesome in order that the mold may be easily modified. In some cases it may be necessary to install a vacuum connection to insure the evacuation of entrapped gas.

TRANSFER TUBE (LOADING SPACE REQUIRED)

Assuming you have established the volume of the part to be molded, the number of cavities, the press size, and the transfer plunger size from the previous press capacity data, the following information will be helpful in calculating minimum depth of loading tube for a given mold design.

A. Volume of one part _____ cu in.
B. Volume of (*No.* of cavities) _____ cu in.
C. Volume of runners, cull _____ cu in.
   (A or B + C) net volume _____ cu in.
D. Gross volume = net volume (C) $\times$ (material bulk factor, either for
   loose powder or preforms _____.
E. Area of transfer plunger _____.
F. Depth of chamber required _____.

$$F = \frac{D}{E} + 2 \text{ in.}$$

Two inches is an arbitrary figure to be used as a safety factor in loading the molded material into a tube which will permit entry of the plunger before the mold transfer pressure is applied so that the material entrapped

Area of one part = _____ sq in.
A. Area of one part (___) × No. of cavities (____) = Total
   cavity area _____ sq in.
B. Area of _____ in. diameter transfer plunger = _____ sq in.
C. Area of runners = _____ sq in.
D. Total area (A + B + C) = _____ sq in.
E. Force of (see E′) _____ in. top cylinder = _____ lb
F. Force of (see F′) ___ in. diameter clamp cylinder = _____ lb
G. Transfer pressure (no reducer) (see G′) = _____ psi
H. Transfer pressure at balance point of (see H′) lb clamp = ____ psi
E.′ Area of top cylinder × line pressure psi (−10%)
F.′ Area of clamp cylinder × line pressure psi (−10%)
G.′ E/B
H.′ F/D

Fig. 7.11. Transfer pressure information.

does not escape. It will be found frequently that mold fabrication details such as upper cavity height, retainer and press platen plate thickness, etc. will provide adequate length for the volumetric needs of the transfer tube. This secondary check against the formula will insure proper loading space in the transfer tube.

Volume calculations essential to insure proper space in the transfer tube for loading the proper weight of preforms must be predicated on their size *after* preheating. Preforms expand considerably in diameter and thickness during preheating.

TRANSFER PRESSURE

The information in Fig. 7.11 is useful in calculating the transfer pressure. It is self-explanatory, and upon completion of your study, you may refer to the "press capacity" (Fig. 7.9) to determine type of equipment for which the mold may be designed.

RUNNER SYSTEMS

Information and test data to date make it impossible to offer specific information on design and size of runner systems for all materials and conditions. General practice and broad experience have resulted in the following techniques:

Main runners are trapezoidal in shape from ¼ to ⅜ in. wide and ⅛ to ¼ in. in depth, with the auxiliary runners being made approximately 30 percent smaller in area.

Many molders have gone to full round runner systems using the ¼ to ⅜ in. diameter for main systems with auxiliary runners approximately ½ of this volume.

Runners are generally cut into the ejection side of the mold, with the ejectors also located on the runners. In the case of full round runners, it is best to design hold down ribs to permit positive ejection of runners and parts.

Runners are cut into the "cull" area of the loading chamber to permit immediate filling of the cavities. Length of runners should be kept short for better material flow and speed of transfer time in filling the cavities. This can be accomplished by grouping the cavities completely around the transfer chamber, instead of laying out two single rows of cavities.

Figure 7.12 shows a 24-cavity mold layout for a piano sharp with 16 runners located directly from the cull and 4 runners feeding 2 cavities each

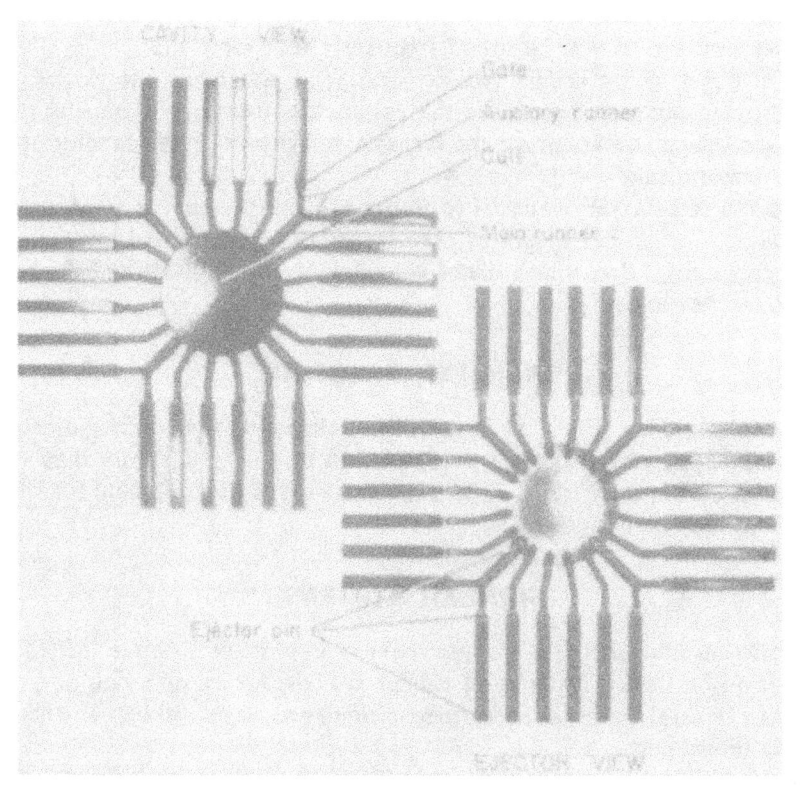

FIG. 7.12. This well designed 24-cavity transfer mold runner and ejector system illustrates many essential design features. Note the short runners for minimum flow and runner take-off from the cull. (*Courtesy Tech Art Plastics Co., Morristown, NJ*)

in order to obtain uniform filling and comparable quality in the finished parts. In this fashion the runner system is kept to a reasonable length for better manufacturing latitude, and for ease of removing the parts from the mold in one "shot." Note ejectors on the runner and cull.

Large area and sectional parts require larger runner systems and parts made of impact or filled materials may require much larger systems to permit the fillers and resin to flow into the cavities with minimum loss of strength.

GATING TRANSFER MOLDS

Gating also presents another problem in designing. It is impossible to state specific design and size information, and we are again dependent on years of experience and practice to offer design guidance that works. These comments apply also to the gating of injection molds for thermosets. Normal practice is to position the gate at the thickest section of the part.

Sizes of the gates depend on the material being molded—general-purpose materials require the smaller areas, i.e., 1/16 to ⅛ in. wide × 1/64 to 1/16

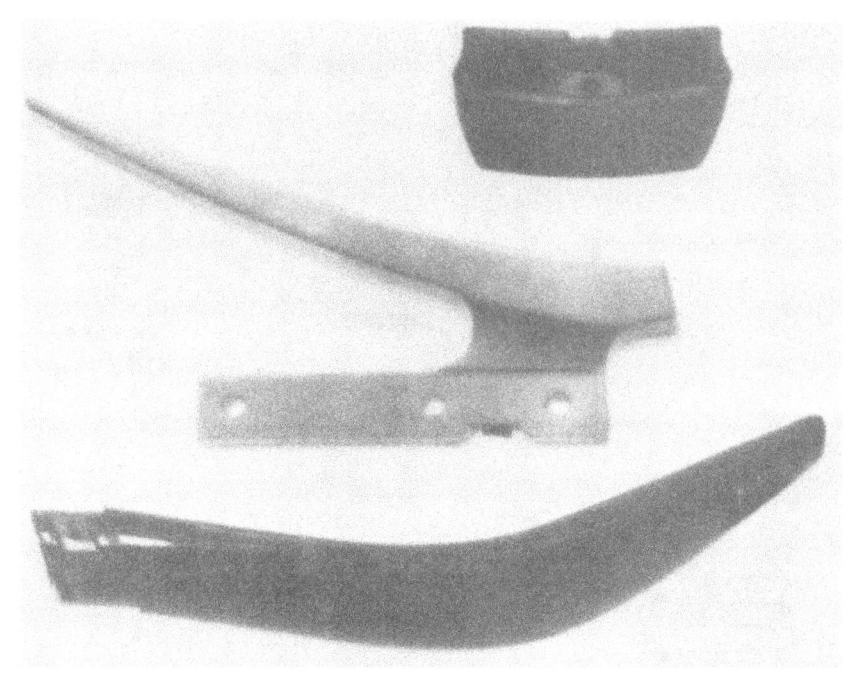

FIG. 7.13. Molded parts showing typical depressed gating where a broken and otherwise unfinished gate slug will not be objectionable.

in. deep. Filled and impact grade materials require larger areas, i.e. ⅛ to ½ in. wide by 1/16 to ¼ in. deep. Materials having long fillers generally require a boss molded in front of the gate area to facilitate the finishing of the part by the removal of the depression or voids that are caused by the "tearing" of the fillers when the runner is removed.

Gate removal and final finishing must be considered seriously because of their potentially large effect on the product cost. Gates that are broken only are the least expensive, and a product may often be designed to permit the gate to be at a depressed area, as shown in Fig. 7.13, on the side of a part. When this inset gate is broken off, the extension will not protrude beyond the outside contours of the part and interfere with fit to adjacent parts. Gates should be located, when possible, on a hidden area of the product.

Figures 7.14 through 7.22 show the various types of gating that are generally used today.

Large complicated parts may be transfer molded as shown in Fig. 7.23. This part is molded in diallyl phthalate glass-filled material, and due to the long glass fillers, it is necessary to use a large runner and gating system. Such gating is needed for some glass filled compounds.

This particular design required a material charge of approximately 1350 grams with a 5-in.-diameter transfer plunger, and fan-type of runner and gating system from 1¾ to 3½ in. × ¼ in. deep. This system permits a good

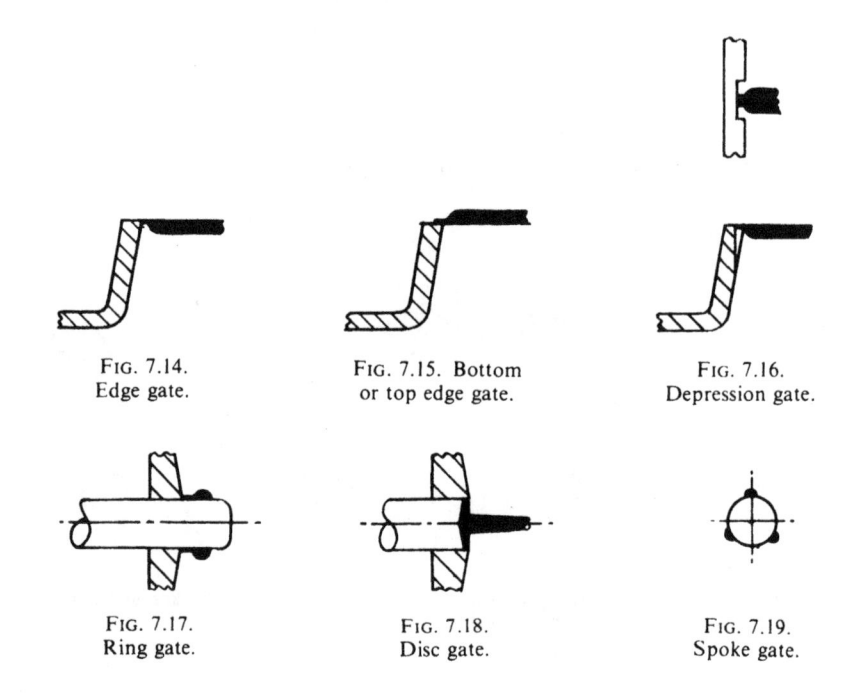

Fig. 7.14.
Edge gate.

Fig. 7.15. Bottom
or top edge gate.

Fig. 7.16.
Depression gate.

Fig. 7.17.
Ring gate.

Fig. 7.18.
Disc gate.

Fig. 7.19.
Spoke gate.

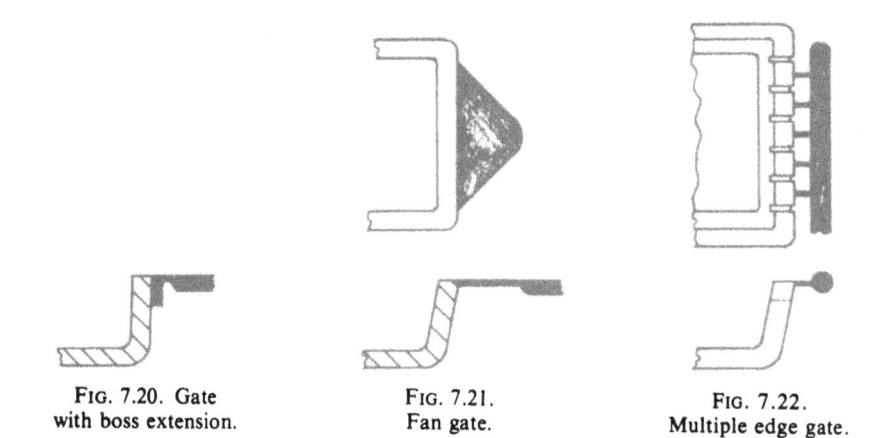

FIG. 7.20. Gate
with boss extension.

FIG. 7.21.
Fan gate.

FIG. 7.22.
Multiple edge gate.

flow of the material from the pot into the cavities and around the many delicate molding pins and cores.

Large moldings of this type require higher costs in degating operations and in some cases it is necessary to use special diamond cutting wheels.

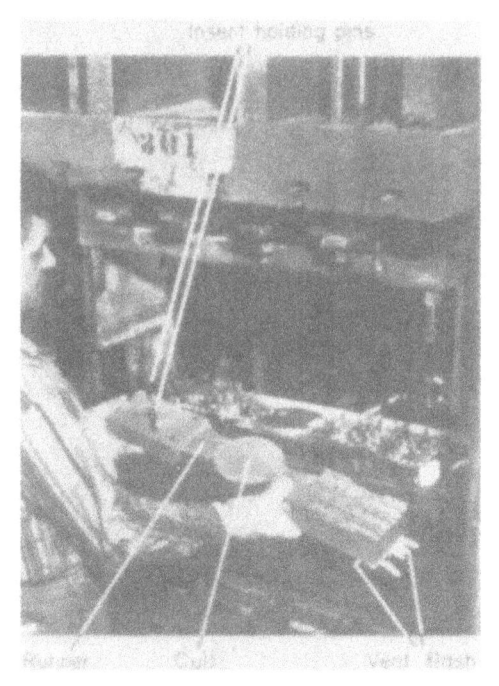

FIG. 7.23. View of 2-cavity circuit-breaker case and cover showing enlarged fan-gating system. Note the insert holding pins in the left piece and the vent flow material at the right. (*Courtesy Tech Art Plastics Co., Morristown, NJ*)

PLUNGER TRANSFER MOLD (EXAMPLE) OF BASE AND COVER

Fig. 7.24A-B shows a fuse case and cover molded of general-purpose phenolic material.

The product drawing for these parts is shown in Fig. 7.25A-B. Note

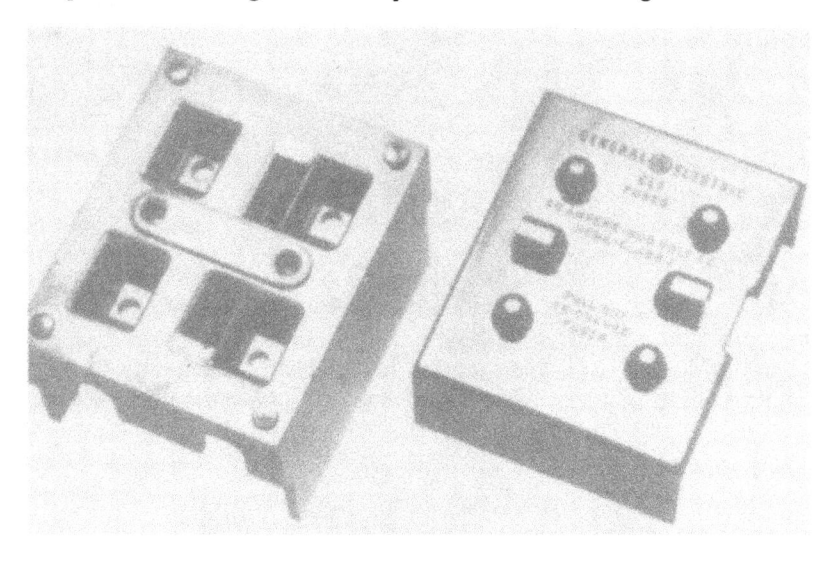

Fig. 7.24A. Fuse case and cover to be molded of general-purpose phenolic compound.

Fig. 7.24B. Under side of case and cover.

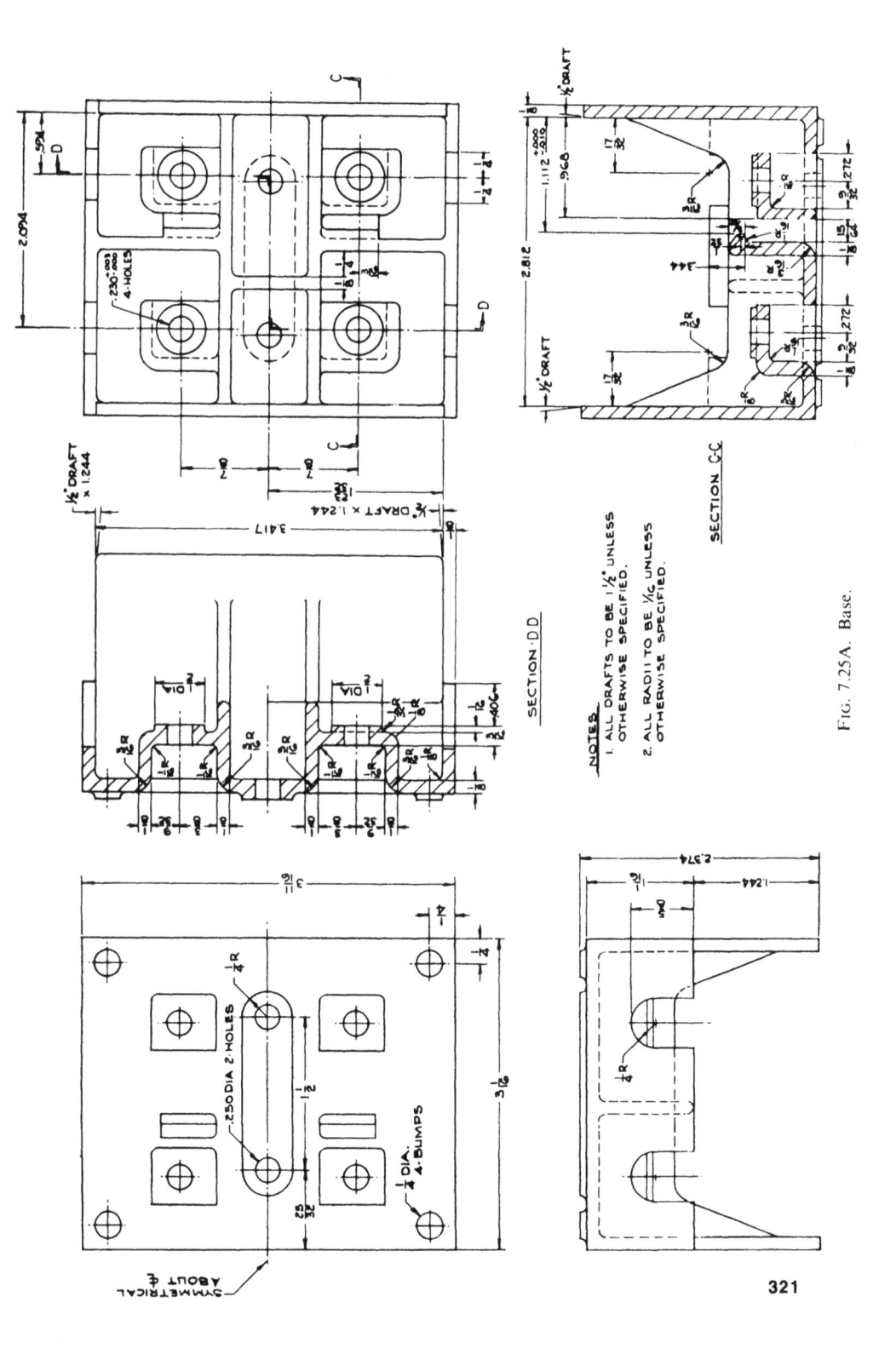

FIG. 7.25A. Base.

NOTES

1. ALL DRAFTS TO BE 1½° UNLESS OTHERWISE SPECIFIED.

2. ALL RADII TO BE 1/16 UNLESS OTHERWISE SPECIFIED.

SECTION-DD

SECTION-CC

SYMMETRICAL ABOUT ℄

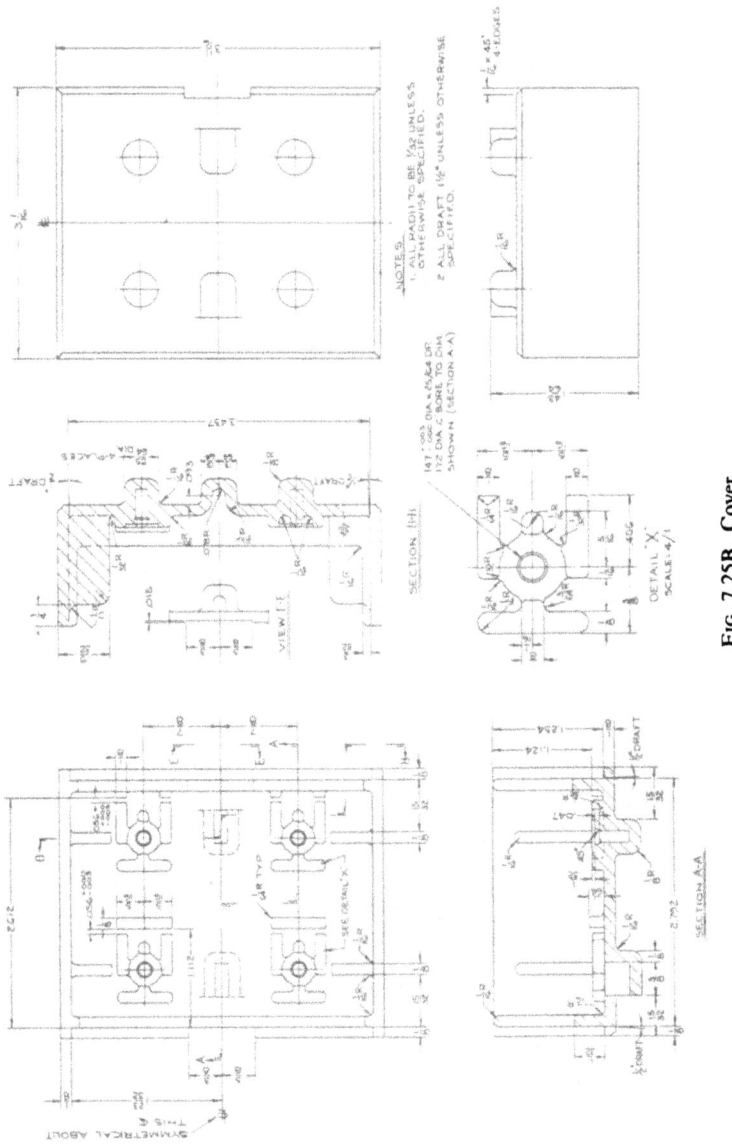

Fɪɢ. 7.25B. Cover.

PLASTICS MOLD ENGINEERING

Area of part   11.293 _____ sq in.

A. Area of 2 parts   22.586 sq in.
B. Area of 2⅝ in. diameter plunger   5.412 sq in.
C. Area of runners   1.732 sq in.
D. Total area   29.730 sq in.
E. Force of 7½ in. diameter top cylinder   99,400 lb
F. Force of 14½ in. diameter clamp cylinder   371,500 lb
G. Transfer pressure (no reducer)   18,365 psi
H. Transfer pressure at balance
   point of _____ lb, clamp   12,495 psi
E.' Area of top cylinder × 2250 psi
F.' Area of clamp cylinder × 2250 psi
G.' E/B
H.' F/D

FIG. 7.26. Pressure work sheet—2-cavity base and cover.

ample tapers, sharp corners removed, simplicity in forming side holes, and gating location.

Figure 7.26 is the pressure work sheet which was used to begin a tooling design. Note that the transfer pressure available is more than that required for the material and it was reduced during molding operations by adjusting the transfer plunger line pressure.

The profile sectional view is presented in Fig. 7.27. Note the uniform spacing of cartridge heaters, uniform ejection system with self-contained unit for press setup arrangement, ample support pillars, ease of venting, and over-all ease of machining. The mechanical strength is built in the mold design for long production life.

Figure 7.28 is view "A" of the assembly drawing and shows the top half assembly of the mold. This view is self-explanatory. Note the ease of machining, good construction strength, tapped eye bolt hole for easy tool room or press handling, temperature control for thermocouple, gating design, and press mounting slots. Figure 7.29 is view "B" of the assembly drawing and shows the lower half assembly of the mold. Noteworthy is the uniform ejection system, well-balanced support pillar areas, eye bolt tapped holes for ease of handling, return pin locations, heater and thermocouple holes, press mounting slots for ease of machining, and good mechanical support.

Figure 7.30 is the end view showing the detail of the lower ejection assembly and mounting slots for installation in the press. Figure 7.31 shows the completed mold.

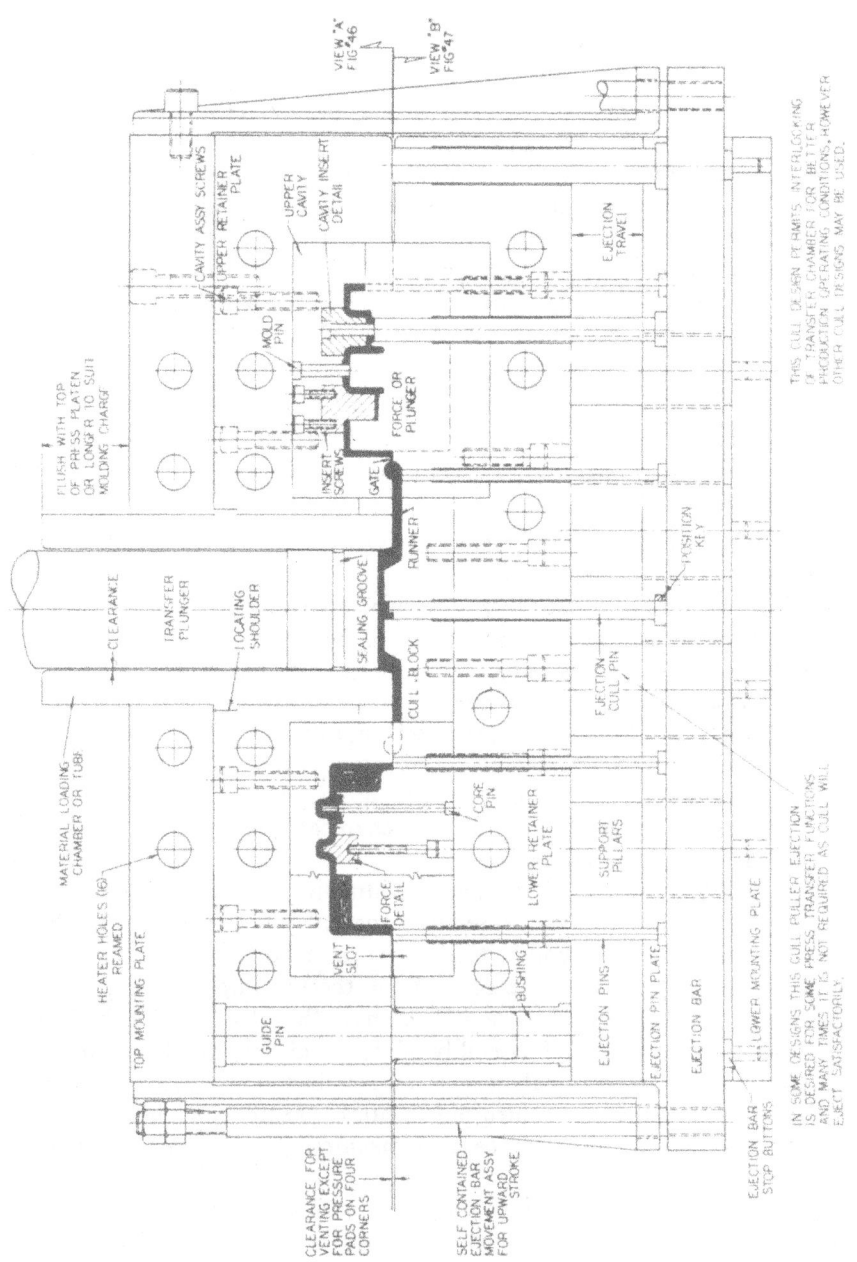

FIG. 7.27. Plunger transfer mold base and cover. Profile sectional view.

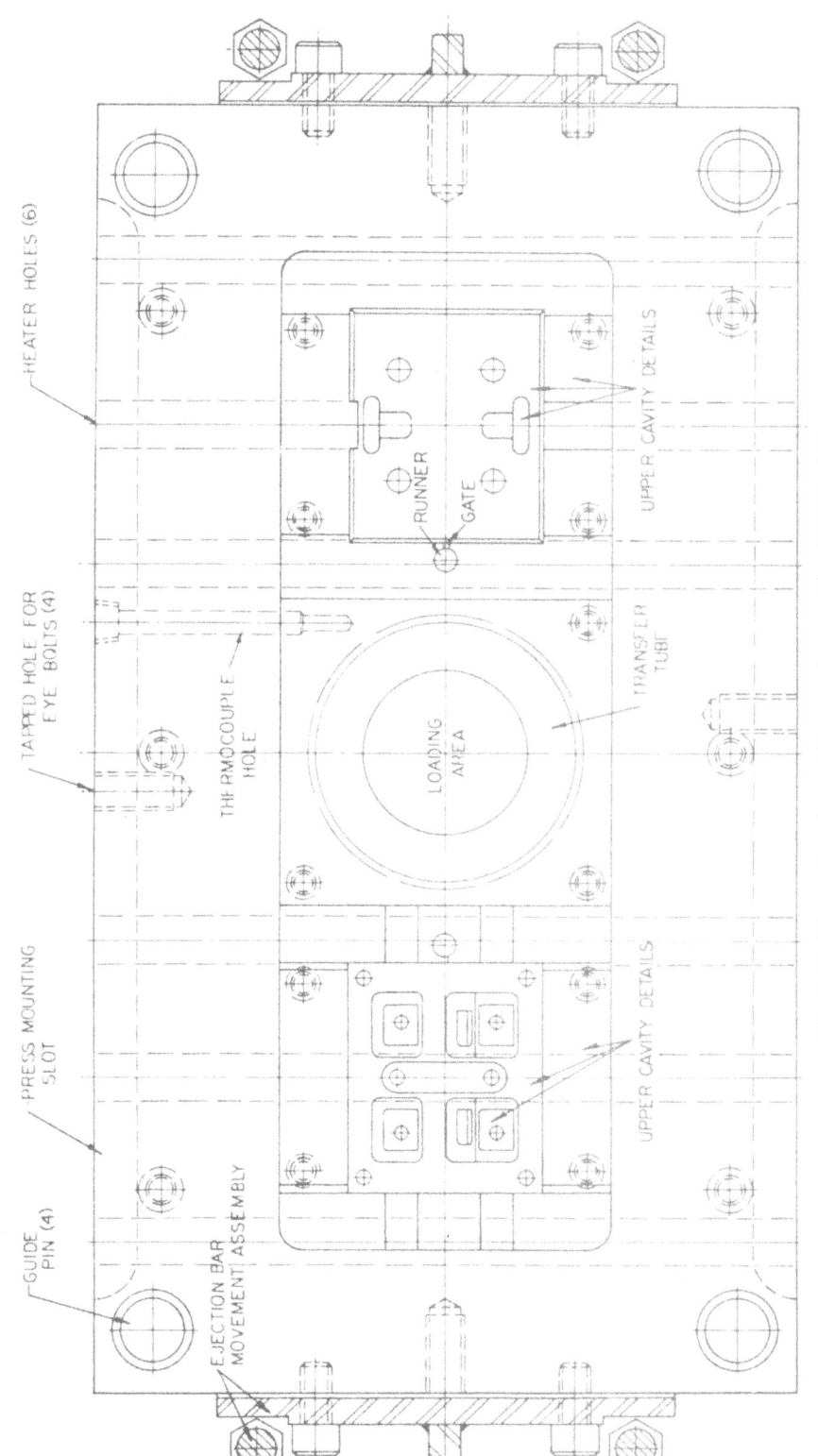

Fig. 7.28. Plunger transfer mold base and cover. View "A" Fig. 7.27.

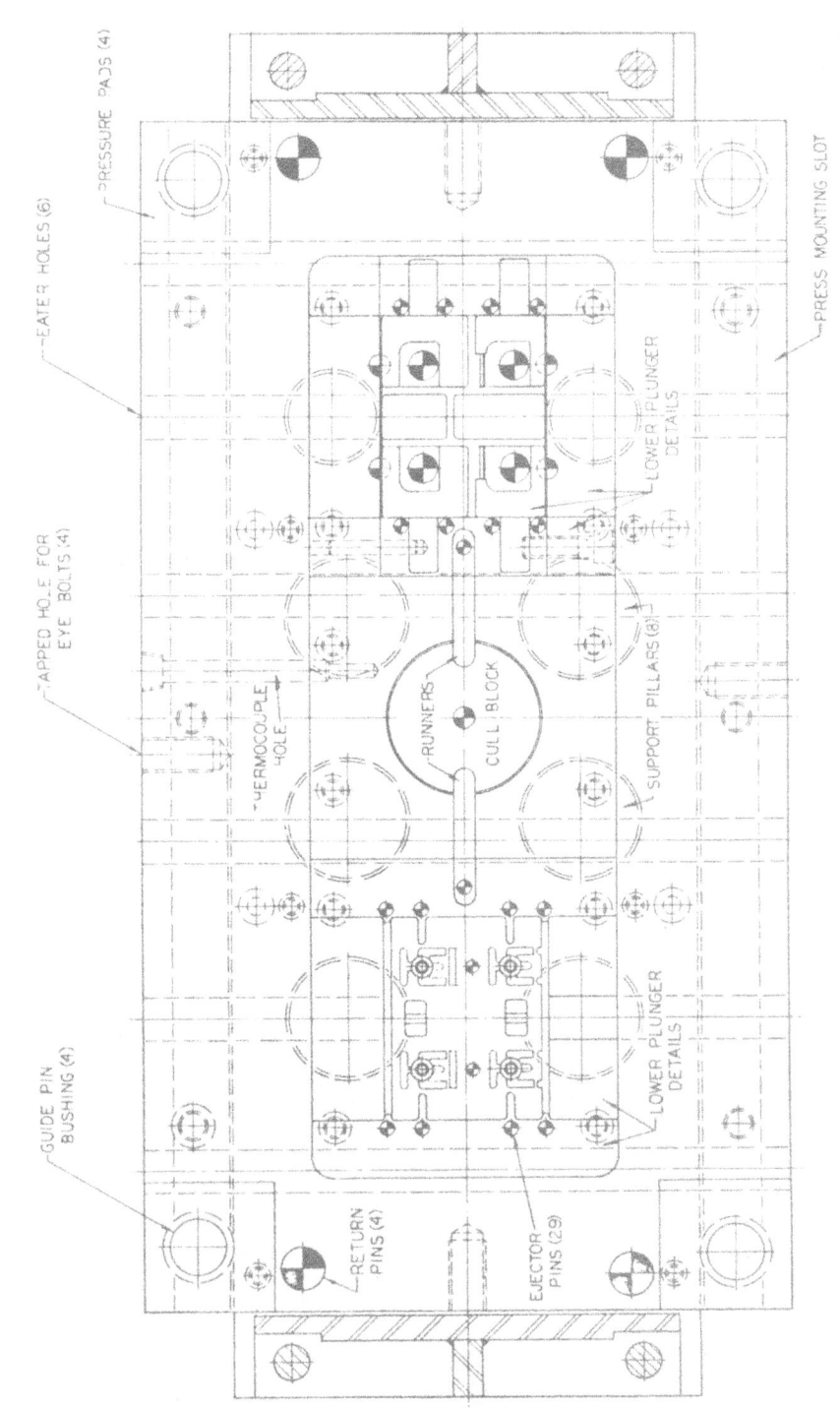

Fig. 7.29. Plunger transfer mold base and cover. View "B" Fig. 7.27.

PRESSURE PADS (4)

HEATER HOLES (6)

PRESS MOUNTING SLOT

LOWER PLUNGER DETAILS

TAPPED HOLE FOR EYE BOLTS (4)

THERMOCOUPLE HOLE

RUNNERS

CULL BLOCK

SUPPORT PILLARS (8)

GUIDE PIN BUSHING (4)

LOWER PLUNGER DETAILS

RETURN PINS (4)

EJECTOR PINS (29)

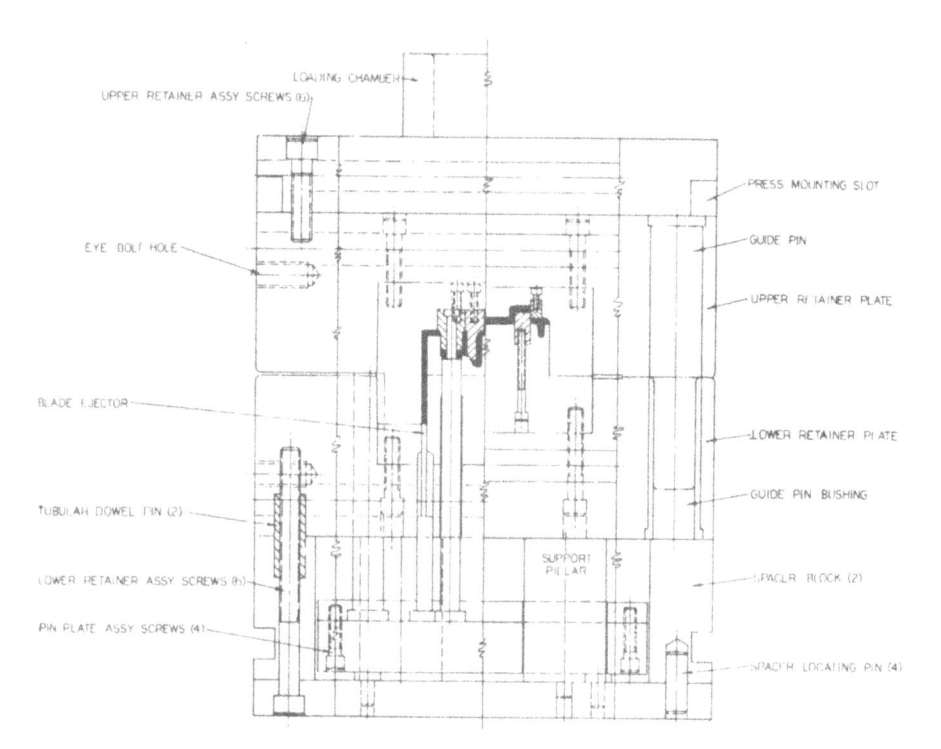

FIG. 7.30. Plunger transfer mold base and cover. End view.

FIG. 7.31. Two-cavity combination plunger-transfer-mold for fuse case and cover.

Types of steel, hardening, finishing requirements, venting clearances, etc., are generally the same as used in conventional-type transfer and compression molds.

Where possible, runner system and mold gating areas should be made of replaceable mold inserts for easy maintenance since the areas wear through long use and abrasion. Mold design consideration and construction must be executed with regard to ease of repair and maintenance through the medium of mold inserts, standard size ejectors, and mold parts. However, as a word of caution, when mold cavities or plungers are fabricated with sections, be sure that you have clearly defined the mold fitting area of the flash line to avoid future assembly or design appearance problems when such assembled parts are on an appearance or mating surface.

STANDARD MOLD FRAMES FOR TRANSFER MOLDS

Many molders have developed their own standard frame designs, whereby the frame is mounted into the press with a master ejection setup and a retainer for mold cavities and plungers.

Commercial standard mold frames are available from several sources. Figures 7.32 and 7.33 show one type of these unit frames. Figure 7.34 shows mold units which are made to fit these frames. Various sizes are offered and

FIG. 7.32. Standard mold frame for transfer mold. (*Courtesy Master Unit Die Products, Inc., Greenville, MI*)

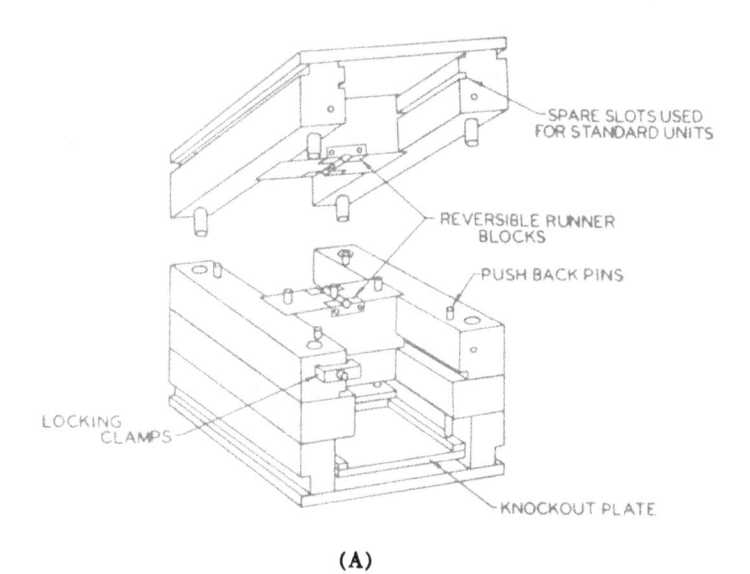

(A)

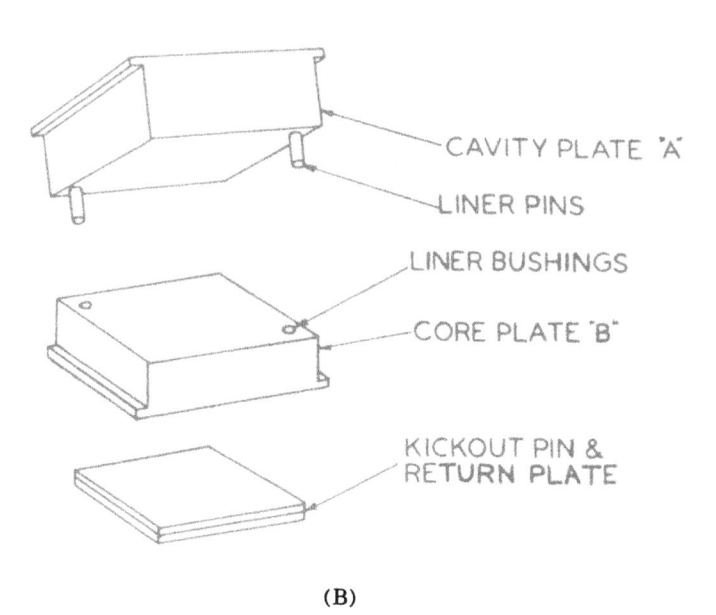

(B)

FIG. 7.33. (A) Master die set. (B) Master unit die insert. (*Courtesy Master Unit Die Products Inc., Greenville, MI*).

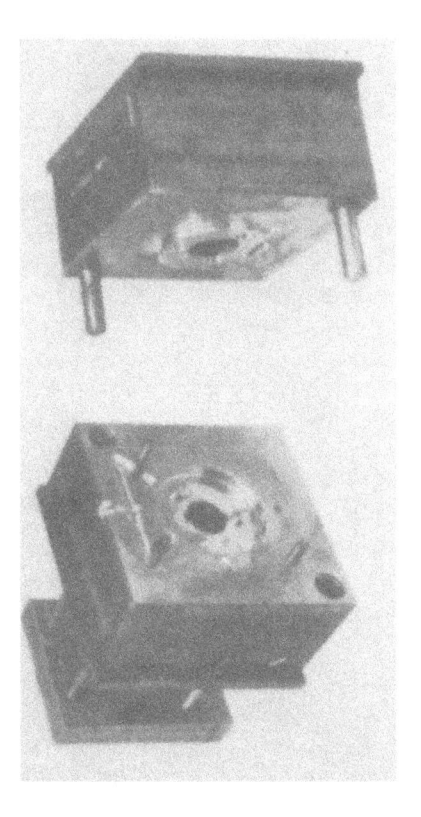

FIG. 7.34. Single-cavity mold unit for assembly in standard frame for transfer molds. (*Courtesy Tech Art Plastics Co., Morristown, NJ*)

considerable savings are encountered in design, mold construction, and production installation time.

Basically, the unit consists of the cavity and plunger, cut solid into the unit; many sets are installed with an auxiliary ejection assembly which is controlled through the master frame set up in a press. Different parts may be molded at one time in separate units as long as the molding material is the same for both parts.

INJECTION MOLDING OF THERMOSETTING MATERIALS

Molded thermosetting products produced by the injection (automatic transfer molding) process offer a cost savings and greater production capacity. Undoubtedly this process will supersede the older transfer processes for most jobs. The following data are given to describe the in-line process. A typical press is shown in Fig. 7.35. An important gain from injection molding results from the fact that smaller and less costly molds can be used to obtain the desired production capacity.

FIG. 7.35. Typical thermoset injection molding machine used for materials with short filler fibers. (*Courtesy Stokes Div., Pennwalt Corp., Philadelphia, PA*)

The thermosetting injection machine is similar to the thermoplastic injection machine with the exception of the heating barrel design, reciprocating screw configuration, nozzle and press controls. Machines are available with interchangeable components to facilitate injection molding of thermosets and thermoplastics. The screw design differentials are shown in Fig. 7.36. Material temperature in the plasticizing chamber is much lower for thermosets than for thermoplastics. Mold temperatures for thermoplastics are often much lower than those used to cure the thermosets. Special thermosetting compounds formulated for injection must be used.

An example of these temperature differentials is given below.

Thermosetting phenolic	Thermoplastic polycarbonate
130 to 240° F	Barrel Temperature 530 to 570° F
330 to 410° F	Mold Temperature 180° F

Experience has proved that thermoset moldings having heavy or ribbed sections will cure faster than the time needed to cool-harden a similar thermoplastic part.

Thermoset injection, as compared with other thermosetting processes,

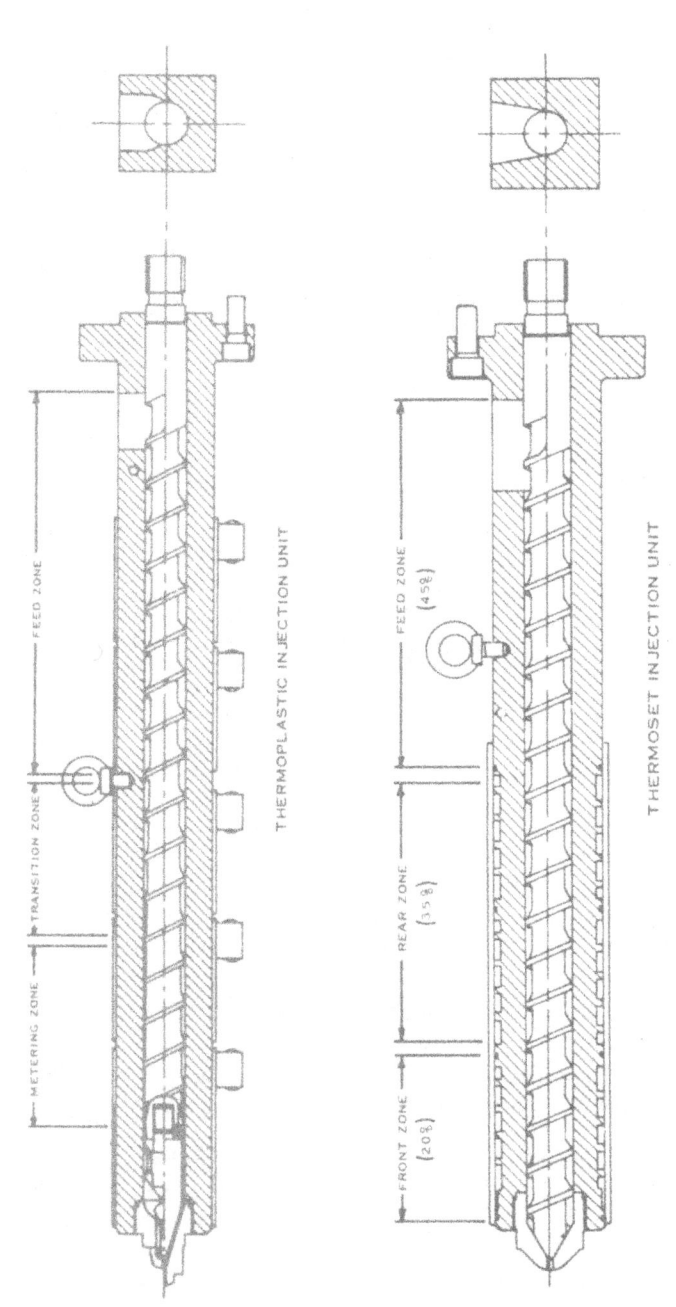

FIG. 7.36. (Courtesy Ingersoll Rand Co., IMPCO Division, Nashua, NH)

has resulted in "net cure" reductions as follows: These studies were made on a piece with a 0.250 wall section in four different molding systems.

1. Cold powder compression—60 sec net cure time. Compression
2. Radio frequency preheat   40 sec net cure time. Compression
3. Radio frequency preheat   30 sec net cure time. Transfer
4. Thermoset injection   24 sec net cure time. Injection

The foregoing analysis may be used only as a guide factor in the process selection. Noteworthy is the fact that the net cure time has decreased 60% since the start of thermosetting molding with contemporary materials. Injection molding of thermosets when used for product designs having heavier wall sections show far greater reduction of molding time; recent analysis of comparable molding cycles have demonstrated time reductions of 27 to 125%. See Fig. 7.37.

Another most desirable gain from thermoset injection is the elimination of preforming, preheating, and subsequent material handling. Preforming is costly, requires production control for inventories of the various sizes and preform materials. In the injection process, the bulk powder is loaded directly into the press hopper.

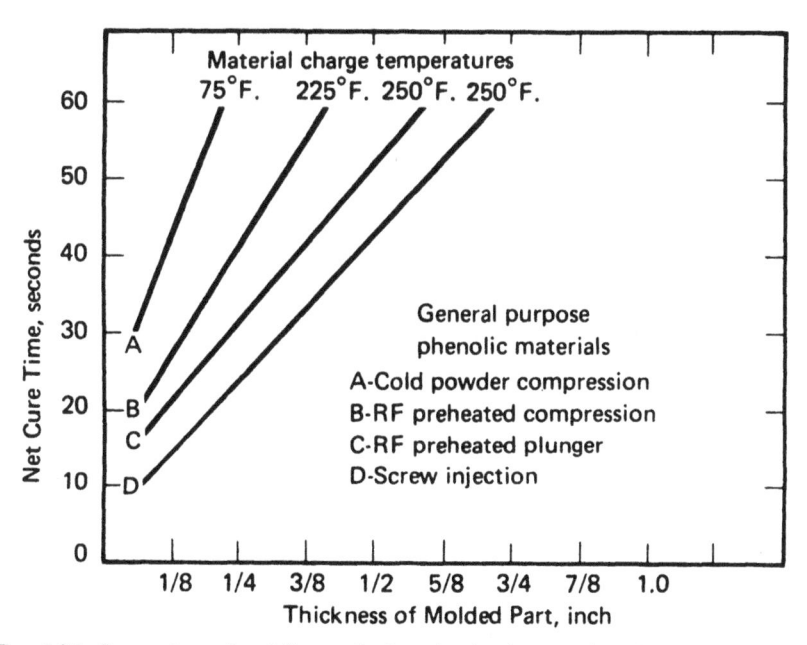

FIG. 7.37. Comparison of molding methods: reduction in cycle times from use of preheated charges.

Experience now predicts that all contemporary methods for the molding of thermosets will continue to be used in the foreseeable future. The mold designer and product engineer must select the right process for each job within the limits of press availability and the economics. Caution must be exercised, since the screw injection of thermosets with abrasive fillers will cause screw and mold wear; Runner and gate wear may indicate the need for very hard alloy gating areas. Presses are available for vertical or horizontal injection in various sizes.

Injection modules as shown in Fig. 7.38 are available for the conversion of existing transfer and compression presses to thermoset injection. The forgoing systems, eliminate the airborne dust that is so objectionable in

FIG. 7.38. Extruder module for screw preheat system which may be used for existing compression or transfer molds. (*Courtesy Alliance Mold Co. Inc., Rochester, NY*)

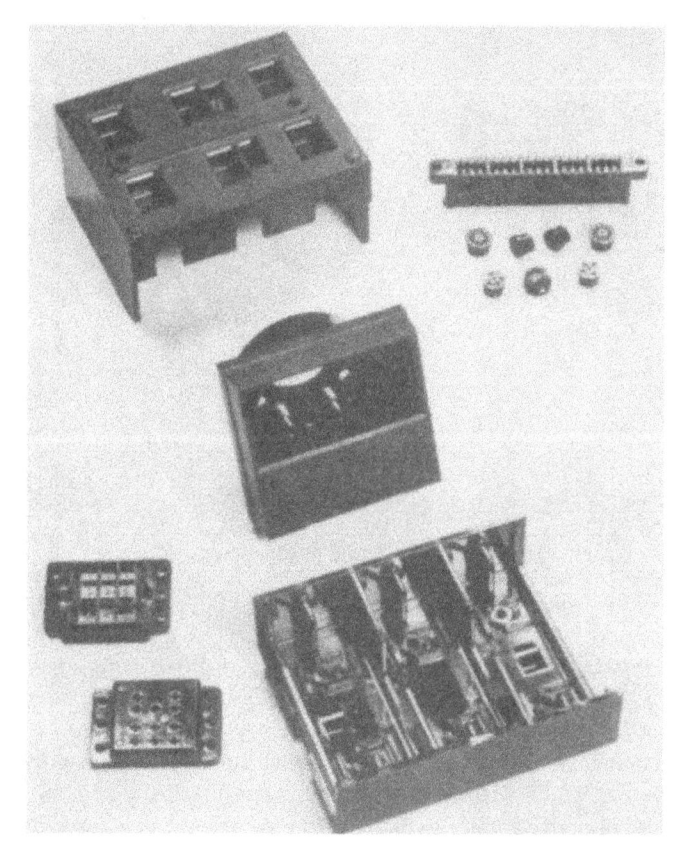

FIG. 7.39. Thermoset moldings from injection molding process. (*Courtesy Tech-Art Plastics Co., Morristown, NJ*)

the older transfer and compression molding processes. Typical injection molded thermosetting products are shown in Fig. 7.39.

Glass polyester premix and BMC materials have led to the development of coaxial plunger injection machines and dual plunger machines for injection molding with minimal fiber degradation as shown in Figs. 7.40, 7.41, and 7.42. Products molded by this process exhibit uniform consistency and are able to retain maximum fiber strength. Fig. 7.43 illustrates the type of mold often used for this work.

Cold Manifold Thermoset Injection Molds.

Thermosetting compounds are also injection molded by reciprocating screw injection machines in molds that are designed to keep the material below the

FIG. 7.40. Glass polymer premix and BMC molding press. (*Courtesy Litton, New Britain Plastics Machine Div., Berlin, CT*)

curing temperature until it is in the cavity. The two general types of such molds are "warm runner" and "cold manifold." Such molds are sometimes called *insulated runner molds.*

Some insulated runner molds for short run jobs use a construction as shown in Fig. 7.43 with considerably enlarged sprue and runner systems that are insulated with air gaps from the cavities. Such molds require considerable patience in starting to gain the thermal balance needed to prevent the compound from setting up in the runners or sprues. Latches are provided to lock the two halves of the runner plate together during normal operation and to permit easy removal of the sprue and runners in the event of accidental hardening in the sprue/runner system. The cold manifold design for fully automatic injection molding of thermosetting compounds offers many advantages and is recommended for all molds that are to run in reasonable or large volume.

Runnerless molding of thermosets has an advantage for long running jobs because of the material savings. Few thermosetting materials can be salvaged by grinding and blending after the cure. Conventional three-plate molds, as depicted in Fig. 7.43A for thermosets, waste the material in the sprue, runner, and gate. Cold manifold molds may cost 10 to 20% more than the simple three-plate mold, and are justified only by the material saved. Cold manifold molds built at a $3000 to $4000.00 premium have saved $10,000 to $20,000 per year in material. A simple check to determine the

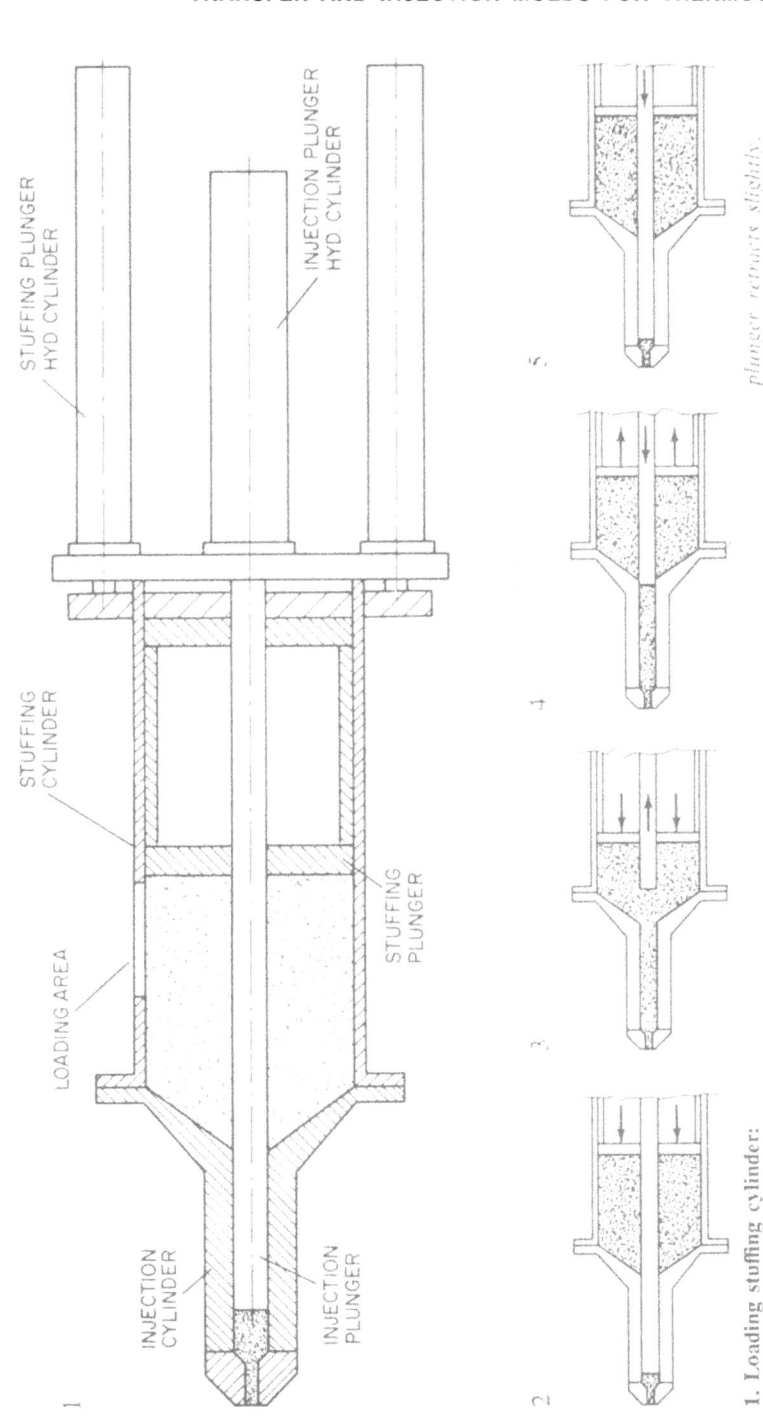

1. Loading stuffing cylinder: injection plunger forward, stuffing plunger retracted.
2. Compressing; stuffing plunger advances, compacting premix to stuffing plunger advances to preset mechanical stop.
3. Loading injection cylinder: injection plunger retracts, fill injection cylinder.
4. Readying for shot: injection plunger advances, packing injection cylinder; stuffing plunger retracts slightly, decompressing stuffing cylinder.
5. Full shot: injection plunger advances, filling mold.

Fig. 7.41. Schematic—Injection molding press with two in-line plungers for glass polyester premix and BMC materials. (*Courtesy Litton, New Britain Plastics Machine Div., Berlin, CT*)

FIG. 7.42. Thermosetting polyester premix injection molded with minimum fiber breakage by the use of a stuffer (top cylinder) and a piston plunger to force the compacted material through the heating barrel into the mold. Screw plasticization is not used because it breaks the glass fibers and strength is lost. (*Courtesy Hull Corporation, Hatboro, PA*)

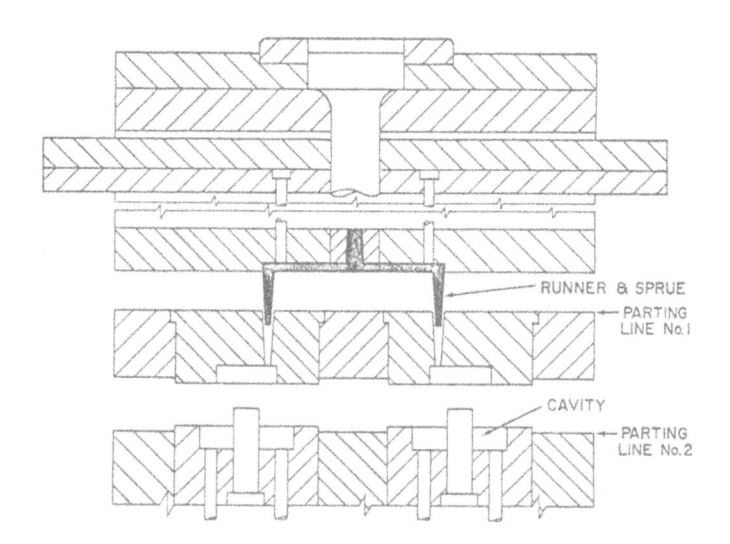

FIG. 7.43. Depicts the conventional 3-plate injection mold as used for short run injection molding jobs. In this case, the sprue and runners harden and are ejected after each shot. This basic design is modified as shown in Fig. 7.43A—warm runner or runnerless mold.

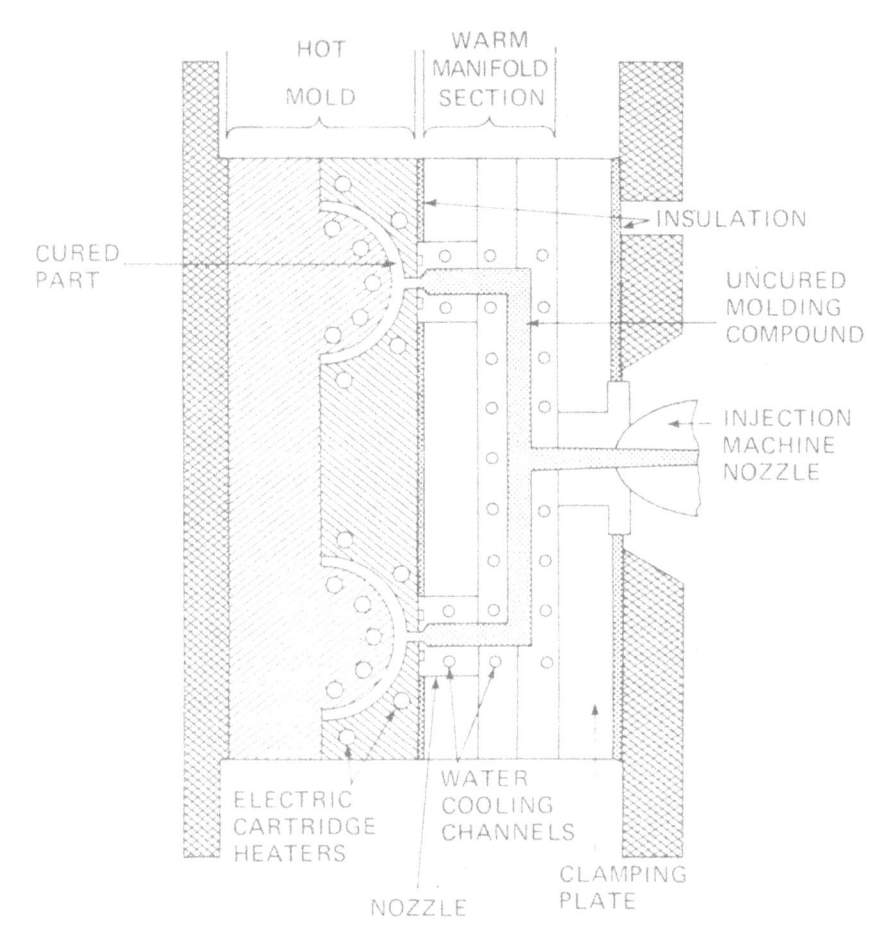

FIG. 7.43A. This schematic mold design illustrates the warm runner or runnerless mold. Note the enlarged runner and sprue areas.

justifiable premium for cold manifold molds is to calculate the value of the material lost in the runners in the simple three-plate mold.

The patents* on such molds will provide very instructive study and care must be exercised in their use.

A typical cold manifold mold design is shown in Fig. 7.44. Note that an insulating barrier is placed between the cool manifold section and the cavities. This insulation may be high temperature thermoset laminate, transite

*Cold manifold mold Patents: 3,591,897—3,499,189 and 3,189,948, owned by Litton Industries and 3,819,312 by Stokes Div.

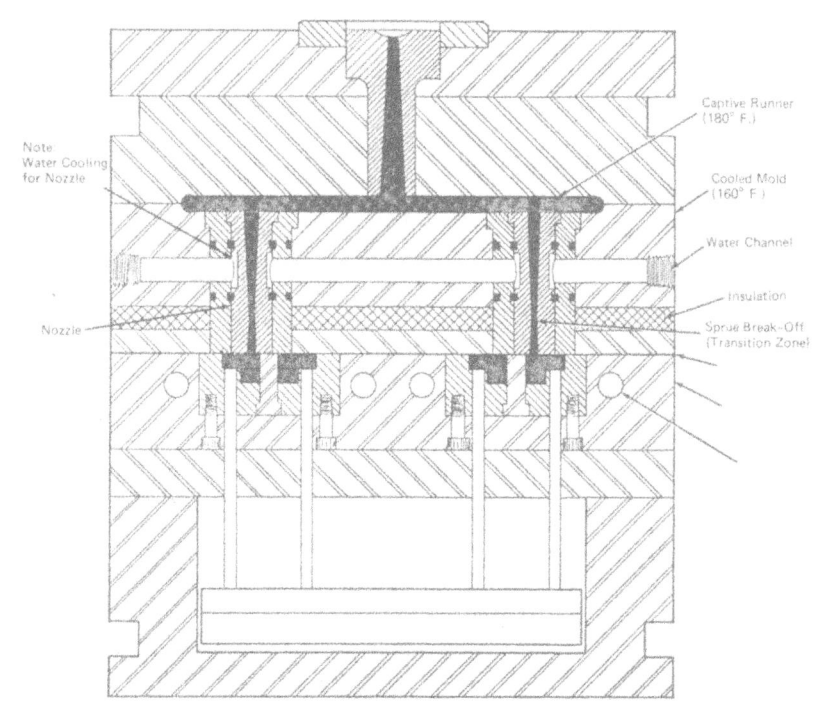

FIG. 7.44. Typical cold manifold design

asbestos, or glass-bonded mica. Glass-bonded mica is preferred when the insulation must maintain parallelism and dimensional control since it may be ground with the same precision and dimensional stability as steel and perform up to 750° F. It will be seen also that the sprue section, as well as the runner section of the manifold, is cooled so that the cure will take place outside the manifold components. Independent nozzles may be used or, one nozzle may serve a cluster of cavities.

Another type of cold manifold mold is shown in Figs. 7.45 and 7.46. In this design, which was patented by the Stokes Division, the mold frame is similar to the conventional injection mold frame. Runner and sprue sections are insulated from the cavity components with suitable thermal control to prevent precure of the materials.

Balanced runner systems in the manifold and cavity layout are mandatory; material must reach each cavity in the same transportation time. Off-center gating is used as shown in Fig. 7.47 to equalize and minimize the flow distance in the cavity.

The nozzles or sprues must be provided with sensitive temperature control since material in the entry half must remain plastic while the material hardens

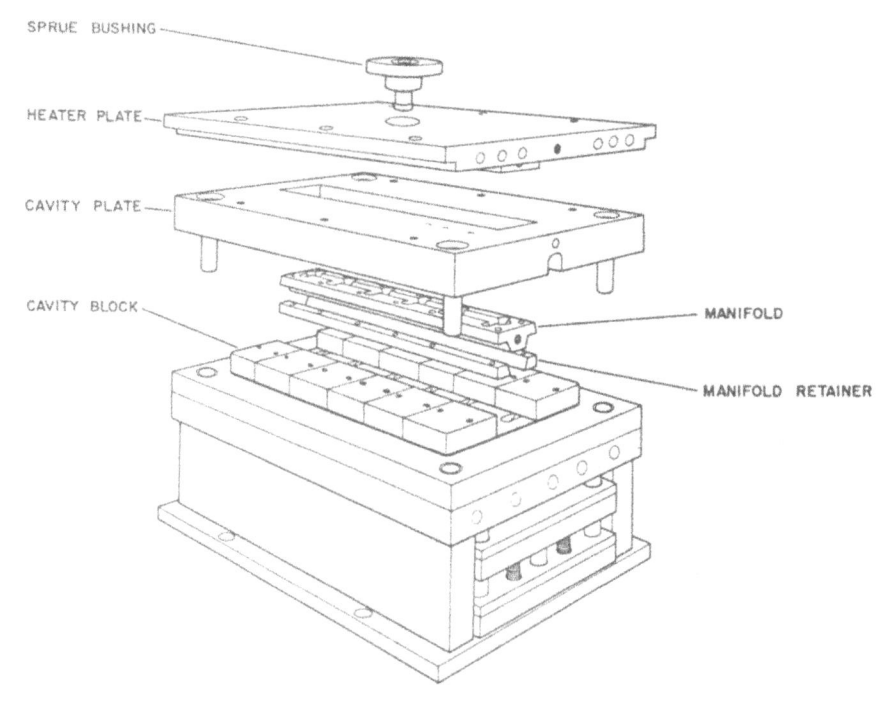

FIG. 7.45. Exploded view of a Stokes cold runner manifold. (*Courtesy Stokes Div., Pennwalt Corp., Philadelphia, PA*)

FIG. 7.46. Manifold of Stokes design cold runner mold as depicted in Fig. 7.44.

Fɪɢ. 7.47. Instrument housing with distributive gating.

in the exit half. These nozzles or sprues may feed directly into a heavy section of the part or feed a cluster of cavities with tunnel or edge gating. The cluster runners in the cavity section should be as small as possible to save material. Nozzles must be insulated from hot cavity plates. This is accomplished with a series of air gaps and water cooling channels. Water coolant control capability must be ±5°F or better.

Direct, edge and tunnel gating are being used. Carbide inserts are recommended since the filled thermosets are abrasive and gate wear must be anticipated. Gate locations take into consideration the usual problems of weld lines, venting, product appearance, gate removal cost, etc. Fan gates will break off satisfactorily for many applications; round gates must be cut where appearance is consequential. Tunnel gate details as illustrated in Fig. 7.48 are excellent where applicable. Direct sprue gating into a central disc, as shown in Fig. 7.49, is reminiscent of gating with the older integral transfer mold. Table 7.1 explains the strength loss that may be expected in various gate/runner areas and serves as a guide in gate designing. Tables 7.2, 7.3 and 7.4 depict the comparable cure times and the shrinkage of typical cold manifold formulations of phenolic compound. Manufacturers of thermosetting compound have done an excellent job in the preparation of special formulations that facilitate trouble-free injection molding of thermosets. Basically, these thermosets must have long flow duration so they retain their plasticity for the extended period of time that they are in the manifold. Additionally, these injection thermosets must cure rapidly at the "set" temperature in the mold to gain the large resulting economies.

Small gating areas result in additional frictional heating of the material as it enters the cavity, resulting in minimum cures. Tunnel gating for thermosets is most desirable to achieve automatic operation of molds with minimum product disfiguration. The success of tunnel gating in thermosets is dependent on the angle, location, ejector pins, and the hot flexure of the material being molded. The following data on tunnel gate design are adapted from a paper by Stephen H. Bauer of Hull Corporation entitled, "Tunnel Gating Thermosets."

The angle of the tunnel gate is quite important. Various tunnel angles have been evaluated. As shown in Fig. 7.48B, there are two angles with which

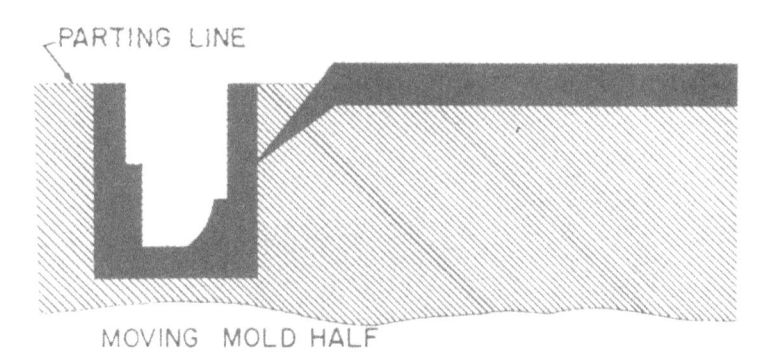

A. TUNNEL GATE

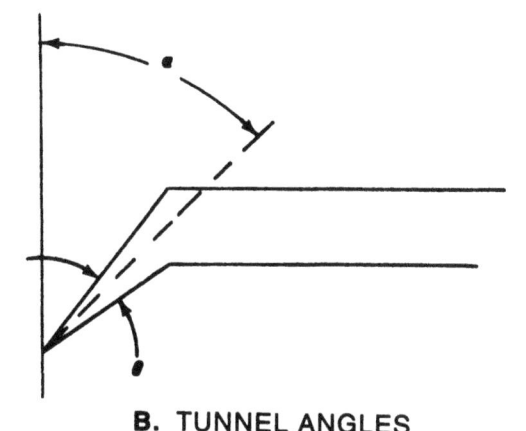

B. TUNNEL ANGLES

Fɪɢ. 7.48. Tunnel gating. (*Courtesy of Hull Corporation, Hatboro, PA*)

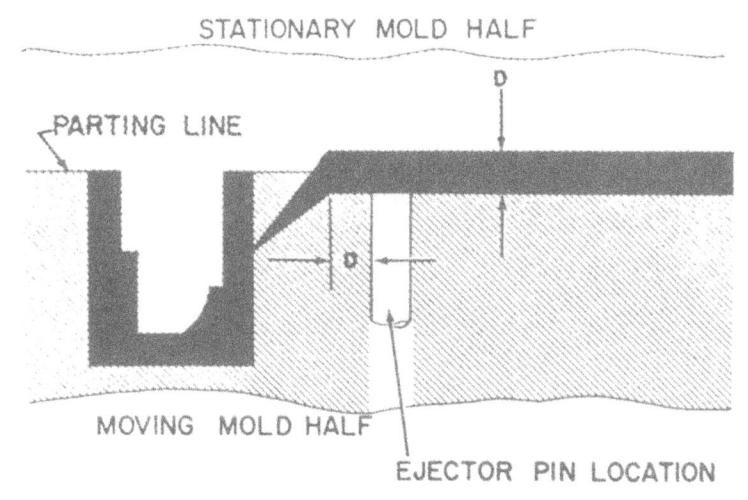

C. TUNNEL GATE

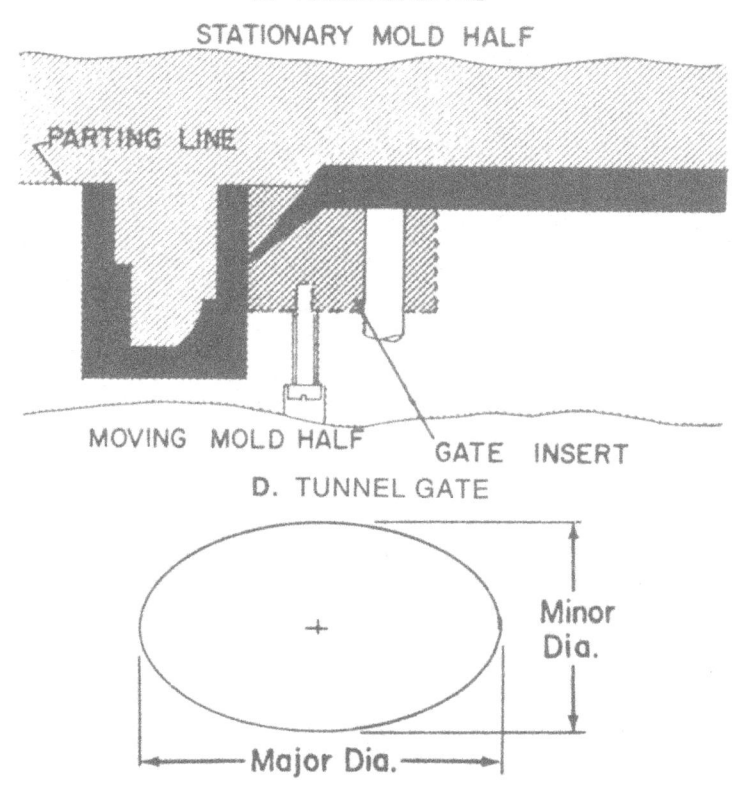

D. TUNNEL GATE

E. ELLIPTICAL GATE—created by tunnel gate intersecting a flat wall in the part.

Fig. 7.48. (*Continued*) Additional tunnel gating types.

TABLE 7.1.   Physical Properties of Injection-Molded Parts Related to Runner and Gate Area for General-Purpose Phenolics.

	Runner and Gate Area, sq in.			
	0.055	0.061	0.116	0.122
Properties	Values			
Injection pressure, psi	1200	950	875	700
Izod impact strength, ft-lb/ in of notch	0.26	0.26	0.26	0.27
Drop-ball impact strength, in.	10.0	10.2	10.7	11.2
Flexural strength, psi	9100	8200	9600	8900
Molding shrinkage, mils/ in.	10.9	10.5	10.1	9.6
Cure speed, sec	45	50	45	50

(*Courtesy General Electric Co.*, *Pittsfield, MA*)

TABLE 7.2   Cure Time.

Cold Manifold Mold		Conventional Mold	
Material	Min. Cure	Material	Min. Cure
P4000-CH	40 sec	P4000 Flow M	40 sec
P4300E-CM	45 sec	P4300E Flow M	40 sec

TABLE 7.3.   Shrinkage.

Cold Manifold Mold			Conventional Mold		
Material	Shrinkage $\rightarrow$	$\downarrow$	Material	Shrinkage $\rightarrow$	$\downarrow$
P4000-CH	.008	.008	P4000 Flow M	.010	.008
P4300E-CM	.008	.007	P4300E-M	.009	.006

$\rightarrow$ Shrinkage perpendicular to direction of flow.
$\downarrow$ Shrinkage in line of flow.
(*Courtesy General Electric Co.*, *Pittsfield, MA*)

to be concerned. First is the centerline angle of the tunnel (α) and the second is the included angle (θ). If the centerline angle is too shallow, the tunnel will not eject; if the angle is too deep, construction of the mold is difficult. When the tunnel becomes too long, its cleanliness becomes a problem. Success has been achieved with centerline angles ranging from 25 to 30° and with included angles ranging from 30 to 35°. Different materials may require different angles and much is to be learned by trial.

TABLE 7.4. Rigidity.

	Cold Manifold Mold		Conventional Mold	
	P4000-CH		P4000 Flow M	
	Cure	Rigidity	Cure	Rigidity
	40 sec	32 mils	40 sec	35 mils
	45 sec	24 mils	45 sec	31 mils
	50 sec	20 mils	50 sec	20 mils
	P4300E-CM		P4300E Flow M	
	45 sec	38 mils	40 sec	17 mils
	50 sec	33 mils	45 sec	12 mils
	55 sec	24 mils	50 sec	7 mils

(Courtesy General Electric Co., Pittsfield, MA)

The location of the runner ejector pins is quite important in relation to size and type of runner. The ejector pins are located on the runner in a position that allows it to flex during ejection. If the ejector pin is located too close to the tunnel, the material will not flex during ejection and it will shear and stay in the tunnel, blocking the next shot. If the ejector pin is placed too far from the tunnel, there might not be enough ejector force to eject the tunnel from the tunnel cavity. When ¼ in. full round runners in various mold designs are used it has been found that ejector pins should be located at a distance from the end of the runner equal to the diameter of the runner,

FIG. 7.49. Cosler wheel with direct disc gating.

as shown in Fig. 7.48C. In general the diameter of the ejector pin should be equal to that of the runner.

It is necessary to have good hot shear strength so that the material is not sheared and torn during the ejection cycle. Hot strength in a given thermoset is a function of the material characteristics, flow and percentage of cure. The molding material must have good flexural strength since it is necessary to bend and flex the runner and the tunnel to remove them from the cavities.

The gate design must include a careful blending of the runner into the tunnel. Thermosets do not flow well when abrupt changes in direction of material flow are required. This blending generally requires additional hand finishing of the mold but it is a worthy investment. The size of the gate orifice is a most important consideration. The orifice must be large enough to allow the material to flow into the cavity but at the same time it must be small enough to shear during ejection. The location of the gate on the part should be into a heavy, nonobstructed section to allow the material to flow readily into the cavity. As a thermoset flows through a restricted orifice, there is tremendous heat build-up in the material. This heat build-up is primarily a function of the frictional work being put into the material as it flows through the orifice. This heat build-up facilitates rapid cycles through high temperature crosslinking of the material in the cavity. At the same time it must be remembered that the frictional heat input is coming from frictional contact with the mold surface with the expected wear. All tunnel gates for thermosets must be readily replaceable as depicted in Fig. 7.48D. A steel such as "Stellite" and "Feratec" should be used for gating such sections. Best results have been obtained in gaining effective sprue removal by draw polishing and chrome plating the gating area.

The interchangeable insert should be constructed with two things in mind: (1) the ease of gate insert removal, and (2) the ghost line the insert will leave on the part since the gate insert is a part of the cavity. This leads to the necessity for locating the gate insert in a section of the mold that is duplicated easily.

At this stage of development it is necessary to determine by experimental work the size of the orifice for each new product design and material to achieve the best results. Small gates permit fast cures and easy ejection. Small gates wear more rapidly from the frictional work. Small orifices can build up stresses in the product which may show up as warpage, dimensional instability, and stress cracking. Excessively large gates lead to longer molding cycles while awaiting gate freeze-off. The intersection of the conical gate with the flat wall of the part creates an eliptical orifice. Users of successful tunnel gated molds today report oval orifices of .050 and .070 in. minor diameter as depicted in Fig. 7.48E.

Injection speed control of the thermoset injection press is quite important.

Tunnel gated molds are run slower on the injection part of the cycle than edge gated molds to minimize gate wear and precuring from the excessive frictional heat. Hydraulic ejection is recommended for tunnel gated molds to gain control of speed and force of ejection.

Accurate control of manifold and nozzles temperatures is essential and here the thermal control devices developed for the thermoplastics will be adequate. The nozzles for most thermosets must be maintained within a 155 to 185° F range. Best results are achieved with separate controls for each cavity or each cluster of cavities. One innovation makes use of a complex series of water flow dividers and flow meters to insure accurate temperature for each cavity and nozzle.

RUNNERLESS INJECTION-COMPRESSION MOLDS*

Before studying this particular section, you should have studied *compression*, transfer, and the previous part of this chapter on *injection molds for thermosets*. Otherwise, some of the terminology and description may not have much meaning for you as we describe and illustrate the patented system known as *runnerless injection-compression molds* (hereafter abbreviated as RIC-molds).

As the name implies, this design is a combination of several other types of molds and parting line requirements. Leon G. Meyer (see footnote) describes it: "It combines the molded part properties of compression molding with the speed of processing in an injection machine. We have added to this the dimension of completely runnerless, sprueless molding, hence the title *runnerless injection-compression*." Please note that coverage in this text is intended for wider dissemination of the technique, as well as urging the obtaining of a license for RIC$^{(tm)}$.

The RIC-mold is not suitable for all types of parts, nor is it likely to replace some of the compression, transfer and injection mold design concepts now in use for many years. The RIC-mold is indicated under conditions requiring some or all of the following criteria:

1. High production using fast cycles (under 10 seconds reported).

*This section written by Wayne I. Pribble and based on information supplied by Leon G. Meyer, and incorporating a report by Robert Q. Roy. It is used by permission of Occidental Chemical Corporation, DUREZ$^{(tm)}$ Resins and Molding Materials, North Tonawanda, NY. We are grateful to Leon G. Meyer for a review for publication.

U.S. Patents originally issued on RIC-molds and process are: US-4,238,181, Dec. 9, 1980; US-4,260,359, Apr. 7, 1981; US-4,290,744, Sept, 22, 1981; US-4,309,379, Jan. 5, 1982. Seven other U.S. Patents have been issued, and two european patents have been allowed.

For a complete list of patents and for license information, write to: Occidental Chemical Corporation, DUREZ Resins and Molding Materials, P.O. Box 535, North Tonawanda, NY 14120.

2. Dimensional stability of molded part (equivalent to dense compression molded parts).
3. Gates already sealed off and eliminates a removal operation. However, the savings in gate removal is partially offset by the need for a secondary operation to remove the parting line flash, just as is needed in compression molded parts.
4. Injection equipment available and sequenced for RIC-molding.
5. Uses vertical parting of mold (not recommended for insert molding but is recommended for automation).
6. Uses less tonnage for molding and clamping (larger part in same press, or a lesser tonnage press needed for the same part in straight injection mold).

From these characteristics, it should be evident that a RIC-mold will produce molded parts with a more uniform density and provide more uniform shrinkage, resulting in good dimensional control of the part. In addition, the RIC-mold lends itself well to automation of the molding process, when combined with a computerized process control unit for the injection machine.

The first requirement in operating an RIC-mold is to have available an injection machine which has been electrically and mechanically modified to provide a molding sequence of: partially close; inject material; fully close; cure; open and discharge; repeat. Because such press sequencing is readily available as a retrofit, or can be provided as an option by most press manufacturers, there is no need to elaborate here. The balance of this description will refer to the *mold design* and the modifications or combinations needed to achieve the desired result. To achieve a basic understanding of what end result is to be achieved, we can describe the process as differing from a "normal" injection cycle only because the mold is not fully closed at the moment of injecting the plasticized material. Under "normal" injection molding conditions, a partially closed mold would result in a heavily flashed mold, no density in the part, and, in general, what is known as a "mess" in any molding operation (a condition to be avoided at all costs!!). Thus, a mold resembling a *positive mold* used in compression molding becomes the "container" for the just-injected material. The use of screw injection allows the metering of a quite precise volume of plasticized material into this container. As soon as the low density fill is in place, the mold is *fully closed*, and the final action is the same as in compression molding. The "low density fill" is sometimes referred to as "injection-purging of preplasticized material through a manifold system."

The three basic types of RIC-molds are;

1. Single cavity molds for large parts, this injects directly into the cavity through a manifold system as an extension of the screw plasticizer.

2. Multiple cavity molds for small parts will use the flash-type parting line with two or more cavities in a subcavity arrangement. The material is injected through the manifold and into the confined space of the well and cavities. Final closing of the mold is a straight compression molding technique.
3. A multiple manifold system is used for multicavity molds for large parts, multicavity molds for deep draw parts, or multigrouping of subcavities utilizing the capability of a larger capacity press.

Design Criteria for RIC-Molds

For illustration purposes, we have chosen two of the three types to briefly discuss and point to the critical items requiring consideration. We are sure the reader will understand that the exact and precise details are patented and that the patent holder willingly supplies all of their know-how accumulated since the filing in 1978 (in exchange for the nominal license fee, of course). The direct injection type should be self evident, thus we will illustrate the second and third type listed above.

Figure 7.50 illustrates the multiple cavity for small parts centrally gated on the outside bottom of a cup-shape. The mold is in the partially closed position and is ready to receive the low density fill. The numbers refer to items in the description of Patent 4,238,181, and we will refer to only a few listed in that description. The manifold is essentially an assembly of plates (105 and 107) and bushings (21, 113, and 115) utilizing material channels (19, 109, 111, 117, and 119) into the cavity area (121 and 123). This assembly (bracket 103) is isolated from the hotter mold by means of insulating boards (131). Air space (125) is used to reduce heat conduction from the plate to the secondary sprue. Note that the sprue-gate segment (117 and 119) will remain attached to the molded part at ejection, and would require a secondary operation to remove same.

A study of Fig. 7.51 will disclose a manifold system similar to that of Fig. 7.50. However, this design provided for an edge-gate with squeeze-off blocks (not numbered) to eliminate the gate as it is shown in the fully closed, compression molding, part of the operation.

Another type of manifold is illustrated by Fig. 7.52. The need for maintaining the fluid state of the preplasticized material in the manifold requires a manifold bushing (17) with a controlled heat input through channel 33 connected to 35. This RIC-mold is also shown in the fully closed, compression molding, part of the operation.

With these preliminary descriptions as a basis, permit us to point to several points to be considered in designing RIC-molds, but neglecting the technical details which the liscensor would supply.

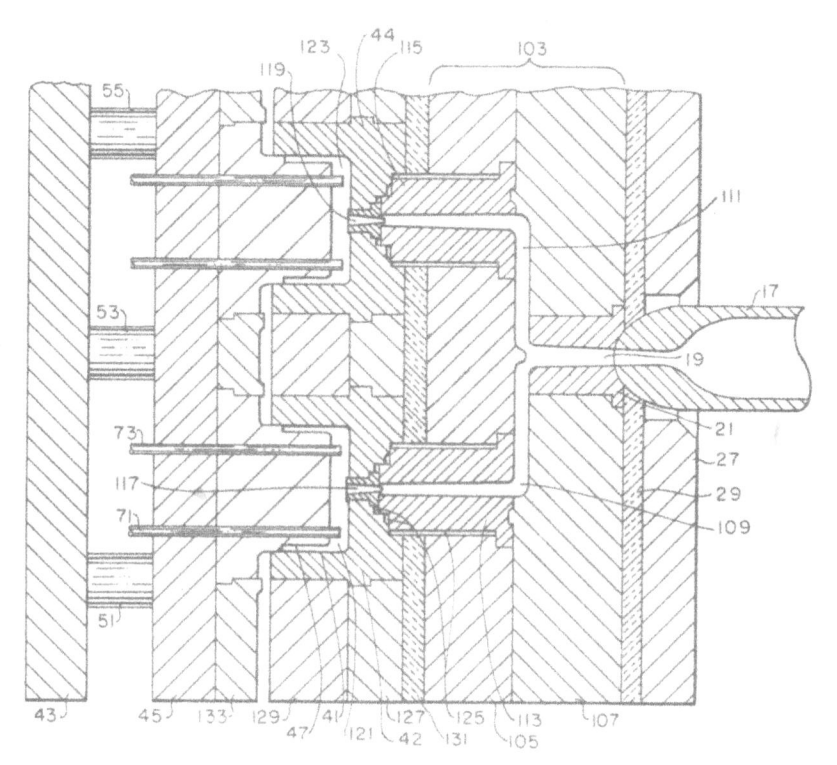

FIG. 7.50. Runnerless injection-compression mold with direct gate into cavity. From U.S. Patent 4,238,181. (*Courtesy Occidental Chemical Corporation, North Tonawanda, NY*)

1. To calculate the press tonnage, use the molding pressure range, as is supplied along with the specifications on the material to be used. Molding temperatures and pressure ranges are pretty well established and available on material specification sheets. Use the pressure and temperature values given for *compression molding*.

2. Generally, the shrinkage value given (cold mold to cold piece) for compression molding will be slightly on the high side by .001 or .002. The reason is because the *density* of the molded part will be approximately 0.8% greater than an equivalent compression molded part. Note that "normal injection molding" *increases* the shrinkage.

3. Be sure to allow for the *flash* resulting from the RIC-mold process. Of course, flash allowance is only made when land areas around the cavities (as in a sub-cavity gang RIC) is a fact. If the force entry into the cavity is fully positive, allowance for flash is *not* required. In any case, and with any standard injection grade of material, an allowance of .008 is a good beginning point. A record of actual flash on each

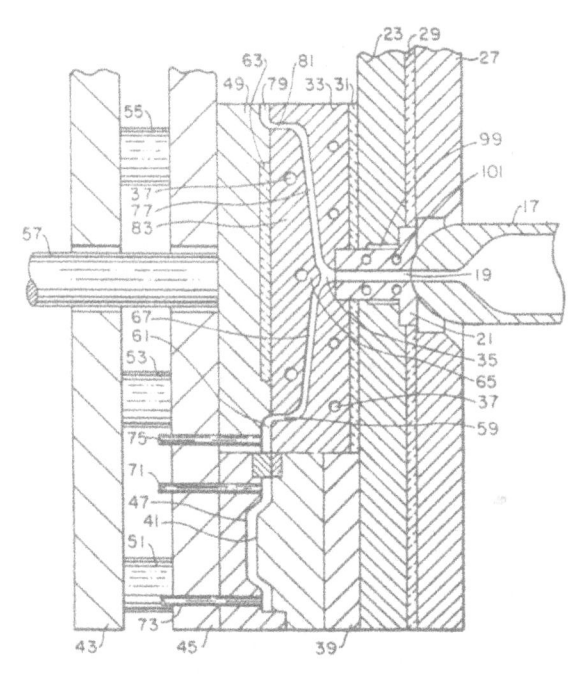

Fɪɢ. 7.51. Runnerless injection-compression mold with manifold and edge-gate. From U.S. Patent 4,309,379. (*Courtesy Occidental Chemical Corporation, North Tonawanda, NY*)

mold should be made at the time of the initial sampling of the RIC-mold.

4. As stated in item 3, any standard injection grade of thermosetting material can be processed using RIC-molds. This includes the fiber-filled impact grades, which have slightly higher impact values from RIC-molds than they have from compression molds. This fact is particulary important because "normal injection" greatly *reduces* the impact strength of fiber-filled materials.

5. As mentioned in item 2, the *density* value will increase by approximately 0.8% (8 tenth of 1%). The calculations of purchased volumes may need to take this into account. This fact is also relative to the pricing of the product, because the reduction in sprue, runner and flash allowance is affected by this increase, slight though it may be.

6. Item 5 mentions the "sprue, runner and flash allowance." Reports of users of RIC-molds reveal that this allowance is in the 5 to 6% range. This is a marked reduction compared to the flash allowance on compression molded parts, or the "throw-away" sprues and runners of "normal" injection molding.

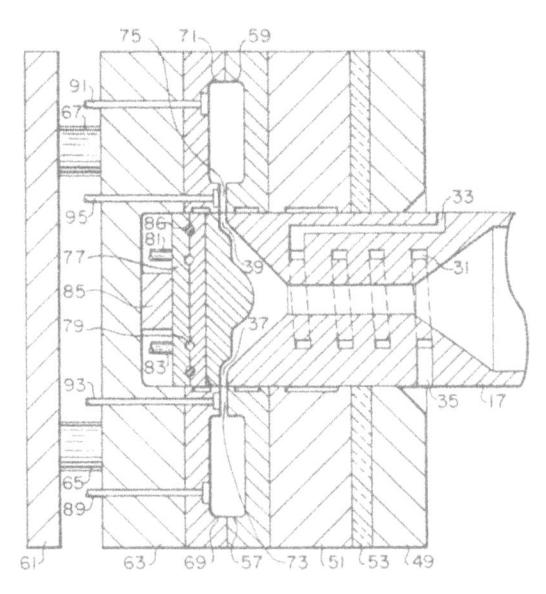

FIG. 7.52. Runnerless injection-compression mold with heated manifold. Mold fully closed. From U.S.Patent 4,260,359. (*Courtesy Occidental Chemical Corporation, North Tonawanda, NY*)

7. Finally, in the event that the designer is called upon to *convert* an existing compression, transfer or injection mold, he must remember that *there are no short cuts*. The RIC-mold technique is quite flexible, but it will not be very forgiving in the event of inadequate temperature control, inadequate isolation of differential temperature zones (as from manifold to mold cavity area), or inadequate attention to some critical dimension where "cut-and-try" was used to achieve a precise dimension in the original process.

ENCAPSULATION

Thermoplastics and thermosetting materials are used for encapsulation (potting) of electronics components as shown in Fig. 7.53. Much of this work started with the thermosets because of their better thermal endurance and their ability to be processed with semiautomatic operations. Some of this work is done with thermoplastics to obtain their better electrical properties; in either case the molds are quite similar from the point of material entry into the mold. The encapsulation process is used to form a chassis or package of plastics compound designed to fix and retain relationship between

FIG. 7.53. Encapsulated thermosetting parts. (*Courtesy Pennsalt Chemicals Corporation, Philadelphia, PA*)

various individual product components. The molds for common encapsulation programs make use of an insert/product holding section that is loaded between the mold halves and becomes a part of the mold during the "shot" (see Fig. 7.54). Thus, the inserts are loaded in the removable frame which indexes with one half of the mold and becomes a third or middle plate during the cycle. After the molding cycle is complete, the tray is removed from the press with all of the molded pieces.

The tray design is dictated by the inserts that are to be included; small pins in the frame are used to locate accurately the position of the inserts (see inserts Fig. 7.55). When possible, the inserts are interconnected parts produced in a single stamping with indexing holes. Clearances are provided so that another stamping operation after molding will separate the pieces. In practice, two trays are used, one out of the press where finished parts are removed and replaced with virgin inserts ready for the next "shot."

Additional data are given in Fig. 8.75A, page 429; concerned with propor-

FIG. 7.54. Illustrates the insert loading rack used in an encapsulation process. *Encapsulation* or potting of electronic devices is often done by transfer molding. In this work, removable tray molds are designed to facilitate loading of the electronic components during the "cure" portion of the cycle. (*Courtesy Hull Corporation, Hatboro, PA*)

tional gating as used with the thermoplastics to "center" inserts that must be free to float in the mold and be encapsulated with a uniform wall of plastics.

Proportional Gating

Some delicate components are encapsulated through proportional gating regulated by fluidic beam deflection amplifiers* to control the incoming

*Capsonic Group (designers of high speed encapsulation systems) Elgin, IL.

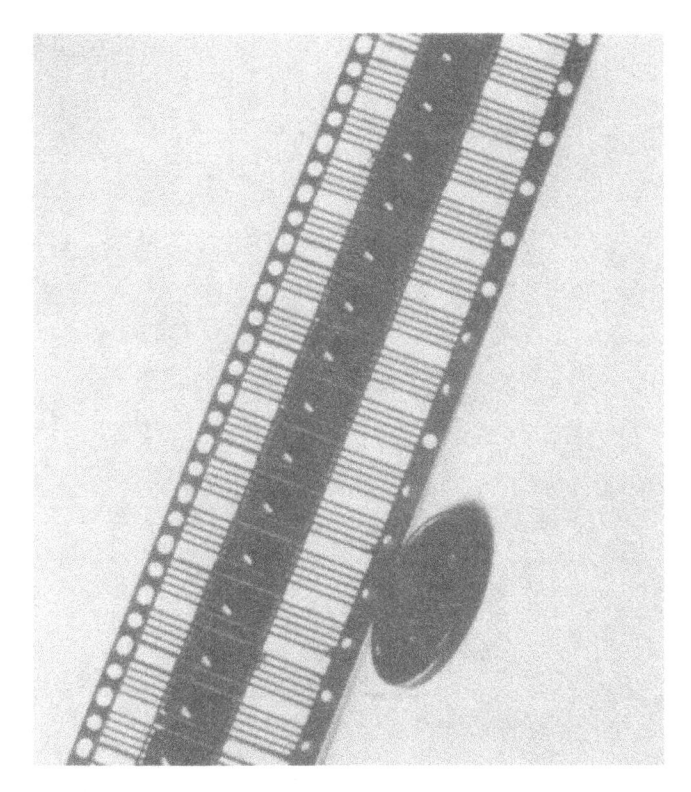

Fᵢɢ. 7.55. Depicts an insert stamping with locating holes as used for encapsulation. A trimming operation after molding separates the pieces. (*Courtesy Hull Corporation, Hatboro, PA*)

plastics melt. The flyback transformer as shown in Fig. 8.75A is excapsulated without the necessity for rigid locating positions or areas for the insert. This proportional gating system uses two incoming streams of plastics which counteract each other to centralize the position of the insert. Fluidic beam deflection amplifiers are used to control the flow and distribution of the melt. The amplifiers utilize the outflow of vent gas to control the incoming material flow. Operation of this system is shown in Fig. 8.75B. Proportional gating is used to encapsulate delicate items such as diodes, triacs, diacs and numerous other sensitive solid state devices.

REFERENCES

Bauer, Stephen H., *Tunnel Gating Thermosets*. Hatboro, PA: Hull Corporation, 32nd Antec, 1974.

Bauer, Stephen H., *Runnerless Injection Molding of BMC Polyester Compounds*, Hatboro, PA: Hull Corporation.

Butler, J., *Compression and Transfer Molding of Plastics*, New York: John Wiley, 1960.
Designing inlet channels of thermoset injection molds, *Plastics Engineering*, Sept. 1974.
DuBois J. H. and John, *Plastics*, 6th Ed. New York: Van Nostrand Reinhold.
DuBois, J. H. and Pribble, W. I., *Plastics Mold Engineering*, 1st Ed., Chicago: American Technical Society, 1946, 3rd Ed. New York: Van Nostrand Reinhold, 1978.
Duda, Edward F., *Warm Manifold and Runnerless Injection Molding*. North Tonawanda, NY: Durez Div., Occidental Petroleum.
Duda, Edward F., *Can Runnerless Injection Molding Reduce Cost?* North Tonawanda, NY: Durez Div., Occidental Petroleum.
Grigor, John M., *Runnerless Molding of Thermosets for Automation*. Litton Industries.
Mali, David T. J., Large parts polyester injection molding; up to the learning curve, *Plastics Technology*, p. 83, Apr. 1982.
Olmstead, B. A., Injection molding thermosets, *Modern Plastics Encyclopedia*, 1983-84 Ed., p. 268.
Perras, Henry A., *Design and Construction of Runnerless Molds for Thermoset Materials*, Oslo & Whitney, Inc.
Prost, Henry, Hyperthermal—runner system: New, faster way to mold thermosets, *Plastics Techology*, May 1981.
The reciproscrew injection machine for thermoset processing, *SPE ANTEC Reprint*, May 1967.
Rosler, Robert K., Encapsulation of semiconductor devices, *Plastics Design Forum Focus Issue*, p. 49, Apr. 1981.
Sors, Laszlo, *Plastics Mould Engineering*, Oxford: Pergamon Press, 1967.

Chapter 8 / Injection Molds for Thermoplastics

Revised by S. E. Tinkham and Wayne I. Pribble

Injection molding of thermosets and thermoplastics is the fastest growing element of the molding industry.* New materials are being developed continuously; these and modified materials greatly enlarge the market for new plastics products. Molding machinery, mold engineering, product design, methods engineering, and automation have also been developed at a fast pace to keep up with materials developments. The mold designer must follow all of these developments in order to expand his understanding and experience.

INJECTION EQUIPMENT

The reciprocating-screw injection machine has made the original plunger-type press obsolete in most cases, and it has been a tremendous help in expediting the growth of injection molding and the use of molded products.

The most widely used contemporary injection molding equipment includes the following basic types:

(1) The plunger-injection press utilizes a heating chamber and a plunger operation to force material into the mold and is explained in Fig. 8.1.

(2) The reciprocating-screw press,** often identified as an in-line press, utilizes a reciprocating screw to move and melt the granules of material as they are milled by the screw and passed through the heated injection cylinder. Most of the melting is achieved by mechanical working of the molding compound. Upon melting, the material builds up in front of the screw, forcing it to retract. At this point the screw stops and becomes the plunger, moving

---

*See Chapter 7 for injection molding of thermosetting materials.
**Reciproscrew®—Egan Machinery Co., Somerville, NJ.

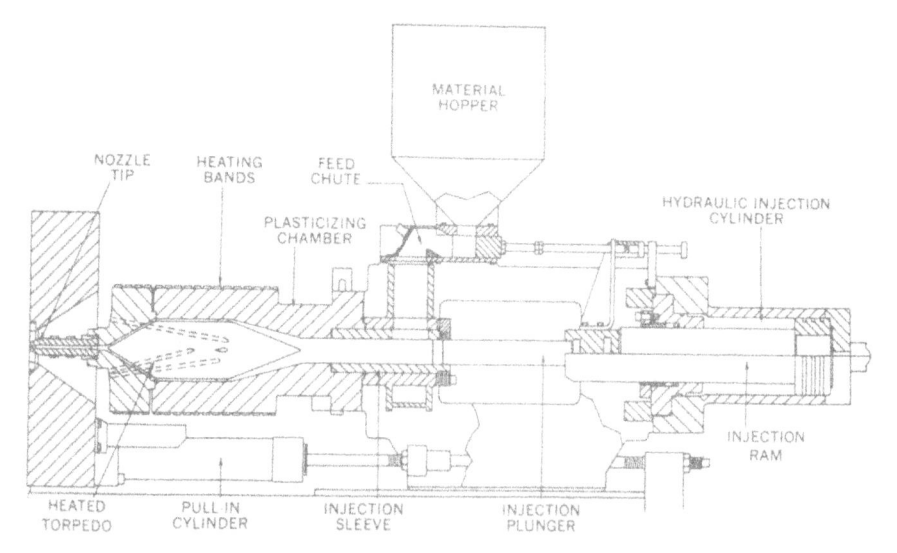

FIG. 8.1. Schematic view of conventional plunger-injection press.

ahead to push the plasticized material into the mold. Machines of this type are illustrated by Figs. 8.2, 8.3 and 8.4.

(3). The two-stage screw press in most cases uses a fixed screw to plasticize the plastic granules and force the molten compound into a holding chamber from which it is transferred by a plunger into the mold, as shown in Fig. 8.5.

(4) The rotating spreader, an alternate to Type 1 above, is driven by a shaft causing it to revolve within the heating chamber, independent of the injection ram and thus to melt the plastic granules. Final filling of the mold is accomplished through the continuing movement of the injection ram, similar to the action in Fig. 8.1.

An injection press consists of the clamping or movable end of the press which is moved by a hydraulic system, or by a toggle clamp* arrangement actuated by a hydraulic system; the stationary end of the press provides nozzle protrusion and retention of the fixed half of the mold plus the plasticizing and material feed units.

Presses are available with horizontal or vertical movement. Vertical movement presses are particularly desirable for insert or loose coring types of molding. The moving half of the press contains ejector or knockout systems for the most commonly used mold operations.

In addition to the mechanical motions, ciamping, and plasticizing functions, the press is equipped with numerous valves, timers, heating controls,

*See Fig. 2.1, p. 22.

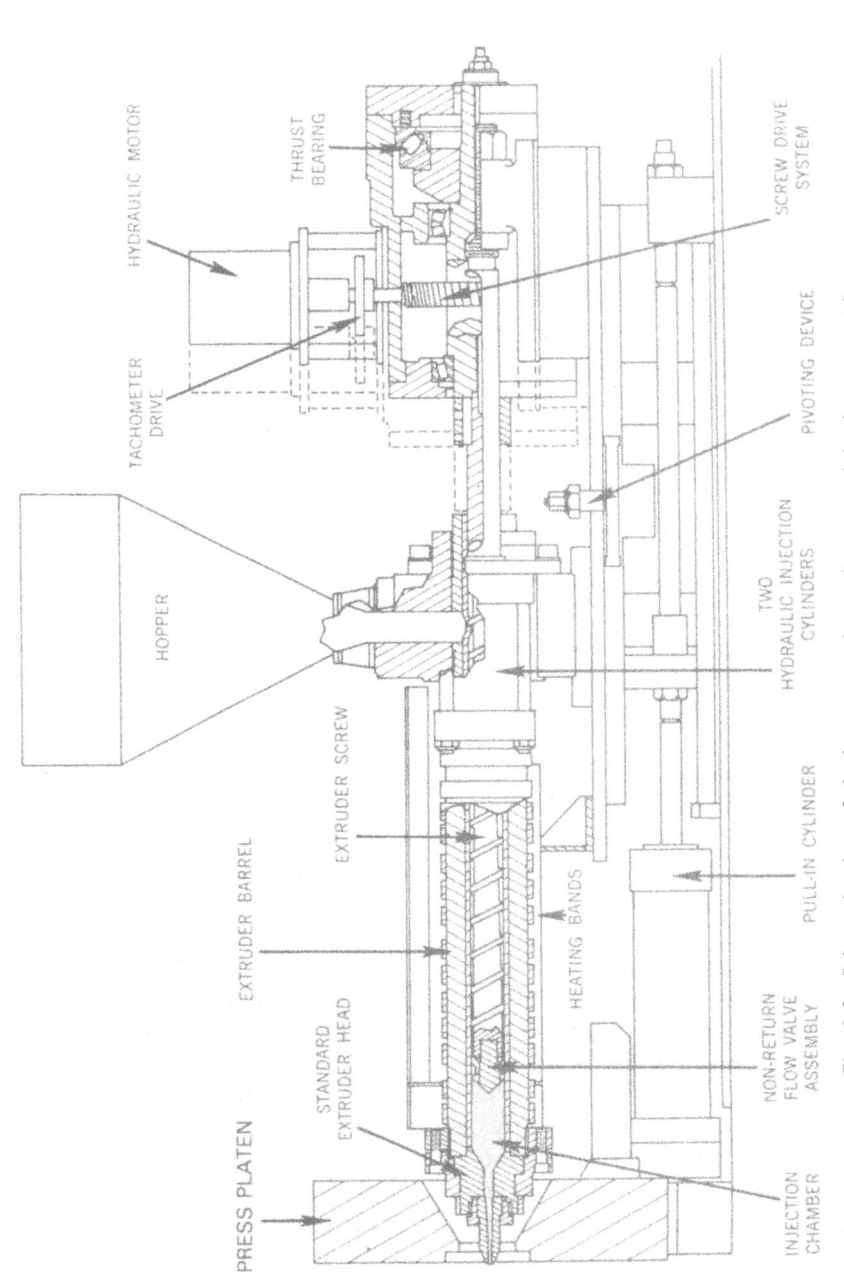

Fig. 8.2. Schematic view of single-stage, reciprocating screw injection press. (*Courtesy HPM Corp., Mt. Gilead, OH*)

Fig. 8.3. Injection molding machine, 200 ton clamp equipped with a 14 ounce injection end. (*Courtesy HPM Corp., Mt. Gilead, OH*)

and safety features for semiautomatic and completely automatic molding. It is designed and constructed to work continuously on fast cycles.

Mold designs must be achieved that meet the press requirements, material flow characteristics, speed, and cooling needs of the desired cycle.

Capacities of injection presses are rated in ounces of polystyrene molded per cycle, or in cubic inches of material displacement. Caution must be used in converting the rated machine capacity to equivalents for the material that is to be used. This is a density differential correction.

In order to understand the remainder of this chapter it is suggested that the reader study the fundamentals as presented in Chapters 2 and 5. Basic design data, terminology, and elements of mold design are presented in these chapters. A review of Chapter 7 on transfer and injection-thermoset mold design will also be helpful.

Fig. 8.4. 2500 ton injection molding machine. (*Courtesy Ingersoll-Rand Plastics Machinery, IMPCO Division, Nashua, NH*)

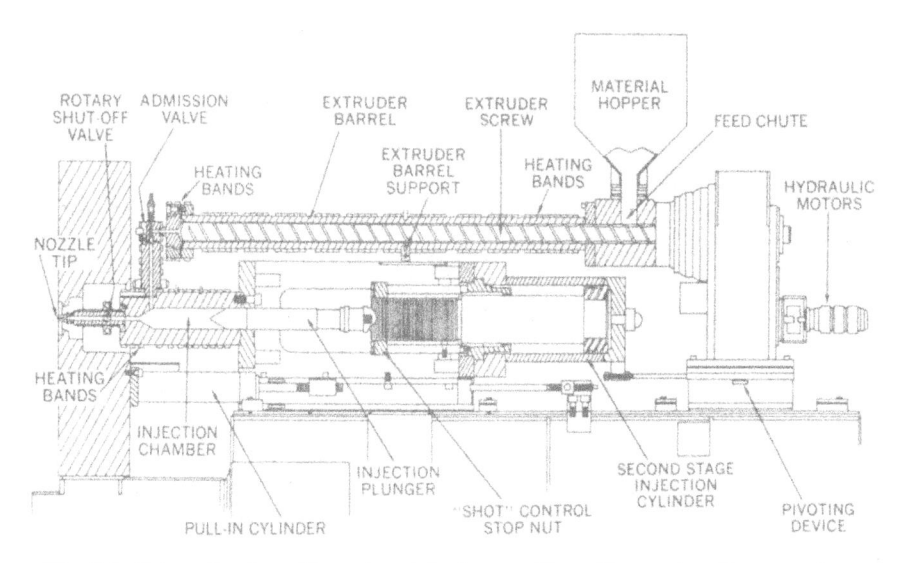

Fig. 8.5. Schematic view of two-stage injection machine with one stage for screw for pre-plasticizing material and second stage for plunger to fill the mold. (*Courtesy HPM Corp., Mt. Gilead, OH*)

PROJECTED AREA PRESS CAPACITY

Capacities of injection presses vary with the product designs and material being molded. The generally accepted practice is to use 2½ to 3 tons psi for plunger-injection presses, and 2 to 2½ tons psi for reciprocating-screw presses. These standards are based on 20 thousand psi of injection pressure.

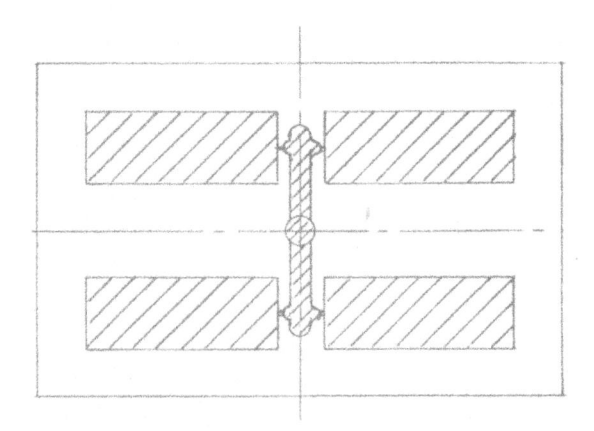

FIG. 8.6. Projected area is the total area of moldings, plus runner and sprue system, at the parting line of mold. The total shaded area above is projected area.

This pressure density is predicated on the projected area of the molding and all runner systems. See Fig. 8.6 for a definition of projected area.

Variations in the above pressure factors on specific molded product designs should be reviewed with materials and equipment manufacturers. Their recommendations will assist in the tool designing calculations.

Clamping pressures must be sufficient to keep the mold closed during the filling or injection phase of the cycle. The size and sectional area of molded parts, runner areas and length, mold and cylinder temperatures, and size of gating all affect the injection pressure required.

The function of the injection mold is to shape the part and confine the molten plastics material under pressure until it is sufficiently rigid to permit ejection. It must perform this function repeatedly in continuous production on minimum cycles without sticking, distortion, wear, component breakage, or excessive maintenance. The mold must also provide a rapid and efficient transfer of heat out of the plastics material.

DESIGN CONSIDERATIONS

Chapter 5 shows an extensive engineering and design check list which should be reviewed at this time and also in conjunction with every mold design undertaken.

Table 8.1 is included to show the type of information needed by each designer. This is not an exhaustive list nor is it typical of any particular molding plant. Table 8.2 applies only to a 6-oz Lester model. To design a mold that will operate in any given press, you need to complete a similar layout giving all the pertinent data about the press, including the information in Tables 8.1 & 8.2. The primary purpose of Table 8.2 is to show clamp bolt hole and ejector hole locations rather than to describe the same in a chart.

TYPES OF MOLDS

Injection mold designs differ depending on the type of material being molded, necessitating various gating and ejection principles to meet the application with maximum economy. Production requirements, product life, and allowable product cost factors will dictate the size of the mold, amount of mechanization, and the efficiency that will be required in cycling.

The most commonly used designs are as follows:

(1) *Two Plate.* Mold cavities are assembled to one plate and forces to the other plate, with the central sprue bushing assembled into the stationary half of the mold, permitting a direct runner system to multiple-cavity molds or direct center gating to individual-cavity molds. The moving half of the mold normally contains the forces and the ejector mechanism, and in most

TABLE 8.1. Typical Injection Machine Data Chart Needed to Design a Mold to Fit.

Press Nos.	Press Model	Capacity oz.	Max. Proj. Area	Clamp in Tons	Screw or Plgr.	Clamp Motor	Injection Press. psi	Hyd. Line Press.
3198	L-100 Lester	2/3	35/40	100	Plgr.	25	20,400	2,000
1	L-100 Lester	2/3	35/40	100	Plgr.	25	20,400	2,000
12	L-125 R5 Lester	6	40/50	125	Scr.	25	19,000	2,000
13	L-125 R5 Lester	6	40/50	125	Scr.	25	19,000	2,000
14	L-125 R5 Lester	6	40/50	125	Scr.	25	19,000	2,000
15	175 New Br.	10.3	88	175	Scr.	30	19,300	2,000
16	175 New Br.	10.3	88	175	Scr.	30	19,300	2,000
8	L-150 BR5 Lester	13	65/70	150	Scr.	25	20,000	2,000
9	L-150 BR5 Lester	13	65/70	150	Scr.	25	20,000	2,000
10	L-250 R52 Lester	13	100/125	250	Scr.	30	20,000	2,000
11	L-250 R52 Lester	13	100/125	250	Scr.	30	20,000	2,000
4	3214 Lombard	12/16	110/130	325	Plgr.	30	20,000	2,000
	200 Arburg	2	18	38	Scr.	6-1/2	18,800	1,350
	200 S Arburg	2	18	38	Scr.	6-1/2	18,800	1,350
	Minirounder Arburg	1/2	4-1/2/6-1/4	13	Plgr.	5	9,000	1,420
A	C 4/B Arburg	1/3	4-1/2/6-1/4	5-1/2	Plgr.	Air	4,400–13,000	100 psi Air
B	C 4/B Arburg	1/3	4-1/2/6-1/4	5-1/2	Plgr.	Air	4,400–13,000	100 psi Air

designs, the runner systems. This is the basic design for injection molding and all other designs are developed from this fundamental design which is illustrated in Fig. 8.7.

(2) *Three Plate.* The introduction of another intermediate and movable plate, which normally contains the cavities for multiple-cavity molds, permits center or offset gating of each cavity from the runner system which

Four-Way Valve	Min. Mold Hite	Max. Mold Hite	Mold Open Stroke	Ejec. Stroke	Clearance Between Rods H × V	Platen Size H × V	Tie Rod Size
No	4	15	8-3/4	0–2-1/2	12-1/2 × 10-1/2	19 × 17	2-1/2 φ
No	4	15	8-3/4	0–2-1/2	12-1/2 × 10-1/2	19 × 17	2-1/2 φ
Yes	4	15	8-3/4	0–2-1/2 0–3 & H	12-1/2 × 12-1/2	19 × 19	2-1/2 φ
No	4	15	8-3/4	0–2-1/2 0–3 & H	12-1/2 × 12-1/2	19 × 19	2-1/2 φ
No	4	15	8-3/4	0–2-1/2 0–3 & H	12-1/2 × 12-1/2	19 × 19	2-1/2 φ
	6	14	14-3/16		15-1/4 × 15-1/4	23-1/2 × 23-1/2	3φ
	6	14	14-3/16		15-1/4 × 15-1/4	23-1/2 × 23-1/2	3φ
No	9	12	6/9	0–3	12 × 14	18 × 20-7/8	3 × 4-5/32
No	9	12	6/9	0–3	12 × 14	18 × 20-7/8	3 × 4-5/32
Yes	7″ with 3″ spacer	22	8/16	0–6-1/4	20 × 17-3/8	27-1/2 × 29-3/8	3-3/4 × 6
Yes	7″ with 3″ spacer	22	8/16	0–6-1/4	20 × 17-3/8	27-1/2 × 29-3/8	3-3/4 × 6
Yes	8	22	8-1/2/20		20-1/4 × 20-1/4	32 × 32	4φ
No	5.9	7-7/8	7-7/8		7-7/8	max. mold 7-7/8 × 9-7/8	
No	5.9	7-7/8	7-7/8		7-7/8	max. mold 7-7/8 × 9-7/8	
No	4.4	6-1/2	5-1/4		4	max. mold 4 × 5	1-3/8 φ (2)
No	2-15/16	4-1/3	3-5/32			max. mold 3.9 × 4	
No	2-15/16	4-1/3	3-5/32			max. mold 3.9 × 4	

connects to the central sprue bushing. This design is widely used, and in many cases, it is necessary to use multiple-sprue pullers for efficient operation. Three-plate mold design is illustrated in Figs. 8.8, 8.107, 8.108, 8.109 and 8.110.

(3) *Loose Details.* Threads, inserts, or coring which cannot be produced by normal operation of the press are often processed by separate mold details

TABLE 8.2   Schematic Press Layout Showing Typical Information Needed About the Presses in Your Plant.

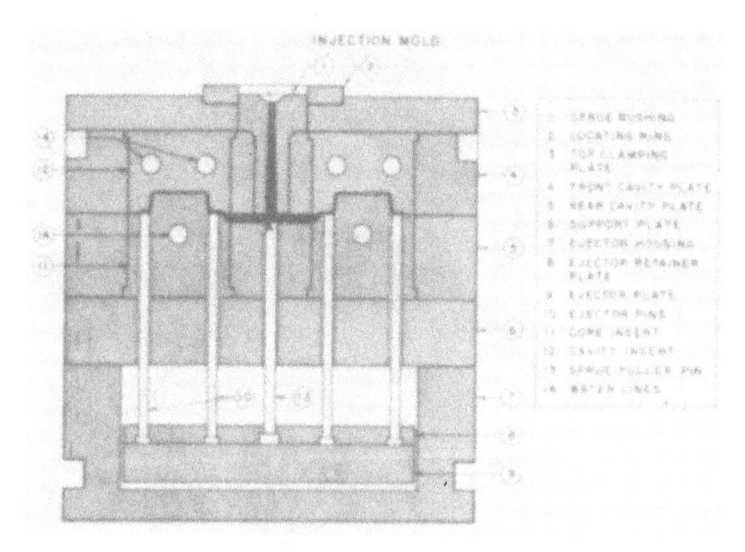

Fig. 8.7. Illustrates the various components of a typical two-plate injection mold used for injection molding. (*Courtesy Dow Chemical Co., Midland, MI*)

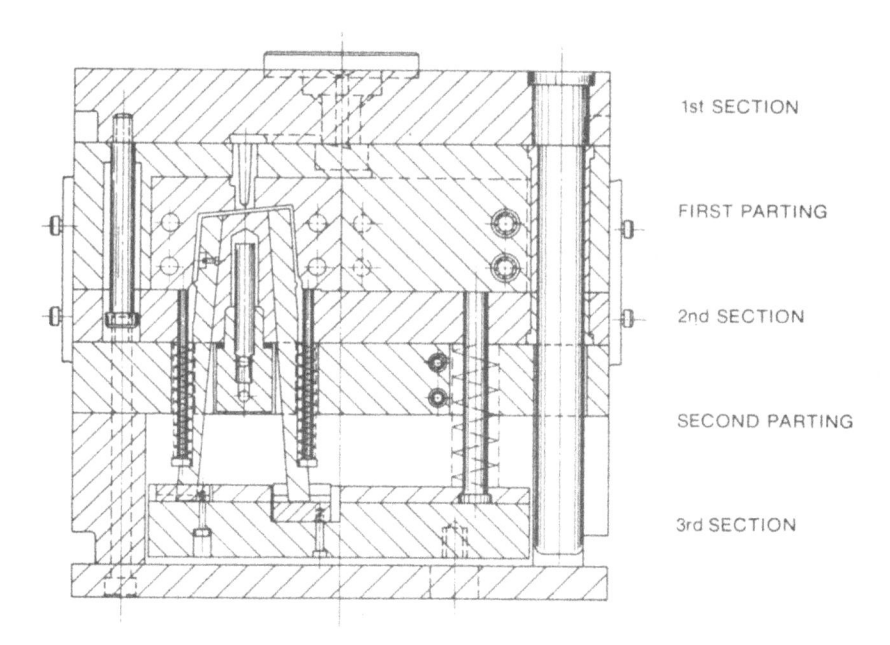

Fig. 8.8. Typical three-plate injection mold. (*Courtesy Marland Mold Co., Inc., Pittsfield, MA*)

Fig. 8.9. Two-cavity loose detail injection mold. Note the two mold details which are ejected with the molded piece (top). (*Courtesy Tech Art Plastics Co., Morristown, NJ*)

which are ejected with the part and removed by hand or with a disassembly fixture after cycling. This practice is often used for experimental or small production requirements to minimize the more costly semiautomatic mold details. Loose cores must be designed to permit foolproof location in the mold and thus to prevent mold damage. Such molds are shown in Figs. 8.9 and 8.88.

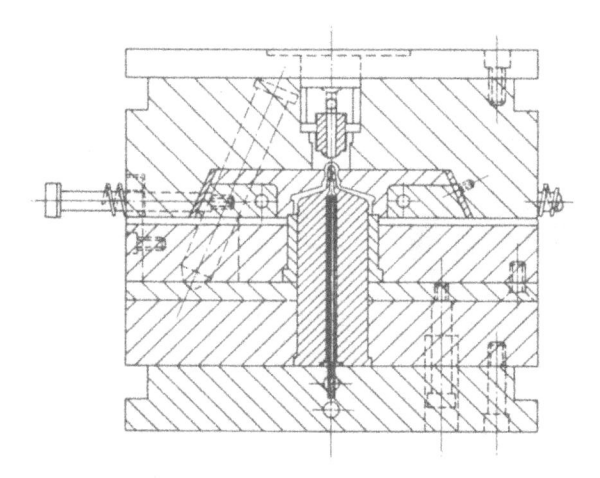

Fig. 8.10. Typical design for horizontal core movement. (*Courtesy Marland Mold Co., Inc., Pittsfield, MA*)

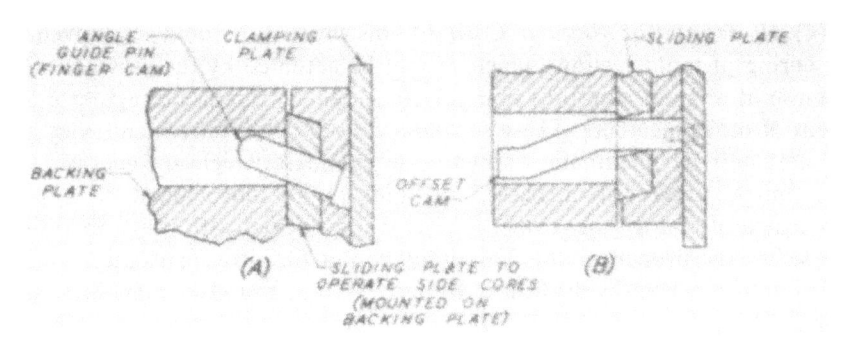

FIG. 8.11. Angle guide pins or finger cams (A) and offset cams (B) are frequently used to operate side cores.

Fig. 8.12. Two cavity semiautomatic injection mold with side core actuated by hydraulic cylinder and loose details for molding metal inserts in position during molding. (*Courtesy Tech Art Plastics Co., Morristown, NJ*)

(4) *Horizontal or Angular Coring.* This practice permits the movement or coring of mold sections which cannot be actuated by the press, through the use of angular cam pins that permit secondary lateral or angular movement of mold members. These secondary mold detail movements may also be actuated by pneumatic or hydraulic cylinders which are energized by the central press system, cams, solenoids or an independent air supply. Sequence of control movement is always interlocked with press operations for safety and proper cycling. This design is used for intricate product production requirements. Typical mold designs are illustrated in Figs. 8.10, 8.11 and 8.12.

(5) *Automatic Unscrewing.* Internal or external threads on product designs requiring large volume and low production costs are processed from molds that incorporate threaded cores or bushings actuated by a gear and rack mechanism, and moved by a long double-acting cylinder sequentially timed within the molding cycle as shown by Figs. 8.13 and 8.36. Various other types of movements may be used, i.e., electric gear motor drives or friction type of mold wipers actuated by double-acting cylinders engaged

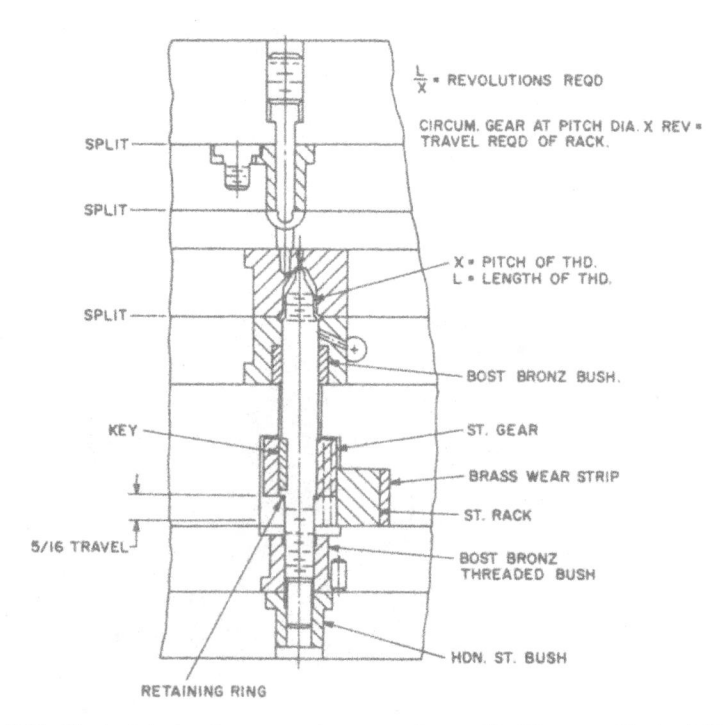

Fig. 8.13. Typical design for automatic unscrewing mold. (*Courtesy Marland Mold Co., Inc., Pittsfield, MA*)

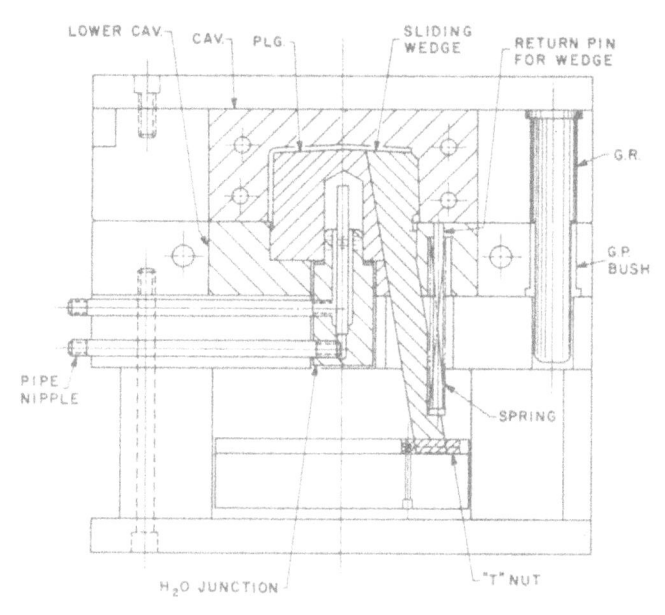

LOWER CAV. CAV. PLG. SLIDING WEDGE RETURN PIN FOR WEDGE
G.R.
G.P. BUSH
PIPE NIPPLE
SPRING
H_2O JUNCTION
"T" NUT

Fig. 8.14. Typical design for rising cam, sliding wedge, and ejector angular movement. (*Courtesy Marland Mold Co., Inc., Pittsfield, MA*)

against the perimeter of a molded part. In all unscrewing molds, it is good practice to use large bearing guidance and self-lubricating wear strips for rack movements. Polytetrafluoroethylene is commonly used for wear strips. Collapsing cores as shown by Figs. 8.99 and 8.100 are an alternative to automatic unscrewing.

(6) *Rising Cam or Ejector Angular Movement.* This design is used to mold undercuts on the interior of parts. Angular movement of the core through the ejection travel permits release of the metal core from the part as illustrated in Fig. 8.14.

(7) *Ejection on Nozzle Side of Mold.* This design is used when it is necessary to have gating and ejection on the nozzle, or fixed half of the mold as shown in Fig. 8.15. The ejection bar is moved by chain pulls.

FLOW RESISTANCE IN THE HEATER SYSTEMS

A very important aspect of hot runner molding is the design of flow channels in such a way that rheological principles are observed, Fig. 8.20. In a study of comparison between a hot edge gate system and the more widely used heated probe-type hot runner system, it became evident that the pressure drop in the heated probe-type manifold, flowing through an annulus, is

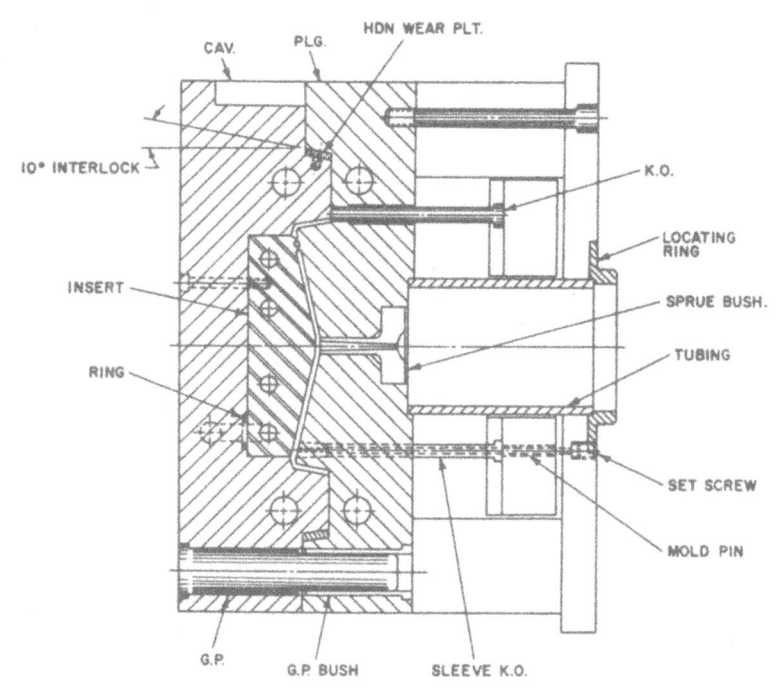

Fig. 8.15. Typical design for injection mold, with ejection nozzle side of mold. (*Courtesy Marland Mold Co., Inc., Pittsfield, MA*)

almost twice as much as the pressure drop flowing through a cylindrical bore. Laminar flow is more prevalent in annulus runners.

HOT MANIFOLD SYSTEMS FOR THERMOPLASTICS

The ideal injection molding system delivers molded parts of uniform density free from runners, flash and gate stubs from every cavity, leaving no runners, sprues, etc., for reprocessing (thus resulting in large material savings). Hot runner and insulated runner systems have been devised to achieve this goal; they are also called *hot manifold* molds. In changing over from the earlier style runners in three-plate molds, the plate that delivers the fluid plastics is properly called a *manifold*. In effect, hot runners are the extension of the heated machine nozzle in the mold.

Alternatively, the cold runner type of mold used for thermosets is properly called a *cold manifold* mold (Chapter 7) since, in this case, the manifold temperature must be kept below the curing temperature of the material.

Hot runner molds require more time to design, and require greater expense to manufacture than conventional molds. The production cost of the molded products is very greatly reduced; other gains are increased produc-

Fig. 8.16. Hot runner mold with hot tip bushings and manifold system. (*Courtesy Incoe Corp., Troy, MI*)

tion, no regrinding of scrap and the material is not contaminated. The gate mark (vestige) is limited to a small dot.

In consideration of the high cost of runnerless molds, it is desirable to consider the alternatives offered by the several systems, by various zone and entry locations to the manifold and by the type of tip. The design selected must facilitate use of the basic runner and gate system components for subsequent mold cavities that can run in this same base and system at a later date. Specialists are available with stock components that are very versatile and economical. Standard or custom made manifolds as shown in Fig. 8.16 can be had with various port or entry locations.

One novel procedure employs a cast hot manifold and achieves the effects of direct gating in multi-cavity molds with integral heaters for the sprue plus hot edge or valve gating. This system is described in the gating section and depicted by Fig. 8.17.

Other forms of hot manifold molding make use of a hot runner plate (Fig. 8.18A and 8.18B) with cartridge heaters and with electrically heated exit nozzles gating directly into the part. An alternate is the insulated runner mold (Fig. 8.19) that makes use of very large area runners in the third plate. In this case the "A" plate and the back-up plate are latched together. Material flowing into the runners hardens as it contacts the runner surfaces but continues to flow or "tunnels" through the center of the runner. In this case, the outer hardened core of material insulates the central fluid plastics as it

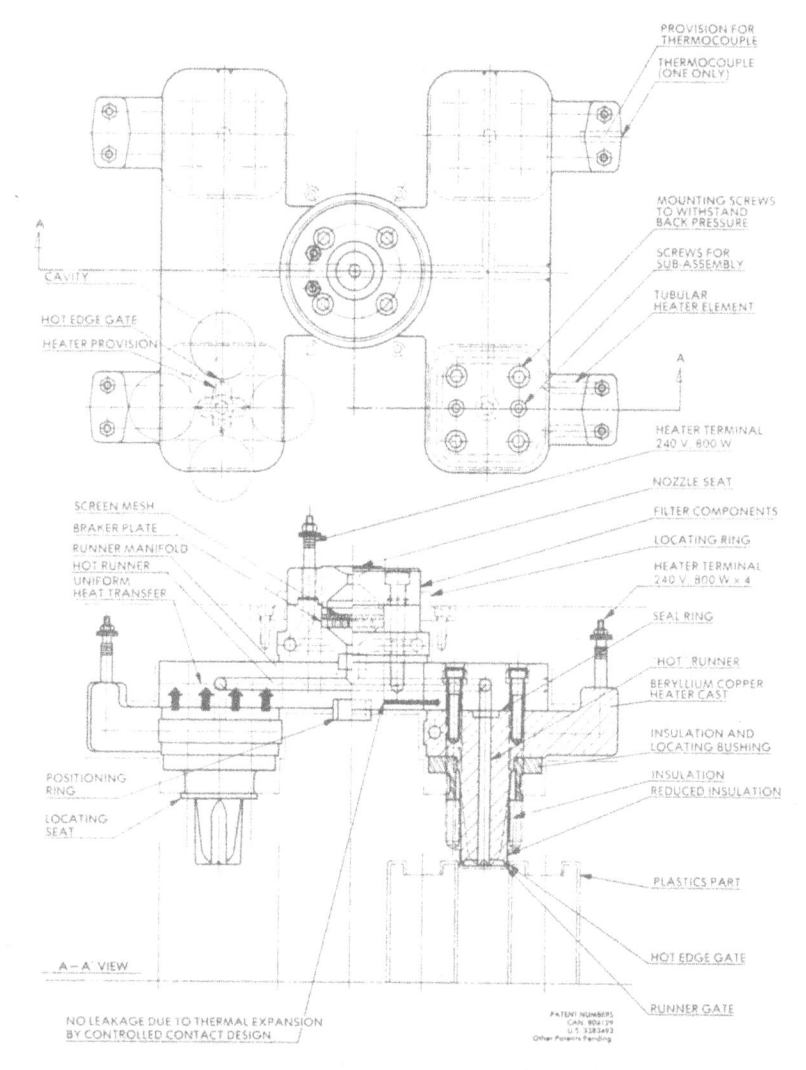

Fig. 8.17. Typical arrangement for 16 cavities with "H" type manifold using heater cast and component screen pack heater cast. (*Courtesy Mold Masters, Ltd., Georgetown, Ontario*)

moves into the cavity. Such molds use probes or hot tips (Fig. 8.19) at the cavity gate point to prevent freeze-up in the gate and to permit slower cycling speeds. Insulated runner molds are simpler than other hot manifold types and the lowest in cost.

Another definition for a *hot runner mold* is a mold utilizing electrical heating elements in hot tips at the cavity gate points, in addition to heaters in the manifold. An *insulated runner mold* is a mold utilizing electrical heating elements in hot tips at the cavity gate points in conjunction with a

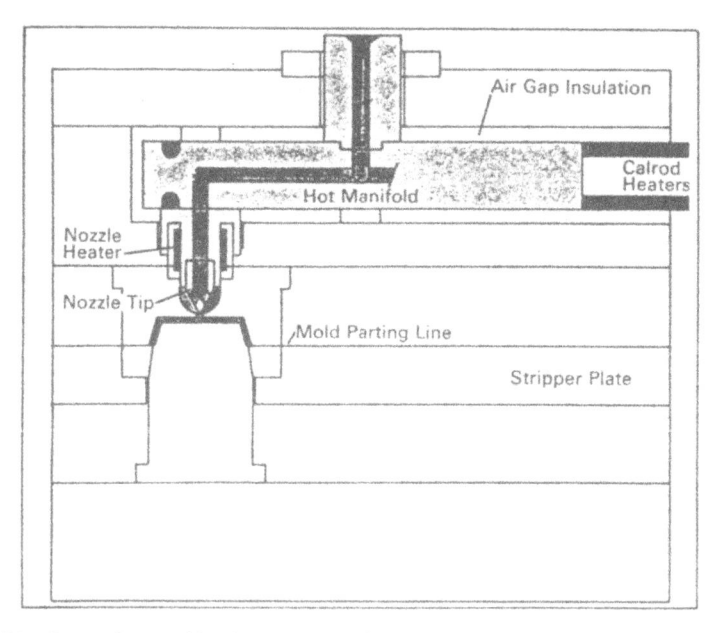

Fig. 8.18A. Runnerless molds—hot runner mold. (*Courtesy Husky Injection Molding Systems, Ltd., Bolton, Ontario*)

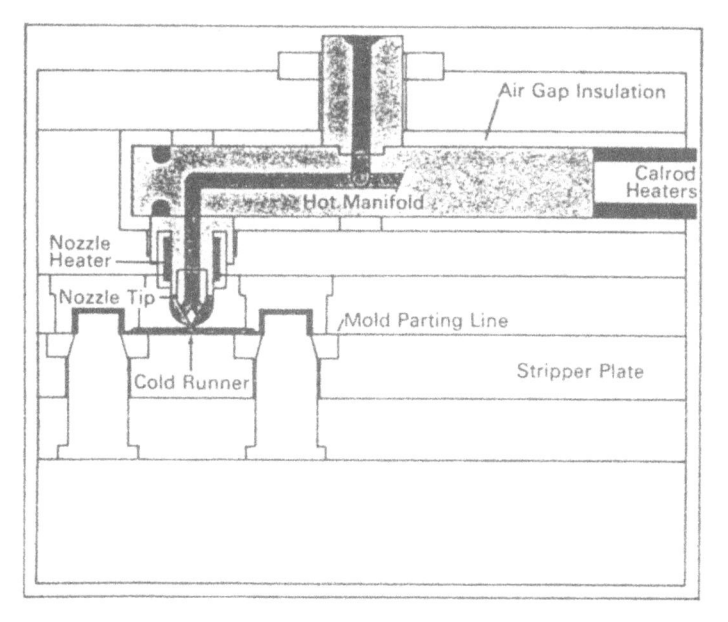

Fig. 8.18B. A combination of hot runner with short cold runners. Greatly reduces volume of cold runner molded per shot. Cold runner can be self-degating from parts. Cold runner may feed 4 or 6 cavities. (*Courtesy Husky Injection Molding Systems, Ltd., Bolton, Ontario*)

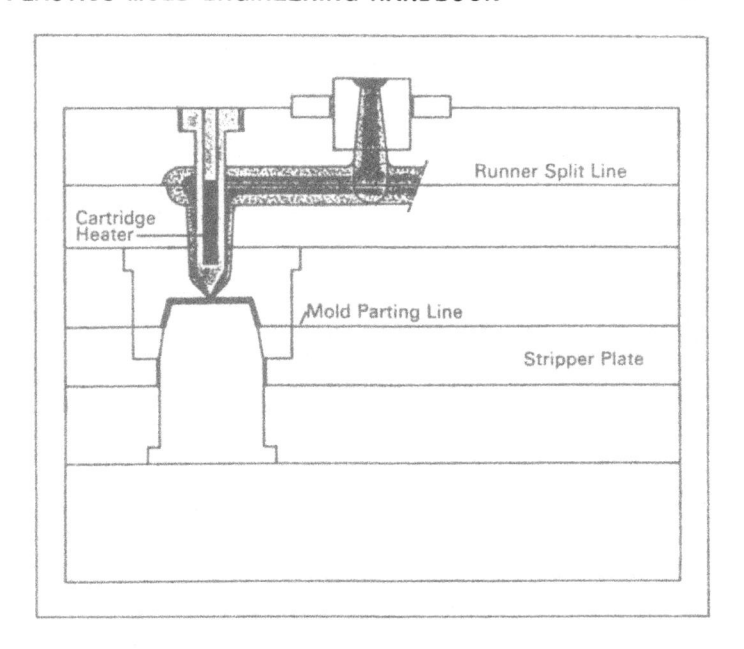

Fig. 8.19. Insulated runner mold. May be used without heaters. Relies on plastic as an insulator to keep runner from freezing up. Mold must be split to remove frozen runner. (*Courtesy Husky Injection Molding Systems, Ltd., Bolton, Ontario*

colder manifold section. These basic hot manifold systems without refinement introduced a variety of problems: material drooling or freezing in the gates, balancing flow, packing, thermal control, contaminents clogging nozzles and freeze-up in the runner system. All hot runner systems should have strainers to stop foreign material that could plug the nozzles. This problem is minimized when virgin material only is used.

A little discussed question that merits consideration as a basic issue is: how is the heat distributed in the material as it enters the cavity? There are two basic concepts of this heat transfer process that are to be considered: (1) When the internal heat from the probe is transferred into an annulus type runner. (2) When the heat transfer source is external and the runner configuration is that of a small cylindrical rod. The internally heated probe system is most widely accepted and is dominant in contemporary runnerless molds. The simple cylindrical system is much more costly and its advantages are not as obvious (Fig. 8.20). Annulus type runner systems require careful balancing and sophisticated heat controls. Cylindrical runner configurations do not require close thermal control at every gate.

Heat pipes* are expected to improve the heating and cooling of the plastics by flowing the plastics over a tube. A heat pipe consists of a closed envelope,

*Heat Pipes-Hughes Thermal Products, Torrance, CA and Noren Products, Inc., Redwood City, CA 94062.

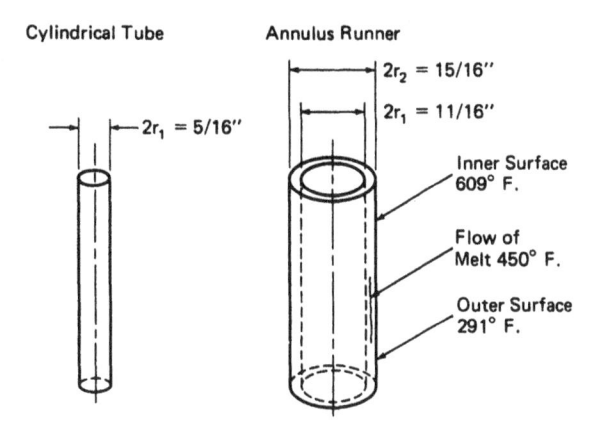

Runner	Equation	Pressure Drop
Cylindrical tube	$\Delta P = 8\mu LQ/\pi r^4$	470 psi/in.
Annulus runner	$\Delta P = 8\mu LQ/\pi(r_2^2 - r_1^2)\,((r_2^2 + r_1^2) - (r_2^2 - r_1^2)/\ln(r_2 + r_1))$	780 psi/in.

Fig. 8.20. Pressure drop in the hot runner—flow in a cylindrical tube and annulus path. (*Courtesy, Mold Masters, Georgetown, Ontario*)

usually a metal tube, containing a capillary-wick structure and a small amount of vaporizable liquid. It functions on the same evaporation-condensation principle as is used in closed-cycle heating and cooling systems. Heat energy from the source is transferred by conduction from end to end through the container wall where the fluid vaporizes. Vapor flows through the core to the condenser where the vapor condenses and returns through the wick by means of capillary action. It is expected that current research on this plan will improve the thermal efficiency of molds.

Many other considerations will dictate the mold design and the higher cost methods merit a careful analysis. One important consideration is to achieve the minimum thermal gradients in the melt as it enters the cavity. Different layers of temperature cause a laminar condition that can introduce warpage weeks after molding.

Essential to good mold design for hot manifold operation is a balanced runner system and calculations that will give effective cooling of the cavity clusters. Thermal insulation for the hot manifold is achieved by the inclusion of air gaps and minimum contact areas between hot and cold portions of the mold. Doubling the air gap spacing increases the thermal insulation by a factor of eight. The mold components must not be contacted in any way by the hot runner plates except for the minimal support areas essential for separation of the sections as depicted in Fig. 8.16. Transite asbestos, laminated plastics and glass-bonded mica are often used to minimize thermal

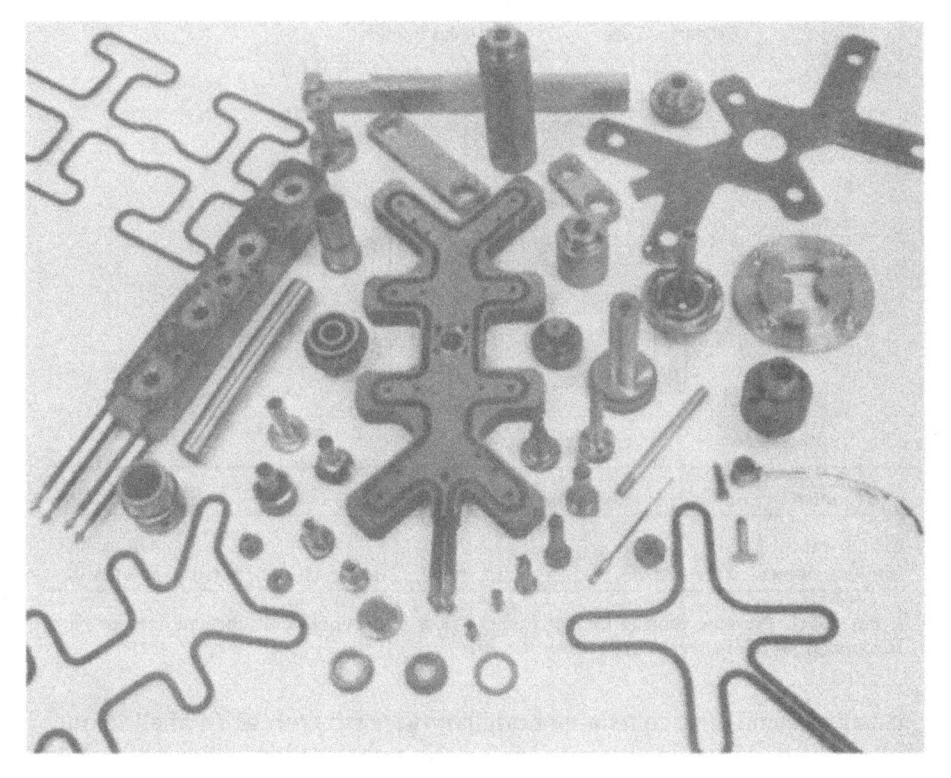

Fig. 8.21. Some standard mold components for hot manifold molds. (*Courtesy Husky Injection Molding Systems, Ltd., Bolton, Ontario*)

transfer from the hot to cold components. Glass-bonded mica* has an advantage since it can be ground to absolute flatness and parallelism the same as steel and it has total dimensional stability up to 700° F. Allowance must be made for differential expansion between the hot manifold element and the colder mold components. This is commonly accomplished by an arrangement of sliding mold sections between the manifold plate and cluster sprue plates. Flat ground surfaces provide the essential sliding surfaces or a piston-cylinder arrangement may also be used as shown in Fig. 8.22. Some type of probe or hot tip bushing is used with annulus gates to maintain temperature at the exit from the manifold. Manifold and tips are heated to maintain barrel temperature in the compound.

The minimal manifold area follows the cavity layout pattern and the manifold is cut to the minimum contour that will contain runners and heating elements as shown in Fig. 8.21. Fundamental manifold design considerations include heaters that parallel the internal runner passages as shown in Fig. 8.21. Thermocouples to maintain desirable temperatures are essential.

*Mykroy Ceramics Company, Ledgewood, NJ.

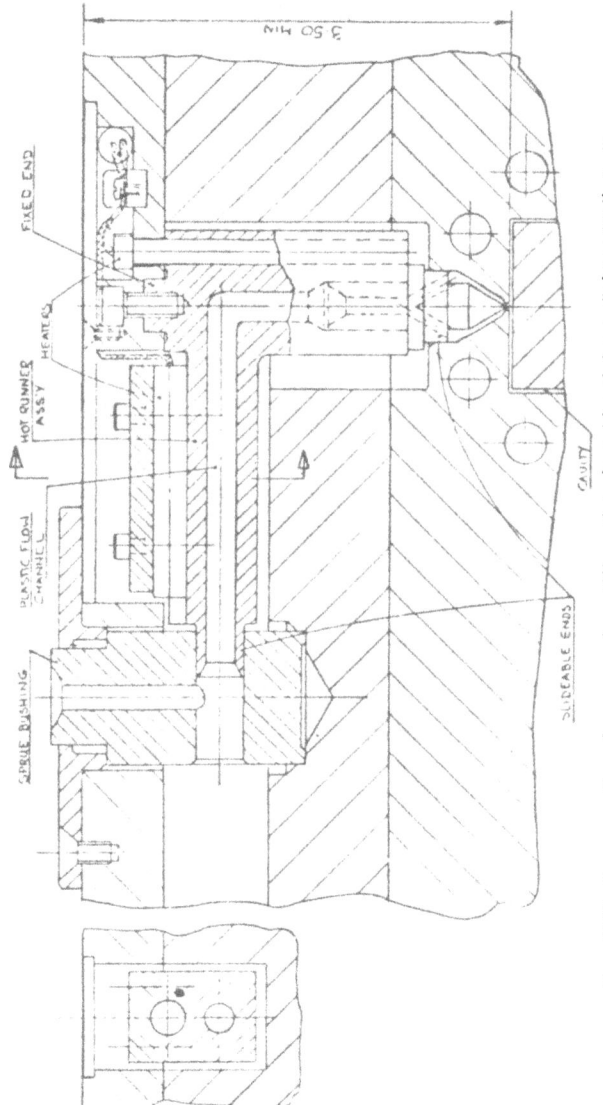

FIG. 8.22. Compensation for expansion differentials is achieved by a "piston/cylinder" arrangement. (*Courtesy DISCO Limited*)

Since the hot tips are indexed positively by the cavities which are at lower temperature than the manifold, they must be free to move in compensation for their differential expansion. A positive seal must be maintained between the bushing and the manifold. In the Disco PW hot runner system, Fig. 8.22, the hot runner is free to expand in the directions from the fixed end by sliding in the bores of the sprue bushing or nozzle and in the cavity bore. The slidable ends are positively sealed by specially designed metal seals.

In some cases the distribution runner clusters may require external heating from a controlled source.

Hot tips, heated probes, hot sprue bushings and heater casts are important to improved hot manifold molding. By heating at the sprue tip in a heat controlled gate, the melt time is extended and material will not freeze in the sprue. Sprues and probes are heated by being enclosed in a cast beryllium copper body or by the use of an internal heating element; in some cases a thermocouple is included also. A variety of sprues, probes, hot tips and thermocouples are shown in Figs. 8.23, 8.24, 8.25, 8.26, 8.27, and 8.28. Figure 8.29 depicts a comparison of the several systems.

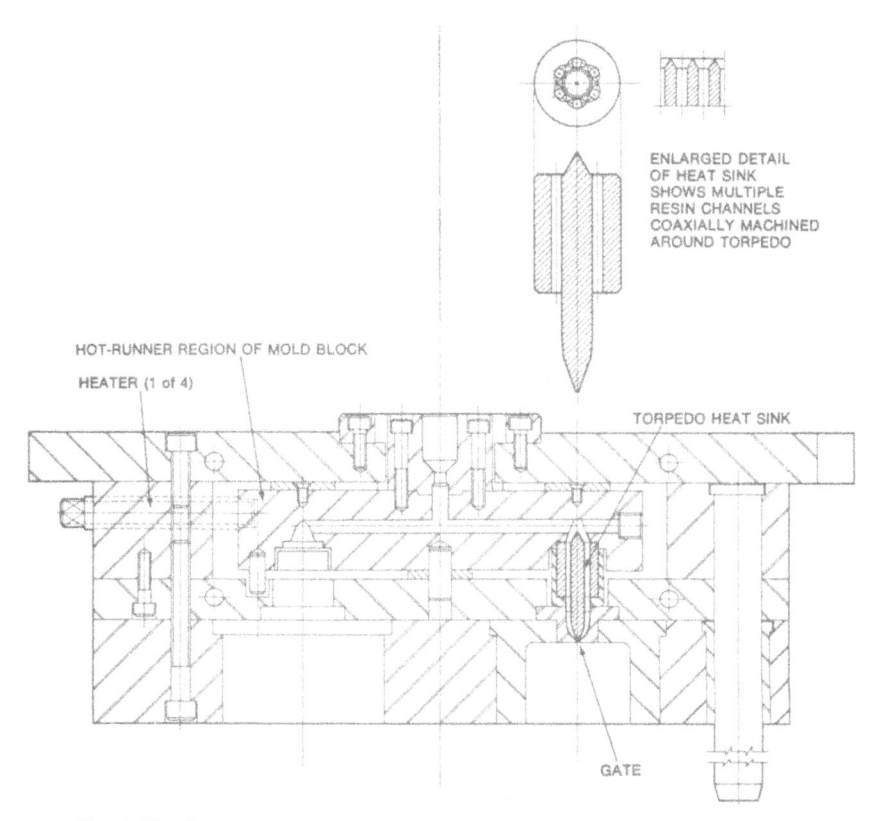

ENLARGED DETAIL OF HEAT SINK SHOWS MULTIPLE RESIN CHANNELS COAXIALLY MACHINED AROUND TORPEDO

HOT-RUNNER REGION OF MOLD BLOCK

HEATER (1 of 4)

TORPEDO HEAT SINK

GATE

Fig. 8.23. Cross section of a mold, showing hot runner and torpedo heat sink.

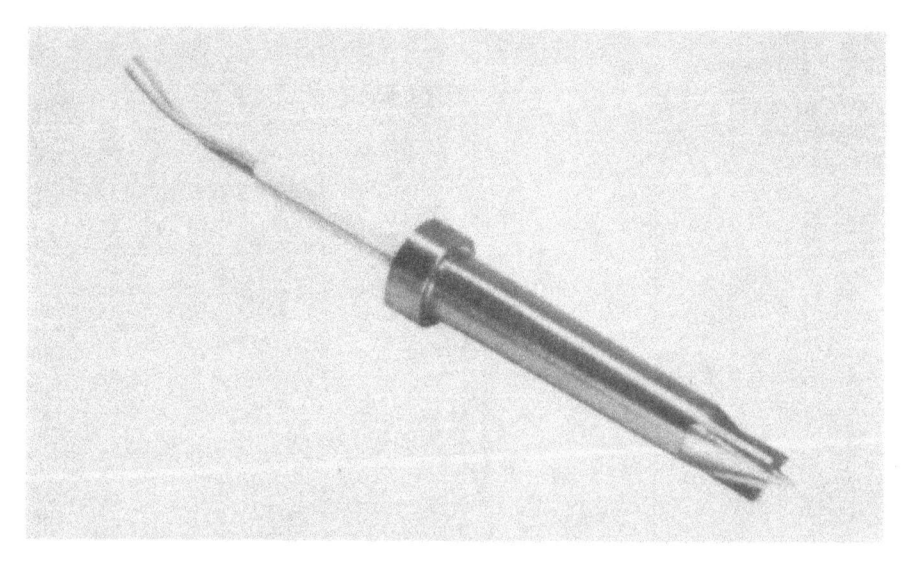

Fig. 8.24. The tip configuration of the D-M-E Auto-Fixed™ Probe prevents both lateral movement and thermal expansion of the tip into the gate. A replaceable thermocouple-cartridge heater permits close temperature control. This tip fixes the probe "tip to gate" relationship. (*Courtesy D-M-E Corp., Madison Heights, MI*)

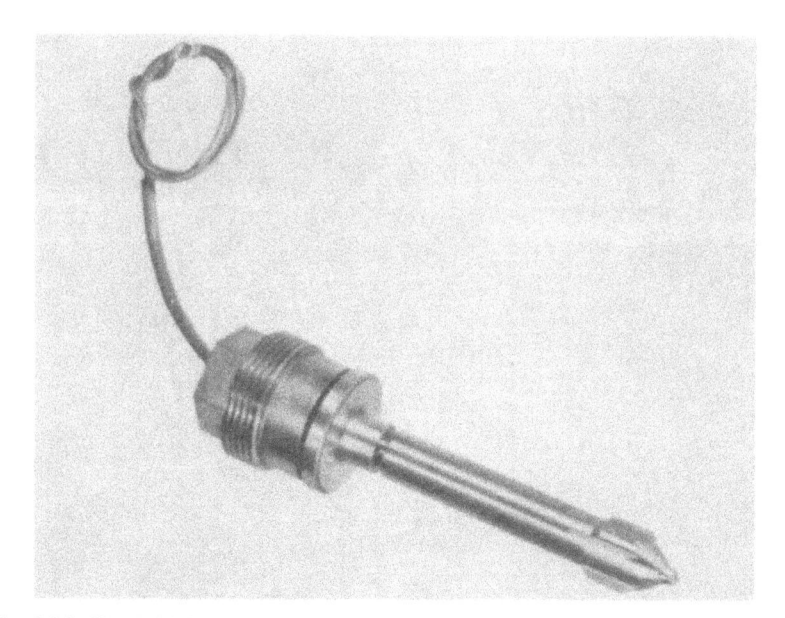

Fig. 8.25. The D-M-E Auto-Shut™ Probe provides positive drool-free shut off at the gate. Injection pressure forces the probe tip back; when pressure ceases, Belleville™ spring washers move the tip forward for positive shut off. A replaceable thermocouple-catridge heater permits close temperature control. (*Courtesy D-M-E Corp., Madison Heights, MI*)

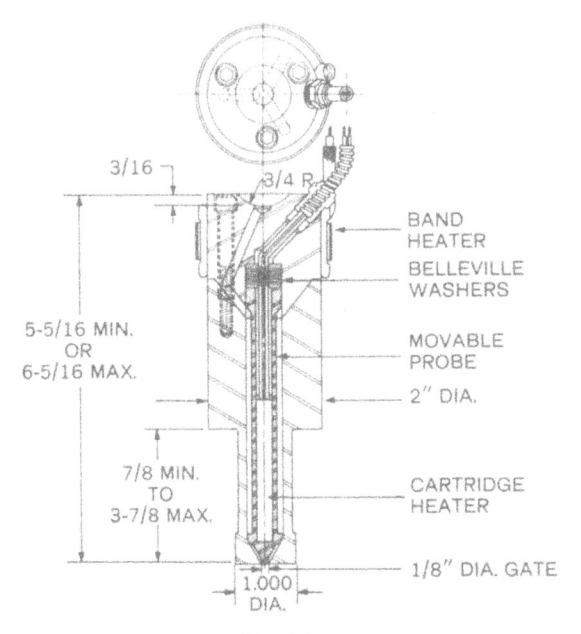

FIG. 8.26.

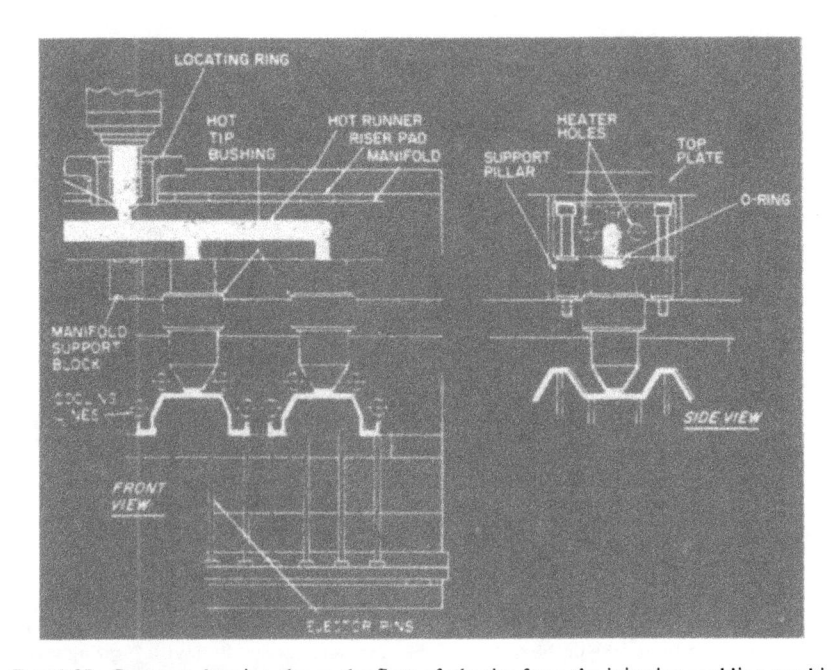

FIG. 8.27. Cutaway drawing shows the flow of plastics from the injection molding machine nozzle through the manifold and hot tip bushing and into the mold.

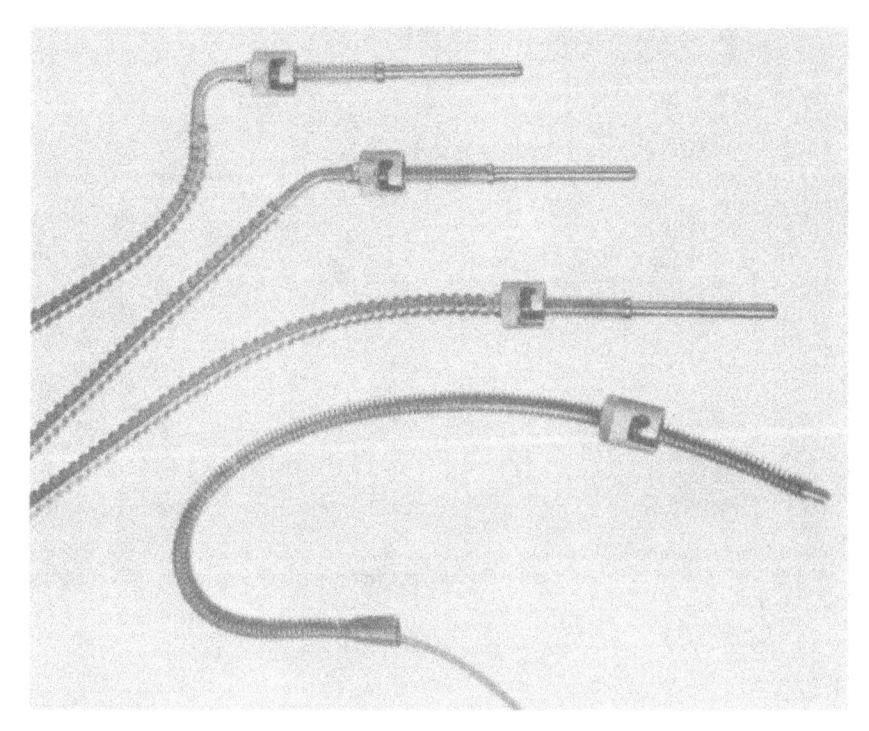

FIG. 8.28. Runnerless molding demands accurate temperature sensing and a variety of standard thermocouples are available. (*Courtesy DM & Company*)

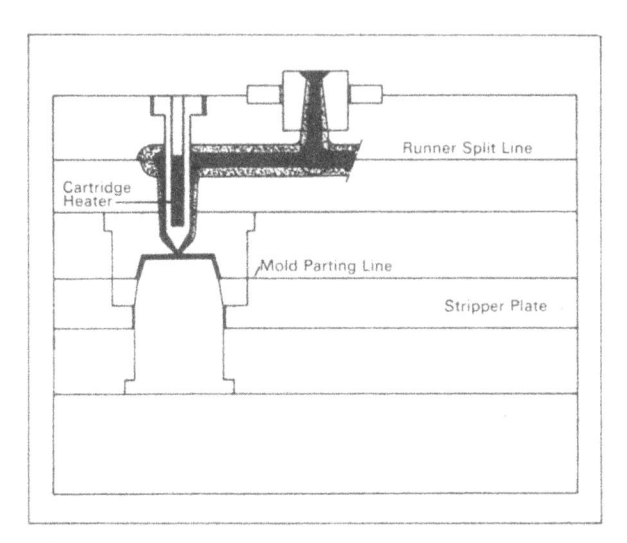

Fig. 8.29A. Insulated runner mold. May be used without heaters. Relies on plastic as an insulator to keep runner from freezing up. Mold must be split to remove frozen runner.

(*Fig. 8.29A,B,C,D,E Courtesy Husky Injection Molding Systems, Inc., Bolton, Ontario*)

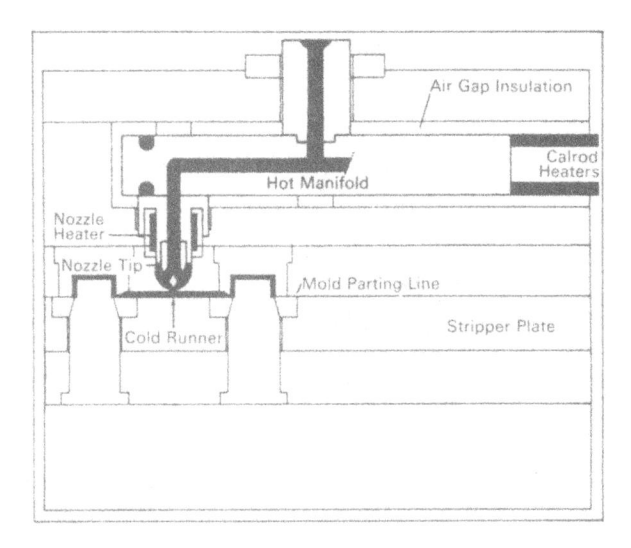

FIG. 8.29B. A combination of hot runner with short cold runners. Greatly reduces volume of cold runner molded per shot. Cold runner can be self-degating from parts. Cold runner may feed 4 or 6 cavities.

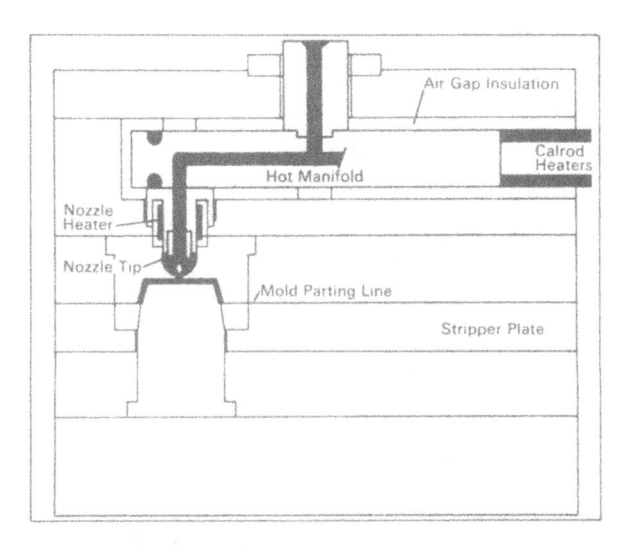

FIG. 8.29C. Hot runner, "runnerless" mold.

Hot Runner Valve Gate

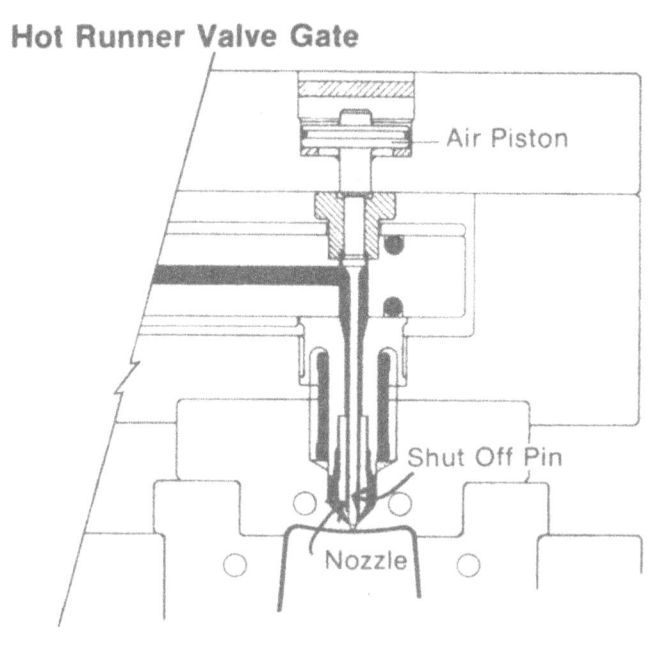

Fig. 8.29D. **Hot Runner Valve Gate.** A variation of standard hot runner design which provides a pin shut off in each nozzle tip. Practically eliminates gate mark.

Hot Runner Edge Gate

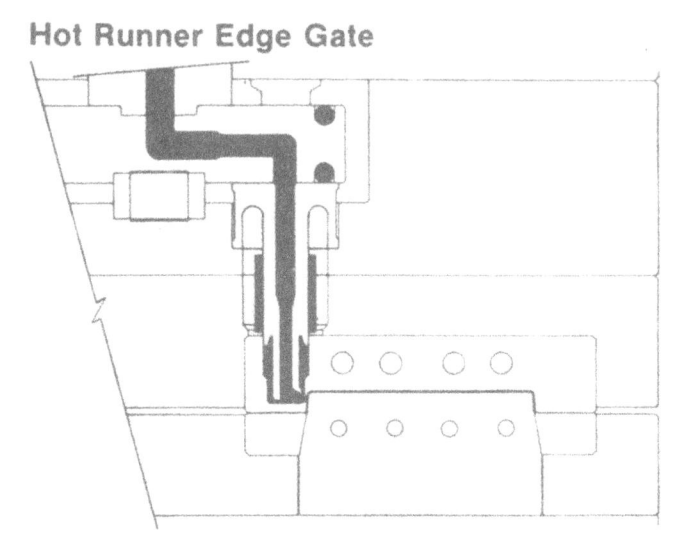

Fig. 8.29E. **Hot Runner Edge Gate.** Used where a gate mark on the outside top surface is not permissible. The hot gate is in the side of the part and is sheared off when the mold opens.

The "Italian" Sprue

The device illustrated in the mold shown in Fig. 8.23 is a system that combines both possible worlds. The runner is heated electrically, but a nonelectric means keeps the material from freezing off at the gate. Instead of utilizing a probe externally heated by means of electrical heating elements, the design of the nozzle incorporates what is in effect a *torpedo*. The entrapped heat of the plastic heats the torpedo, which in turn heats the plastic at the localized gate area so that it does not freeze off. The torpedo helps equalize the heat within the nozzle, both throughout the material cross-section and between front and back of the nozzle.

Cartridge heaters are normally used to supply heat to some manifold designs and the following guidelines are the result of several studies:

1. Use the lowest wattage heaters possible since they offer the longest life. Best results are obtained when they are operated between 30 and 70% of their rated output.
2. Select the largest heater diameter that is practical. For a given design, heater performance and longevity are generally determined by its watt density (watts per square inch of heater surface area versus its physical fit in the mold for heating. By the use of larger heaters with greater surface area, the fit is slightly less critical and the larger heaters have better characteristics as illustrated in Table 8.3. A ground O.D. to insure a tight fit is the best practice for heaters that are not cast in beryllium copper.
3. Make sure that all heaters, contacts and mold contact areas are clean; the presence of foreign material between the heater and the mold component will create a hot spot and cause premature heater failure.
4. Extra care must be exercised when new heaters are installed or when the mold has been in storage for a considerable period to bake out the moisture that has accumulated. A bakeout for several hours at 10 to 20% reduced power is recommended. A bakeout is always necessary before sealed terminals are applied.
5. Caution must be exercised in the mold design to permit an orderly arrangement of the wiring when such is necessary.
6. Ample space must be provided for hookup. This includes a check for tie-bar press clearance and other adjacent auxiliary components.
7. All heater wires must be identified for proper zone control prior to press installation. It is near impossible to identify multiple heater leads after the mold is in the press.
8. Evaluate the special cast-in beryllium copper heating elements for hot runner gating since they greatly simplify wiring, minimize burnouts and put the heat where it is needed.

Several patents have been issued or are pending on hot manifold molding which merit further study. E. R. Knowles patent No. 2,237,263, Dec.

TABLE 8.3. CARTRIDGE HEATERS FOR HEATING MOLDS
[by insertion into holes]

HOW TO SELECT HEATER SIZE:

1. Establish maximum temperature at which part will operate.
2. Calculate total wattage needed to produce this temperature.
3. Establish the diameter and length of the Heater that is best suited to the dimensions of your part.
4. Estimate the number of HEATERS needed for even heat. Divide total wattage required by the number of Heaters to determine wattage rating of each unit.
5. Determine the watt density of the size selected.
6. Determine fit.
7. Use graph below to make certain that the watt density you have established does not exceed the maximum allowable watt density of the cartridge.

Read along the part temperature curve to the horizontal line representing the fit you have established. From this intersection point read down to the Watt Density Scale. This is the maximum watt density you should use. A higher rating would shorten cartridge life. A lower rating prolongs cartridge life.

8. If you find that watt density is excessive you can correct in three ways: (1) Use a tighter fit (2) Use more or larger heaters (3) Use lower wattage. (In this case allow for longer heat up time.)
9. Tight fits are achieved by grinding OD of cartridge and reaming the hole size accurately.

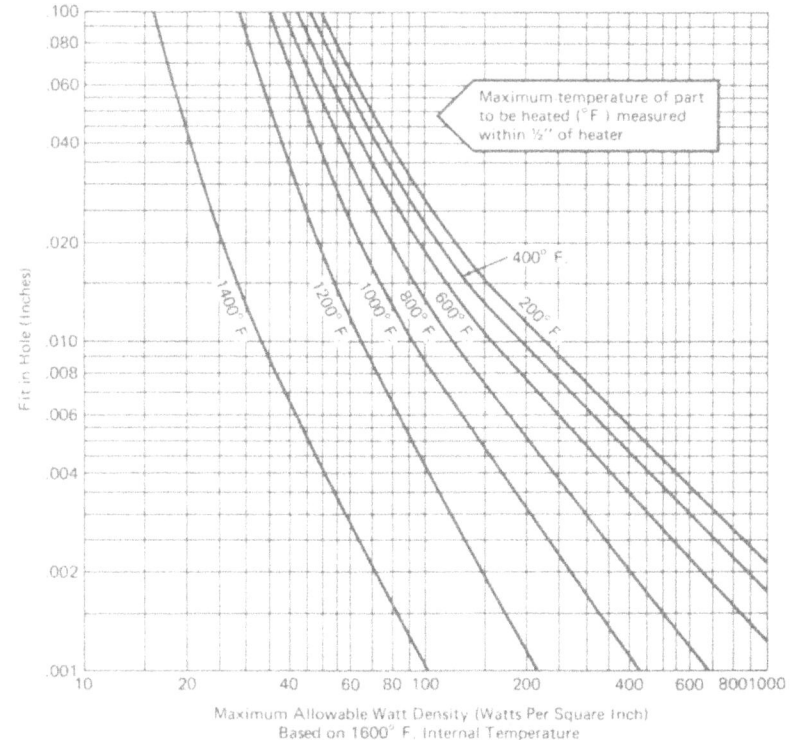

Table 8.3. *(Graph courtesy Watlow Electric Mfg Co., St. Louis, MO)*

31, 1940; Patent No. 3,093,865, June 18, 1963 covers the work of D. L. Peters and John M. Scott Jr. at Phillips Petroleum Company. Mold Masters Ltd. patent Nos. 3,530,539, 3,822,856 and 3,806,295 cover the work done by Jobst U. Gellert on hot edge and valve gating with cast manifold. The "Italian" sprue (Fig. 8.23) is reported to have been developed in the United States; Pyro Plastics are said to have used hot manifold molds with valve gating in the late forties. Columbus Plastics and Milton Ross were other pioneers. One vendor now offers the Driscoll Sprue with their own valve gates. The internally heated system was perfected at Stokes Trenton. Sam Son Molds, Inc. has a proprietary hot manifold type with a patented self-sealing feature and single control.

Completely integrated control panels of the closed loop-on/off type and the steady current adjustable type using triacs are available for a variety of applications. Temperature sensing probes, Fig. 8.28, have been developed to insure accurate reporting of thermal changes and immediate correction. Typical of the many sophisticated control systems is a solid state expandable modular electronic power controllers shown in Fig. 8.30. It is a fully time-proportioning closed-loop feedback system capable of excellent control over a wide range of operating conditions. This control is obtained by sensing the temperature at the required point by means of a thermocouple and then applying the exact amount of power called for to

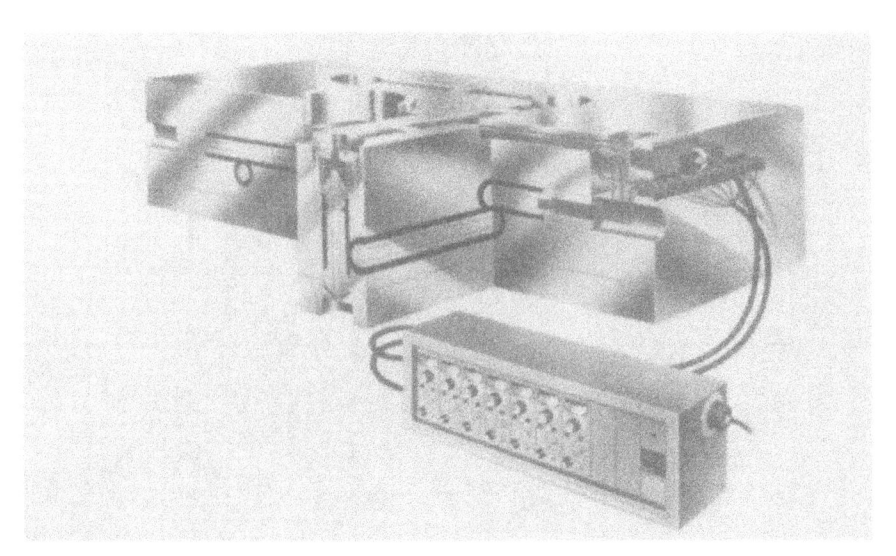

FIG. 8.30. The distributor system of a D-M-E runnerless mold employs distributor heater tubes located within distributor bores of a larger diameter. Plastics melt flows around the tubes, which contain cartridge heaters with integral thermocouples. The D-M-E electronic controls continually monitor and maintain melt temperatures within the entire mold. (*Courtesy D-M-E Corp., Madison Heights, MI*)

FIG. 8.31. Where precise temperature control is not required, the D-M-E phase fired power controls permit accurate control of the voltage supplied to the heating elements. Vernier control permits accurate adjustments at low voltage settings. (*Courtesy D-M-E Corp., Madison Heights, MI*)

maintain a pre-set temperature. The duration time that power must be applied is automatically sensed and continuously adjusted as required to maintain very close limits in properly designed system. Included features are ± 1° control, calibrated deviation meter, automatic/manual control in the event of thermocouple failure coupled with automatic shut-down in the event of such failure. Zero crossover switchover switching for extended heater or load life is used. These and similar units are supplied by several vendors.

Another useful type of control is the electronic power controller depicted in Fig. 8.31. This unit is designed to provide current control for any system that requires a very accurate and reliable variable power source for manifold, cartridge, tips, nozzles, probes, heater casts and insulated runner heaters.

Table 8.4 and 8.5 summarize the accumulated contemporary data on hot manifold/runnerless molds.

TYPES OF OPERATION

A balanced runner system is essential for successful operation of this principle. Experience has also indicated that the insulated runner system is generally satisfactory where cycles do not exceed 1 minute and where the part is large enough to allow rapid displacement of material in the runner system. The volume of the part should be about three times the volume of

TABLE 8.4.   Summary of Manifold Runnerless Molding Systems.

Types of Runners	Advantages	Disadvantages
Insulated runner no heat	• Simplest in Design • Fast Color Changes • No Controls • Runs Heat-Sensitive Materials	• Impractical on small parts or longer cycles • Freezes up on delay • Requires runner removal to start • Leaves largest gate vestige
Insulated runner heated probes	• Fewer controls than full hot runner • Smaller gate vestige than Insulated Runner • Most efficient type if there are frequent color changes	• Same as above except will tolerate somewhat longer delays and has less gate vestige
Hot manifold center gate	• Fewer freeze-up problems • Easier start-up than insulated type	• More controls • Longer warm-up time • Tends to undermine mold support similar to ejector box • Longer time to change colors
Hot edge gate	• Good for parts not suitable for center gate • Only 1 control for small molds • Least vestige of any system	• Overheat on delay • Uneven cooling around cavities • Most delicate at gate
Internally heated hot runner	• Starts up most quickly and at normal operating settings • No overheat or freeze off on delay • Less energy use than hot manifold • Most efficient type for high production 1 color	• Most prone to blocked gates • Requires more controls than other types • Takes longest time for color changes

(*Courtesy Stokes Trenton, Inc., Trenton, NJ*)

TABLE 8.5. Comparison of Various Runnerless Systems.

Criteria	Ins. Runner	Ins. with Hot Tip	Hot Manifold	Hot Edge Gate	Int. Ht'd Hot Run
Gate vestige	D	B	B	A	B
Frequency-Gate freeze off	C	B	A	B	A
Ease of freeing frozen gate	C	B	A	A	A
Ease in freeing blocked gate	A	A	C	C	C
Can shut off one or more cavs.	X	C	A	X	A
Start-up time	C	C	B	B	A
Waste of plastics material	C	B	B	A	A
Simplicity	A	B	C	B	C
Power consumption	A (=0)	B	C	B	B
Number of controls	A (=0)	C	C	B	C
Usable with closed-loop control	X	C	B	A	C
Run small light weight parts on slow cycle	X	C	A	B	A
Run large heavy parts on long cycle	X	C	A	B	A
Overall efficiency*	D	C	B	A	A

A = best   B = good   C = fair   D = poor   X = impractical

*Depends on application but assumes high volume of parts.

(*Courtesy C. C. Davis Jr., Stokes Trenton, Inc., Trenton, NJ*)

the runner system. However, where cycles are longer than 1 minute and when the item weight is small, it has been found desirable to use auxiliary heat, not only in the mold, but in the runner system itself.

Molds are designed for use in the following areas of production:

(1) Hand or "V" molds, shown in Fig. 8.32, consist of loose cavity and force plates which are fitted into "V" nests or other types of holders mounted into the press. The loose members of the mold are removed from the press with the parts after cycling and are separated by hand or fixture external to the press. This design permits low tooling costs for experimental or small-scale production. The unit cost is high due to the extended cycling time and hand labor.

(2) Standard frame and unit-type molds are used quite extensively, with many molders employing their own custom-designed frames or one of the many commercial types available as shown by Figs. 8.33, 8.34 and 8.35. The master frame, attached to the press, contains the sprue bushing, retainers, cooling or heating channels, and a master ejection system. The cavity and force areas available for the master frame are of a predetermined dimension for each frame size, and provide an auxiliary ejection system. This product, which is identified as a component or unit, is assembled into the master

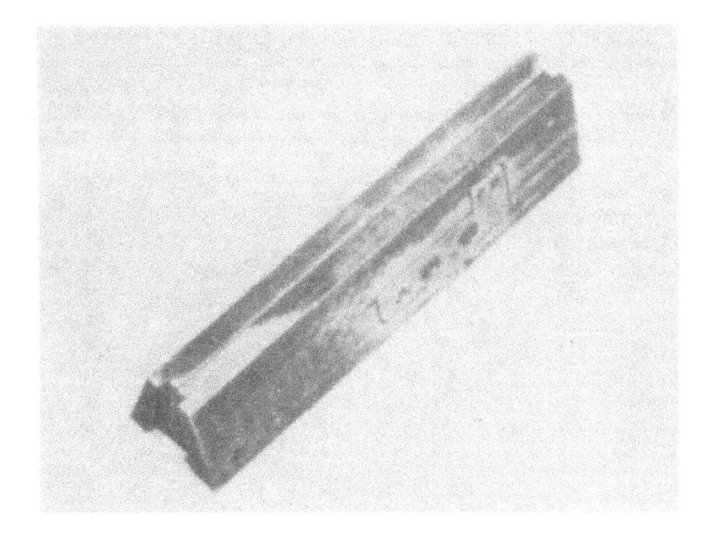

(A)

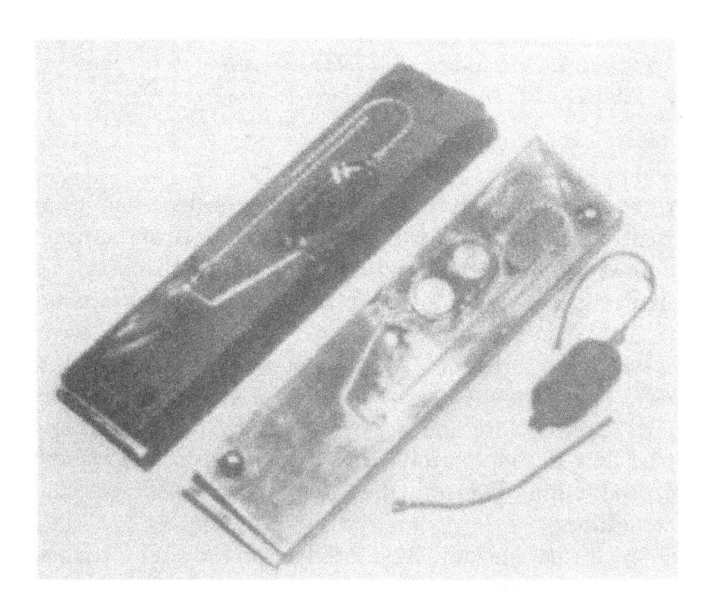

(B)

Fig. 8.32. (A) This view shows exterior shape of a typical "v" bar mold. The mold is placed in a "V" shaped mold block in the press for molding. (B) Interior view of "V" bar mold showing cavity and force construction for molded part illustrated in lower right corner. Cavity is shown on left; force in the middle.

FIG. 8.33. Mold components for this reproduction of a 100-year-old plastics **daguerreotype** case are mounted in this master unit die frame. These mold cavities were cast from the original plastics pieces as models. (*Courtesy Plastics World, Boston, MA*)

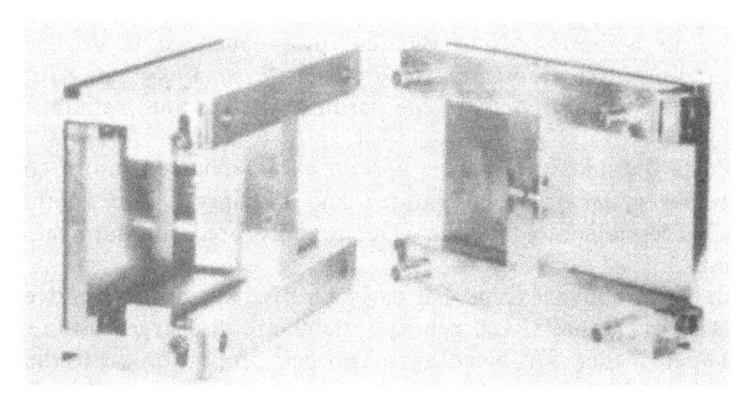

FIG. 8.34. Master unit die frame (11½ in × 16 in.) for standard unit inserts 7 in. 7 in. (*Courtesy Master Unit Die Products, Inc., Greenville, MI*)

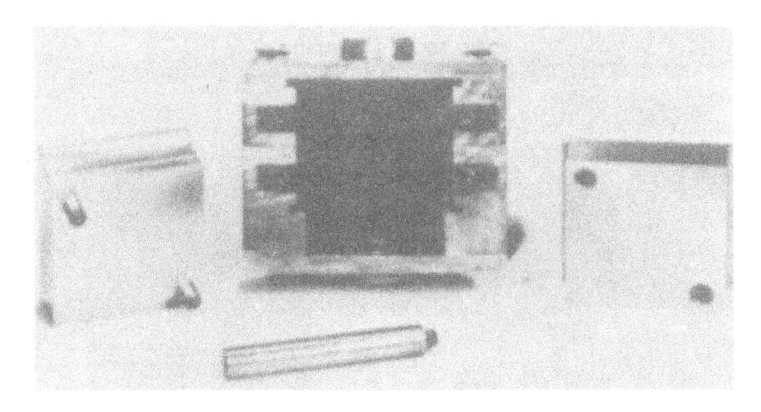

FIG. 8.35. Master unit die frame and single-mold insert for Arburg 1/3-oz. press (*Courtesy Master Unit Die Products, Inc., Greenville, MI*)

frame for production. Setup time is at a minimum and it permits semi-automatic operation for production of the molding, resulting in a lower parts cost and minimum tooling investment. Multiple and combination types of parts may be molded concurrently if the same material and color are required. If the standard frame is large enough to permit a variety of cavities, runner and gate shut-offs may be employed to keep the tooling investment low and to facilitate balanced production.

(3) Semiautomatic Molding. The mold is fastened to the stationary and moving half of the press, and ejection takes place upon opening the press. The operator discharges the parts and recycles the press as illustrated in Fig. 8.36.

(4) Automatic Operation. This process is the most efficient since the press is operated by predetermined cycling elements and is continuously repetitive in operation. No individual press operator is necessary and completely uniform production quality is achieved. Fully automatic molding gives best quality control. The various types of automatic operations include:

(a) Parts degated from runner system and discharged into a container with runner system pieces separated and discharged into another container. Three-plate mold design is commonly used for automatic mold operation.

(b) Same as above, except that parts are discharged into a conveyor for removal from the press, and runner systems are discharged into a grinder and hopper loader for blending in proper proportion with the virgin material.

(c) Use of the hot runner or insulated type of mold design eliminates the losses and handling of the runner systems.

FIG. 8.36. Semiautomatic press operation, showing operator opening gate on press to remove parts and runner system from a 16-cavity automatic unscrewing mold for an internally
threaded closure. Note the cylinder operated rack mechanism to unscrew threaded cores.
(*Courtesy Wheaton Plastics Co., Mays Landing, NJ*)

(d) Other variations of automatic molding may be used depending on
the product and its production requirements. Some volume products,
which require best quality control of dimension and appearance, will
necessitate a high inspection standard and special packaging, and may be
set up in a two or more press battery with conveyorized movement of parts
to a central inspection station employing one packer inspector only for
all presses. A two-press installation for piano keys is shown in Fig. 8.37.

FIG. 8.37. Automatic molding using two Lester 150-ton 13-oz. reciprocating screw presses with 4-cavity mold operation which drops the degated parts on the inclined conveyor and delivers them to the central inspection station for inspection and packing. (*Courtesy Tech-Art Plastics Co., Morristown, NJ*)

FIG. 8.38. Stokes vertical automatic injection press with swing plunger tray that moves from side to side under the cavity section to permit automatic loading of inserts while other half of tray is positioned for molding and ejection. (*Courtesy Pennwalt Corp., Stokes Div., Philadelphia, PA*)

(e) Special automatic mold operations for products requiring metal insert components may be mechanized as shown in Fig. 8.38. Figure 8.39 illustrates another vertical-operating automatic press, which is suitable for fully automatic injection molding with feedback of runner system. Automatic molding designs require that molds be built of the best materials. They should utilize hardened or prehardened mold retainers, and employ the best possible ejection systems for positive removal of parts and extraction of runners. Automatic ejection is often achieved by the use of mold wipers, as shown in Fig. 8.40, or by the use of an air blast assuring complete removal of the parts, flash and runners at time of ejection. The presses must be set up to utilize safety interlocking controls that eliminate recycling when parts are not properly ejected. A low pressure adjustment on the clamping system that will reduce the clamping pressure to a minimum when pieces are not completely ejected will prevent mold damage. Magnets should be installed in the press hopper to collect foreign metal and thus prevent damage to the mold, press plunger, screw or nozzle.

The mold design must include a positive system of actuating and return

FIG. 8.39. This 15-cavity automatic injection mold is operating in a fully automatic reciprocating screw press with comb to pick up parts and gates separately and discharge the parts into a tote box. The runner system is discharged into a grinder-feeder for blending with virgin material and loading into the hopper. (*Courtesy Rapid Tool and Mfg. Co., Newark, NJ*)

FIG. 8.40. This press-actuated wiper removes the molded piece and the runner from the mold, assuring completely automatic operation with no possible "double-shot."

of the ejector bar and pins. Double knockout or ejector systems may be needed to insure positive ejection.

(5) Two-level Molding (also called "Stack Mold" and "Tandem Mold"). This system has been used extensively in the compression type of molding, and many multi-level molds have been used for injection, compression, and transfer molding. Injection two-level designs have proven to be desirable for parts of small weight and large area when the part configuration permits the mold to be designed in two levels. This is desirable to gain use of full injection cylinder capacity. Figure 8.41 illustrates a typical two level mold design.

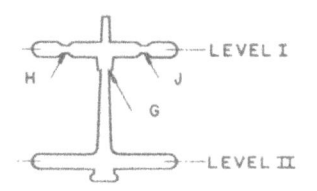

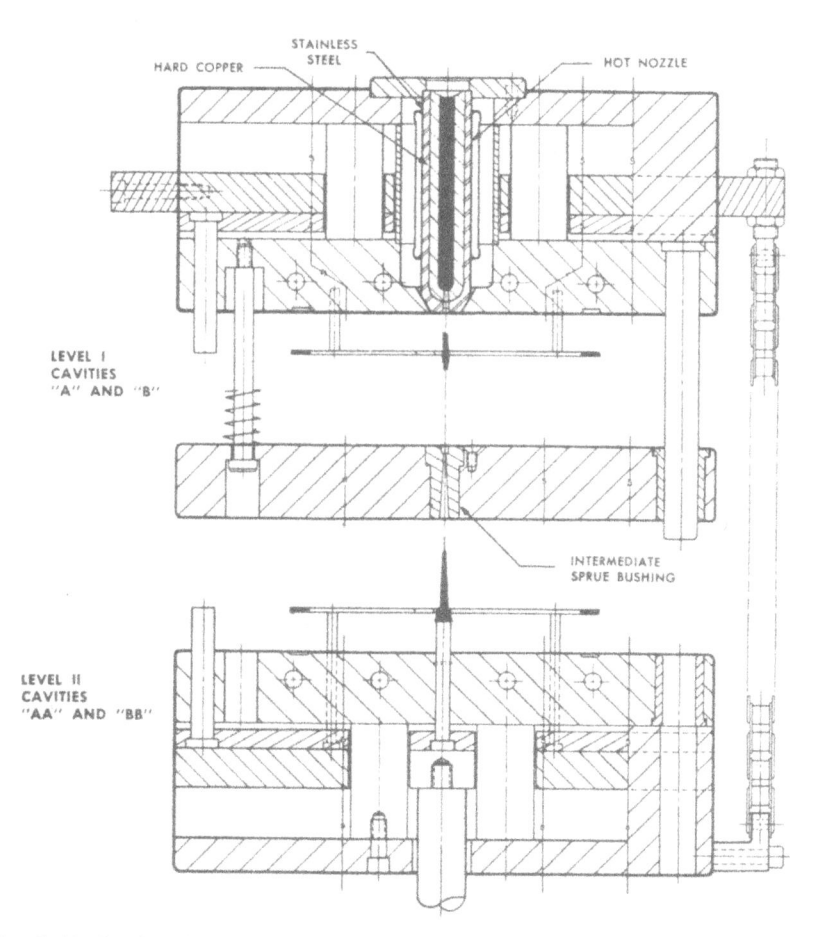

FIG. 8.41. Stack mold which permits two or more large area parts such as picture frames to run in a small press with half the clamping pressure that would be needed to parallel the cavities in one plane. (*Courtesy Interstate Mold & Hobbing Co., Inc., Union, NJ*)

FIG. 8.41A. This is a hot runner stack mold, the upper half the lower half each comprise a complete mold (*Courtesy Stokes Trenton Inc., Trenton, NJ*)

EJECTION OR KNOCKOUT SYSTEMS

Ejector travel must be sufficient to clear the molding from fixed members in the mold. Undercuts or "pick-up ribs" may be machined into mold members to insure that the molded part remains on the ejection side of the mold. Pick-up sizes are determined for the type of the material being used.

Parts may be removed from a mold using the common type of ejector or knockout system.

Ejectors actuated by an ejector bar must contain pushbacks or safety return pins to reposition the ejector pins prior to the start of the injection or mold filling cycle. Many times it is desirable to add positive alignment to the system to gain more mold life and to offer trouble free operation. Fig. 8.42 shows guide pins and bushing assembly in the ejector system. The primary benefit is in injection molds operating in a horizontal position where side thrust is a major factor in ejector pin wear or breakage.

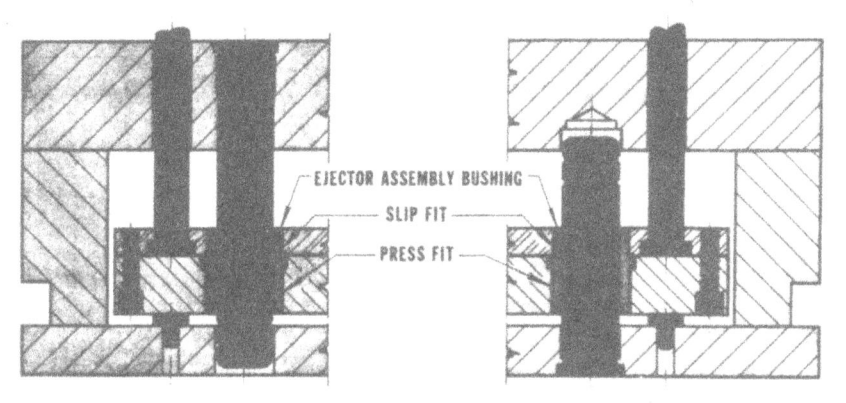

FIG. 8.42. Schematic showing best alignment for ejection travel. Note guide pins and bushings assembled into ejector assembly. (*Courtesy National Tool & Mfg. Co., Kenilworth, N.J.*)

Careful consideration is essential to decisions concerning the number and area of ejectors to be used and the type of system to be employed for the various types of materials. It must be understood that, in most cases, the part being ejected may, to some degree, be soft due to the high temperature at time of ejection. For lowest cost, molding parts are ejected when they are just sufficiently hard to prevent distortion and in many cases, the mold itself is heated to achieve maximum feasible temperature at ejection. Figure 8.43 illustrates desirable knockout pin locations for soft ma-

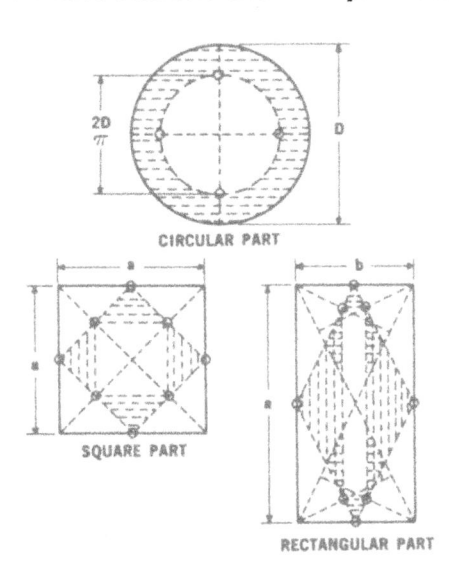

FIG. 8.43. Knockout pin locations. (*Courtesy E. I. DuPont de Nemours & Co. Inc., Wilmington, DE*)
Note: Based on calculation to give minimum bending—shaded areas are preferable ejector pin locations for soft flexible plastics.

terials such as polyethylene. Ejection marks may be styled into the part when desirable by adding design configuration in these areas. This is accomplished by decorating or adding a series of concentric rings on the ejector pin surface.

(1) *Ejector Pin or Blade.* Round or rectangular shaped pins are used as indicated by the available areas for product removal. Pins should be made to the largest possible area for longest tool life. Typical ejector or knockout pins are shown in Fig. 8.44.

(2) *Sleeve Ejector.* A sleeve ejector is used around bosses wherein a center coring pin is employed. It is attached to the lower mounting plate, permitting a sleeve or bushing-type pin to provide uniform ejection around the coring pin. This type sleeve is used for all molding materials. It is shown in Fig. 8.45.

(3) *Stripper Plate or Ring.* This system permits uniform ejection for thin-walled or deep moldings of all types of materials. It is normally as-

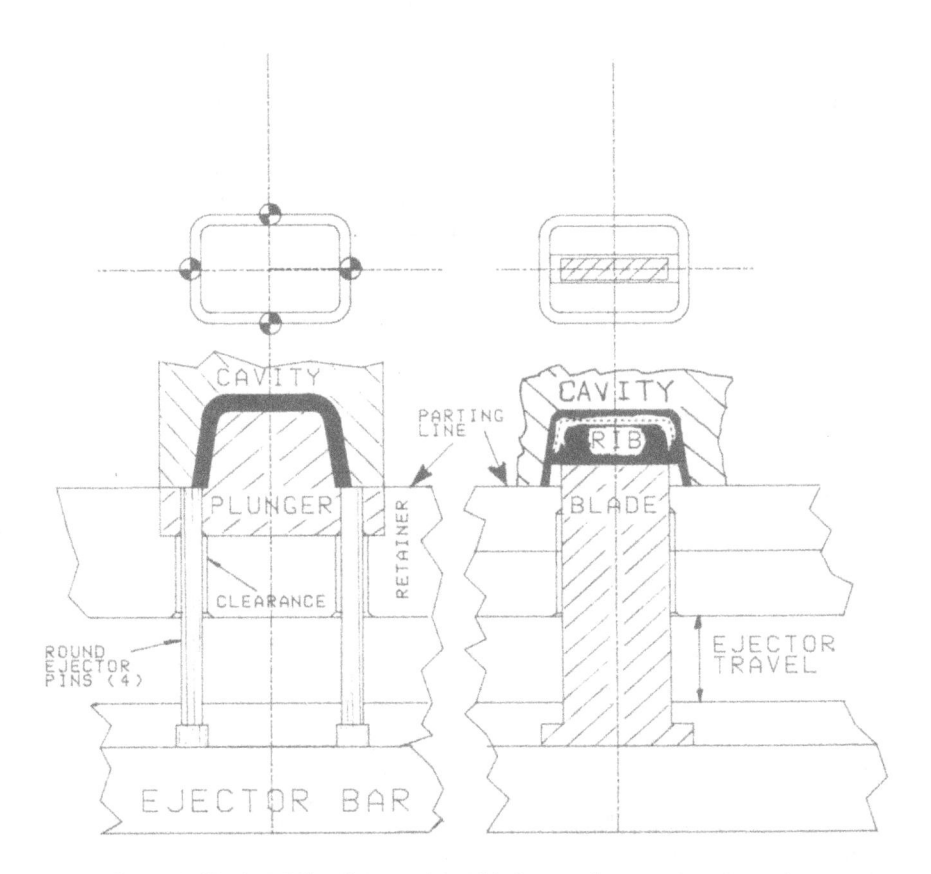

FIG. 8.44. Typical design for round and blade-type ejector or knockout pins.

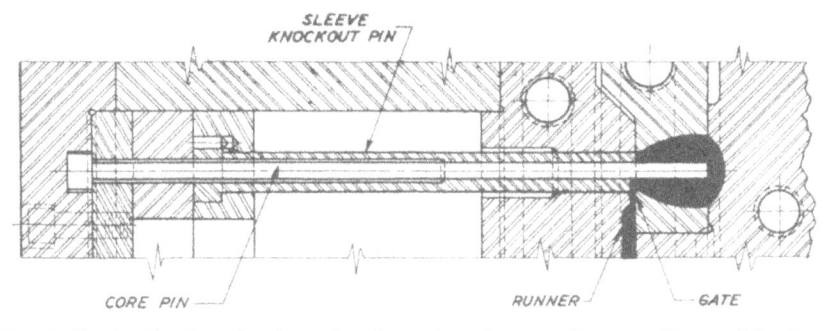

FIG. 8.45. Application of a sleeve knockout pin. (*Courtesy Eastman Chemical Products, Inc., Kingsport, TN*)

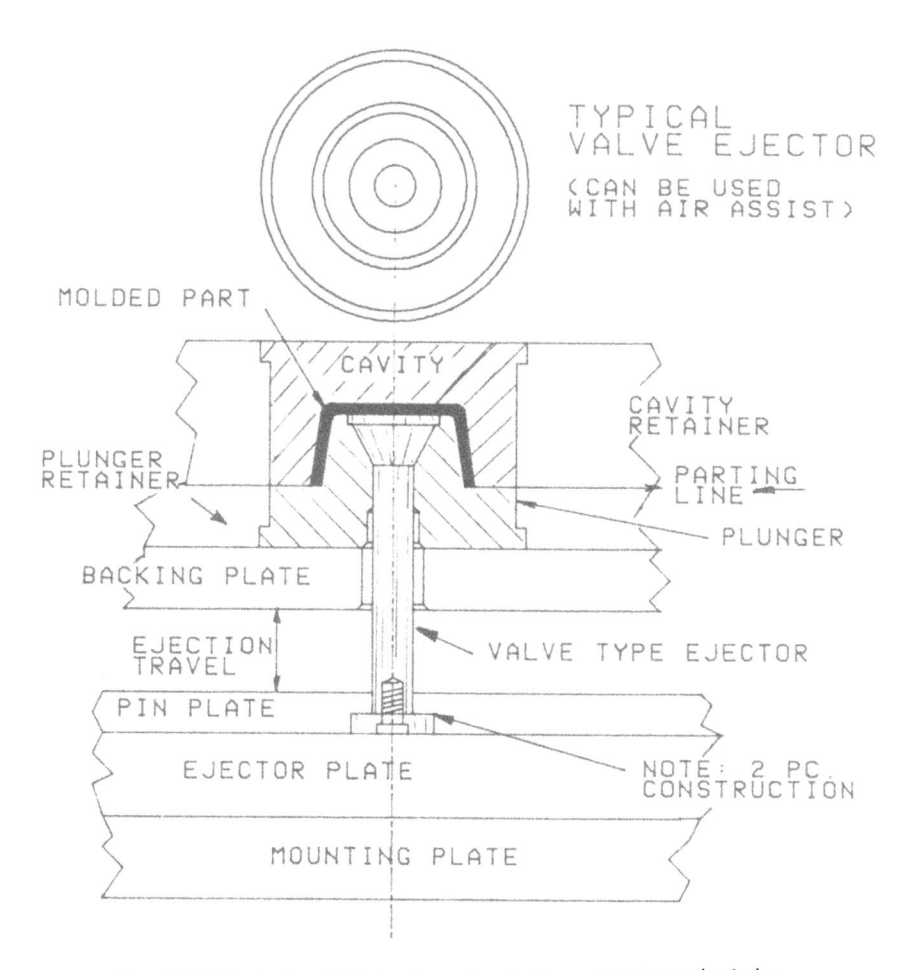

FIG. 8.46. Typical mold design for valve ejector or knockout pin design.

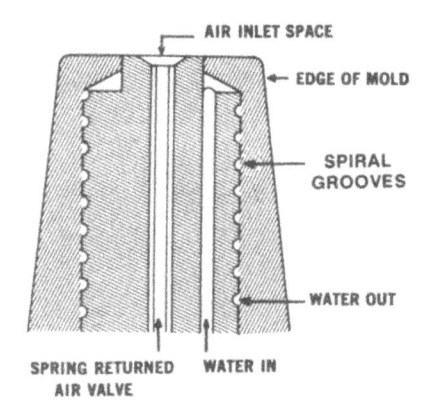

FIG. 8.47. Typical air-ejection valve assembly. (*Courtesy E. I. DuPont de Nemours & Co. Inc., Wilmington, DE*)

sembled into a third plate actuated by the ejector bar. Several figures in this chapter illustrate stripper plate design. Details of the stripper plate design are also covered in Chapters 2 and 6.

(4) *Valve Ejector Pin.* This ejector pin has the shape of a valve and stem which provides a large area for ejection in tooling designs that limit the use of conventional ejectors; it also provides good release and tool strength. It is commonly used for the flexible materials and in many disc-gated molds. Details of the valve ejector pin are shown in Figure 8.46.

(5) *Air Ejection.* High-pressure air is channeled into the coring on the interior of a part, with the port or valve area actuated by a moving pin attached to an ejector bar which, upon opening of the mold, releases the air and forces out the molding. Normally, the valve is spring loaded for the return stroke. Air ejection is commonly used for flexible plastics and deep draw products. A typical design is shown in Fig. 8.47.

(6) *Two-stage Ejection.* Designs requiring thin walls or areas which have undercuts on the interior require one stage of ejection to remove the parts from the cams or the mold force forming the inner wall design and the second stage to remove the part from the mold. This double system permits the part to be freely ejected from the mold. This design may be used for all materials. It is illustrated by Figs. 8.48 and 8.49.

RUNNER SYSTEMS*

Typical runner cross sections are shown in Fig. 8.50.

Full round runners are preferred as they have minimum surface-to-volume ratio, which minimizes heat loss and pressure drop. Trapezoidal

*See also Hot Runner Molds—Figs 8.16 through 8.29.

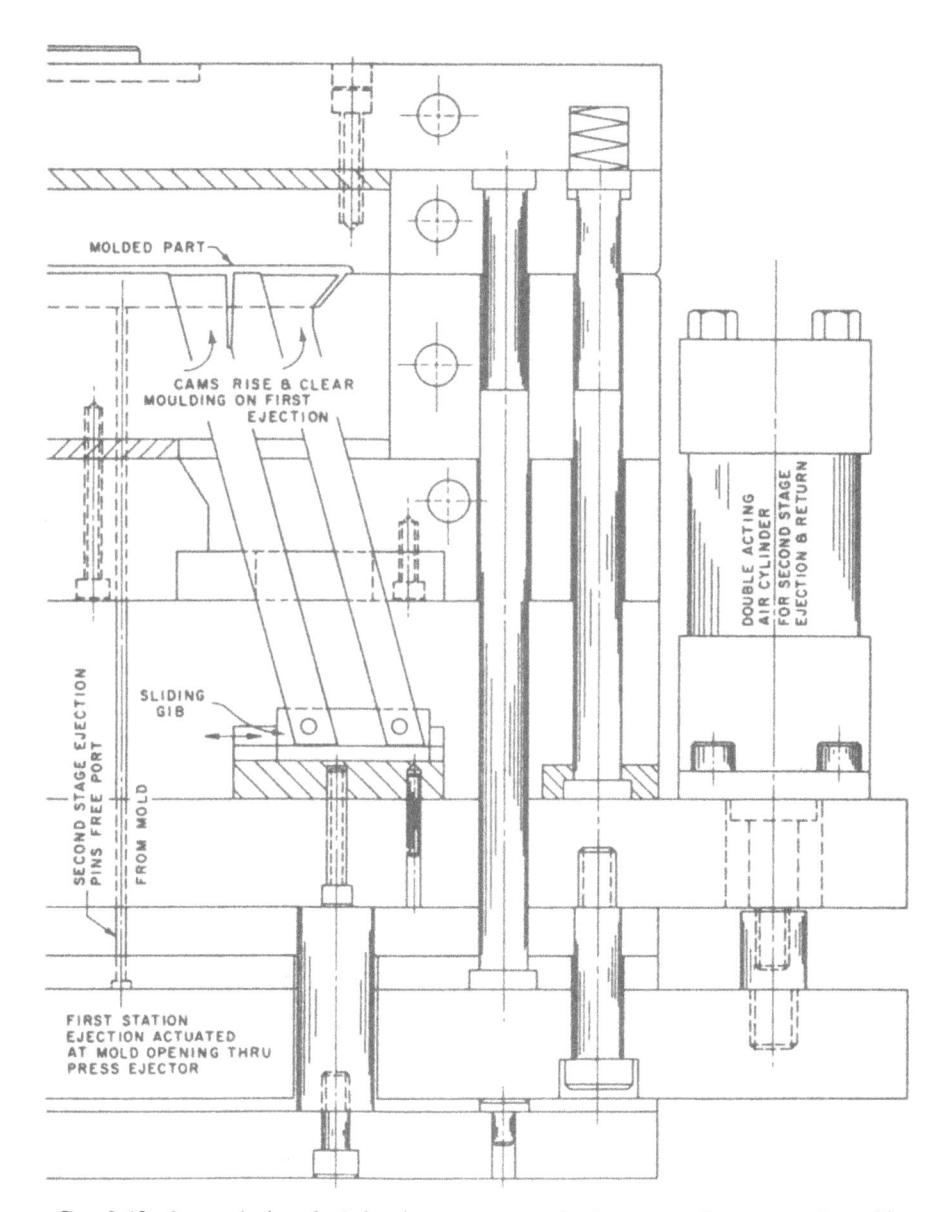

FIG. 8.48. One typical method showing a two-stage ejection system for an organ key with undercuts on interior. Second stage is actuated by double-acting air cylinder. (*Courtesy Marland Mold Co. Inc., Pittsfield, MA*)

FIG. 8.49. This 16-cavity balanced-runner mold has tab gating and 2-station ejection. Hydraulic cylinders actuate the first stage (Fig. 8.48) ejection. (*Courtesy Tech-Art Plastics Co. Morristown, NJ*)

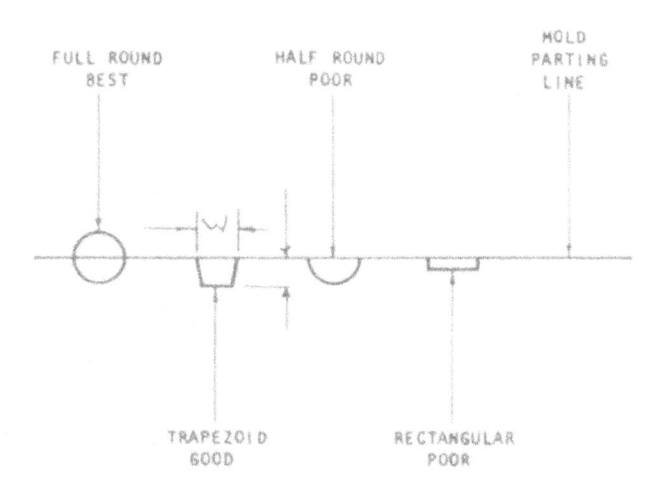

FIG. 8.50. Typical runner cross sections. (*Courtesy Rohm & Haas Co., Philadelphia, PA*)

runners work well and permit the runner system to be designed on one side of the mold. This type of runner is commonly used in three-plate molds where the full round runner may not release properly, and at the parting line in molds with a sliding action where the full round runner would interfere with the sliding movement.

The preferred type of runner layout is known as the *balanced runner*, as shown in Figs. 8.49 and 8.51. The balanced runner permits best uniformity of flow of the material from the sprue into the various cavities since all cavities are at equal distance from the sprue. This design is beneficial for

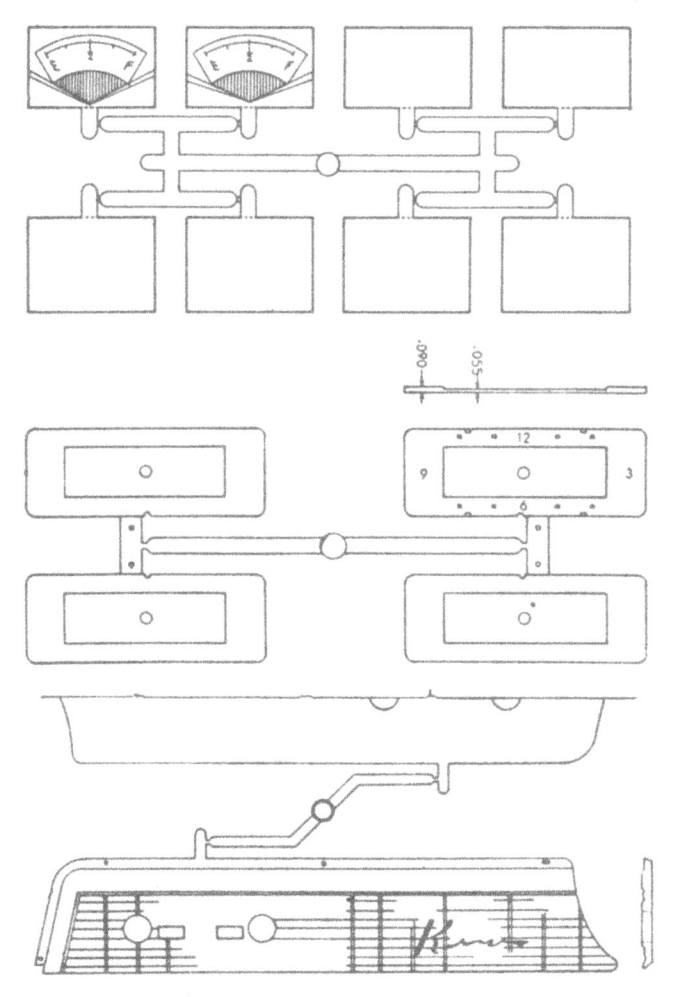

FIG. 8.51. Typical balanced runner designs. (*Courtesy Rohm & Haas Co., Philadelphia, PA*)

all types of materials. The secondary runners are smaller than the main runner since less volume flows through them and it is economically desirable to use minimum material in the runners. Other types of runners are shown in Figs. 8.52 and 8.53.

Restricted runner systems, as shown in Fig. 8.53 are used quite satisfactorily for some acrylic product mold designs. Material is heated by friction as it passes through a restricted area. The restriction is located approximately 2/3 the distance from the sprue to the gate. This design improves the heating and flow of the material as it passes through the runners, and finally provides a rapid pressure drop along with the heat rise, permitting better control of injection pressure without danger of overpacking the cavity. Restricted areas are approximately 25 per cent of the

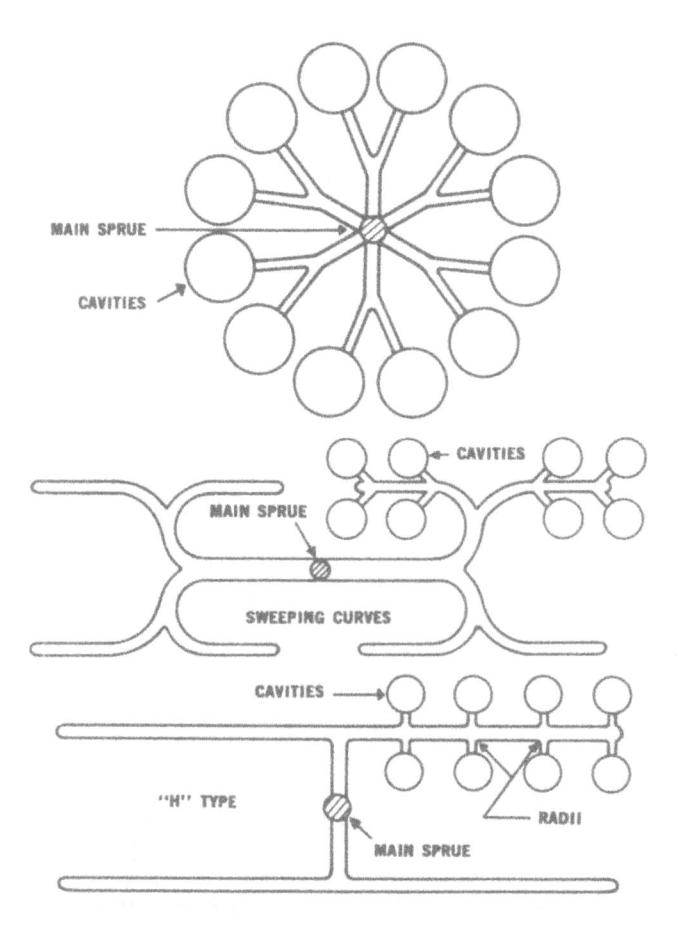

FIG. 8.52. Typical runner designs. (*Courtesy E. I. DuPont de Nemours & Co., Inc., Wilmington, DE*)

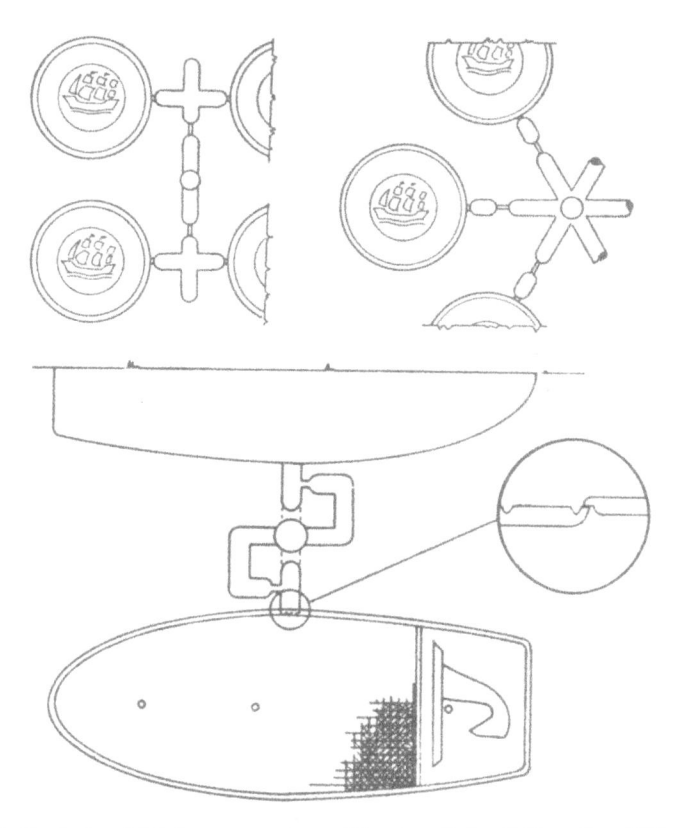

F<small>IG</small>. 8.53. Typical restricted runner systems. (*Courtesy Rohm & Haas Co., Philadelphia, PA*)

runner diameter with a land length of approximately ¼ in. In all cases, it should be smaller than the gate area.

Sprueless-type runner systems are used as shown in Fig. 8.54 with the nozzle extending as close as possible to the main runner, with a sprue height of approximately ½ in., or in special cases the nozzle extends to the runner system. This design reduces the amount of material required, eliminates sprue sticking, increases the mold filling efficiency and improves cycling.

It is not practical to specify accurately the size, shape, and length of runners to be used for injection molds. However, general experience and empirical analysis offer the following suggestion for the design of runners.

Runners must be designed to fill the cavity rapidly, and for easy ejection and easy removal from the molded parts. The surface of runner areas must be polished, for most materials, and it is desirable to have multiple sprue pullers and ejection locations in extended runner systems. (Polyethylene normally does not require as good a surface as most other materials.)

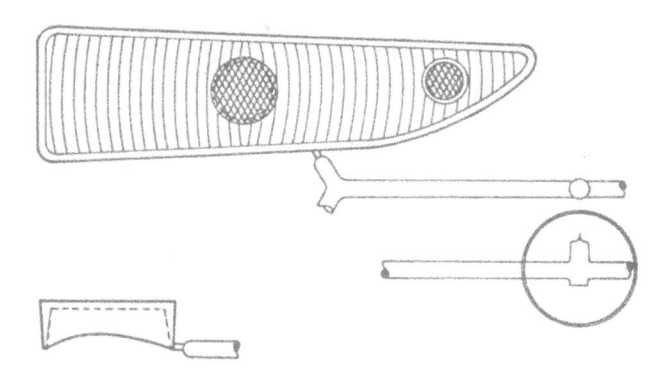

Fig. 8.54. Typical sprueless mold for acrylics. (*Courtesy Rohm & Haas Co., Philadelphia, PA*)

Typical runner diameters for various unfilled materials are given in Table 8.6. (The areas of other type runners should be equal to or greater than these round runner areas.) These approximate sizes are for conventional runner molds and *not* for hot runners or insulated runner molds.

For critical programs and new materials, it is well to ask the materials supplier to review runner sizes and types and to make recommendations. Part designs of small cross-sectional areas and short runner lengths normally will require runners on the low side of the above indicated sizes, and parts of larger cross-sectional areas and nonuniform sections with short or long runner systems will require the maximum diameter. The runner area selected must equal the sprue area to permit rapid flow of the material to the gating area.

The "flow tab" has proven to be quite successful when included in the runner system for controlling quality and dimensional factors in multiple-cavity molds as explained in Figs. 8.55 and 8.56. "The flow tab (Fig. 8.55), extending from a runner, is made by machining a groove about 5 in. long, 0.25 in. wide and 0.035 in. thick. It is marked at $\frac{1}{8}$ in. intervals to indicate the extent of flow into the mold. The operator can check the reading periodically and adjust his machine conditions (usually injection pressure) to keep the flow and pressure in the mold uniform. For example, a burned-out heater band would be indicated immediately by a decrease in flow tab length. In multi-cavity molds, carefully made flow tabs on each cavity can give excellent indications of the relative pressure developed in each cavity."*

"Figure 8.56 demonstrates a fairly simple method of controlling molding quality and size with multiple-cavity molds. It consists of an engraved

*E.I. Dupont de Nemours Co. Inc., Wilmington, DE

TABLE 8.6. Runner Diameters for Unfilled Materials.

Material	Runner Diameter	
ABS, SAN	3/16-3/8	
Acetal	1/8 -3/8	
Acetate	3/16-7/16	
Acrylic	5/16-3/8	
Butyrate	3/16-3/8	
Fluorocarbon	3/16-3/8	Approximate
Impact acrylic	5/16-1/2	
Ionomers	3/32-3/8	
Nylon	1/16-3/8	
Phenylene	1/4 -3/8	
Phenylene sulfide	1/4 -1/2	
Polyallomer	3/16-3/8	
Polycarbonate	3/16-3/8	
Polyester thermoplastic	1/8 -5/16	Unreinforced
	3/16-3/8	Reinforced
Polyethylene—low to hi-density type		
1, 2, 3, 4	1/16-3/8	
Polyamide	3/16-3/8	
Polyphenylene oxide	1/4 -3/8	
Polypropylene	3/16-3/8	
Polystyrene—general purpose medium		
impact–hi impact	1/8 -3/8	
Polysulfone	1/4 -3/8	
Polyvinyl (plasticized)	1/8 -3/8	
PVC Rigid (modified)	1/4 -5/8	
Polyurethane	1/4 -5/16	

'ruler' adjacent to the base of the sprue bushing and is, in effect, a built-in 'pressure gage.' It should be 3½ in. long, ¼-in. wide, and 0.050-in. thick for 'Plexiglas'; 0.030-in. thick for 'Implex.' Numerals are stamped or engraved, 1-15, spaced ¼-in. apart. When the mold is first tried out, careful records should be kept as to the number reached on the gage in relation to part size and quality.

"In production, all parts filled to the correct degree (as determined by the gage) will be uniform in size and shape. The molding machine operator need only look at the number of the gage when inspecting the parts. Molding precise parts such as fountain pen barrels where thread size is so important, and decorated parts where mask fit is vital has proved the value of this simple tool."*

It should be remembered that when the runners systems are not re-

*Courtesy Rohm & Haas Co., Philadelphia, PA.

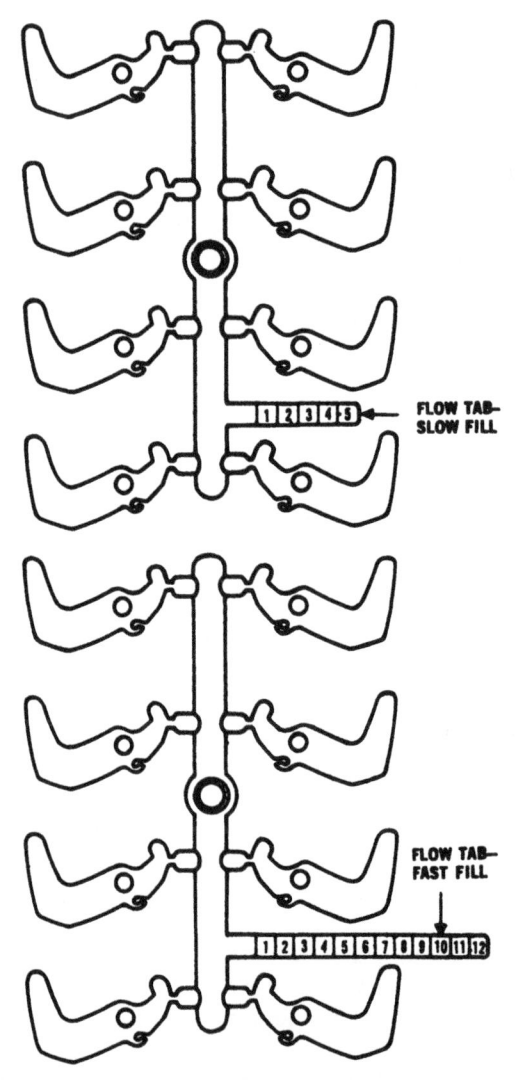

FIG. 8.55. Pressure gage for polyolefin molds illustrating one method of checking suspected mold pressure and fill rate loss because of machine malfunction utilizing a mold with a flow tab machined into it similar to that shown above. The flow tab may also be incorported into any mold for in-process mold pressure recording where molded part shrinkage is critical. (*Courtesy E. I. DuPont de Nemours Co. Inc., Wilmington, DE*)

ground and remolded, the cost of the lost material in a specific job is substantial. The resale value of scrap runners represents only a small fraction of the virgin materials cost. It is often impossible to use any scrap when producing colored or crystal clear moldings for top quality products.

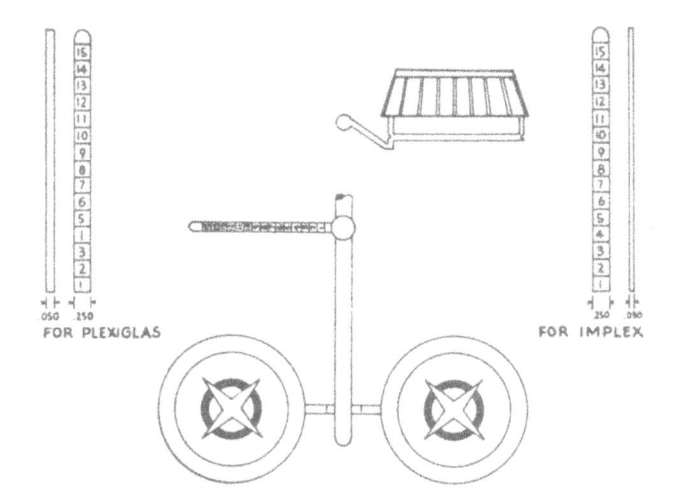

FIG. 8.56. Pressure gage set-up for acrylics. (*Courtesy Rohm & Haas Co., Philadelphia, PA*)

GATING SYSTEMS

Gating is a widely controversial subject, and it is impossible to forecast completely accurate design information. The gate is the opening between the runner system and the cavity. Gates are often initially cut to the minimum size and enlarged as required for concurrent filling of all cavities. Based on empirical studies and experience, the following information can be used as a guide: General design practice to insure uniform ejection is to permit tapered areas on sidewalls of gating areas, with minimum taper required for flexible materials, and greater tapers for rigid and filled materials. The land area (Fig. 8.57) is very critical and is generally .025 to .060 in. in length. Multiple-cavity molds require accurate gating and land area.

Sometimes, multiple-cavity molds with long runner systems require enlarging the gate area in the cavities at the extreme ends of the runners.

Gating sizes shown are approximate and were compiled over a wide range of parts that have been produced as a result of many years of experience. In designing gate areas, consideration must be given to the sprue size and the runner system to facilitate rapid filling of the mold.

Where possible, gate locations should be designed into the part, as shown in Fig. 8.58. This eliminates costly finishing operations and greatly facilitates automatic molding.

Many illustrations are presented here to show the most commonly used types. With the wide range of materials, complex formulations, and fillers available today, it is suggested that actual gate specifications be reviewed

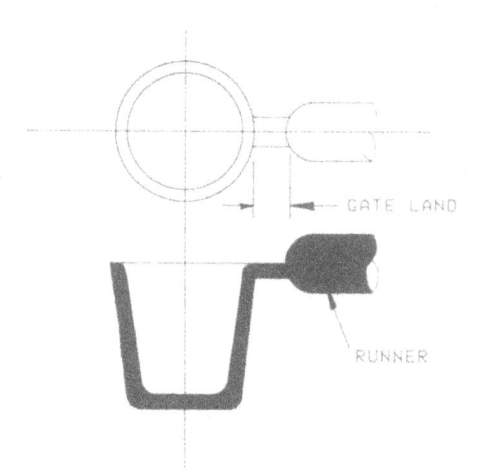

FIG. 8.57. Typical gate land area.

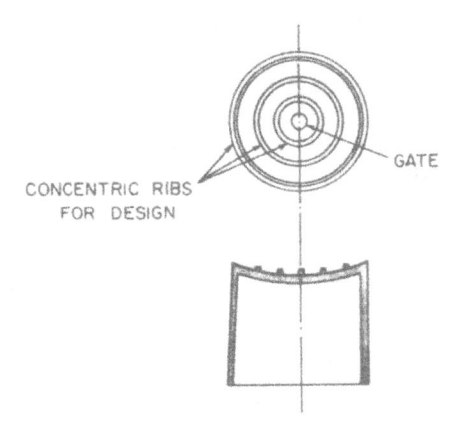

FIG. 8.58. Typical examples of styling gate location into a product design.

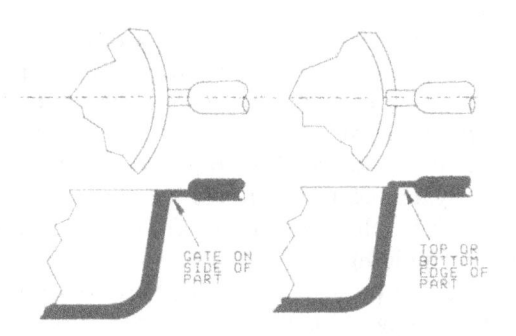

FIG. 8.59. Typical edge gating.

with specific materials manufacturers when there are no comparable experience data available.

Typical gating designs are as follows:

(1) Edge gating, Fig. 8.59, is used on the side, top or bottom of a part. The typical size is 1/64 to ¼ in. deep and 1/16 to ½ in. wide.

(2) Pin point gates, Figs. 8.69 and 8.110 are widely used for many materials and permit automatic ejection from the runner system. Typical sizes are 0.10 to 1/16 in. in diameter.

(3) Disc or diaphragm gate, Fig. 8.60, is used for most materials and in product designs with large cut-out areas. It offers improved molding characteristics and eliminates weld lines. The disc must be removed after molding. The typical size is .010 to .050 in. thick in the gate area.

(4) Tunnel or submarine gating permits automatic degating of the part from the runner system during the ejection cycle. It is widely used for many types of plastics materials. The typical size is .010 to 5/64 in. in diameter. It is tapered to the spherical side of the runner. (See Tab and Plug gating variations, items 11 and 12.)

(5) Spoke or spider gate, Fig. 8.61, may be used for moldings similar to the parts shown in Fig. 8.60. It has the advantage of producing parts

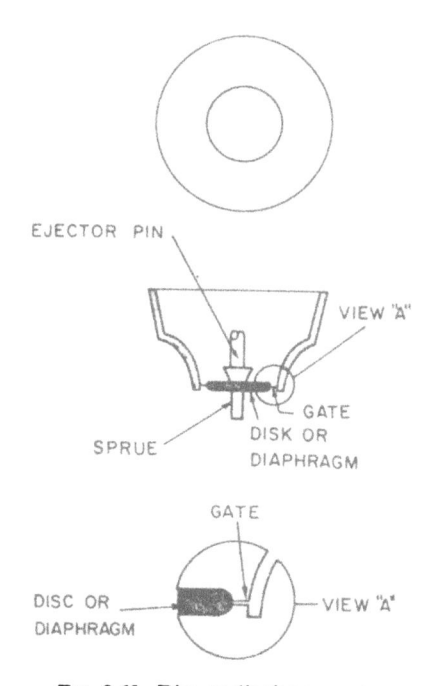

Fɪɢ. 8.60. Disc or diaphragm gate.

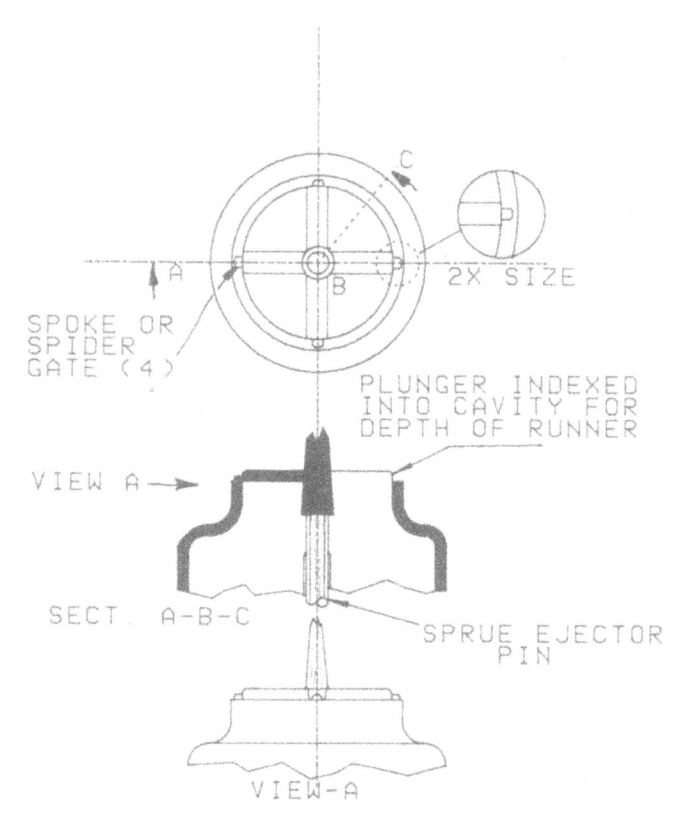

Fig. 8.61. Typical design for "spoke" or "spider" gate.

with a lower degating cost and less material usage. These multiple gates will show weld lines between them. However, the molding will be stronger than with a single side gate. It is desirable to use this gating when center core alignment is critical; it permits good positioning between both sides of a mold. Typical sizes range from 1/32 to 3/16 in. deep × 1/16 to 1/4 in. wide.

(6) Ring gates, Fig. 8.62, are used on cylindrical shapes. The typical size is .010 to 1/16 in. deep. In this case the material flows freely around the core before it moves down as a uniform tube-like extrusion to fill the mold. The optional bleeder permits the first incoming material to move out of the cavity and assists uniformly hot material to enter the cavity.

(7) Center or direct gate, Figs. 8.15 and 8.63, differs from the pin point gate in that it is larger and has the gate extension left on the molding for subsequent removal. It is used for larger moldings and for single-cavity

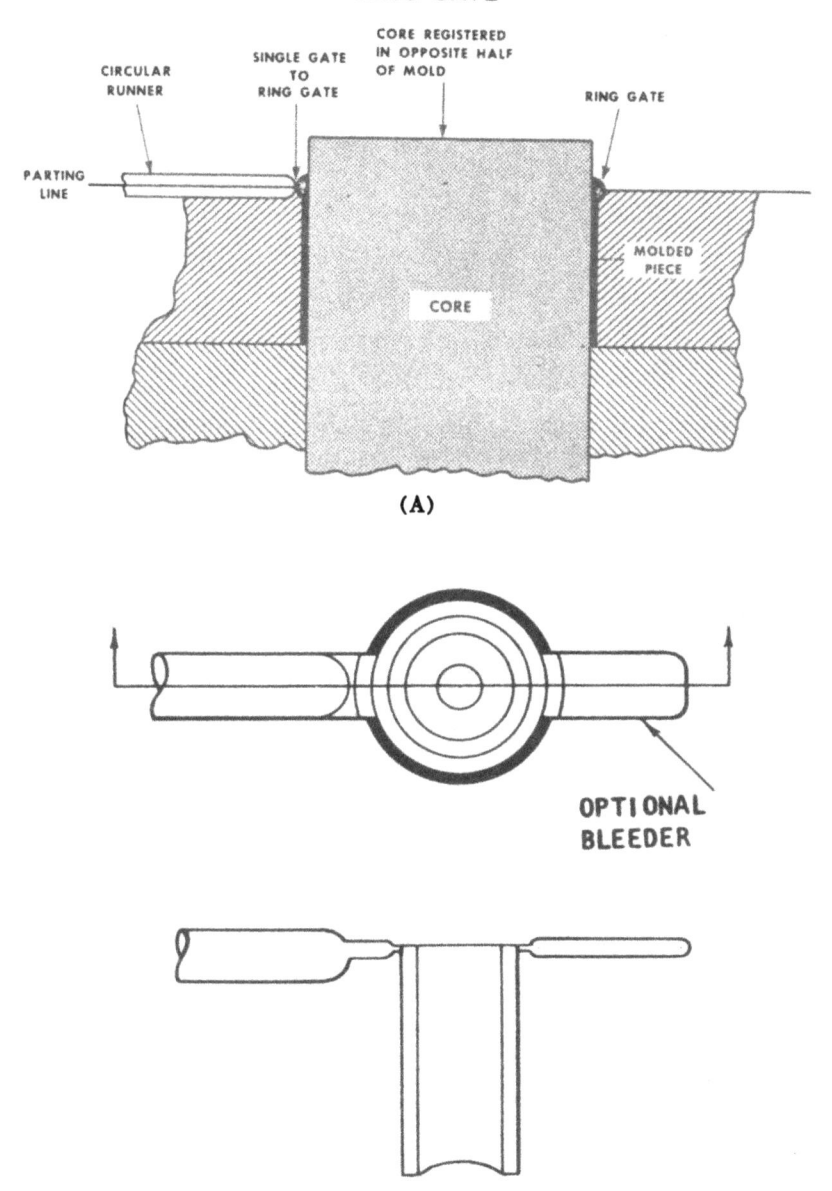

FIG. 8.62. Ring gates. (A) *Courtesy E. I. DuPont de Nemours & Co. Inc., Wilmington, DE.* (B) *Courtesy Rohm & Haas Co., Philadelphia, PA.*

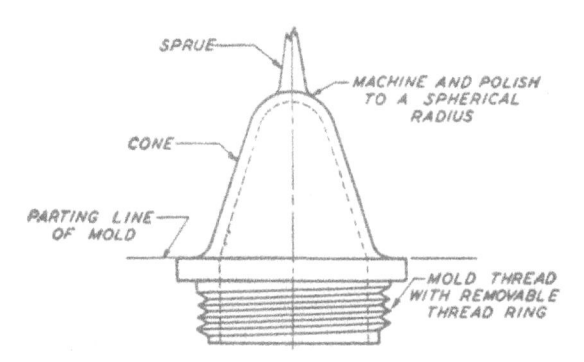

FIG. 8.63. Molded part which uses a direct gate.

designs. Typical size ranges from 1/16 to ¼ in. in diameter and 1/16 to ½ in. long, with a minimum taper of one degree.

(8) Multiple gate, Fig. 8.64, is widely used for all materials when the product design has fragile mold areas with restricted flow, since it permits a better flow and balances the pressure around the fragile mold sections. Typical sizes range from .010 to .040 in. deep and 1/64 to ⅛ in. wide.

(9) Fan gate, Fig. 8.65, permits the material to flow into a cavity through a large area gate and utilizes a very shallow gate depth. It is used on product designs that have fragile mold sections and for large area parts wherein

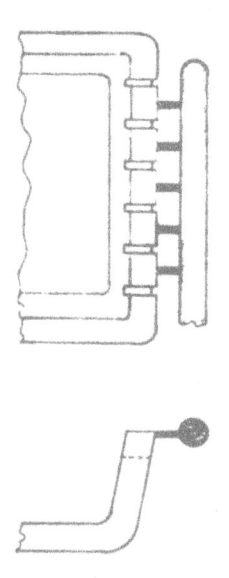

FIG. 8.64. Typical design for multiple-edge gating.

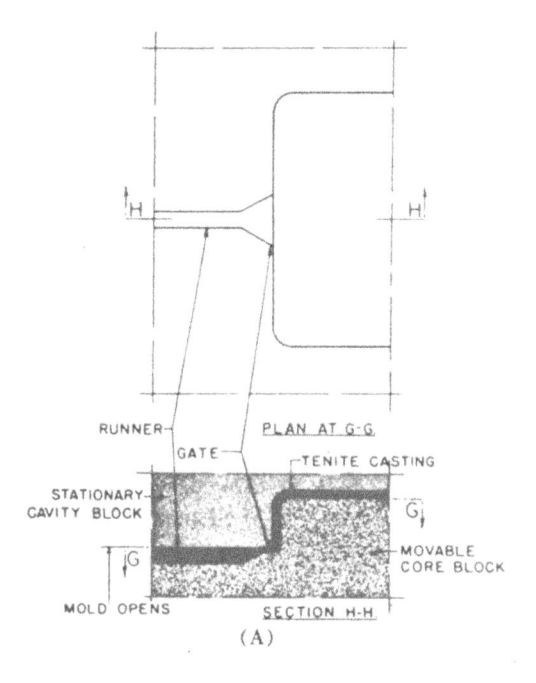

(A)

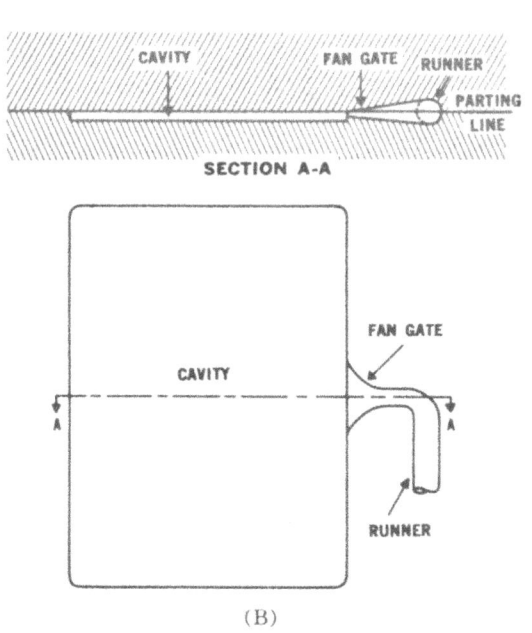

(B)

FIG. 8.65. Typical fan gate designs. (A) *Courtesy Eastman Chemical Products Inc., Kingsport, TN.* (B) *Courtesy E. I. DuPont de Nemours & Co. Inc., Wilmington, DE.*

the material may be injected into the cavity through a large entry area for rapid filling. Typical sizes are from .010 to 1/16 in. deep × ¼ in. to 25% of the cavity length in width of gate.

(10) Flash gate, Fig. 8.66, is used for acrylic parts, and generally for flat designs of large areas where flatness and warpage must be kept to a minimum.

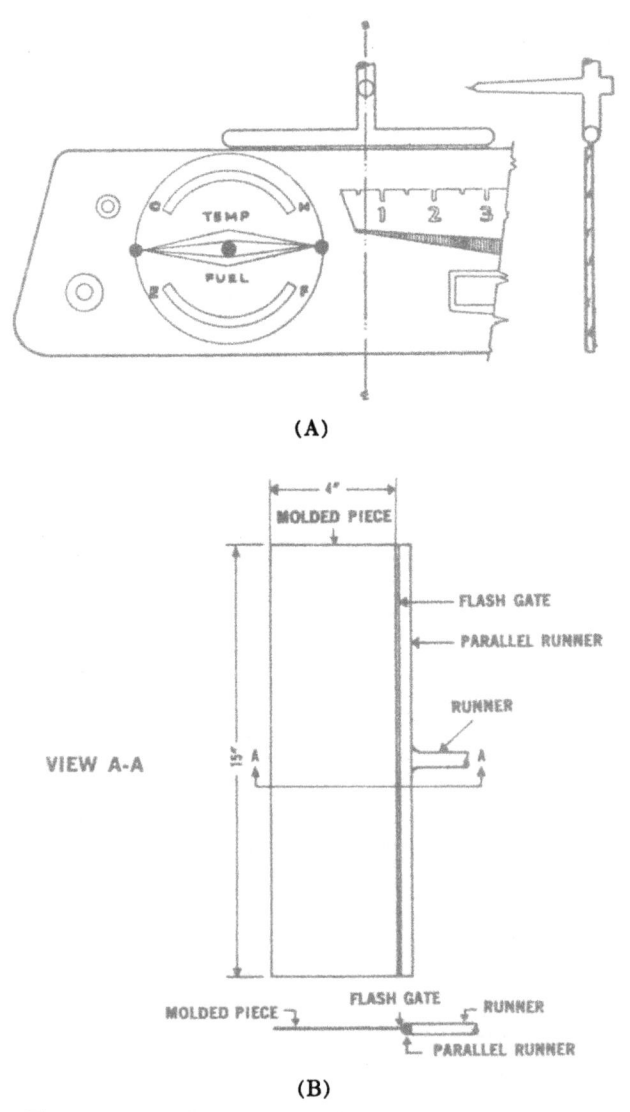

(A)

(B)

FIG. 8.66. Flash-type gates for large-area pieces to help reduce warpage. (A) *Courtesy Rohm & Haas Co., Philadelphia, PA.* (B) *Courtesy E. I. DuPont de Nemours & Co. Inc., Wilmington, DE*

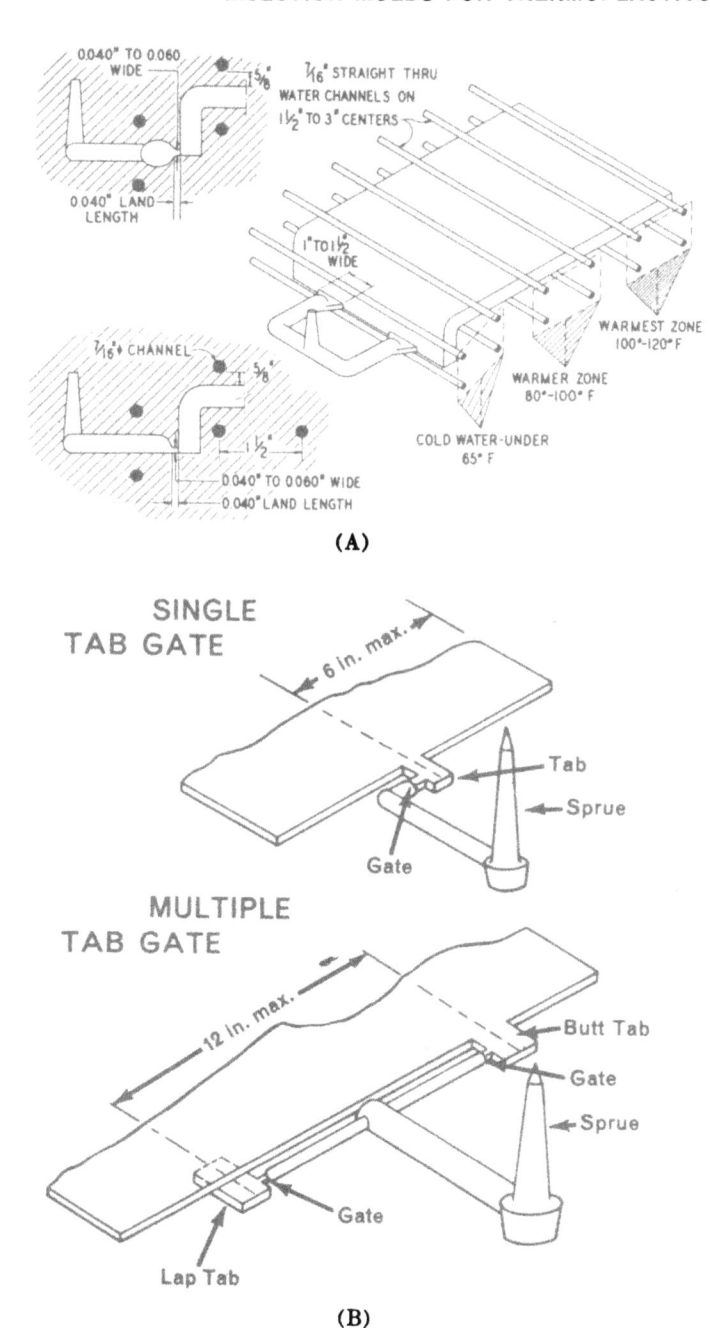

FIG. 8.67. Typical tab gate designs. (A) Tab gating flat item. (*Courtesy Phillips Chemical Co., Pasadena, TX*) (B) Single and multiple-tab gates. (*Courtesy Borg-Warner Chemicals, Inc., Parkersburg, WV*)

Gate size is small (approximately .010—.025 in.) and approximately 25% of the cavity length in width of gate. Land area must be kept small, approximately .025 in.

(11) Tab gating, Figs. 8.67 and 8.68, is used extensively for molding polycarbonate, acrylics, SAN and ABS type materials. The small gate area permits the material temperature to build up through frictional heat, and the tab area forms a well permitting the hot material to impinge against the blank wall of the tab, and finally to fill the cavity with a flat even flow of well plasti-

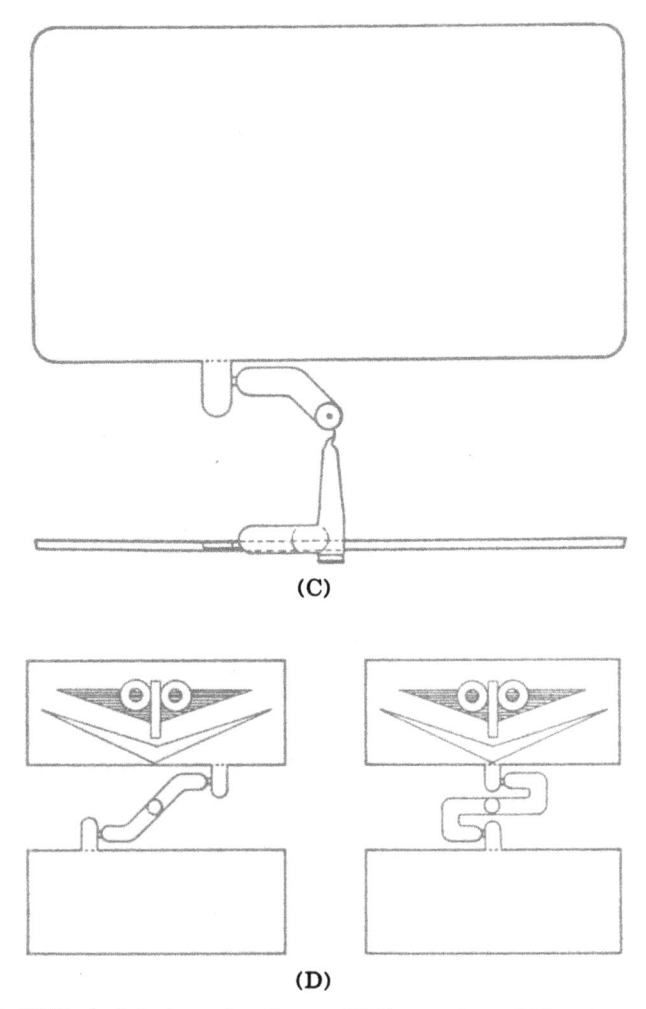

(C)

(D)

Fig. 8.68.   (C) Typical single-cavity tab gate. (D) Two-cavity mold for tab-gated parts. (C) *and* (D) *Courtesy Rohm & Haas Co., Philadelphia, PA.*

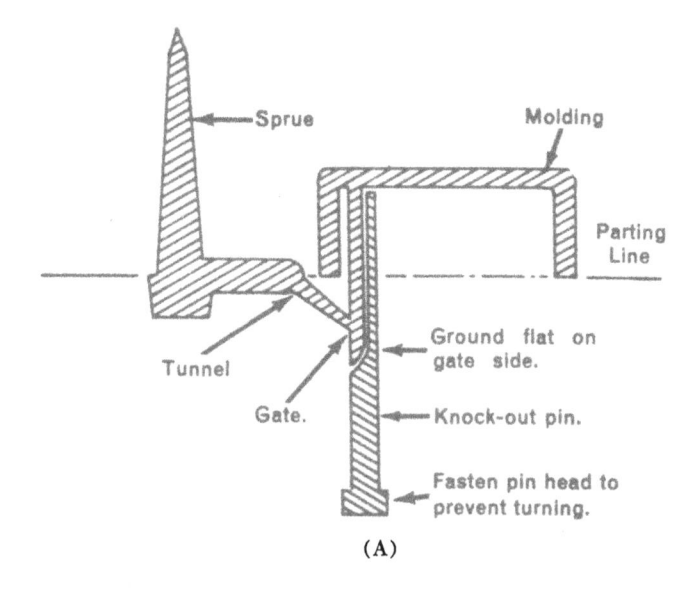

(A)

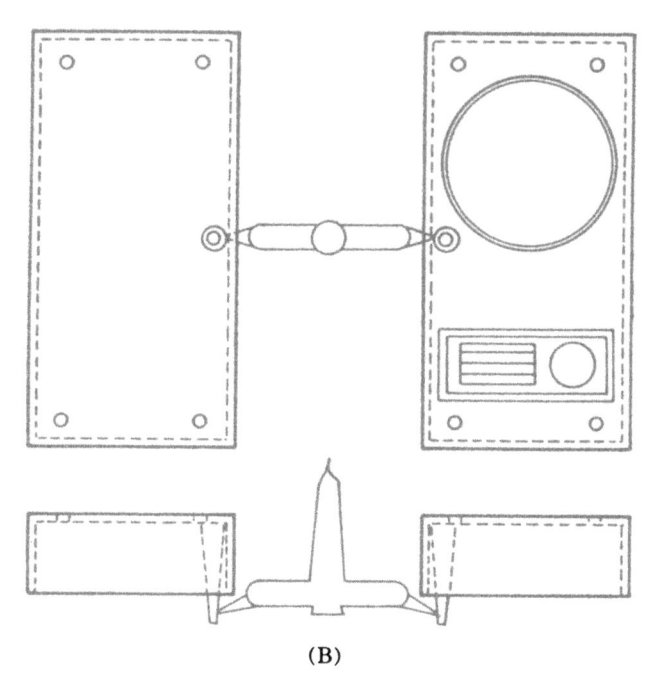

(B)

FIG. 8.69. Typical plug, tunnel or submarine gating system. (A) Tunnel gate into knockout pin. (*Courtesy Borg-Warner Chemicals, Inc., Parkersburg, WV*) (B) Transistor radio cam, plug gated. (C) Variations of plug gate with flat, vertical tab. (B-C) *Courtesy Rohm & Haas Co., Philadelphia, PA.*

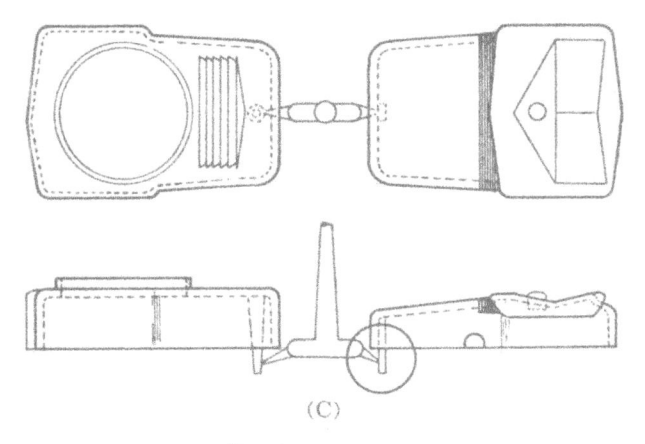

(C)

FIG. 8.69. (Contd.)

cized material. Minimum tab sizes are ¼ in. wide by 75% of the depth in the cavity for the cross section. Gate width, normally is double the thickness of the tab.

Tabs may be horizontal or vertical and, in many cases, the product design may be modified to permit the tabs to be left on the part. However, if not possible, they must be removed after molding by shear or saw cut. Some vertical tab designs permit the use of tunnel or submarine gating. The ejector pin must be located under the vertical tab.

(12) Tunnel or submarine plug gating, Fig. 8.69, is similar to the vertical tab gating system except that normally, a round tapered plug gate extension is used. Typical plug sizes are tapered from ⅛ in. at the small end of the gate to ¼ in. diameter in the cavity area. Submarine or tunnel gating is used with ejectors located under the plug.

(13) See schematic Fig. 8.70 which covers various modifications of gating systems to perform a specific need.

HOT EDGE GATING*

Hot edge gating permits the making of many parts with minimal material loss resulting from drool, string and post-molding gate removal. The solidified gate in this case, serves as a seal and shears clean. Material remaining in the gate/hot runner is ready for the next shot. Figure 8.71 illustrates the operational details that facilitate this simple system. Note that the heater element is cast in the beryllium copper block gaining complete transfer of heat from the heater to the hot runner manifold and the gating area; these

*U.S. Patents 3,530,539 and 3,822,856.

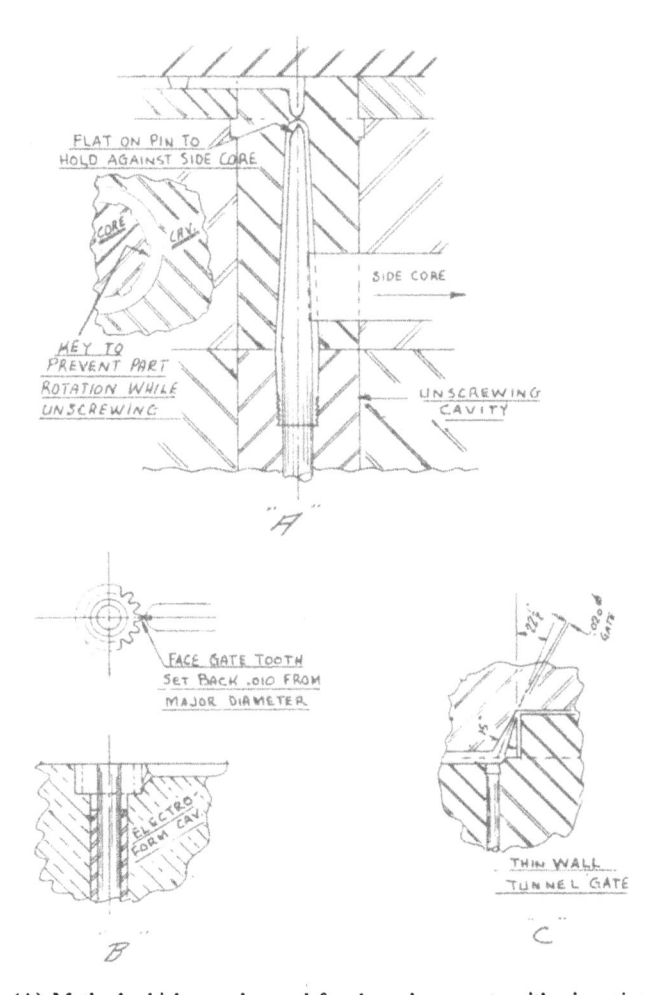

FIG. 8.70. (A) Method which may be used for deep draw parts with pin point gating and revision to center core to direct flow of plastic materials to prevent movement. Also note internal core design with "Keys" to prevent part rotation while unscrewing threaded section.
(B) Good example of gate location for gear tooth to eliminate secondary operations.
(C) Good design for tunnel gating geometry.

heating elements are not likely to burn out when operated at normal voltage. Acetal resins must not be molded in contact with copper without protective plating.

The gate diameter is, in general, 2/3 of the wall thickness and it is located on a vertical or slightly drafted wall on the cavity side of the piece. The molded piece must be held on the ejector side of the mold during the mold opening which shears the gate flush in that initial opening movement.

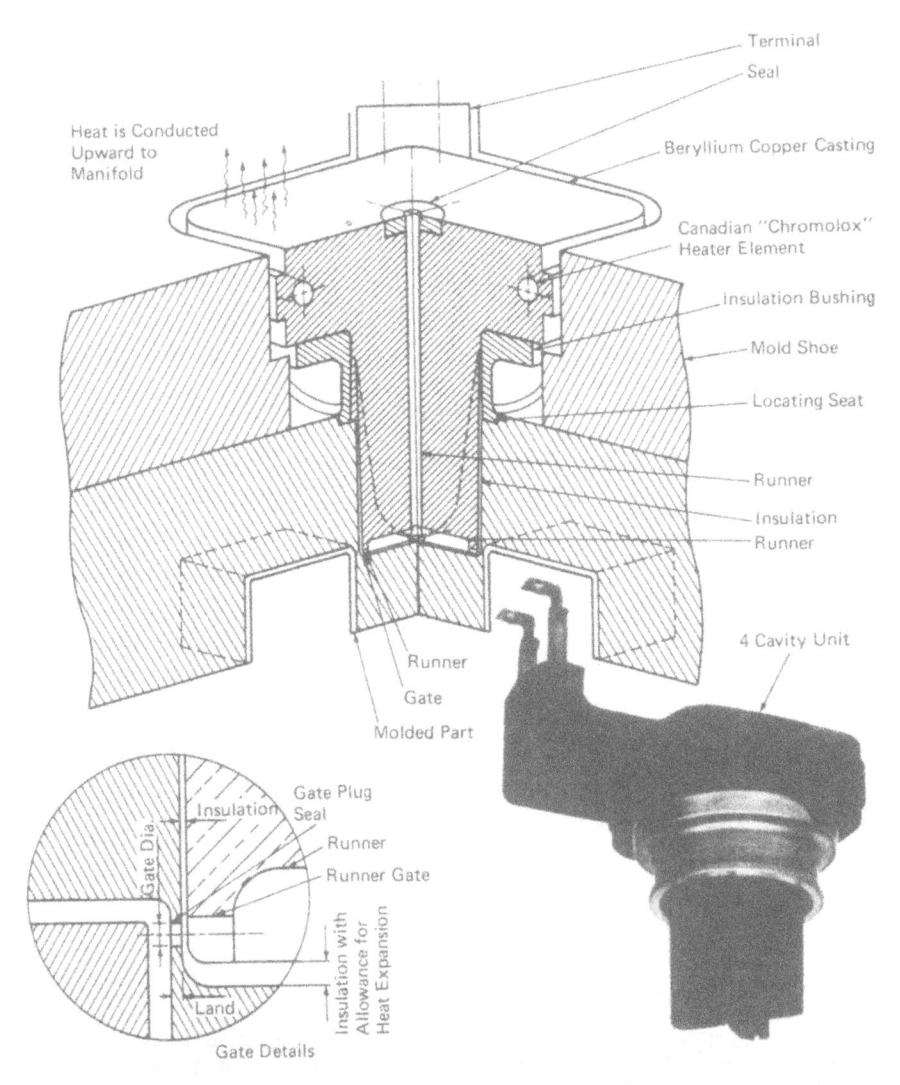

FIG. 8.71. Hot edge gating details. (*Courtesy Mold Masters Ltd., Georgetown, Ontario*)

Note that the gate is in the cooled portion of the mold and the thin insulation of plastics serves as a barrier between the copper casting and the cooled portion of the mold. The thin, tiny, sheared-off plug-seal melts quickly and becomes part of the inflow when the injection starts to fill the next shot. The insulation of plastics between the copper and steel in the gate area in restricted to such a measure that the reduced viscosity of the insulating material

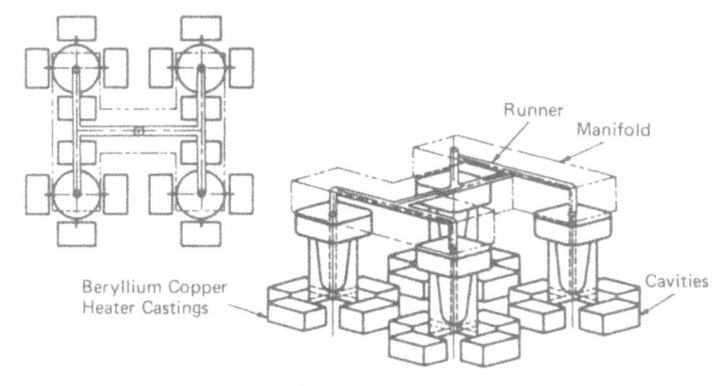

FIG. 8.72. 16 cavities with 4 heaters (H arrangement). (*Courtesy Mold Masters Ltd., Georgetown, Ontario*)

does not allow stale or decomposed material to enter the flow of the incoming material. Very large multiple-cavity molds may be built with this system and the schematic layout of a 16-cavity mold is shown in Fig. 8.72. Other bonus gains are no gate wear and the reduction of core or plunger shift since minimal filling pressures are used. This type of hot runner mold may be miniaturized to include a large number of cavities in a small area fed by a single hot edge gate unit.

VALVE GATING

The valve gating process pioneered by Messrs. Kelly and Seymour have been followed by many mechanical gating developments. The Mold Masters system is a simple and trouble free standard units for the construction of hot-runner valve-gated molds.

This process is illustrated by Fig. 8.73 and makes use of a pressure cast beryllium copper body with cast-in sheathed heating elements. This basic mold construction unit supplies heat to the hot runner manifold, the hot tip and the valve gate. The heated gate tip contains a piston-like valve gate-pin that seats in the hot tip outlet. Injection pressure unseats this gate pin, permitting material to flow into the mold cavities. A built-in rocker arm is actuated by the in-mold piston to close this valve and stop the flow of material into the mold. Thus the material is stopped at the hot tip, leaving no gate stub or drool.

The signal to close the gate is given by the press when the reciprocating screw stops turning. A typical mold assembly is shown in Fig. 8.74.

Valve gated molds must run in presses that permit a feed back into the

FIG. 8.73. This four cavity family mold for shoe trees uses valve gating from the hot manifold casting. The sprues are integral parts of the heater coil. (*Courtesy Mold Masters Ltd., Georgetown, Ontario*)

nozzle when gates are closing. All hot runner molds must include a strainer or filter. Manifold/gating units of this type may be used with a variety of cavities that fit in the standard frame. Gate balancing problems are not encountered. Valve gated molds are recommended especially for hot runner systems molding nylon, polycarbonate and for heavy wall sections. Beryllium copper must be given a protective plating coat if it is used in contact with hot acetal resin.

Proportional Gating

Some delicate components are encapsulated through proportional gating regulated by fluidic beam deflection amplifiers* to control the incoming plastics melt. The flyback transformer as shown in Fig. 8.75 is encapsulated without the necessity for rigid locating positions or areas for the insert. This proportional gating system uses two incoming streams of plastics which counteract each other to centralize the position of the insert. Fluidic beam

*Capsonic Group (designers of high speed encapsulation systems) Elgin, IL.

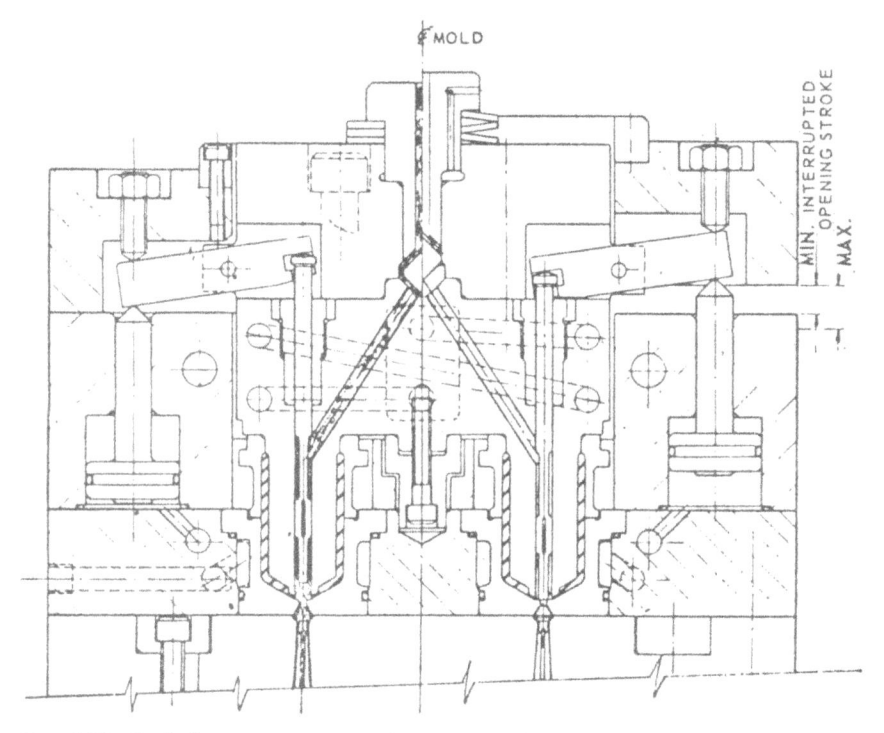

FIG. 8.74. Typical arrangement of Mold-Masters valve-gating system. Left view shows valve open for injection. Right half shows valve closed for ejection.

FIG. 8.75A. Proportional gating of transformer.

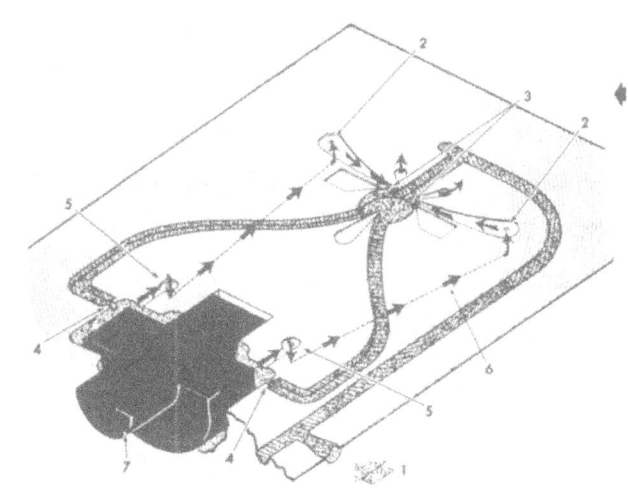

Flow of plastic melt through runner system and flow of exhaust gases through fluidic control. 1. Incoming plastic melt. 2. Control ports (gas input). 3. exhaust gas deflects plastic. 4. Mold gate. 5. Cavity exhaust port. 6. Exhaust gas flowing to control ports. 7. Component mounted in cavity.

Diagram of proportional gating-fluidic control system. At the beginning of the injection cycle, plastic melt flows through runner and is divided at splitter into two runners. More plastic enters gate B than gate A causing uneven plastic distribution in cavity and more exhaust gas to be vented at port B (indicated by heavy arrows). The greater gas flow is returned to input B where it is channeled to the interaction area to proportionally divert melt from runner B to runner A. This action can cycle from side to side as many as 50 times in a 3-second injection cycle.

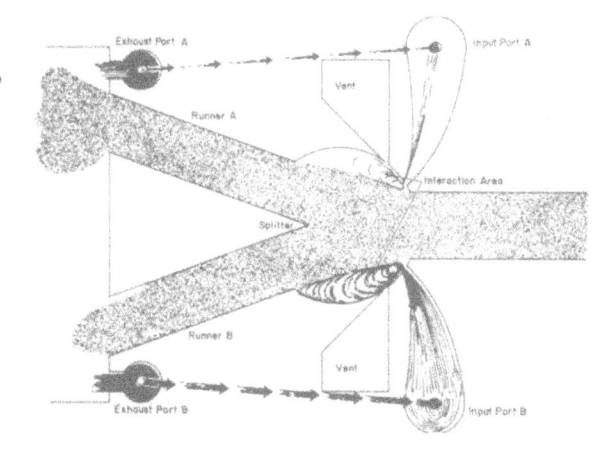

FIG. 8.75B. (*Courtesy Capsonic Group, Elgin, Ill.*)

deflection amplifiers are used to control the flow and distribution of the melt. The amplifiers utilize the outflow of vent gas to control the incoming material flow. Proportional gating is used to encapsulate delicate items such as diodes, triacs, diacs and numerous other sensitive solid state devices.

DOUBLE SHOT MOLDING

Product designs requiring the use of two different colors or two different materials combined into a single molded piece have benefited from special standard mold frames and presses. This molding process eliminates the additional cost of hot stamping and other identity procedures. Special

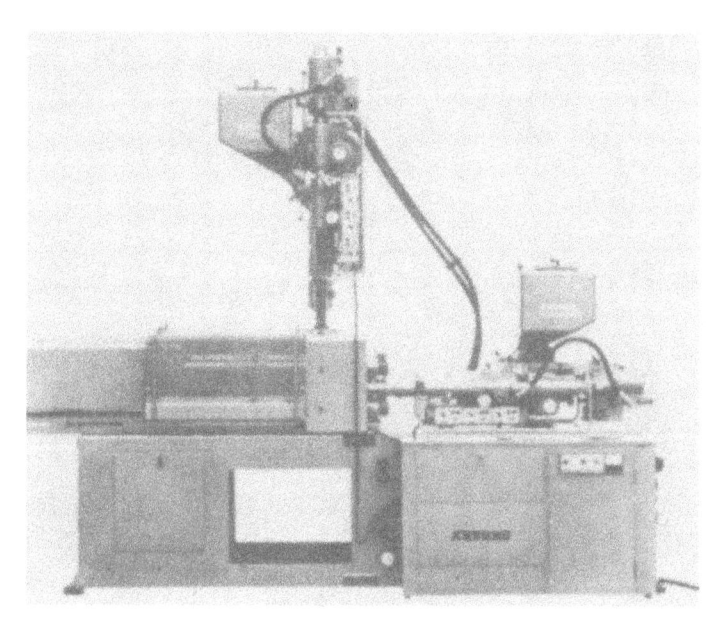

FIG. 8.76. The two color–two material process is made possible by a special machine equipped with two cylinders—one for each color. See also Figs. 8.77A and 8.77B for the indexing mold design.

presses, mold frames and mold designs are available to make these parts at minimum costs.

The special injection molding machine for two-color molding has the conventional horizontal plasticizing cylinder plus another mounted at an angle or in the vertical position, as illustrated in Fig. 8.76. Both plasticizing-injection units are synchronized in the main control. Figure 8.77A shows the design of a special indexing frame for two-color molding and the frame itself is shown in Fig. 8.77B. The mold requires cavities and forces for each color. Upon completion of filling the first cavity, the mold is rotated 180° bringing the first shot position into the second shot location for the second color. Ejection designs are critical for each of the two molding stages. The first shot must be held in position for movement into the second stage position. After mold filling in the second stage, a stripper plate ejection system may be used raising the piece by the second shot area. Figure 8.77C shows the first and second shot moldings as made in Figures 8.77A and 8.77B tooling. Figure 8.78 shows another typical double shot molding.

Other potential economics and utilitarian parts are achieved by double shot molding by using high cost materials for critical sections and low cost materials for the unimportant areas. Another product area for this type of molding results from molding the first shot smaller than the finished

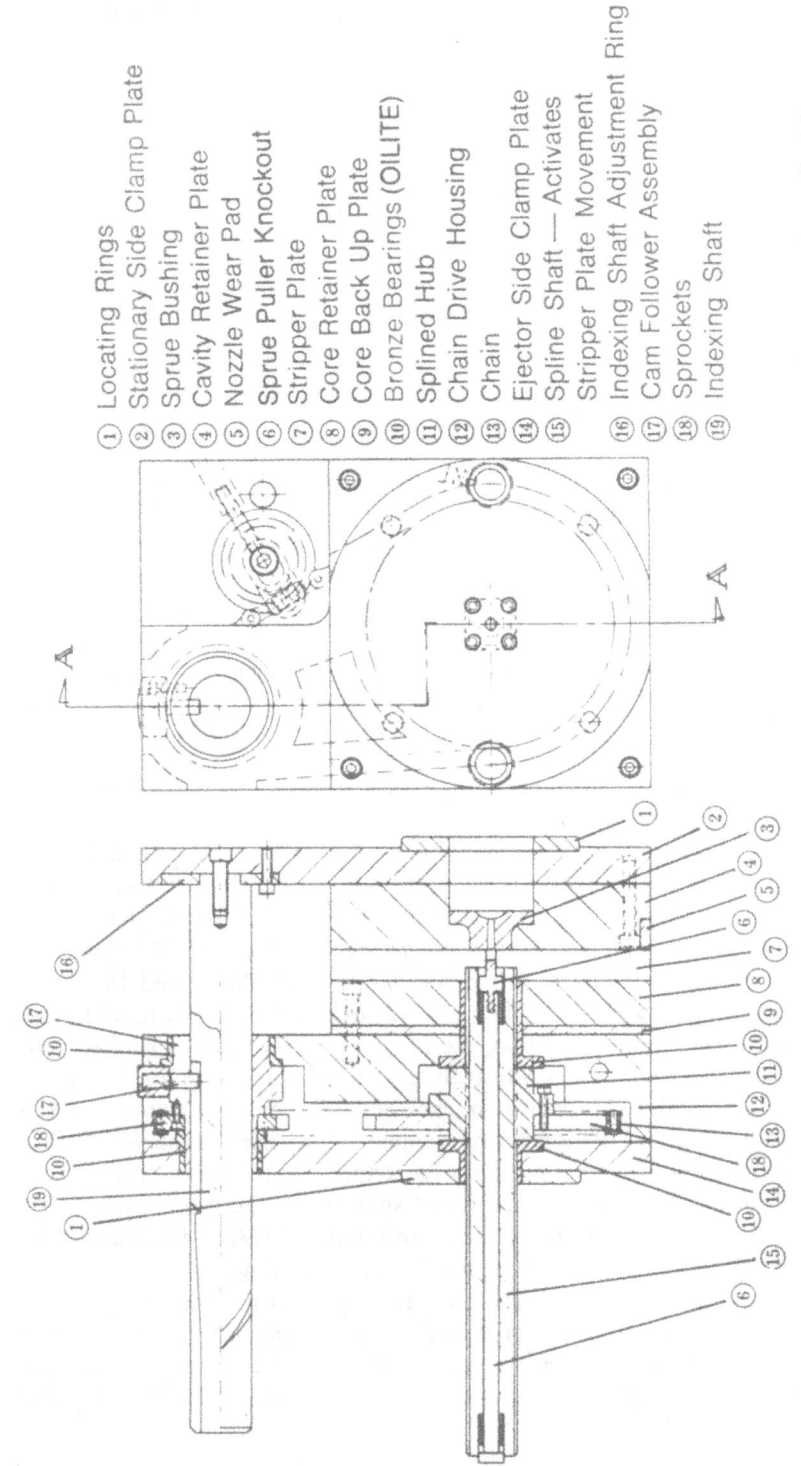

① Locating Rings
② Stationary Side Clamp Plate
③ Sprue Bushing
④ Cavity Retainer Plate
⑤ Nozzle Wear Pad
⑥ Sprue Puller Knockout
⑦ Stripper Plate
⑧ Core Retainer Plate
⑨ Core Back Up Plate
⑩ Bronze Bearings (OILITE)
⑪ Splined Hub
⑫ Chain Drive Housing
⑬ Chain
⑭ Ejector Side Clamp Plate
⑮ Spline Shaft — Activates
    Stripper Plate Movement
⑯ Indexing Shaft Adjustment Ring
⑰ Cam Follower Assembly
⑱ Sprockets
⑲ Indexing Shaft

FIG. 8.77A. Schematic of two-color material standard mold frame. (*Courtesy Master Unit Die Products, Inc., Greenville, MI*)

FIG. 8.77B. Standard two material mold frame, with positive 180° indexing, mechanically rotated by the clamp open stroke. (*Courtesy Master Unit Die Products, Inc., Greenville, MI*)

FIG. 8.77C. First and second shot molding. (*Courtesy General Electric Co., Louisville, KY*)

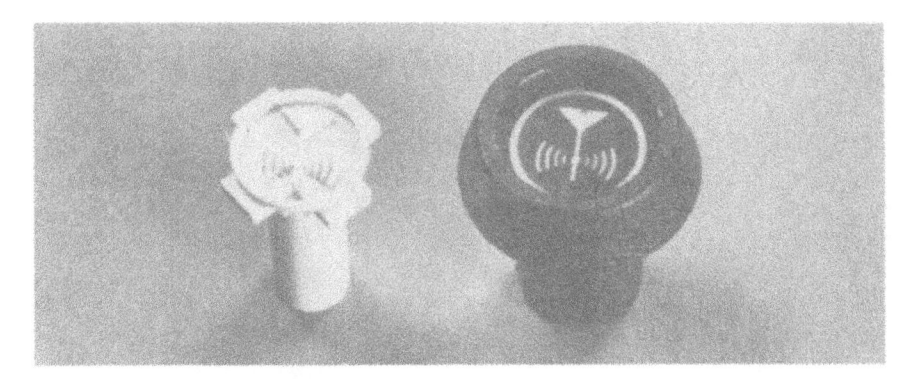

Fig. 8.78. First and second shot molding. (*Courtesy Master Unit Die Products, Inc., Greenville, MI*)

piece. This will permit the first molding to absorb the bulk of the shrinkage, warpage and other dimensional problems.

VENTING

Proper escapement of air and gas in a closed mold is absolutely essential. Poor escapement or venting results in unfilled, weak structural areas, poor appearance, poor ejection, inefficient cycling, and burned material.

Injection equipment is designed for rapid filling of the mold. Trapped air or gas retards the filling, causing the above defects; thus demanding complete and fast mold venting. The faster the anticipated cycling, the more acute this venting requirement becomes.

The mold areas to be vented are located along the parting line at intermittent positions from .0005 to .003 in. deep $\times$ 1/16 to ½ in. wide. Larger channels evacuate the air or gas into the atmosphere. Free escape of the air or gas from the mold is essential; it is of no benefit to release it from the cavity area, and then seal it within the mold. It may be necessary, in some extreme cases, to employ a vacuum reservoir to evacuate the mold cavity prior to injection. Figure 8.79 shows this auxiliary venting which should be used only when absolutely essential.

COOLING

"Cooling," as used in this chapter, means only that the die or mold is colder than the incoming material. Actual mold temperatures vary from 0°F (–18°C) to 800°F (425°C) depending upon the material being used. A check with the material makers processing sheet will tell you whether the mold must

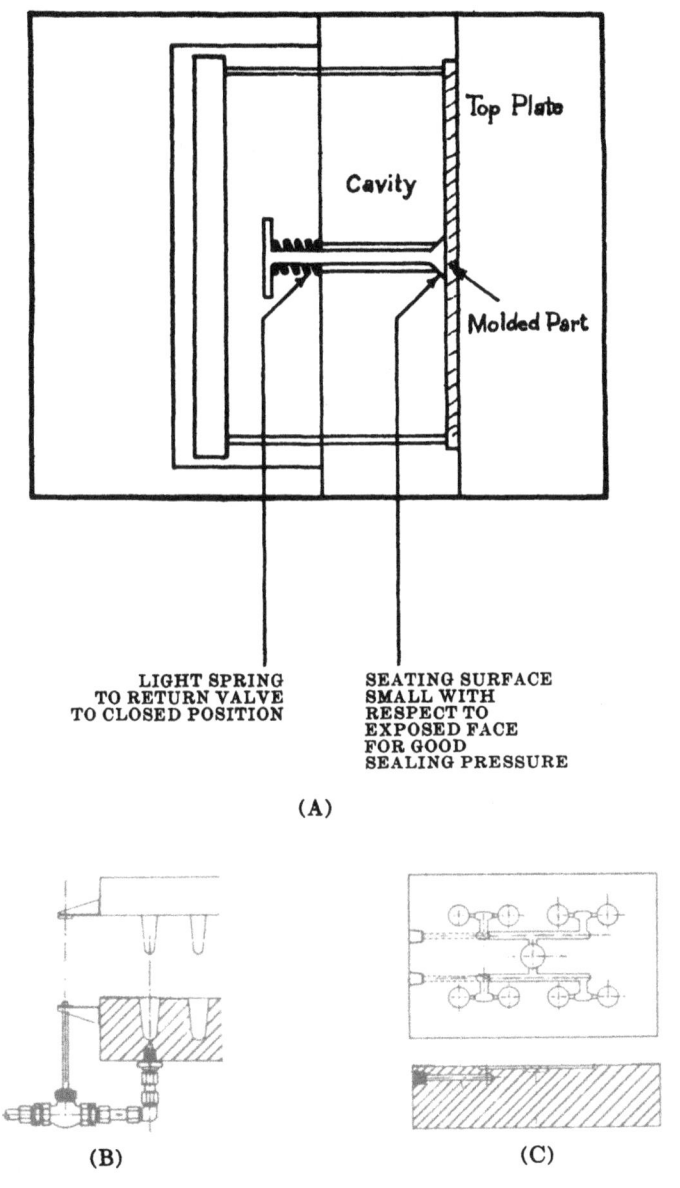

LIGHT SPRING
TO RETURN VALVE
TO CLOSED POSITION

SEATING SURFACE
SMALL WITH
RESPECT TO
EXPOSED FACE
FOR GOOD
SEALING PRESSURE

(A)

(B)

(C)

FIG. 8.79. Typical auxiliary venting methods. (A) Valve installed to break vacuum. (B) Vacuum applied to cavity. (C) Vacuum applied to runners. To mold at a lower pressure with a cooler melt it may be necessary to draw a vacuum on the vents. This procedure can be especially useful when contending with blind or poorly vented cavities where air entrapment is a problem. The vacuum can be applied either to one of the mold cavities or to the runner.

be cooled or heated, and it will also tell the desired temperature range. Molds that run on several materials may need provision for heating and cooling.

Paradoxically, when the material is being "cooled" above the ambient temperature, it may be necessary to include heater elements that will facilitate cooling down from the melt temperature to the "set" temperature. High temperature liquid coolants such as hot water, steam, Dow Therm, oil, etc., are also used to achieve high temperature cooling.

When working within the "cooling" range of cold/hot water, the molds are normally channeled to permit the flow of liquid at various controlled

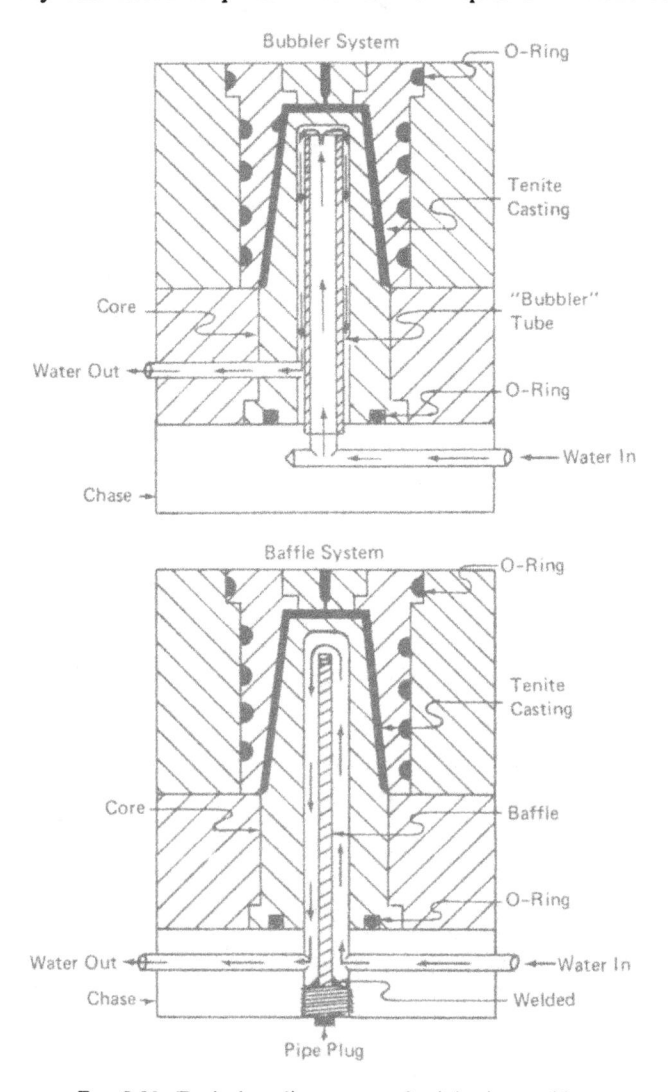

FIG. 8.80. Typical cooling systems for injection molds.

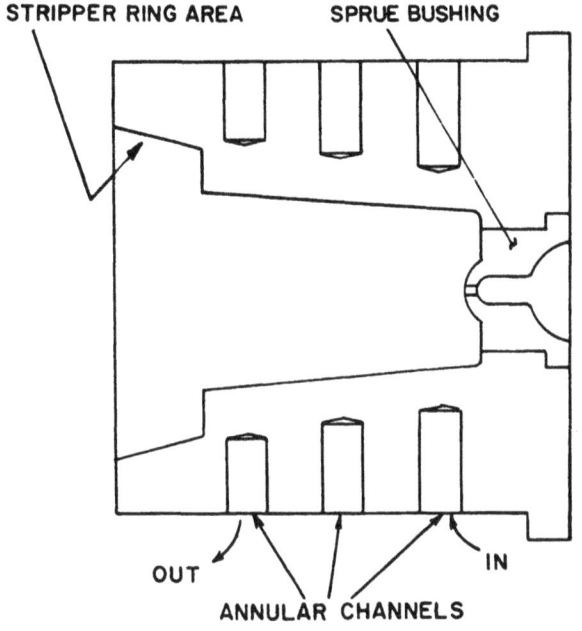

STRIPPER RING AREA   SPRUE BUSHING

OUT   IN

ANNULAR CHANNELS

FIG. 8.81. Typical cooling design. Annular cooling channels around cavity. Water in at sprue bushing end and out near parting line end.

temperatures. Many times the channeling is set up for zone control in various areas of the molds, and in moving members, as well as the areas surrounding the cavities and plungers. Channels must be large enough to permit the rapid flow of the cooling or heating medium. Figures 8.80, 8.81 and 8.82 show typical methods.

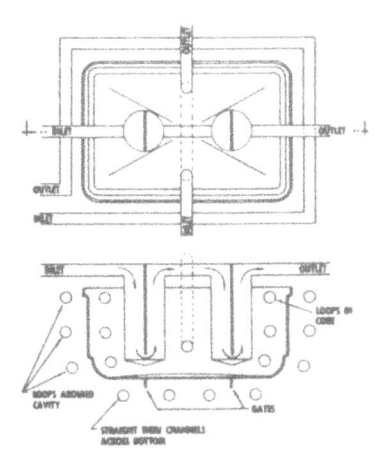

FIG. 8.82. Mold cooled by straight channels and bubbler coolers. (*Courtesy Phillips Chemical Co., Pasadena, TX*)

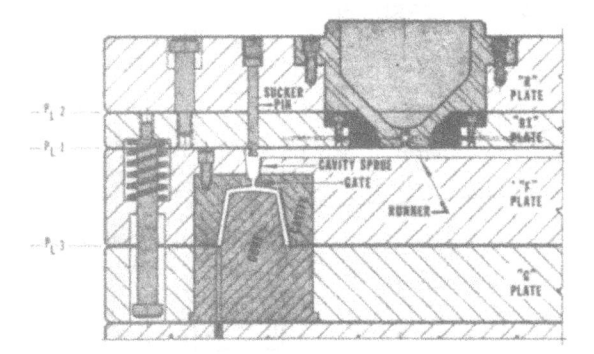

PHASE NO. 1

Mold closed.

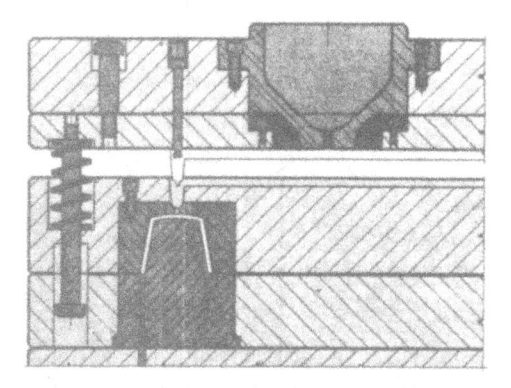

PHASE NO. 2

Mold starts to open with springs assuring first opening between "RX" & "F" plate; RUNNER is pulled from "F" plate and held to "RX" plate with sucker pins.

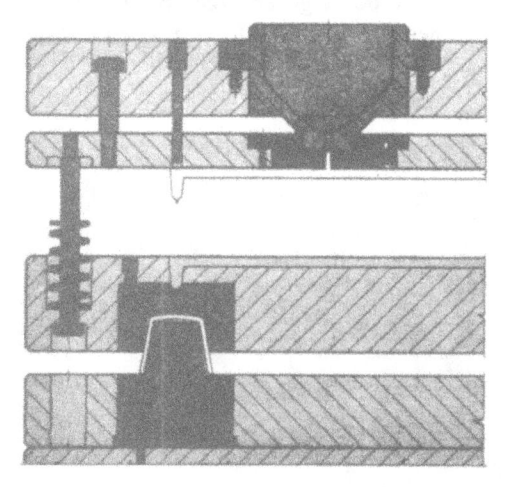

PHASE NO. 3

At further opening of mold, stripper bolts activate "RX" plate which strips RUNNER from sucker pins and sprue from recessed sprue bushing. RUNNER is free to fall clear of mold.

FIG. 8.83. Schematic illustrations depicting operational sequence of a four plate stripper plate mold with recessed sprue bushing are shown here. This bushing permits press nozzle engagement closer to the main runner. Result is better control of molding conditions and less sprue regrind. (*Courtesy National Tool and Mfg. Co.*)

Refer to Chapter 5 for simple formulas for calculating heating or cooling load and flow of temperature control media.

SPRUE BUSHINGS

Sprue bushing diameters and lengths vary with the type of mold design and the press required for a given product design. The sprue diameters must be tapered to permit easy sprue removal and the small diameter must be larger than the nozzle diameter in the press to prevent sprue holding in the nozzle. Typical sprue designs are shown throughout this chapter.

Sprue bushings are normally catalog items and representative catalogs should be collected by the designer. Recessed sprue bushings as shown in Fig. 8.83 permit the use of extension nozzles. An extension nozzle shortens or eliminates the sprue to give improved control of material flow temperatures, as well as to reduce the quantity of regrind material. Extension nozzles are used for either single or multiple cavity molds. Please refer to other sections of the text for additional information concerning heated nozzle bushings, hot sprue bushings, hot tips and hot manifolds.

SPRUE PULLERS

Provision must be made in the mold design to permit pulling the sprue stem out of the sprue bushing and to break it away from the nozzle when the mold opens. This is accomplished by a sprue puller. This system also permits the cold material, which may be formed at the face of the nozzle, to flow into a well, (which is identified as a "cold slug") before the hot material flows into the cavity. Ejection of the cold slug is achieved by means of an ejector pin actuated by the ejector bar.

Cold slug area and density of molded material for large sprue and runner systems should be minimized by adding a coring pin on the sprue puller ejector pin as depicted in Fig. 8.101. Caution must be exercized to keep the pin below the runner system or it will form an obstruction to material

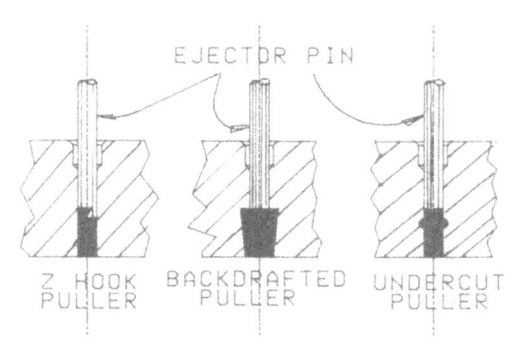

FIG. 8.84. Typical sprue puller designs: A. hook; B. backdraft; C. undercutting. (*Courtesy Phillips Chemical Co., Bartlesville, OK*)

flow. Typical types of sprue pullers are shown throughout this chapter and in Fig. 8.84.

CAVITIES

The cavities of an injection mold may be formed by any of the conventional means, as discussed in Chapter 3. Many molders prefer to have the cavities formed directly in a solid die block, or in several blocks which are later fastened firmly on a suitable mounting plate. This method of making a mold may be objectionable from the tool-maker's standpoint because of difficulty experienced in the alignment of guide pins, dowel pins, etc.

Other molders prefer to have the cavities inset as units in a retainer plate. This permits accurate boring of the holes designed to receive the cavity blocks when round sections are used. The retainer plates are often hardened to resist the abrasive action of the flowing material.

In all cases the sprues, runners and gates should pass over hardened surfaces. For some compounds, this may not be too important, but friction will be reduced, better moldings obtained, and less maintenance required if this procedure is followed.

Chrome plating may be required in some cases where the material has an acid content that would attack ordinary steels or unplated steels. The tool engineer must learn whether any of the materials that may be used are corrosive. Chrome plating of all molds is good practice.

Side Cores

Side cores may be considered as part of the cavity, since they often affect the position of the cavities in the mold. Adequate tapers should always be provided on the core pins. A taper of $\frac{1}{2}°$ per side is considered satisfactory. The minimum taper for good extraction should be 0.005 in. per inch per side. There are, of course, cases where zero draft and negative draft are useful. Parts having negative draft can usually be extracted more easily from injection molds than from compression molds operating on thermosetting material, the reason being the relative softness of the material on ejection.

If possible, the core pins should be pulled before the mold is opened and while the part is still in place. This will minimize distortion. In any case make a careful study of the location of ejector pins with relation to the side core. Be sure the sequential operation provides for noninterference, otherwise, expensive and repetitive damage and broken pins will result. Side cores may be operated satisfactorily in many ways. Devices that are operated by the machine are used, or the pins may be operated externally by electric motors operating through speed reducers. Solenoids, hydraulic or air cylinders, levers, offset cams, angle guide pins or finger cams are commonly used for operation of the core pins. When angle guide pins (Fig. 8.11A) are used, the angle of offset should be less than 30°, with 20° a much safer figure. The pins should be not less than ½ in. in diameter and should be well lubricated to provide long and trouble-free life. On machines which have large holes in the platens, offset cams may extend from one half of the mold through the other half so that side cores or wedges may be operated in either or both halves.

Inserts Used in Injection Molds

Inserts may be molded in place or they may be pressed in after molding. If they are to be molded in, they should be held firmly in the mold. There is great turbulence in the cavity because of the speed with which the material is injected. If inserts or long core pins are used, the gating of the mold should be studied carefully so the first rush of material will not be pushed directly against the insert. The compound should have a swirling action if possible, with the insert located at a relatively dead spot. Inserts should be located near the gate because the last mass of material to enter is the hottest and therefore will knit best around the insert. If the brass color of inserts is objectionable in clear plastics materials, the inserts may be oxidized by baking. Most clear plastics materials will tend to magnify any internal structure or defect.

Loading fixtures for inserts are not often used in the injection process, since the molding cycles are short and it would take the operator longer to load the fixture than it would to load the mold. Of course, if automatic location of the insert is required, then it will be necessary to use a loading fixture of some kind. Certain jobs may be operated economically by using two operators.

Injection mold designs that take inserts should place an ejector pin behind them to avert a tendency of the insert to hang in the mold and distort the piece. Parts requiring inserts may be molded on a semiautomatic cycle, the press itself injecting the material, timing the setting, opening for ejection and remaining open until the operator has removed the parts and loaded the next round of inserts.

DESIGNING INJECTION MOLDS

Standard Mold Frames

The standard frame was mentioned in Chapter 2 in the discussion pertaining to construction of injection molds. Some additional information is given at the end of this chapter and some commonly used sizes and other design details are also presented. There are, of course, cases when it would be im-

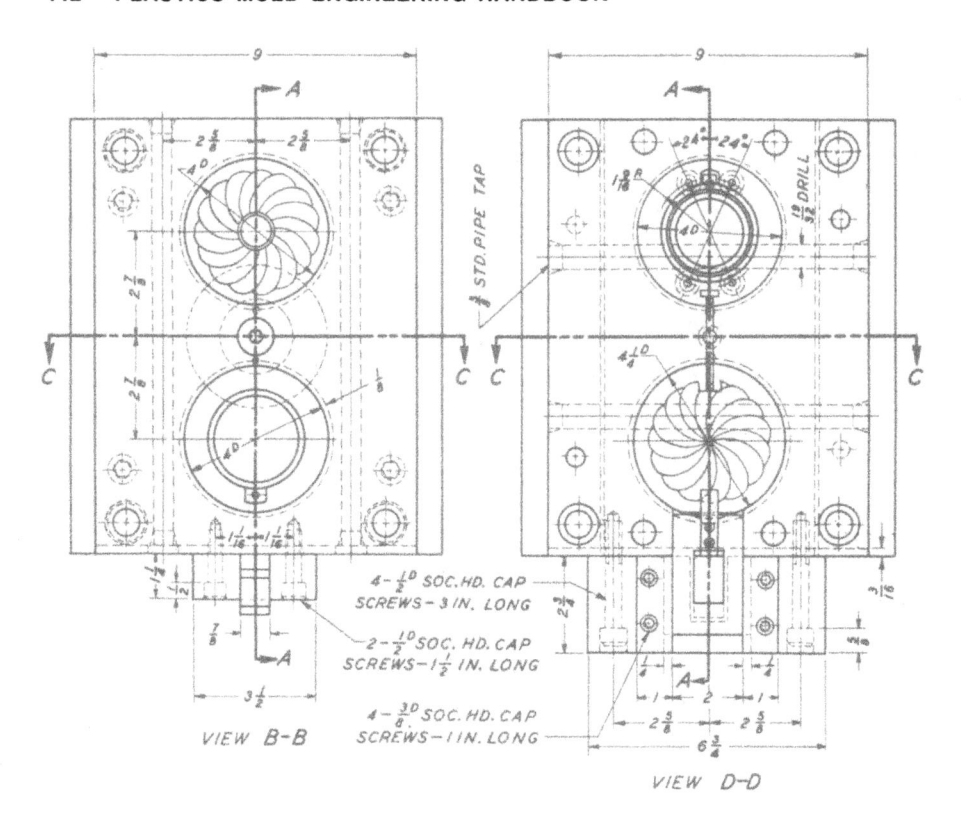

VIEW B-B

VIEW D-D

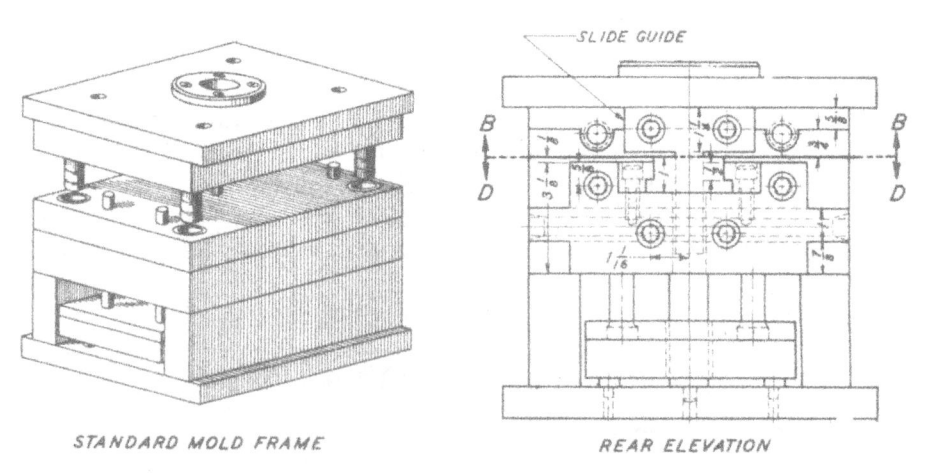

STANDARD MOLD FRAME

REAR ELEVATION

FIG. 8.85. An injection mold for powder compact assembled in a standard frame. (See also Fig. 8.86.) (*Courtesy Eastman Chemical Products, Inc., Kingsport, TN*)

possible to use a standard frame because of conditions surrounding some special feature of design. However, for many jobs, such as the powder compact shown in Figs. 8.85 and 8.86, a standard frame could be used to good advantage. This mold requires a side-pull pin for a slot perforating one side of the lid. The pull pin is operated by an offset cam fastened to the stationary mold section.

In designing this mold, the first thing the designer should do is to make a layout, using a master templet furnished by the manufacturer of the standard mold frame. This master templet offers all the detail of the frame, but it is unnecessary to place the full detail in the layout. The only caution to observe is to make sure that temperature control channels, screw holes, ejector pins, etc., do not interfere with any of the frame dimensions.

The size of the compact is such that a body diameter for the plunger and cavity of $4\frac{1}{4}$ in. will allow for a $\frac{1}{4}$-in. land. An allowance of $\frac{1}{8}$ in. is made for the heel, using a stock diameter of $4\frac{1}{2}$ in. The diameter of the sprue brushing, where it passes through the plunger plate, is 1 in. An allowance of $\frac{3}{8}$ in. from the edge of the sprue bushing to the edge of the plunger, plus the $4\frac{1}{2}$ in. stock diameter, will make a total of $5\frac{3}{4}$ in., center distance of the cavities.

It will be noted that the cavities will be horizontal when the mold is mounted in the press. This also permits use of horizontal runners and provides a more uniform flow of material. Whenever possible, the major portion of the runner and the flow of material should be horizontal. The side-pull pin will also be horizontal. If it were mounted vertically, it is possible that the slide block would fall back to the closed position after the offset guide pin was withdrawn, thereby damaging the mold if, subsequently, it were closed with the pin already in the closed position.

The construction of the side core is interesting. The offset cam is made in a T-shape, as shown in the rear elevation. This permits two socket-head screws to be located $2\frac{1}{8}$ in. apart for minimizing any deflection resulting from the clearance in the screw holes. The slide is designed as a rectangular block having the proper size hole for a sliding fit to the offset cam. The 30° offset in the pin is about the maximum that can be used if efficient operation is to be maintained. If the angle is any greater, the guide pin will tend to move the slide in the direction in which the pin moves instead of moving it at right angles to the direction of the pin. The small tongue that molds the slot in the side of the part is made as a separate unit so that it can be replaced easily in case of breakage. This also gives the heat treater a better chance to harden the tongue properly, since there will be no combination of thick and thin sections. The two No. 10–32 screws are sufficient, as there will be little pull on the tongue.

Note that the cavities oppose each other, that is, one cavity is on the sprue side and the other cavity is on the knockout side. The inside of the

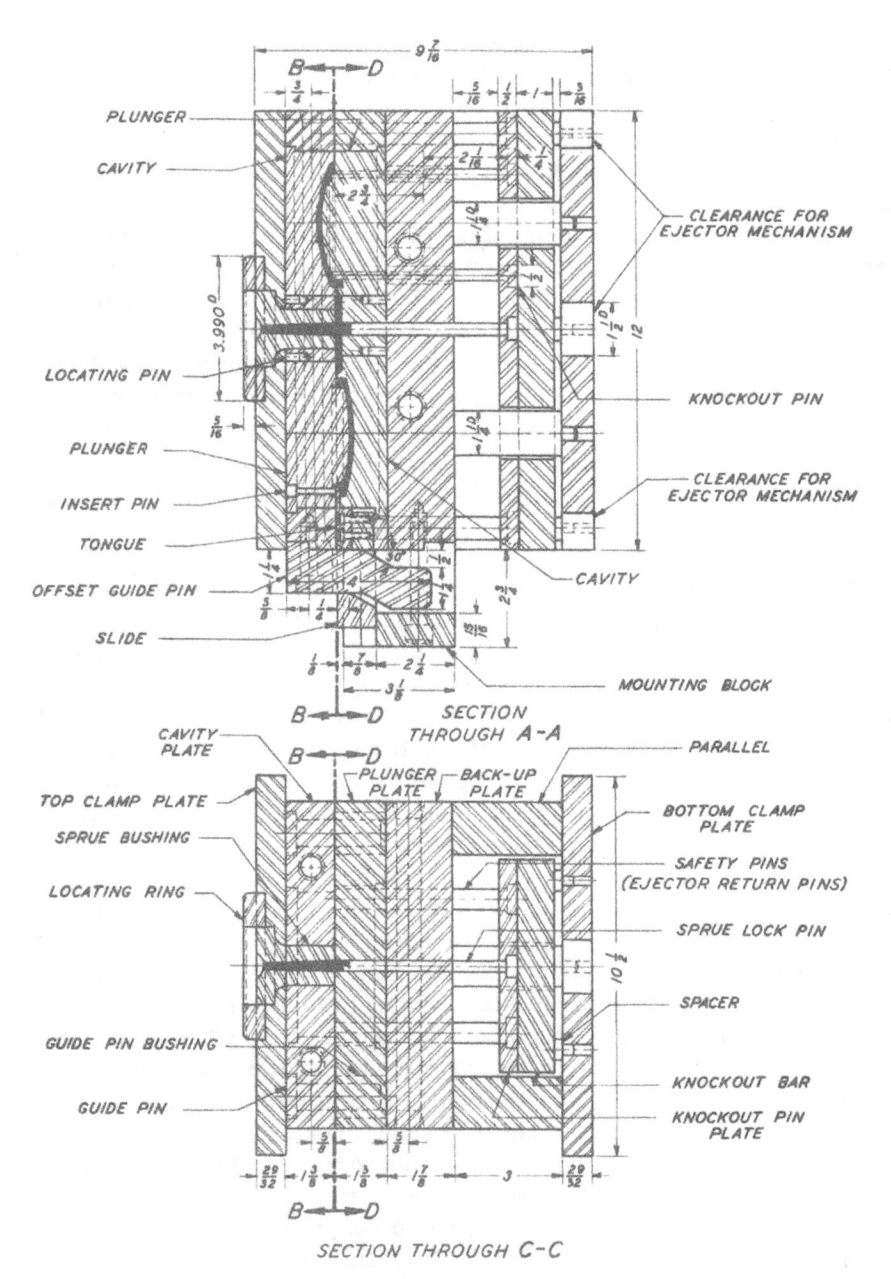

FIG. 8.86. Cross sections of mold shown in Fig. 8.85. Section through A-A shows detail of side-pull pin.

bottom of the compact (upper cavity in Section *A-A*) is more intricate than the other cavity and is expected to stay on the ejector side. This cavity, therefore, is laid out so that ejector pins may be used. The tongue on the one side and the gate on the other will insure that the top of the compact shall remain on the ejector side. The top design is such that no ejector can be used and this means that the runner, sprue and gate must be depended upon to push the molding from the cavity as the sprue lock pin moves forward.

If vents are needed on this mold, they can be placed at the tongue in the one cavity and directly opposite the gate in the other. The ejector pins are located at 24° (and this could be varied slightly) so they will be close to the places most liable to stick when the parts are ejected. Another reason for placing them in this manner is to clear the 19/32 in. drilled hole in the back-up plate. This hole must clear both the dowel pin holes in the frame and the clearance holes for the ejector pins.

In the operation of the mold, the ejector half will move away from the stationary half for a distance of ½ in. At this time, the offset guide pin will come in contact with the angle of the slide, and a continued motion will cause the slide to move away from the cavity for a distance of ½ in., since this is the amount of offset in the guide pin. Continued movement of the ejector half will cause the ejector mechanism (which usually consists of a simple rod which hits against the frame of the machine) to push the ejector assembly forward and eject the parts from the cavity. The sequence of operations is reversed as the mold closes.

Two support pins, or support pillars, are specified. These are placed under the center of each cavity, which also means on the center line of the ejector bar. The clearance hole should be not more than 1/32 in. larger than the support pin if this construction is used. The best construction is to have the support pin hardened and insert a hardened bushing in the ejector bar. This will eliminate any side or end play from the assembly, therefore no lateral stresses will be placed on the ejector pins. Soft support pins will wear quickly and may need replacing, and the holes in the bar may have to be rebored for a fit to the support pins.

The guide pins are almost always located in the stationary half of the mold. This is done so that they will not interfere with free ejection of the parts when a fully automatic cycle is used. If the pins were put in the movable half of the mold, it would be possible for the shot of parts to hang on one of the pins and thus damage the mold when it was closed for the next shot.

Longer life might be expected of this mold if a hardened block were inserted between the cavities for the runners. Runner blocks as described in Chapter 7 are used for this purpose. Many molders construct all parts of their injection molds from oil-hardened tool steel because of the high operating pressures which the molds must withstand.

Rack and Gear Core-Pulling Mechanism

The core-pulling mechanism shown in Fig. 8.87 is of the rack and gear type, which automatically operates as the press platen moves. The plate which holds the gear and pinion is bolted to the movable platen of the

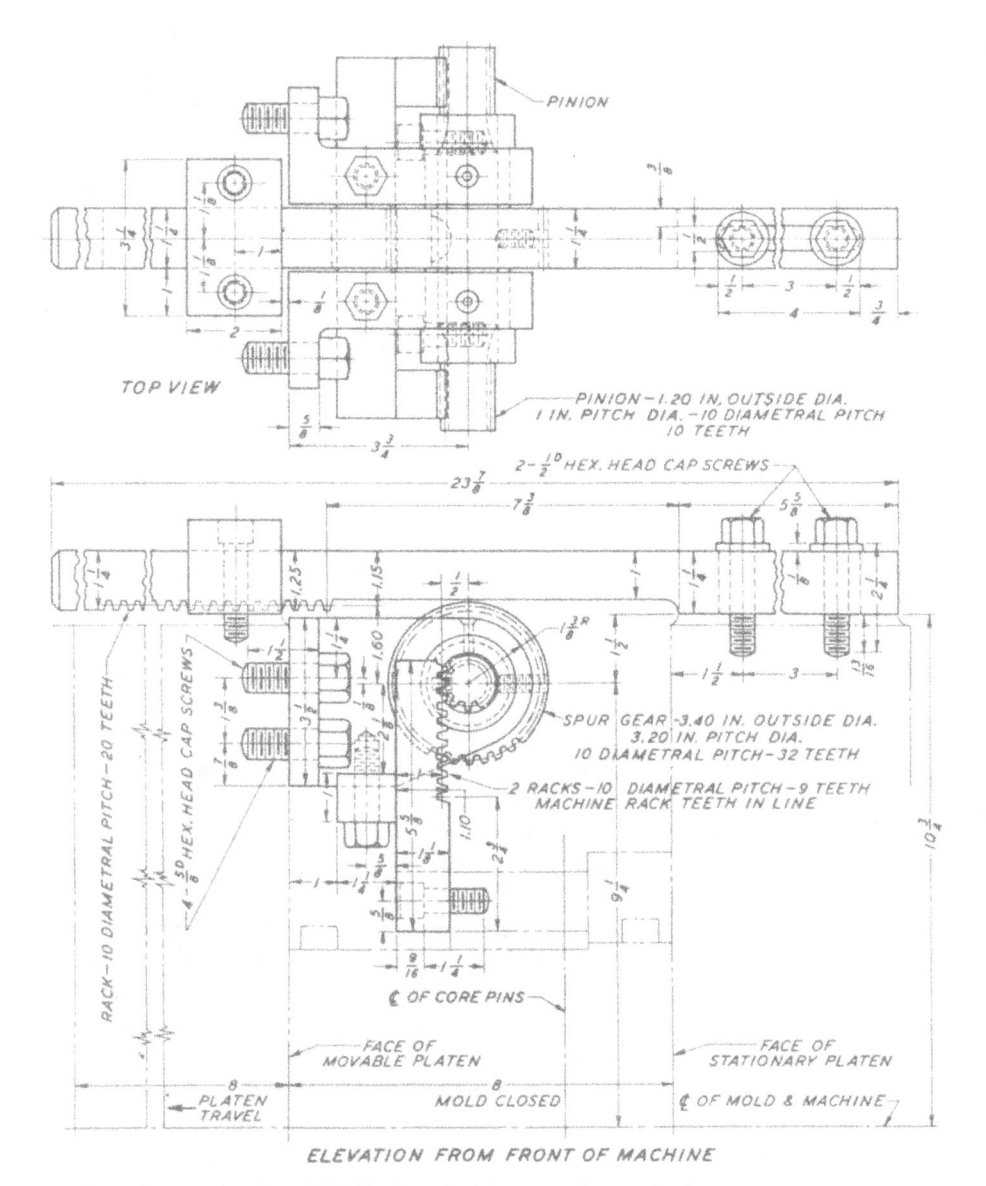

FIG. 8.87. Rack and gear core-pulling mechanism.

machine. The large rack is bolted to the stationary platen and extends over the movable platen. It is guided and held in place on the movable platen by a simple U-shaped block which permits free movement of the rack yet holds it in engagement with the gear. As the press opens, the molded parts containing the core pins move with the movable platen until the large gear comes in contact with the rack. Continued motion causes the gear and pinions to revolve and thus move the double rack vertically. This action pulls the core pins.

It is usually a good practice to have the large rack adjustable so that timing of the core-pulling mechanism may be adjusted to start and end the movement at exactly the right time. If it starts too soon, the core pins will be subjected to excessive motion. If too late, the core pins may not be free from the molded parts before the ejector pins come into play. See also collapsing cores.

Removable Wedge Molds

Injection molds for parts such as spools, or for other parts having undercuts or threads, may require movable mold sections. In cases where removable sections are used, it is usual practice to provide duplicate sections so that one may remain in the mold while the other is on the bench undergoing necessary assembly and disassembly operations.

A single-cavity injection mold for a simple spool is shown in Figs. 8.88, 8.89, and 8.90. A mold of this type could with little more tool work be made a two-cavity mold, and this would afford better balance in the machine. If a single-cavity mold is to be used, the end of the mold opposite the cavity should be made not less than one-half the length of the cavity end. This will relieve some the side strain on the machine.

Overhanging clamping plates are necessary in this case so the clamping slots will come at the proper place to match the mounting holes in the machine. The clamping slot should leave no less than ½ the width of the slot from the edge of the slot to the end of the plate. This provision is made so as to offer sufficient strength to prevent the small ear from bending.

Unscrewing Devices

Various methods are employed for removing threaded parts from injection molds. (See also collapsing cores.) Simple threaded sections may be stripped from the mold sections directly without unscrewing in many cases. Cam or rack and pinion-operated screwing mechanisms are often used. Single-cavity or other simple molds may use hand-operated levers to unscrew the mold from the part. Large external threaded sections are frequently

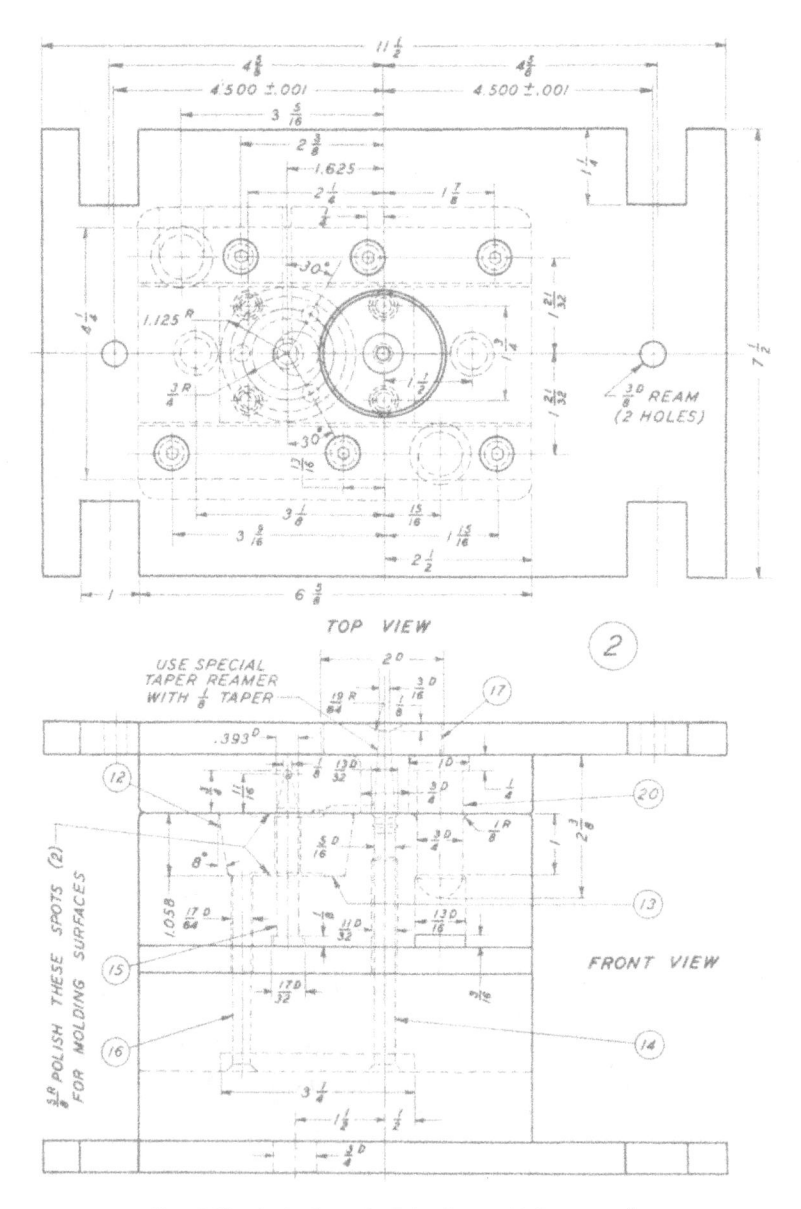

TOP VIEW

FRONT VIEW

FIG. 8.88. A single-cavity injection mold for a spool.

formed by means of a thread ring which is unscrewed from the part after its removal from the mold, as described in Chapter 6. Complicated molds are very costly and their maintenance is high; considerable thought must be given to such designs to insure rugged, long-wearing construction and elimination of breakdowns.

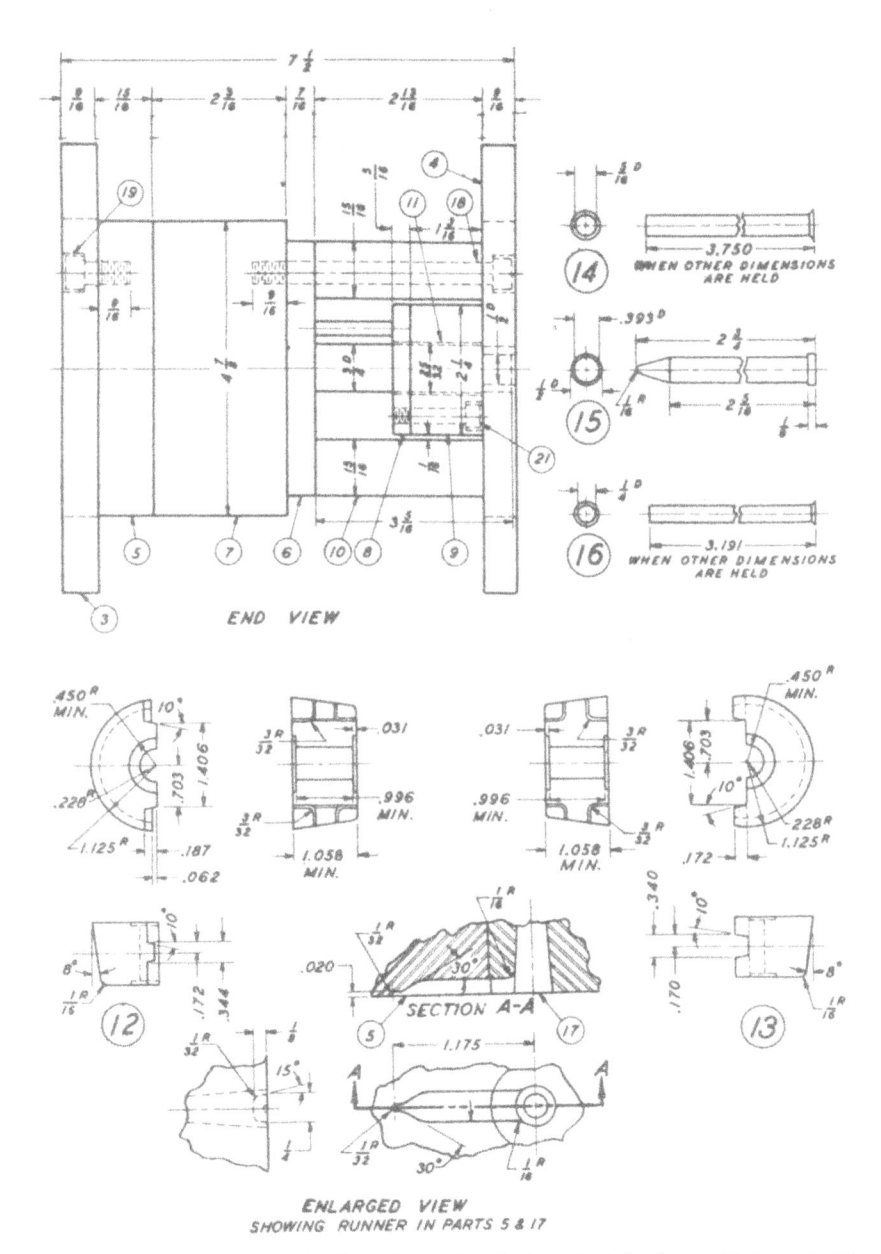

FIG. 8.89. End view and detail of mold parts for single-cavity injection mold shown in Fig. 8.88. Parts 12 and 13 are removable wedges.

1-CAVITY INJECTION MOLD

QTY	MATERIAL	OPERATION	NO.	NAME	D.	L.	W.	T.
4	STEEL	PT. 21 1/4-20 THD.	21	SCREW-SOC. HEAD	1/4	1		
2	ST. TOOL	HARDEN	20	GUIDE PIN	1	2 1/2		
6	STEEL	3/8-16 THD.	19	SCREW-SOC. HEAD		3/4		
6	STEEL	3/8-16 THD.	18	SCREW-SOC. HEAD		3 3/8		
1	ST. TOOL	HARDEN GR.	17	BUSHING	2 1/8	1 5/8		
3	DR. ROD	HARDEN	16	K.O. PIN	1/4	3 3/8		
1	DR. ROD	HARDEN	15	MOLD PIN	3/8	2 7/8		
1	DR. ROD	HARDEN	14	K.O. PIN	5/16	3 7/8		
1	ST. 33I2		13	WEDGE	2 1/2	1 1/2	1 1/4	
1	ST. 33I2		12	WEDGE	2 1/2	1 1/2	1 1/4	
2	ST. C.R.		11	SUPPORT PIN	3/4	3 1/2		
2	ST. MACH.	GRIND	10	PARALLEL	6 3/4	3	1	
1			9	K.O. BAR	6 3/4	2 3/8	1 1/4	
1			8	PIN PLATE	3 3/8	2 3/8	3/8	
1			7	SHOE	6 3/4	5	2 1/4	
1			6	BUTT PLATE	6 3/4	4 3/8	1/2	
1			5	RETAINER	6 3/4	5	1	
1			4	BOTTOM PLATE	11 5/8	7 5/8	5/8	
1	ST. MACH.	GRIND	3	TOP PLATE	11 5/8	7 5/8	5/8	
X	MISC.		2	ASSEMBLY				
	MATERIAL		NO.	NAME	D.	L.	W.	T.

Fig. 8.90. Material list for mold shown in Figs. 8.88 and 8.89.

One of the best ways to learn injection mold design is to study molds which have been operated successfully. Some injection molders prefer the types of jobs that tax their mechanical ingenuity and are well known in the industry for the "trick" molds which they design. It is true, however, that the design of injection molds, as of all other types of molds, must be kept as simple as possible if long, trouble-free operation is to be realized. Any removable parts of the mold should be made foolproof, which is to say, they should be made in such manner that they cannot be inserted in the mold the wrong way.

Some product designs may be suitable for solid-plate mold construction, as shown in Figs. 8.91A and 8.91B. Note that the individual cavities and forces are eliminated since they are cut solid into the movable and stationary retainer plates. This design, when applicable, saves time and money

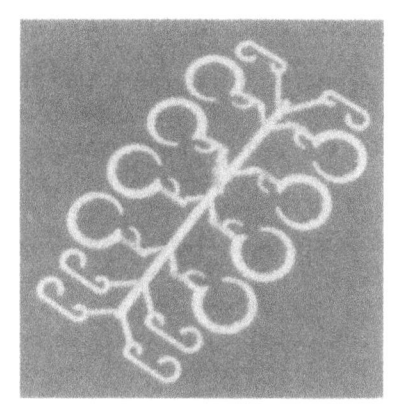

FIG. 8.91A. The cavities and runners for parts such as these may be cut in a single mold plate as illustrated in Fig. 8.91B. Such cavities are often difficult to repair or replace but the original mold cost is quite low as compared with molds which are assembled from single blocks for each cavity—assembled in a mold base.

FIG. 8.91B. The two halves of the mold are set up in a rotary-head milling machine and milled in a single setup. Inset shows finished parts. (*Courtesy Kearney & Trecker Products Corp., Milwaukee, WI*)

since there are no individual mold sets. Caution should be used in predicting the ejection side of the mold and it is frequently necessary to use pick-up ribs as required or provide some mechanical method of holding runner in the runner system and on the ejector side.

STANDARD MOLD BASES*

There are many commercial suppliers today that make and supply standard mold frames and various universal mold components. Some toolrooms and molders use their own standard mold frames. It is good design practice to use available standard mold frames and components, where possible, and thus to benefit in cost, time, quality, and maintenance.

The following illustrations and chart show some typical designs that are widely available. Several manufacturers have prepared catalogs giving complete information on many of the designs. It is good practice to review the availability of specific frame designs and components when starting a mold design and to learn the preference of the tool shop that will build the mold.

Figure 8.92 shows a standard mold base. Figures 8.93A and 8.93B show engineering drawings that may be used (with a considerable saving in time) to lay out a mold design for one type of standard frame. Figure 8.94 illustrates the availability of master layouts as furnished by one supplier, which will save considerable time in designing molds.

STANDARD MOLD BASES AND COMPONENTS

Figure 8.95 contrasts the smallest standard mold base in one manufacturer's line of assembled bases with its largest base. All of the individual components are available as separate "off-the-shelf" items; also available are individual plate and block items ranging up to $45\frac{3}{4} \times 66$ in. By choosing

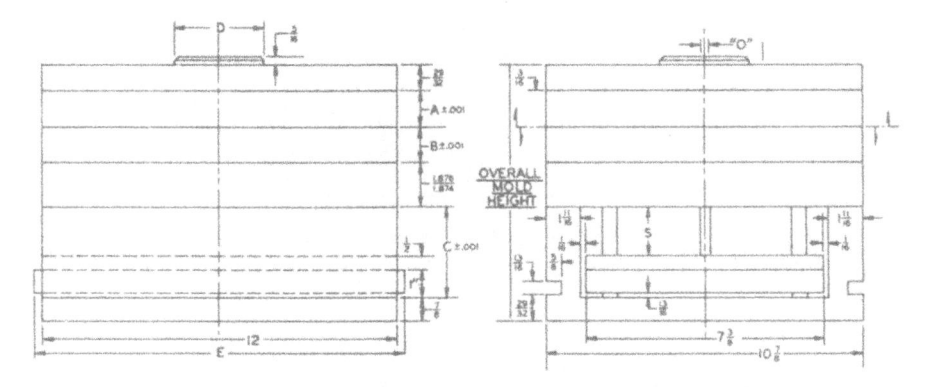

FIG. 8.92. Side view—engineering drawing of mold base to facilitate mold layout. (*Courtesy D-M-E Co./VSI Corp., Madison Heights, MI*)

*(Courtesy D-M-E Corp., Madison Heights, MI)

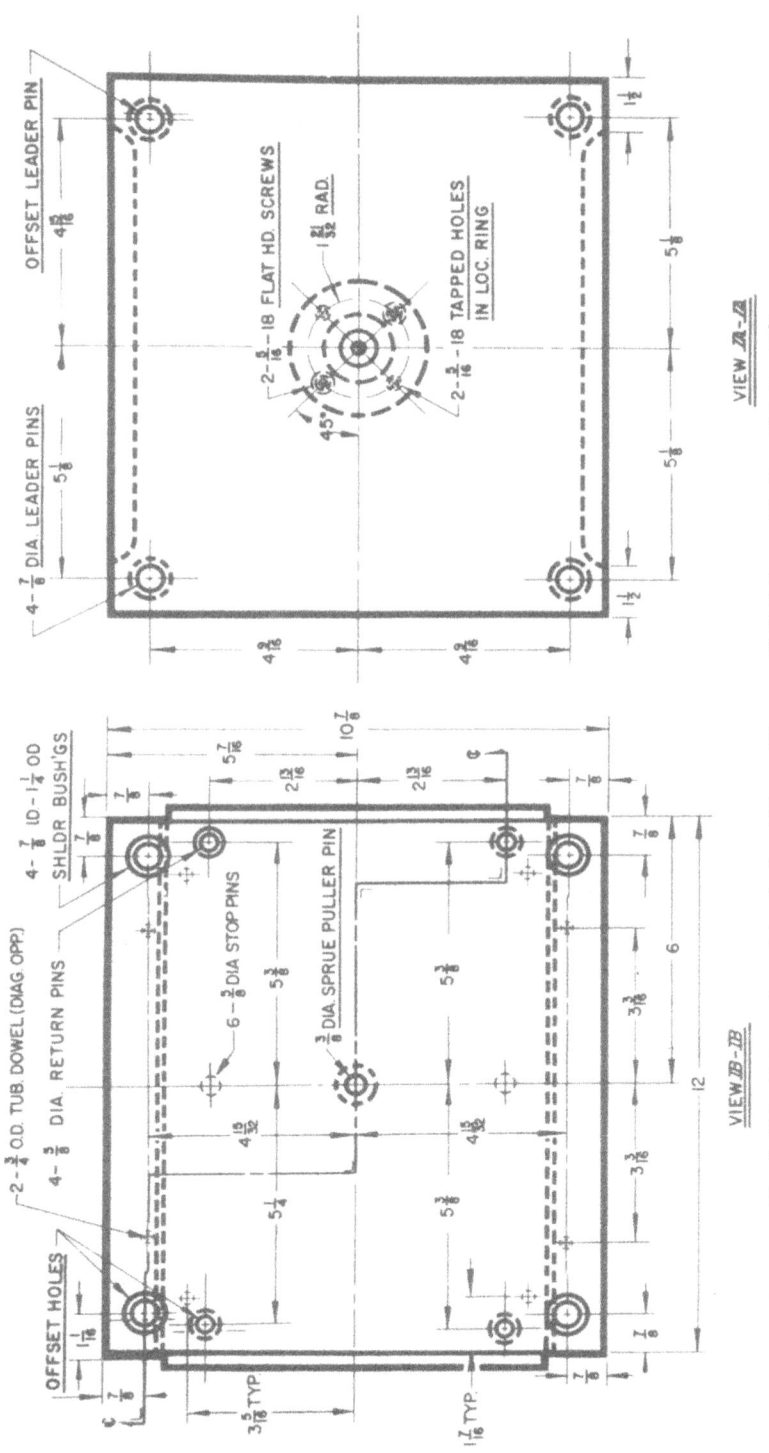

FIG. 8.93A. Bottom view of standard mold base. (*Courtesy D-M-E Corp., Madison Heights, MI*)

453

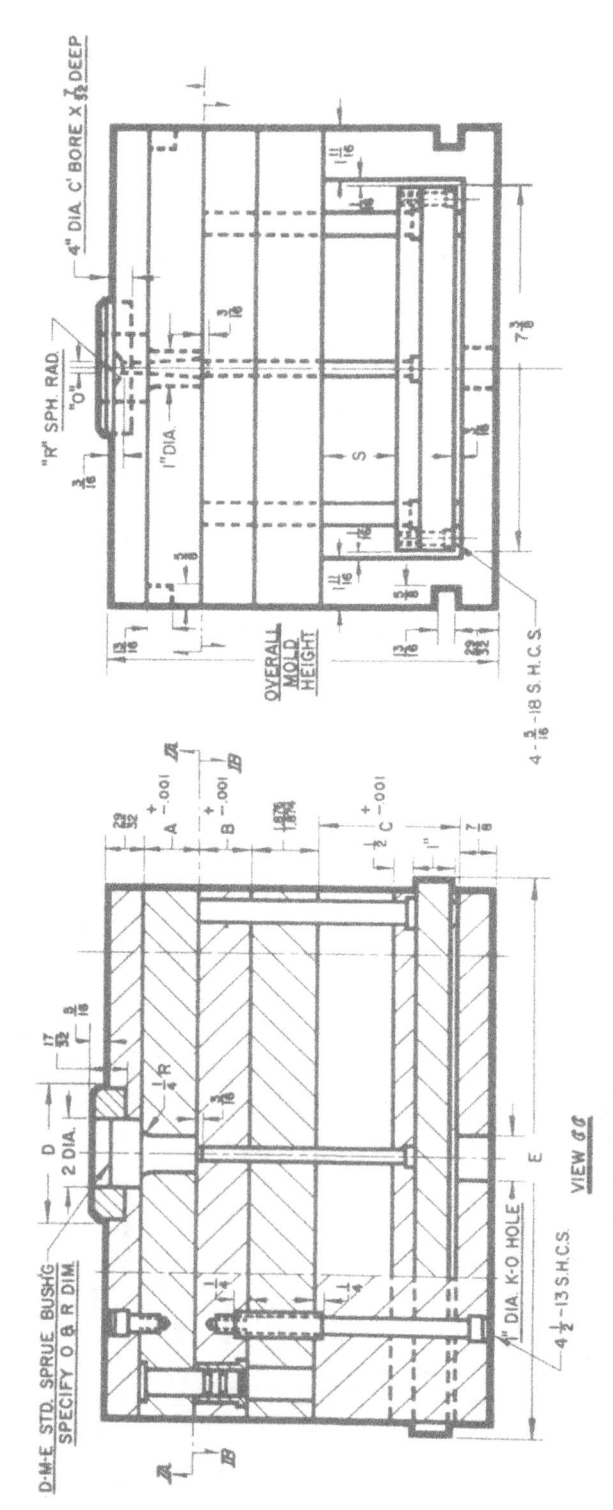

Fig. 8.93B. Section view of one type of standard mold base. (*Courtesy D-M-E Corp., Madison Heights, MI*)

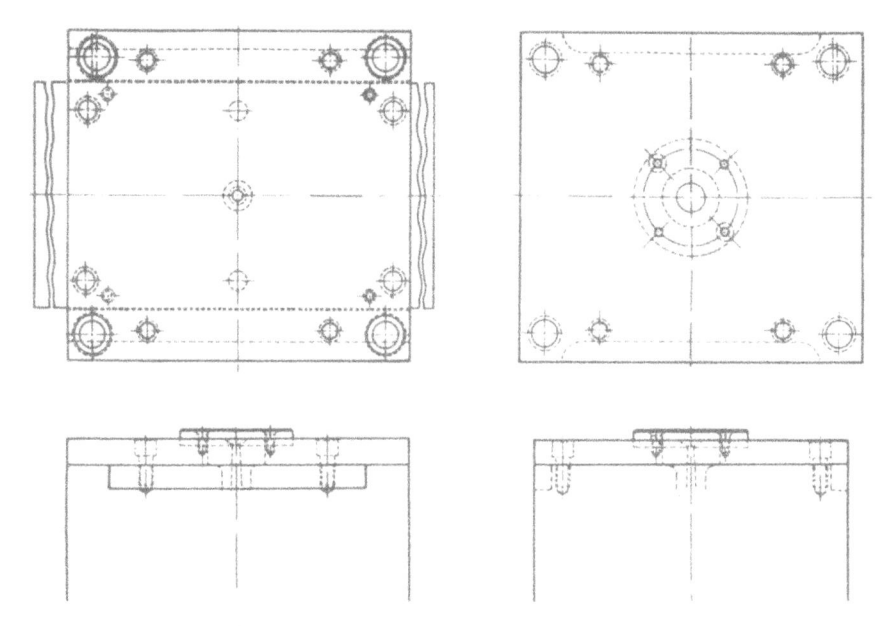

FIG. 8.94. Master drawing layouts can save valuable mold design time. They are reproduced full scale so the designer can draw the cavities and other required parts without the necessity of drawing such details as screws, dowels, leader pins, etc. This eliminates the chance for error in locating the majority of mold parts. The efficient mold designer should have a set of master layouts on hand covering all sizes he is likely to use. (*Courtesy D-M-E Corp., Madison Heights, MI*)

various plate thicknesses, parallel (riser) heights, locating ring sprue bushings, etc., it is possible to have over 12,000 different, but standard mold bases (frames).

Standard mold bases have also been developed for particular and special-purpose injection machines and for quick change operation as follows:

1. Shuttle presses using one upper half and two lower half assemblies. The lower half assemblies "shuttle" alternately to the upper half while the out-of-service lower-half is being reloaded with inserts. The principal application is for insert or wiring harness production.

2. Universal mold adapter plate system where the adapter plate becomes the platen for mounting of a core and cavity unit. This application is for quick change molding and compatability with many different presses.

3. Vertical automatic molding machines where two ejector systems are used.

4. Master mold base for insertion of standard modules of force, cavity and ejector system as shown in Figs. 6.1 and 8.96.

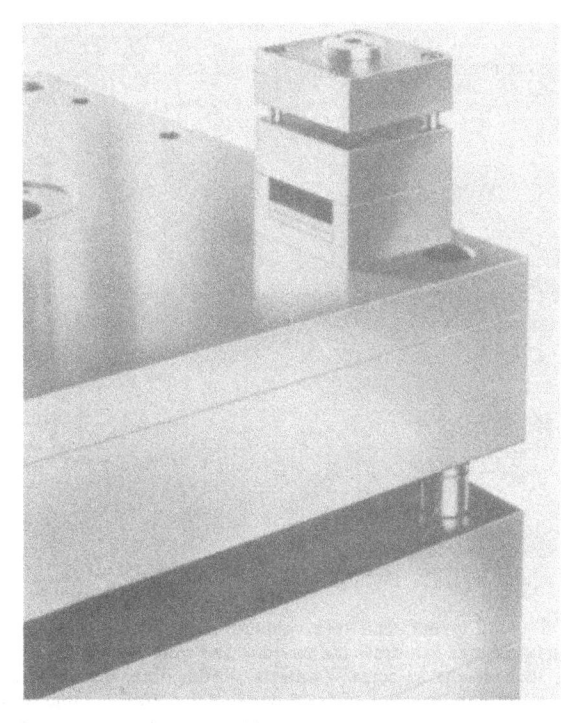

FIG. 8.95. A wide range of sizes of standard mold bases from 3½ × 3¾ to 23¾ × 35½.

Other pre-engineered tooling devices for which you will find vendor catalogs very helpful:

1. Early ejector return systems as illustrated by Fig. 8.97. This application is for a requirement where a cam slide on the ejector side of the mold passes over an ejector pin as the mold closes. The ejector pin must be retracted before the cam slide passes over it to prevent damage. The early ejector accomplishes this sequencing.

2. Latching device for floating and locking plates as shown in Fig. 8.98. This application is in three or four plate top runner molds or where a cam slide must be installed in the upper or force plate.

This particular unit can be used with any plate that must be locked, then floated away. Usually two or more are used on each mold (at least one on each end of the mold). Action and reaction time are controlled by the length of the latch and release bars which are cut to length as specified by the designer.

3. Collapsible cores are an available, off-the-shelf item in diameters from approximately 1 to 3½ in. for closures and other parts requiring

FIG. 8.96. Standard frame. Three plate construction to permit pin point gating in individual mold sets. Frame has capacity for four individual mold sets. Other frames are available for two mold sets. (*Courtesy Master Unit Die Products, Inc., Greenville, MI*)

internal undercuts (usually threads or a snap ring). A collapsing core allows removal of the molded part without any unscrewing mechanism detailed earlier in this chapter. Figures 8.99 and 8.100 show the expanded and collapsed positions. Basically, the collapsible core is a cylinder, slit lengthwise at one end to form tine-like fingers which dovetail adjacently with each other to form a segmented sleeve. A solid pin, through the center of the sleeve, supports the segments against injection pressure.

The desired threads or undercut shapes are ground into the expanded sleeve. The mold shown in Fig. 8.101 is in the closed (molding) position. Opening the mold will pull the molded cap out of the cavity and cause it to

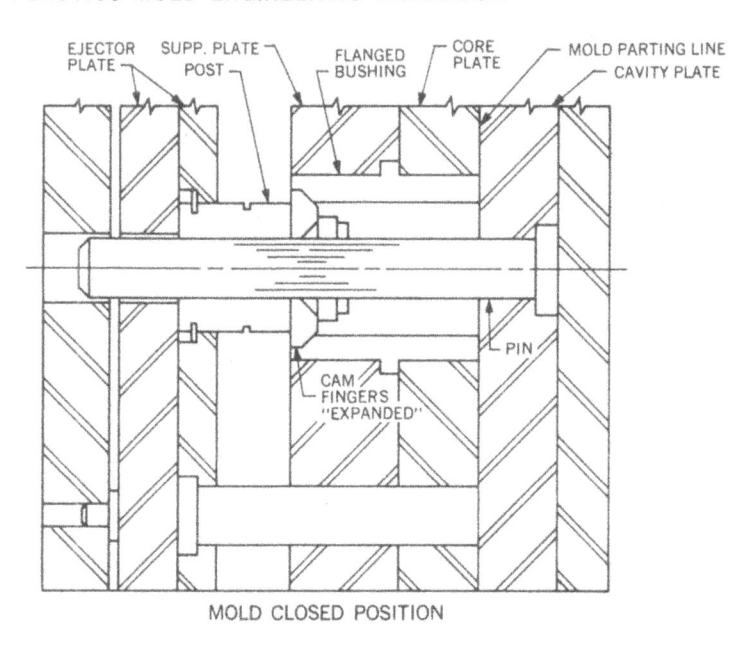

EJECTOR PLATE — SUPP. PLATE — POST — FLANGED BUSHING — CORE PLATE — MOLD PARTING LINE — CAVITY PLATE

PIN

CAM FINGERS "EXPANDED"

MOLD CLOSED POSITION

FIG. 8.97. Early ejector return unit in injection mold assembly. (*Courtesy D-M-E Corp., Madison Heights, MI*)

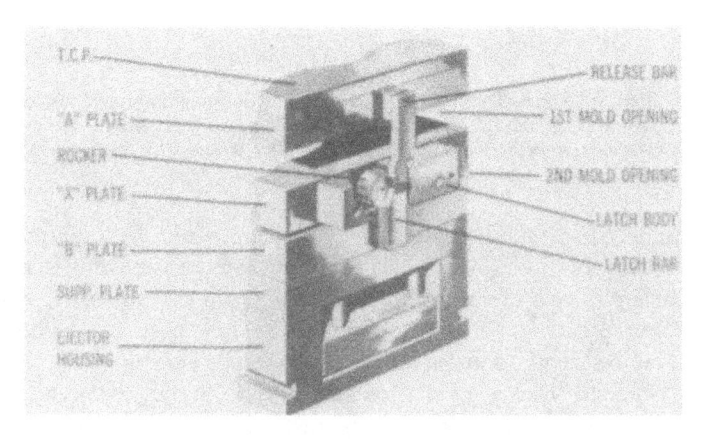

T.C.P. — RELEASE BAR — "A" PLATE — 1ST MOLD OPENING — ROCKER — 2ND MOLD OPENING — "X" PLATE — LATCH BODY — "B" PLATE — LATCH BAR — SUPP. PLATE — EJECTOR HOUSING

TO FLOAT "X" PLATE AWAY FROM "A" PLATE WHILE LOCKING "X" and "B" PLATES

FIG. 8.98. Latch-Lok in mold assembly.

FIG. 8.99. A standard collapsible core—expanded.

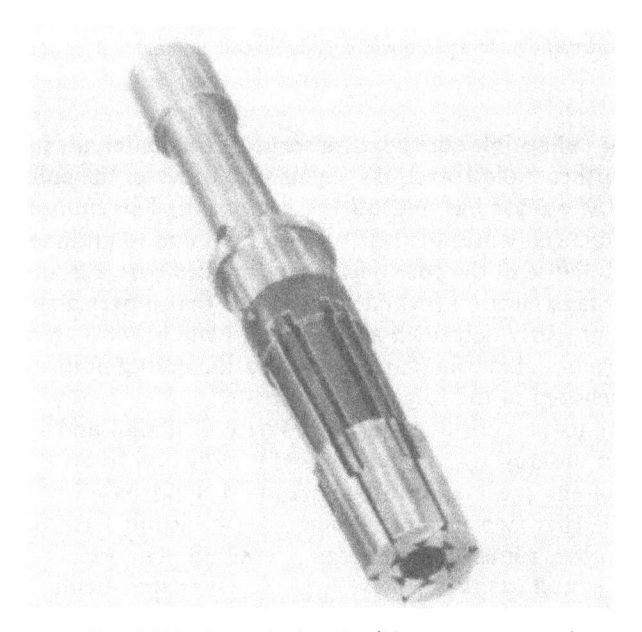

FIG. 8.100. A standard collapsible core—contracted.

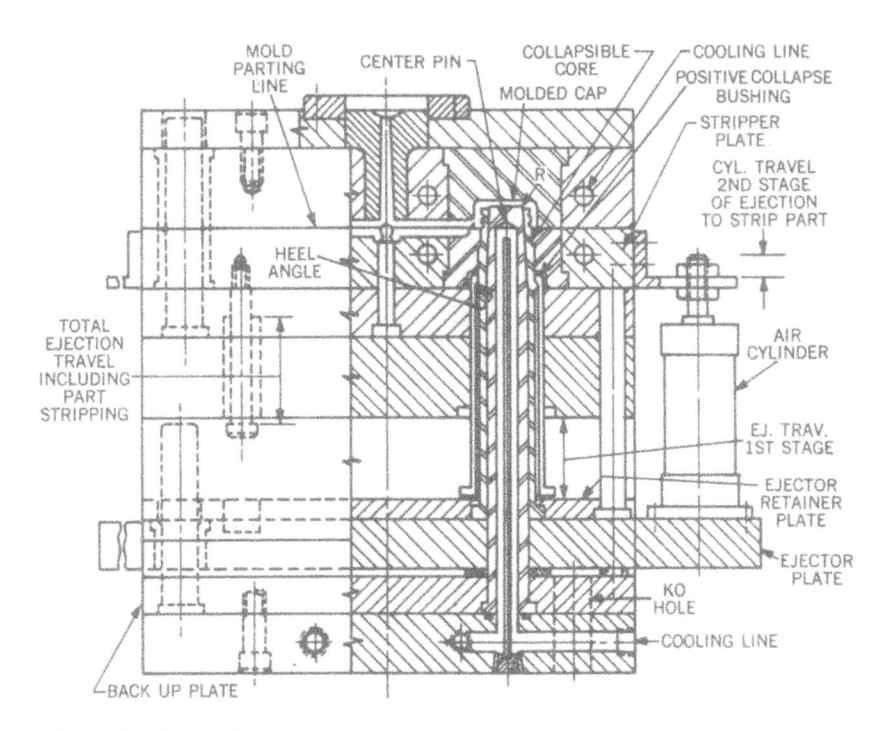

FIG. 8.101. Schematic mold assembly of collapsible core in mold with edge or tunnel-gated cap.

remain on the collapsible core because of the thread undercuts in the molded part. With further mold travel, the segmented sleeve of the collapsible core moves with the ejector bar and off the center pin. This motion allows the tines to contract (collapse) of their own volition due to an inherent spring-like action. Collapse of the core releases the molded part. Ejection is assured by a second stage ejection (cylinder operated) that moves a stripper plate. This mold is properly described as a stripper plate injection mold with collapsible core and submarine gate. Note also the coring point on the sprue puller pin anchored in the force back-up plate.

Advantages to be considered are simplicity of design and specification, elimination of unscrewing mechanism and a reduction in cycle time.

Selection of the most appropriate standard mold base and other components is the first step in mold design after decisions have been reached on cavity number, runner system, gating, ejection plan, etc. In selecting the mold base, it is well to take the next larger size than the calculated minimum because, wall thickness studies, cooling/heating channels, etc. often leave the designer or mold maker short of space to do a good job.

Figure 8.102 shows a product drawing for a vial and cap as designed for

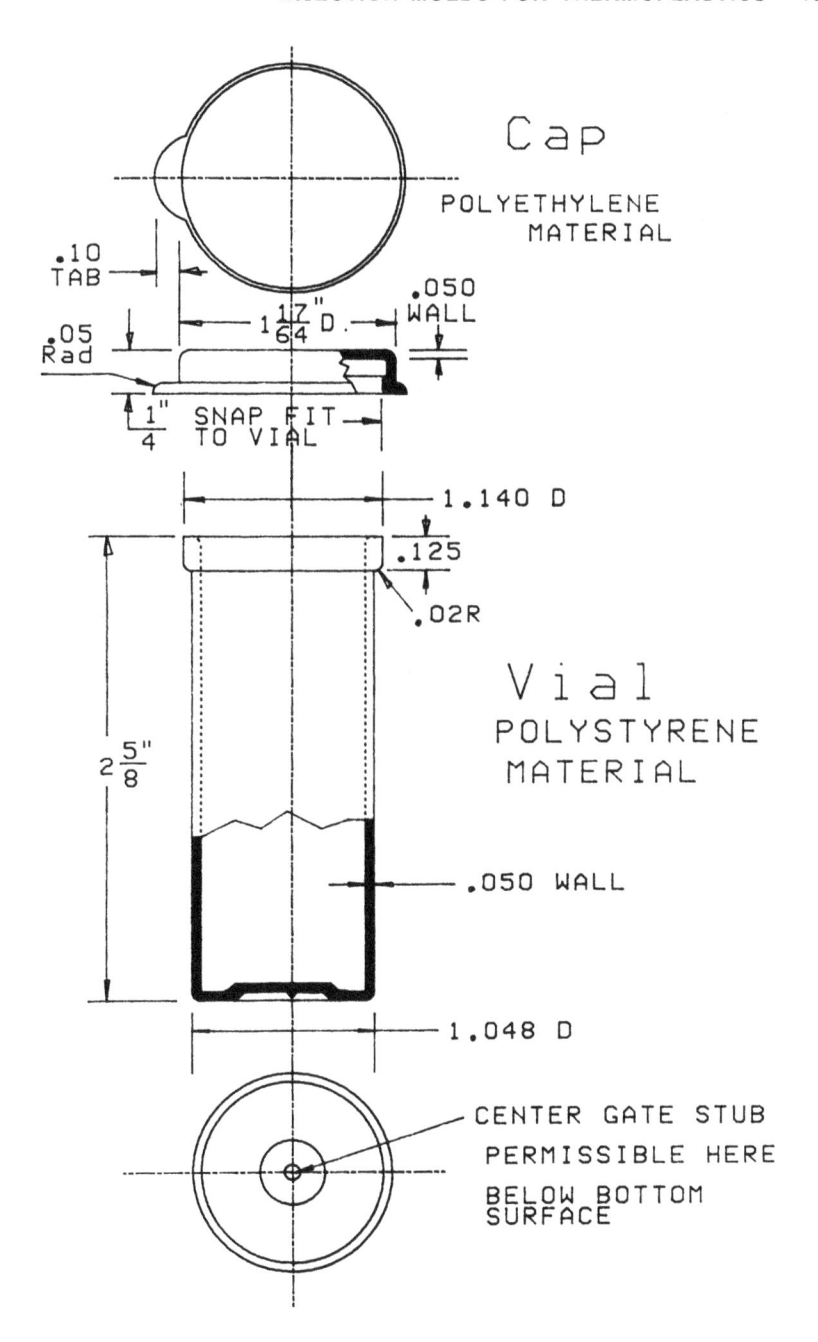

FIG. 8.102. Plastic vial and cap.

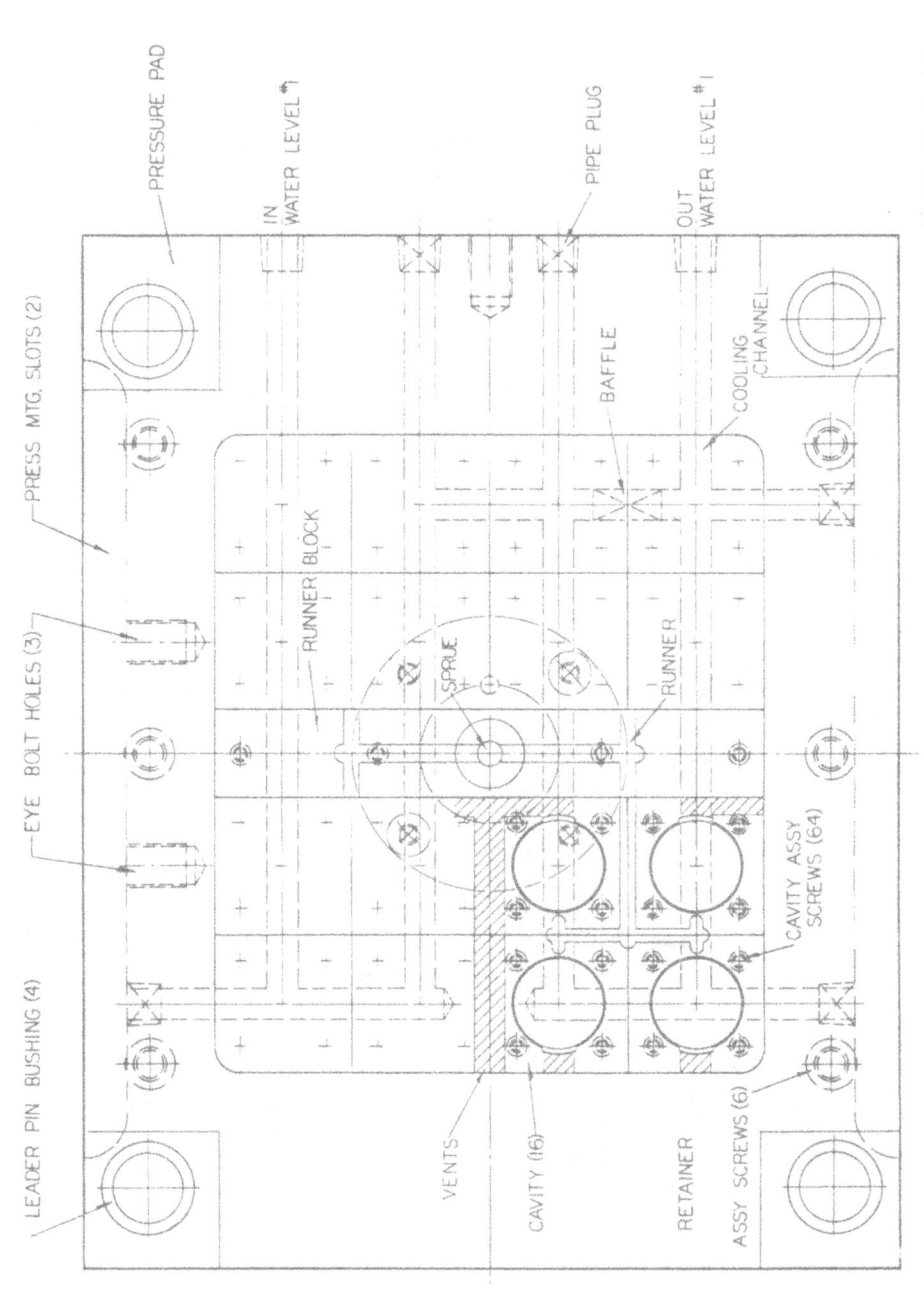

FIG. 8.103. A 16-cavity tunnel-gated stripper-plate mold for cap. Stationary half of mold view "A," as identified in Fig. 8.105.

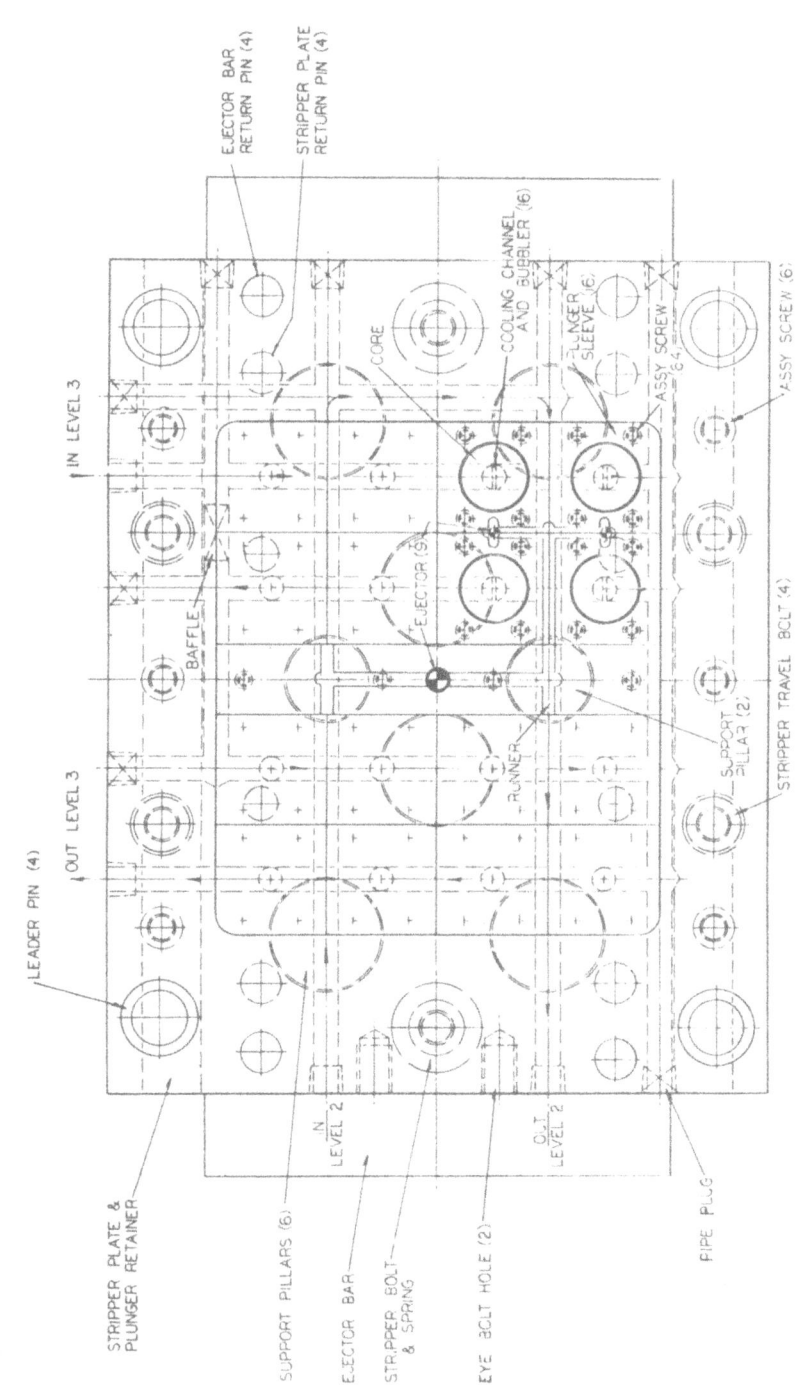

FIG. 8.104. A 16-cavity tunnel-gated stripper-plate mold for cap. Movable half of mold view "B," as identified in Fig. 8.105.

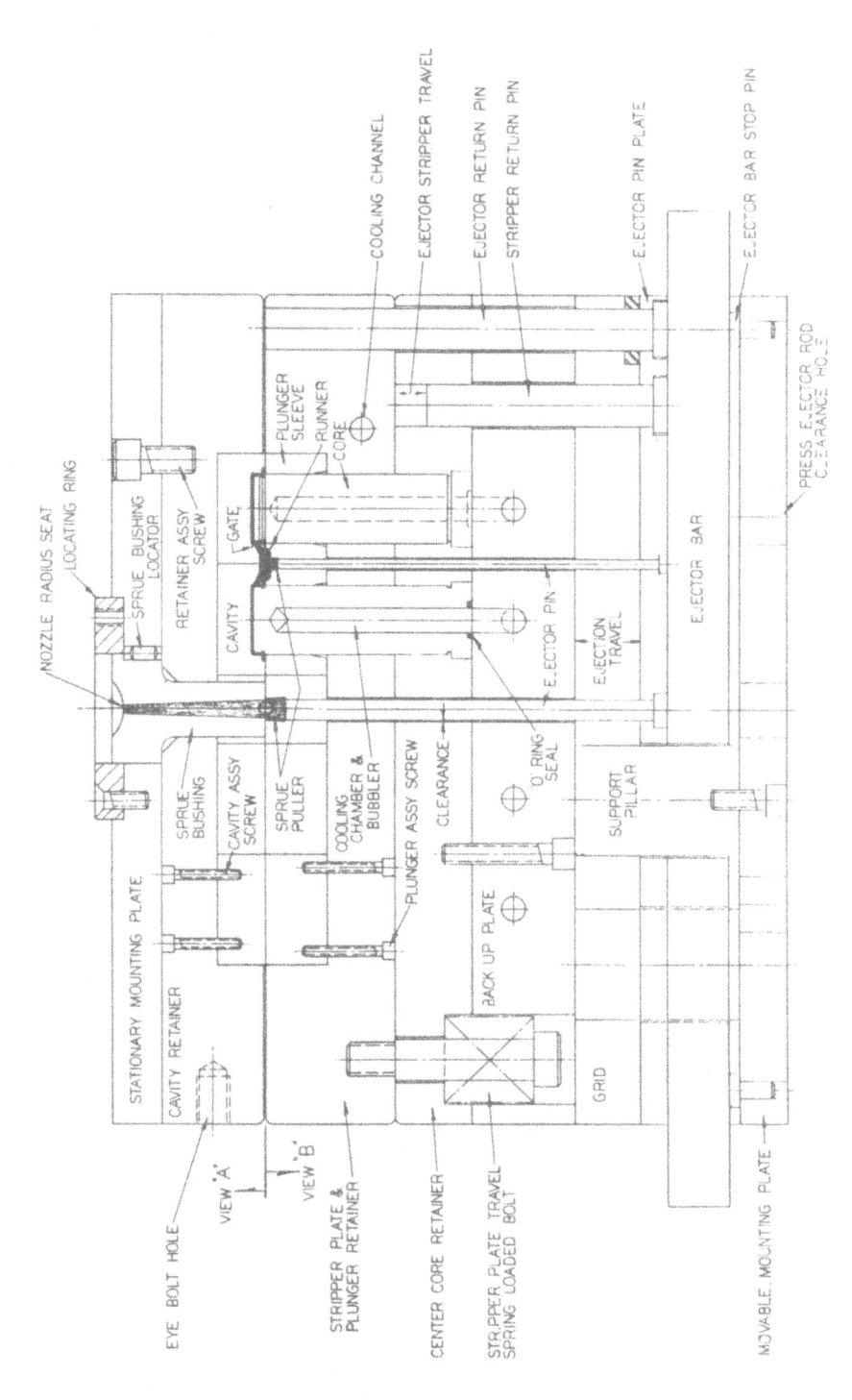

FIG. 8.105. A 16-cavity tunnel-gated stripper-plate mold for cap. Profile view.

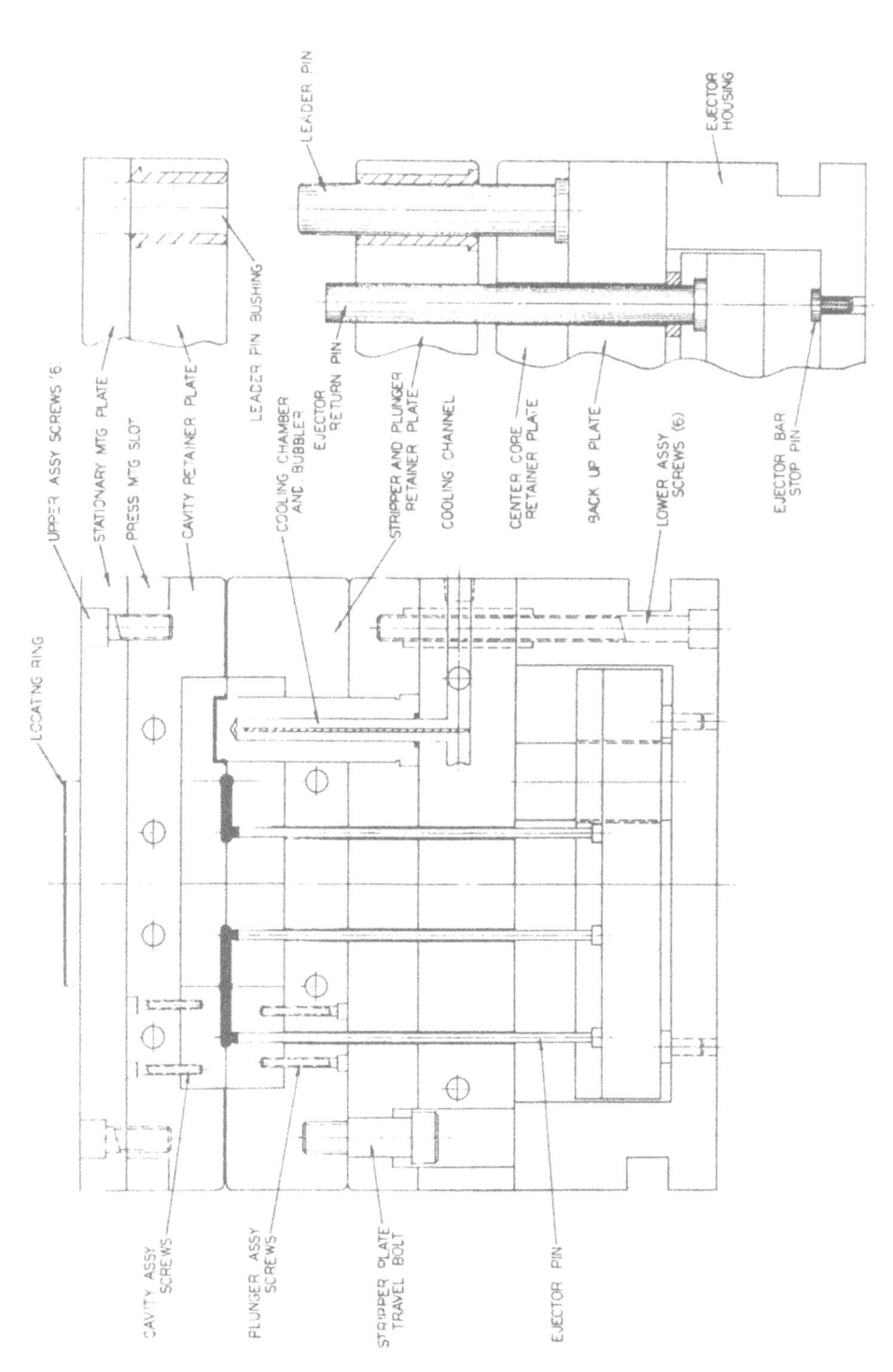

LOCATING RING

UPPER ASSY SCREWS '6

STATIONARY MTG PLATE

PRESS MTG SLOT

CAVITY RETAINER PLATE

LEADER PIN BUSHING

COOLING CHAMBER AND BUBBLER

EJECTOR RETURN PIN

STRIPPER AND PLUNGER RETAINER PLATE

COOLING CHANNEL

CENTER CORE RETAINER PLATE

BACK UP PLATE

LOWER ASSY SCREWS (6)

EJECTOR BAR STOP PIN

LEADER PIN

EJECTOR HOUSING

CAVITY ASSY SCREWS

PLUNGER ASSY SCREWS

STRIPPER PLATE TRAVEL BOLT

EJECTOR PIN

FIG. 8.106. A 16-cavity tunnel-gated stripper-plate mold for cap. Endview. Right-hand views show fully open position of members.

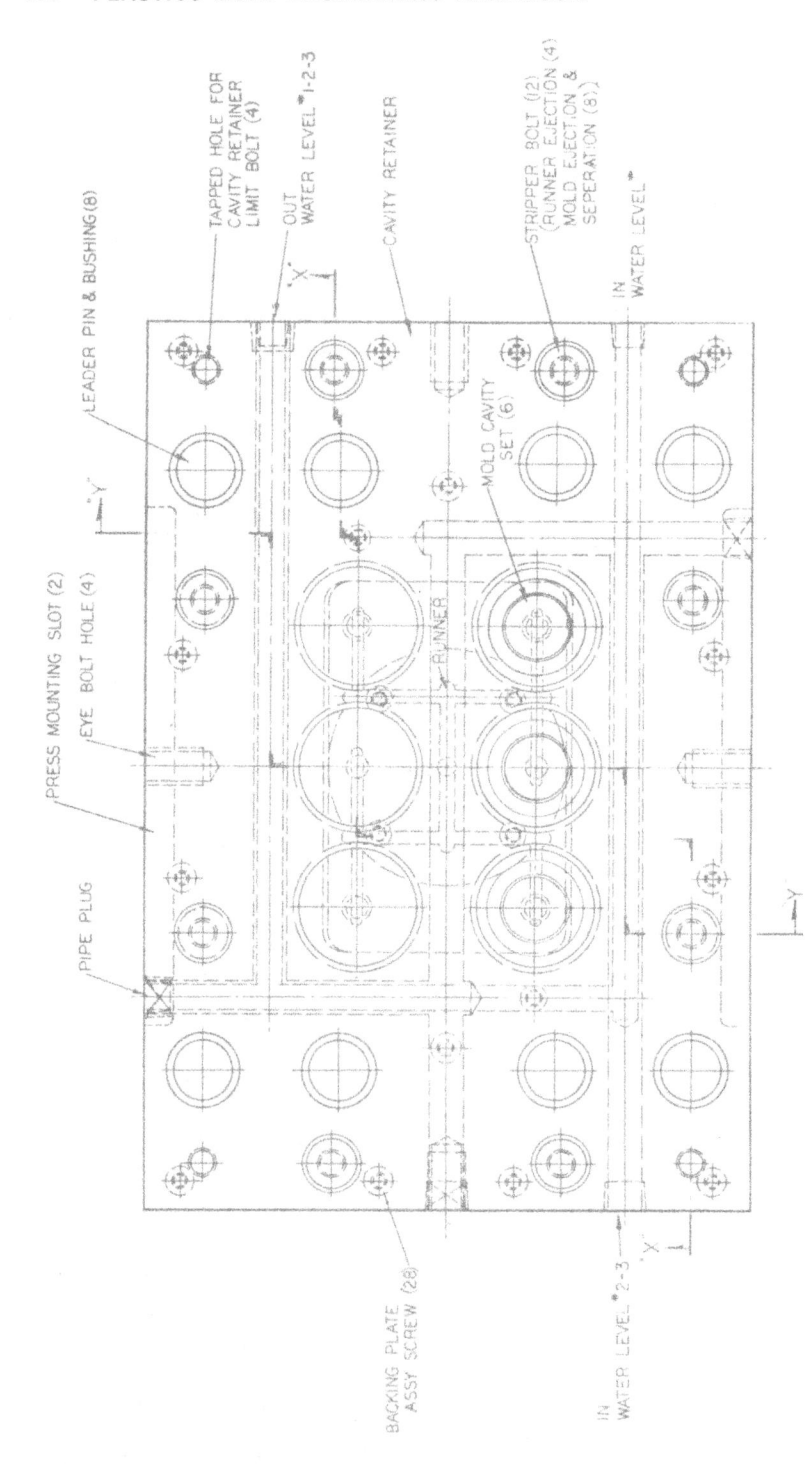

FIG. 8.107. A 6-cavity stripper-plate pin-point center-gated mold for vial View "A," as identified in Fig. 8.110.

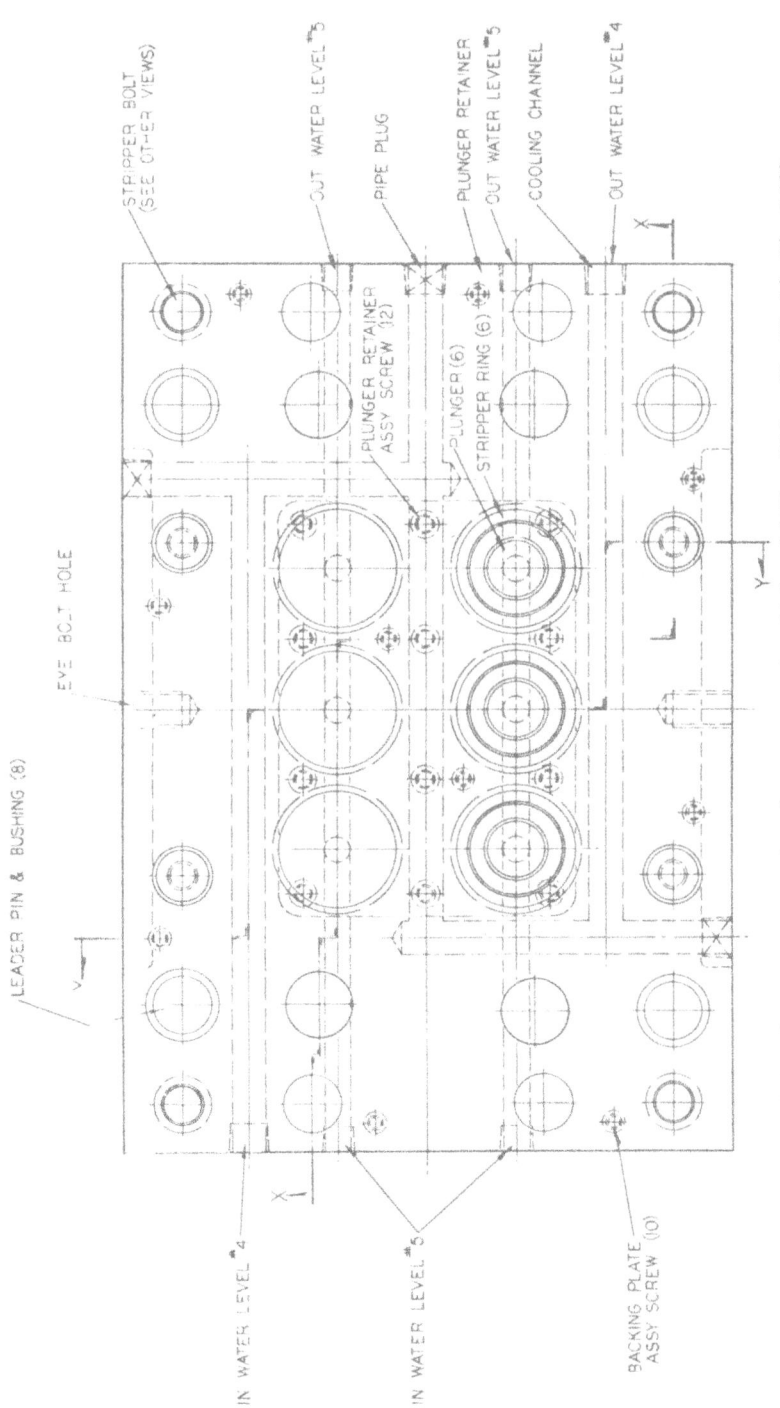

FIG. 8.108. A 6-cavity stripper-plate pin-point center-gated mold for vial View "B," as identified in Fig. 8.110.

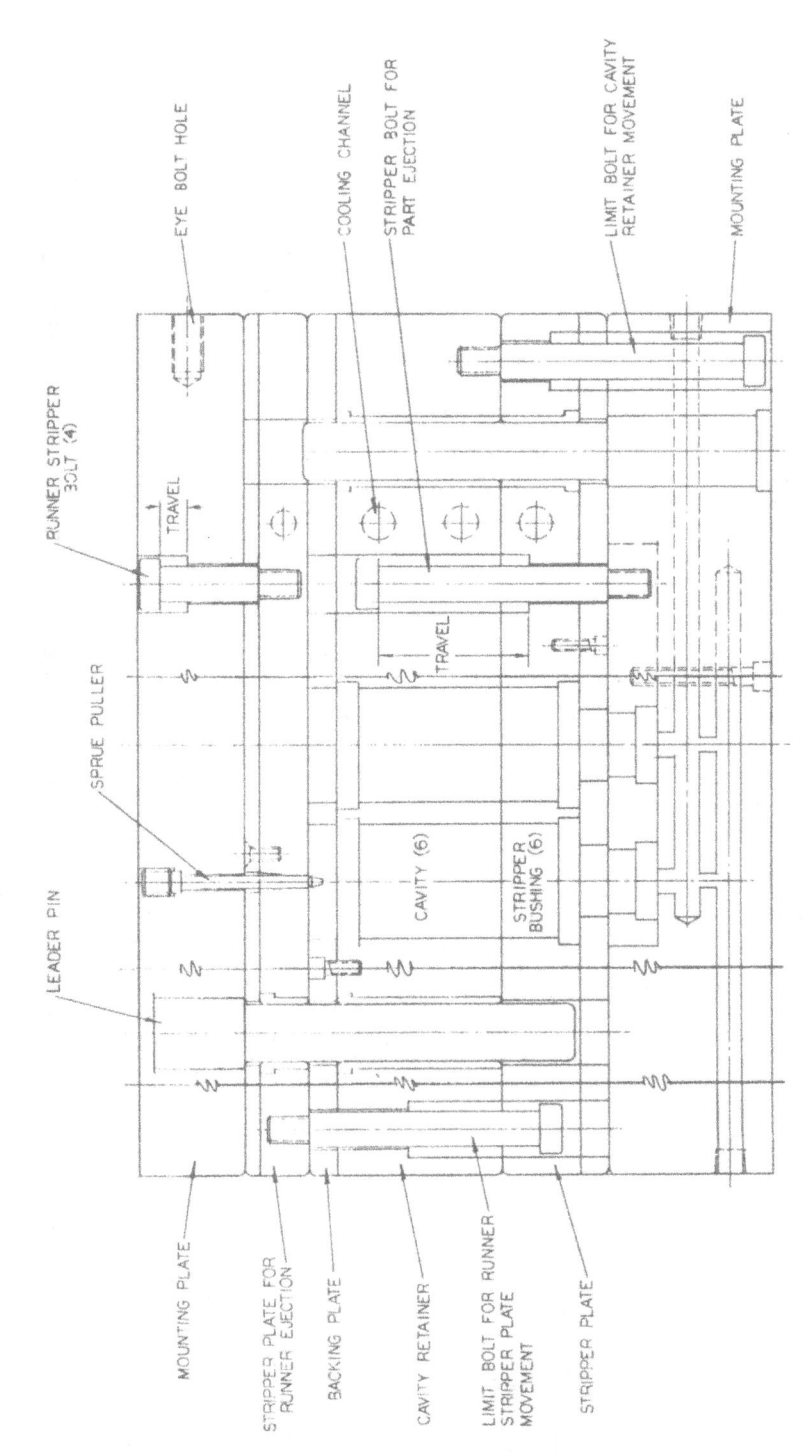

Fig. 8.109. A 6-cavity stripper-plate pin-point center-gated mold for vial Profile view "X-X," as identified in Figs. 8.107 and 8.108.

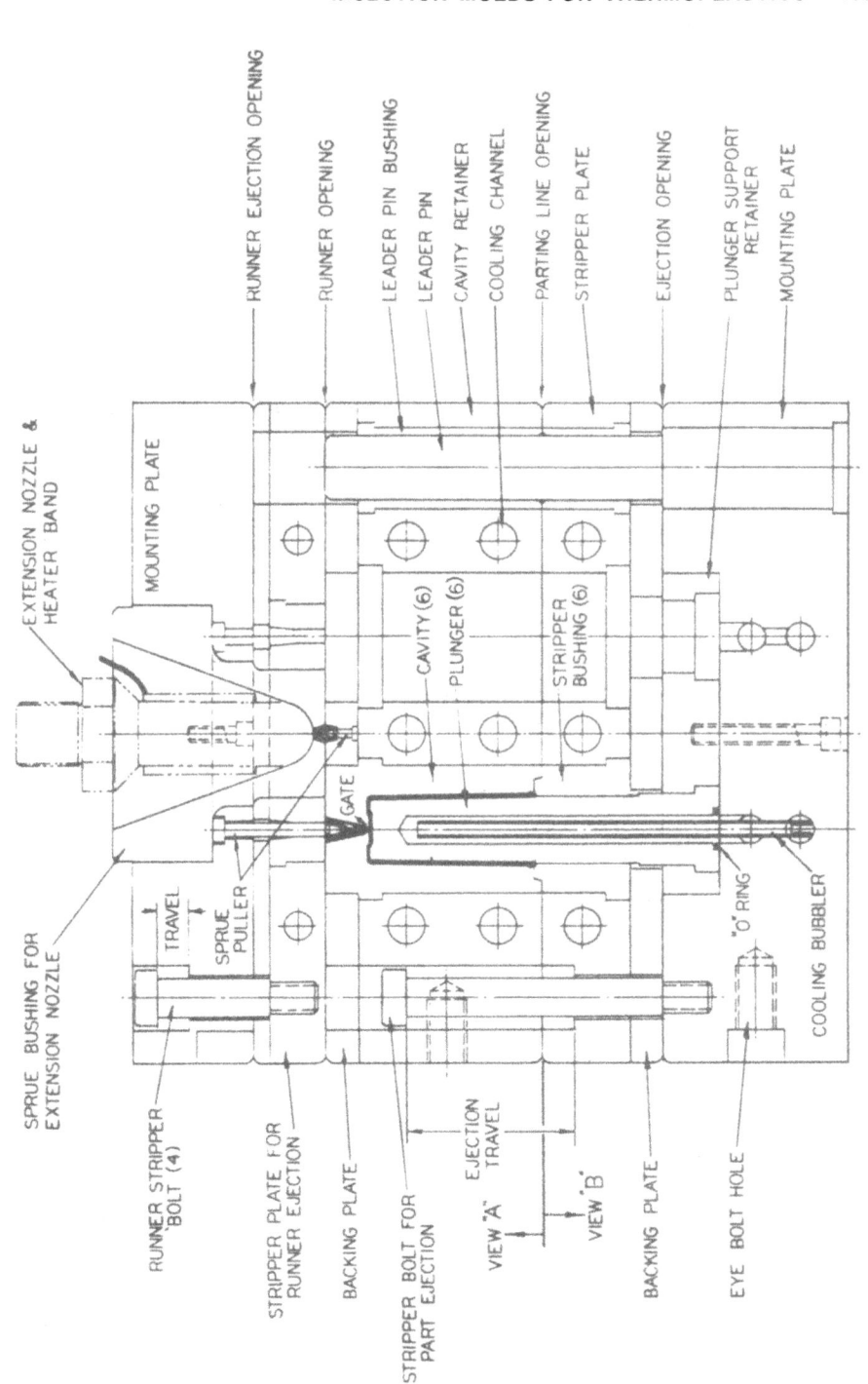

Fig. 8.110. A 6-cavity stripper-plate (pin point) center-gated mold for vial. End view section "Y-Y", as identified in Figs. 8.107 and 8.108.

good molding practice. Figures 8.103, 8.104, 8.105, and 8.106 show the tool design for the cap. Figures 8.107, 8.108, 8.109, and 8.110 show the tool design for the vial. Each drawing is self-explanatory. It is important to note the sound engineering construction of each detail. Molds of this general design have performed well in production, resulting in long production runs with a minimum of maintenance costs.

The mold designer should always be present when each new mold is sampled to insure proper installation, freedom of moving parts, and to learn by experience how to do a better job the next time a similar mold is to be designed.

VERY LARGE INJECTION MOLDS*

We have already described many of the features of large and very large molds. The designer must understand that the line of demarcation between various sizes is defined by the judgment of the designer. For our purposes here, we will consider very large molds as being those in excess of 500 square inches projected area of the part. The ultimate determination for a tool design is the final requirements placed on the molded part. Raw material manufacturers can provide thousands of variations in formulae in an almost unlimited variety of base materials with their fillers and reinforcements. Machinery manufacturers can provide the ultimate in combinations of processing equipment. The very best properties of the molding material or molding equipment can be completely lost or nullified by an incorrectly designed mold.

Because of the previous large volume of data on molds and their individual features, we will confine ourselves to a listing, in this section, of the important features to consider when designing a very large injection mold. First of all, we will list the probable properties of the plastics material which your mold design will be required to mold. Then, we are going to list the characteristics of the required injection machine, as they will differ from the "regular" machines, where preciseness is not the essential requirement. Finally, we will then list mold features which will assure a molded part of the required quality and dimensions. Inasmuch as this is a text on mold designing, we shall concentrate on the items which must be incorporated into a mold to ensure that trouble-free operation is accomplished.

Currently, thinwalling is an apparent contradiction in terms, where increased(ing) thinwalling means decreased wall thickness. European molders

*This section written by Wayne I. Pribble in consultation with Roger Herrick, Michigan Plastics Products Co., Div. JSJ Corp., Grand Haven, MI. Refer also to: Big-part molding gears up for the automotive panel, *Modern Plastics*, Oct. 1986.

are using wall thicknesses of less than 2.5 mm (.098 in.) in some automotive parts. In the USA, one auto maker (Ford Motor Co.) is using 4.0 mm (.156 in.) in the bumper for a 1987 model automobile. The PBT-modified polycarbonate data sheet gives 16 ft-lb/sq-in. notched izod. This must be maintained, because impact strength is critical in an auto bumper. We will discuss the details of this mold in the following section, entitled "Mold Design Requirements." While automobile applications are the projected large volume users of very large moldings, do not overlook the appliance industry, and the sports industry, as users of panel-type moldings. In fact, the largest press now in use is a 10,000 ton model used to mold windsurfer hulls with high-density polyethylene (HDPE) material.

The authors agree with a DuPont representative, as he describes the combinations necessary for proper processing, "the wrong combination of machine, controls, and tooling can turn any good resin into a bad part." Specifically, we will enunciate many of the requirements for the material, machine and tooling involved. Thus, it becomes the mold designer's objective to be sure that the mold tooling is capable of molding good parts.

Plastics Material Requirements

Processability will probably require some tradeoff with physical properties, as it has in the past. The tougher the resin, the more difficult it will be to mold the resin without a loss of properties (particularly impact strength.)

A new-generation of molding materials are on the way. At least six major suppliers are developing on-line paintable thermoplastics. The final properties must approximate the steel or the Sheet Molding Compound (SMC) currently in use. A major problem involves the reduction of the coefficient of expansion to match other parts of an assembled unit.

A surface finish to Class A, as molded, is a prime requirement. "Class A" in automotive terms means no flow lines, no knit lines, no gate marks, high gloss, and paintable with current spray-bake processes.

Heat resistance to 240°F is the current standard for interior and exterior panels. In the engine compartment, heat resistance to 390°F is the current standard. Ease of motion without lubrication may be another requirement along with heat resistance. A good designer will check all of the specifications, including the choosing of the molding material and the specifications to be met.

Maximum leeway on processability conditions of temperature, pressure and flow characteristics are an objective. Unfortunately, it seems that the higher the performance required, the more likely the "window" of acceptable processing temperature becomes proportionally narrower, even as low as 20°F.

The term "energy management" is applied to plastics materials, in the sense of absorbing energy applied from outside of the product, such as impact on steering wheels when resisting crash conditions. The meaning changes when the term is applied to machines, as noted below.

Machinery Requirements

1. A clamping pressure of 5,000 tons or higher is necessary.

2. Process controls capable of close control of processing temperatures and exact sequencing of operations. The control must also provide repeatability of shot weight, cycle times, variable pressures, etc.

3. Screws designed for engineering alloys require low compression ratios and minimum shear of plastics material as it moves through the processing zones. These are also called "polyblend" screws and used with vented short barrels which will allow processing of material without predrying. These screw designs provide "easy" treatment of the plastics material in order to retain its designed properties as it is delivered to the mold.

4. 'Energy management'' is applied to machines in the sense of managing the energy needed to process the material. Most of the energy management of machines consists of systems such as heating or cooling units, and the controls for them, whether applied to machines, preprocessors, or molds.

5. Redesign or retrofitting is required because current large (3000 ton and over) presses were designed for free flowing styrenic or olefinic blends and relatively slow production cycles with reasonably wide product tolerances. The thermoplastic panel press requires rapid movement of heavy-mass platens, precision alignment under high pressure (10,000 PSI clamp on projected area of part) and dry-cycling of under 20 seconds. However, the most critical need is probably the very precise control of hydraulic pressures at given points in the injection cycle to provide the best possible melt at the mold.

Mold Design Requirements

We reemphasize that the critical balance of properties in the raw material can easily be lost in a poorly designed mold. The very best designed machine cannot overcome deficiencies in mold design. Thus, the design of a mold must take into consideration the specific resin to be processed, along with its peculiarities of processing. The mold must also be designed for a specific machine so that the mold can deliver the melt to the mold cavity in as good a condition as it received it from the processor of the machine.

We recommend the use of one of the Computer Aided Engineering (CAE) programs to evaluate the temperature control of molds where the flow of material would exceed 18 in., or where a changing cross section will drastically affect the flow of material. Chapter 14 covers the essential description of one basic program for evaluating the thermodynamics (Heat transfer or exchange) of any size or shape of mold.

Hot runners will be the primary choice for melt delivery systems in the mold. For example, in current production at the Ford Motor Co. is the 14-foot long bumper mentioned above. The material is a PBT-modified nylon (G.E. Co., Xenoy). The presses available are 2,500 to 3,500 ton machines retrofitted with special low-compression screws and high performance process controllers. The average wall thickness is 4 mm (.156 in.). There are the usual requirements of no gate mark, no knit lines and a class ''A'' surface, as molded, for this paintable item.

The mold builder selected a hot runner for melt delivery through a series of valve-gates up to 36 in. in length. Ten locations for valve-gates were selected along the 14-foot length of the molding. By sequentially opening and closing the valve-gate, the melt could be introduced into the mold so that a maximum flow of 18 in. would be required. Thus, a minimum of shear would be generated and a minimum of packing pressure would be required. The principle of sequencing valve-gates provides an almost unlimited length of molded part with melt flow of 18 in. or less in the mold cavity. In other words, as the flow from open valve-gate #1 passes the closed valve-gate #2, valve-gate #1 closes at the exact instant that valve-gate #2 opens. Thus, the melt flows uninterruptedly toward valve-gate #3, but the flow between #1 and #2 is stopped and that part of the mold cavity is completely filled and beginning its cold-set. The open-and-shut sequence must be controlled by a very precise controller as the flow proceeds through the ten valve-gate locations. Obviously, the fill time is the time for flow between any two valve-gates multiplied by the number of valve-gates plus one. The set or cooling time becomes almost the equivalent of molding a small part of 18 in. or less. With this description of a mold design for a particular large part, the designer should be able to visualize other molds and how to reduce the molding problems to a manageable size by consideration of the individual components which make up the system. Once again, we emphasize the need for a designer to be thoroughly familiar with components already available in the marketplace and how to combine them for the upgraded requirements of newer materials and machines. It is seldom necessary to ''reinvent the wheel.''

Where there is a possibility of breaking mold pins during a production run, we recommend a design where such vulnerable pins can be replaced without disassembling a mold that may weigh five or ten tons. This principle is illustrated in the section covering compression molds for FRP materials.

Shrinkage allowances in the direction of material flow may need to be different from shrinkage allowances in the transverse flow direction. An example would be a refrigerator door frame, where a very lengthy flow of material is required to encircle a 36 in. by 60 in. frame. Shrinkage in the direction of flow may turn out to be approximately twice the shrinkage in the transverse direction. Our recommendation to use one of the CAE programs to analyze large molds is also a recommendation to use it for this point regarding shrinkage. Any good CAE program should be able to predict shrinkages, flow patterns, residual stresses, etc. In passing, we might note that a hot runner mold with sequential delivery (as described earlier) will undoubtedly reduce the overall shrinkage control to that needed for the smaller section fed by each valve-gate.

Temperature control of the mold, including the hot runner and valve-gate system, is an essential ingredient of the mold design. A CAE program will provide information on heating or cooling channels and their orientation and connnections. Lacking a CAE program, the designer should organize temperature control channels in the general order of the coldest control media being introduced near the hottest plastics material. Directed flow is necessary when cooling is involved. Uniformity of temperature is also critical. Thus, a large part would have at least two circulation systems starting with the coldest media nearest to the sprue and progressing toward the outer end of the mold. A general principle is to provide at least 20% excess cooling or heating, especially when a CAE program is not available.

The ejector bar in these very large injection molds can weigh several hundred pounds. Most of the very large molds will be operated in a horizontal position. This means that gravity tends to bind all of the ejector pins because of the weight of the ejector bar and the ejector pin plate. Thus, it becomes necessary to provide support and guidance for the bar. We recommend regular guide pins and bushings spaced similarly to the guide pins and bushings between the mold halves. In all cases, we recommend grease fittings which are easily accessible while the mold is installed in the molding machine. In addition, we recommend temperature control channels in the ejector bar to preclude the possibility of differential expansion creating binding of the ejector pins at the molding surface. Binding or misalignment of ejector pins will create excessive wear in the ejector pin hole and will soon result in unacceptable flash at the molding surface. It should not be necessary to point out, but we will say it anyway, that ejector pins should never be located on an ''appearance'' surface. Some current engineering materials for injection require mold temperatures in the 600°F range. A 600°F mold with a room temperature ejector system and pins spaced 60 in. or more apart can create major problems of differential expansion and create operational problems that a good mold design would have avoided.

It is not likely that a novice will be designing a very large injection mold as the first of his designs. However, part of the training of a mold designer should be devoted to participating in evaluation sessions relating to the mold designs of others. There is no substitute for practical experience. In any case, we believe that the most important of these recommendations is this one. With at least six of the largest material suppliers and with at least six of the press manufacturers pouring monumental amounts of money into molds, materials and equipment, it is a foregone conclusion that even the statistics presented here as very large injection molds will be considered run-of-the-mill applications by 1990. This writer has consulted others well versed in the field of very large moldings. The reader should consult other designers and find out what problems they may have had with their previous mold designs. Then, avoid those problems. It is frequently more important to know what to avoid than it is to know what has been successful. There may be hundreds of ways to do-it-wrong, and only a few good ways to do-it-right. Consult the mold maker experienced in building very large injection molds and ask, "How would you design this particular mold?" This writer's experience is that all of the really knowledgeable people in mold design or mold making are ready and willing to share their expertise. Obviously, material manufacturers and machine manufacturers have a vested interest in molds that "work the first time in the press." Make your design one of them!

REFERENCES

Bernhardt, Ernest C., and others, Computer-aided engineering is ready for injection molding, *Plastics Engineering*, Jan. 1984.

Bernhardt, E. C., ed., *Computer-Aided Engineering for Injection Molding*, SPE sponsored, New York:Hanser/MacMillan, 1983.

Groleau, R. J., Intelligent molds, *Modern Plastics*, p. 73, Nov. 1984.

Moldflow cuts Buick parts cycle time 30%, *Modern Plastics*, p. 32, July 1984. Moldflow Australia, Bridgeport, CT (license fee).

Mold-Master System for Valve Gating, Georgetown, Ontario, Canada:Mold-Masters Limited (1975 and later).

Mold Standardization, A Modular Approach, Bolton, Ontario, Canada:Husky Injection Systems (1975 and later).

Plagate Systems, Tokyo, Japan:Saito Kohki Co., Ltd.

Precision molding via heat pipes, *Modern Plastics*, p. 28, Mar. 1984.

Ross, M. I., Mold design for automatic runnerless molding, *SPE Journal*, June 1976.

Sensitip (tm) Runnerless Technology, Elmhurst, IL:Elmhurst Manufacturing Co., Inc., 1984.

Shaw, R. J., *Transisprue (tm) Technical Data*, Brookfield, WI:R. J. Shaw Products Corp. (1983).

Shaw, Richard J., A temperature regulated sprue can solve tough molding problems, *Plastics Engineering*, May 1983.

Smith, George E., *New Runnerless Injection Molding Systems*, Chatworth, CA:Spears Systems, 1984.

Sneller, Joseph A., Big-part molding gears up for the automotive panel, *Modern Plastics*, Oct. 1986. (Includes bibliography of suppliers).

Sneller, Joseph A., Runnerless tooling: A fast and reliable way to mold any material, *Modern Plastics*, p. 64, October 1983, (Includes bibliography of suppliers).
Wood, Richard, Large-part injection molding; Developments in Europe, *Plastics Machinery and Equipment*, Apr. 1984.

SUGGESTED FOR FURTHER READING

Lachowecki, Walter W., Molding reinforced plastics: Tips for thin wall parts, *Plastics World*, June 1983.

Chapter 9 / Cold Mold Design

Leon R. Egg

Cold molded materials are formed or pressed in a mold at room temperature and subsequently cured in ovens. Hence the term *cold molding*.

In the early nineteen hundreds, when this process was in its infancy, the molder compounded his materials according to his own formulas. He then processed them by proprietary methods.

To insure that molds were built to suit the peculiarities of his compounds, he maintained his own mold shop, with personnel trained by the metalforming and die-casting industries. The molds were built from the ground up, using special castings, etc., for each individual mold. There was little or no standardization.

The advent of commercial suppliers of standard frames and other components, and of specialized mold shops, changed this practice. The designs became more standardized and more economical to build.

The early compounds used bituminous and portland cement binders, with some use of phenolics. Later melamine binders made their appearance. At the present time the bituminous binders are being phased out, due to fire hazards in their preparation and restrictions as to air pollution.

The fact that these materials have poor flow, and must be distributed in the mold as uniformily as possible prior to compressing, becomes an important consideration in the design of the mold. The molds are of fully positive design.

For the most part, molds are of the semiautomatic type, being securely bolted in the press. The presses are generally hydraulic, although toggle presses equipped with oil cushions have been used. Punch presses have been used for some jobs.

The pressing or molding operation is fast, as the cure time is not a factor. The charge is weighed before placing into the mold cavity and distributed

477

properly before closing the mold. The design of some parts lend themselves to the use of a "rake" or mechanical slide for loading the mold (Fig. 9.1). In these cases the cavity controls the volume and weight of the material used. It therefore becomes necessary for the material to be of a free-flowing granular type. A typical cold mold is shown in Fig. 9.2.

With parts of fairly simple design, the cavity can be filled with a slide which moves forward to fill the cavity and at the same time pushes the molded part off the mold into a receptacle in the front of the press. This device can be incorporated in the pressing action of the press, thus making the molding operation completely automatic. It is necessary only to fill a hopper in the rear of the press and remove the molded parts from the front.

After the part is formed, the press is opened, the part is removed and carefully placed on trays in a "soft" condition, to be taken to the oven for curing.

Because of this "soft" condition, the area of the lower pad or piston should be the same as, or as near as possible to, the area of the cavity.

For pieces having undercut sections, the mold is designed with wedges or splits. In some cases these wedges are removed from the mold with the molded part, and released from the part by hand. Other wedges or splits are so hinged that they can be opened manually while still in the mold and then the molded part removed. This type of split can be made self-opening, facilitating the removal of the part.

Variation in the amount of the charge for each pressing will introduce a difference in the thickness or build-up dimension. A tolerance of plus or

FIG. 9.1. The material is raked into cavities from hopper at back of press for a volume load in this four-cavity cold mold.

FIG. 9.2. A cold mold in its press. (*Courtesy Garfield Mfg. Co., Wallington, NJ*)

minus 1/64 in. must be allowed on the thickness dimension of cold mold parts.

Through holes are produced by mold pins secured in the lower or bed plate, extending to the top of the jacket or cavity. The upper ends of these pins are provided with a generous radius to facilitate entry into the plunger which telescopes over the pins as the mold closes. The top plunger should be provided with clean-out holes running horizontally, front to back of the mold. As these clean-out holes take care of the excess material flowing up around the pins, they should be smooth and polished, and readily cleaned by the operator.

The plunger plate, which serves to hold the plunger in position, should be at a minimum distance of 1 in. from the top of the jacket when the mold is in the closed position. This distance is maintained by small pads fastened to the under side of the plunger plate. This is done for two reasons: first as a safety consideration, as it minimizes the danger of catching the operator's hands between plates as the mold closes; second, the plunger is prevented from decending too far into the cavity, causing damage to both plunger and cavity.

Holes which run at right angles to the main vertical holes, present some difficulty in molding. They should be eliminated if possible or modified to permit the use of a cross slide mechanism. As a last resort, drilling after curing may be the answer.

All vertical walls, with the exception of the perimeter wall, should have approximately a 2 degree taper. As the outer walls are formed by the jacket, it is economical in mold construction to omit any taper. This is also true of the vertical pins as the part is easily stripped off these pins because of the large area of the lower pad.

Cold-molded parts are pressed to shape only. Since they are "soft" when extracted from the mold, it is not practical to use small knockout pins such as are used for the hot-setting materials. Large ejector pads are commonly used to distribute the ejection force over a sufficiently large area, thus insuring that the piece will not be distorted. The design of cold-molded parts must be such that the outline of the pad or extraction member will not be objectionable. Ejector pads should cover most of the area at the bottom of the cavity. In many cases it is possible to use the entire bottom of the cavity as an ejector as shown in Fig. 9.3. The extraction or ejection bar must offer sufficient "lift" to raise the entire piece above the top of the mold, where it can be handled easily and removed without distortion. Shown in Fig. 9.4 is a schematic diagram of a typical cold-mold design.

For parts having depressions or projections on the under surface, i.e., the surface carried on the ejector pad, auxiliary ejector pins are used. These pins are made with a shoulder at the upper end. The surface of the ejector pad is counterbored to take this shoulder. The smaller portion of the pins extends below the bottom surface of the ejector pad a short distance. These pins are located at positions advanageous to the depressions or projections.

When the ejector pad is raised to its full height, the lower end of the pins projects below the pad, leaving a space between the end of the pin and the bed plate of the mold. A plate or fork is inserted into this space and the ejector pad lowered slightly. As this is done the pins are held up by the plate or fork, so that the piece rests on the top of the pins when the pad is lowered. The molded part can then be removed with minimum distortion. Figures 9.5, 9.6, and 9.7 illustrate this action.

FIG. 9.3. The ejector plate in this single-cavity mold covers the entire bottom surface of the molded part.

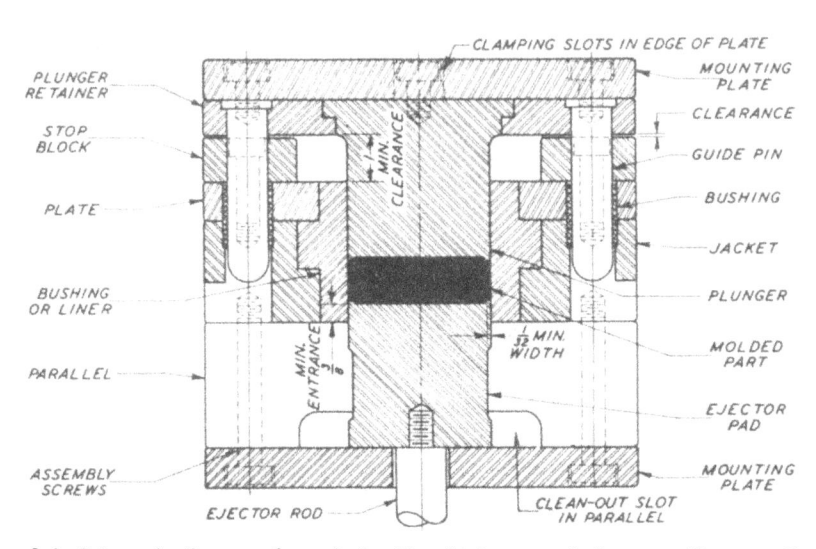

FIG. 9.4. Schematic diagram of a typical cold mold for a round, flat part. (*Courtesy Garfield Mfg. Co., Wallington, NJ*)

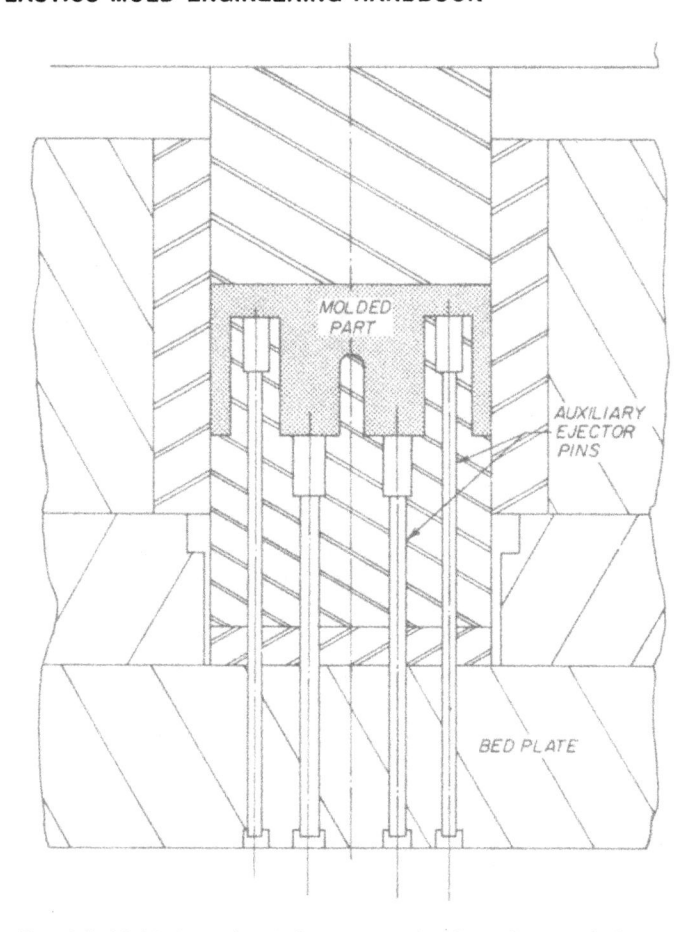

FIG. 9.5. Molded part just before top pad is raised after part is formed.

A further refinement of this method is illustrated by Fig. 9.8. In this method the shoulder at the upper end of the ejector pins is omitted, as the lower ends of the pins are secured in a floating plate shown in the diagram. The mold proper is raised sufficiently above a clamping plate, to accommodate the ejection mechanism. The accelerated cam device, now furnished by commercial suppliers, is mounted above the clamping plate on a plate free to move up and down actuated by a long rod. The lower end of this rod is connected to the lower ram of the press. The upper end is secured in the bottom of the pad. As indicated in the diagram, when the ram moves the rod upward, the pad, as well as the plate supporting the cams, also rises. When the molded part is slightly above the top surface of the cavity, the cams strike pins in

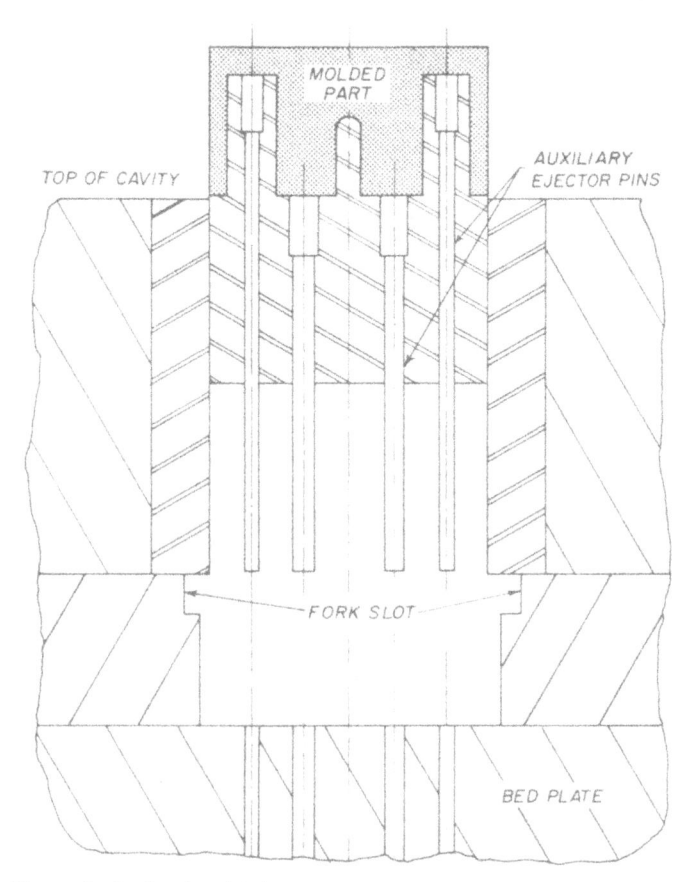

FIG. 9.6. Top pad raised and molded part raised on lower pad above the top of cavity.

the under side of the bed plate, causing the cams to lift the ejector pin plate. This action further raises the molded part above the cavity on the ejector pins where it can be easily removed. On the return stroke of the lower ram, the ejection mechanism (cams, pins, plates, etc.) is returned to the proper molding position. This method eliminates the "fork," which has to be handled by the press operator, thus shortening the time of ejection and removal.

Because of uncontrollable variables found in the processing of cold-molded parts considerable dimensional allowance must be provided. The "build-up" tolerance has been previously discussed. The other dimensions should follow the tolerance schedule shown in Table 9.1.

High carbon and high chrome content steels are used extensively in the

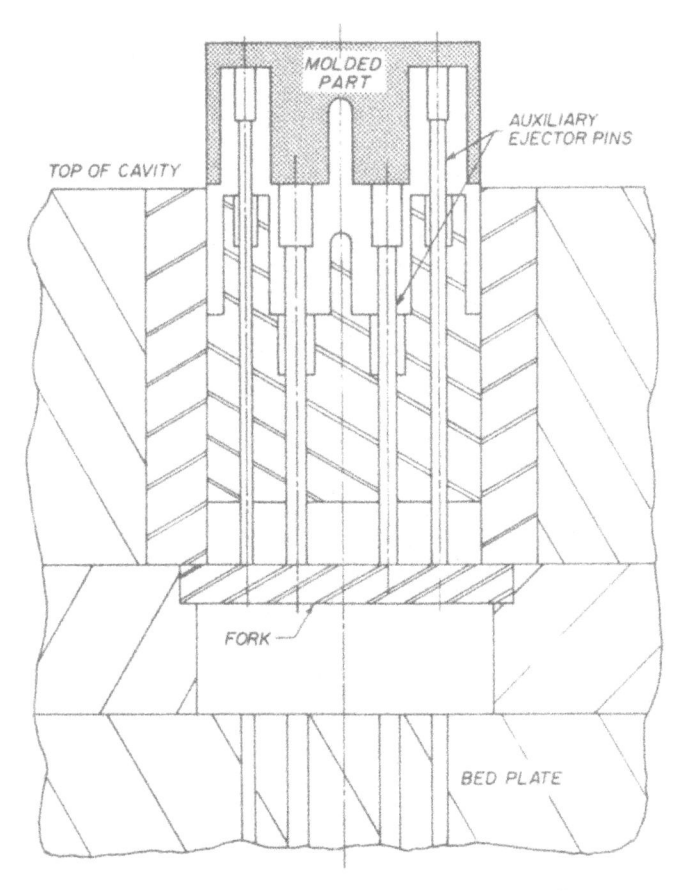

FIG. 9.7. Lower pad dropped so that molded part rests on the auxiliary ejector pins ready for removal of part.

TABLE 9.1. Cold-Mold Tolerances.

Nominal Dimension (in.)	Tolerances (Plus or Minus) (in.)	Nominal Dimension (in.)	Tolerances (Plus or Minus) (in.)
¼	.009	5	.023
½	.011	6	.028
1	.014	7	.032
2	.016	8	.036
3	.018	9	.040
4	.020	10	.048

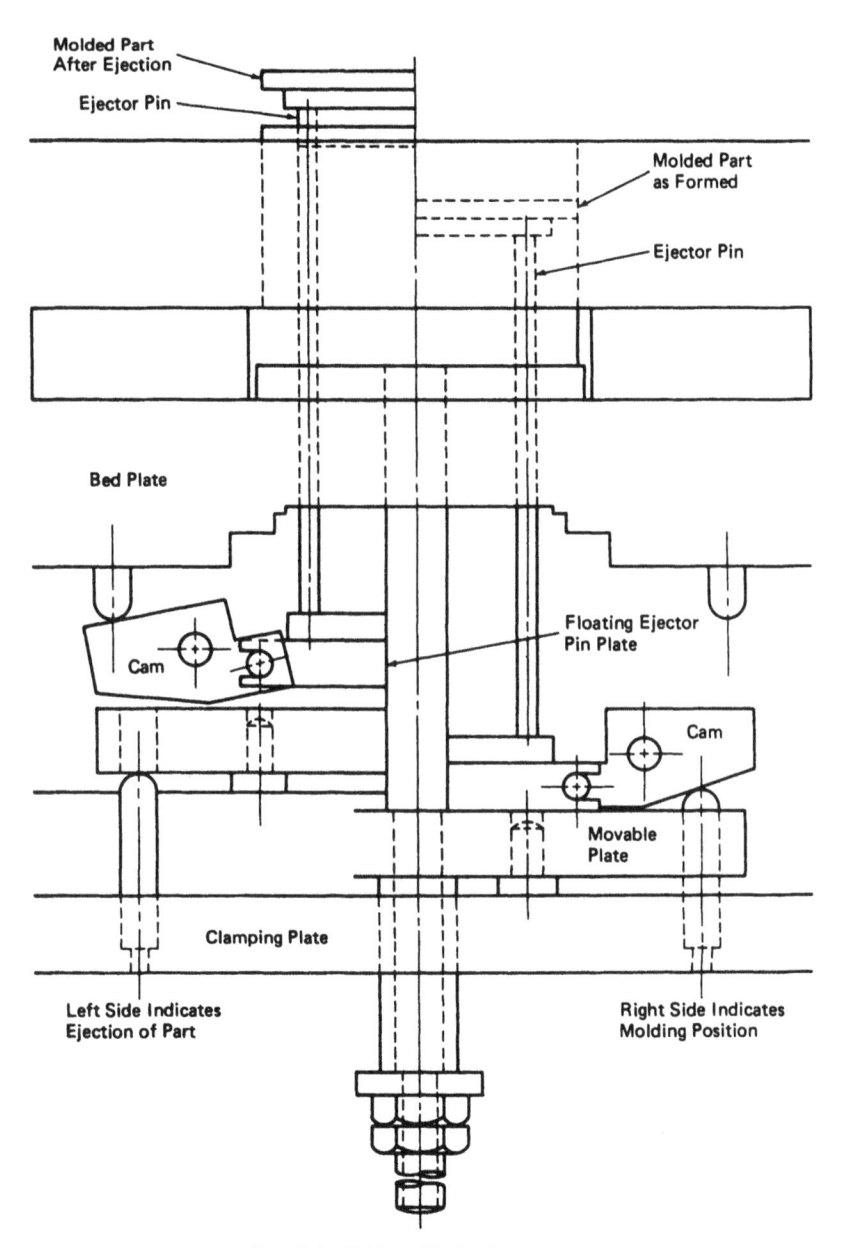

Molded Part
After Ejection

Ejector Pin

Molded Part
as Formed

Ejector Pin

Bed Plate

Cam

Floating Ejector
Pin Plate

Cam

Movable
Plate

Clamping Plate

Left Side Indicates
Ejection of Part

Right Side Indicates
Molding Position

FIG. 9.8. Cold mold ejection system.

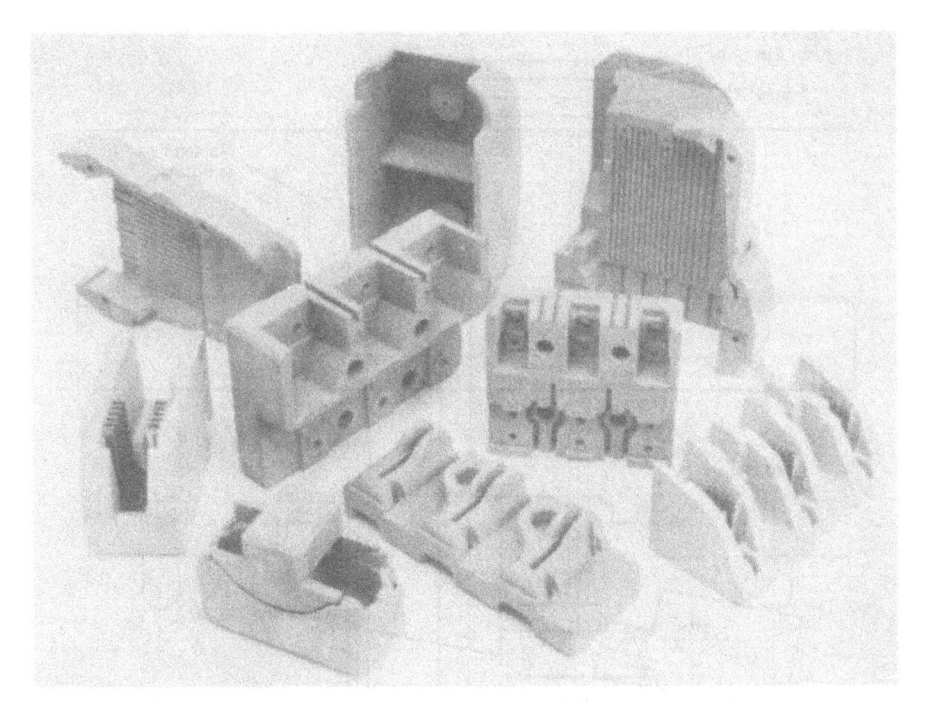

FIG. 9.9. Typical cold molded products. (*Courtesy Garfield Mfg. Co., Wallington, NJ*)

construction of cold molds. The chrome nickel steels provide needed strength to withstand the pressures used in cold molding. As cold molding compounds are highly abrasive, it is advisable to use steels which offer maximum wear resistance. Chrome plating the molding surfaces is extensively used for its wear resistance and better release of the part from the mold. Molds should be designed in such a way that they can be disassembled easily for repair and replacement of worn parts.

Designs of cold molds vary to some degree to suit the molding material used and considerable proprietary art is exercised in the cold mold shops. Typical cold mold products are shown in Fig. 9.9.

REFERENCE

DuBois and John, *Plastics*, 6th Ed., New York: Van Nostrand Reinhold.

Chapter 10 / Extrusion Dies and Tools for Thermoplastics

Revised by Sidney Levy

Statistics for the plastics industry show that the major portion of plastics materials pass through an extruder at some stage in processing enroute to the final product. The extruded material may be only the raw material for another molding process, such as injection molding, blow molding, compression molding, or another extrusion. Either thermoplastic or thermosetting material can be and is processed through an extruder at some stage of producing molding material. However, this chapter will be confined to the extrusion of finished products with the provision that a particular finished product may be an intermediate step to still other finished products, for example, extruded film that eventually becomes a trash bag, a wastebasket liner, or a sandwich bag.

The extrusion process is fundamentally a simple process, especially when we describe it as "the forcing of a softened plastic material through a shaped channel in a die, and through the die orifice." In practice, the use of sophisticated tooling makes extrusion one of the most versatile of the many methods for producing plastics products.

A constant cross section is the primary characteristic of an extruded shape. By the nature of the process, the length of the cross section is limited only by the auxillary equipment available to store the output in a manageable form. For example, wire coated by extruding a plastic coating can be thousands of feet long on one take-up reel. Other products are cut to length by cutters moving at the same rate as the moving extruded section. There are tooling variations (dies or auxillary equipment) that make it possible to create periodic changes in cross section and, thus, extend the range of products which can be produced by extrusion. Typical extruded products include sheet, rod, tube, film, wire jacketing, and profile shapes. The dies and the take-off tools for each of these products are described in this chapter. It is also possible to combine plastics with other material while extruding the plastics. Such combinations as plastic coated paper, plastic coated cloth, lamination of multiple layers of plastics (including more than one variety of plastic in the various layers) are examples of combinations.

One or more of the references at the end of the chapter details the auxillary equipment. Other references detail the extruder as a machine, and include the geometry of the screw(s). Thus, there is no need to discuss that particular material in this text. This chapter is intended to provide the basic principles of designing *an extrusion die* (previously referred to as the "orifice"), which is always located at the exit end of the extruder barrel. The rotating screw propelling the plastic melt is contained in a tube called a *barrel*. Our definition of *tooling* includes the die, any post-shaping units, handling equipment, and a suitable melt delivery system. The *melt delivery system* includes all of the system components preceding the die or the orifice.

The usual melt delivery system is a single-screw extruder. Its function is to supply a continuous nonfluctuating stream of plasticized material at the correct temperature and pressure to the die. The single-screw extruder is used for over 75% of all extrusion. The remainder is done through twin-screw or other types of machines designed to perform on some specific plastic material, such as PVC.

Inasmuch as the product made by extrusion results from several tooling elements and pieces of equipment, operations are separated into several categories. One category is called "standard products," and consists of such items as sheet, film, tube and pipe, rod, monofilament, and wire coating. A second category is profile shapes and specialty extrusions. The third category is a "catchall" miscellaneous group of operations, which includes the use of extruders in conjunction with injection molding, blow molding, and compounding.

This text will discuss some examples from the first two groups. It will also illustrate the approach to tooling and the manner in which the dies and tools are designed for high quality and high rates of production. Because each material and each product has specific melt requirements, the extruder screw becomes an important part of the total design requirements for an extruder and its related melt conditioning equipment.

GENERAL CONCEPTS OF DIE DESIGN

An extrusion die is a shaped orifice made in a block of steel or other suitable structural material that is wear- and corrosion-resistant. The function of the die is to receive the melt stream emerging from the extruder screw and to reshape it to the required form. In some instances, this is the final shaping operation, and in others, post-die shaping equipment completes the shape to the required form. An example of post-die forming is the three-roll stack used to size and cool the plastic sheet coming from the die.

Designing dies requires an accurate knowledge of flow characteristics of the particular plastic material to be processed. The designer must determine the precise shape of the flow channel(s) in the die, and the exact shape of the exit orifice. Unfortunately, the amount of data available to a designer is quite limited. In fact, much of it is still considered proprietary information by die-makers or by extruder operators. Fortunately, the authors have persuaded some of these craftsmen to permit publication, for the first time, of many of the basic principles developed over many years. The information in general availability is flow data for the material to be extruded through the die. Inasmuch as many errors result from "overlooking the obvious," we recommend that each designer collect all possible information concerning the various plastics materials for which die designs may be required.

Further, the designer must glean a knowledge of *how* plastic flows under pressure. From these two pieces of information—the characteristics of material to be processed and how it flows—the die passage design can be tentatively generated so that the relative flow from each portion of the die lips (orifice) corresponds to the extrudate dimension required at that particular point.

In order to accomplish the exact dimension, it becomes necessary that the pressure drop along the flow path(s) in the die be equalized. The result of a "correct" design will be precise relative rates of flow from all regions of the orifice of the die.

ANALYTICAL RELATIONSHIPS: FLOW RATES VERSUS PRESSURE

Analytical relationships for flow rates versus pressure, which apply to the flow of plastics materials, are shown in Table 10.1. However, this table is a strict application to Newtonian fluids, where the resistance to flow is independent of the shear rate, and the viscosity is a constant—a condition that does *not* prevail for polymers. At higher shear rates, the polymer molecules (and short fibers in filled plastics) become aligned, and the resistance to flow is reduced. Aligned molecules and/or fibers, coupled with reduced resistance to flow, translates to a *lower* viscosity. In order to use the mathematical approach, it becomes necessary to adjust for this reduction in viscosity as the shear rate increases. The viscosity of the plastics material is also temperature-dependent. The flow through the die causes shear in the material and generates frictional heat. The amount of heat generated must be considered in the design of the die. High shear regions in the die will result in lower viscosity, which in turn will result in higher relative flow rates of the plastics material.

TABLE 10.1. Flow Rate Relationships in Die Channels.

Simplified Analytical Approach
to Die Design

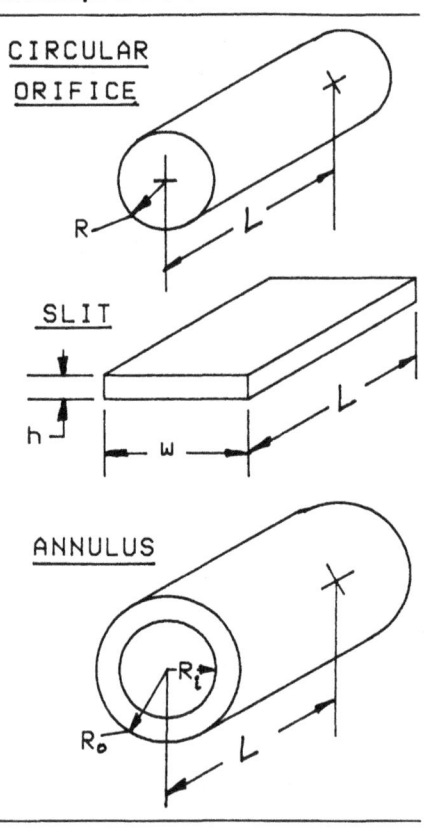

CIRCULAR
ORIFICE

The flow of melt under pressure through a die may be calculated most simply by using the Newtonian flow equation:

$$q = \frac{kp}{\mu}$$

where k is defined as follows for the geometric shapes shown at right:

SLIT

$$k = \frac{\pi R^4}{8L} \quad \text{circular orifice}$$

$$k = \frac{wh^3}{12L} \quad \text{slit}$$

$$k = \frac{\pi(R_0 + R_i)(R_0 - R_i)^3}{12L} \quad \text{annulus}$$

ANNULUS

and where all dimensions are in inches, and

p = pressure drop through the die
μ = viscosity of melt flowing through the die

Shear rate for round channel = $4q/\pi R^3$

Shear rate for flat channel = $6q/wh^2$

q = throughput in cubic inches per seconds.

Data on the previously mentioned criteria is reported in the literature on curves such as those shown in Fig. 10.1. If sufficient data are available on a specific material, it is possible to use the equations in Table 10.1 to accurately define the shape of the flow channel required to produce the desired extrudate. Computer Aided Design (CAD) programs are becoming increasingly available to perform analysis for die design. Most of the CAD programs currently listed in trade literature are for sheet and tubing products. Even computer analysis needs evaluation and modification based on prior experience. Analytical judgment is also required in using the simplified relationships between channel depth, channel length, and flow rate as a guide to the "correct" design. Practical experience is most needed when designing a die for profile shapes where the flow can be quite complex.

Another complicating factor in die design is *die-swell*. When a plastic material emerges from the die lips (orifice), the material swells to a dimen-

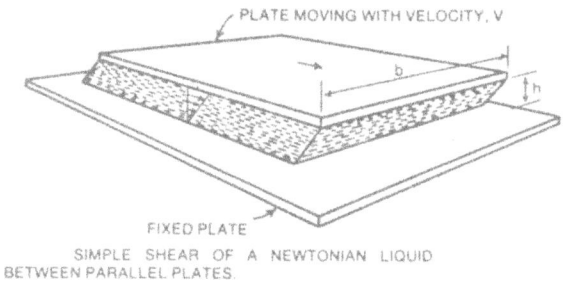

SIMPLE SHEAR OF A NEWTONIAN LIQUID
BETWEEN PARALLEL PLATES.

VINYL RESINS   B.F. GOODRICH CHEMICAL COMPANY
GEON * 8700A
(RIGID POLYVINYL CHLORIDE)

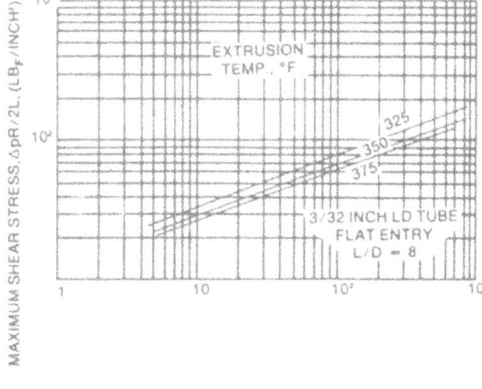
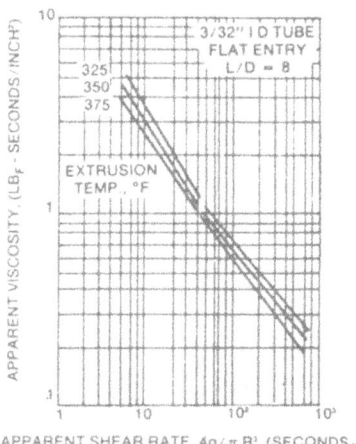

FIG. 10.1. Analytical relationships for flow rates versus pressures. Application is to Newtonian fluids and is not directly applicable to polymers. (*Courtesy B. F. Goodrich Geon Vinyl Div., Cleveland, OH*)

sion larger than the die opening. The swell occurs as a result of energy stored in the plastic material, i.e., *elastic energy*. The amount of die-swell will vary from one material to another, and will also vary with changes in the operating conditions for the extruder (the speed of screw rotation, temperature of the barrel or screw, backpressure, etc.). Depending on the material and the melt conditions, die-swell will range from less than 5% to over 100% of the orifice dimension. Once again, we repeat that there is no substitute for practical experience. Estimating of die-swell resolves itself to an "educated guess."

The design procedures use these relationships: flow will vary as the cube of the channel depth (*D*) and inversely as the channel length (*L*) with any given upstream pressure. These relationships are used in many ways, ranging from the "rule-of-thumb" calculations of the flow character of a particular die channel to the exacting computer-generated design analysis that defines the shape of the die opening, the orifice dimensions, and the precise shape of the interior flow channels in the die.

To illustrate these basic design concepts, we now proceed to the tooling for a few specific and typical products to show how die design develops, and how dies are operated in conjunction with size and shape control devices to produce products to required sizes and specifications.

SHEET-DIE DESIGN

Sheet plastics are made in thicknesses ranging from 0.010 in. to 1.0 in. or more, and in widths of 120 in. or more. The die orifice is a slit of the appropriate height and width (plastic sheet is nearly always extruded in a horizontal position—thus *height* refers to the *thickness* of the sheet). The design effort is directed toward producing a suitable internal channel inside the die to achieve "proper" distribution of the melt, entering from the screw-tip, to the die-lips in order to produce a uniform thickness of material exiting from the die-lips and along the full width of the die. (*Note*: We use the term "die-lips" when discussing "sheet" whereas we use "orifice" when referring to "shaped" output.) To assist with the regulation of material flow, the die is equipped with adjustable restrictions along its width. These restrictors permit adjusting for spot changes in sheet thickness by changing the flow pattern in the die. In most sheet operations, the output of the die is fed to a three-roll calender stack where the entering nip-roll spacing controls the final thickness of the sheet. Without the restrictors, as mentioned above, the excessive variations along the width results in the nip-roll forcing the plastics material sidewise instead of stretching it. The side movement of the plastic produces an unacceptable "smearing" of the sheet and could result in reduced physical properties.

The simplest manifold in use with sheet dies is the T-shape shown in Fig. 10.2. The melt enters from the extruder screw and on the leg of the "T." The melt is then distributed to both sides on the cap of the "T." The downstream side of the main channel in the manifold is open, but with a restriction, to feed the melt to the die-lips.

The T-shape manifold may be an erratic performer in sheet extrusion because of the variable pressure drop along the main manifold. The rate at which the material flows is a function of the pressure drops along the flow path. Thus the longer flow path to the end of the "T" manifold results in a lesser flow of material as it approaches the ends of the manifold. Correction of this unequal flow is usually achieved by a narrow opening at the center of the die and a gradual increase to the wider opening at each end of the die.

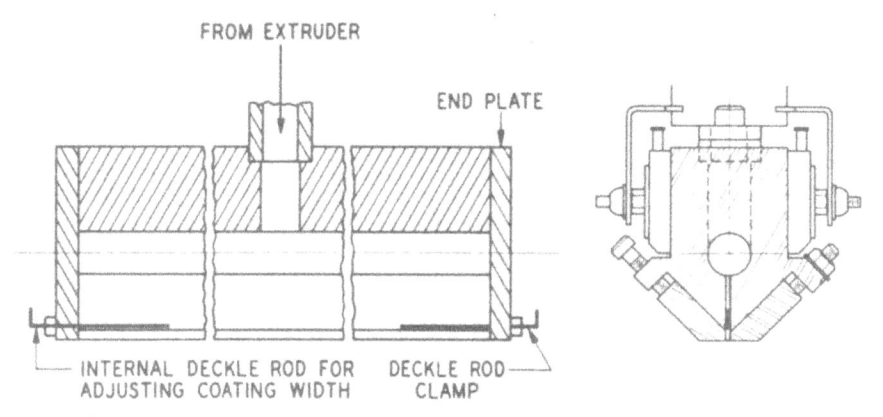

FIG. 10.2. T-type coating die with width adjustment. (*Courtesy Phillips Chemical Co., Pasadena, TX*)

Choke elements (restrictors) are needed and used to help balance the flow to the die lips.

The "coathanger" configuration of Fig. 10.3 is the most frequently used design for the extrusion of sheet products. In this design, a triangular pre-land restriction is introduced into the flow path. The distributor manifold is shaped like a "coathanger," thus the descriptive name. The plastic material first flows into the manifold behind the pre-land section. Then the plastic melt moves across the pre-land in the direction of the die-lips. The pre-land

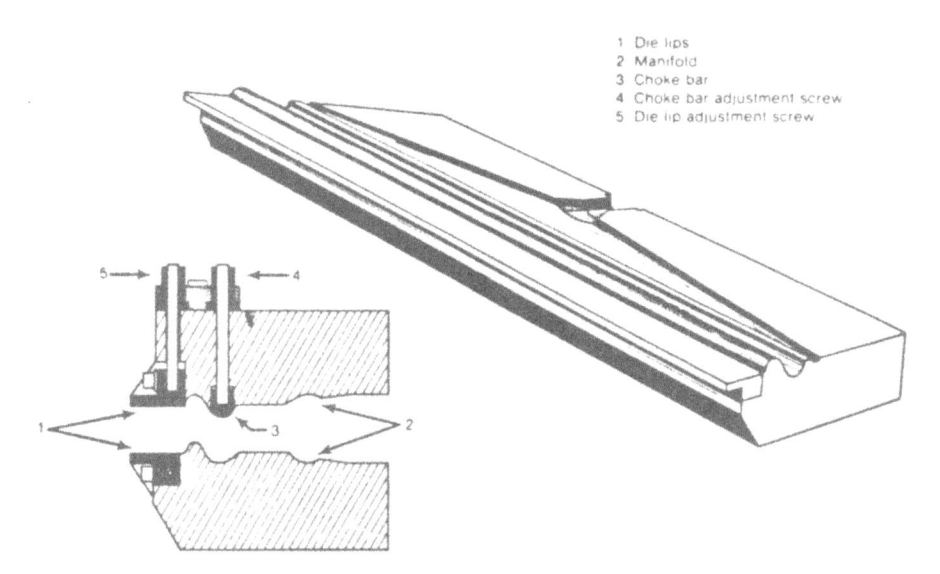

FIG. 10.3. Coat-hanger manifold sheet die. Lower half and a cross section.

distance to the die-lips is longest near the center of the die. It decreases to its shortest distance at each end of the die. The pressure drop across the land is equalized to a nearly constant pressure as the material emerges from the die-lips. By appropriate selection of the angles of the pre-land, and the gap in the pre-land area—usually about two to three times the die opening—to match the polymer rheology, a balanced flow to the die-lips is achieved. The "coathanger" will give better results than the T-shape described in the previous paragraph. The selection of the "coathanger" becomes more attractive as the sheet thickness increases and as the sheet width increases. One of the references details methods for calculating the channel in the "coathanger" manifold.

Fig. 10.4 illustrates a die equipped with a choke bar assembly used to compensate for flow variations that may exist across the width of the die. The final adjustment for extruded sheet thickness is made with the "die blade adjustment screws" to distort the "adjustable die blade." "Distort" means the adjustable die blade may not be straight, and it may not be parallel to the fixed die blade.

Two additional refinements in the design of dies for sheet extrusion are used when a material has a complicated melt rheology, and when the material is highly sensitive to shear. One of these modifications is the tapering of the manifold diameter to a smaller diameter as the distance from the melt entry point increases. Tapering provides a variable pressure drop condition that improves the distribution of the melt along the die width. The second modification is to create the manifold as a curve rather than a straight line.

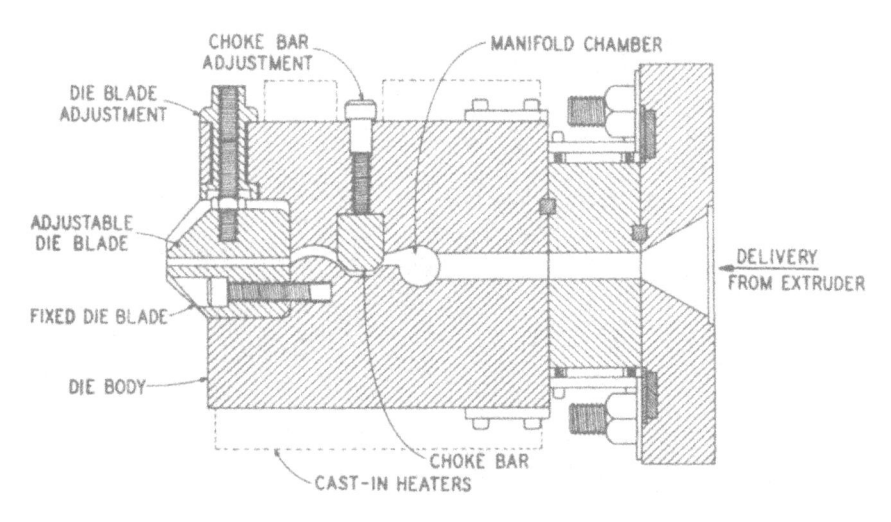

FIG. 10.4. Cross section of sheet extrusion die. (*Courtesy Phillips Chemical Co., Pasadena, TX*)

A curved manifold will result in a curved pre-land barrier. Choosing the proper curve, and perhaps including taper at the same time, can enable a closer match between the pressure drop and the melt rheology. Sheet dies with curved and tapered manifolds are called "fishtail" dies because of the resemblance of the manifold to a fishtail shape. Another of the references at the end of this chapter describes the finite element flow analysis used to determine the shape of the manifold, and the size and shape of the pre-land area as applied to a specific extrusion material.

An improved, but more expensive, method for making the die-lip adjustment is illustrated by Fig. 10.5. A flexible upper lip is used instead of the adjustable die blade of Fig. 10.4. The upper die-lip has a very thin section capable of being deflected by the downward adjustment of the screws along the top side. The "normal" condition of the top lip is flat and parallel to the bottom lip, and the adjusting screws prevent the upward movement of the lip when it is under pressure from the extrudate. The major advantage of this design is the lack of a joint where the hot melt can be trapped and decompose. A secondary advantage is the ease of adjusting the die opening to control the extrudate thickness along the width of the sheet.

Figure 10.6 illustrates how the choke bar can be eliminated and replaced by a flexible pre-land section of the die. A flexible pre-land section is particularly useful when materials highly sensitive to variations in heat, such as rigid PVC, are being extruded. The pre-land space is adjusted by increas-

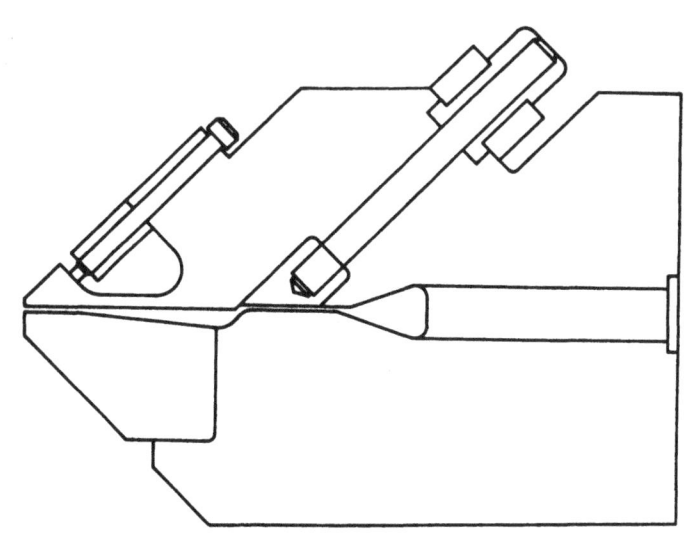

Fig. 10.5. Flexible lip sheet die with 45° choke bar. (*Courtesy Extrusion Dies, Inc., Chippewa Falls, WI*)

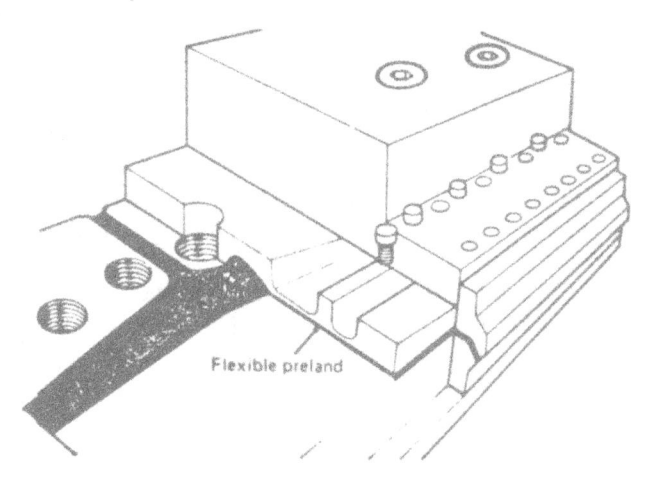

FIG. 10.6. Internal construction of fishtail sheet die with flexible preland.

ing or decreasing the pressure of the individual screws along the width of the die.

A clear line of demarcation between the tooling and its operation is lacking, as we consider the many innovative ways in which adjustment can be made by increasing the sophistication of the individual elements of the die or its controls. For example, Fig. 10.7 illustrates the AutoFlex® die design. The die bolts, which press on the flexible lip of the die, are fixed at the head-end. Electrically heated "bolt-heater blocks" are applied to the indi-

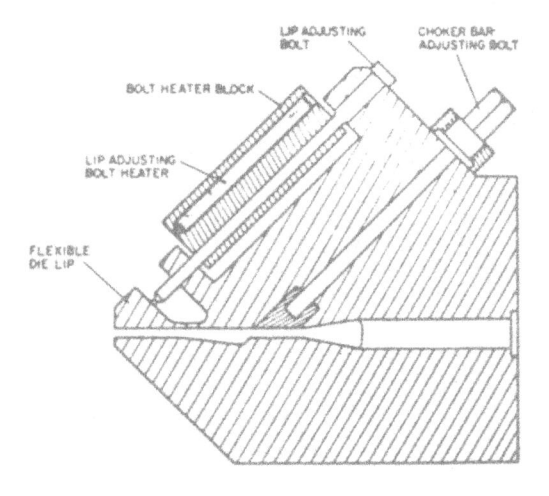

FIG. 10.7. Auto-Flex® sheet die. The die opening is controlled electronically by adjusting heat input to the die adjusting bolts.

vidual bolts. Each of the electric heaters is controlled by a closed-loop control activated by the sheet thickness emerging from the die. Such controls allow the expansion or contraction of the pressure bolt to continuously and automatically adjust the die opening to compensate for a variation in sheet thickness at any given point along the entire die width. Close control of the thickness of any sheet coming from the die results in close control of the product dimensions.

The sheet dies are equipped with heaters to heat the entire die to operating temperatures, and to maintain the dies at a given temperature. Figure 10.8 illustrates the application of cartridge heaters usually employed for this purpose. Most catalogs of suppliers of electric heater elements also contain the formulae for calculating heat losses, watts per sq. in., and other information needed to determine the ''correct'' size, placement, and wattage needed to raise the die to operating temperature in some ''reasonable'' time and to maintain a given temperature without either the die overheating or the electric heaters excessively on-off cycling. The authors recommend that each designer study the catalog formulae and follow the suppliers' recommendations.

An extruder for thick or wide sheets is quite large equipment. Obviously a die made of steel and 10 ft. wide (long) is a heavy piece of equipment. Dies are bolted or taperlocked to the end of the barrel. Most dies are also equipped with a support structure to carry some of the die weight and relieve the stress on the main extruder. The die designer will probably be called upon to design some of this auxillary structure.

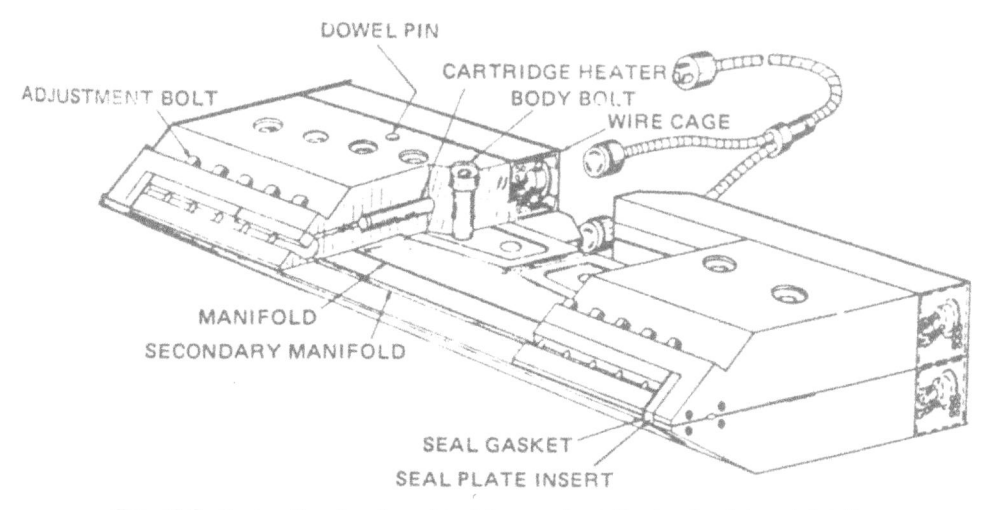

FIG. 10.8. Perspective drawing of coat-hanger sheet die showing internal details.

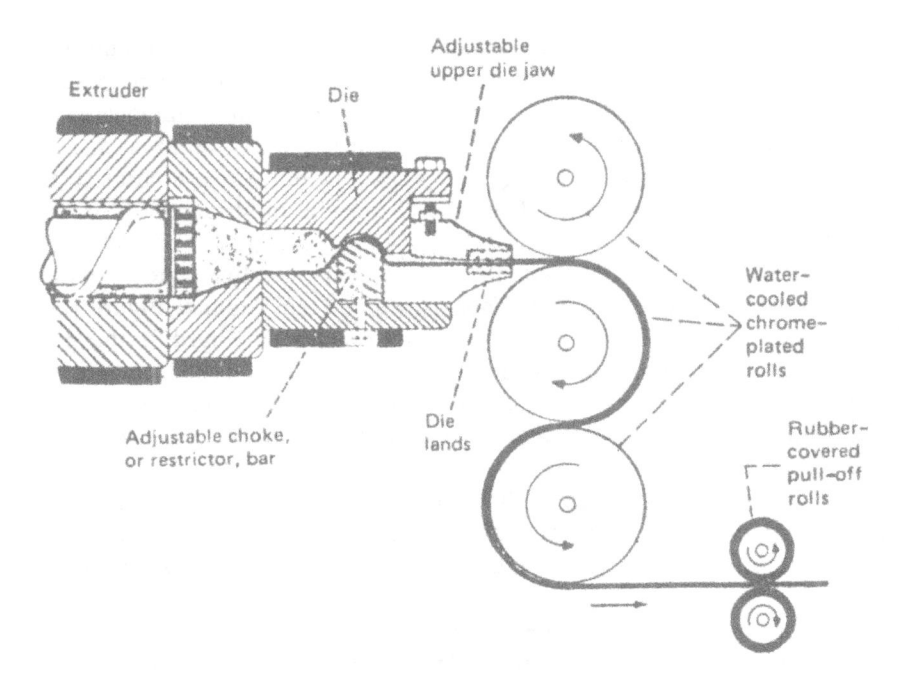

Fig. 10.9. Cross section of sheeting die and take-off unit for extruded sheet.

We have previously mentioned the "three-roll stack" as part of the take-off mechanism for extruding sheet stock. Figure 10.9 illustrates a typical unit as part of a production line. The three-roll stack performs at least three functions in producing the final sheet. First of all, the top rolls act as a calender to reduce the extrudate to the "correct" thickness. A portion of the thickness reduction is also enabled by control of the rotational speed of the rolls. At the same time, reduction is taking place, the sheet is being cooled. Roll temperature is controlled by the circulation of a controlled temperature fluid through the rolls. The functions of a three-roll stack are calendering, pulling (or stretching), and cooling. Designing and Building three-roll stacks is an art all its own. The die designer should have a good working knowledge of the limitations (or capabilities) of the three-roll stack. Most of the newer lines will be equipped with computerized control of roll speed, temperature, and spacing adjustment; and include computerized control of the extruder processing conditions of temperature, screw speeds, die adjustment, and so forth.

BLOWN-FILM DIES AND TOOLING

The basic die used in making blown film is a tubing die that produces a thin-wall tube. After the thin-wall tube is extruded, it is inflated by air pressure to form a large-diameter bubble of much lesser wall thickness, hence the name "blown film". The tooling used is the tubing die, the venturi support with its cooling ring, and the cooling tower with its nip rollers. The latter is to hold the bubble shut (but not "sealed") to contain the air pressure while pulling the bubble away from the die opening at several times the speed of extrusion. The tubing is expanded from 5 to 15 (or more) times the original diameter with a resulting wall thickness which usually ranges between $\frac{1}{2}$ mil (0.0005 in.) to 10 mils (0.010 in.) but *not* in the same bubble! Any one film must be *uniformly* blown, or else disaster lurks in the bubble or in the next product using the film. Because of the very large expansion factor (thinning of the tube wall), the design of the tubing die becomes critical in several aspects, as described hereafter. Any variation in output around the extruded tube will be amplified by the blowing process, and will produce an out-of-specification product. Even though the material may be "useable" with wall variation(s), it is almost certain that the variable wall thickness will cause problems in the winding and handling of the film.

The primary requisite for a die for blown film is that it must make a tube with a uniform wall. A tube with a uniform wall will expand to make a uniform film. The die construction or design may have any one of several variations based upon the inlet feed point as well as the manner in which the melt is spread around the bushing. Tube sizes range from approximately 1 in. to 100 in. or more. Tube-wall thickness ranges from approximately 0.015 in. to approximately 0.075 in. Exactly which diameter to choose, and exactly which wall thickness to choose depend on the blow-up ratio and on the final film thickness to be produced. The designer should be well aware of the engineering process used to determine the tube diameter, the wall thickness, and the blow-up ratio. We suggest that working with an experienced engineer is probably the best way to develop this awareness. As of this edition, most of the specific knowledge is jealously guarded as "proprietary information." Thus this text confines itself to detailing the principles of design.

Figure 10.10 illustrates one of the simpler variations. The melt is fed from the bottom (*note*: vertical extrusion), and the inner mandrel is supported by a "spider" element. The spider legs are a "necessary evil" because they support and center the spreader, but they also act as an unwanted restriction to flow. Their restriction can produce thin spots in the film unless counteracting effects are introduced. To minimize the restricting effects, the flow path through the die is equipped with several reservoir manifolds in the flow path. These manifolds eliminate the pressure drop caused by the spider legs. They also eliminate the thin spot(s) in the extruded tube. This design technique is adequate when materials with a relatively high melt index are being processed. For plastics with a low melt index, another design will be more suitable, as illustrated in the following examples.

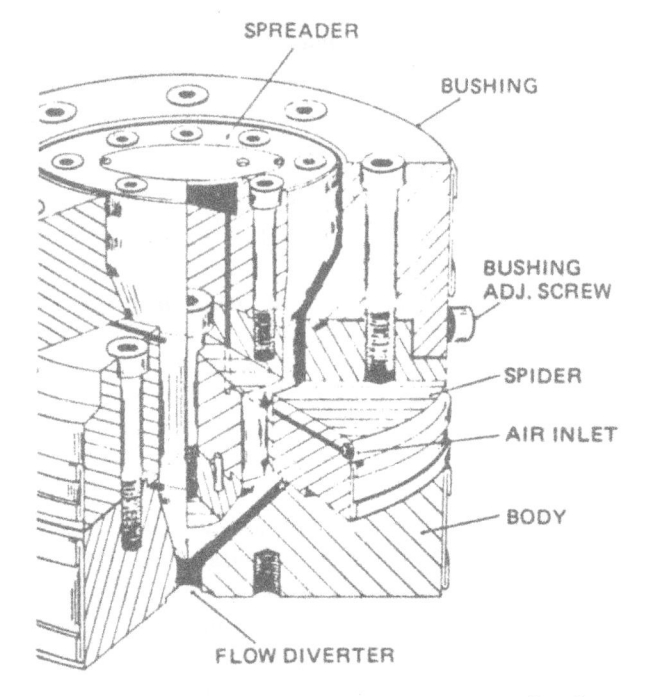

FIG. 10.10. Johnson bottom-fed spider-style blown film die.

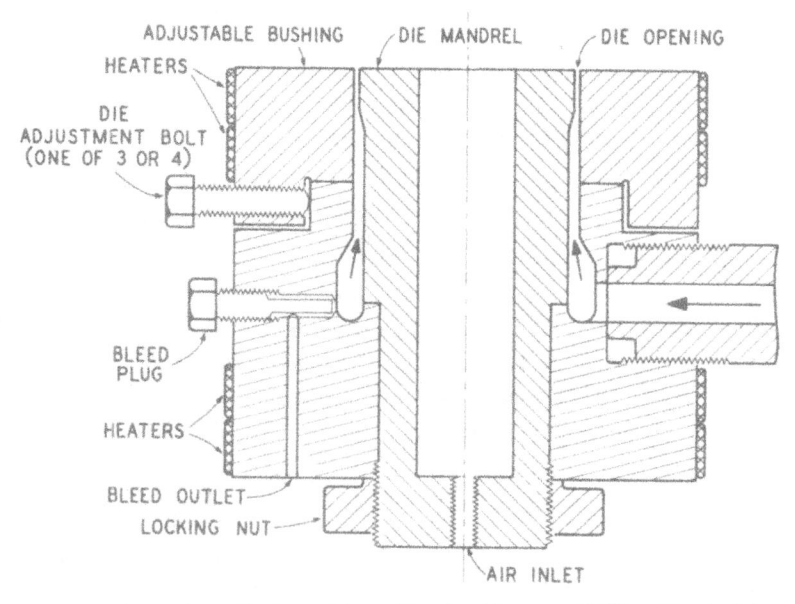

FIG. 10.11. Side-fed manifold-type blown film die. (*Courtesy Phillips Chemical Co., Pasadena, TX*)

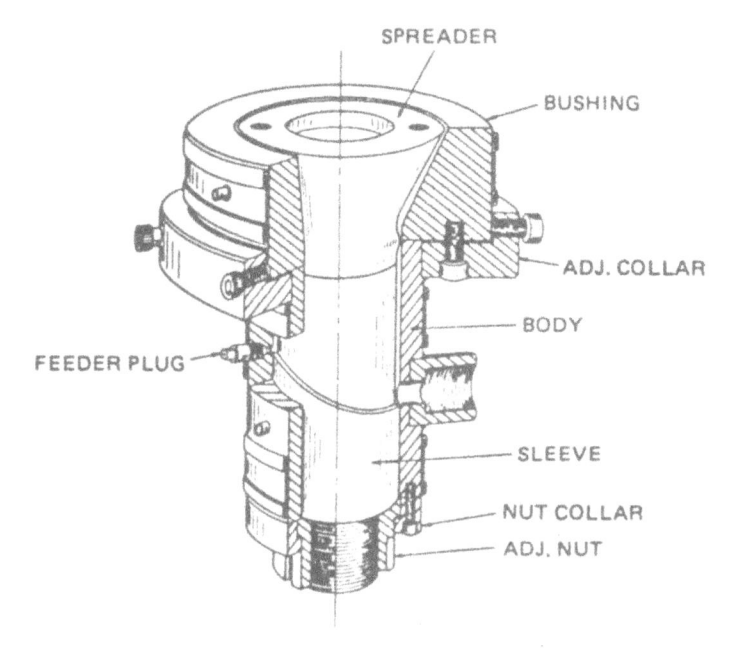

Fɪɢ. 10.12. Johnson side-fed blown film tubing die.

Figure 10.11 illustrates a "side-feed manifold die," where the mandrel is supported on one end and the enlarged manifold is used to conduct the resin completely around the mandrel to maintain a more uniform melt temperature and to level out any flow differences. Figure 10.12 shows a Johnson-type side-fed die in which the developed design of the deflector unit produces a uniform flow of the plastics melt. Still another variation appears in Fig. 10.13, in which the bottom feed is followed by a spiral-type mandrel. In this die, the mandrel is made with a spiral channel (much like a square thread on a bolt) that has a decreasing depth of spiral as the melt flows toward the die opening. Note that some material will flow over the lands while other material flows around the spiral channel at the same time to give a uniform distribution around the mandrel. Each of these designs is useful with a particular plastics material, melt range, and range of tubing sizes. In general, the larger the diameter of the tube, and the thinner the tube wall, the more sophisticated must be the conduction of the melt from the die entry point to the exit point of the extrudate.

SELECTION AND DESIGN CRITERIA

The selection of the appropriate flow-control arrangement is the first step in the design. Ring reservoirs are then incorporated in the flow path to mini-

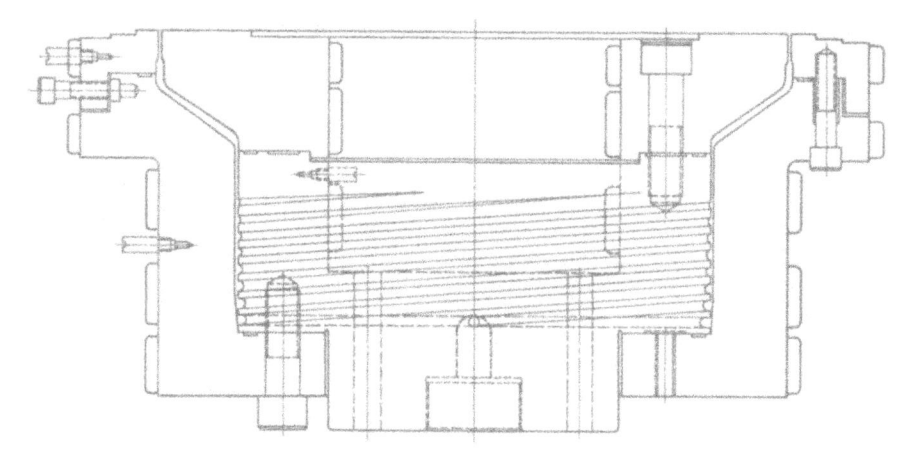

Fig. 10.13. Bottom-fed, spiral mandrel, tubular film die. (*Courtesy Egan Machinery Co., Somerville, NJ*)

mize the effects of any unbalanced flow in the feed region. The ultimate design objective is to provide a flow-path configuration that will result in the plastic melt approaching the die opening with as uniform a flow as possible. Blown-film dies are equipped with a centering arrangement for either the mandrel or the bushing to insure that the slot in the die is uniformly round and that it remains round under the pressures of extrusion. In some of the larger dies, the bushing or the mandrel may become out-of-round. To offset this possibility, the die bushing is designed with a thinned-lip section and adjusting screws to allow adjustment of the die opening so the extruded tube has a uniform wall.

The controlling factor in the design of blown-film dies is the requirement for uniformity in the wall of the extruded tube. A parallel requirement is uniformity in the melt characteristics of the plastics material being processed. The exact thickness of the tube wall is not critical, inasmuch as the blow-up ratio and the pull-speed are the primary controls for the film thickness. For the same reason, die-swell is not considered in the blown-film-die design. Fig. 10.14 illustrates the Johnson-type cooling ring and venturi unit. Film must not be brought into contact with a cool or cold surface while the film is still hot from the extruder. The cooling effect of expanding compressed air is utilized by this unit as it surrounds the just-blown bubble. The venturi unit also acts as a centering device and a support for the film in the critical first stage of cooling. However, each material and each gauge of film has its own peculiar cooling and handling character for which a cooling ring and venturi unit must be designed.

Computer Aided Design (CAD) is frequently used to design tubing dies for blown-film. It is true that the basic design is relatively simple. However,

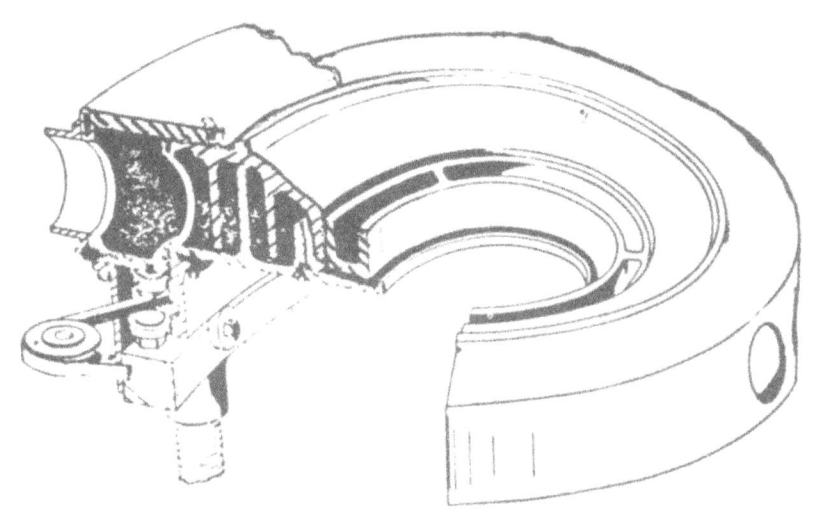

FIG. 10.14. Johnson-type cooling air ring and venturi unit.

the very large dies and the very thin tube walls introduce complexities in the design. Critical elements become heater placement, heater wattages, temperature control, and even the stresses and/or distortions created by the heating of massive die sections.

PIPE AND TUBING DIES

The distinction between *pipe* and *tubing* is somewhat arbitrary, particularly in relation to extruded plastics products. The principle distinction is the flexibility of the product. Pipe is usually *rigid*, whereas tubing is generally a *flexible* product. In any case, the designation *pipe* is used to identify those products which conform to standard pipe sizes and wall thicknesses according to American Standards Association (ASA) standards. These standards are available in any mechanical engineering handbook.

All pipe, whether made of plastics or any other material, is made to ASA standards, and is generally supplied in cut lengths for convenience in shipping and handling. However, plastics tubing would generally be supplied in a coil or roll form unless it were also of the rigid variety. Throughout this chapter we will use *tubing* to refer to pipe or tubing because a pipe is also a tube, but a tube may or may not be a pipe within the standards definition. Regardless of the precise distinction, an extrusion die design follows the same principles for either pipe or tubing.

The dies used for pipe and tubing are annular opening structures with a *bushing* that defines the outside diameter, and a *mandrel* or pin that forms the inside diameter. The bushing and the mandrel are supported coaxially by one of several different arrangments or methods, and the annular die space is fed with melted plastic (melt) by one of several feed arrangements. The choice of support and choice of feed arrangement are both dependent on the material being processed and the size of the pipe or tubing.

Products in the pipe and tubing categories range from catheter tubing with diameters from 0.050 in. to 0.100 in. and wall thickness of 0.005 in. or less, to heavy wall pipe with diameters over 60 in. and a wall thickness of 2 in. or more. Tubing, 1 to 3 in. in diameter and with walls of 0.005 in. to 0.010 in. is widely used as a package for consumer products, such as hair brushes, novelities, and toys.

Probably the most widely used die design is the straight-through flow die, in which the mandrel is supported by a spider-unit. Figure 10.15 illustrates one configuration of a spider-unit die. One leg of the spider has a hole drilled from compressed-air passage to the mandrel. Introducing air pressure prevents a vacuum for forming inside the tube as the tube moves away from the die. A vacuum inside the tube could cause the hot tube to collapse or change size. The spider legs present a localized restriction to the uniform flow of the plastic melt. Part integrity could be affected; thus countermeasures are indicated. Usually a circular manifold region is cut out of the die, and beyond the spider legs in the direction of melt flow. This manifold is used to balance the flow and minimize the effects of the spider-leg restrictions. Beyond the spider-leg limitations, this type of die is one of the simplest to construct and the least trouble to operate. Most of the dies are streamlined to minimize pressure drop and to prevent stagnation of the plastic melt. Either the bushing or the mandrel is designed to be adjusted with centering screws, to produce a uniform wall in the extruded product.

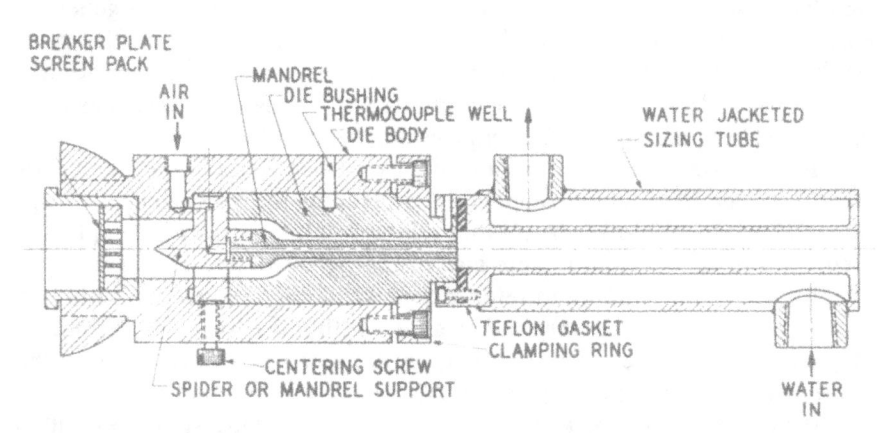

FIG. 10.15. Spider-type tubing die with sizing tube. (*Courtesy Phillips Chemical Co., Pasadena, TX*)

Determination of the "correct" streamlined shape becomes the major problem in design of the flow channel in a tubing die. The "natural" streamline varies from one plastic material to another, and the entry angles into the die may be only 20 degrees during the processing of polyethylene, but may be over 150 degrees when rigid PVC or acrylic is being processed. There is some information in the trade literature concerning the shaping of the flow channels, but the bulk of streamline design is based on someone's experience. In the text which follows, the design criteria are enumerated. Specific information is regarded as proprietary by most processors, thus an experienced engineer is undoubtedly the best source for the working knowledge needed by the die designer. Some Computer Aided Design (CAD) programs are becoming available for designing for optimum flow, serving best as a supplement to the empirical methods that determine the most nearly correct internal configuration for die.

The straight land length in the tubing die usually ranges from 10 to 20 times the wall thickness (L/D ratio). For very large tubes, or tubes with thick walls, the 10- to 20-L/D ratio becomes impractical. As a rule, the L/D ratio *decreases* as the wall thickness *increases*. The designer must calculate the mandrel size and the bushing size necessary to produce the required profile for the tubing. In practice, the tubing is stretched by the take-off mechanism in amounts varying with the need for sizing the final product. For example, a cooling tank used for vacuum calibration may require only 5 to 15% drawdown to the size of the entry hole into the finishing tank. On the other hand, a drawdown of 1000 to 2000% may be used in the production of small-diameter flexible PVC tubing. Such a large drawdown is performed because it is more economical to make larger tubing and draw to a smaller diameter than to move plastics material directly through a small die. In other words, 1000 feet of drawn product can be produced in the same amount of time one foot of small-diameter tube can be extruded.

When tubing is to be free-drawn, it is essential the mandrel and bushing sizes allow for die-swell and that the ratio between the inner diameter (ID) and the outer diameter (OD) is correct at the point where the tubing leaves the extruder die. Drawdown is a proportional effect. Thus the ratio will be preserved and the drawdown to a specific OD will also determine the wall thickness. Occasionally, there is an anomalous drawdown characteristic to a specific material. Such a characteristic will quickly become apparent, and the mandrel or the bushing, or both, must be adjusted to accomodate this peculiarity.

When tubing is sized in a vacuum calibrator, the internal size of the cooling mandrel will determine the OD. The wall thickness can be somewhat controlled by the pulling (stretching) rate of the take-off mechanism. The range of adjustment is limited by two factors: (1) excessive pulling will detach the tubing from the cooling mandrel; and (2) inadequate pulling will

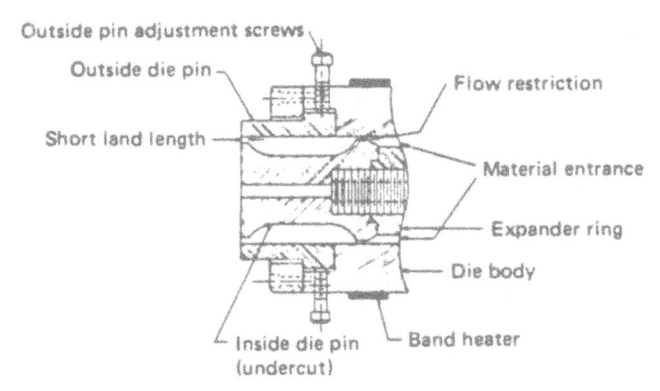

FIG. 10.16. Cross section of die designed for thin wall tubing. (*Courtesy Eastman Chemical Products, Inc., Kingsport, TN*)

cause a material pile-up at the entry into the vacuum tank. The best design will account for die-swell effects, voiding the need for the adjustment of pulling speed to control tubing-wall thickness.

The extrusion characteristic of certain materials and the elimination of the effects of material weld in the spider region make necessary the use of a variety of configurations for the tubing die. Figure 10.16 illustrates a design for producing thin-wall tubing. It has a long pre-land section and a short die land to premit better control of wall thickness. Figure 10.17 illustrates a crosshead support die, which allows access to both ends of the mandrel. This construction is useful which the interior mandrel requires heating or a cooling mandrel is used to size the interior of the extruded tube. The design includes a reservoir manifold to minimize the effects of material flowing around the mandrel and knitting on the far side. (poor knitting or welding

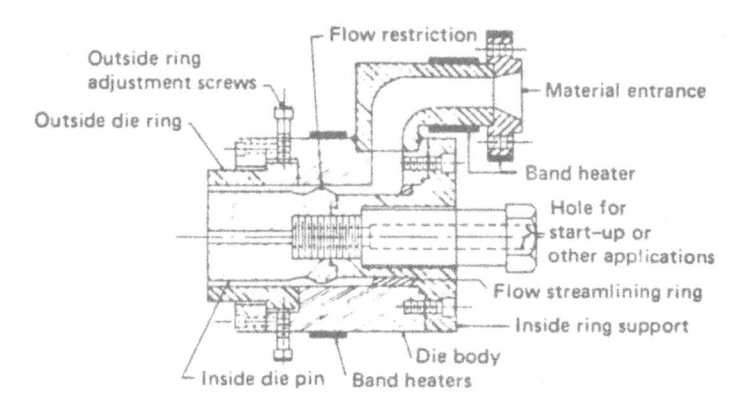

FIG. 10.17. Offset tubing die. (*Courtesy Eastman Chemical Products, Inc., Kingsport, TN*)

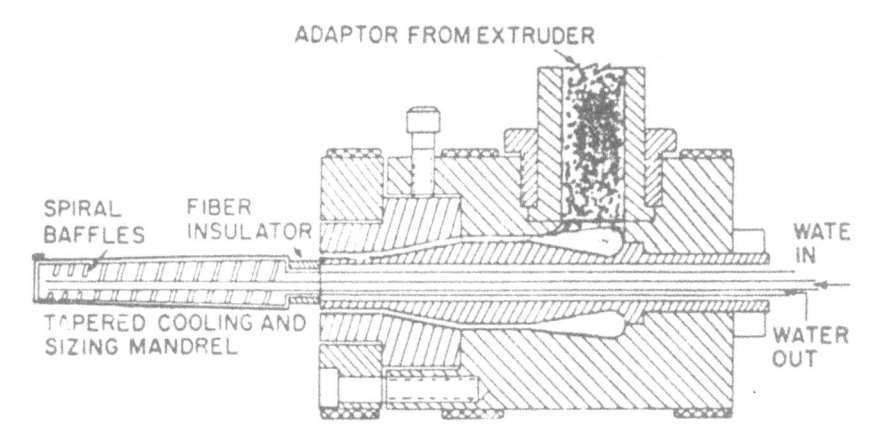

FIG. 10.18. Tubing extrusion die equipped with internal cooling and sizing mandrel. (*Courtesy Union Carbide Corp., Danbury, CT*)

can cause a weak spot in the extruded tube). Figure 10.18 shows another design, which utilizes a carefully designed flow diverter to minimize problems related to a poor weld in the flow pattern.

Figures 10.19 and 10.20 illustrate two more of the variations sometimes useful in the production of tubing. Figure 10.19 is a die similar to a blown-film die in which the helical feed unit acts as a mechanism to distribute the material melt uniformly around the mandrel. Figure 10.20 illustrates a basket support die in which the mandrel is supported by perforated tube section connected to the die structure. Both of these designs avoid the restrictions of the spider-unit and its resulting weld problems, and are particularly in-

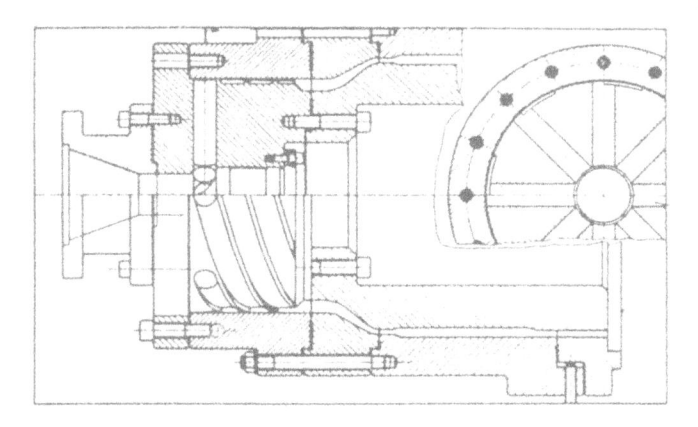

FIG. 10.19. Die with helical channel type mandrel made by Reifenhäuser. (*Courtesy Industrial & Production Engineering, Vol. 3, p. 78, 1980*)

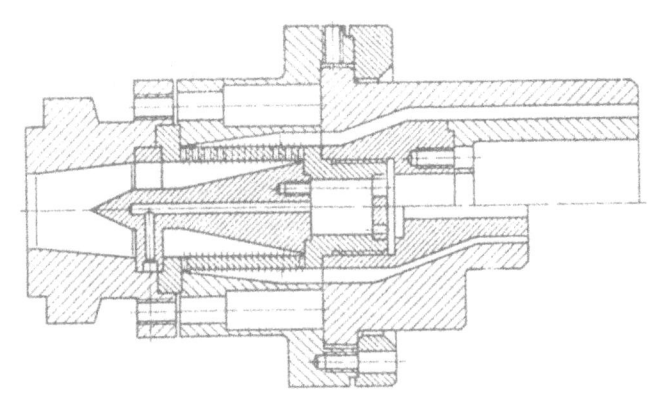

FIG. 10.20. Die with screen cylinder-type mandrel support. (*Courtesy Industrial & Production Engineering, Vol. 3, p. 78, 1980*)

dicated for extruding materials, such as polypropylene, into large-diameter pipe.

Band heaters are usually used to heat the body and the bushing of the tubing die. The designer should select heater bands that will completely surround the die to assure uniform heating without any cold spots. A gap in the heater band will probably lead to weak spot(s) in the wall of the extruded tube.

The extrusion of tubing usually requires some form of calibrator for sizing the OD or ID. A calibrator is usually unnecessary if a flexible material, such as plasticized PVC, is to be extruded. The most common calibrator unit is the vacuum sizing tank illustrated in Fig. 10.21. The extruded tube moves through a cooling mandrel that has a series of rings to control the OD of the

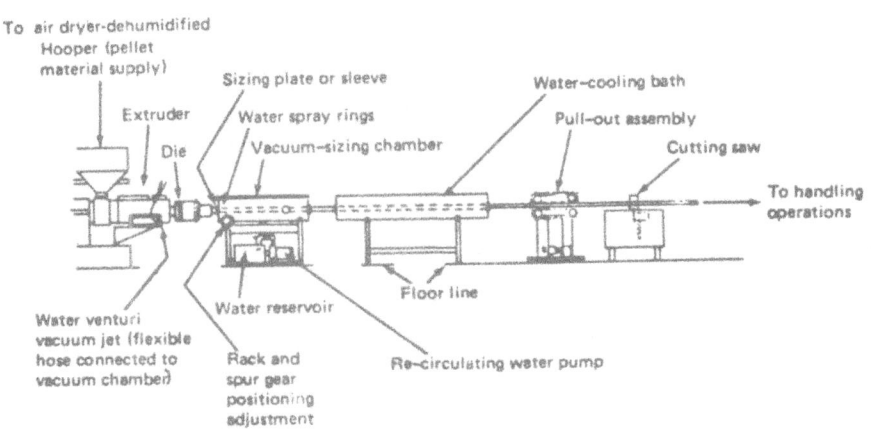

FIG. 10.21. Sketch of tubing extrusion line using vacuum sizing chamber.

tube. The mandrel is inside a tank that can be filled with water after the tubing passes through the entry bushings. After the water level is raised over the tube, a vacuum of 5 to 10 in. Hg is applied to draw the tubing to the mandrel rings. The water cools the tube as the mandrel rings size the OD. Note that the shrinkage allowance is approximately the same allowance as would be used if the tube had been injection-molded.

Another type of calibrator is the internal cooling mandrel, shown in Fig. 10.18. The internal cooling mandrel is used only with a crosshead die because the mandrel extends past the die orifice and requires the coolant to pass through the die head. The extruded tubing is drawn down to the mandrel, where it cools and sets. This type of calibrator is used only with thin-walled tubing because of limited cooling ability, and because take-off speed must be closely controlled.

Figure 10.15 illustrates still another form of calibrator, sometimes used where extrudate contact with water is undesirable. This is a *dry calibrator*, which utilizes a vacuum chamber surrounded by a second chamber with internal passages for the cooling medium. The rate of heat transfer is considerably less than direct contact with the cooling medium permits; thus the dry calibrator is indicated for those materials that cannot be cooled rapidly (sometimes cooled *shock cooling*).

Tubing dies, particularly the smaller ones, are bolted or taperlocked to the extruder barrel. The designer must be sure that the die matches the machine configuration. The larger dies may require die carts or other means for supporting, moving, and handling. Shop practice is the controlling factor and the designer should maintain a file covering these practices.

CROSSHEAD COVERING DIES

The primary use of crosshead covering dies is to apply insulation to wire and cable. Other uses include the covering of metal tubing or wood dowel stock for use as towel bars or similar items. It is also used to cover, or jacket, woven metal hose that may be exposed to abrasion. The item to be covered must be fed to the crosshead die in continuous lengths of uniform cross section. By ''continuous lengths,'' we also mean shorter lengths fed in close sequence. In any case, the item to be covered passes through the die on its way to being covered by the extruded tube of plastic. Figures 10.22 and 10.23 illustrate the two basic concepts of crosshead covering dies. At first glance, they will seem identical; the key difference is in the die bushing.

Figure 10.22 shows the ''tube-on'' process, with the die orifice *larger* than the cable or wire. Intimate contact of tube and wire occurs after the tube leaves the die orifice coaxially with the wire to be covered. The take-

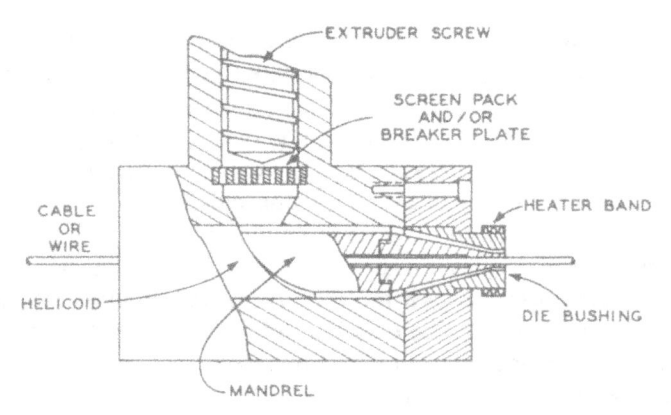

FIG. 10.22. Typical crosshead extrusion with "tube-on" die assembly. (*Courtesy Phillips Chemical Co., Pasadena, TX*)

off mechanism creates a reduction in tube size by the amount of the draw-down. The extrusion rate and the wire pull rate must be independently variable and carefully controlled to provide conformance of the tube ID to the wire OD. In another version of this die, a vacuum is applied at the entry point of the wire to aid in shrinking the tube against the wire. Control of drawdown is critical to providing firm contact between tube and wire.

Figure 10.23 is the "pressure-covering" process, with the wire directly in the melt stream and forming the inside of the covering without any need for drawdown. The wire is coated by the plastic under pressure. The pressure-die operation differs from the tube-on-die operation in that the wire, in the former, actually pulls the plastics material through the die. The wall thickness resulting from the pressure process is controlled by the die orifice

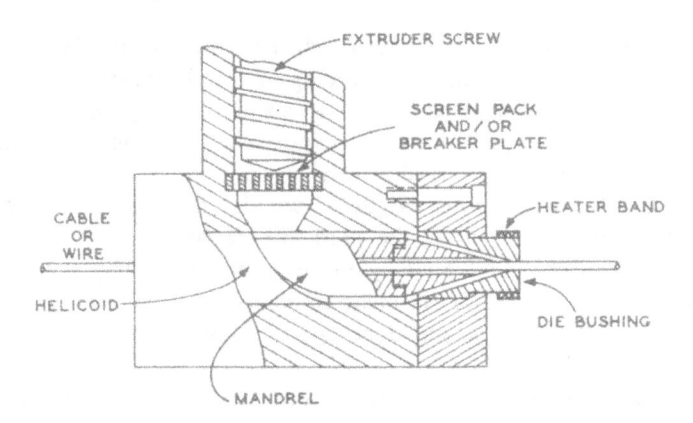

FIG. 10.23. Typical crosshead extrusion with "pressure-type" die assembly.

diameter. The pull rate through the die does have some effect; in fact, excessive pull will cause the coating to break. A "correct" design will provide the desired coating with a minimum of draw.

The tube-on method is preferred in those cases in which contact of the product with the hot melt in the crosshead could damage the product. An example is a nylon overcoating on polyethylene. Most primary insulation is applied with the pressure method to assure intimate contact between the wire and covering. Any voids may result in corona or other electrical problems. Coating for substrates, such as wood dowels or metal tubing, is done by the pressure-covering method. Good adhesion is the object in covering these types of substrates.

Two design problems arise in the design of crosshead dies. One problem is how to determine the "correct" size of the die orifice. In the case of the tube-on method, the design procedure is identical with that used for regular tubing. Allowance for die-swell must be made in either the tube-on or the pressure method. Otherwise, the final wall thickness will be incorrect. In the case of pressure-covering dies, the flow through the die orifice is complicated by the moving wire that pulls the plastic through the die orifice at a greater rate than would otherwise occur. Thus, a balancing of the pressure flow with the pulling effects of the wire must be considered in determining the proper size of the orifice. A study of Caswell and Tanner will give an understanding of the proper approach to the design of such orifices.

The second design problem is the crosshead itself. Flow within the crosshead is complicated because the plastic material makes a right-angle turn in the crosshead at the same time that it passes over the crosshead pin. The crosshead must not have any regions where the plastic can stagnate. By the same token, the design must minimize the weld effects on the covering. Finite element analysis of the flow in the crosshead is commonly used to determine the interior configuration.

Figure 10.24 illustrates the result of finite element analysis of a particular crosshead. Several references at the end of this chapter elaborate on the details of this procedure.

A practical test of the smoothness of flow and streamlining can be done by starting with a clean die and extruding a small amount of any color available. Then, without any cleaning of the crosshead, change to a contrasting color and extrude a small amount. Remove the cull from the die; the locations at which the first color remains stagnated in the cull will clearly show where further streamlining must be effected.

Heat for the crosshead and die is usually provided by cartridge or band heaters. The specific design of the crosshead will dictate the size and type of heaters to use.

The dies are usually bolted directly to the crosshead. The crosshead is

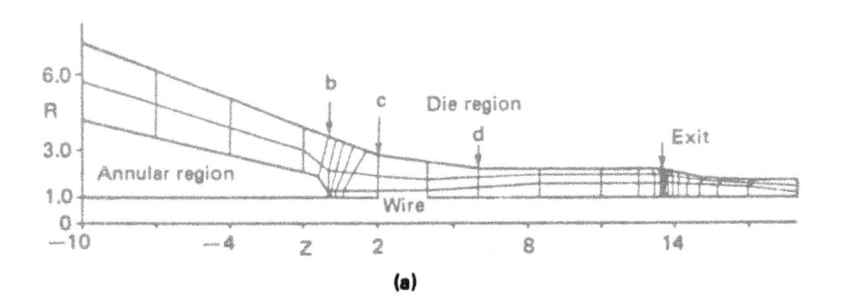

(a)

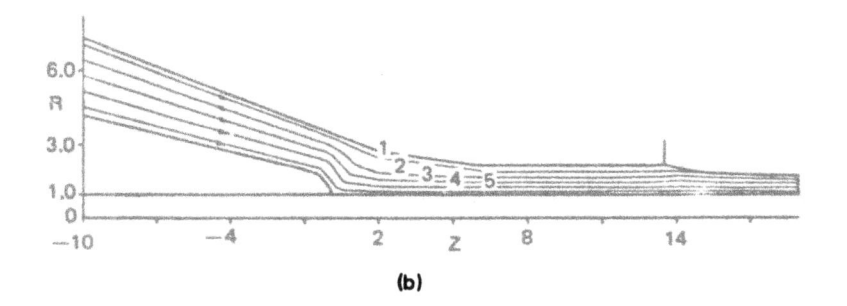

(b)

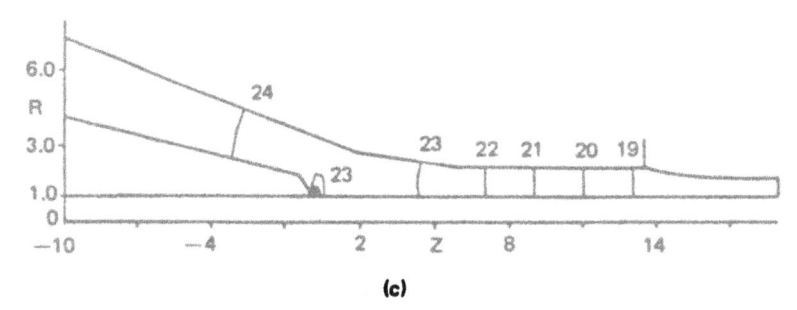

(c)

FIG. 10.24. (a) Finite-element grid, (b) streamlines, and (c) pressure contours for wire-covering die designed by finite-element analysis. (*Courtesy Polymer Engineering and Science, Vol. 18, p. 420, April 1978*)

then attached directly to the machine-barrel flange by bolts or by some sort of taper-clamp arrangement. The designer should maintain a file from which to draw the specific details as needed. The designer may also be called upon to design or specify the wire-feed mechanism as well as some type of temperature conditioning unit. The wire-feed mechanism and the take-off mechanism will depend on the wire size(s) being handled and the rate of production. Capstans are frequently used for tension control on wire-coating lines. Computer control and synchronization of production extruders and associated equipment is creating an increasingly competitive environment.

ROD AND MONOFILAMENT DIES

The only distinction between *rod* and *monofilament* dies is the diameter of the finished extrusion. The division point is arbitrarily set at 0.075 in.; anything smaller is *monofilament* and anything larger is *rod*. The die orifice is a round hole in a die bushing. Diametral control of the extrusion is by (1) the die orifice and (2) the drawdown ratio. Generally, dies for rod are single-opening dies. However, there are a few multi-orifice dies used for the smaller sizes of rod. In contrast, dies for monofilament are nearly always multi-orifice, with the orifices arranged in a circular pattern. The circular form is frequently called a "spinnerette" because of common usage to produce monofilament for textile applications. Another arrangement might be side-by-side and fed by a manifold.

The principal problem encountered in the design of multi-orifice dies is the requirement for balancing the flow to several openings so that uniform-diameter filaments will result. In the circular arrangement, balancing the flow is relatively simple because the flow paths are of equal length. The only other requirement is to streamline the channel(s) to avoid material hang-up or stagnation.

The side-by-side arrangement uses an approach manifold much like the manifold in a sheet-die, as discussed earlier in this chapter. A coathanger section will distribute the melt uniformly to the die orifices. The final adjustment is made by moving the individual orifice bushings in or out to increase or decrease the restriction of the local flow.

Drawdown of monofilament is usually by several hundred percent and in a controlled-temperature oven. A device called a *Godet Stand* utilizes drum-type pullers to control the relative rate of drawdown. A major reason for this procedure is that orienting filaments by drawdown results in increasing strength and stiffness of the material by a factor of 20 to 30. For example, low-density polyethylene as extruded has a tensile strength of approximately 2500 psi, but when the filament is oriented by drawdown, the tensile strength increases to over 60,000 psi. Reference to any good text on polymer chemistry will explain this phenomenon.

COEXTRUSION

Coextrusion is a modified extrusion process in which two or more plastic materials are extruded simultaneously and immediately combined to make the final product in one pass through the extrusion die. The major application is in producing sheet stock, but the method can be applied to film, tubing, wire-covering, and other operations. The principles discussed here apply to sheet extrusion.

An example of the utility of the process is in the extrusion of sheeting to be used in thermoforming food containers. A low-cost core material can be used to start the sandwich. Then, the side that will come in contact with food is formed with a material approved by the Food and Drug Administration (FDA). At the same time, the opposite surface, which is nearly always the outside surface of the thermoformed part, is formed with a material that can be easily decorated with ink printing, hot stamping, or other suitable means. Inasmuch as FDA-approved materials are usually priced higher per pound than either of the other materials, the saving is substantial when one considers millions of thermoformed parts, such as containers for butter, margarine, jams, cheese spread, and so forth.

Coextrusion requires an extruder or a pumping unit for each material to be extruded. The melt streams are fed to a single die, where they are combined into a single melt stream. Two different systems are in use for sheet extrusion, and the dies are different for each of these systems. The "feedblock" system, shown in Fig. 10.25, uses the feed-block, consisting of the manifold, the feedport module, and the transition, to combine two or more melt streams. From the point of combining the melt streams, the fishtail-type of sheet die is used to pass the material into a "standard" sheet where the layers have been spread to make the multilayer product. Figure 10.25 oversimplifies the feed-block, and it is actually as shown in the schematics of Fig. 10.26. These illustrations show only two layers, but there can be as many layers in the final sheet as there are feedports. The thickness of each layer in the final sheet is proportional to the thickness of its particular feed-block. This approach has the advantage of using the design described under *Sheet-Die Design*, as the basic die design. Thus, any flow problems are

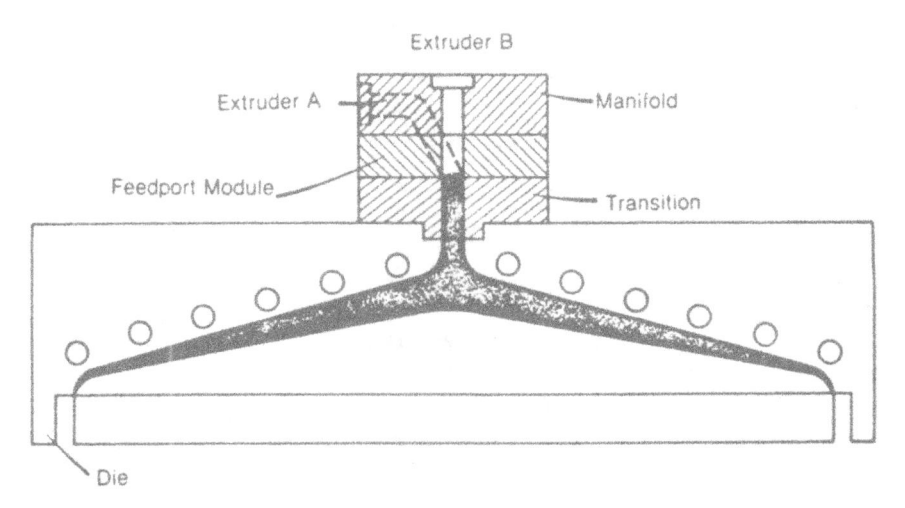

Fɪɢ. 10.25. Schematic of fishtail-type sheet die with feed block for coextrusion.

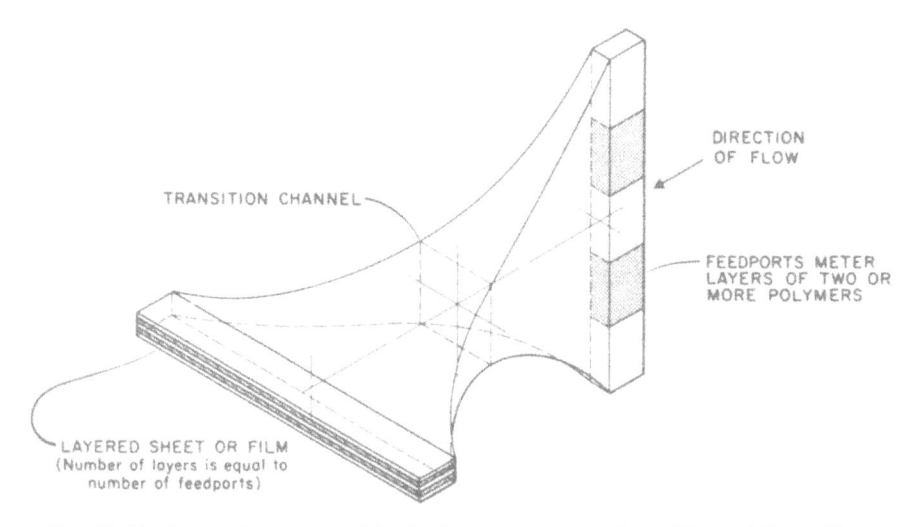

FIG. 10.26. Schematic diagram of the feedback arrangement for making multilayer film or sheet using a single manifold die. Two or more polymers are fed through individual feedports that distribute individual layers to their desired location. (*Courtesy Dow Chemical Co., Midland, MI*)

concentrated in the feed-block(s). Careful and thoughtful design is needed to achieve constant layer thickness across the width of the extruded sheet. However, the limitations of the feed-block system is that it requires plastics materials with quite similar rheologies for successful coextrusion.

Where the feed-block system is not desirable or feasible, the second system for coextrusion utilizes a separate manifold, and a separate flow path, for each material on its way from the extruder to the die lips. The separate paths converge just before the die lips as shown in Fig. 10.27. Obviously, a different set of dies is required for each sheet design, with the result that die costs are quite high compared to feed-block systems. In general, the "multi-manifold" system is required when the dissimilarity of rheology of the resin in each layer is sufficient to preclude the use of the feed-block system.

Blown film, cast film, tubing, and wire-coverings are coextruded by the use of appropriate die design and structure. Figure 10.28 illustrates a die for coextrusion into a blown film configuration. Figure 10.29 shows the schematics of a die for coextrusion of pipe or tubing. The general principles applying to coextrusion are universal to the end product. Thus the design of a coextrusion die is controlled by the designer of the end product by way of an attempt to devise the "ideal" material for the particular application of plastics material. Each die designer should, by this time, have process sheets from each of the raw material suppliers. In every case, and prior to starting

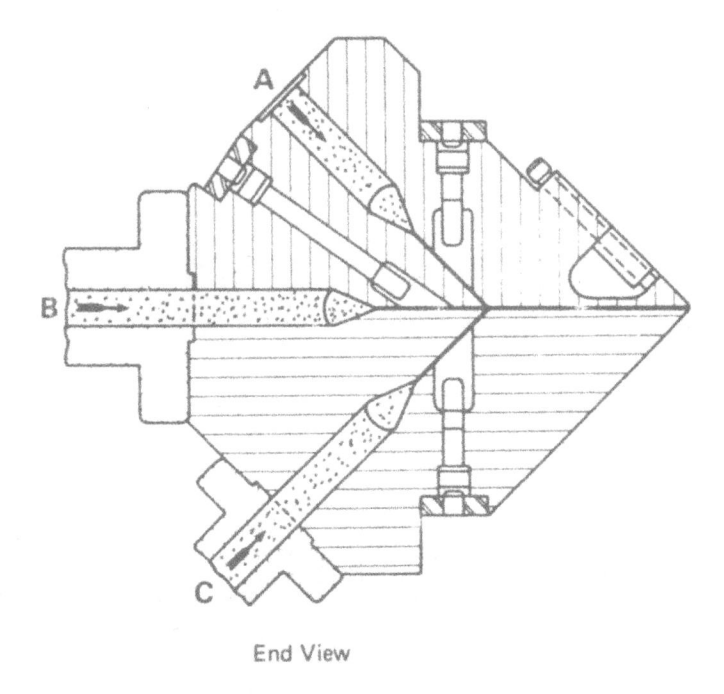

End View

FIG. 10.27. Coextrusion die with separate manifolds (A, B, C) for three different resins.

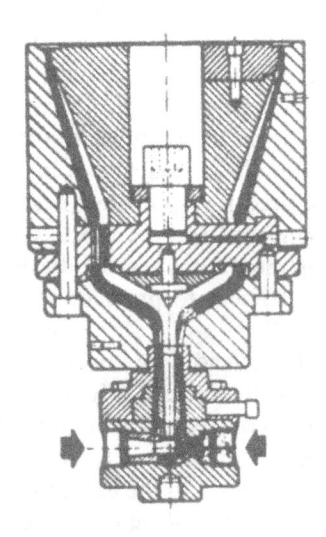

FIG. 10.28. An extrusion die for coextrusion of film. One port needed for each material. Two ports shown.

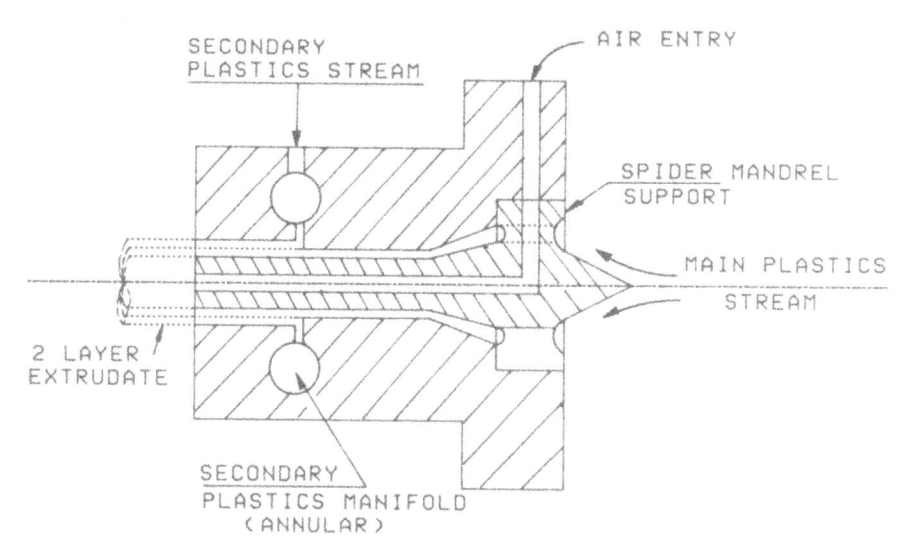

SECONDARY
PLASTICS STREAM

AIR ENTRY

SPIDER MANDREL
SUPPORT

MAIN PLASTICS
STREAM

2 LAYER
EXTRUDATE

SECONDARY
PLASTICS MANIFOLD
(ANNULAR)

FIG. 10.29. An extruder die for coextrusion into pipe or tubing. Ports for two materials shown.

a design, we recommend that the designer review all available information on die design, and processing of the specific materials.

DIE CONSTRUCTION

The dies used for the production of sheet, film, pipe, and other products, as previously described, are usually made from alloy steel. Conventional machining processes, such as turning, milling, grinding, and other machining methods, are utilized as described in Chapter 3. The use of electrical discharge machining (EDM) is increasing in popularity. Either the conventional EDM, using a carbon or copper electrode, or the wire EDM, using a moving wire, may be used, depending upon the die section or the contour required. For example, the complicated fishtail die manifold shapes can be more easily fabricated by EDM than by profile or numerical control (NC) milling. The previous statement is particularly true if stainless steel or an exotic alloy is used for the die material. The wire EDM method is most widely used to cut the dies for profile shapes. Wire EDM eliminates the need to grind and fit die sections. We will discuss machining with EDM in greater detail in the following section on profile extrusion dies.

Some extrusion dies are produced by precision casting or electroforming. These methods are generally limited to dies which must necessarily be made of an exotic material that *must* be cast, or because of the difficulty of machining the particular shape required. Electroforming is most useful in making very fine detail that cannot be produced by machining techniques. Extrusion dies are normally made from steel. The range of types is from mild steel to the 400 series of semistainless steel to high-alloy stainless steel, such as the 300 series and the hardenable 17–5 PH types. In some cases, Hasteloy or Z-Nickel are used to meet some special requirement for corrosion or high-heat resistance. The choice of material for the die is dependent on structural strength requirements, the corrosion resistance needed, and the exposure to heat in processing. In the final analysis, personal choice and successful shop practice should be considered by the designer when specifying the material for the die. Inasmuch as high-alloy steels are high-priced and, in many instances, not as strong as tool steels, dies can be made using an inlay of high-alloy steel in a mild-steel body. The net result is a lower overall cost, which is shown in Table 10.2. When you also consider that some modern tools may weigh several thousand pounds, and where the ease of machining the lower-alloy steels will represent a considerable savings, bimetallic dies become even more economical.

Every designer should be concerned with the fabrication of dies that conform to design drawings, and dies that will produce a product to specification. The advice of experienced tool makers should be sought to be certain a design does not present unusual difficulty in machining or finishing a die. In view of the present state of the art, in which it is difficult to obtain precise information on die contouring, we thereby recommend as a practice, close contact between the designer and the toolmaker. Quite a lot of die detail has, in the past, been left to the toolmaker's knowledge and/or intuition. With the use of CAD and finite element analysis now available to compute die passages, it becomes important to avoid the former intuitive approach to construction. Careful craftsmanship and attention to detail are key elements to the proper functioning of tools. Inspection (meaning measurement)

TABLE 10.2. Bimetallic Die Construction Versus Dies Made in Total from Various Metals.

Material Description	Prices of Dies Made from Various Metals[a]	Bimetallic Die Selling Price[a]
4150 Steel, chrome plated	$13,300	$
316 S.S.	19,000	14,200
420 S.S.	21,000	14,800
15-5 PH S.S.	33,300	16,800
17-5 PH S.S.	25,650	16,400
Hastelloy C	46,600	19,200
Duranickel	42,300	18,300

[a]Prices are dependent on numerous options and represent prices effective 1986 (*Courtesy Egan Machinery Co., Somerville NJ*)

of tools to ascertain conformance to design is an important part of the design effort.

PROFILE EXTRUSION DIES

Profile extrusions are extruded shapes that are not described within any of the standard shape designations, for example, sheet, rod, tubing, film, and so forth. Profile shapes cover an almost unlimited range of possibilities, some of which are shown in Figs. 10.30 and 10.31.

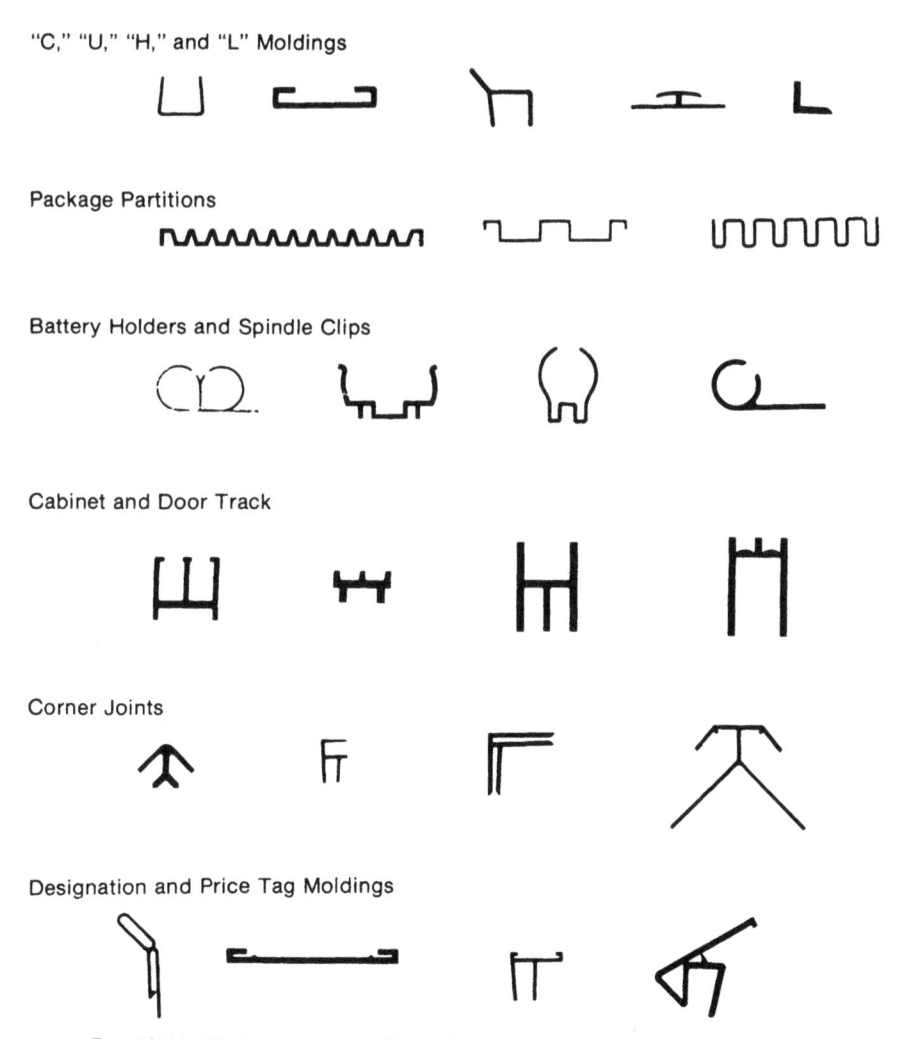

"C," "U," "H," and "L" Moldings

Package Partitions

Battery Holders and Spindle Clips

Cabinet and Door Track

Corner Joints

Designation and Price Tag Moldings

Fig. 10.30. Typical commercially available uniform-walled profile extrusions.

Extruded Handles

Corner Joints

Name Plate Holders

Terrazzo and
Floor Strips

Extruded Lenses

Miscellaneous Profiles

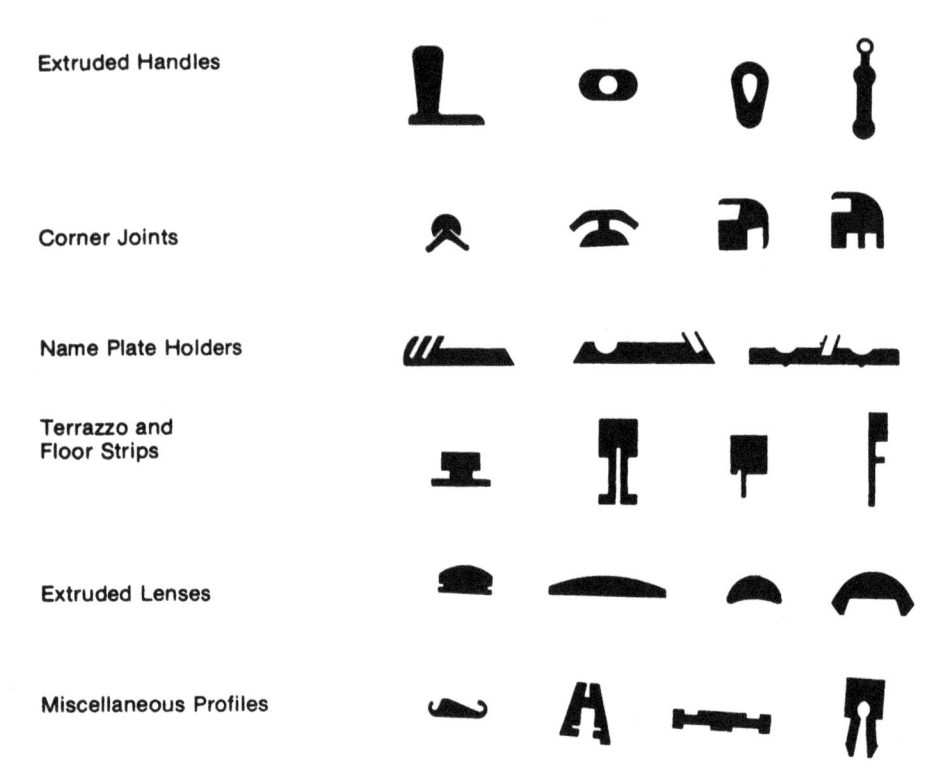

FIG. 10.31. Typical commercially available heavy section and non-uniform-walled profile extrusions.

The dies used to shape these extrusions, and the auxiliary tooling used for control of the part dimensions and shape, use the same basic flow concepts as previously described. However, the flow in the profile die is considerably more complex, and the design procedure thus becomes more complicated.

There are two categories of profile dies. Each category refers to the flow of the material as it leaves the die. One category is the one-D (one-dimensional flow) which is *all* flow in the direction of the extrusion. The second category is the two-D (two-dimensional flow) in which there is a considerable amount of crossflow in the die. The first category (one-D) consists of a variety of "slit" dies for which the aspect ratio (width vs. thickness) is 10:1, or more, and where the overall thickness of the extrusion is less than 0.100 in. (2.5 mm). Whatever the type of material and its rheology, this ratio will insure that the flow is in the extrusion direction (assuming that the part has essentially the same wall thickness throughout). If the wall thickness of any section is increased, or the aspect ratio decreased, or the wall thickness is not uniform, then there will be *some* crossflow and the die

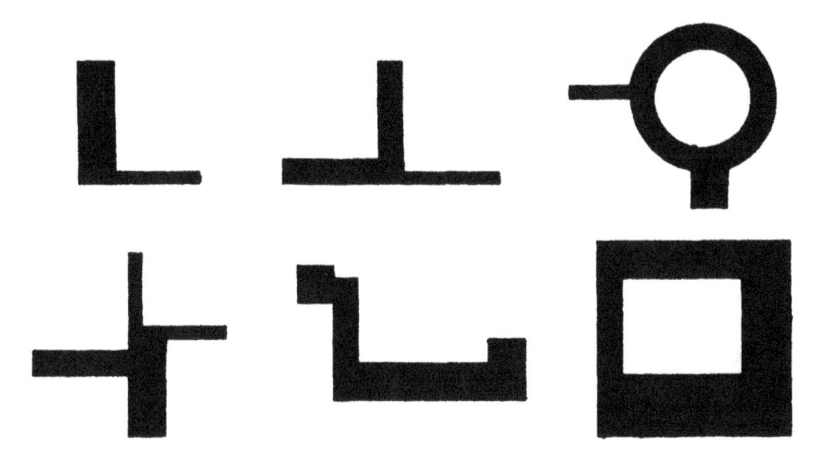

Fig. 10.32. One-dimensional shapes made from combinations of slot sections.

becomes a two-D die. A round die is also a one-D die, but not in the same sense as a one-D die of the rectangular strip, because other considerations must be incorporated into the round profile (see the discussions on ''Pipe and Tubing Dies,'' and ''Rod and Monofilament Dies,'' previously).

A study of Figs. 10.32, 10.33, and 10.34 will provide a concept of one-D and two-D shapes. Fig. 10.35 illustrates a specific one-D die, and we will step through the design procedures as it applies to this profile. The shape to be considered is an angle. The plastic material might be rigid PVC or ABS. Either of these materials behave well in profile extrusion. There is enough available flow data to make a reasonably good ''first design'' which would require very little correction after the first sample run. If the die were made to the exact shape of the part, the extruded profile would be incorrect because of *die-drag* and *die-swell*. Figure 10.36 illustrates the founded corners and the fillet result of an ''uncorrected'' die. The corners at the end of each

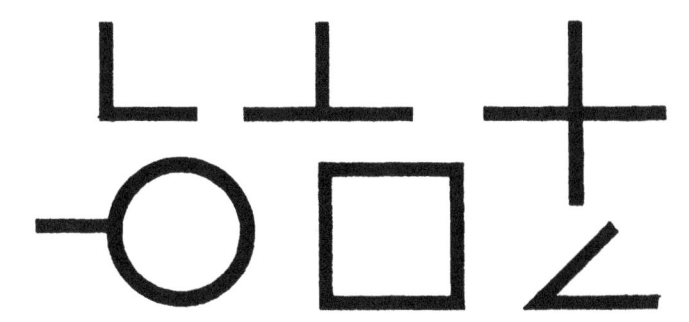

Fig. 10.33. Variable wall thicknesses made from combinations of strip die shapes.

FIG. 10.34. Solid profile shapes which will use two-dimensional flow of extrudate.

leg are rounded because of *corner-drag*. The right-angle corner will be rounded externally and internally. The inner corner rounding is a result of reduced drag (greater flow) as compared to the straight-wall section. The outer corner is rounded by the same effect as that which produces the rounding at the end of each leg, namely, corner-drag (reduced flow).

Refer again to Fig. 10.35, where we have shown how the die is ''corrected'' by changing the land lengths to alter the flow pattern. The corners, where there is corner-drag, are enlarged for a part of the land length in order

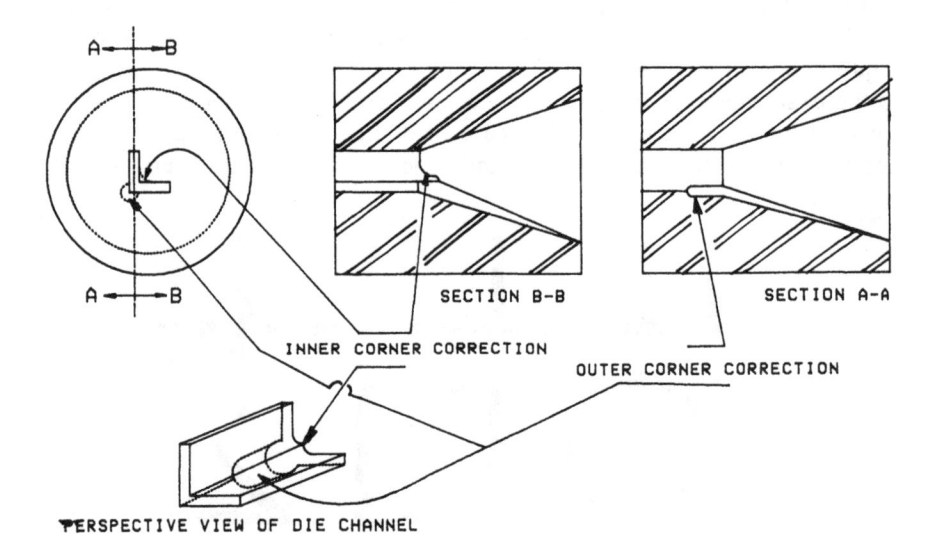

FIG. 10.35. Extrusion die corrections for angle profile having unequal legs.

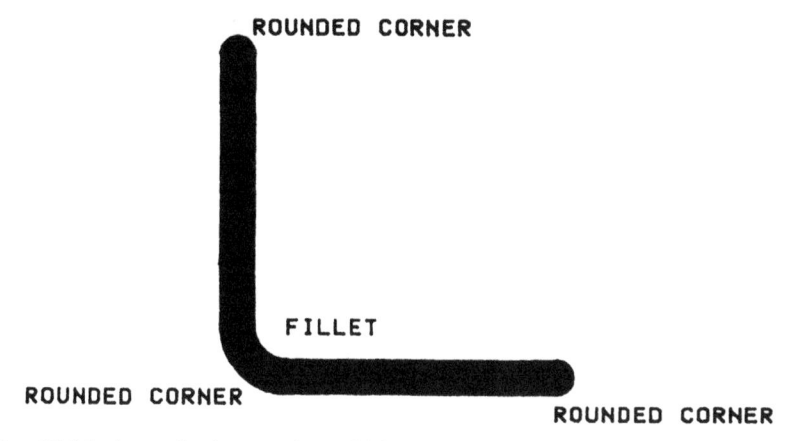

ROUNDED CORNER

FILLET

ROUNDED CORNER

ROUNDED CORNER

FIG. 10.36. An angle shape as it would be produced from an uncorrected die. See Fig. 10.35 for corrections.

to increase the flow and fill the corners. The inner corner, which has become rounded because of too much flow of material, is restricted by extending the land length to decrease the flow of plastic melt.

Typically, die land lengths range from 10 to 20 times the channel depth (thickness of the extrusion), which corresponds to an L/D ratio of 10:1 to 20:1. The correction outlined above results in a change in the ratio of L/D by some fraction of the original land length. Most extrusion shops normally have developed data on flow rate vs land length for a particular design configuration. By examination of this data, the corner cutaway can be estimated. When rigid PVC compounds are being extruded, the corner cut will be about $\frac{1}{8}$ in. in diameter and extend for 30% of the land length as it is applied to the ends of the legs. The corner cut for the right-angle will usually be $\frac{1}{4}$ in. in diameter and extend for 25% to 40% of the land length. The exact percentage will depend on the thickness of the extrusion. The formula is: *the thicker the extrusion, the greater the percentage of land-length reduction*. The extension or lengthening of the land length on the inside corner follows the same formula except the land length is increased.

It should be apparent that the design procedure requires the generation of flow data for the specific material. Ordinarily, test dies will be operated under a variety of conditions of temperature, pressure, and speed just to generate useable data for specific profiles. One such curve is shown in Fig. 10.37. Figure 10.38 illustrates the configuration of the test die and related equipment. Each designer must recognize that the information generated by a particular business entity is considered proprietary and highly confidential. With the advent of computer design and analysis, much of this information will become more generally available. In any case, the designer *must*

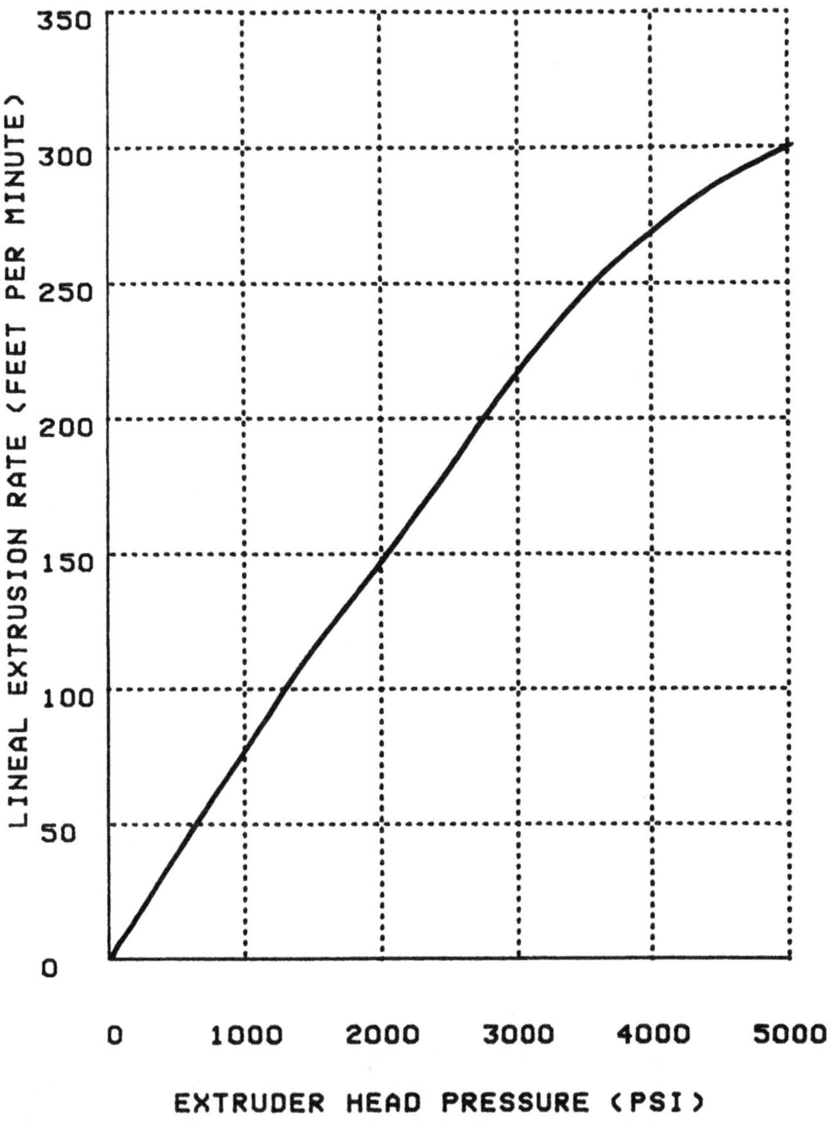

FIG. 10.37. Chart of lineal rate of extrusion from a .020 in. × .050 in. slit die using flexible PVC material at 310° F die temperature.

have this information to produce an effective and practical die design. Our purpose here is to develop the principles without revealing specific data, or betraying confidences.

Two-dimensional flow (two-D) dies are more complicated to design simply because the crossflow(s) acts to alter the flow through a die in an even

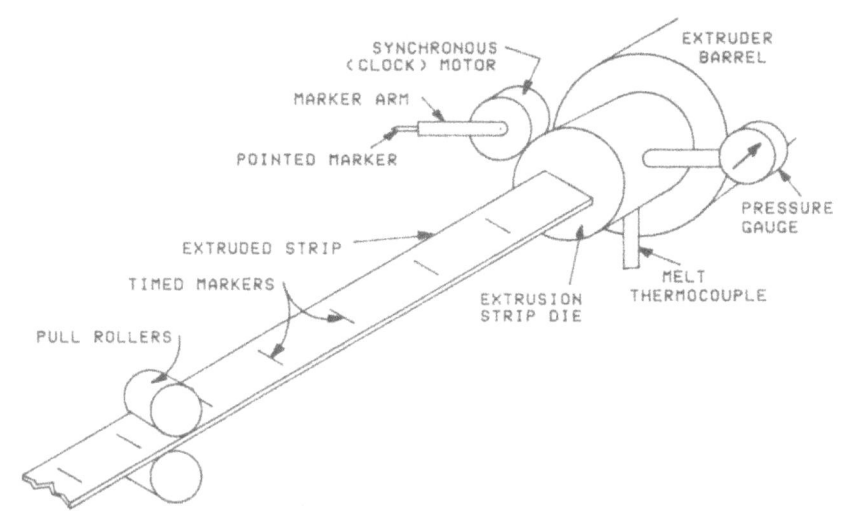

FIG. 10.38. Schematic diagram of extrusion rate tester.

more drastic manner than do the corner effects just described. The unequal-leg angle shown in Fig. 10.39 is an example of such a die. Adjusting for the disproportionate flow through slits of different sizes is done by enlarging the flow channel in the thin leg and by using a *flow-block* or restriction to

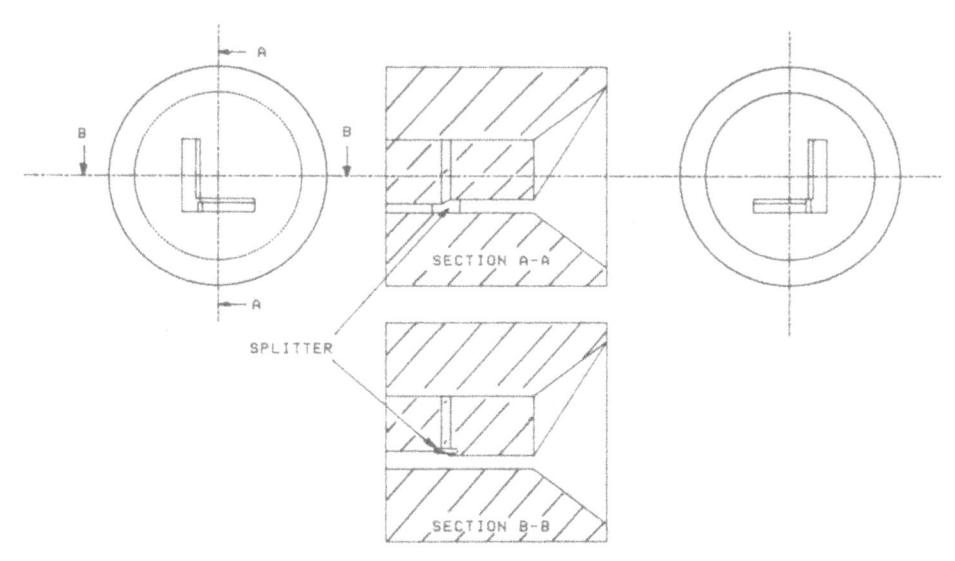

FIG. 10.39. Extrusion die design for angle profile with unequal legs. Note "splitter" to allow reaction as separate legs.

reduce the flow in the thicker leg. The relative flow rate in each section of the die must be proportional to the wall thickness. The length of the die lands can be calculated from the same flow-rate data used for the one-D dies.

The corrections for the end of the leg are generally similar to those used in the previous example. The angle-corner corrections are more difficult and less predictable because of the crossflow possible at the corner. Quite frequently, a cut-and-try approach may be needed. However, the use of a *flow-splitter* (barrier), as shown in Fig. 10.39, will simplify the problem. The advice to use a flow-splitter is contrary to previous advice to avoid splitting flow patterns. In this case, the flow-splitter will allow the two legs to react as separate strips that are rejoined after passing the splitter and before entering the final section of the die lands. A splitter length of two to three times the thickness of the leg is sufficient to give a good weld of the material in the melt stream. With this technique, the inside and outside corrections for the angle corner can use the same formula as is used for the end of the leg. The restriction on the crossflow provided by the splitter will result in an ''almost sharp'' corner on the very first cut of the die. Thus, very little subsequent correction will be needed.

Die-swell complicates the design of profile dies. Die-swell adds to the extrudate dimension and in all directions. The dimensions of the profile can be adjusted by control of the drawdown ratio, but this control will not eliminate the effects of die-swell. The die-swell curve shown in Fig. 10.40 for a particular PVC material shows a 9% to 18% increase in diameter depending upon the shear rate. The shear rate, in turn, is dependent on the channel depth. For the slit die, such as the angle, the shear rate is much less across the width than it is in the thickness simply because of the dimensional difference and its effect on the shear rate. The rule-of thumb formula is: *Die-swell will add a constant amount to each dimension of the part shape as it exits from the die orifice.* ''Constant amount'' translates to a ''constant envelope.'' It should be clear that drawdown is proportional to each dimension. Thus, drawdown will alter the shape of the extrusion. The designer must calculate (or estimate) the die-swell and make allowance for it in the orifice shape in order that drawdown will bring the extrusion to the exact size and exact shape required.

Die-design procedures result in approximations that are highly dependent on the plastics material to be processed and the rate at which it is extruded. Changing the material grade, or changing to a different melt index, will result in a change in the shape of the extrudate. Changing the rate of extrusion will also alter the shape simply because delivery rate is shear-dependent. Altering the overall shear rate will alter the rates of flow through the die passages with a resulting change in the profile shape as it leaves the die

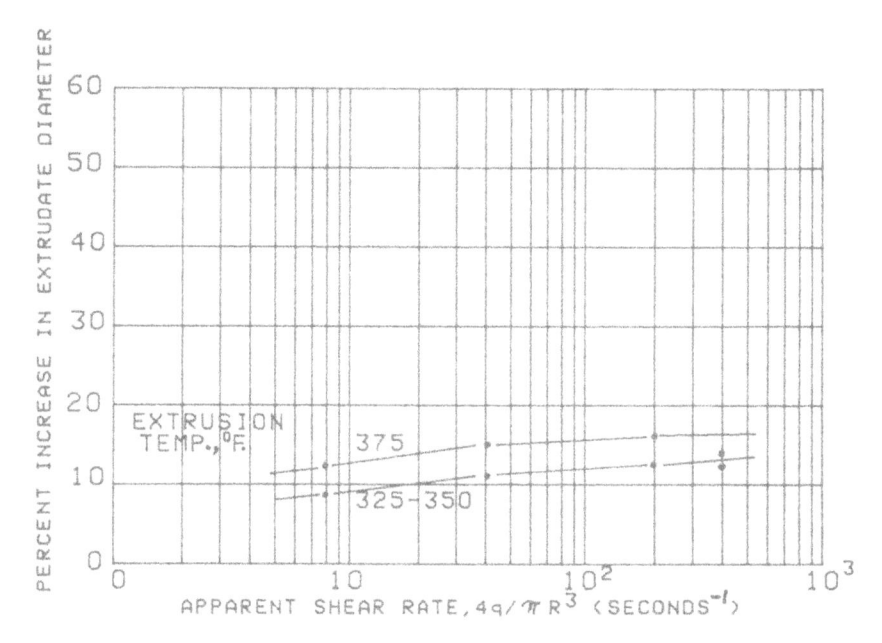

FIG. 10.40. Graph of die-swell for a 3/32 in. ID tube, flat entry into die. L/D ratio = 8.

orifice. The designer needs to communicate to the extruder operator the operation parameters that have been anticipated in the design of the die.

The only known method to determine the shape a die will produce is to test the die in actual use with the equipment to be used in production, and under the production conditions of time, temperature, and pressure. Even small differences between machines can and will affect the profile. In any case, the final correction to a die should be attempted only after production conditions and equipment are finalized. Final corrections include changing die land length, changing land depth, and/or changing the shape of the orifice to correct for die-swell and drawdown effects.

Figure 10.41 is a chart of the most commonly used plastics material and the relative ease of extruding them into profile shapes. Inasmuch as some materials are highly sensitive to temperature and pressure changes, we recommend choosing the most tractable material sufficient to fill the application requirements. Difficult shapes in difficult materials should be attempted only when ample funds and time are available, or when the dynamics of the end use leave no other alternatives.

We turn now to other considerations in the design of profile dies. One important consideration is the approach section to the die orifice. In some instances, with thermally stable plastics materials and for short runs, the die plate is attached to a die holder and run in this manner. A considerable

Material	Thixotropy	Melt viscosity	Temperature setting range	Frictional heat	Melt elasticity
Rigid PVC	1	1	3	4	4
Flexible PVC	2	2	3	2	3
Acrylic	3	2	3	3	6
Modified PPO	2	3	2	2	5
ABS	3	3	3	3	4
Impact styrene	4	4	3	2	4
Polycarbonate	4	3	4	4	4
Thermoplastic polyester	4	4	2	3	4
Polypropylene	6	4	3	2	4
High-density polypropylene	6	5	5	2	2
Polyethylene	7	5	6	2	4
Polyamides	8-9	7	3	2	3

FIG. 10.41. Chart for low-melt-index materials and the characteristics affecting profile extrusion. Numbers represent relative difficulty involved in extruding profiles of the listed materials. Scale is 1 to 10, with the lowest number being easiest to extrude. (*Courtesy Plastics Machinery & Equipment*)

amount of material may be trapped in the corners where the die plate attaches to the holder. Figure 10.42 shows the die-entry profiles for three popular materials.

Obviously, the best practice is to use a streamline approach. The shape of the streamline approach section will vary with the plastics material to be used, and to some extent is dependent on the rate of extrusion. Calculation of the exact curve can be difficult, if not impossible.

The pragmatic approach is illustrated by Fig. 10.43. The die is run with the particular plastics material, and the die cull will self-autograph. Observation of the color change caused by decomposition of material such as PVC is one method. Material that changes color when subjected to protracted heating is another method.

A third, and probably the easiest method, is to use two colors of the same material, run each through the die in sequence, then remove the cull and cut it in half along the axis. This sectioned cull will show quite clearly the exact area of stagnation. Machining an approach section to this same contour will result in a very close match from the breaker circle to the orifice shape.

The use of streamlining will increase throughput, decrease the probability of degradation, reduce power requirements, and provide best control of the profile shape. In passing, note that a nonmoving layer of plastics material is an excellent heat insulator. For evidence, refer to "Hot Runner Molds" in Chapter 8.

Most dies are provided with either band or cartridge heaters and with thermostats to control the die temperature. The choice of heating method depends upon the die and the profile. Heater wattage should be high enough

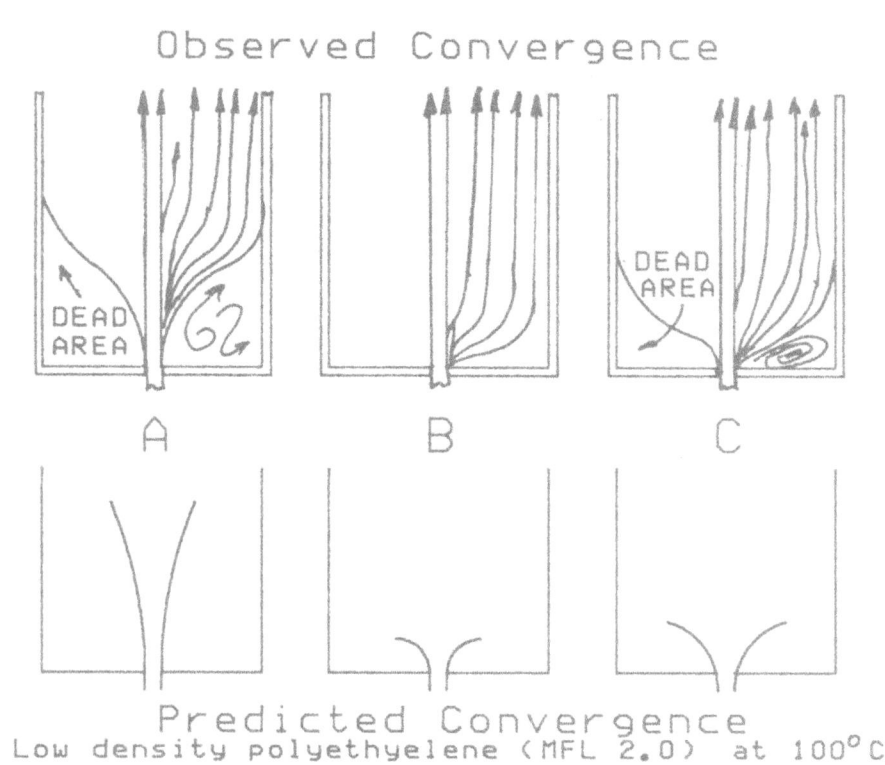

Observed Convergence

A   B   C

Predicted Convergence

A Low density polyethyelene (MFL 2.0) at 100°C.
B Extrusion grade polypropyelene at 230°C.
C Molding grade acrylic at 230°C.

FIG. 10.42. Die entry flow for shear rate 300 sec⁻¹ in a circular die. Observed convergence is symmetrical about center line of the die. (*Courtesy Polymer Engineering and Science, January 1972, Vol. 12, No. 1*)

to bring the die up to operating temperature in one to two hours. The oversizing of heaters will result in excessive cycling during the production run, when maintaining temperature is the only requirement. In general, the die will be approximately the same temperature as the melt passing through it. It is also possible to provide some control of profile shape by the use of localized heat effects.

Other tooling involved in profile extrusion is the downstream equipment needed to cool the extrudate and to maintain the profile as the material cools and sets. A puller is used to draw the extrudate as it emerges from the die. It is also directly involved in the sizing of the profile. In some cases, the downstream equipment is used to reshape the profile to final requirements. One unit used to retain part shape as the material cools is the *vacuum sizer* (calibrator) of the type shown in Fig. 10.44. The profile is held against the

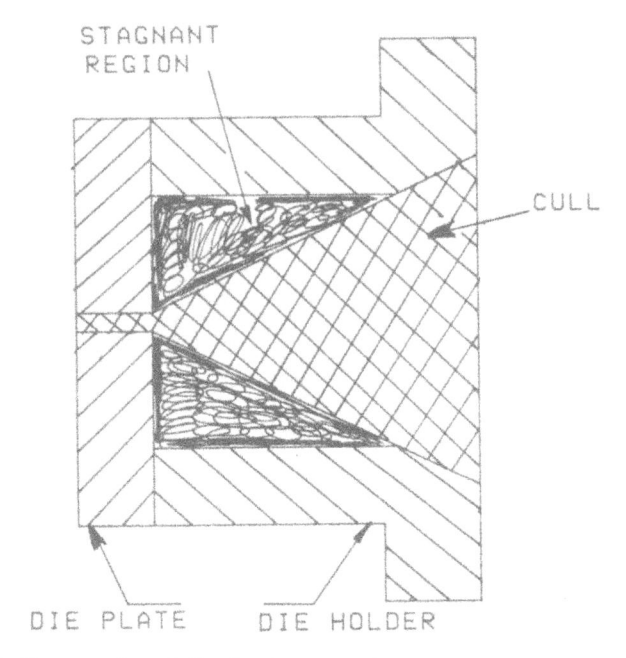

FIG. 10.43. Cross section of die showing streamline flow in the cull. Good die design should eliminate all stagnant region(s).

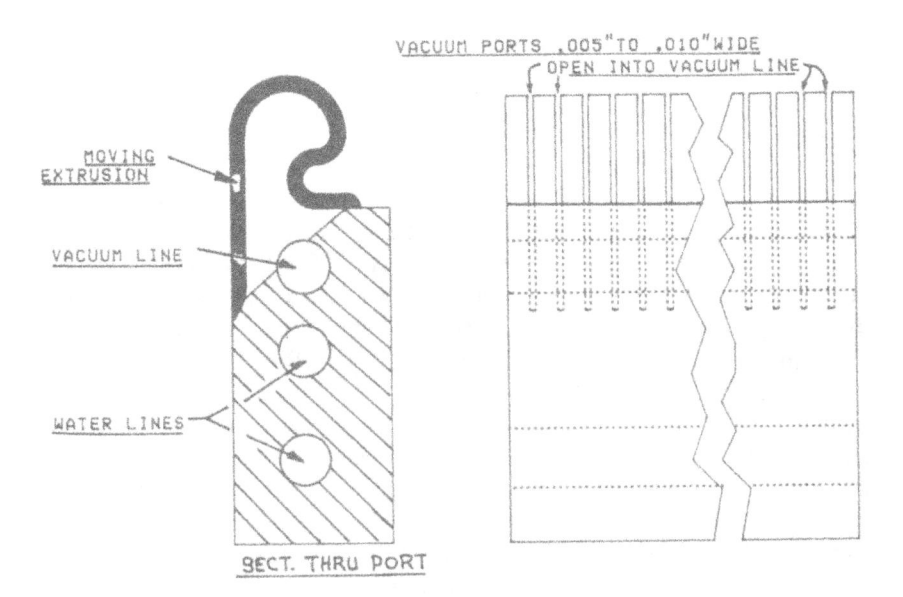

FIG. 10.44. Vacuum sizer for controlling one side (inner side) of extrudate.

shaped and cooled mandrel as the profile "slides" along the length of the mandrel. The mandrel shown contacts only the inside surface of the profile. However, it could also envelop the profile and cool it from the outside. This type of sizer cannot simultaneously contact both inside and outside of the profile. Nor will it be suitable if the shrinkage of the profile will "lock" the profile to the mandrel. Figure 10.45 illustrates three acceptable units and one which will "lock" the profile. Sizers vary from 24 to 60 in. in length. The longer units are usually used as two or more sections, and made to dimensions that allow for profile shrinkage. Aluminium is the material most frequently used, because of its rapid transmission of heat. A low-friction coating, such as Teflon® (TFE), is usually applied to the sizer to facilitate the "sliding" of the profile. Some sizers are made of chrome-plated brass. The shape of the sizer must be carefully calculated and accurately made because the sizer determines the final shape of the profile.

Figures 10.46 and 10.47 show downstream equipment for supporting the profile shape to prevent it from collapsing before it can set (become rigid). Initial cooling may be created by blowers directing cold (cool) air over the hot profile. Quicker cooling is also achieved by sophisticated units using a combination of venturi air cooling units and water fog sprays.

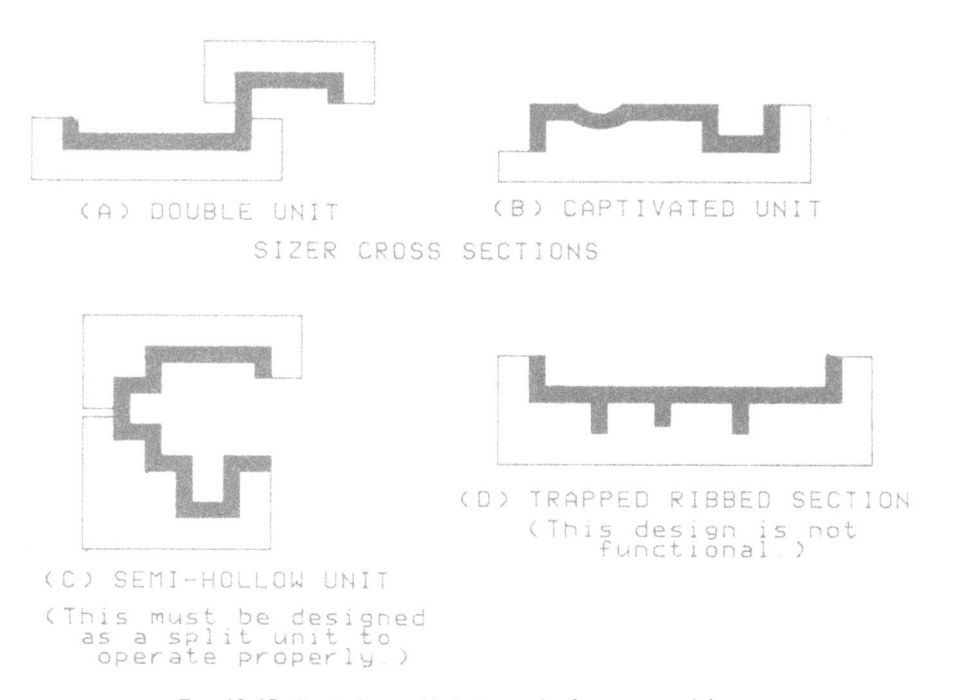

FIG. 10.45. Typical one sided sizer units for vacuum sizing.

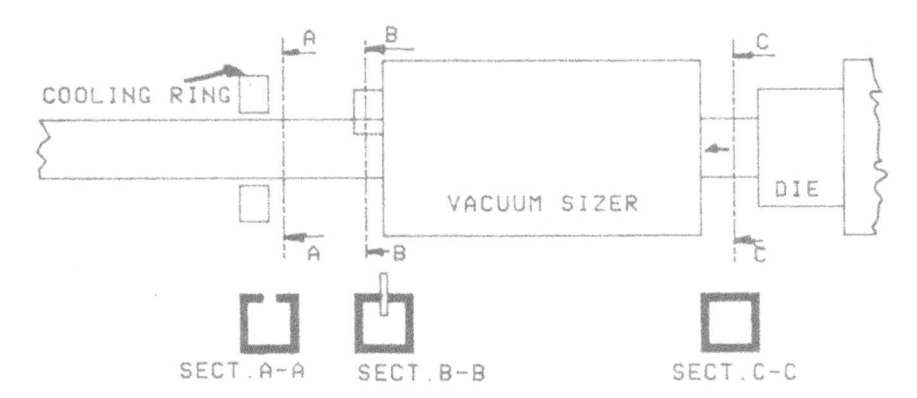

Fig. 10.46. Post die shaping technique using vacuum sizer and cooling ring(s).

A recent development of increasing usage is profile extrusion as a coextrusion of two or more materials in one profile. Two (or more) dies sequentially or coincidentally form the sections of the coextruded profile. The adhesive effect of hot contact is the usual method of holding the materials together. In some cases, mechanical interlocking is used, or else a separate interlayer of adhesive is coextruded to join the materials. Figure 10.48 illustrates a coextrusion die to produce a two-material channel. Rigid PVC and flexible PVC are the most used combination. However, the coextrusion of profiles has also been extended to other materials. Reference to the discussion of coextrusion above will provide additional information.

Punching, reforming, or other operations are sometimes performed on-line with the extrudate. These operations are not usually regarded as a die designer's planning or designing responsibility. The tooling involved for stamping or reforming on-line with the extruder is similar to that used for

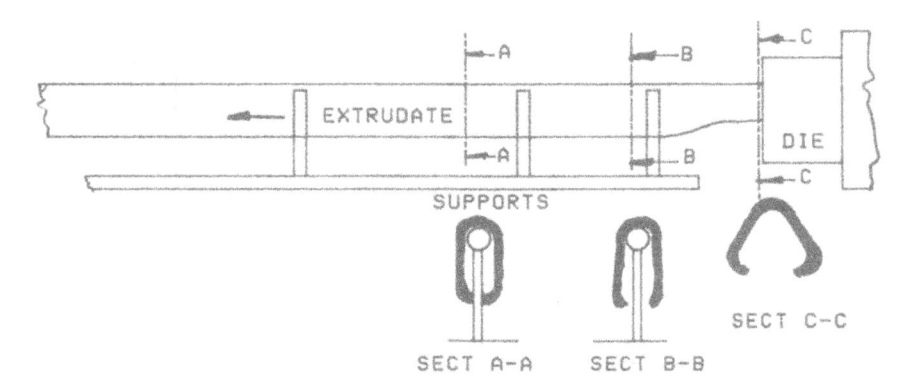

Fig. 10.47. Post die shaping using support forms. Applicable to rigid extrusion materials.

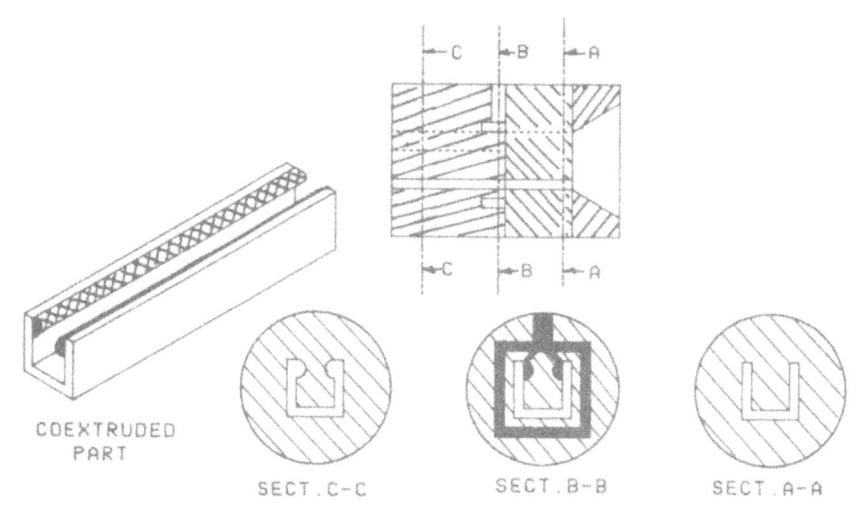

Fig. 10.48. Die made in three sections for ease of building and cleaning (when necessary). Used to coextrude two different, but compatible, materials.

stamping and reforming operations when done separately. The principal difference is the length of time that can be allowed for contact with a moving extrudate. The usual solution is to move the press or reforming unit in synchronization with the take-off equipment. Even so, the shorter the contact time between one punch and the material, the less the criticality of synchronization.

MATERIALS USED IN PROFILE DIES

Profile dies may be made from any one of several materials, just as the sheet, rod, and tube dies use various materials. Semi-stainless or stainless steel in the 410 series or the SAE 4140 type is typical. Some dies are made of 18-8 or 17-5PH steels when long runs with corrosive plastics materials are expected. The dies are frequently plated to improve corrosion and abrasion resistance. Hard nickel or chrome is the most commonly used plating material. Specialty coatings, such as Neodex® TFE-impregnated porous electroless nickel helps prolong die life. Tufram®, a TFE-sealed hard anodize coating over aluminium, is frequently used on vacuum sizer mandrels to improve wear resistance and to provide increased lubricity.

Fabrication of profile dies uses conventional machining as well as EDM techniques. Figure 10.49 shows a two-part die that is shaped by grinding the two halves, matching the halves, and assembling with screws. The present trend is to cut the die orifices by using wire EDM. This makes it possible to cut long die channels and still allow die material for corrections after the

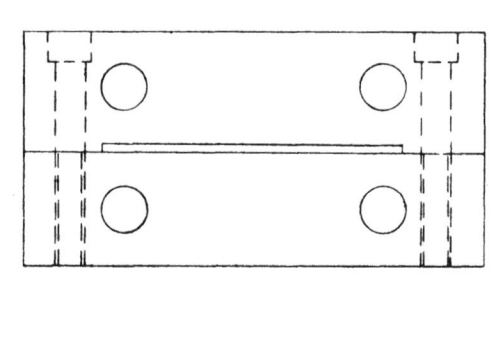

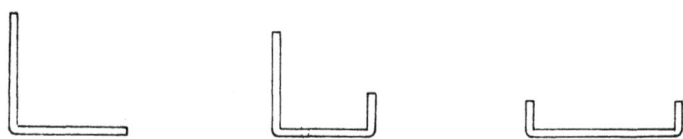

FIG. 10.49. Tape profile die (top) from which profile shapes are formed by takeoff mechanism. Two piece die shown, but can be made in one piece by wire EDM cutting of orifice. (*Courtesy, Anchor Plastics Co., Long Island, NY*)

sample run. Always size the die channels to allow for the *removal* of metal. Great advances have been made in the art of ''putting-on-metal,'' but it is always more economical to ''take-off-metal.'' Wire EDM equipment of today is quite capable of channels up to 8 in. long, and areas of 10 in. by 30 in. Not many profile dies exceed this range.

The options for machining the contours of the approach sections are limited only by the equipment available to do it. Contour milling, with or without numerical control, is readily available. Generating a carbon master by using the signature-cull method, then using EDM equipment to ''burn the shape'' is common practice. The designer should be well-versed in the available options in general shop practice and in the shop of the die maker.

The references contain a wealth of information for the designer who is interested in additional reading to learn about a particular subject in depth. This chapter is intended as a guide to well-designed dies for the unlimited variety of shapes and sizes which can be created by extrusion.

REFERENCES

Barney, J. J., Trends in Extrusion Die Design, *Plastics Design and Processing*, p. 12 (Feb. 1974).

Caswell, B., and Tanner, R. I., Wirecoating Die Design Using Finite Element Analysis, *Polymer Engineering & Science* 18(5) (April 1978).

Cogswell, F. N., Converging Flow of Polymer Melts in Extrusion Dies, *Polymer Engineering & Science* 12(1) (Jan. 1972).

DuBois, J. Harry, *Plastics History—USA*, Boston, Cahners Books, 1972.

DuBois and John, *Plastics*, sixth revised edition, New York, Van Nostrand Reinhold, 1981.

Gutfinger, C., Boyer, E., and Tadmore, Z., An Analysis of a Crosshead Die with the Flow Analysis Network (FAN) Method, *Polymer Engineering & Science* 15(5) (May 1975).

Han, C. D., and Rao, D., Studies on Wire Coating Extrusion, *Polymer Engineering & Science* 18(10) (Oct. 1978).

Levy, Sidney, Maximizing Flow through Profile Dies, *Plastics Machinery & Equipment* 7(4) (April 1978).

——, Melt Rheology: Its Effect on Profile Extrusion Dies. *Plastics Machinery & Equipment* (Sept. 1978).

S. Levy P.E. & Associates, Report on Plastics Profile Tooling and Technology, Fresno, Ca., May 1979.

Matsubara, Y., Design of Coat Hanger Sheeting Dies Based on Ration of Residence Times in Manifold and Slot, *Polymer Engineering & Science* 20(4) (Jul. 1980).

Schrenk, W., Multilayer Film from a Single Die, *Plastics Engineering*, p. 65 (Mar. 1974).

Veazey, Barnette, and Pate, Optimizing Extrusion of LLDPE film, *Plastics Technology*, p. 33, (Sept. 1984).

Wood, Richard, Advanced Die Designs for Coextrusion. *Plastics Machinery & Equipment* (May 1984).

Chapter 11 / Blow Mold Construction and Design

Robert S. Musel, Brooks B. Heise,
and Robert D. DeLong

PLASTIC BLOW MOLDING PROCESSES

Many commercially available processes and machines are used for plastic blow molding; each one has its advantages and limitations, as indicated in the descriptions which follow. The mold design for any given blown product will be governed by the process and machine selected. When the molder has more than one process or machine size available, the selection of the machine and mold size will be governed by various factors, including blown product size and configuration, material to be blown, production requirements, quality requirements, manufacturing cost estimates, and investment funds available. Figure 11.1 shows a variety of packaging containers. The variety of industrial containers is illustrated by Fig. 11.2.

Continuous Extrusion Process

One or more hot plastic tubes (parisons) are extruded continuously at the thickness required to manufacture the blown product to the desired weight and wall thickness. During extrusion, the parison thickness may be varied (programmed), as shown in Fig. 11.3, to control the uniformity of the wall thickness in the blown product. Blow-mold halves close around and pinch the parison and then must be quickly moved away from, or along with, the continuously extruding parison; the machines commercially available employ various kinds of mobile mechanisms, discussed below, to move and operate the molds.

Wheel Machine. The molds are mounted on a vertical wheel turning at a speed which is synchronized with the speed of parison extrusion (Fig. 11.4).

FIG. 11.1. Blow molded bottles used for packaging consumer products. (*Courtesy Kautex Machines, Inc., (Krupp-Kautex source) Linden, NJ*)

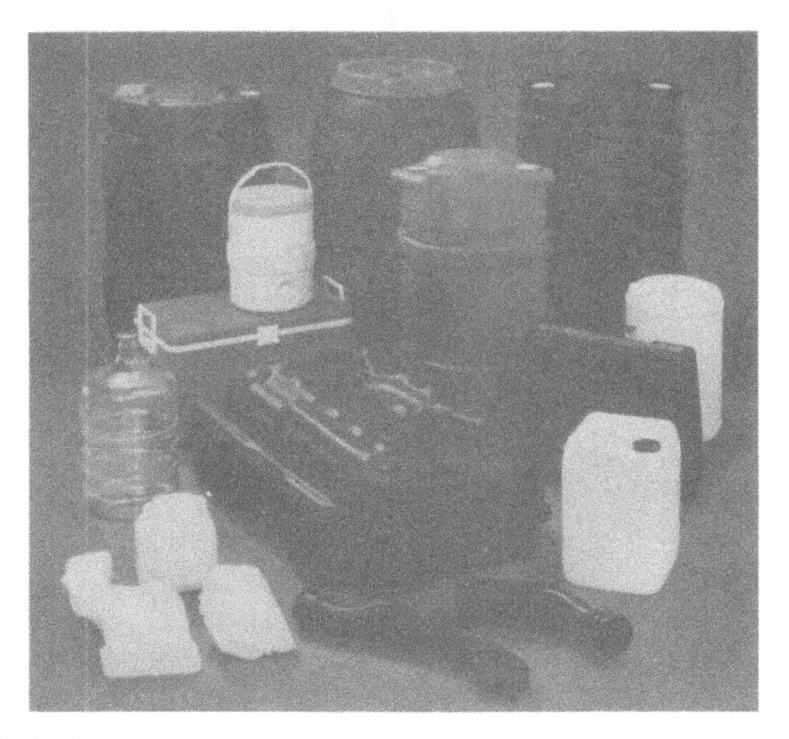

FIG. 11.2. Blow molded industrial containers. (*Courtesy Hoover Universal, Plastic Products Division, Ann Arbor, MI*)

As the mold halves are moved into position on either side of the parisons, they are sequentially closed, the parisons are pinched at both ends, and the product is blown, cooled, and removed from the machine. Generally, a secondary trimming operation is required to remove the excess plastic from both ends of the product.

The wheel machine is used primarily for high volume production of containers, in a size range of 8 ounces to 1 gallon for packaging household chemical, food and beverage, and automotive products.

Shuttle Machine. The mold halves are mounted on a mobile clamp carriage which is synchronized to shuttle the mold halves laterally, or angularly, or in an arc to the parison-extruding position (Fig. 11.5). The mold halves close and pinch the parisons at the bottom end, the clamp carriage is moved to a blowing position, a blow pin is inserted into the parison, and the product is blown, cooled, and removed from the machine. Shuttle machines are available with dual mobile clamp carriages which shuttle alternately from dual blowing positions to a central parison-extruding position (Fig. 11.6).

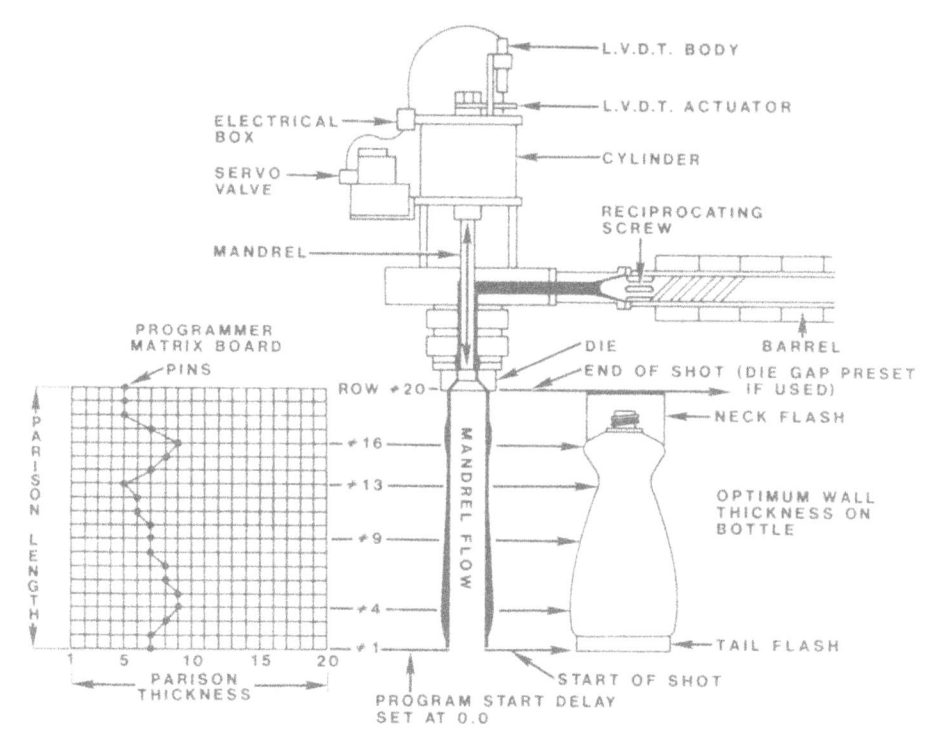

FIG. 11.3. Schematic of parison programming for variable wall thickness. (*Courtesy, Hoover Universal, Plastics Machinery Division, Manchester, MI*)

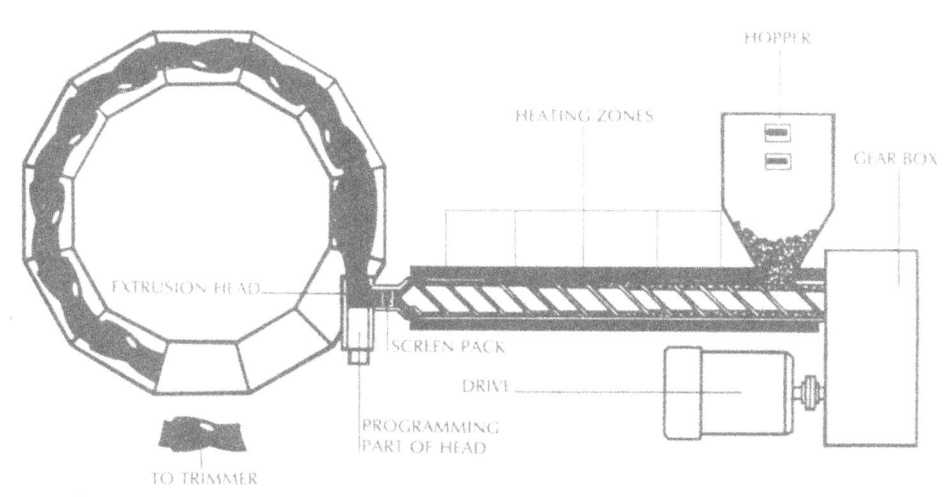

FIG. 11.4. Wheel machine allows high volume and extra cooling time. (*Courtesy Graham Engineering Corp., York, PA*)

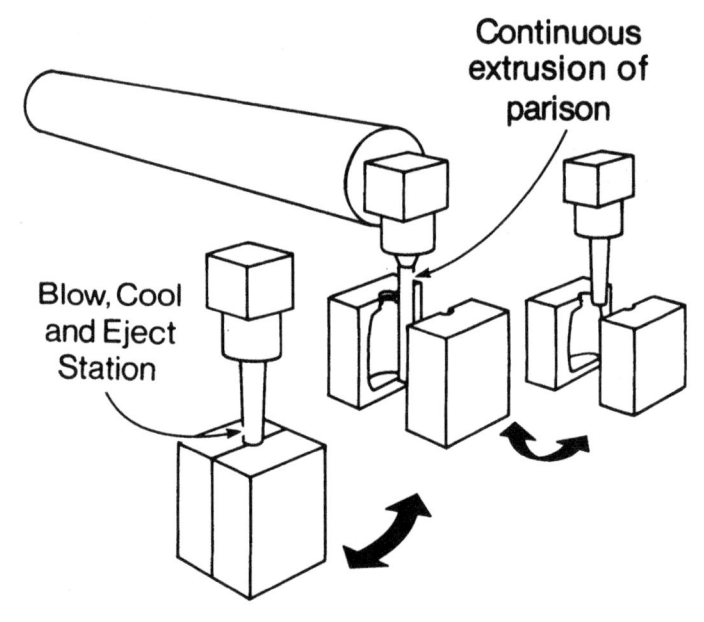

FIG. 11.5. Central parison with dual blowing positions (shuttle machine). (*Courtesy Hoover Universal Plastics Machinery Div., Manchester, MI*)

FIG. 11.6. Shuttle machine with dual mobile clamp carriage. (*Courtesy Bekun Plastics Machinery, Inc., Williamston, MI*)

The shuttle machines are widely used because of the great variety of blown products which may be manufactured, in a size range from a few ounces to 5 gallons. Productivity can vary from a minimum on a single-parison, single-carriage machine to a maximum on a multiple-parison, dual-carriage machine.

Intermittent Extrusion Process

The intermittent process is very similar to the continuous extrusion process, except that the parisons are extruded intermittently to accommodate different kinds of operation mechanisms.

Accumulator Machine. One or more accumulating chambers are intermittently filled with material by an extruder and the material is rapidly forced from the chamber by a ram in parison form. The mold halves are in position to receive the parison as it is extruded, and they immediately close when the parison reaches the required length. While the product is being blown and cooled, the accumulating chamber is being filled again for the next molding cycle. When the product has cooled sufficiently, the mold is opened, the product removed, and another parison is extruded for the next cycle. The machines may be equipped with parison programming for wall-thickness control. Some machines utilize a stationary clamping mechanism as shown in Fig. 11.7, while others are equipped with single and dual mobile clamp carriages.

The accumulator machines are primarily used for the manufacture of large containers and other large products in a size range of 2½ to 800 gallons for the chemical, industrial, automotive, and toy markets.

Reciprocating-Screw Machine. Material is intermittently accumulated in the extruder barrel at the front end of the extruder screw. By reciprocation of the screw, the material is forced from the barrel in the form of parisons. As in the accumulator machine, the molds are in the parison-extruding position as the parisons are extruded. Reciprocating-screw machines can be equipped to extrude 2 to 12 parisons simultaneously (Fig. 11.8). They are primarily used for the manufacture of containers in a size range of 8 ounces to 5 gallons for the dairy, food, automotive, and household chemical markets. The most popular product application has been thin-wall and light-weight 1-gallon milk bottles.

Extrusion Process Variations

Many of the extrusion process machines are capable of manufacturing a great variety of blown products through the use of optional equipment and cleverly designed molds, as shown in the following discussion.

FIG. 11.7. Blow molder and mold for plastic fuel tank. (*Courtesy Kautex Machines, Inc. (Krupp-Kautex Source), Linden, NJ*)

FIG. 11.8. Reciprocating-screw blow molder producing one gallon bottles. (*Courtesy Hoover Universal Plastic Machinery Div., Manchester, MI*)

Neck Calibration. After the parison has been cut off, the blow pin is plunged into the end of the parison to simultaneously (1) force the hot material into the neck ring threads, (2) form the inside diameter and top surface of the neck, and (3) pinch off the excess plastic above the neck of the container (Fig. 11.9).

Handleware. Containers with integral handles are formed by using a thin-wall parison which is large enough in diameter to pinch out the handle area as the mold closes and flattens the parison (Fig. 11.10).

Pinched Neck. Containers having a ratio greater than 3/1 between the body width or diameter and the neck diameter, are formed from a parison having a diameter greater than the neck diameter. When the mold is closed, the excess plastic will be pinched off at the neck ring and from a section of the container shoulder (Fig. 11.11).

Captive Air. Blown products, without a conventional neck or any other opening, are formed by pinching one end of the parison as the mold closes, and, after being blown, the other end of the parison is closed and sealed by sliding pinchers in the mold (Fig. 11.12).

3-STEP CALIBRATION

1. MOLD BODY
2. BLOW PIN
3. STRIKER PLATE
4. CUTTING SLEEVE
5. STRIPPER PLATE
6. STRIPPER PLATE HOLDER

| NECK CALIBRATION BOTTLE BLOWING | 1st. STEP NECK FINISH SEPARATION FROM THE TOP FACE OF BOTTLE NECK | 2nd STEP FOLLOWING MOLD OPEN, BOTTLE DROP BLOW PIN LIFTS TO MIDPOINT ALLOWS CARRAGE TO SHUTTLE TO POSITION | 3rd STEP BLOW PIN LIFTS TO UPPER POSITION TO STRIP NECK FLASH |

FIG. 11.9. Schematic of neck calibration (3 step). (*Courtesy Hoover Universal, Ann Arbor, MI*)

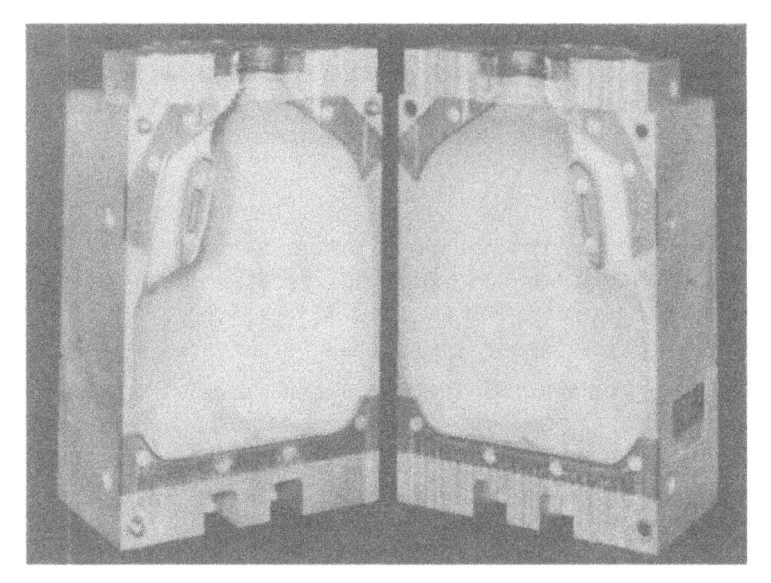

FIG. 11.10. Pinch out for handle of one gallon bottle mold. (*Courtesy Hoover Universal Plastic Products Div., Ann Arbor, MI*)

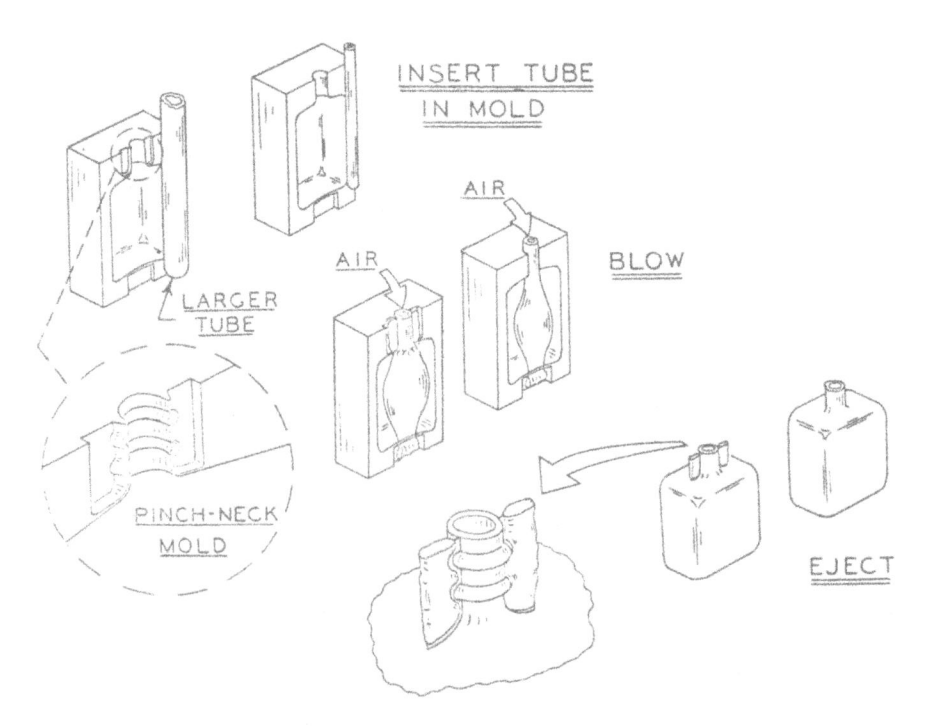

FIG. 11.11. Pinch-neck and "regular" neck processes. (*Courtesy Monsanto Chemical Co., Hartford, CT*)

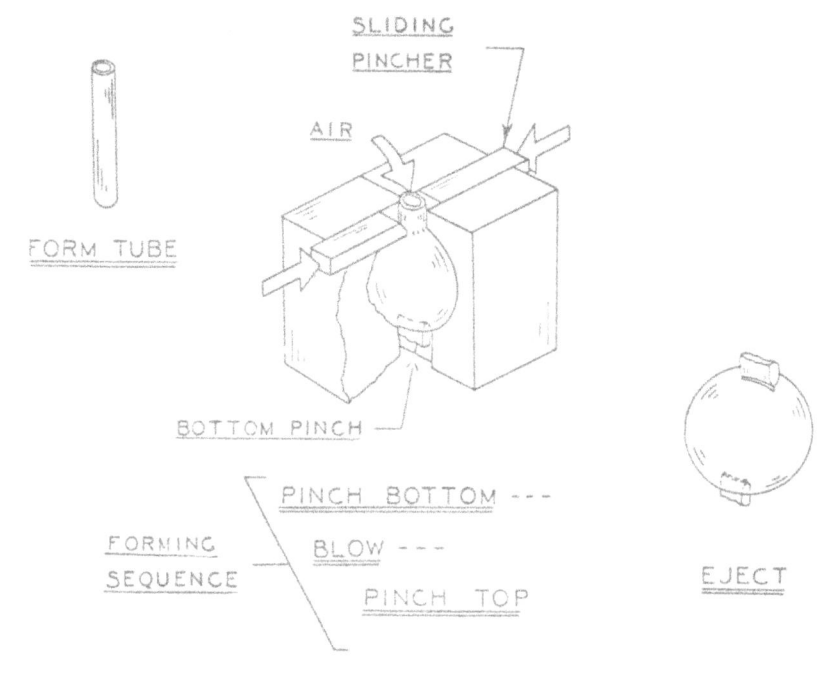

FIG. 11.12. Captive air tooling. (*Courtesy Monsanto Chemical Co., Hartford, CT*)

Form-Fill-Seal. Containers are formed conventionally. Before the mold is opened after forming, the container is filled with the liquid to be packaged. The filling operation is through the blow pin in each cavity. When filling is complete, the neck end of the container is pinched and sealed (Fig. 11.13).

Formed Neck Process

In this process, the neck of the container is precision-formed by pressing the neck ring and integral blow pin against the extrusion die and extruding the plastic material into the interior space. After the neck is formed, the neck ring and blow pin move away from the extrusion die while additional material is extruded at a controlled speed for the section of the parison which will form the body and base of the container. In one variation of the process, a short blow pin is used and the parison is pinched off by the mold halves which close on it (Figs. 11.14A and 11.14B). In another variation, a long blow pin is used which supports the entire length of the parison; when the desired parison length has been extruded, it is cut off and the parison is rotated 180° by the blow pin to the blow mold (Fig. 11.15). Commercially available machines are not yet in wide use, but proprietary machines employing the first variation of the process have been used for many years.

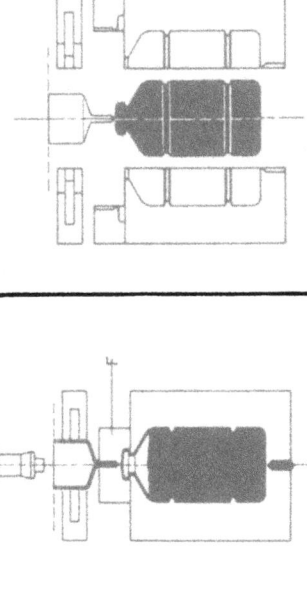

1. The thermoplastic (A) is continuously extruded in a tubular shape. When the tube (or parison) reaches the proper length, the mold (B) is closed and the parison is cut off at (C). The bottom of the parison has been pinched closed and the top is held in place with a set of holding jaws (D). The mold is then transferred to a position under the blowing and filling nozzle.

2. The blow-fill nozzle (E) is then lowered into the parison until it forms a seal with the neck of the mold. The bottle is formed by blowing filtered compressed air into the parison, expanding it out against the walls of the mold cavity. The compressed air is then vented from the bottle and a metered amount of product is forced into the bottle thru the fill nozzle. When the bottle is filled, the nozzle is retracted to its original position.

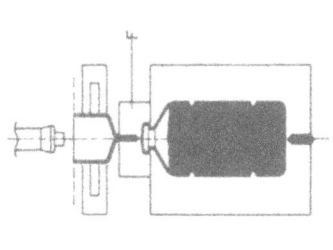
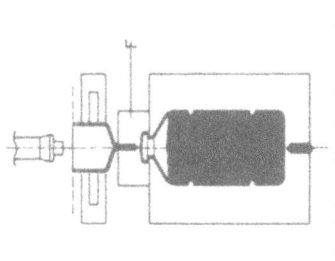

3. At this point in the cycle, the length of parison between the top of mold and the holding jaws is still semi-molten. Separate sealing molds (F) close to form the cap and hermetically seal the bottle.

4. After the bottle is sealed the molds open. The finished bottle, completely formed, filled, sealed and bottom de-flashed is then conveyed out of the machine.

Fig. 11.13. Schematic of mold, fill and seal in one operation of four steps. (*Courtesy Automatic Liquid Packaging, Inc., Elk Grove, IL*)

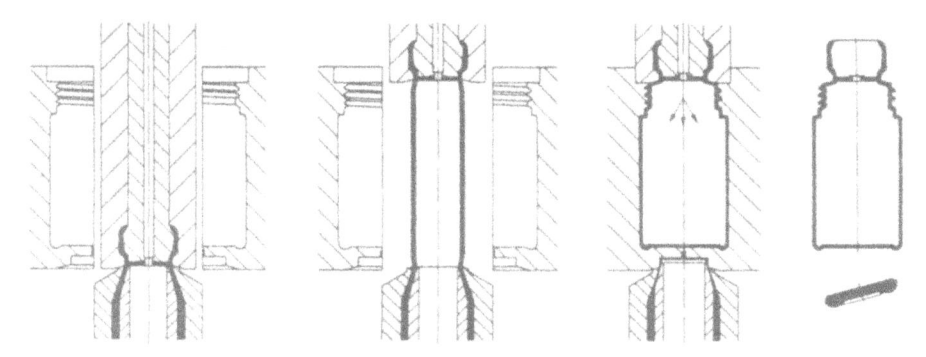

FIG. 11.14A. In injection stretch-blow of a roll-on bottle, the ball holder and the inner-cover section are injected to size without scrap. The parison is then drawn through a calibrated nozzle. The hollow body is blown and the bottom is trimmed.

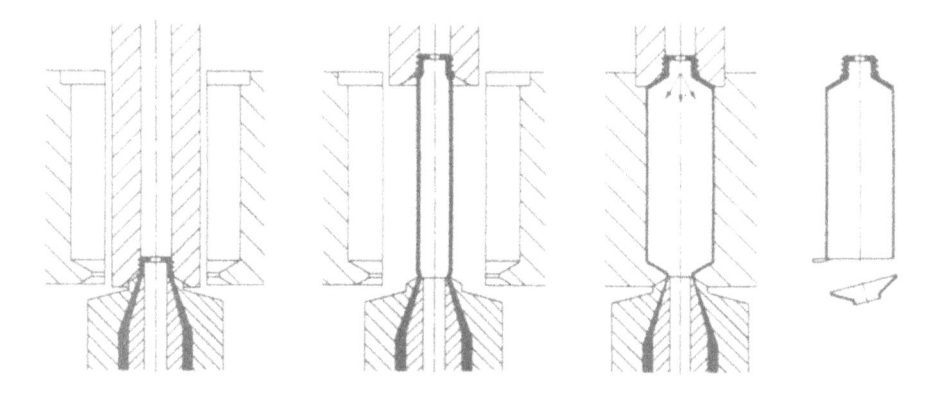

FIG. 11.14B. A collapsible tube can be produced in similar manner, with the neck injected to size, the parison drawn through a calibrated nozzle, the hollow body blown and the bottom cut off. (*Ossberger-Turbinefabrik Source*)

Injection-Blow Process

The injection-blow process combines the function of blowing with the precision of injection molding; it requires a parison-forming mold with integral neck ring, multiple core rods (blow pins) and a blow mold. A core rod is moved into the parison mold and material is injected into the mold to form a precision parison with a fully formed neck. Within certain limits, the parison can be shaped internally by the core rod, and externally by the parison mold, to control the uniformity of the wall thickness of the blown product. While the body of the parison is still hot and plastic, the core rod with the parison on it is lifted from the parison mold and then transferred to the blow mold. Air is introduced through the core rod to blow the parison to the final

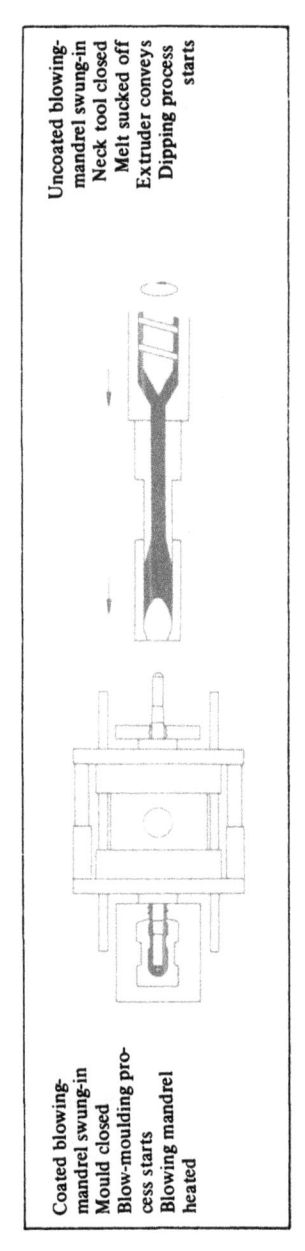

Uncoated blowing-
mandrel swung-in
Neck tool closed
Melt sucked off
Extruder conveys
Dipping process
starts

Coated blowing-
mandrel swung-in
Mould closed
Blow-moulding pro-
cess starts
Blowing mandrel
heated

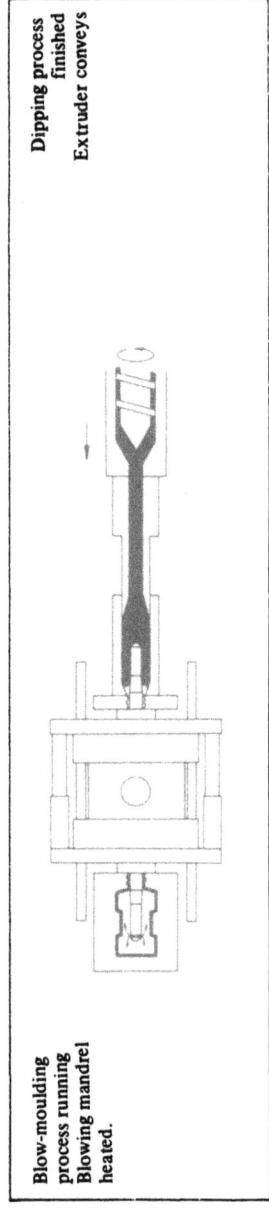

Dipping process
finished
Extruder conveys

Blow-moulding
process running
Blowing mandrel
heated.

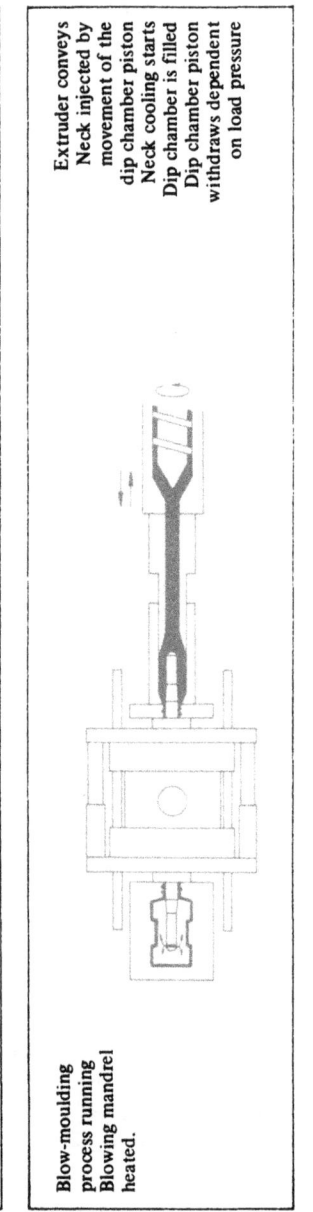

Extruder conveys
Neck injected by
movement of the
dip chamber piston
Neck cooling starts
Dip chamber is filled
Dip chamber piston
withdraws dependent
on load pressure

Blow-moulding
process running
Blowing mandrel
heated.

Blow-moulding process running Blowing mandrel heated.

Dip chamber filled Extruder stops Release of load pressure by movement of dip chamber Coating starts

Blow-moulding process is finished Venting starts Mould opening starts Blowing mandrel heating finished

Coating finished Knife cuts between preform and melt Neck tool opens

Mandrel carrier rotates Bottle is being stripped off

Melt suck-off starts Extruder conveys

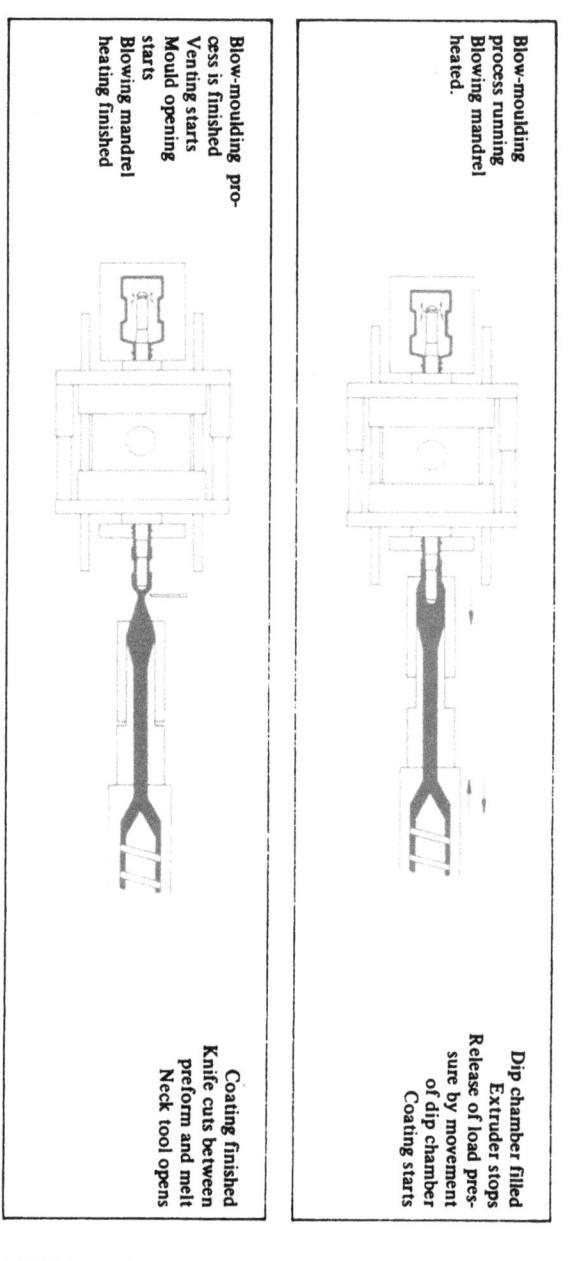
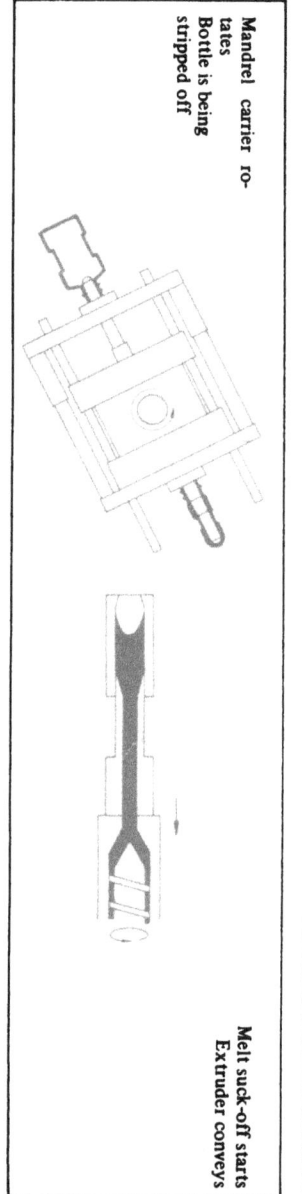

FIG. 11.15. Patented dip-blow molding for small hollow articles. (*Courtesy Staehle Maschinebau GmbH, Dieselstrasse 35, FRG*)

form of the product. After cooling, the product is removed from the core rod as a completely finished article without the need for secondary trimming operations.

The injection-blow process is used to manufacture products requiring close control of dimensions, weight, capacity, and material distribution.

There are three main types of injection-blow machines commercially available. They are discussed below.

Three and Four-Station Rotary Machines. The parison mold and blow mold are mounted in a fixed radial position with respect to each other. The core rods are mounted in a head which indexes intermittently to transfer the core rods from the parison station to the blow station and then to an ejection station (Fig. 11.16). The four-station machine has an extra station, after the ejection station, for conditioning (heating or cooling) the core rod before it is transferred into the parison mold (Fig. 11.17).

The rotary machines are the most widely used of the injection-blow machines because of their versatility in producing precision containers in a size range from a fractional ounce to 64 ounces.

Two-Station Machine. In comparison with the four-station rotary machine, the ejection and conditioning stations have been eliminated. Ejection of the finished product occurs as the blow mold is opened. Conditioning in a two-

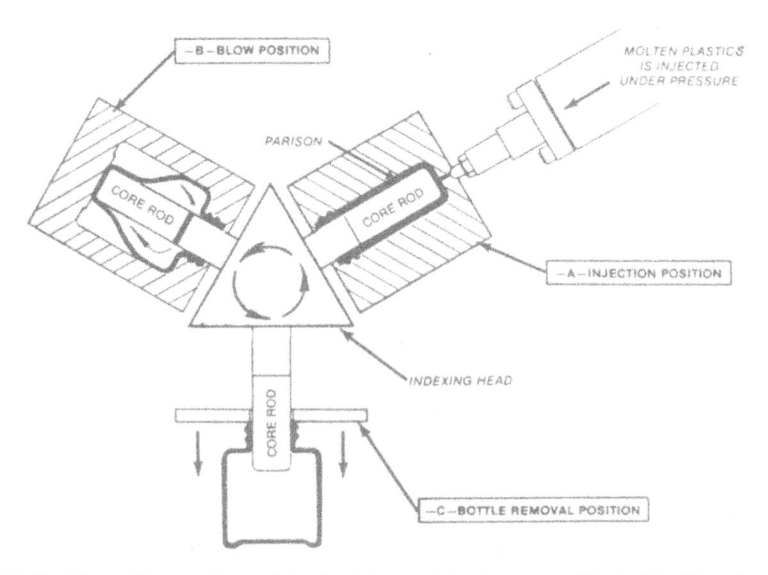

Fig. 11.16. Typical layout for an injection-blow mold. (*Courtesy of Rainville Div., Hoover Universal, Middlesex, NJ*)

FIG. 11.17. Four station rotary machine. (*Courtesy General Machinery Div. of Wheaton Industries, Millville, NJ*)

station machine is normally accomplished by the circulation of temperature-controlled fluid through the core rods.

The two-station machines are used primarily for producing specialized containers for paint or food products.

Adaptive Tooling. Various kinds of systems have been designed and built to adapt standard injection molding machines for blow molding. Such systems generally find application in injection molding companies that want the flexibility to occasionally use their standard machines as blow molders.

Stretch-Blow Process

The stretch-blow process is used to manufacture plastic containers which generally are lighter in weight and have greater clarity and improved physical properties, compared to containers manufactured by other blow-molding processes. In stretch-blow molding, the container material is biaxially-oriented by stretching and blowing the parisons at a critical temperature, which is different for each of the materials commonly used (Polyethylene terephthalate (PET, also PETP), Polyvinyl chloride (PVC), and Polypropylene (PP)). The parisons, after being formed, are temperature-conditioned, stretched longitudinally up to twice their original length by mechanical means, and then blown in a blow mold to the container shape. Stretch-blow machines and molds are more complex and more costly than those used in other blow-molding processes and, as a consequence, the stretch-blow process tends to be limited to high-volume production.

The machines commercially available vary in the means used to form, temperature-condition, stretch, and move the parisons from station to station in the machine, some of which are shown below.

Single-Stage Machine. In a single-stage machine, all of the steps involved in producing a fully formed, biaxially oriented container occur in continuous sequence in a single machine.

In one kind of machine, the parisons are precision-formed in a fashion similar to the injection-blow process. After the parisons have cooled to a point at or below the temperature at which orientation occurs, they are removed from the parison tooling and transferred to a conditioning station where they are heated and/or allowed to equilibrate to their critical orientation temperature. The parisons then are moved to the blow station, where they are held by the neck finish, stretched mechanically by rods inserted into the open necks, and immediately blown to the shape of the container. In one type of machine, the sequence of operations occurs in a rotary fashion (Figs. 11.18 and 11.19), and in another type the operations are carried out in linear movements (Figs. 11.20 and 11.21). Such single-stage machines are being used to produce clear, light-weight PET containers in a size range of ½ to 2 liters for packaging carbonated beverages.

Fig. 11.18. Single stage stretch-blow injection machine with four station carousel. Primary use is for PET bottles of limited production volume. (See Fig. 11.19) (*Courtesy Cincinnati Milacron, Batavia, OH*)

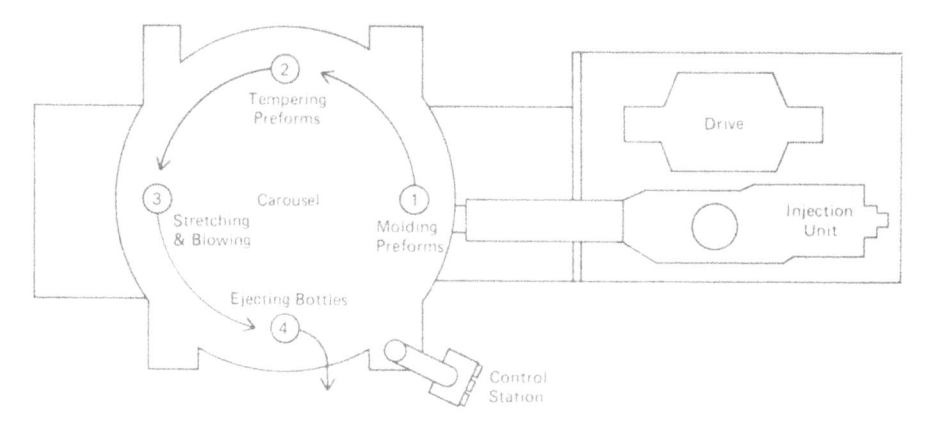

FIG. 11.19. Schematic of machine shown in Fig. 11.18. (*Courtesy Cincinnati Milacron, Batavia, OH*)

Another kind of single-stage machine is similar to a continuous-extrusion shuttle machine except that a preform station and mold have been interposed between the parison extruding and blowing stations (Fig. 11.22). The operational sequence is shown in Fig. 11.23 and typical tooling is shown in Fig. 11.24. This type of machine is used primarily to produce containers from PVC or other heat-sensitive materials, in a size range of ⅓ to 2 liters.

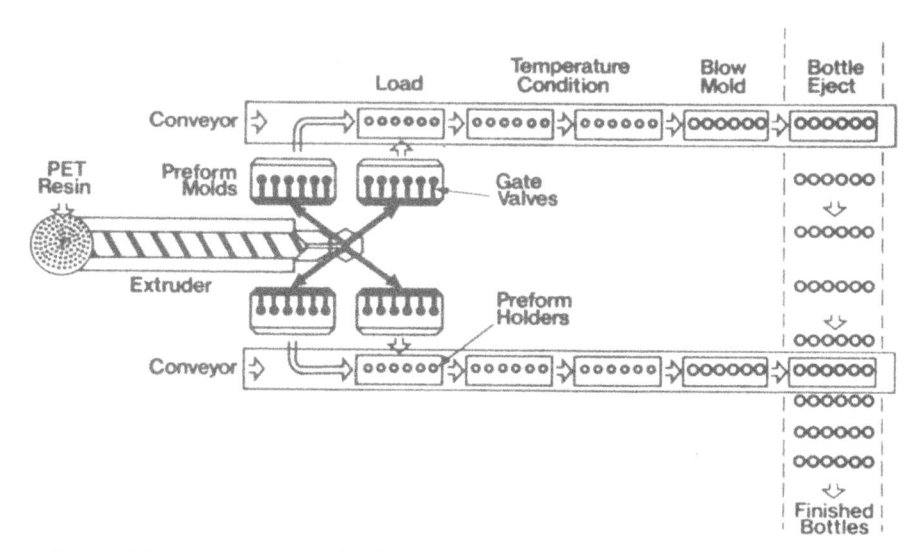

FIG. 11.20. Continuous extrusion into four preform molds discharging to two conveyors. Processing is identical for both conveyor lines. Specific design for 1/2 liter PET beverage bottles. (*Courtesy Van Dorn Plastic Machinery Co., Strongville, OH*)

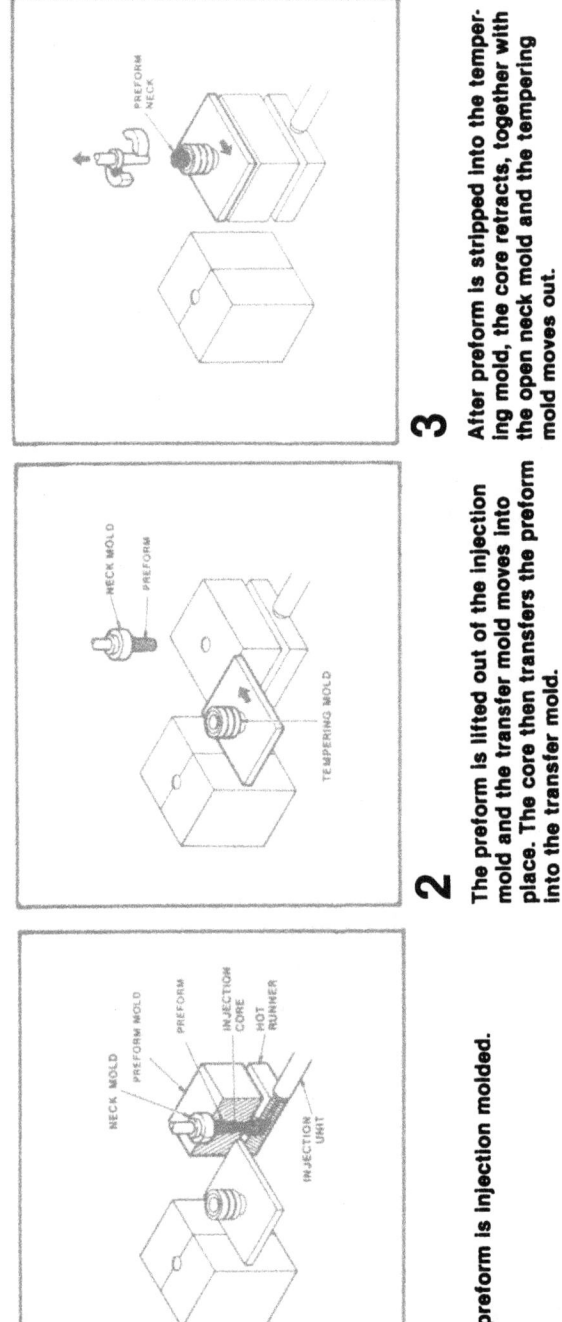

3

After preform is stripped into the tempering mold, the core retracts, together with the open neck mold and the tempering mold moves out.

2

The preform is lifted out of the injection mold and the transfer mold moves into place. The core then transfers the preform into the transfer mold.

1

The preform is injection molded.

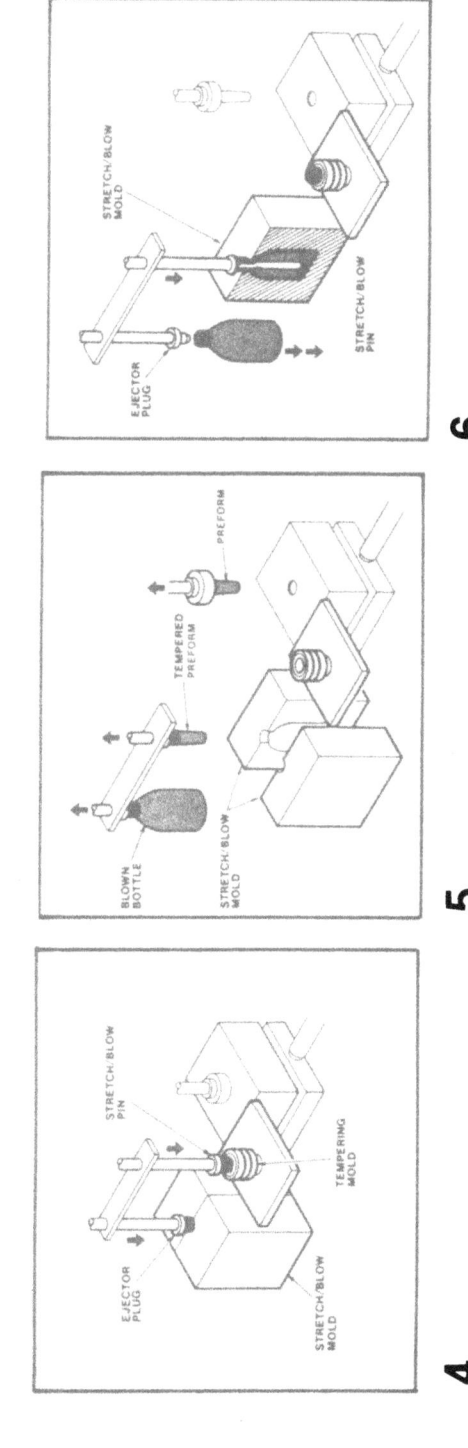

4

The transfer/tempering mold carries the preform to the tempering station. Grippers hold the necks for transfer to a next tempering station or to the stretch/blow mold.

5

The preform is ready for the stretch/blow mold while a finished bottle is transferred to the removal station.

6

The stretch/blow pin stretches the preform the full length of the blow mold and blows it into final shape while the ejector plug discharges the previously finished bottle.

FIG. 11.21. Injection stretch-blow schematics (six steps). (*Courtesy tpT Machinery Corporation, Norwalk, CT*)

FIG. 11.22. Single-stage stretch-blow molding machine with a preform station between the parison extruding and the blowing station. (*Courtesy Bekum Plastics Machinery, Inc., Carlstadt, NJ*)

Two-Stage Machine. In two-stage processing, parisons are formed in a first-stage machine which is normally off-line from the second-stage machine where the containers are blown. Parisons are formed either by injection molding, using specially designed runnerless molds, or by extrusion molding in a fashion similar to pipe extrusion (Fig. 11.25). Injection-molded parisons are precise by nature with fully formed necks, whereas extruded parisons are essentially short pieces of pipe which require secondary forming at both ends. The fact that parison forming is independent from parison blowing allows for the possibility of molding parisons in one location and transporting them to another location for blowing.

The parisons are normally at room temperature when they are fed by automatic unscramblers into the second-stage blowing machine. The parisons pass through a chamber where they are gradually heated to the orientation

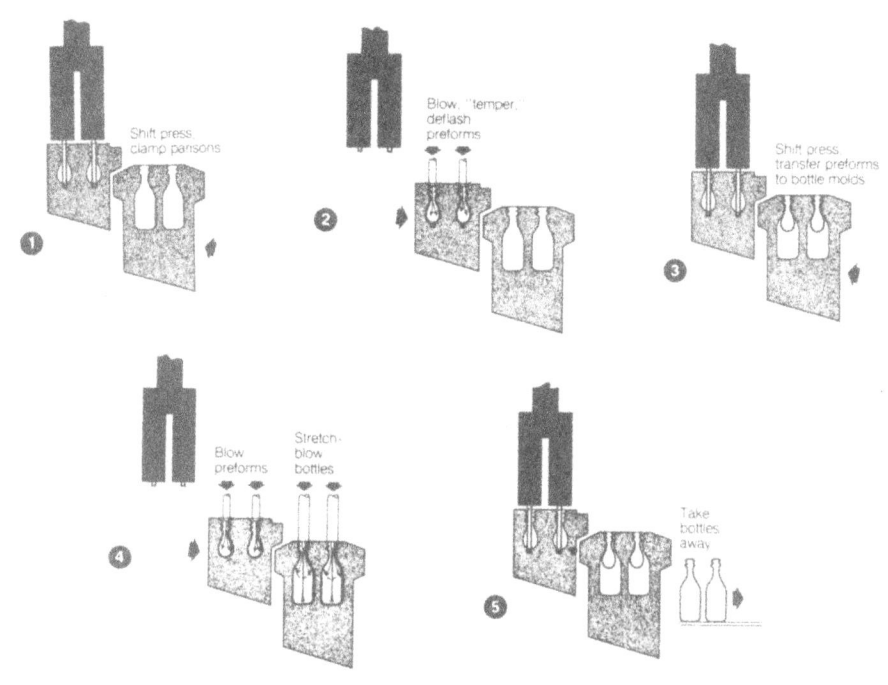

FIG. 11.23. Schematic operation sequence of stretch-blow machine shown in Fig. 11.22. Note shuttle transfer. (*Courtesy Bekum Plastics Machinery, Inc., Carlstadt, NJ*)

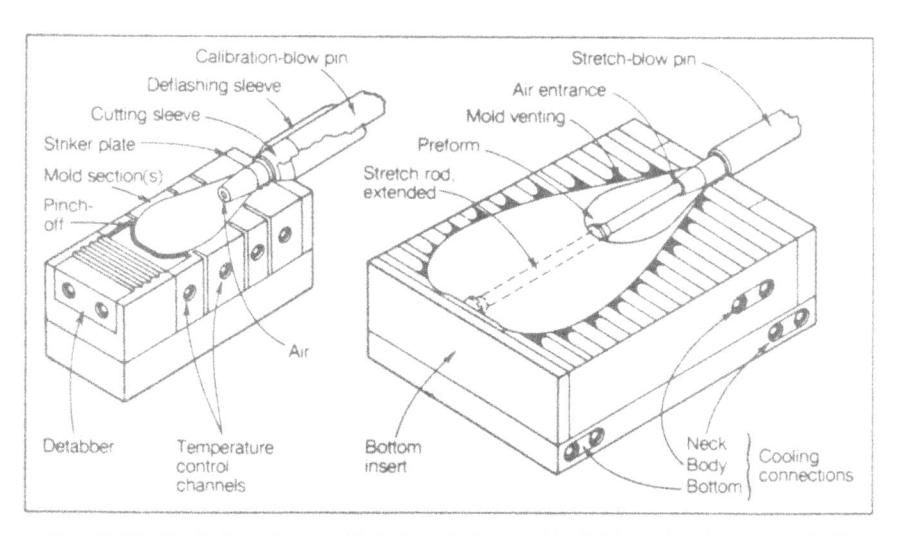

FIG. 11.24. Typical preform mold (left) and blow mold (right) used with machine in Fig. 11.22 and schematics of Fig. 11.23.

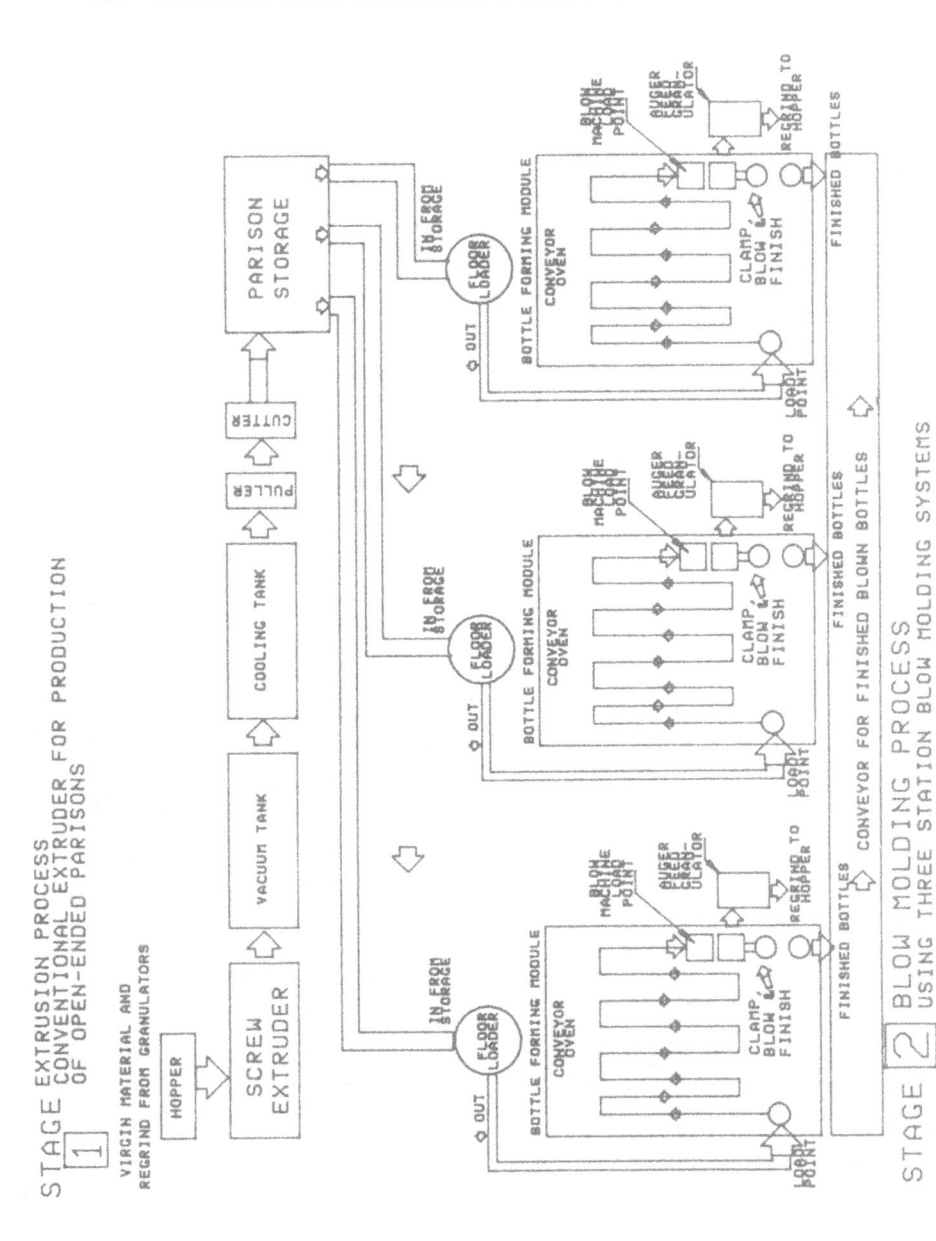

Fig. 11.25. Schematic of two-stage process where parisons are independently produced and stored. Later, parisons are reheated, then blown to bottle shape desired. (*Courtesy Deacon Plastics Machine, Inc., North Adams, MA. for process*

temperature required for the material being processed. The parisons generally rotate as they pass through the chamber, and the chamber is usually equipped with multiple heat zones to optimize the temperature of the parisons.

In one kind of second-stage machine, which uses extruded parisons, the heated parisons are transferred from the heat chamber to the blow station, where one end of the parison is reformed by neck rings which close around it, and by a core (swage) which compression-forms the neck finish (Fig. 11.26); the other end of the parison is gripped by a picker which mechanically stretches it to a length sufficient for the blow-mold halves to close around it and pinch the end shut (Fig. 11.27). This type of machine is being used to produce relatively clear PP containers in a size range of 4 to 48 ounces for packaging sterile medical solutions, household chemicals, and food products.

In another version of the second-stage machine, which uses injection molded parisons, the parisons are restrained at one end by a bead at the base of the neck finish, stretched mechanically by rods inserted into the open necks, and then immediately blown to the shape of the container (Fig. 11.28). This version of the second-stage machine is widely used to produce clear, light-weight PET containers in a size range of ½ to 2 liters for packaging carbonated beverages.

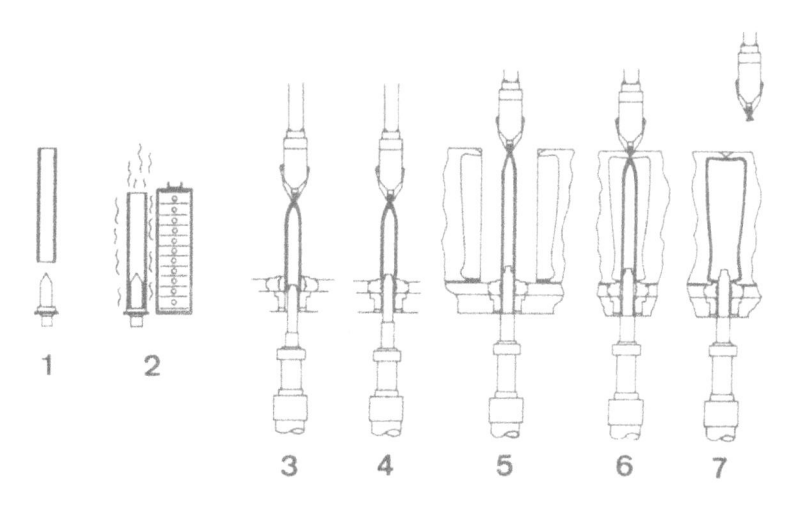

FIG. 11.26. Two-stage stretch-blow schematic. (*Courtesy Deacon Plastics Machine, Inc., North Adams, MA*)

Fig. 11.27. Two-stage stretch-blow tooling. (*Courtesy Deacon Plastics Machine, Inc., North Adams, MA*)

Multiple-Layer Processes

Multi-layer barrier containers have reached the market only recently. Their market success indicates they will be used in a wide range of sizes for packaging edible foods and other products which heretofore were not considered suitable for packaging in plastic containers. Many types of plastics materials are used in the manufacture of barrier bottles. For food packaging, the following properties are needed in some proportion depending upon the exact application: Clarity and low color; impact strength; hot-fill resistance; retortability; water barrier, oxygen barrier, or solvent barrier.

Fig. 11.28. Stretch-blow using injection-molded parison, restraining bead at base of neck, stretch and then blow. (*Courtesy Cincinnati Milacron, Batavia, OH*)

Obviously, multi-layer containers cost more than single-layer containers because they are more difficult to manufacture. Cost becomes secondary when the properties of a barrier layer are needed to minimize permeation (passage) of oxygen and moisture into a container. Also, when it is necessary to prevent escape of the product ingredients outward through the container sidewalls. Costly, but superior, barrier materials (such as Polyvinylidene chloride (PVDC) or Ethylene-vinyl alcohol (EVOH)) are combined in thin layers, usually .004 to .006 in., with heavier structural layers of lower cost materials (such as High density polyethylene (HDPE) or Polyphenylene oxide (PPO)) to form a bottle wall of the desired thickness (usually in range of .032 to .045 in.) In some applications, adhesive tie-layers are used to prevent delamination of the barrier layer and the structural layer. The precision of the shot size, the temperature of cavity and core pin, and injection pressure determine the positioning of the barrier layer(s) in relation to the structural layers. Thus, the principal concern of a tool designer is adequate control of temperatures of various tool components whether in the actual mold, in the manifold, in the die head, or in the feedblock. Temperatures used in the layer dies and feedblock are:

> For the EVOH barrier approx.   425°F.
> For the PVDC layers approx.   375°F.

For PET or polycarbonate   550°F.
For the polyetherimides approx. 700°F.

Other engineering material combinations may require temperatures to 800°F for processing. All of these temperatures may be required in the same feedblock or die head at the same time and just a short distance from each other. Thus, careful design is required to maintain the heat isolation.

Cooling time of the hot thermoplastic material generally determines the minimum cycle time. Thus, not only temperature control of individual parts of each mold become important, but also the ability to move heat from one place to another within the mold. Fast Heat Co., Inc., (Elmhurst, IL) and others are providers of heat transfer devices, and the designer should have a file of catalogs on these devices.

A small number of prominent molders have developed proprietary blow molding equipment for the manufacture of multi-layer containers. However, several types of machines are now commercially available. Others are likely to appear as the barrier container market grows, and the technology of manufacturing multi-layer containers becomes less expensive and less sophisticated. Machines currently used for multi-layer containers are basically the same machines described earlier in this chapter. However, major adaptations make them usable for producing multi-layer parisons.

Continuous Extrusion Machines. The shuttle and the wheel types of machines have been adapted so that multiple screw-extruders feed into a common extrusion head to continuously form a parison tube having three, five, seven, or more concentric layers of materials. Figures 11.29A and 11.30 illustrate the modified machines. All processing generates some scrap material. Multi-layer scrap is difficult to recycle because it is a mixture of materials which cannot be separated. Some applications will allow this mixed scrap to be reused as one of the noncritical layers. Careful control of reprocessing of scrap is required because reprocessing into the wrong layer of a container could be a calamitous situation.

Proprietary wheel-type machines have been used for the majority of barrier bottles manufactured to date. However, wheel-type and shuttle-type are now commercially available. The choice of type of machine basically depends on the rate of production needed, or the flexibility. Choose the wheel machine for high-volume production. Choose the shuttle machine for short production runs where a single parison tube is adequate. Shuttle machines, with their greater flexibility, will become more competitive with the wheel machine in the near future when extruder heads are available to simultaneously extrude two or more multi-layer parisons.

Coinjection-Blow Machines. Nissei ASB Co. (College Park, GA), offers the only coinjection stretch-blow machine commercially available today. It

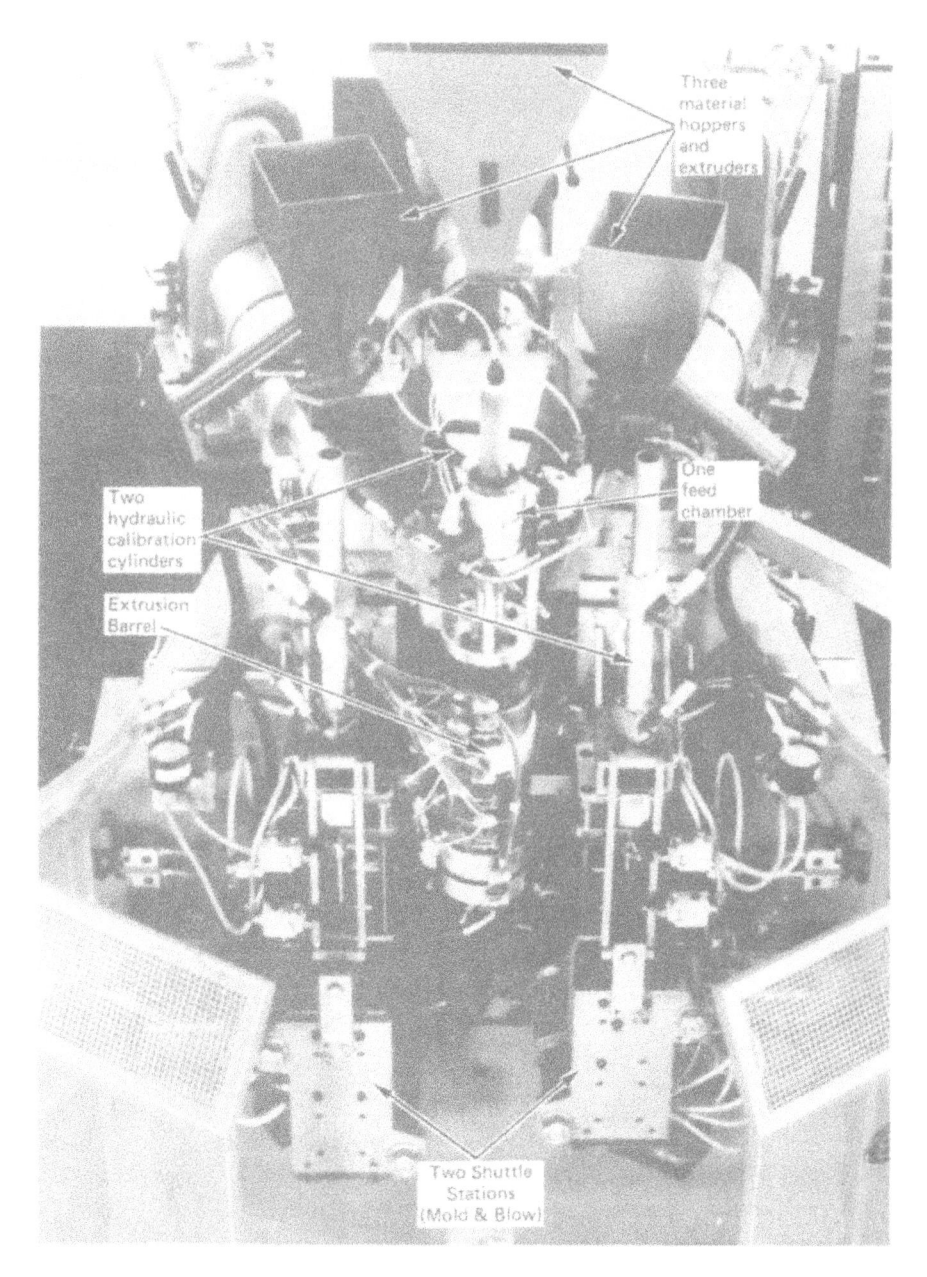

FIG. 11.29A. Continuous extrusion shuttle machine with 3-layer coextrusion. Hydraulic calibration, automatic neck and bottom waste removal. (*Courtesy Bekum Plastics Machinery, Inc., Williamston, MI*)

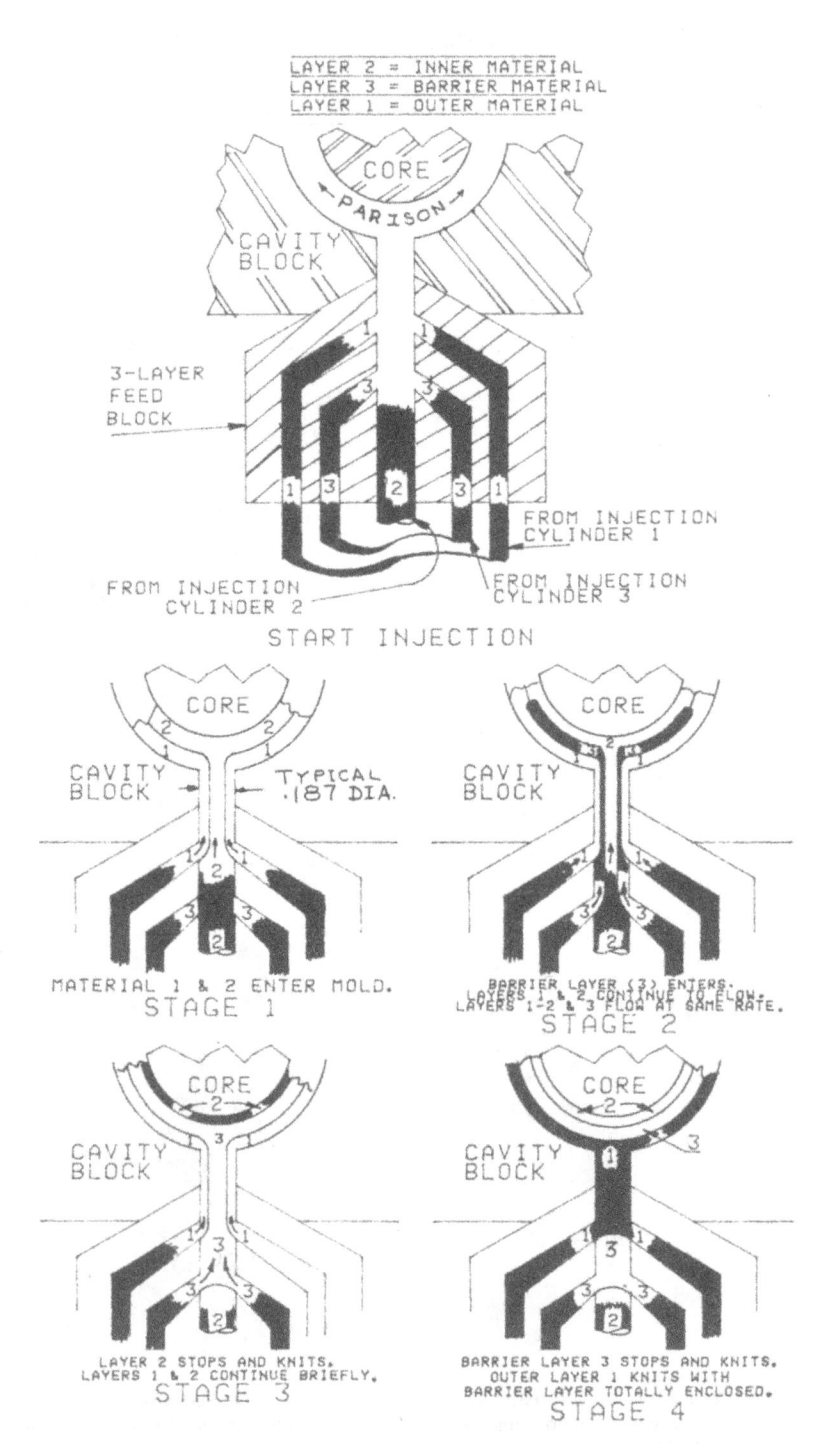

Fig. 11.29B. Multi-layer parison demands precise injection sequence to ensure total encapsulation of barrier material layer. (*Courtesy American Can Co., Barrington, IL*)

is used to manufacture narrow-neck bottles. The machine is similar to their standard stretch-blow machine, but it is equipped with two extruders. One extruder is for the structural material (such as PET) and the other extrudes the barrier material (such as EVOH, or nylon). A special manifold accepts the feed from the two extruders, but keeps the material separated until they are sequentially combined at parison nozzles. The PET material is the first material to be injected into the parison mold. It is followed immediately by the barrier material which flows into the hot interior section of the parison wall, between its cooling interior and exterior walls, to form a three-layer parison with the barrier material sandwiched between the PET layers. An adhesive layer is not used in this process where only three layers are the end result. Heat fusion is relied upon to produce a wall which will not delaminate in service. Coinjection-blow molding has been used with molds up to ten cavities. More cavities are theoretically possible, but costs and molding problems increase in some geometric ratio after the economical point.

A proprietary process and machine (American Can Co., Barrington, IL and reported in Feb. 1985, *Plastics World*) has been developed to produce wide-mouth can-shaped containers to package retortable food products. ("Retortable" means food subject to heat of cooking or sterilization after packaging). The process involves a five-layer tubular parison having an inner and an outer layer of polyolefin, two tie (or adhesive) layers, and an EVOH barrier. The tie-layers incorporate a desiccant to absorb moisture out of the EVOH barrier as the container cools after the retort operation. EVOH loses oxygen barrier properties as it absorbs water moisture, thus, the desiccant restores the barrier to full effect. A study of Fig. 11.29B will reveal the precise timing sequence needed to ensure total encapsulation of the barrier material. It also shows only a three-layer operation, but it should not be difficult to visualize the tie-layer introduced to both sides of the barrier layer as it is injected. One advantage of coinjection-blow molding is the elimination of scrap formed at the neck and base of containers blown by other processes.

Designer involvement in coinjection-blow molding includes:

1. Design of a variety of container shapes.
2. Design of cavities and cores for the molds.
3. Design of nozzles and determination of passage sizes and configuration.
4. Knowledge of plastics materials and their properties as used in this process.

Incidentally, a CAD system capable of three-dimensional solid modeling will be of great assistance in depicting the final container in all its peculiar shapes and geometry. The same CAD program would then generate the CNC instructions for accurate cutting of cores and cavities.

Another proprietary injection-blow process is illustrated in Fig. 11.30. In this three-stage process, the first stage is the extrusion of a multi-layer sheet of material; the second stage is the thermoforming of the sheet into tubular liners having the same shape as the core rods; in the last stage, the liners are automatically fed onto the core rods of the injection-blow tooling. Structural plastic material is then fed into the parison mold with the result of a precision parison, lined with one or more barrier materials. Blowing to size and shape follows the same process as unlined parisons. There is a definite

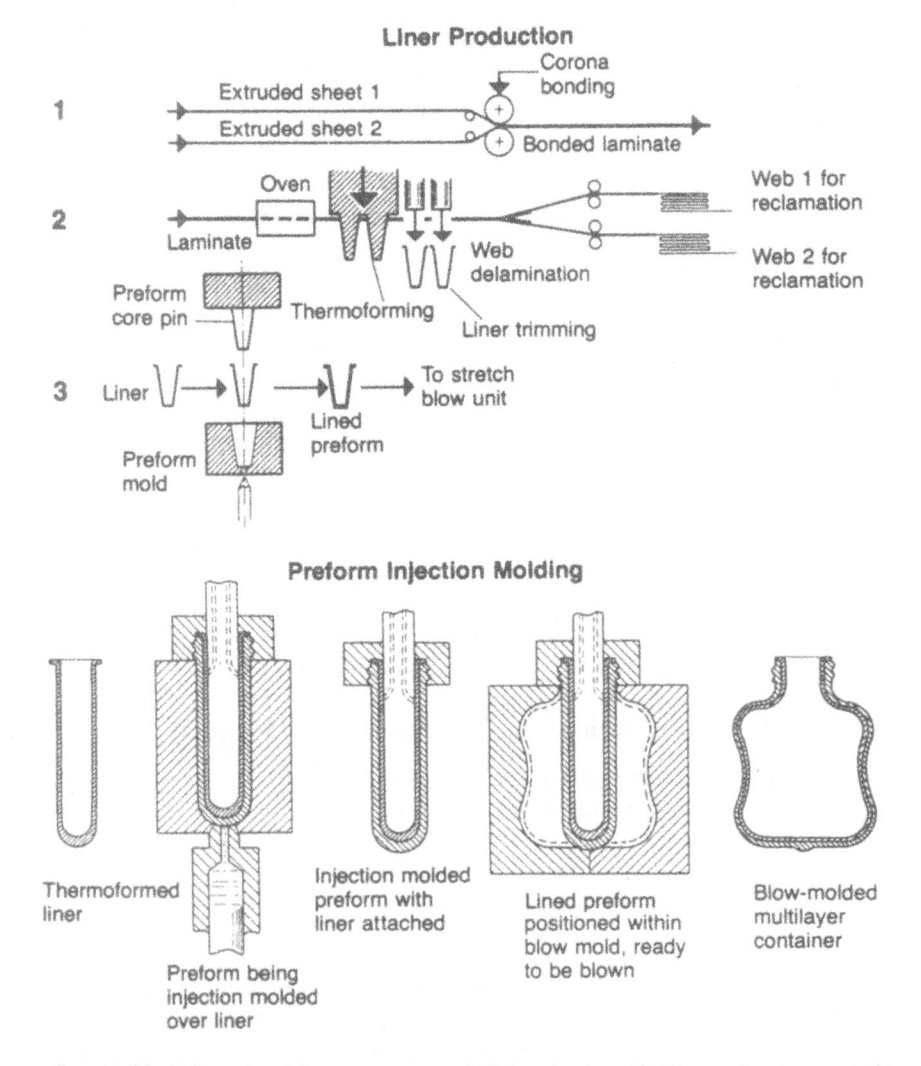

Fig. 11.30. Schematic of 3-stage process of (1) Lamination; (2) Thermoforming; and (3) Preform injection and blow. (*Courtesy National Can Corporation, Bethel, CT*)

limitation on the ratio of length-to-diameter in the thermoforming stage. This information is considered proprietary as of this publication, therefore, we can only alert the designer to a limitation and suggest consultation with "experts" prior to commiting to a size and shape of container.

Two-Stage Machines. Multi-layer containers can be blown on commercially available two-stage stretch-blow machines. A multi-layer parison is extruded in the form of pipe, then cut to the desired length. Subsequently, the parison is reheated and presented to the blowing machine. Obviously, some special technique is required to reform and weld the bottom end of an extruded parison just to assure that the barrier material layer is totally buried and continuous throughout the container. Injection molding of a shaped multi-layer parison seems probable by the publication date of this text. We would be pleased if one of our readers gained enough information from this text to qualify him or her as the designer of the injection mold for a multi-layer parison.

Enhancement of Barrier Properties. The barrier properties of plastic containers are enhanced in ways other than multi-layer processing.

In a process developed by Air Products and Chemicals, Inc. (Allentown, PA), polyolefin containers are blown by a mixture of fluorine with other gasses (instead of compressed air). A standard extrusion blow molding machine is modified to handle the gas mixture. Fluorine is toxic and requires adequate safety precautions against accidental release and injury to operators. During the blowing cycle, the fluorine makes contact with the inner surface of the container and chemically changes the surface to decrease permeation of the container to such nonpolar liquids as gasoline, xylene, and toluene.

In a different process, polyolefin containers are blown by any conventional process. Then, sulfonic agents are used to post-treat and chemically change the inner surface of the container. One application of this process is in the production of automotive fuel tanks.

A third process, used principally in food and beverage packaging, used PVDC or EVOH as a coating material. The coating is applied by spraying, or dipping, or roller coating, in a secondary operation using commercially available equipment.

A fourth kind of coating uses ion deposition of metallic silicon nitride in a thickness of about 1300 Angstrom units (the Angstrom unit is named after Anders J. Angstrom, and is equal to one ten-billionth of a meter. Thus, 1300 Angstrom units equals thirteen one hundred-millionths of a meter). This coating is invisible to the naked eye.

All of the four processes just mentioned have a single purpose: increasing resistance to the permeation of the container by oxygen and carbon dioxide. Every designer must bear in mind the long-term needs of the product and the process. In the case of food and beverage items, a shelf life of two years or more is not uncommon.

Mono-Layer Barrier Process

A single layer of material having the same or similar properties as the multi-layer materials would drastically reduce the cost of containers with low permeability. Material manufacturers are developing such materials for extrusion, or injection, on standard blow molding machines. The materials will be available in the form of homopolymers, copolymers, and alloys of two or more polymers. The blending technology is under development. Through the use of compatabilizing agents, interpenetrating networks, and other techniques, structural and barrier properties of materials will be enhanced to meet the requirements for each end use. Costs of the materials will undoubtedly follow a downward curve as the market volume increases. A projection of the final cost of a container will reveal the intersection point at which mono-layer and multi-layer barrier containers are identical. From that cost point, the public will benefit from the lower cost of machines, simpler processing technology, and more efficient recycling of scrap.

E. I. DuPont De Nemours and Co. has patented and commercialized a process, using a special (amorphous) nylon material (Selar RB), to improve the barrier properties of containers used for household, agricultural, and industrial chemicals. Four to eighteen percent Selar RB material is blended with polyethylene. The blend is processed through an extrusion blow molding machine, but uses a special technique to minimize material shear. The result is a laminar dispersion of discontinuous, but overlapping platelets of nylon in the base polyethylene. This structure inhibits the permeation of volatile hydrocarbons from the packaged products.

A few of the many other blends and alloys in development are combinations of: PET/copolyester, PET/nylon, and EVOH with various base polymers. Various barrier polymers and copolymers are currently available. Others will be commercialized in the near future in this quite volatile market. Watch for developments in materials such as acrylonitriles, polyacrylic imides, and amorphous nylons.

EXTRUSION BLOW MOLD MATERIALS DESIGN AND CONSTRUCTION METHODS*

The blow mold has undergone many changes in its design evolution, but its basic requirement of good heat-transfer properties and resistance to wear

*This section written by Brooks B. Heise, Heise Industries, Inc., West Berlin, CT.

remain of maximum importance. These two factors must be given major consideration in the design of blow molds.

Although many metallic and nonmetallic materials have been used with varying degrees of success, this presentation will consider only metallic materials as the best production material for blow molds. The most commonly used materials are aluminum, beryllium copper and zinc, and their applicable properties are listed in Table 11.1, and rated for those characteristics of greatest interest to a manufacturer and user of blow molds. A discussion of their characteristics follows to clarify the reasons for the arbitrary ratings of excellent, good, fair and poor.

Aluminum

High thermal conductivity, light weight, ductility, and excellent machining characteristics make aluminum an obvious choice for many blow mold applications. While Table 11.1 lists the thermal conductivity of pure aluminum, the alloying elements usually added to enhance its ability to be cast and machined do not seriously impair its heat conductivity. Wear resistance has been rated as poor, but some of the high strength aluminum alloys such as 6061 and 7075 have been used successfully in those areas of a mold which come in contact with the plastics and are subject to wear.

Minor repair of dents or nicks to mold cavity edges can usually be made by peening or by mechanically moving material into depressions. The amount of material movement is limited by its ductility which varies with the different alloys of aluminum. Where mold construction conditions permit, some repairs can be made by inlaying pieces of similar aluminum and mechanically fastening them in place. This type of repair is more desirable than welding in that no heat is involved to impair the strength of the aluminum. For major repairs, aluminum can be successfully welded, but as a result of this process the aluminum becomes annealed thus losing much of its hardness and compressive strength. It is for this reason that the repairability of aluminum is classified as fair in Table 11.1.

In consideration of aluminum's low density of .097 lb/cu in. and its price of $3.50 to $5.00 per lb in 1986, it is one of the lowest cost blow mold materials.

Beryllium Copper

Beryllium copper is chosen as a mold material where the prime considerations are hardness, high wear resistance, high thermal conductivity, resistance to corrosion, and ease of repairability. Copper, when alloyed with a very small percentage of beryllium to gain hardness and castability, loses

TABLE 11.1.   Properties of Blow Mold Materials.

Material	Thermal Conductivity	Wear Resistance	Ability to be Cast	Ability to be Repaired	Density lb/cu in.	Ability to be Machined and Polished	Approx. Cost per Pound—1986
Aluminum	.53	Poor	Fair	Good	.097	Excellent	$3.50–5.00
Beryllium copper	.15–.61[b]	Excellent	Good	Good	.129–.316[b]	Fair	$6.70–8.00
Zinc	.27	Fair	Excellent	Good	.217	Excellent	$ .40– .50[c]

[a]Calories per square centimeter per second per degree centigrade.
[b]Data per Brush Wellman Co.
[c]Data per American Metal Market.

some thermal conductivity and machinability. As noted in Table 11.1 thermal conductivity can vary from .15 to .61 cal/sq cm/sec/°C. Typically, beryllium copper having 2% beryllium, .5% cobalt, and the balance copper will show a conductivity of .15 in the cast state. Heat treating this same material and increasing its hardness from Rockwell B55-70 to C37-42 can improve its conductivity to .31. By reducing the beryllium content to .6% and increasing the cobalt content to 2.60%, the conductivity of cast material can range from a low of .25 to a high of .61 when undergoing a change in hardness from Rockwell B30-45 to B90-100.

A compromise formulation containing approximately 2.0% beryllium is usually chosen when casting blow molds of beryllium copper. If mold inserts or cavities are to be machined directly from bar stock, the percentage beryllium usually used is approximately 1.65.

With material cost in 1986 ranging between $6.70 per lb for casting ingot to $8.00 per lb for high density stock, beryllium copper is the most expensive mold material presently used for blow molds. Oftentimes, to keep costs at a minimum, beryllium copper is used only in those areas of a mold where its most desirable properties are required. Usually these are in the neck and base areas where the plastics material is squeezed between the mold halves. Even if cost is of no object, the material weight can sometimes eliminate use of beryllium copper for mold construction because of mold machine limitations or because of in-plant handling considerations.

Its ductility, as in the case of aluminum, permits minor repairing by peening. Inert gas welding is usually employed for major repairs. While some annealing takes place as a result of welding, the serviceability as a blow mold material is not jeopardized.

While the tendency of beryllium copper to work harden and its toughness make it difficult to machine, adherence to proper cutting angles, coolants, etc., can facilitate this problem. Booklets describing best machining practices are available from the manufacturers of beryllium copper.

Zinc

Zinc or the many commercial formulations of zinc alloy have been widely used for molds, primarily for the ease with which they can be cast. Although its thermal conductivity is lower than that of aluminum or beryllium copper, the assurance of sound castings precludes the possibility of reduction in conductivity due to porous sections. Porous castings must be avoided for best thermal transfer. It is therefore possible to use zinc for molds having widely varying thickness of wall section which are not practical with other basic materials that can easily develop porous sections in the heavy areas.

The wear resistance of zinc is only fair. This characteristic probably compares equally with aluminum, such as 2024. Molds have been found to show consequential wear in the areas where the plastics material is squeezed between the mold faces. The resistance of zinc to wear does not compare favorably with that of beryllium copper. Wear such as described above is normally corrected by removing a minimum amount of material from the mold parting faces. Material may be added to the mold to facilitate repairs by standard gas or Heliarc welding and can be displaced by peening or staking.

Zinc has the lowest price per pound, and even though its density runs a close second to beryllium copper, its net cost for mold work is the most favorable. A lower mold cost is also reflected by the ease with which this material can be machined. As in the case of beryllium copper, weight is an important factor that must be considered in comparison with aluminum.

Steel

Steel has not been a popular material for extrusion blow molds due to its relatively low thermal conductivity and its poor resistance to corrosion. For injection blow molds, however, a variety of prehardened and stainless steels are used quite extensively. A complete discussion of steels for this application is covered in Chapter 4.

EXTRUSION BLOW MOLD FABRICATION

The two most widely used methods of extrusion blow mold constructions are by machining or casting. The choice of method is dependent on factors such as, number of cavities, mold material, intricacy of design and accuracy of cavity sizes. Generally, if many cavities are involved, and the material is one that lends itself to casting, and cavity size tolerances are not critical, then casting produces the lowest cost mold. Some designs, due to the intricacy of detail, are easier and therefore less costly to produce by casting rather than machining and vice versa.

The mold material to be used is probably the most important factor in selecting the fabrication process. Zinc, which is readily cast, can produce excellent results by gravity casting over a metal master made from mild steel. The detail is sharp and the accuracy of cavity sizes is as good as any casting technique providing times and temperatures during casting are closely controlled. If a ceramic master is used then both detail and accuracy are sacrificed.

Aluminum as noted in Table 11.1 is difficult to cast and only certain alloys are castable. Generally alloys that are high in silicon are best suited for casting, while a high strength alloy such as 7075 cannot be cast and must be machined. The best detail and accuracy in an aluminum casting can be obtained by pressure casting and die casting wherein a hardened metal master usually made from an H-13 steel is used. Neither of these casting methods has been popular because of the ease with which a higher strength alloy can be machined with the inherent accuracy of cavity sizes in this cutting process. Gravity casting of aluminum over ceramic has been limited primarily to large mold cavities of the 5 gal. variety and up. Even in these castings the aluminum foundry will core out a portion of the casting to permit as nearly uniform thickness of wall section as possible in an attempt to keep porosity at a minimum.

Beryllium copper, like aluminum, can best be pressure cast to prevent porosity in molds of varying cross sections. A hardened metal master of H-13 steel is usually used, but under repeated heating and cooling the H-13 becomes annealed. Subsequent re-hardening of the master can cause distortion and a change in size which affects the size and accuracy of the ultimate cavity. Machined beryllium copper has produced the highest quality blow mold.

Regardless of whether a mold is machined or cast, shrinkage factors must be considered. The machined mold cavity as shown in Fig. 11.31 must allow for shrinkage of the plastic used plus any other shrinkages which will take place in the subsequent post-molding operations on the blow molded part. For a cast mold, the above shrinkages must be considered together with an allowance for the amount the metal will shrink during casting. Once all shrinkage factors have been predicted, the pattern is displaced to check its size before metal is either cast or cut. This check verifies the mathematical computations used to determine the pattern size initially and prevents possible costly errors in the actual mold construction.

Manufacturing the Blow Mold

While the hand duplicator or copy milling machine illustrated in Fig. 11.31 is still the backbone of the mold-making industry because of its ability to cut just about any shape with a high degree of accuracy, it is slow and requires a highly skilled operator in order to obtain the optimum quality and accuracy of cut. This hand duplicator has given way to the automatic duplicator. Even more recently, it has given way to the numerically-controlled (NC) or computer numerically-controlled (CNC) machine, because automatic duplication or computer control will increase productivity as well as provide more accurate contours needing less handwork to finish to final surface and dimension.

The automatic duplicator, Fig. 11.32, with its multiple heads, can cut as many duplicates as there are cutting heads. The path of the cut is controlled by a stylus which traces over a pattern of the desired shape. The stylus is connected, in turn, to a sensing device which will transmit to the hydraulic or electronic control units the information they need to direct the machine

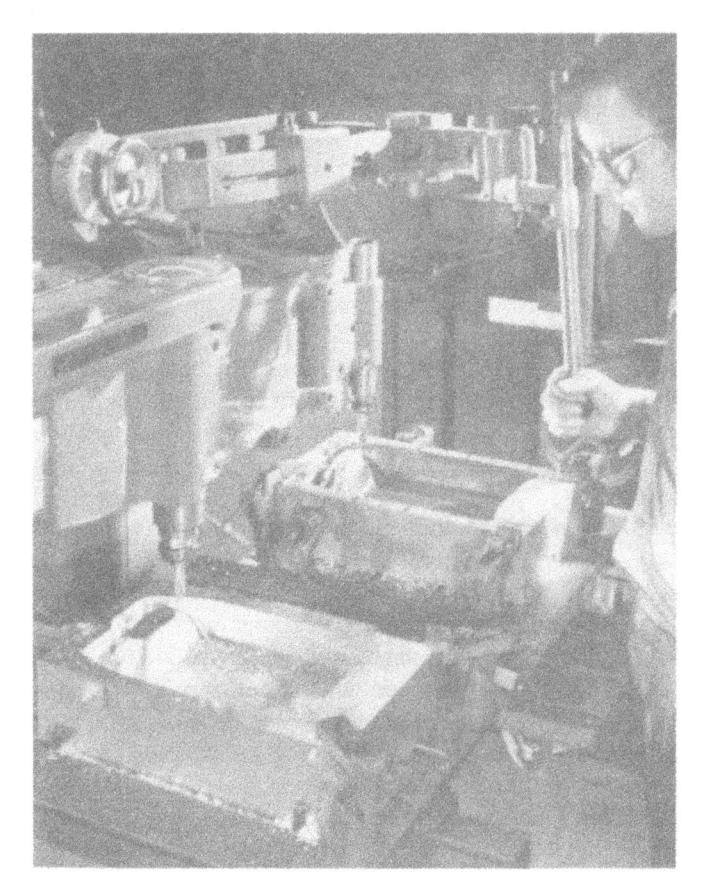

Fig. 11.31. Typical blow mold construction. The cavity at left is being duplicated from master form at right. (*Courtesy Heise Industries, Inc., East Berlin, CT*)

to cut the correct shape of the reproduction. In all instances, some movement of the stylus is required to transmit a change in direction to the tool path. This fact, together with the necessity for the stylus diameter to be larger than the cutter diameter, eliminates this method of duplication when fine detail is required.

Numerically controlled machinery has been available for a number of years for two- and two-and-one-half axis machining. Now these machines have added controls with their own computer to enhance their flexibility of operations. The most recent technological improvement has been the use of computer graphics to program or direct the operations of the NC machines. This recent technology is the so-called CAD/CAM system (CAD is the acronym for Computer Aided Design; CAM is the acronym for Computer

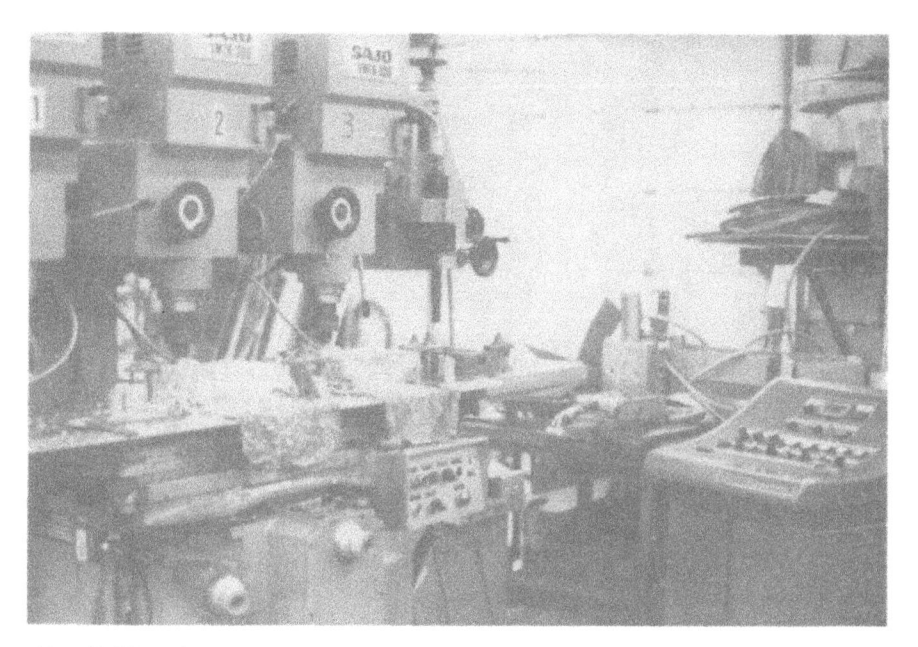

Fig. 11.32. A 3-spindle automatic duplicator. (*Courtesy Heise Industries, Inc., East Berlin, CT*)

Aided Manufacturing). In reality, the mold designer uses a CRT (cathode-ray tube) on which to draw what would normally be drawn with pencil and paper. Figure 11.33 illustrates this CRT technique. That which is drawn on the CRT can be reproduced at *any* desirable size or scale to make the familiar "tracing" associated with drafting. In the case of CAD, this reproduction is called "plotting." The important point here is this: *whatever shape can be drawn or modeled on the CRT can be reproduced by the NC or CNC machine within the normal limitations of tool geometry.* (Refer to Fig. 11.34.) It should be obvious that the potential for the CAD/CAM system of mold-making is enormous. The overall cycle time to make a mold can be greatly reduced by eliminating the need for a three-dimensional pattern, by lessening the handwork on a mold, and by increasing the accuracy of the duplication. Many NC machines performing multiple tool operations (using automatic tool changers) are capable of unattended operations on a 24-hour basis. However, someone who knows the "shut-down procedure" should be nearby, for that rare occasion when "automatic" is not good enough to prevent catastrophe.

We digress here to make a very important point about "pencil and paper." At some point in every design, it is essential for the eye to see and the mind to comprehend the *actual size of an object.* For example, it is

FIG. 11.33. Cathode Ray Tube (CRT) and keyboard of a Computer Aided Design (CAD) system. (*Courtesy Heise Industries, Inc., East Berlin, CT*)

possible to represent, on a 12-in. CRT, the scale representation of a 10 ft mold. This means that a plate of 12-in. thickness will appear *on the CRT* approximately ¾ in. in dimension. Try looking at a picture drawn at ¾ in. and then try to convince your mind it is *actually* 12 in. We do not wish to belabor this point except to recite an engineer's observation: "That injection machine (manufacturer nameless) was the most *overdesigned* machine he had ever seen. No one had bothered to lay out on a piece of paper the *actual thickness* of the platens. They were almost twice as thick as they needed to be!" By the same token, a .005 in. section can be "zoomed" to fill the 12-in. CRT. The deception to the eye and mind is still there, but in the opposite manner.

At the present time, the technology for modeling nongeometrical (meaning irregular) three-dimensional shapes has not reached the point where *all* such shapes can be accurately described. Those areas on a model which are now blended by the skilled pattern or moldmaker present the greatest challenge to manufacturers of CAD/CAM systems. To overcome this limitation, some CAD/CAM manufacturers offer a digitizing option to generate the data required to establish a tool path for the NC machine. This digitizing option requires "scanning" a model with a stylus similar to that used on the automatic duplicator. The stylus movement is then translated to "digital code" which the NC machine understands and can translate into movement of a cutter or movement of a table, or both simultaneously.

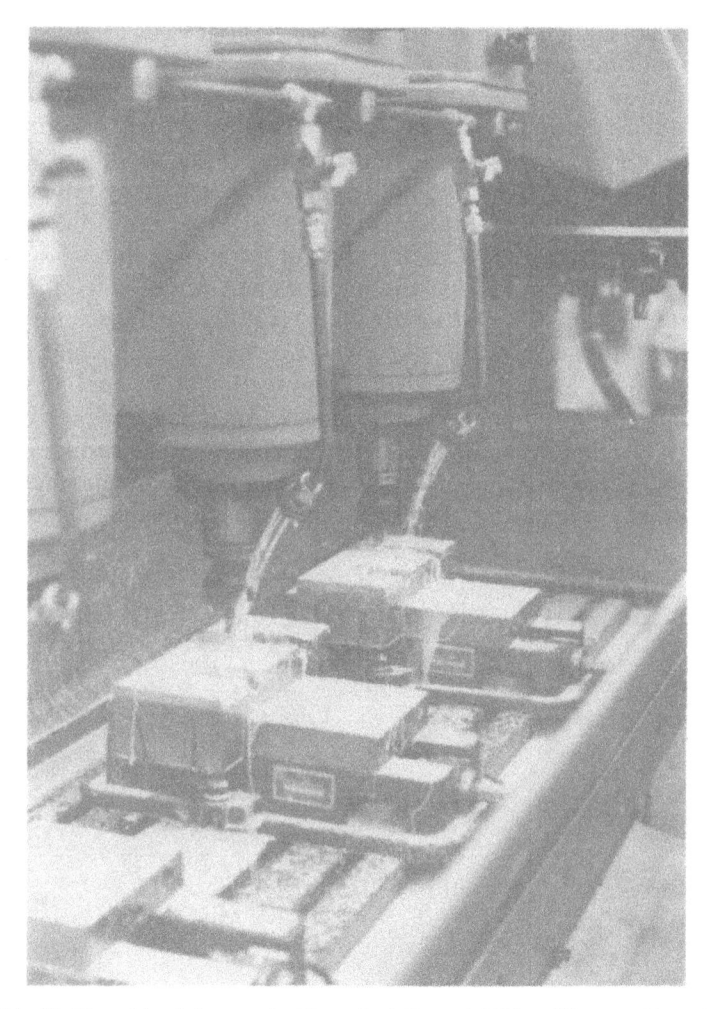

FIG. 11.34. Mold cavities being cut by Numerical Control (NC) milling machine. (*Courtesy Heise Industries, Inc., East Berlin, CT*)

If the makers of NC equipment have their way, all hand duplicators or automatic duplicators will become obsolete. It is true that usage of the less sophisticated machinery will decline in the future. However, bear in mind the areas of the world in which the "industrial revolution" is still to come. The authors believe that this text can educate our successors, even while we address our contemporaries.

With the completion of the mold cavities, they are fully machined on those surfaces which must mate with inserted parts and are sized to fit the particular machine for which the mold is intended. The mold halves (for a two-part mold) then have dowel pins and bushings added to assure proper alignment of the cavity.

Simultaneously with the machining of the cavity halves, inserts to form the neck and base ends of the cavity are being completed. Inserts serve two purposes (Fig. 11.35). First, they include areas that are likely to wear and can therefore be replaced without reworking or scrapping the entire mold. Second, this provides an opening to the cavity that permits the mold maker to match up accurately the cavity halves at the parting line.

Before assembly of inserts, cooling lines are added to the mold. The use of drilled lines and milled channels (Fig. 11.36) is usually employed to carry the coolant. It is desirable to maintain a large area through which the coolant flows and at the same time keep the cross section of the cooling stream at a minimum. The latter requirement keeps the amount of cooling medium to a minimum and assures a rapid change of coolant over the surfaces that are to be cooled. The consistency of cross-sectional area through which the coolant flows is also important. With an increase in the area the coolant velocity

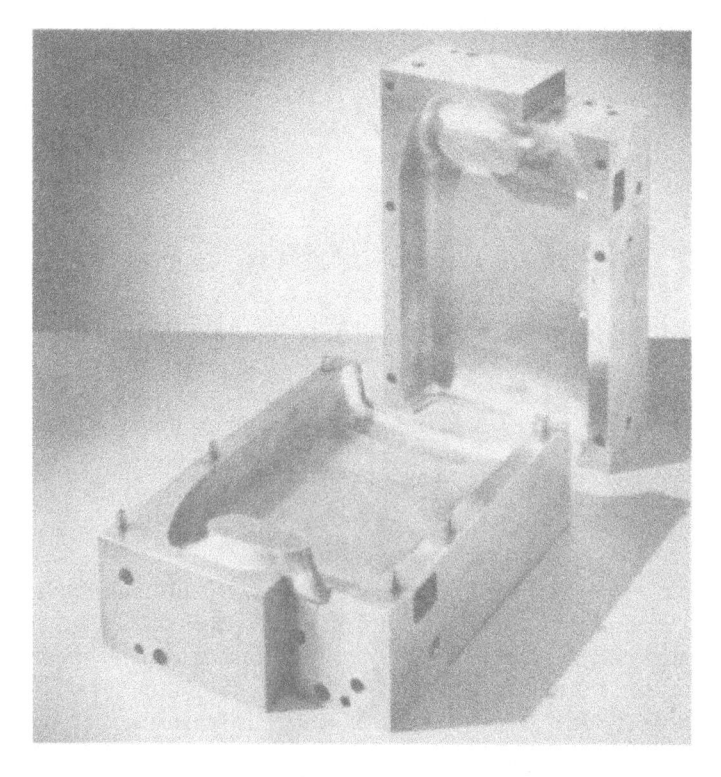

FIG. 11.35. Completed mold bodies with areas machined to accept inserts. (*Courtesy Heise Industries, Inc., East Berlin, CT*)

decreases, developing a hot spot in the mold because the heat is not being conducted away as rapidly as from other areas in the mold.

Drilled lines are usually added to all inserts making up the mold assembly. All parts of the mold are normally cooled independently to afford the blow molder greater flexibility to his molding cycle through varying coolant temperatures.

With the assembly of inserts and matching of halves mentioned above, relief is milled in the mating mold faces to permit displacement of the excess plastics material, as depicted in Fig. 11.37. The depth of the relief is critical and dependent on the weight of the article molded. If the relief is too shallow, the mold will not close completely and poor seam lines will appear on the molded article. If the relief is too deep the cold mold will not be in intimate contact with the excess plastics material and remove its heat. When this happens the article is removed from the mold, retaining hot pieces of attached excess material. These, in turn, can adhere to and spoil other articles.

Because it is so difficult to accurately determine the depth of pinch relief prior to molding, the use of a blown pinch has been employed. By cutting the relief relatively deep the excess plastic can be blown by either bleeding off some of the major blowing air or providing a separate air source. The heat in the excess material is now rapidly removed by touching the cold mold face on the outside and by the cooling air on the inside.

FIG. 11.36. Typical blow mold construction. Bottom view of insert showing milled cooling channels. (*Courtesy Heise Industries, Inc., East Berlin, CT*)

FIG. 11.37. Typical blow mold construction. Completed mold with inserts fastened in place and showing pinch relief being cut. (*Courtesy Heise Industries, Inc., East Berlin, CT*)

The final operations in making a blow mold consist of attaching back-plates, water testing all cooling systems for leakage, and blasting the cavity surface with an appropriate abrasive material. The mold cavity is usually given a rough finish except for those items requiring a high surface gloss. This rough surface vents the air trapped between the plastics and the cavity surface during the blowing cycle. Without this rough mold surface or other venting means, the resultant surface on the blow article would be rough and pock marked from air entrapments.

Nearly all molds in use today consist of two halves matching on a single flat surface. There are, however, molds having irregular parting lines and more than two movable pieces. This type of mold becomes necessary when the part to be molded has reentrant surfaces such as shown in Figs. 11.38 and 11.39. While multiple-action molds increase the design possibilities of blown articles, the cost of tooling can be considerably higher. Only the economics of a specific job can dictate the type of tooling.

DESIGN OF EXTRUSION BLOW MOLDS

Standard data on blownware thread designs (bottle finishes) are given at the end of this chapter. Data on extrusion dies are given in Chapter 10.

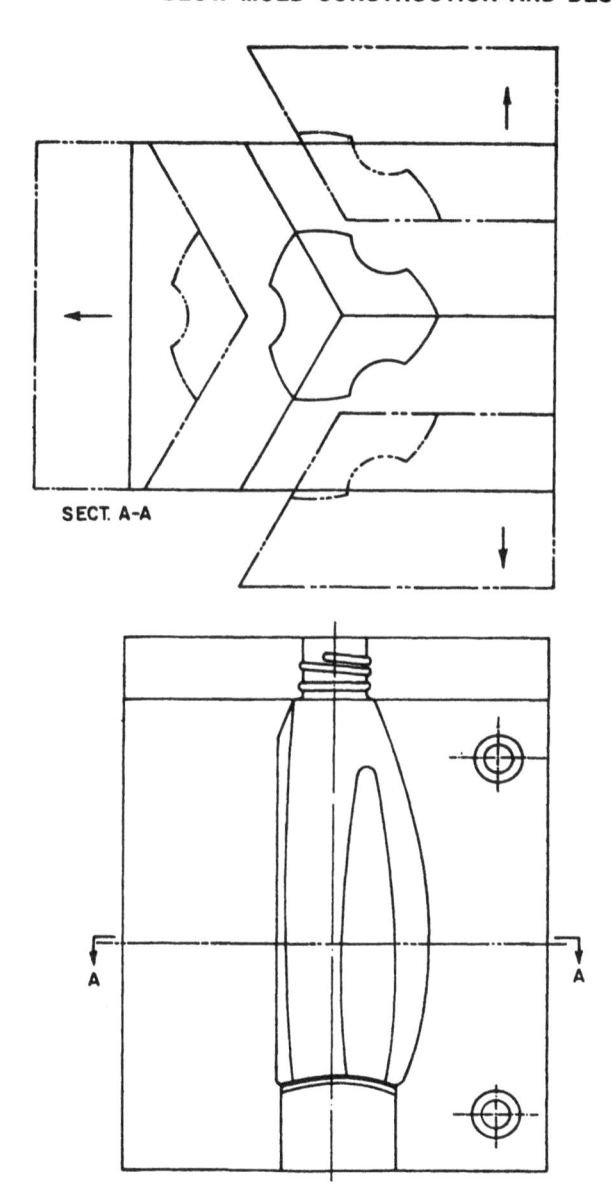

SECT. A-A

FIG. 11.38. Mold operation for bottle with reentrant contour. Dashed lines show movement of three parts in direction of arrows to release reentrant sides. (*Courtesy Heise Industries, Inc., East Berlin, CT*)

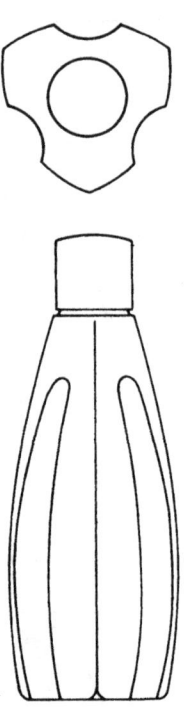

FIG. 11.39. Bottle to be made in reentrant mold shown in Fig. 11.38.

The actual design of a blow mold will be guided by the standard data available to adapt the mold to the machine and the applicable process. The mold cavities will be replicas of the exterior surface of the blownware product with the essential material shrinkage added to the product dimensions. A typical mold is shown in Fig. 11.40.

The blow mold designer's initial consideration is the location of the mold parting line. This is determined first by the shape of the article which must be molded so it can be ejected from the mold, and secondly, by its appearance. The mold parting line should be positioned so that there is no tendency for the molded piece to "lock" in the mold. Most thermoplastics used for blow molding are sufficiently hot and pliable at the time the mold opens to permit some negative relief or hook. Care must be exercised in the amount of hook used, for an excessive amount will cause the finished part to stay in one mold half. While some form of manual or automatic ejection can be employed, parts with excessive hooks are subject to severe surface scuffing. Intentional hooks can sometimes be designed into tooling advantageously. For example, opposite and opposed hooks which create a force couple at the time of mold opening will spin the molded article and allow it to eject freely.

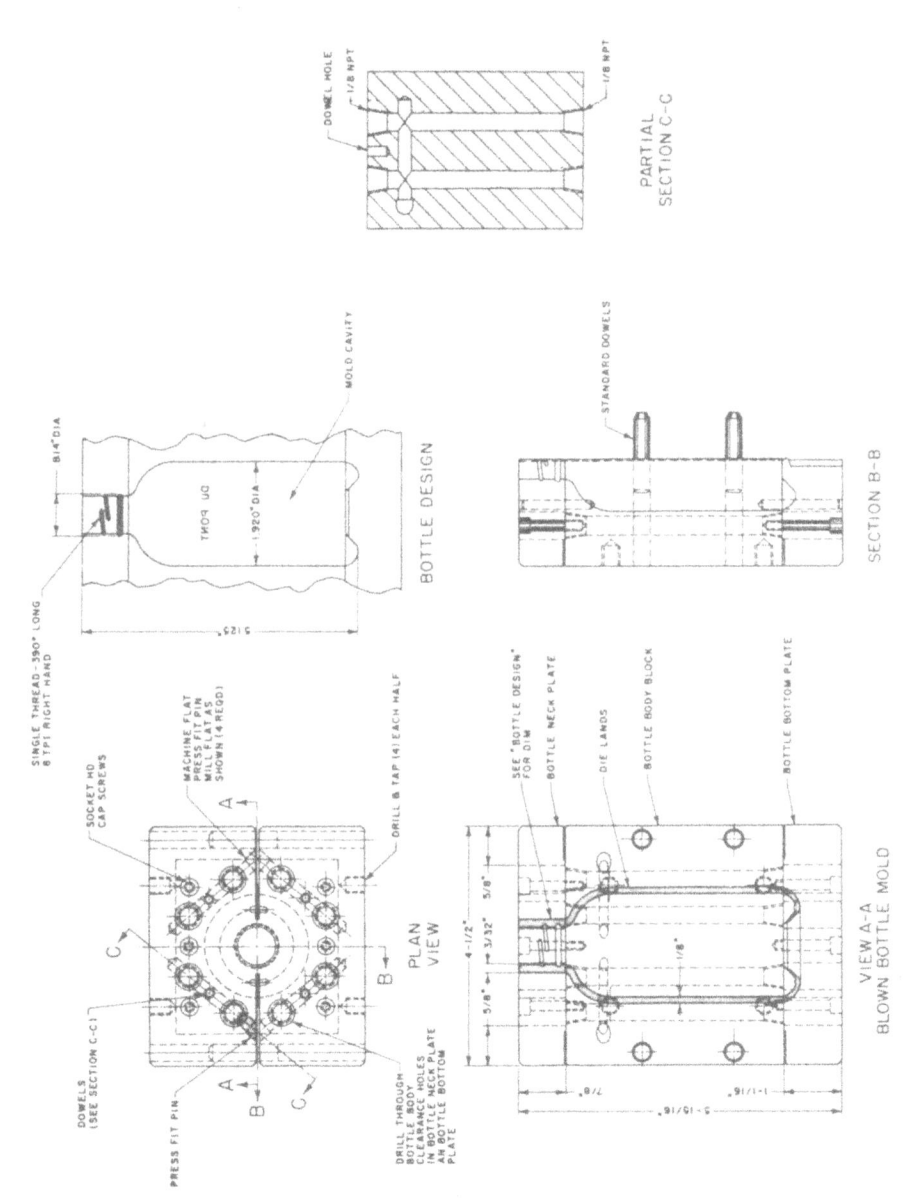

FIG. 11.40. Typical blown bottle mold design. (*Courtesy E. I. DuPont de Nemours & Co. Inc., Wilmington, DE*)

The mold parting line should be hidden in the container design if at all possible. It should also be kept off surfaces which will subsequently be decorated with silk-screen, hot stamp, or paper labels. Any one of these labels will show some irregularity or flaw caused by the uneven surface associated with the mold parting line. The quality and speed of molding should not be sacrificed for appearance design. On a square bottle, for example, it might appear advantageous esthetically to hide the parting line on two corners by having it run diagonally across the bottle. A closer analysis will show the hidden parting line is not worth the irregular wall distribution which will result from the pinch-tube method. The base corners perpendicular to the parting line will be much thinner than those on the parting line, and will cause bottom distortion through uneven cooling.

With the parting line established, the extremities of the mold are selected within the limitations of the blow molding machinery on which the mold will be used. At this time the cavity is positioned most advantageously relative to the extruded parison. For most symmetrical items the cavity and parison centerlines coincide and are parallel. There are, however, times when the wall distribution of the final molded part can be improved by tipping the mold cavity centerline relative to the parison centerline.

In most cases, inserted pieces are designed for the neck and base areas of the mold as noted previously. The neck insert cavity component is a replica of the finished product in the neck area, and in addition, carries a configuration above the top of the finished neck suitable for the particular blowing machine. Generally this configuration provides a reduction from the neck diameter to a small diameter which fits the blow pin on the blowing machine. Some blowing machines have a movable blow pin that moves into the neck end of the blown object after the mold closes and completely finishes the part to size in this area. This operation eliminates subsequent trimming operations. For this finished neck process in the pinch-tube method of blow molding, a hardened piece must be added to the neck insert to prevent excessive wear from the movable blow pin (see Fig. 11.41). If the extruded tube from which the article will be blown is larger in diameter than the diameter of the neck insert cavity, then a pinch relief must be provided for the excess plastic which will flash (see Fig. 11.37). With a pinched neck, threads, if used, can be made either continuous or interrupted. The interrupted thread does not cross the parting line, leaving a smooth surface from which to remove the residual plastic flash. The disadvantage of this type of thread is the reduced engagement possible between a molded article and its closure which, in some circumstances, might cause faulty sealing. Where maximum closure torque and superior sealing properties are desired, the thread is made continuous and the pinch relief is cut around each thread at the parting line.

FIG. 11.41. Typical blow mold construction. Hardened striker plate. (*Courtesy Heise Industries, Inc., East Berlin, CT*)

The neck end of any blown piece usually has heavy wall sections thus making it desirable to provide a maximum amount of cooling in the neck insert. Cooling channels having a large surface area and small cross-section area are the most desirable, as noted previously. Coolant inlets and outlets are located for accessibility within the design of the blow molding machine. For a minimum number of fittings and hoses, it is sometimes preferable to connect two or more parts of the mold in series rather than have individual inlet and outlet lines connected to the mold body, neck, and base inserts.

Another effective method of minimizing connections is by the use of a mold back plate which serves as a manifold. The coolant then enters the back plate at one end and enters each of the mold parts through some suitable seal such as an "O" ring. The coolant is then discharged through separate holes in a similar manner.

The base insert carries the pinching edges which seal the end of the tubular parison, and in the usual container, the base indentation or pushup. The base insert should be sufficiently high to accommodate the variations in parison length associated with variations or surging within the extruder.

For molds utilizing some mechanical means of removing the base tab formed by the pinch tube process during the molding cycle, the height of the base insert is kept to a minimum. This minimum height reduces the frictional area over which the tab must be pulled by the removal mechanism. Like the neck insert, the base insert is contacting thick and hot sections of plastic. The cooling channels should therefore be kept close to the parting surface which is the area needing greatest heat absorption.

With heavy sections of plastics material, much distortion can take place during cooling. For this reason allowances must be made for defects which might affect the function of the blown article. For containers having a push-up, the height of the curvature must be increased to allow for its sinking. A correction is also usually incorporated that will insure stability of the container on its base. This latter correction consists of cutting three or four deeper sections in the base inserts that will provide projections on which the container will ultimately set.

The mold body is usually the portion that carries some means of maintaining alignment of the mold halves. A minimum of two sets of hardened dowel pins and bushings can be used for this purpose. For mold materials such as zinc and aluminum, a shoulder type pin and bushing is desirable to prevent them from pulling out of the mold. Aluminum and zinc have so little elasticity that a press fit cannot be relied upon to hold pins and bushings in place. The size of the pin and bushing should be determined by the operation of the blow molding machine. Some machines operate with platen bushings having large clearances and the mold pins and bushings are expected to align the platens at the time the molds close. Obviously, this puts considerable strain on the mold alignment, and the pins and bushings should be large. For machines having well supported platens, pins measuring $\frac{3}{8}$ to $\frac{1}{2}$ in. in diameter can be used; the larger pin being used for larger molds. The neck and base inserts are attached to the mold body with suitable fastenings and means of alignment registration. Dowel pins can be used for parts alignment, and are most suited for inserts which will not be interchanged regularly. For ease of interchanging inserts a register diameter is recommended.

The same principles of cooling channel size and location mentioned for inserts apply for the mold body.

Many different designs have been used in the pinch-relief sections. Two typical ones are shown in Fig. 11.42. Design A is probably the one most widely used for polyethylene. In some instances, however, where the mold must pinch on a relatively thin portion of the parison and next to this pinch-

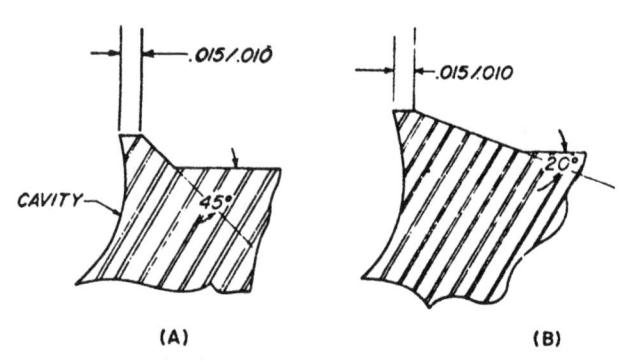

(A)   (B)

Fig. 11.42. Pinch relief sections.

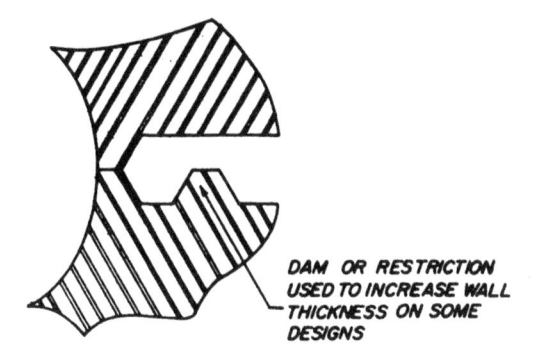

DAM OR RESTRICTION
USED TO INCREASE WALL
THICKNESS ON SOME
DESIGNS

FIG. 11.43. Dam or restriction restricts flow of escaping material at flash line.

ing edge the parison must expand a large amount, the plastic will thin down and may even leave a hole on the parting line. This defect is sometimes seen near the finger hole on containers having handles. The shallower angle of 20° has a tendency to force plastic to the inside of the blown part and increase the wall thickness at the parting line rather than pushing the excess material back into the pinch relief. Another method used for increasing the wall thickness at the parting line employs a restriction or dam in the pinch relief similar to that shown in Fig. 11.43.

In attempting to minimize the residual flash left after trimming, the pinch land is sometimes made sharper than .015/.010 in. Figure 11.44 illustrates a typical design. It should be noted, however, that pinch lands of .003 in. width should only be used on hard, tough mold materials such as beryllium copper.

For certain cavity shapes additional venting may be necessary in addition to that noted previously. In these cases vent holes and/or slots are added.

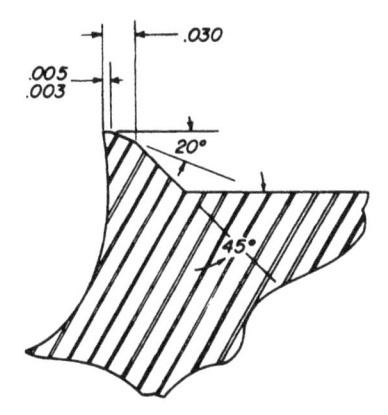

FIG. 11.44. Design often used to minimize residual flash.

Sharp concave depressions or grooves in a cavity will trap air at their deepest point. In these areas, holes ranging in size from .005 to .009 in. in diameter are drilled, permitting the air to escape to the outside of the mold. If air entrapments occur at the parting line, then simple slots approximately ½ in. wide and .002/.0015 in. deep are milled across the entire face of one mold half at frequent intervals.

Needless to say, the cost of drilling holes .005 to .009 in. in diameter can be very great and they tend to become clogged during molding so their use is usually kept to a minimum. If sharp, square-faced, letters are required on a finished blow part, the engraving in the cavity must be vented. With some letters of the alphabet requiring as many as six vent holes to provide proper reproduction, the cost of drilled vent holes could become prohibitive. As an alternative, the mold can be inserted with a porous metal piece in the area to be engraved. Trapped air can then vent through the engraving by way of the porous material and some suitable connecting passage to the outside of the mold. Not only can this porous metal method of venting be less costly than drilled holes, but with proper selection of particle size used in the porous metal, the vent openings will not show on the finished plastic article. This is not necessarily true of drilled holes. With a practical minimum of approximately .005 in. diameter, drilled holes could show a projection on the finished part depending on temperature and type of plastic being blown. The disadvantage of a porous metal insert would be visual; a line would always show at the matching surfaces of the mold and insert. In some instances, this match line can be hidden in the design of the blown article.

The mold design discussion thus far has been concerned with the pinch-tube method of blow molding. Most of the design considerations can, however, be applied to all methods of blow molding. For the neck ring process, a wear-resisting and self-registering pocket must be provided in the mold to accept accurately the neck ring on each cycle.

With the continuous tube process, air must be introduced through the side of the parison rather than through its center. This is necessitated by the fact that the parison is pinched shut at both the top and bottom of the mold. This is done to allow separation of one article from another at the completion of molding. Air is usually injected by means of a hollow needle which pierces the parison. Thus the neck insert design must provide for this needle together with some means of actuating it, such as an air cylinder. The neck insert would also carry a pinching edge to close off the upper end of the parison. With the horizontal continuous tube process, ejector pins are usually required to free the finished part from the lower mold half.

In addition to providing pincher slides and their means of actuation in the trapped air process, much consideration must be given the actual opposing pinching faces on the slides. A seal, trapping the air within, can be effected either by the meeting of two pinching surfaces or by crowding the parison into a cavity which is small enough to collapse the parison and cause it to seal itself. Pinching edges must be designed to provide sufficient wall thicknesses at the parting line. Otherwise there is danger of the trapped air bursting the blown article at a thin spot as it expands due to the heat of the plastics material before the article can be cooled.

INJECTION BLOW MOLDING DESIGN AND CONSTRUCTION*

The design of tooling for injection blow molding as discussed herein is intended for equipment utilizing the horizontal rotary index method of core rod transfer. This principle has been the most successful to date and is used by over 90% of the machines now in service. Figure 11.45 shows the mold and machine.

FIG. 11.45. Operational principle of the injection blow molding process. (*Courtesy Rainville Div. of Hoover Universal, Middlesex, NJ*)

*This section by Robert D. DeLong.

Principle of Operations

As seen in a simplified tooling layout (Fig. 11.16) two sets of molds are required, along with three sets of core rods. To insure alignment and speed mold change-overs, each set of molds is mounted on its own mold base or die shoe. In the first or parison mold, melted resin is injected over the core rods, filling the cavities and fixing the weight of the container. At the same time the neck finish is molded completely. This set of core rods carrying the preformed parisons is rotated to the second or blow station where they are enclosed within the blow molds and blown to the desired shape. The blown containers are then rotated to the stripping station for removal, oriented if desired. Since there are three sets of core rods, all three operations are simultaneously conducted by sequential rotation of the core rods.

Mold Design

The constraints of container length, diameter, and number of cavities are imposed by the physical dimensions of the machine at hand. Parison molds are normally mounted so as to place their centroids of projected area (Fig. 11.46) directly under the centerline of clamp pressure. The distance, then, from the face of the rotating head to the center of pressure will determine the maximum container length. Machines are available to cover the range of ½ to 12 in. length. The maximum container diameter is about ¼ in. less than the mold opening, to allow a minimum of ⅛ in. top and bottom clearance for the blown container to rotate from the blow station to the stripper station. Present machines use from 4 to 6 in. mold opening, depending on the size.

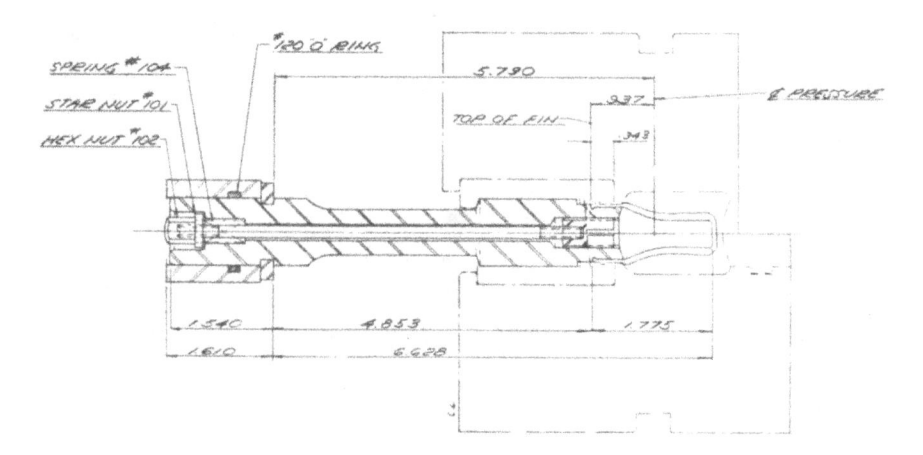

FIG. 11.46. Core rod assembly. (*Courtesy Captive Plastics, Inc., Piscataway, NJ*)

The maximum number of cavities is normally limited by the clamp tonnage of the parison mold and the material being molded. The following clamping requirements are suggested as *minimums*:

LDPE-2800 psi/in.^{2} of projected area
HDPE-3200 psi/in.^{2} of projected area
SAN and P/S-3800 psi/in.^{2} of projected area
PP-4100 psi/in.^{2} of projected area
P/C-4400 psi/in.^{2} of projected area

The nature of the injection-blow process places some limitations on the shapes of the containers which can be made. Injection-blow molding is well suited to wider mouth bottles and conventional shapes. Because the core rod is a cantilever beam, its L/D ratio should be 12/1 maximum to prevent deflection during injection. A 32 oz. cylinder round style container requires a core rod about 9 in. long. Specified with a 24 mm finish, the L/D ratio is an unworkable 14/1. Moving up to a 28 mm or 33 mm would reduce the L/D to a manageable ratio.

Swing weight becomes a consideration with containers over 32 oz. capacity that have wide necks. The core rod for a 48 oz. capacity container with a 110 mm finish, shown in Fig. 11.47, weighs about 9 lb. Mounted at the extremes of the swing radius, they impose severe inertial loadings on the transfer mechanism, as indexing takes place in about 1 sec.

One advantage of injection blow is the diametrical and longitudinal programming of the parison by shaping the parison mold, the core rod, or both. This two-dimensional programming becomes especially advantageous in the production of oval containers. Up to an ovality (container width/container thickness) ratio of 1.5/1 quite satisfactory containers can be blown from circular cross-section parisons. Up to ovality ratios of 2.2/1 can be handled with oval cross-section parisons. Generally the ovalization is done to the parison mold; the core rod remains round. Extensive parison ovalizing, perhaps above 35%, can lead to selective fill during injection and result in visible knit lines in the finished bottle. Under these circumstances it will be necessary to ovalize *both* the core rod and parison cavity to obtain the desired distribution in the finished container. Some provisions to prevent the core rod from rotating must then be incorporated. Ovality ratios above 3/1 are not suggested. The center core rod in Fig. 11.47 shows an ovalized core rod. The preferred blow ratio is between 2 and 3/1, measured as the average parison OD against the bottle OD. This normally will yield an ideal parison thickness of from .120 to .180 in. with predictable expansion characteristics. Increasing the blow ratio by reducing the core rod diameter, reduces the projected area, often enough to allow another cavity to be fitted. Unavoidably, the parison thickness must be increased, to maintain a constant

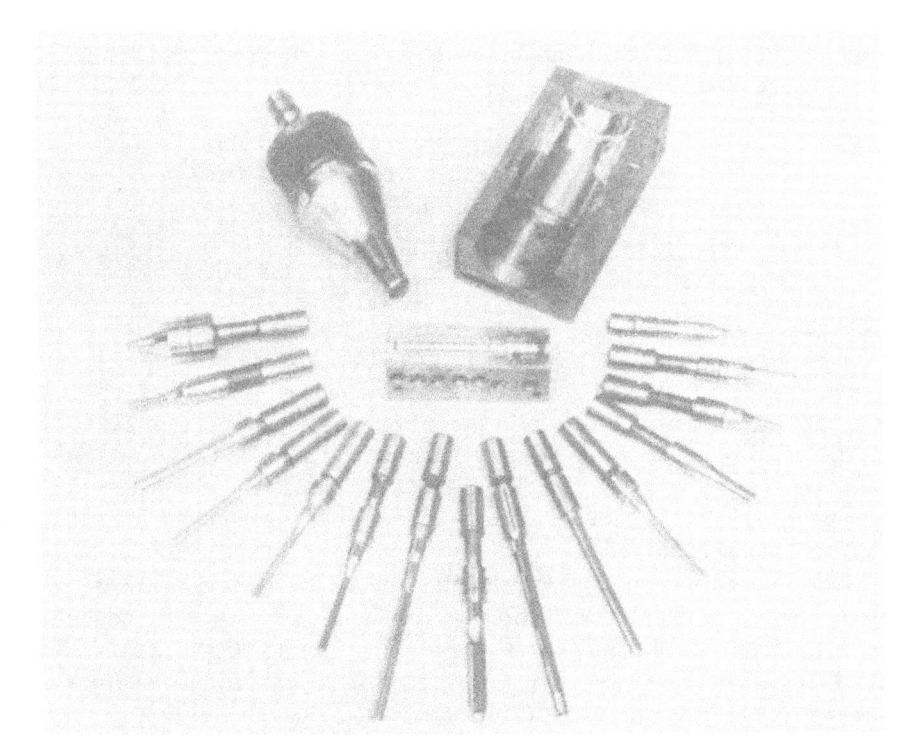

FIG. 11.47. Core rods, parison mold, and blow mold. Upper left shows 48 oz.—110 mm core rod. Lower center blade shape core rods illustrate ovalization and means of preventing rotation. (*Courtesy Rainville Div. of Hoover Universal, Middlesex, NJ*)

weight, and experience has shown parison thicknesses above .225 in. to be unpredictable in their blowing characteristics.

Alignment of Components

Examination of Fig. 11.45 suggests some of the alignment problems encountered in assuring that the core rods are concentric to their respective cavities. It is usual practice to utilize individual cavity blocks and locate these via keyways on the die shoes. Set-up then is simplified as it involves changing only the die shoe with the mounted cavities as a unit. The die shoe is located from a single center keyway mating with the machine; radial spacing from the center of rotation is adjusted as required by shims. Figure 11.48 illustrates a 4-cavity mold with die shoe mounting.

Initial core rod/mold alignment is achieved by the index mechanism of the machine. Upon mold closing, the index head is guided by guide pins to further refine the mold/core alignment, with final alignment by the core rod

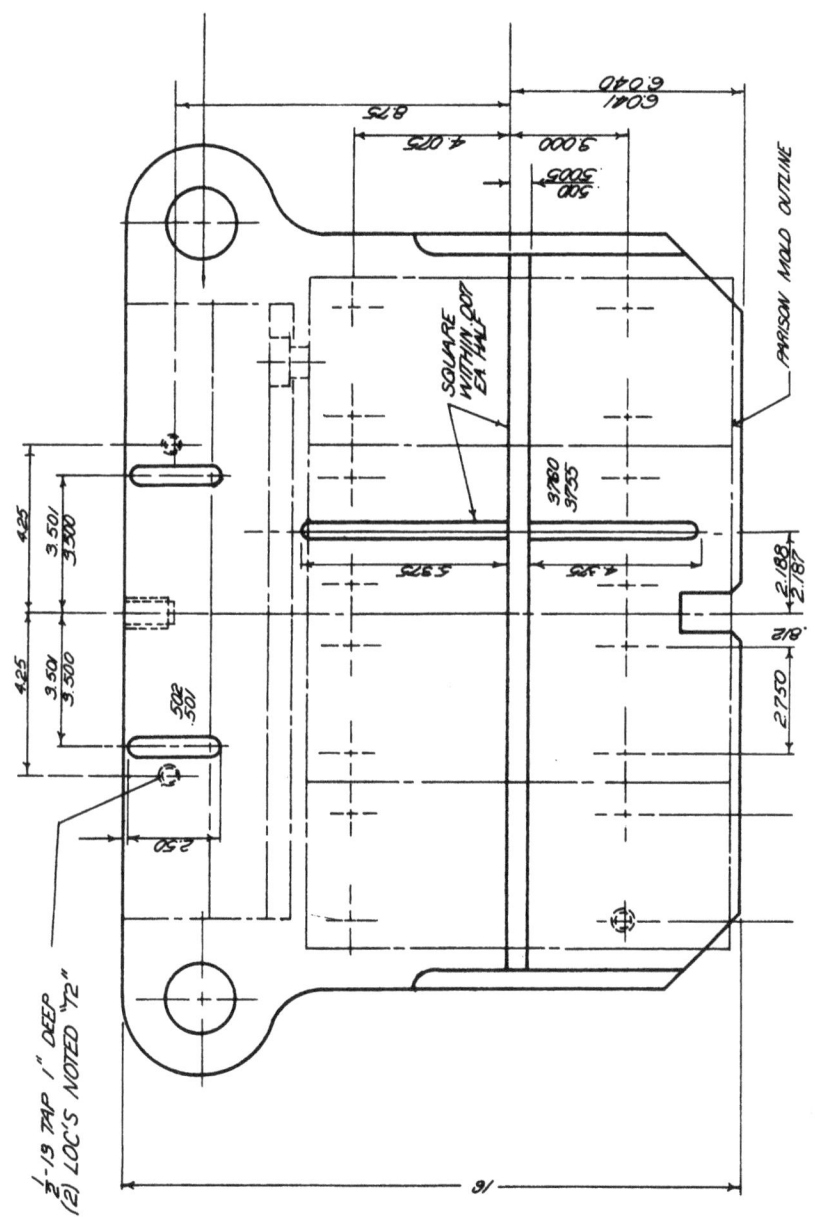

Fig. 11.48A. Parison mold, lower die shoe. (*Courtesy Rainville Div. of Hoover Universal, Middlesex, NJ*)

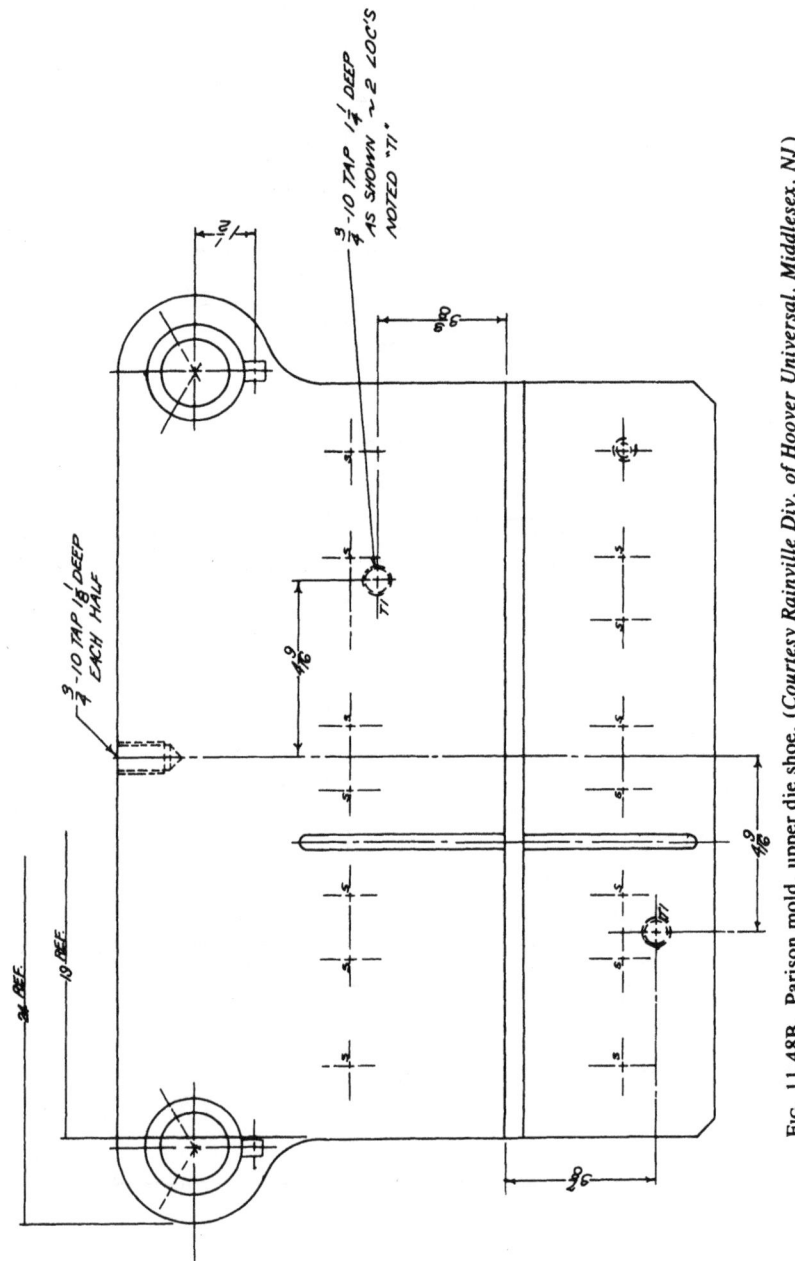

Fig. 11.48B. Parison mold, upper die shoe. (*Courtesy Rainville Div. of Hoover Universal, Middlesex, NJ*)

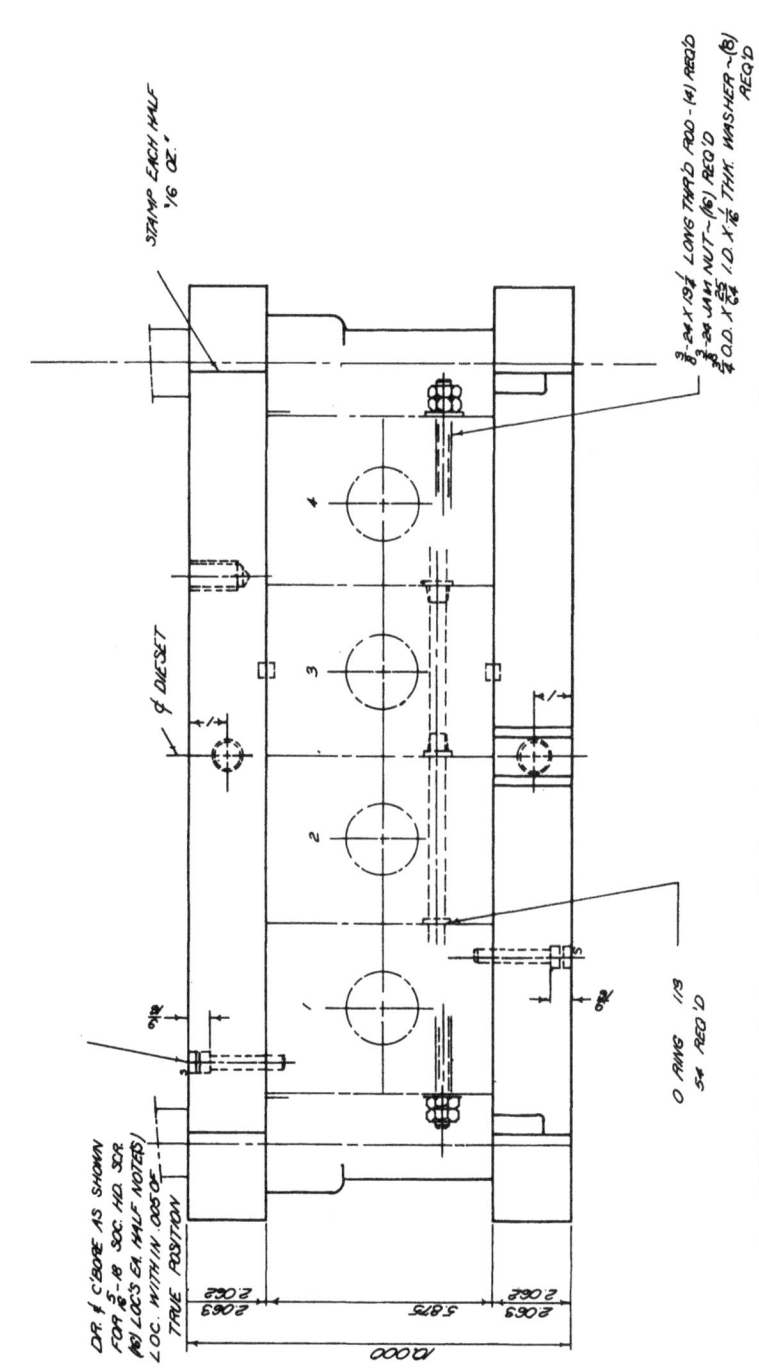

FIG. 11.48C. Parison die shoe assembly, front view. (*Courtesy Rainville Div. of Hoover Universal, Middlesex, NJ*)

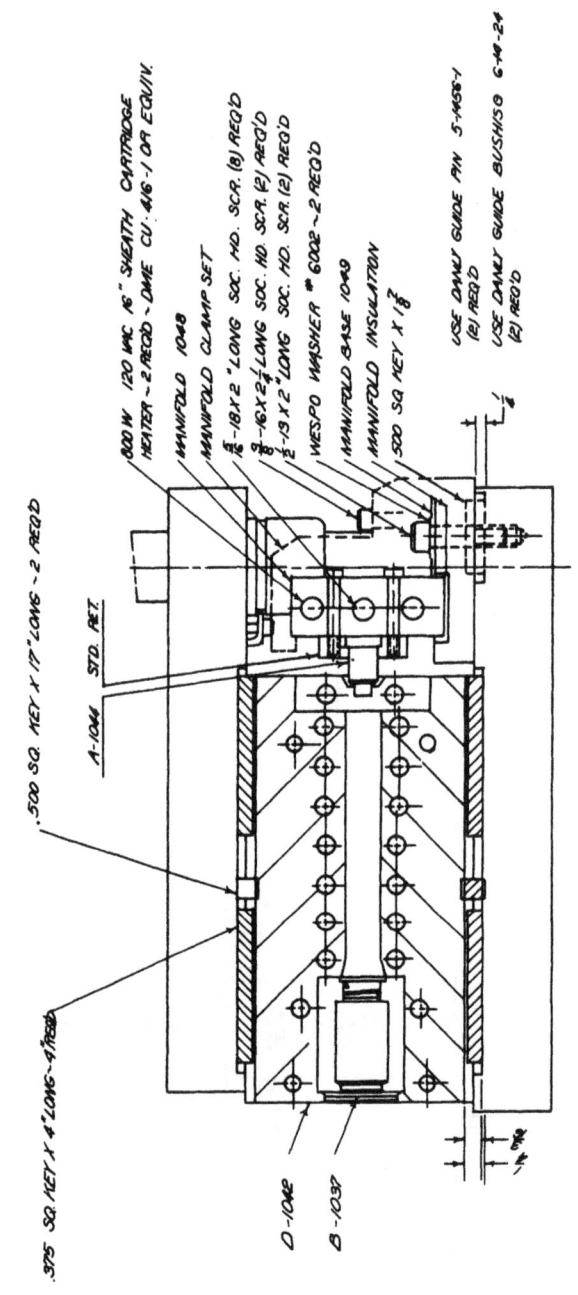

FIG. 11.48D. Parison mold assembly-side view. (*Courtesy Rainville Div. of Hoover Universal, Middlesex, NJ*)

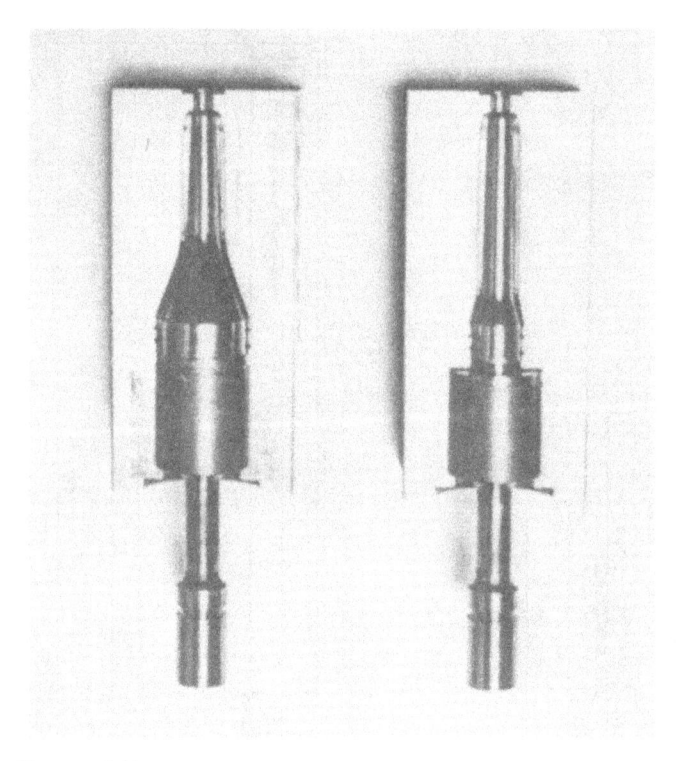

FIG. 11.49. 28mm and 48mm core rods are shown nested in injection mold halves. (*Courtesy Captive Plastics, Piscataway, NJ*)

shank nesting into its seat in the mold. This shank is a precise (.0000–.0005 in.) fit to its nest. Figure 11.49 shows 28-mm and 48-mm core rods nested in their respective parison mold halves. The core rod mounting ends are clearanced .004–.006 in. to "float" in the rotating head to compensate for thermal expansion differences between the hot (200–300°F) parison mold and the cold (40–60°F) blow molds.

Core Rod Design and Construction

Core rods are typically hardened steel (AISI L6-RC 52-54) polished, and hard chromed. Location of the air entrance valve is preferable at the shoulder, but often is moved to the tip when core rod L/D exceeds 6/1, due to mechanical considerations. Figure 11.50 details a typical core rod as shown in Fig. 11.47. See also Figs. 11.55 and 11.56 for core rod details.

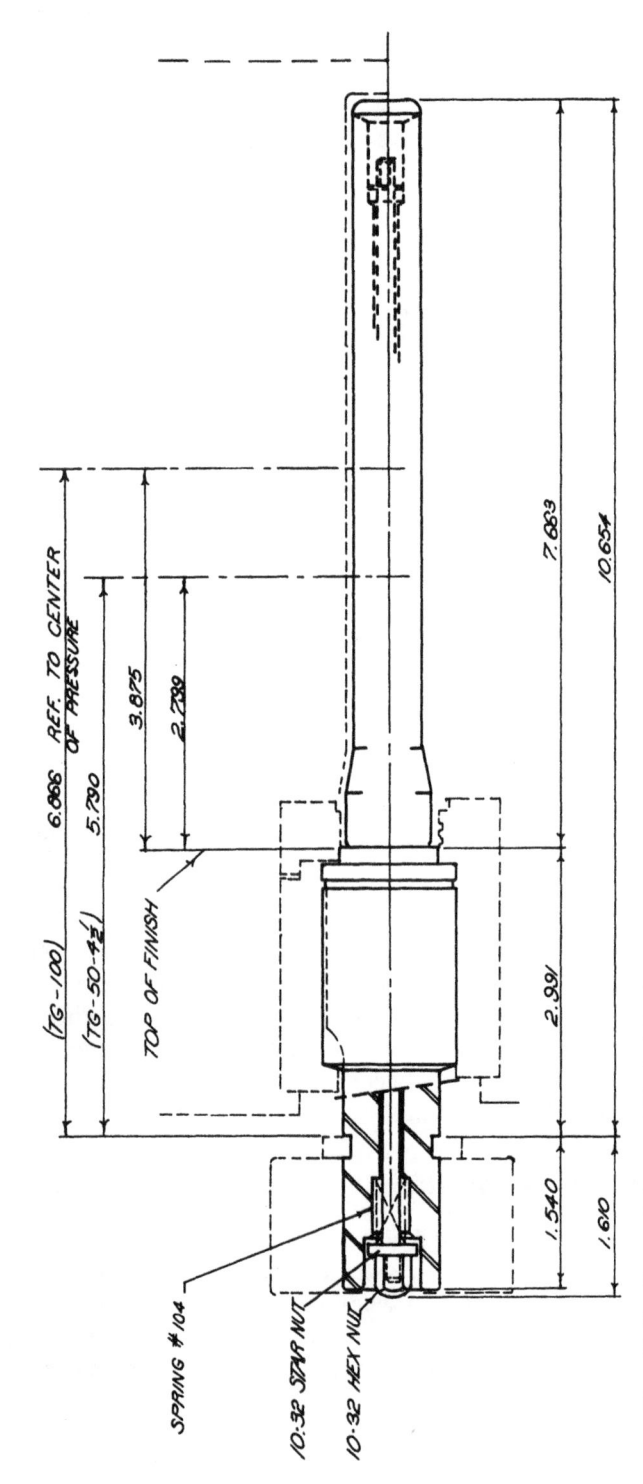

Fig. 11.50. Core rod for assembly in injection mold for a 16 oz. alcohol bottle. (*Courtesy Rainville Div. of Hoover Universal, Middlesex, NJ*)

Normal construction of the core rod is solid, although coring out to reduce mass may be used in larger sizes. Where space permits, some means to enhance heat transfer may be employed, such as aluminum or beryllium copper inserts. Heat pipes (Chapter 8) are being installed in the center of some core rods to obtain faster cycles and close temperature control. Examination of Fig. 11.47 will reveal one or two annular grooves on the core rod near the seating shank. These grooves (.004-.010 in. deep) perform a dual function. They stabilize the parison against elastic retraction during transfer from the parison mold to the blow mold which would result in thread mismatch. They also seal the bottle against excessive air loss during blowing, hence the name, blow-by grooves.

Parison Molds

Individual parison mold blocks are normally fabricated from 1017 HRS or P-20 prehard steel for use with polyolefin molding resins. Molds for rigid resins, such as styrenics, nitriles, carbonates, etc., are usually made of an oil-hardening steel (A-2 or equivalent), rough machined and hardened to RC 40-45. In both cases final polishing is usually followed by hard chroming of the molding surfaces, although this is not required for polyolefin molds.

Cooling lines for control of cavity temperature are placed close to the surface as shown in Fig. 11.51. The use of O-rings betweeen adjacent cavities is standard practice.

No venting is used on the parison mold parting line. If required, the venting is over the core rod and out the back of the mold.

Neck rings are separate inserts, as shown in Fig. 11.57 or Fig. 11.50. Construction is normally of oil-hardening steel (A-2 type) with hardening to RC 54-56 for all resin types.

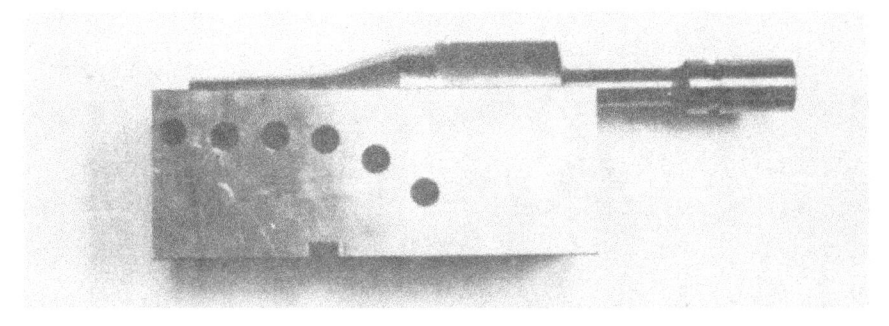

FIG. 11.51. Side view of a parison mold showing cooling holes. (*Courtesy Captive Plastics, Inc., Piscataway, NJ*)

Nozzles seat directly into the parison bottom mold, with the top half of the parison mold clamping and unclamping with every cycle. Mold flash, if it occurs, is most probably at the nozzle, with potential nozzle or nozzle seat hobbing. Practice varies widely on nozzle hardness, but replaceable nozzle seats as shown in Fig. 11.48 D are in wide use. This eliminates the need for enlarging the nozzle seat with time and stocking nozzles of various diameters.

Blow Molds for Rigid Plastics

Blow molds for rigid plastics such as the styrene compounds, nitriles, etc., are of turned or hobbed steel. Generally, an air-hardening steel is preferred, such as A-2 type. Hardening is normally to RC 45-50, followed by polishing and hard chroming. Unless the container is virtually flat-based (.030 in. push-up or less) then a retractable base push-up must be used. Figures 11.52 and 11.53 detail construction of current design practice in this regard. Neck rings are also of hardened steel, polished and chromed. Overall shrink factor for the non-crystalline resins is .005 in./in. in both directions.

Blow Mold Design for PVC

Even though PVC is a rigid plastic, some change in mold design must be made to accommodate its special characteristics of corrosiveness and very poor heat transfer.

Current practice is to use suitably hardened BeCu blow molds. Aluminum molds will rapidly deteriorate; chrome plate will peel from steel molds because the steel corrodes under the chrome; and stainless steel imposes a time penalty on cycles because of its inferior heat transfer properties.

The alignment of a blow mold for PVC is critical because of its low strength when subjected to the notched Izod impact test. Any misalignment

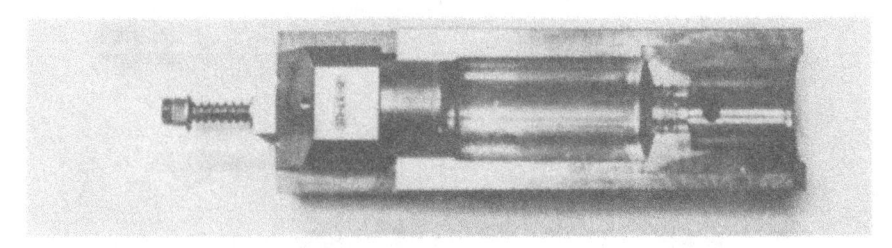

FIG. 11.52. Top view of blow mold showing moving base push-up. (*Courtesy Captive Plastics, Inc., Piscataway, NJ*)

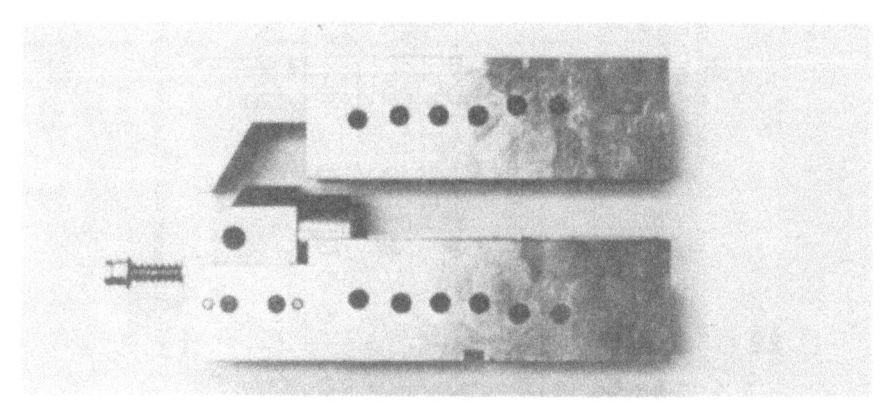

FIG. 11.53. Side view of blow mold with cam operated push-up. (*Courtesy Captive Plastics, Inc., Piscataway, NJ*)

between the mold halves will create a step-notch on the molded part, which will adversely affect the drop impact strength of the container. For this reason, block-style molds are often used for PVC instead of the individual cavities employed in molds for other plastics.

Molds for PVC parisons are usually made from 400 series stainless steel and hardened, because it has proven more satisfactory than chrome-plated steel. Conversely, stainless-steel core rods have proven less satisfactory than tool-steel core rods which have received nickel/chrome plating or just nickel plating.

PVC shrinkage is relatively low (.006–.007 in./in.) and is quite predictable. To eliminate sinks in the thread profile, the thread is narrowed to create a more uniform cross-section. "L" style threads are to be avoided if possible. The explanation for this ability to narrow the thread profile is that thread profiles for plastic bottles were set in the early 1960s with polyethylene in mind. The higher-strength plastics, such as PVC, do not require such massive profiles.

The manifold contour must cause the melt to be distributed uniformly. The critical elements in the design are flow pattern, temperature, and residence time of the melt. For ease of manufacture and for ease in disassembly for cleaning, manifolds are split into sections. The split can be either vertical or horizontal to give access for the cutting and shaping of the flow channel(s). Figures 11.54A, 11-54B, and 11.54C illustrate a horizontal split manifold with flow channel shaped by corner radii to avoid a "dead area" where the melt can stagnate, degrade, and burn. For additional information regarding design of manifolds, refer to Chapter 10 on extrusion dies.

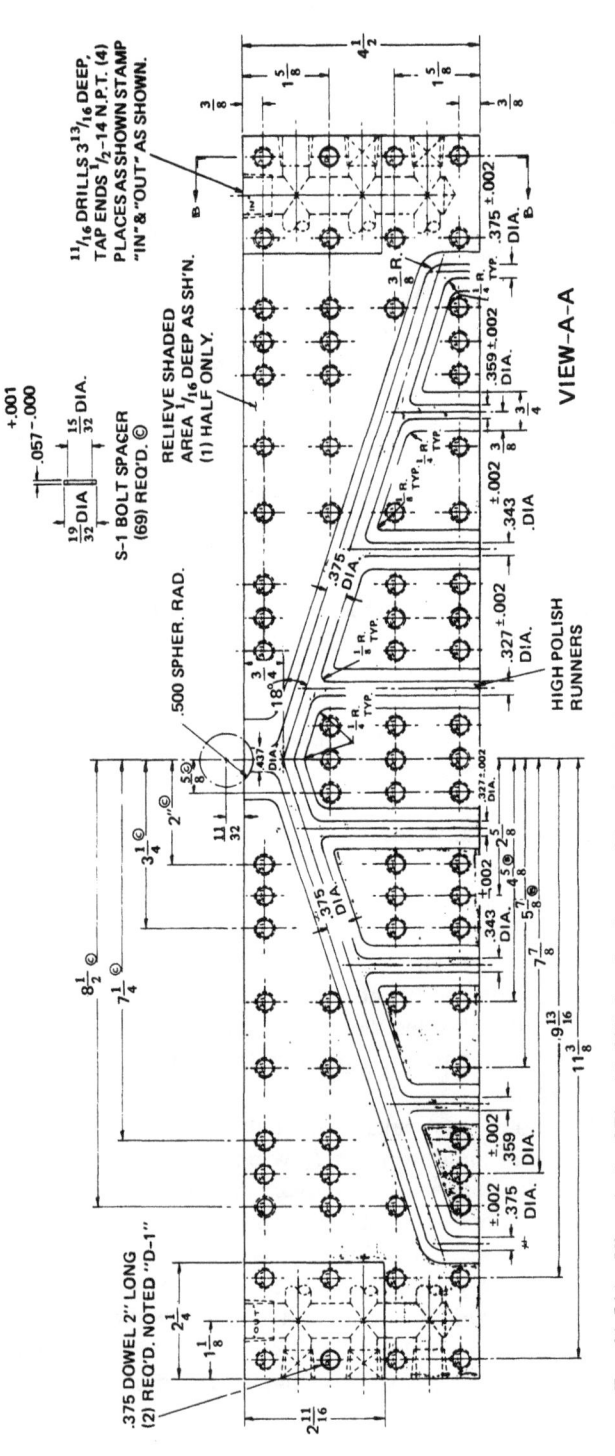

FIG. 11.54A. View A–A of Fig. 11.54B showing layout of 8 cavity, 2.625 in. centerline manifold for .375 in. diameter rod. (*Courtesy Captive Plastics, Inc., Piscataway, NJ*)

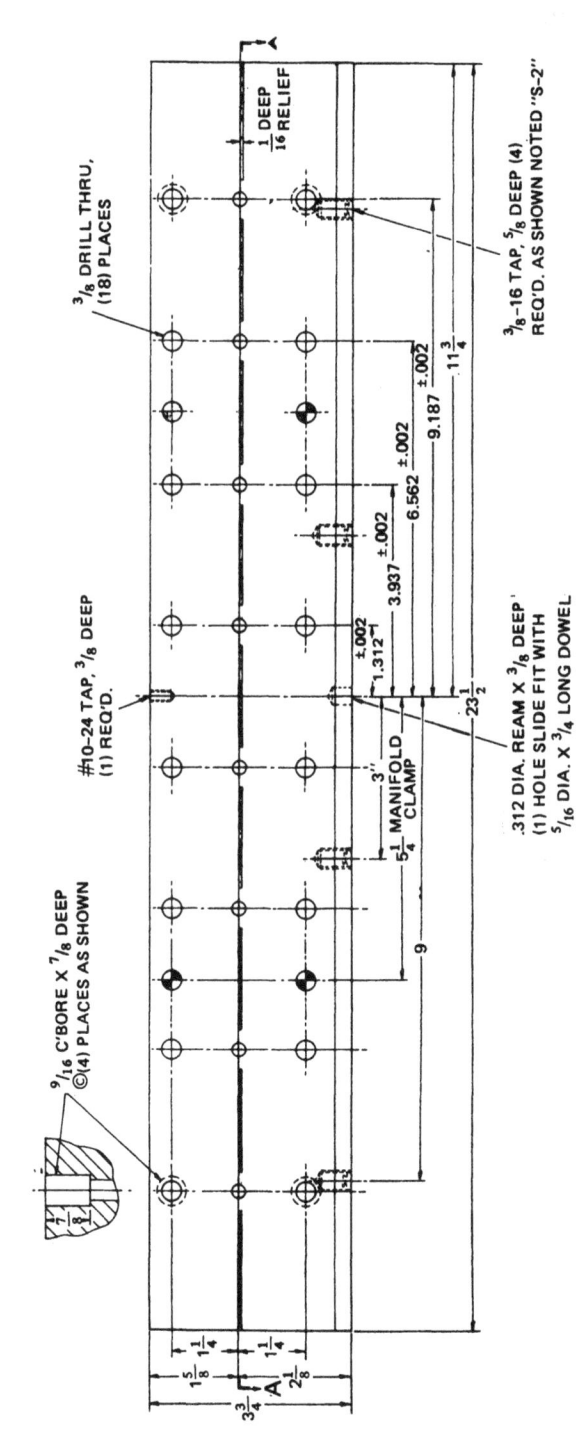

FIG. 11.54B. Front view of 8 cavity manifold for .375 in. diameter rod. (*Courtesy Captive Plastics, Inc., Piscataway, NJ*)

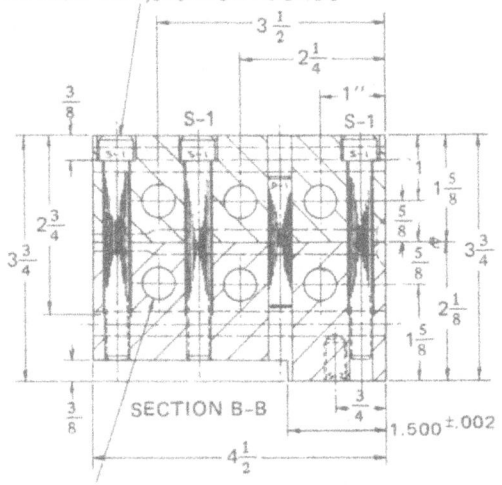

Fig. 11.54C. Cross section of 8 cavity manifold (refer to Figs. 11.54A and 11.54B). (*Courtesy Captive Plastics, Inc., Piscataway, NJ*)

Blow Molds for Non-Rigid Plastics

These usually are 7075 T6 aluminum with turned or pantographed cavities. Hobbed aluminum cavities are used, but must be carefully stress-relieved and retempered. Cast aluminum blow molds are rare, although cast kirksite molds are in limited use. Molds are vapor-honed or sand blasted to promote venting. Grit is seldom coarser than 180, in contrast to extrusion blow molding practice. Parting line vents for both rigid and olefin resins are .001 to .002 in. deep. No provision is required for a retractable base in the blow mold for the non-rigid olefins. Neck rings and base inserts are usually aluminum. A clearance of .003 to .004 in. per side between the parison and blow mold thread diameters is normally allowed. Shrinkage factors are somewhat different than those used with extrusion blow molding, with .012 in./in. used for length and .022 in./in. used for diameter.

Figures 11.55 through 11.59 have been included here to show the actual detail dimensions of some of the smaller parts of the blow mold sections.

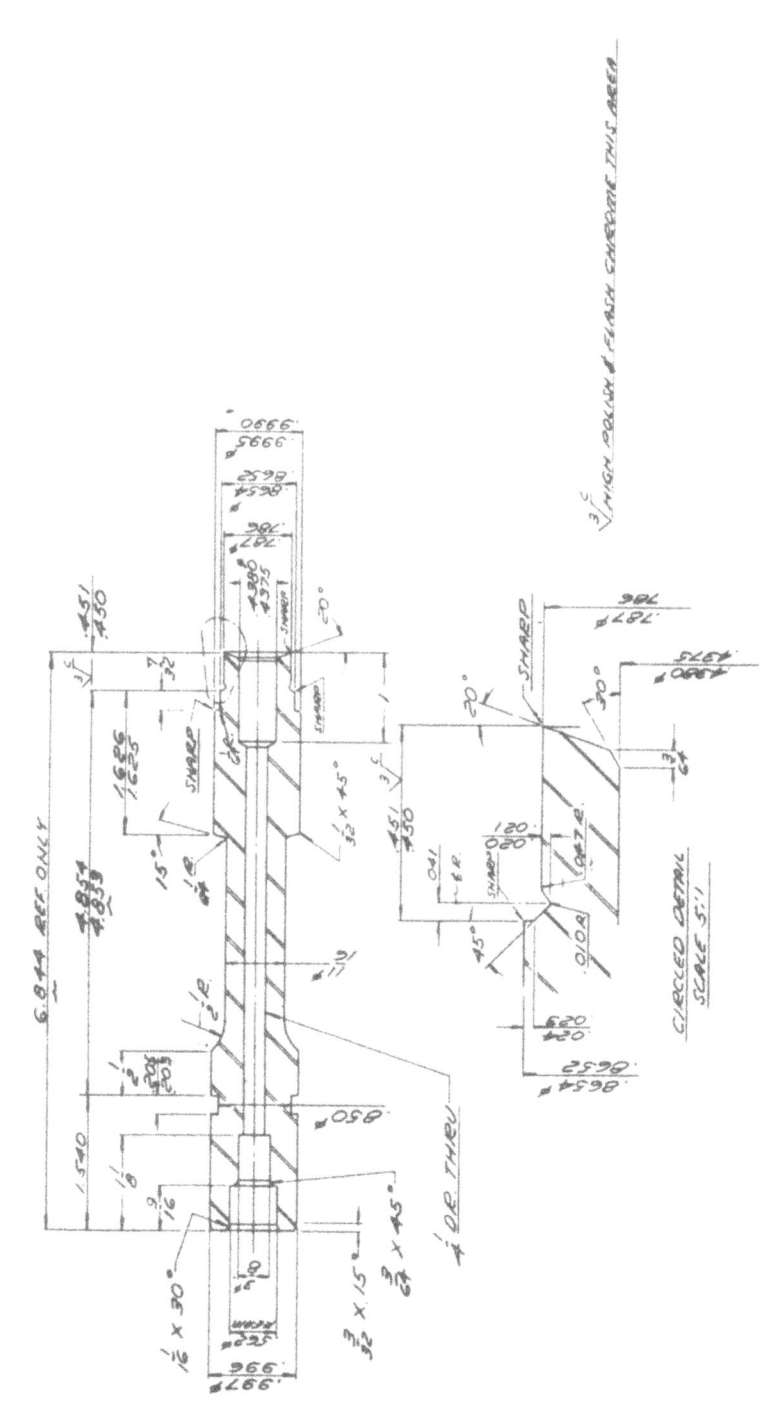

FIG. 11.55. Core rod material: AISI-L6 Tool stl. HDN: R/C 52-54. (*Courtesy Captive Plastics, Inc., Piscataway, NJ*)

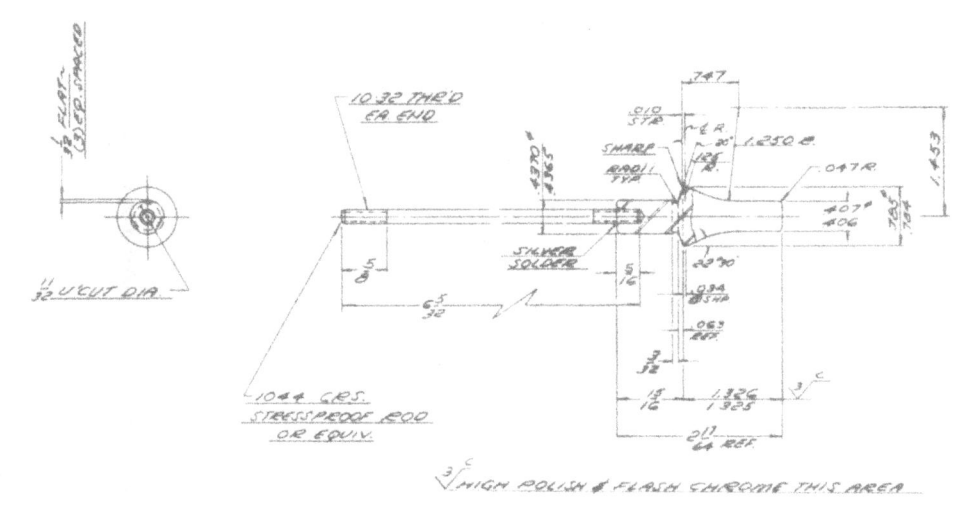

FIG. 11.56. Core rod head. Material: AISI-L6 Tool stl. HDN: R/C 52-54. (*Courtesy Captive Plastics, Inc., Piscataway, NJ*)

Figure 11.55 is the *core rod* made of AISI-L6 tool steel and hardened to RC 52–54, high polish in the thread area and chrome-plated.

Figure 11.56 is the *core rod head*, which is assembled in Fig. 11.55 as shown in Fig. 11.45. It is also made of AISI-L6 steel, hardened, polished, and chrome-plated the same as shown in Fig. 11.55.

Figure 11.57 is the *parison neck ring* made of AISI-A2 tool steel, hardened to RC 52–54, vapor-blasted and chrome-plated in the thread area. Note the single thread (one turn) and the exact and precise specifications of the thread dimensions. We also note the 0.0005 in. tolerances on the critical dimensions which must fit another part of the mold. Thus we reiterate that, even though bottle dimensions do not seem to be exact, the *molds* which make bottles must be quite precise and well-made.

Figure 11.58 shows a *blow neck ring* made of 7075-T6 aluminum. Note that the dimensions of the neck ring are only. 003 to .004 in. larger than the parison neck ring.

Figure 11.59 shows a *bottom plug* made of 7075-T6 aluminum with a vapor blast finish on the spherical surface which will form the "push-up" on the base of the blown bottle. Note that the push-up is .096 in. Bottle numbers and cavity numbers will usually appear on the push-up surface and are specified as "reverse" or "left-hand" characters.

The authors are indebted to Captive Plastics Inc., of Piscataway, New Jersey, for supplying these proprietary drawings of a production tool in order that the design procedure could be well-detailed.

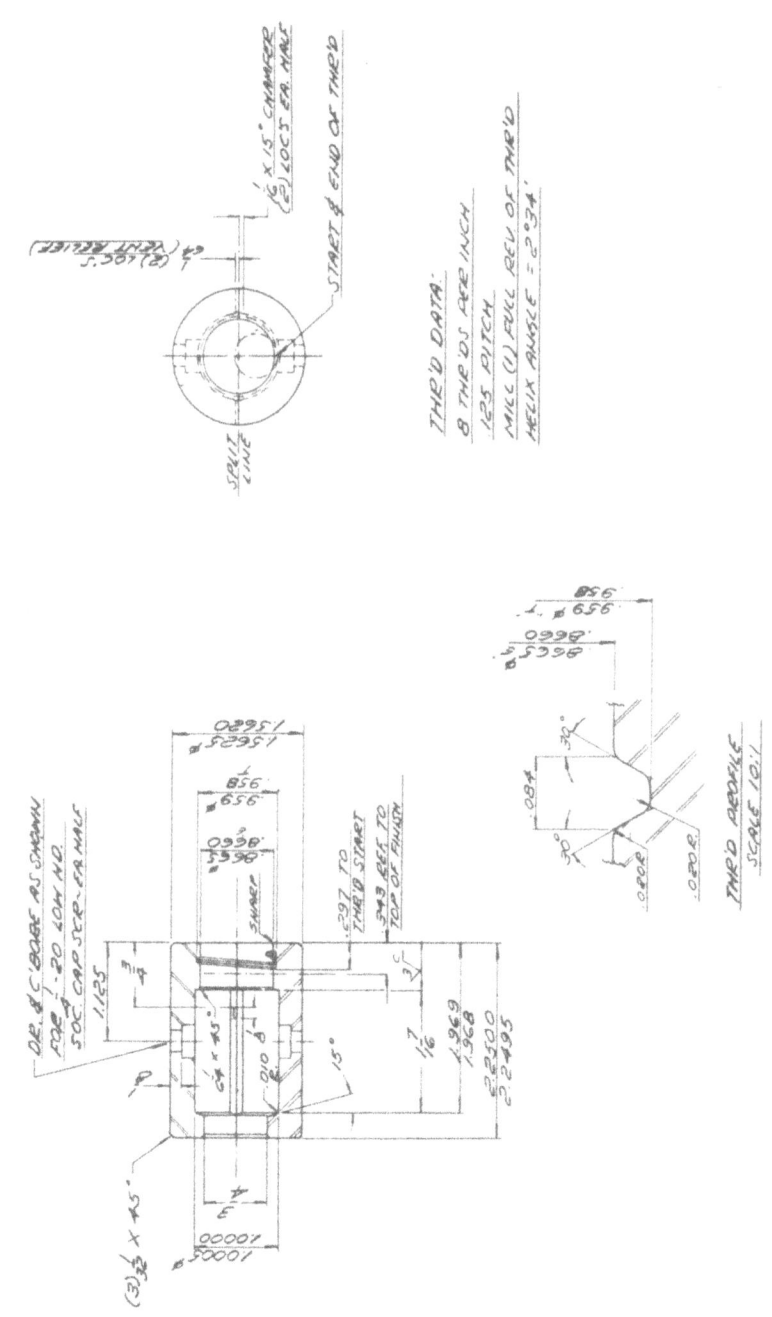

FIG. 11.57. Parison neck ring (detail) Material: AISI-A2 Tool stl. HDN: R/C 52-54. (*Courtesy Captive Plastics, Inc., Piscataway, NJ*)

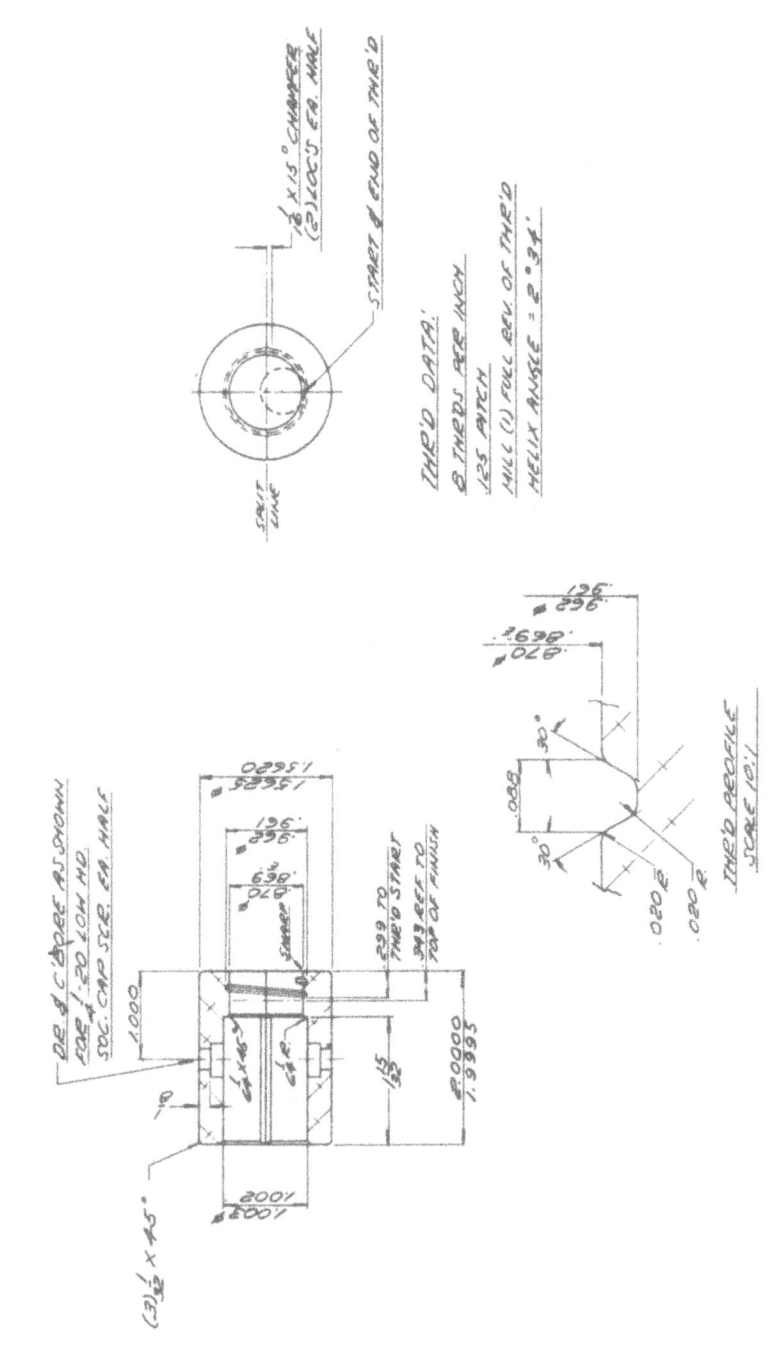

FIG. 11.58. Blow neck ring (detail). Material: 7075-T6 Aluminum. (*Courtesy Captive Plastics, Inc., Piscataway, NJ*)

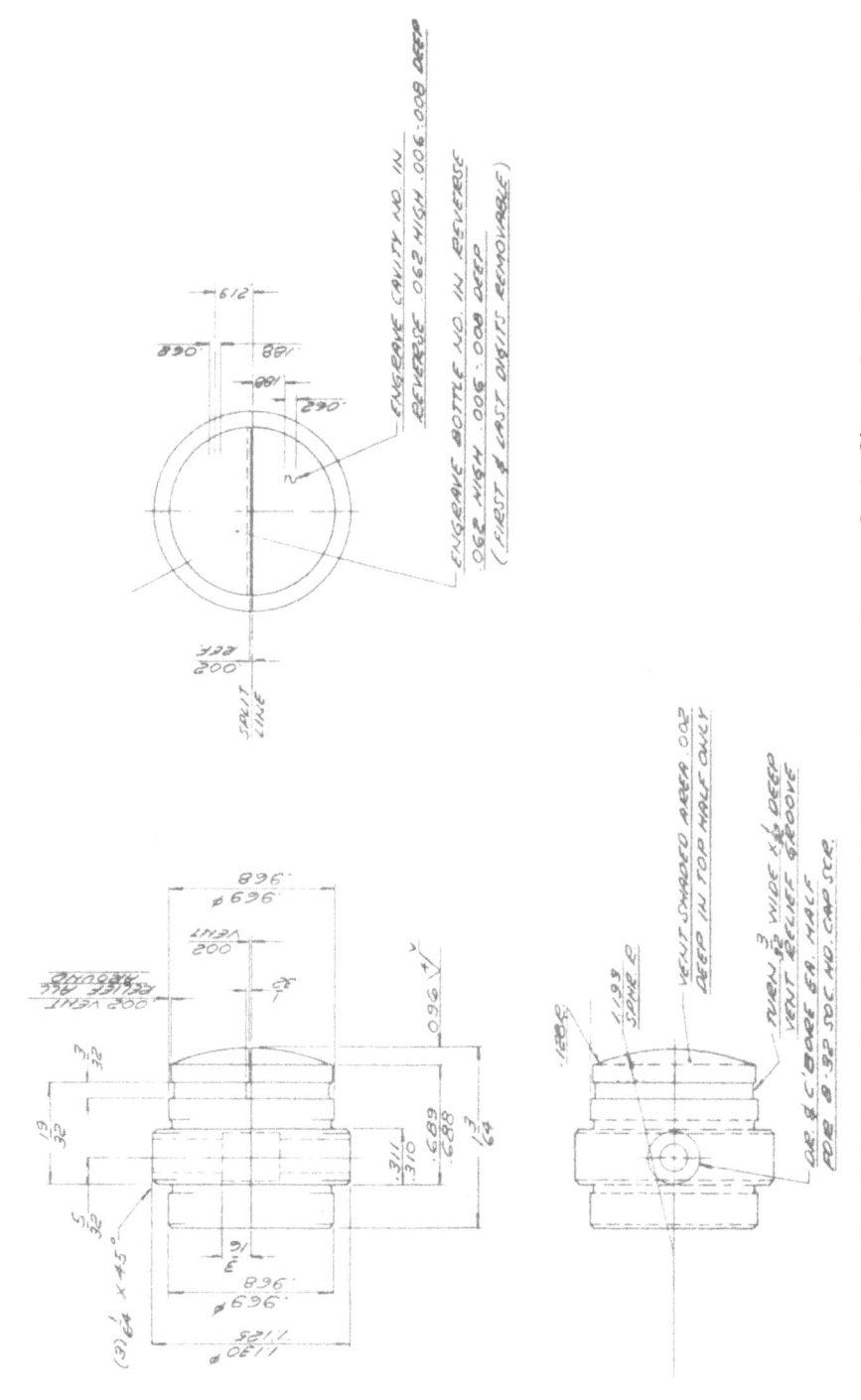

FIG. 11.59. Bottom Plug (detail). Material: 7075-T6 Aluminum. *(Courtesy Captive Plastics, Inc., Piscataway, NJ)*

BOTTLE FINISH*

Neck Finish Specifications

The effectiveness of a plastic bottle as a container depends on the creation of a seal between the bottle and its closure to keep the contents from escaping, while permitting easy opening and resealing of the closure. To achieve a proper seal requires compatibility between the mating bottle and closure threads and other points of engagement; that is, compatibility requires standardization. When plastic blow molding began, standard neck finish specifications developed by the glass industry were used, but subsequently the plastics industry adopted specifications developed by The Plastic Bottle Institute, a division of The Society of the Plastics Industry, Inc. See Fig. 11.60 for bottle-finish terminology; Tables 11.2 and 11.3 for bottle capacity and body dimensional tolerances; and Figs. 11.61, 11.62, and 11.63 for the three most commonly used finish specifications: SP-400, SP-410, and SP-415. Other standard finish specifications are available from The Plastic

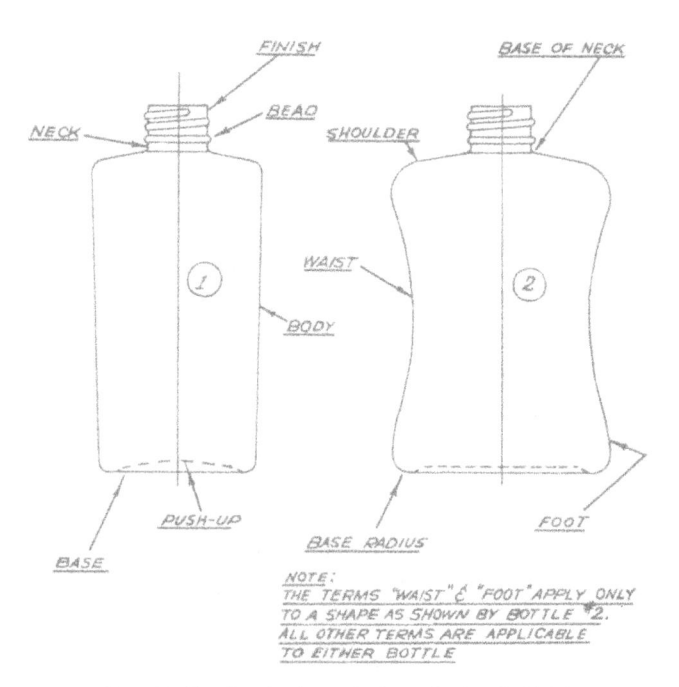

FIG. 11.60. Standard terminology for bottle finish.

*This section by Robert J. Musel, Musel Enterprises, Inc., Pleasantville, New Jersey.

Bottle Institute. Having a copy of their technical bulletin PB1-2 on hand will be very helpful.

Although standard specifications are widely used, container manufacturers produce many nonstandard neck finishes to mate with specialized closures and dispensing fitments. Such nonstandard finishes have proliferated during recent years to meet packaging requirements for (1) cost reduction by using snap-on and linerless closures; (2) improved assembly of fitments which lock in place to dispense products by drop, stream, or spray; (3) consumer convenience by dispensing through the closure; and (4) child and pilfer resistance in accordance with government safety regulations. Specifications for such nonstandard finishes tend to be proprietary to each container manufacturer and are not generally available for publication.

It should be noted that the ability of certain plastic blow molding processes (such as injection-blow) to mold neck-finish configurations and close tolerances (not possible in glass), and the unique properties of plastics used in containers and mating fitments, have provided package designers with great freedom and latitude to develop new package forms.

CONTAINER TERMINOLOGY

Many of the terms referring to various sections of blown plastic containers have been adopted from glass terminology. The following are definitions of the terminology used in Fig. 11.60.

Neck: The section of the container above the shoulder, where the cross-sectional area is smaller than the body of the container.

Base of Neck: The point where the neck meets the shoulder of the container. The base of neck (B.O.N.) often is a reference point for measuring container fill level.

Finish: Shaping of the neck section with a thread form or other configuration suitable for attaching a closure.

Bead: An enlarged diameter of the neck finish used for various purposes: aesthetics, stabilization of the closure, and as a locating point for secondary machining of the neck.

Shoulder: The section of the container where the body decreases in size to meet the neck. The shoulder surface generally is a radius shape.

Body: The section of the container which houses the usable contents.

Waist: A narrowed section of the body.

Foot: The section of the container where the body decreases in size, as a contoured shape, to meet the base surface. The shape may be a simple radius (base radius), or a more complex contour to improve the strength of the container.

SP-400 FINISH FOR

FINISH IDENT	T				E				H			
	MAX INCH	MIN INCH	MAX mm	MIN mm	MAX INCH	MIN INCH	MAX mm	MIN mm	MAX INCH	MIN INCH	MAX mm	MIN mm
18	.704	.688	17.88	17.47	.620	.604	15.75	15.34	.386	.356	9.80	9.04
20	.783	.767	19.89	19.48	.699	.683	17.75	17.35	.386	.356	9.80	9.04
22	.862	.846	21.89	21.49	.778	.762	19.76	19.35	.386	.356	9.80	9.04
24	.940	.924	23.88	23.47	.856	.840	21.74	21.34	.415	.385	10.54	9.78
28	1.088	1.068	27.63	27.13	.994	.974	25.25	24.74	.415	.385	10.54	9.78
30	1.127	1.107	28.62	28.12	1.033	1.013	26.24	25.73	.418	.388	10.62	9.85
33	1.265	1.241	32.13	31.52	1.171	1.147	29.74	29.13	.418	.388	10.62	9.85
35	1.364	1.340	34.64	34.04	1.270	1.246	32.26	31.65	.418	.388	10.62	9.85
38	1.476	1.452	37.49	36.88	1.382	1.358	35.10	34.49	.418	.388	10.62	9.85
40	1.580	1.550	41.13	39.37	1.486	1.456	37.74	36.98	.418	.388	10.62	9.85
43	1.654	1.624	42.01	41.25	1.560	1.530	39.62	38.86	.418	.388	10.62	9.85
45	1.740	1.710	44.20	43.43	1.646	1.616	41.81	41.05	.418	.388	10.62	9.85
48	1.870	1.840	47.50	46.74	1.776	1.746	45.11	44.35	.418	.388	10.62	9.85
51	1.968	1.933	49.99	49.10	1.874	1.839	47.60	46.71	.423	.393	10.74	9.98
53	2.067	2.032	52.50	51.61	1.973	1.938	50.11	49.22	.423	.393	10.74	9.98
58	2.224	2.189	56.49	55.60	2.130	2.095	54.10	53.21	.423	.393	10.74	9.98
60	2.342	2.307	59.49	58.60	2.248	2.213	57.10	56.21	.423	.393	10.74	9.98
63	2.461	2.426	62.51	61.62	2.367	2.332	60.12	59.23	.423	.393	10.74	9.98
66	2.579	2.544	65.51	64.62	2.485	2.450	63.12	62.23	.423	.393	10.74	9.98
70	2.736	2.701	69.49	68.60	2.642	2.607	67.11	66.22	.423	.393	10.74	9.98
75	2.913	2.878	73.99	73.10	2.819	2.784	71.60	70.71	.423	.393	10.74	9.98
77	3.035	3.000	77.09	76.20	2.941	2.906	74.70	73.81	.502	.472	12.75	11.99
83	3.268	3.233	83.01	82.12	3.146	3.113	79.96	79.07	.502	.472	12.75	11.99
89	3.511	3.476	89.18	88.29	3.391	3.356	86.13	85.24	.550	.520	13.97	13.21
100	3.937	3.902	100.00	99.11	3.817	3.782	96.95	96.06	.612	.582	15.54	14.78
110	4.331	4.296	110.01	109.12	4.211	4.176	106.96	106.07	.612	.582	15.54	14.78
120	4.724	4.689	119.99	119.10	4.604	4.569	116.94	116.05	.700	.670	17.78	17.02

THREAD CROSS SECTIONS

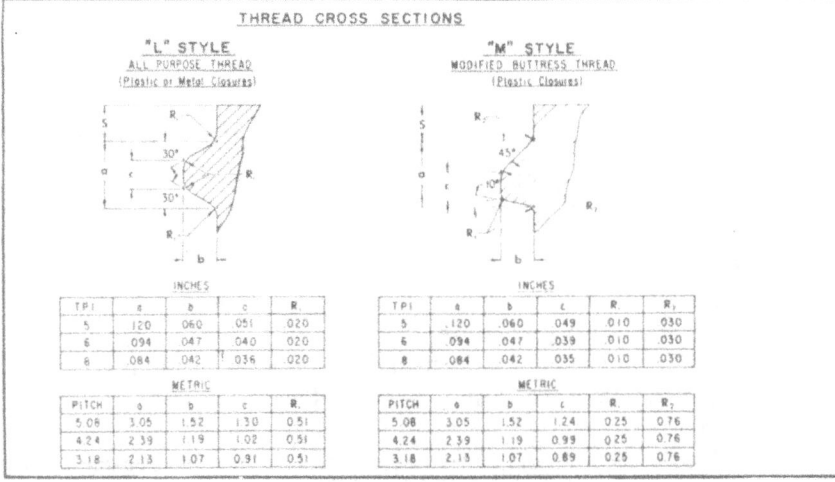

"L" STYLE — ALL PURPOSE THREAD (Plastic or Metal Closures)

"M" STYLE — MODIFIED BUTTRESS THREAD (Plastic Closures)

INCHES

TPI	a	b	c	R
5	.120	.060	.051	.020
6	.094	.047	.040	.020
8	.084	.042	.036	.020

TPI	a	b	c	R	R₁
5	.120	.060	.049	.010	.030
6	.094	.047	.039	.010	.030
8	.084	.042	.035	.010	.030

METRIC

PITCH	a	b	c	R
5.08	3.05	1.52	1.30	0.51
4.24	2.39	1.19	1.02	0.51
3.18	2.13	1.07	0.91	0.51

PITCH	a	b	c	R	R₂
5.08	3.05	1.52	1.24	0.25	0.76
4.24	2.39	1.19	0.99	0.25	0.76
3.18	2.13	1.07	0.89	0.25	0.76

Notes:

1. Dimension H is measured from the top of the finish to the point where diameter T, extended parallel to the centerline, intersects the shoulder or bead [see note #9]
2. A MINIMUM OF 1 — full turn of thread shall be maintained.
3. Contour of bead, undercut or shoulder is optional. [see note #9]
4. Unless otherwise specified, I min. applies to the full length of the opening.
5. Concentricity of I min. with respect to diameters T and E is not included in this standard. I min. is specified for filler tube only.
6. T and E dimensions are the average of two measurements across the major and minor axis. The limits of ovality will be determined by the container supplier and container customer, as necessary.

FIG. 11.61. Table: SP-400 finish for plastic bottles. (*Courtesy the Plastic Bottle Institute, A Division of The Society of the Plastics Industry, Inc., Technical Committee*)

PLASTIC BOTTLES

MAX. INCH	MIN. INCH	MAX. mm	MIN. mm	MIN. INCH	MIN. mm	HELIX ANGLE	CUTTER DIA INCH	CUTTER DIA mm	T.P.I.	PITCH INCH	PITCH mm
.052	.022	1.32	0.56	.325	8.25	3° 30'	.375	9.52	8	.125	3.18
.052	.022	1.32	0.56	.404	10.26	3° 7'	.375	9.52	8	.125	3.18
.052	.022	1.32	0.56	.483	12.27	2° 49'	.375	9.52	8	.125	3.18
.061	.031	1.55	0.79	.516	13.11	2° 34'	.375	9.52	8	.125	3.18
.061	.031	1.55	0.79	.614	15.59	2° 57'	.500	12.70	6	.167	4.24
.061	.031	1.55	0.79	.653	16.59	2° 51'	.500	12.70	6	.167	4.24
.061	.031	1.55	0.79	.791	20.09	2° 31'	.500	12.70	6	.167	4.24
.061	.031	1.55	0.79	.875	22.22	2° 21'	.500	12.70	6	.167	4.24
.061	.031	1.55	0.79	.987	25.07	2° 9'	.500	12.70	6	.167	4.24
.06	.031	1.55	0.79	1.091	27.71	2° 0'	.500	12.70	6	.167	4.24
.061	.031	1.55	0.79	1.165	29.59	1° 55'	.500	12.70	6	.167	4.24
.061	.031	1.55	0.79	1.251	31.77	1° 49'	.500	12.70	6	.167	4.24
.061	.031	1.55	0.79	1.381	35.08	1° 41'	.500	12.70	6	.167	4.24
.061	.031	1.55	0.79	1.479	37.57	1° 36'	.500	12.70	6	.167	4.24
.061	.031	1.55	0.79	1.578	40.08	1° 31'	.500	12.70	6	.167	4.24
.061	.031	1.55	0.79	1.735	44.07	1° 25'	.500	12.70	6	.167	4.24
.061	.031	1.55	0.79	1.853	47.07	1° 20'	.500	12.70	6	.167	4.24
.061	.031	1.55	0.79	1.972	50.09	1° 16'	.500	12.70	6	.167	4.24
.061	.031	1.55	0.79	2.090	53.09	1° 13'	.500	12.70	6	.167	4.24
.061	.031	1.55	0.79	2.247	57.07	1° 8'	.500	12.70	6	.167	4.24
.061	.031	1.55	0.79	2.424	61.57	1° 4'	.500	12.70	6	.167	4.24
.075	.045	1.90	1.14	2.546	64.67	1° 1'	.500	12.70	6	.167	4.24
.075	.045	1.90	1.14	2.753	69.93	1° 9'	.500	12.70	5	2.00	5.08
.075	.045	1.90	1.14	2.918	74.12	1° 4'	.500	12.70	5	2.00	5.08
.075	.045	1.90	1.14	3.344	84.94	0° 57'	.500	12.70	5	2.00	5.08
.075	.045	1.90	1.14	3.737	94.12	0° 51'	.500	12.70	5	2.00	5.08
.075	.045	1.90	1.14	4.131	104.93	0° 47'	.500	12.70	5	2.00	5.08

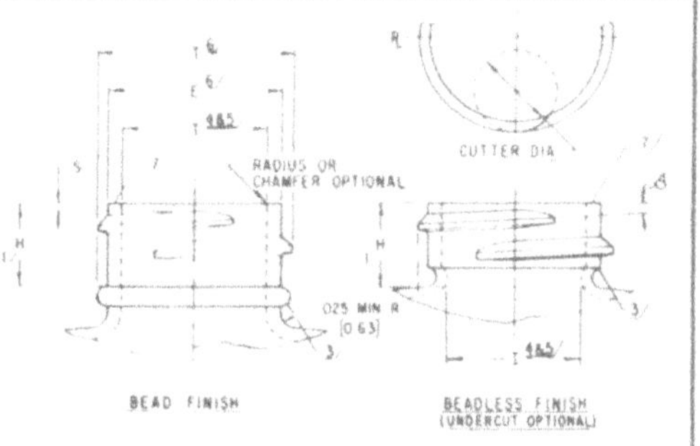

RADIUS OR CHAMFER OPTIONAL

CUTTER DIA

.025 MIN R [0.63]

BEAD FINISH

BEADLESS FINISH (UNDERCUT OPTIONAL)

REVISIONS

The Plastic Bottle Institute
A Division of The Society of the Plastics Industry, Inc.

TECHNICAL COMMITTEE

FIG. NO. 1
REDRAWN

DRAWING NUMBER
SP 400

7. Consideration must be given to the sealing surface width for the sealing system being used.

8. Dimensions in [ ] are in mm.

9. Many child-resistant closures rely on dissimilar simultaneous motions for their proper function. The SP-400 finishes do not necessarily provide physical space for these motions. A special min. and max. H dimension or, as an alternate, a max. bead diameter must then be specified. In addition, to avoid closure contact between adjoining packages during shipment, the bottle diameter should be greater than the largest diameter of the closure. Since these limiting specifications vary between types of child-resistant closures, any user should obtain them from the specific closure manufacturer.

10. Finish to be specified as follows: Thread style, finish identification and drawing number. Example: M28SP400.

To the best of our knowledge the information contained herein is accurate. However, The Society of the Plastics Industry, Inc., assumes no liability whatsoever for the accuracy or completeness of the information contained herein. Final determination of the suitability of any information or material for the use contemplated, the manner of use and whether there is any infringement of patents is the sole responsibility of the user.

SP-410 FINISH FOR

FINISH IDENT.	T				E				H				L	
	MAX. INCH	MIN. INCH	MAX. mm	MIN. mm	MAX INCH	MIN. INCH	MAX. mm	MIN. mm	MAX INCH	MIN INCH	MAX. mm	MIN. mm	MIN. INCH	MIN. mm
18	.704	.688	17.88	17.47	.620	.604	15.75	15.34	.538	.508	13.66	12.90	.361	9.17
20	.783	.767	19.89	19.48	.699	.683	17.75	17.35	.569	.539	14.45	13.69	.361	9.17
22	.862	.846	21.89	21.49	.778	.762	19.76	19.35	.600	.570	15.24	14.48	.376	9.55
24	.940	.924	23.88	23.47	.856	.840	21.74	21.34	.661	.631	16.79	16.03	.437	11.10
28	1.088	1.068	27.63	27.13	.994	.974	25.25	24.74	.723	.693	18.36	17.60	.463	11.76

THREAD CROSS SECTIONS

"L" STYLE
ALL PURPOSE THREAD
(Plastic or Metal Closures)

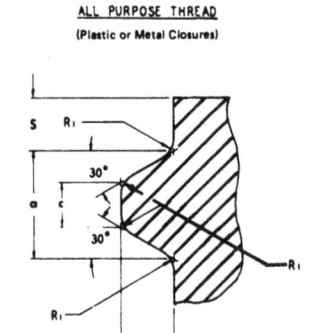

"M" STYLE
MODIFIED BUTTRESS THREAD
(Plastic Closures)

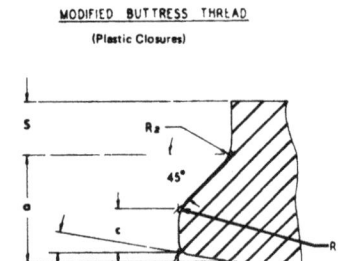

INCHES

T.P.I.	a	b	c	R₁
6	094	047	040	020
8	084	042	036	020

METRIC

PITCH	a	b	c	R₁
4.24	2.39	1.19	1.02	0.51
3.18	2.13	1.07	0.91	0.51

INCHES

T.P.I.	a	b	c	R₁	R₂
6	094	047	039	010	030
8	084	042	035	010	030

METRIC

PITCH	a	b	c	R₁	R₂
4.24	2.39	1.19	0.99	0.25	0.76
3.18	2.13	1.07	0.89	0.25	0.76

Notes

1. Dimension H is measured from the top of the finish to the point where diameter T, extended parallel to the centerline, intersects the shoulder.
2. A MINIMUM OF 1-1/2 Full turns of thread shall be maintained.
3. Contour of bead, undercut or shoulder is optional. If bead is used, bead dia. and L MINIMUM must be maintained.
4. Unless otherwise specified, I min. applies to the full length of the opening.

FIG. 11.62. Table: SP-410 finish for plastic bottles. (*Courtesy the Plastic Bottle Institute, A Division of The Society of the Plastics Industry, Inc., Technical Committee*)

PLASTIC BOTTLES

S				I 4&5/		W 3/		HELIX ANGLE β	CUTTER DIA.		T.P.I.	PITCH	
MAX INCH	MIN INCH	MAX mm	MIN mm	MIN INCH	MIN mm	MAX INCH	MAX mm		INCH	mm		INCH	mm
.052	.022	1.32	0.56	.325	8.25	.084	2.13	3°30'	.375	9.52	8	.125	3.18
.052	.022	1.32	0.56	.404	10.26	.084	2.13	3°7	.375	9.52	8	.125	.3 18
.052	.022	1.32	0.56	.483	12.27	.084	2.13	2°49'	.375	9.52	8	.125	3.18
.061	.031	1.55	0.79	.516	13.11	.084	2.13	2°34'	.375	9.52	8	.125	3.18
.061	.031	1.55	0.79	.614	15.59	.094	2.39	2°57'	.500	12.70	8	.167	4.24

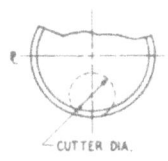

CUTTER DIA.

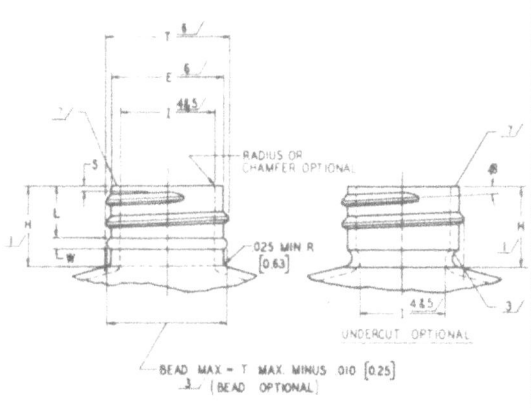

BEAD MAX = T MAX. MINUS .010 [0.25]
3/ (BEAD OPTIONAL)

REVISIONS

The Plastic Bottle Institute
A Division of The Society of the Plastics Industry, Inc.

TECHNICAL COMMITTEE

FIG. NO. 2
REDRAWN 7-4-76

DRAWING NUMBER	REV
SP-410	

5. Concentricity of I min. with respect to diameters T and E is not included in this standard. I min. is specified for filler tube only.
6. T and E dimensions are the average of two measurements across the major and minor axis. The limits of ovality will be determined by the container supplier and container customer, as necessary.
7. Consideration must be given to the sealing surface width for the sealing system being used.
8. Dimensions in [ ] are mm.
9. Finish to be specified as follows: Thread style, finish identification and drawing number. Example: M28SP410.

To the best of our knowledge the information contained herein is accurate. However, The Society of the Plastics Industry, Inc., assumes no liability whatsoever for the accuracy or completeness of the information contained herein. Final determination of the suitability of any information or material for the use contemplated, the manner of use and whether there is any infringement of patents is the sole responsibility of the user.

SP-415 FINISH FOR

FINISH IDENT.	T				E				H				L	
	MAX. INCH	MIN. INCH	MAX. mm	MIN. mm	MAX. INCH	MIN. INCH	MAX. mm	MIN. mm	MAX. INCH	MIN. INCH	MAX. mm	MIN. mm	MIN. INCH	MIN. mm
13	.514	.502	13.06	12.75	.454	.442	11.53	11.23	.467	.437	11.86	11.10	.306	7.77
15	.581	.569	14.76	14.45	.521	.509	13.23	12.93	.572	.542	14.53	13.77	.348	8.84
18	.704	.688	17.88	17.47	.620	.604	15.75	15.34	.632	.602	18.05	15.29	.429	10.90
20	.783	.767	19.89	19.48	.699	.683	17.75	17.35	.757	.727	19.23	18.47	.456	11.58
22	.862	.846	21.89	21.49	.778	.762	19.76	19.35	.852	.822	21.84	20.86	.546	13.87
24	.940	.924	23.88	23.47	.856	.840	21.74	21.34	.972	.942	24.69	23.93	.561	14.25
28	1.088	1.068	27.63	27.13	.994	.974	25.25	24.74	1.097	1.067	27.86	27.10	.655	16.64
33	1.265	1.241	32.13	31.52	1.171	1.147	29.74	29.13	1.289	1.259	32.74	31.98	.772	19.61

THREAD CROSS SECTIONS

"L" STYLE

ALL PURPOSE THREAD

(Plastic or Metal Closures)

"M" STYLE

MODIFIED BUTTRESS THREAD

(Plastic Closures)

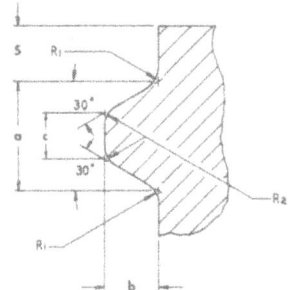
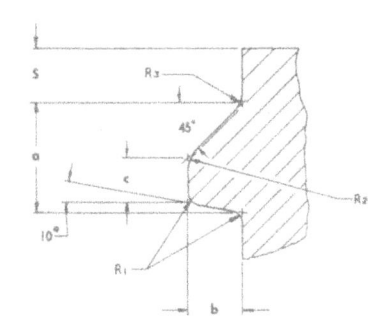

INCHES

T.P.I.	a	b	c	R₁	R₂
6	.094	.047	.040	.020	.020
8	.084	.042	.036	.020	.020
12	.045	.030	.011	.015	.005

METRIC

PITCH	a	b	c	R₁	R₂
4.24	2.39	1.19	1.02	0.51	0.51
3.18	2.13	1.07	0.91	0.51	0.51
2.11	1.14	0.76	0.28	0.38	0.13

INCHES

T.P.I.	a	b	c	R₁	R₂	R₃
6	.094	.047	.039	.010	.030	.030
8	.084	.042	.035	.010	.030	.030
12	.051	.030	.016	.010	.008	.020

METRIC

PITCH	a	b	c	R₁	R₂	R₃
4.24	2.39	1.19	0.99	0.25	0.76	0.76
3.18	2.13	1.07	0.89	0.25	0.76	0.76
2.11	1.29	0.76	0.41	0.25	0.20	0.51

Notes:

1. Dimension H is measured from the top of the finish to the point where diameter T, extended parallel to the centerline, intersects the shoulder.
2. A MINIMUM OF 2 — Full turns of thread shall be maintained.
3. Contour of bead, undercut or shoulder is optional. If bead is used, bead dia. and L MINIMUM must be maintained.
4. Unless otherwise specified, I min. applies to the full length of the opening.

FIG. 11.63. Table: SP-415 finish for plastic bottles. (*Courtesy the Plastic Bottle Institute, A Division of The Society of Plastics Industry, Inc. Technical Committee*)

PLASTIC BOTTLES

S				I 4&5/		W 3/		HELIX ANGLE θ	CUTTER DIA.		T.P.I.	PITCH	
MAX. INCH	MIN. INCH	MAX. mm	MIN. mm	MIN. INCH	MIN. mm	MAX. INCH	MAX. mm		INCH	mm		INCH	mm
.052	.022	1.32	0.56	.218	5.54	.045	1.14	3°11'	.375	9.52	12	.083	2.11
.052	.022	1.32	0.56	.258	6.55	.045	1.14	2°48'	.375	9.52	12	.083	2.11
.052	.022	1.32	0.56	.325	8.25	.084	2.13	3°30'	.375	9.52	8	.125	3.18
.052	.022	1.32	0.56	.404	10.26	.084	2.13	3°7'	.375	9.52	8	.125	3.18
.052	.022	1.32	0.56	.483	12.27	.084	2.13	2°49'	.375	9.52	8	.125	3.18
.061	.031	1.55	0.79	.516	13.11	.084	2.13	2°34'	.375	9.52	8	.125	3.18
.061	.031	1.55	0.79	.614	15.59	.094	2.39	2°57'	.500	12.70	6	.167	4.24
.061	.031	1.55	0.79	.791	20.09	.094	2.39	2°31'	.500	12.70	6	.167	4.24

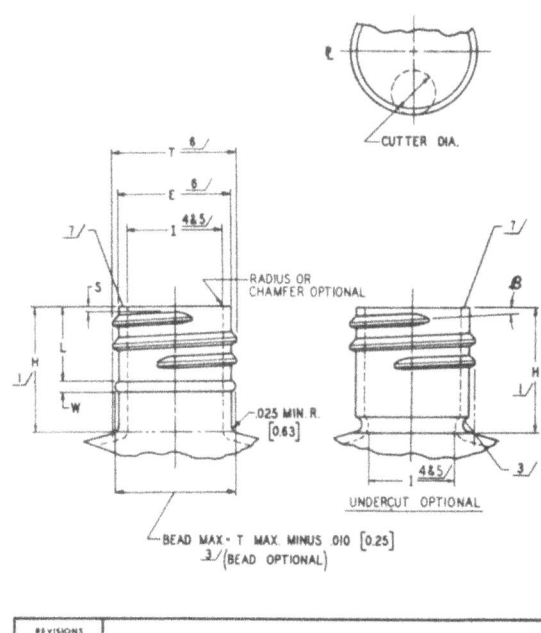

The Plastic Bottle Institute
A Division of The Society of the Plastics Industry, Inc.

TECHNICAL COMMITTEE

REVISIONS
A 10-30-80

FIG. NO. 3
REDRAWN 7-4-78

DRAWING NUMBER	REV
SP-415	A

5. Concentricity of I min. with respect to diameters T and E is not included in this standard. I min. is specified for filler tube only.
6. T and E dimensions are the average of two measurements across the major and minor axis. The limits of ovality will be determined by the container supplier and container customer, as necessary.
7. Consideration must be given to the sealing surface width for the sealing system being used.
8. Dimensions in [ ] are in mm.
9. Finish to be specified as follows: Thread style, finish identification and drawing number. Example: M28SP415.

To the best of our knowledge the information contained herein is accurate. However, The Society of the Plastics Industry, Inc., assumes no liability whatsoever for the accuracy or completeness of the information contained herein. Final determination of the suitability of any information or material for the use contemplated, the manner of use and whether there is any infringement of patents is the sole responsibility of the user.

Base: The bottom surface, on which the container stands upright.

Push-up: A section depressed upward within the base surface. The depression assures that the container will stand upright without rocking on projections caused by parison pinch lines, gate marks, engraving, coding, or other interruptions in the base surface.

Tables—Capacity and Tolerances

TABLE 11.2. Capacity Tolerances

Fluid Ounces				Milliliters			
Bottle Overflow Capacity		Tolerance (Plus or Minus)		Bottle Overflow Capacity		Tolerance (Plus or Minus)	
	less than	0.75	0.05		less than	22	1.5
0.75	and less than	1.2	0.07	22	and less than	35	2.0
1.2	and less than	1.6	0.08	35	and less than	47	2.5
1.6	and less than	2.1	0.10	47	and less than	62	3.0
2.1	and less than	2.8	0.12	62	and less than	83	3.5
2.8	and less than	3.9	0.14	83	and less than	115	4
3.9	and less than	5.4	0.17	115	and less than	159	5
5.4	and less than	7.4	0.20	159	and less than	218	6
7.4	and less than	9.8	0.24	218	and less than	289	7
9.8	and less than	13	0.30	289	and less than	384	9
13	and less than	18	0.37	384	and less than	531	11
18	and less then	26	0.44	531	and less than	767	13
26	and less than	37	0.51	767	and less than	1092	15
37	and less than	51	0.68	1092	and less than	1505	20
51	and less than	72	0.81	1505	and less than	2125	24
72	and less than	98	1.01	2125	and less than	2892	30
98	and less than	119	1.30	2892	and less than	3512	38
119	and less than	139	1.50	3512	and less than	4103	44
139	and less than	160	1.80	4103	and less than	4723	53
160	and less than	180	2.00	4723	and less than	5313	59
180	and less than	210	2.20	5313	and less than	6199	65
210	to 5 gals.		1% of Capacity	6199	To 18.89 liters		1% of Capacity

Note 1. If it is desired to specify in cubic centimeters, convert to cubic centimeters by multiplying milliliters by 1.0017 at 73.4°F (23°C).

(Courtesy of the Plastics Bottle Institute, div. of The Society of The Plastics Industry, Inc., New York, N.Y.)

TABLE 11.3.   Body Dimensional Tolerances

Range of Specific Dimensions		Height Dimensions		Width & Depth Dimensions	
Inches	mm	Inches	mm	Inches	mm
0 up to but not incl. 1	0 to 25.40	.030	0.76	.030	0.76
1 up to but not incl. 2	25.40 to 50.80	.030	0.76	.050	1.27
2 up to but not incl. 4	50.80 to 101.60	.050	1.27	.060	1.52
4 up to but not incl. 6	101.60 to 152.40	.050	1.27	.080	2.03
6 up to but not incl. 8	152.40 to 203.20	.060	1.52	.090	2.29
8 up to but not incl. 10	203.20 to 254.00	.060	1.52	.110	2.79
10 up to but not incl. 12	254.00 to 304.80	.080	2.03	.120	3.05
12 up to but not incl. 15	304.80 to 381.00	.090	2.29	.150	3.81
15 up to but not incl. 18	381.00 to 457.20	.110	2.79	.150	3.81

(*Courtesy of the Plastics Bottle Institute, div. of The Society of the Plastics Industry Inc., New York, N.Y.*)

REFERENCES

Barrier-bottle technology advances on all fronts, *Modern Plastics*, p. 600, Aug. 1986.

Blow molding, *Modern Plastics Encyclopedia*, 1984–85, 1985–86 Eds.

Blow molding unveiled for oriented PP bottles, *Plastics Technology*, Dec. 1985.

Coextruded packaging excitement spurs machinery, material advances, *Plastics Technology*, Dec. 1985.

Coextrusion coming on fast, *Plastics Technology*, p. 104, May 1985.

DuBois and John, *Plastics*, 6th Ed. New York: Van Nostrand Reinhold Co., Inc.

Equipment choices widen for making coex containers, *Plastics World*, p. 42, Feb. 1985.

In-mold labeling: Next value-added for blow molders, *Plastics World*, Aug. 1984.

Kuelling, W. and L. Monaco, Injection blow molds, here's how to build them, *Plastics Technology*, June 1975.

Multilayer bottles mix functionality, economy, *Package Engineering*, Mar. 1982.

O-I (Owens-Illinois) bets on the barrier bottle, *Plastics World*, p. 46, Feb. 1985.

Molds for Reaction Injection, Structural Foam and Expandable Styrene Molding

Chapter / *12*

Revised by Wayne I. Pribble

Reaction injection molding is also called liquid injection molding, high pressure impingement molding and reaction liquid impingement molding with acronyms RIM, LIM and RLIM. These systems are now being used for shoe components, furniture, industrial parts and automobile body components, etc. as shown in Fig. 12.1. In this process, the two components of a resin such as urethane or a synthetic rubber are metered carefully and mixed at very high pressure in a mixing chamber (Fig. 12.2) prior to injection into the mold where a fast thermoset cure is achieved.

Many of the polymers such as caprolactam, silicone, epoxy, nylon and polyester are multicomponent reactive liquids that may be modified for the reaction injection process. Reaction injection molding is particularly suited for large area pieces such as auto body parts, furniture, housing sections, etc. because of the fast cycle and the low mold clamping pressure required. This system reduces greatly the capital investment required for very large parts. For example, a 5,000 ton conventional injection press costs approximately $3,000,000. The rear fender of an automobile would need a 12,000-ton press costing $6,000,000 to $10,000,000 if molded by the conventional high pressure injection process; with reaction injection molding (RIM), the equipment cost might only be one tenth of the conventional machine cost.

The mold clamping unit in the RIM press is usually designed (Fig. 12.3) so that the mold may tilt 300° about its axis to facilitate filling, venting and foaming. During injection, the mold is tilted into a favorable position that will eliminate surface defects such as bubbles and insure a complete mold

FIG. 12.1. Reaction injection molded (RIM) body component molded in a 1.5 minute cycle.

fill. Mold orientation during fill is a cut-and-try process to find the most favorable position. The low cost of RIM molding machines is the result of the low pressures that are used. Molds and presses are usually designed for 600 tons but 300 tons is the actual molding pressure commonly used at this time.

Molds for reaction injection molding are commonly of the clam-shell type to facilitate easy removal of the part (Figs. 12.4 and 12.5). The best material for production molds is steel, with finely polished and plated surfaces. Aluminum, electroformed nickel, beryllium copper or kirksite may be used. Nonporous molding surfaces are essential. Mold making and general construction are similar to that for structural foam molds except for the hinging components and locking devices.

Mold cut-off surfaces must provide a tight seal to minimize flash. Venting is extremely important and the vent design must contemplate easy vent-flash removal. Conventional ejector pins are useless although large area ejector pads may help with some designs. Vacuum part unloaders plus mold pad ejectors are superior to the manual stripping historically used by the rubber industry. Close thermal control is essential for the mold cavities since a constant temperature somewhere between 70 and 90° F must be maintained for RIM integral skin and rigid products. Water lines drilled in mold components work best with a circulating water cool-

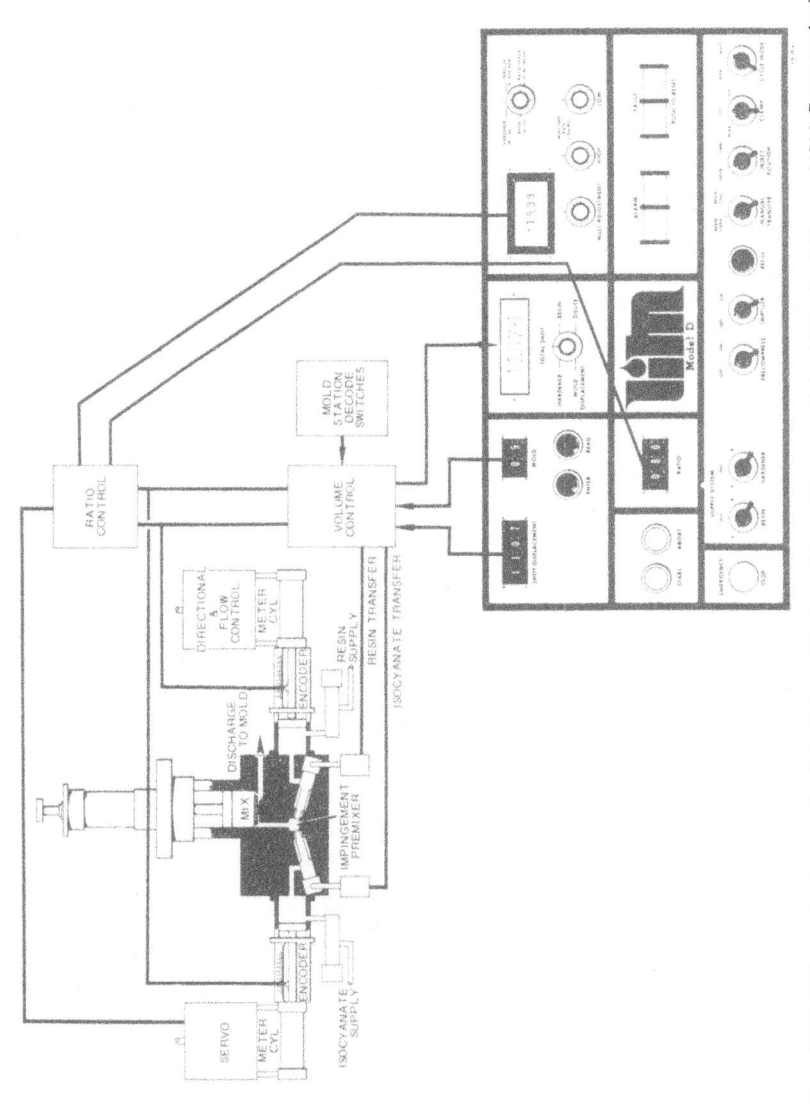

FIG. 12.2. Schematic of a reaction injection molding machine. (*Courtesy United Chemical Group, USM Corporation*)

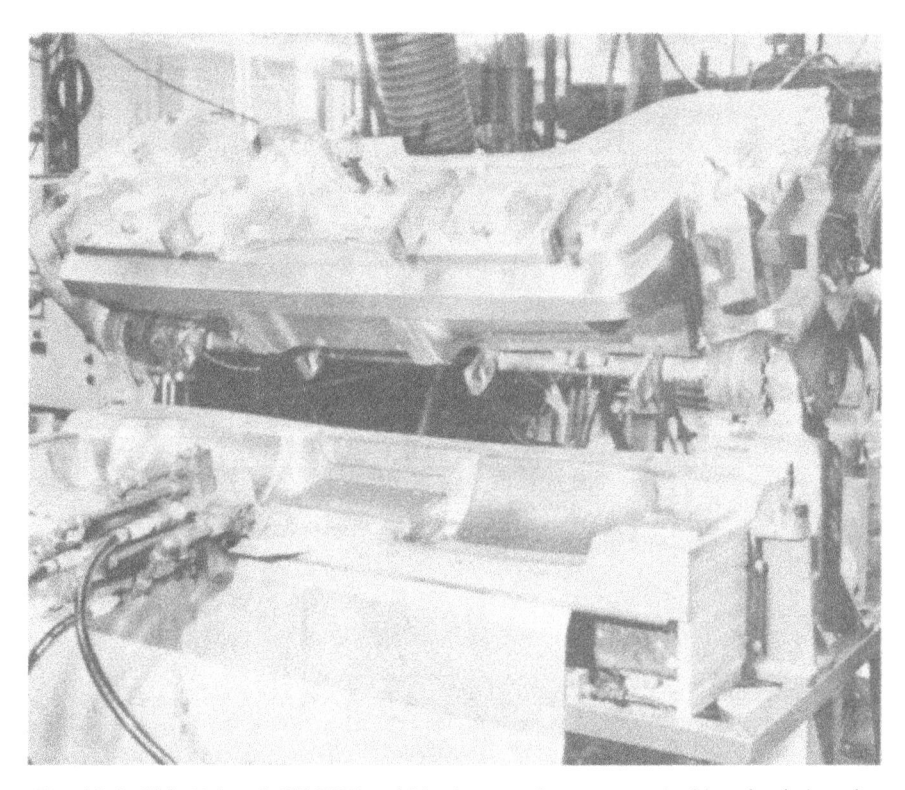

FIG. 12.3. This "clamshell" RIM mold is shown as it rotates on the hinged axis into the clamp position. The gating area can be noted in the front upper center. The injection is in lower front center. (*Courtesy Union Carbide Corp., Danbury, CT*)

ant. The clamp is often assisted in the large molds by solenoid or pneumatic latches that lock the halves together prior to injection.

Gate design is very critical, it must be adequate to permit laminar flow into the mold, as illustrated by Fig. 12.6. An after-mixer, sometimes called a static mixer, illustrated in Figs. 12.7 and 12.8, is built into the mold, integral with the gate, to insure complete mixing before the compound enters the mold. These after-mixers are designed to effect reimpingement and turbulent mixing of the previously mixed material components. This type of static mixer is contingent on velocity control of the material as it flows through such a labyrinth design. This after-mixer is a part of the sprue and runner system and comes out as a cull with them. Turbulence cannot be permitted in the mold.

The mechanical details of the mold design are controlled by the press design and the machine builder will provide guidance on these details. The mold designer must have a real understanding of the material, the ma-

Fɪɢ. 12.4. Shown here is the RIM molded piece being removed from the mold. The rectangular slots in the projecting ears in Figs. 12.3 and 12.4 receive the clamping wedges or latches. (*Courtesy Union Carbide Corp., Danbury, CT*)

chine, the anticipated cycles and volume requirement before undertaking the design of a reaction injection molding mold design. A very simple mold of low cost materials and fabrication will be suitable for small volume jobs if the molding surfaces can be finely polished and plated. Long runs will justify the best mold that can be built.

MOLDS FOR STRUCTURAL FOAM PRODUCTS*

Structural Foam

Structural foam molding is a process that allows the molding of thick sections with a minimum of *shrinkage*, *sinks*, or *warpage*. The process uses standard molding materials into which a *foaming agent* has been introduced,

*This section for Structural Foam Products was rewritten and updated by Wayne I. Pribble in consultation with John Waanders, DeKalb Molded Plastics, A JSJ Corporation Company, Butler, IN. Also after review of The Society of the Plastics Industry, Inc., manual entitled *Structural Foam*, and a review of publications of molding material suppliers. Every designer of structural foam molds needs these publications as part of his library.

FIG. 12.5. RIM mold of Fig. 12.3 shown here in the clamped position ready for the injection through the front center hoses. (*Courtesy Union Carbide Corp., Danbury, CT*)

FIG. 12.6. In this RIM molding, the incoming material flows through the large runner section at right and enters the mold through the reducing section in the center and into a fan gate (at left). A laminar flow results. Gate thickness ranges from $3/64$ in. to $3/32$ in. (*Courtesy Mobay Chemical Corp., Pittsburgh, PA*)

Fɪɢ. 12.7. One type of after-mixer is shown in this RIM mold where the entering material is mixed a second time before it enters the runner. (*Courtesy Mobay Chemical Corp., Pittsburgh, PA*)

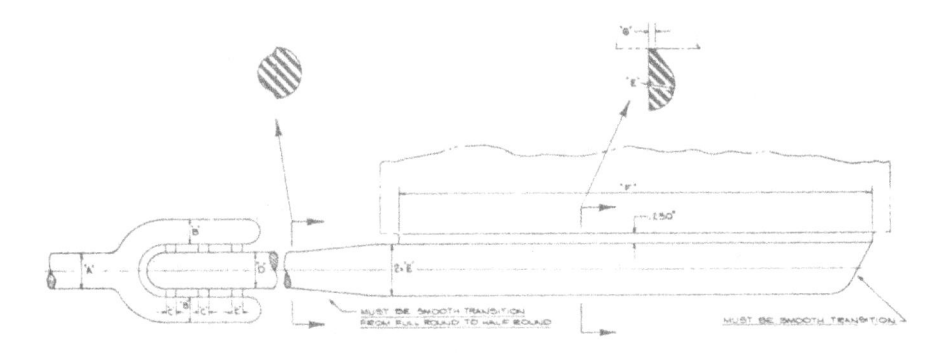

Fɪɢ. 12.8. Depicts another type of commonly used after-mixer built into the mold at the point of material entry system. These after-mixers cause reimpingement and turbulent mixing as the material enters the runner. (*Courtesy Mobay Chemical Corp., Pittsburgh, PA*)

either as a premix or as a direct injection into the melt stream. The result, when the mix is injected into a mold, it becomes a structure with a cellular interior and a solid skin. The interior pressure of the foaming agent offsets the normal tendency of a heavy section to sink toward the center of a heavy wall or boss. Sinks in plastic parts are just another example of the ancient principle, "Nature abhors a vacuum." To offset the tendency to sink, the interior pressure of the foaming agent fills, with a cellular structure, what would otherwise be a void or a sink. Thus, it becomes possible to mold structural foam parts having a wall thickness of several inches.

History and Development

Much of the development of structural foam in the U.S.A was in the New England area, but there was also parallel activity in the Midwest beginning in the early or mid-1960s. Because of the past tendency to be secretive in the plastics industry, it becomes difficult to state precisely who was the first person to conceive of eliminating sinks in such items as furniture parts or window frames by throwing a handful of ordinary household baking powder into the hopper with the resin. The Society of the Plastics Industry, Inc., did create a Structural Foam Division about 1971, from which we deduce a five to ten year prior development period. The names of Shell Chemical Co., Union Carbride Corp., Borg-Warner Corp., and others are prominent in the early developments.

For historical reasons, we mention the various licensed processes instituted from time to time. Most of these processes were not successful in the market place. In fact, some in the industry believe many of the patented processes hindered the growth of the industry rather than helping it. As practiced today, structural foam molding can be done by any injection molder. However, those specializing in the field realize they must be prepared to make very large parts, beyond the capability of the average injection molding plant, or else produce structural foam moldings more economically by the use of specialized machines. The special machines are designed to take advantage of the low cavity pressures, to offset the long cycle times and/or to improve surface finishes. Special machines for structural foam have the common characteristics of a relatively large platen, large shot size, and low clamp tonnage as compared to conventional injection molding machines. The special machines to which we will refer later include, single nozzle machines, manifold with multiple nozzle, rotary table with up to ten mold stations, shuttle mold, shuttle injection unit, and co-injection machines.

The foaming agent can be an inert gas, such as nitrogen, or a *chemical blowing agent* (CBA). Historically, ordinary household baking powder was probably the first of the CBA's, and is probably still being used today. In household use, baking powder incorporated into a cake mix will reach a conversion temperature in the oven and change into a gas. This gas expands and causes the cake to become cellular while retaining its moist characteristics (that is, if the baker is fortunate and the cake doesn't ''fall''). Thus, a CBA is activated by the heat of the injection cylinder and the melt must be quickly injected into the mold where the expansion can take place.

With this rather sketchy developmental history of the process, let us proceed to examine the details of machines, processes and mold design.

Structural Foam Molding Machines

Machines built primarily for structural foam molding were initially similar to ''normal'' injection machine in layout, appearance and operation, but were characterized by lighter construction than the conventional machine and therefore, on the basis of shot weight capability, were cheaper. However, a recent development is the use of counterpressure within the mold. This development has increased the pressures required to keep the mold closed during the injection cycle with a resulting increase in overall clamp tonnage. In general, structural foam machines are of the reciprocating screw type using either a single-stage or two-stage technique. It is most economical to build very large injection units with separate extruders and injection chambers. In most designs, the screw is mounted above or alongside the injection cylinder, and most of them are electrically driven by DC drive motors for precise control of rotation speeds. Figure 12.9 shows the layout

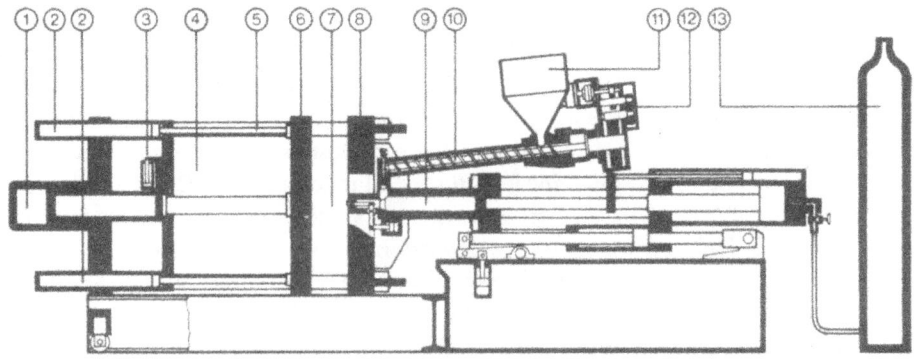

1. clamp cylinder	6. movable mold platen	10. screw preplasticizing unit
2. cylinders for platen movement	7. hydraulic nozzle shut off valve	11. hopper
3. locking plate with cylinder	8. stationary platen	12. screw drive (electrical or hydraulic)
4. clamp unit	9. melt cylinder	13. gas-oil accumulator
5. piston rod for platen movement		

FIG. 12.9. Schematic of a structural foam molding machine. (*Courtesy Krauss-Maffei Corp., Grand Rapids, MI*)

of a typical machine. Valves of various types are used to allow transfer of the melt into the injection cylinder, from which it is injected into the mold.

Several machines have been built where there is more than one injection unit for the very large press. These injection units can be used separately to make two or more parts at the same time, and, if desired, using different materials. They can also be used together, using the same material and in the same cycle, to make one large part requiring the shot capacity of the two or more units.

Sandwich System

The I.C.I. *bi-component* or *sandwich system* is not in common use in the U.S.A., but it merits description on the basis of its place in the history of developments in structural foam molding. This system permits a part to have a skin of one material and a core of another material, or else a solid skin with a foamed core of the same material. The process is characterized by (1) parts of excellent surface finish without sink marks opposite ribs or bosses; (2) a sharp division between core and skin; (3) the ability, as previously mentioned, for the core and skin to be of different materials. For example, a hard and tough but comparatively expensive polymer can enclose a foamed core of less costly material.

Co-injection Systems

A system currently in use to accomplish an objective similar to the sandwich is called *co-injection*. At last count, there were about eight co-injection machines (all made by Battenfeld) in the U.S.A. The principle of co-injection is to inject two polymer formulations sequentially from separate injection units into a mold through a common sprue. The injection machine used for co-injection is specially designed, having two injection units and equipped for effecting sequential changeover from one injection unit to the other during a single shot. The press must also have means for controlling the sequence of mold opening and mold closing movements as well as the operation of core movement mechanisms to allow for expansion of the melt. Partial opening is covered later under the section label "USM Process."

Co-injection machines are about 50% more expensive than conventional machines of comparable size. The sequence of molding operations is shown in Fig. 12.10. Stage 1 shows the mold closed, with both injection units charged, both screws retracted, and the changeover tap shut. In Stage 2, the tap is open to the first unit, which has injected a partial charge into the mold. Stage 3 involves opening the tap to the second unit, injecting the polymer, then forcing the polymer to the extremities of the mold to form a uniform layer of the first polymer without bursting through it. The mechanism is

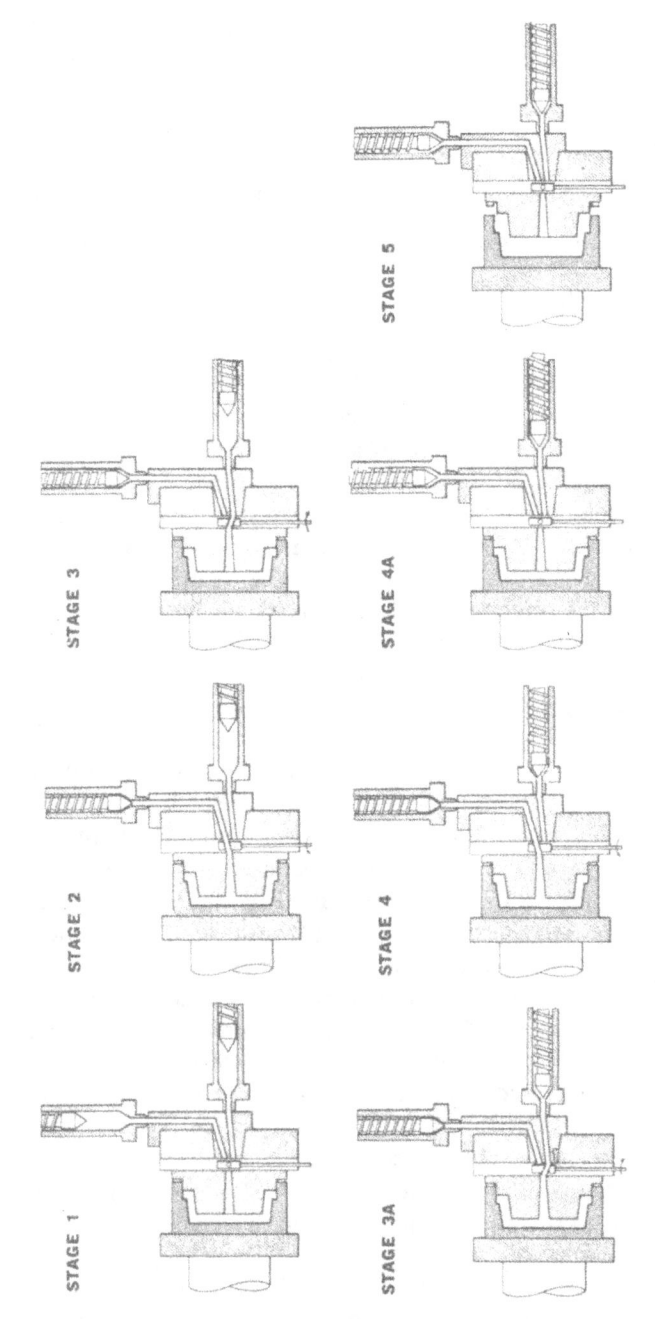

Fig. 12.10. Stages in mold filling in the co-injection structural foam molding process, where two components are sequentially injected to make a single product.

analogous to blowing up a balloon; the first polymer (the balloon) provides the package for the second polymer, providing the rate of filling and the volume of the second polymer (the air) is not excessive. This system works with two-polymer charges provided the conditions of polymer temperature, mold temperature, injection rate and other variables are well controlled.

In Stage 4, a sufficient amount of first polymer (the package) is injected into the cavity to clear the sprue of material. (Without this small additional charge, the foam core would be revealed on removal of the sprue, and residual foamable polymer in the tap would contaminate the skin of the next molding.)

Having filled the mold, closed the tap (Stage 4), and held clamp pressure, the interior material will foam and give a uniform foam structure within a thin skin of solid material (Stage 5). Incidentally, Fig. 12.10 also shows the USM process of opening the mold slightly to give more room for foaming of the core material.

Union Carbide System

The system developed by Union Carbide Corp., sometimes termed the *low-pressure* or *short-shot system*, uses a stationary (fixed) single-screw plasticizing extruder for producing the foamable melt in conjunction with an accumulator and a hydraulically operated injection and mold clamp unit. Union Carbide Corp. patents were held invalid about 1976 and *low-pressure system* is now the accepted descriptive term. The processing technique is shown diagrammatically in Fig. 12.11.

The blowing agent, usually nitrogen, is injected directly into the melt in the extruder barrel. The melt passes from the extruder into the accumulator where it is maintained at sufficient pressure and temperature to prevent foaming until the predetermined charge is collected. A changeover valve is then operated and the charge injected rapidly by a hydraulic ram via multiple nozzles. Inasmuch as the material is a thermoplastic, cooling of the material is required to "set" the material so the mold can be opened, the part ejected, and the mold reclosed while the accumulator is filled for the next shot. In some cases the extruder could operate continuously, but current practice is to stop the extruder when the accumulator charge is collected rather than attempt an extruder screw speed to exactly match the amount of melt required in a given amount of time.

In the low-pressure process, pressures within the mold range between 200 and 800 psi. Thus, relatively low clamp tonnages are required, permitting the molding of components of large projected area on machines with large but comparatively lightly built clamp units. The use of the multiple nozzle system ensures complete and rapid filling of large molds, or, if desired, permits the simultaneous molding of a number of smaller components in a

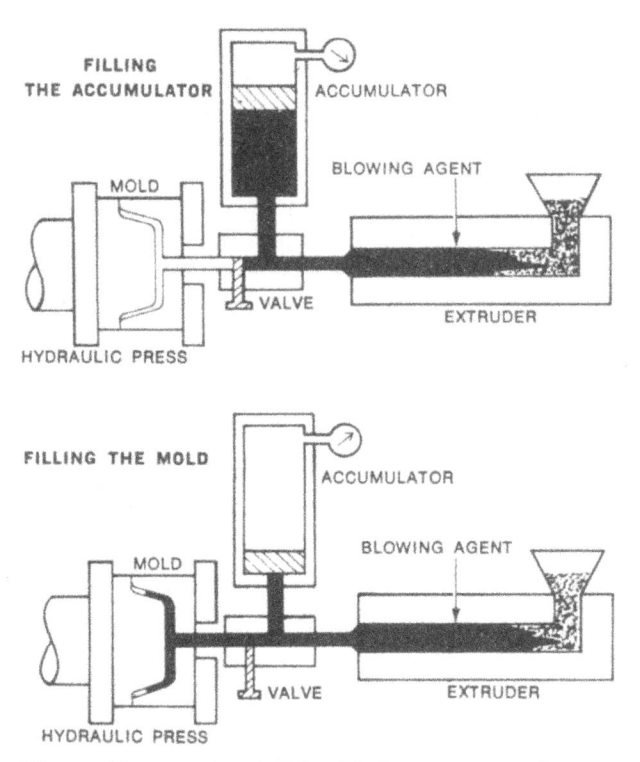

FIG. 12.11. Diagram illustrates the principle of the low-pressure or short-shot process originally developed by Union Carbide Corp., for structural foam moldings.

single shot. Sometimes this is accomplished by *sequential valve-gating* of multiple cavity tooling. (Refer to Fig. 12.22.) These multiple moldings from the same shot can vary in size because the melt flow through each nozzle can be individually controlled. Incidentally, one advantage of sequential valve-gating is the ability to move knit lines out of critical areas by the simple process of increasing or decreasing the valve-gate opening of any valve-gate with respect to any other valve-gate in the same mold. Obviously, such individual adjustment affects the flow pattern within the mold. Various configurations of plasticizing cylinder, accumulator, and injection methods are available from the machine suppliers. A mold designer will normally be required to design a mold for a specific machine already in place. Complete data about the machine must be available to assure ''fit'' of the mold.

The TAF System

The rapid increase of structural foam molding that began with the use of conventional machines and spawned the low pressure, short-shot and sandwich-molding processes encouraged machine designers to seek a system which would give the design flexibility of the former coupled with the surface quality associated with the latter.

In the TAF process, developed by Ashai-Dow Ltd. flexibility is limited by the necessity of expanding the mold, or by withdrawing some of the melt as it expands. Even though the TAF process is not in use in the U.S.A. (to the best of this author's knowledge) we will give a brief description here because large, thick walled parts can be produced and very low densities achieved without deterioration in the mechanically smooth, optically homogeneous surface provided by the sandwich process. The process is commonly designated *high-pressure* or *full shot*, but the pressures generated during the cycle are much lower than those associated with conventional injection molding. Clamp pressure ranging from 600 to 1500 psi may be required. Average foam densities are in the 0.7 range, while cycle times compare with the low-pressure process.

In practice, the foamable melt is plasticized under closely controlled temperature and pressure using a high L/D ratio screw (in the order of 28 : 1), which also provides a close control of viscosity. The melt is injected to completely fill the mold and, after a pause to allow the molten polymer to freeze at the skin, inserts in the mold are retracted, increasing the capacity of the cavity and thus permitting foaming of the core, as shown in Fig. 12.12.

In the TAF process, the cooling time is governed by all those normal factors of mold temperature, material temperature, plasticizing temperature, part surface relative to volume, blowing agent gas pressure, and, of course,

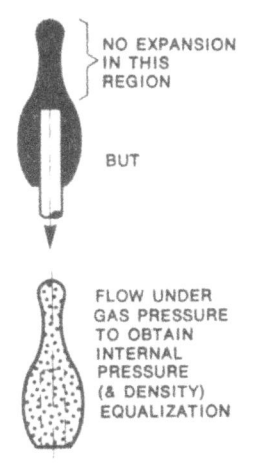

FIG. 12.12. TAF principle of insert retraction for increasing the cavity volume and allowing structural foaming. A good surface finish is the main objective. (*Courtesy Siemag*)

wall thickness. A curve of the relationship between wall thickness and cooling time in the mold results in a parabolic curve.

Wall thickness is initially *much lower* and the density *much higher* than the final thickness and density. Thus, cooling is rapid in the inital stage of molding. In a test, the following times were recorded and will serve as guidelines for similar moldings in any of the processes described:

Wall thickness in mm (in.):

Initial	Final	Cooling time in seconds
7 (.27559)	14 (.55118)	180
12 (.47244)	25 (.98425)	270
35 (1.37795)	120 (4.72440)	420

USM System

The USM system, developed by the USM Corp., operates somewhat similarly to a conventional reciprocating screw injection machine for the plasticization and injection of the foamable polymer, as shown in Fig. 12.13. A full shot of foamable melt is injected into a closed mold and, after freezing of the skin, a mold section is moved (withdrawn) to allow expansion of the melt. In some cases, the mold may be partially opened to allow the expansion.

Figure 12.13(a) shows the mold closed and the plasticized material ready to be injected by forward motion of the screw. In Fig. 12.13(b), the nozzle

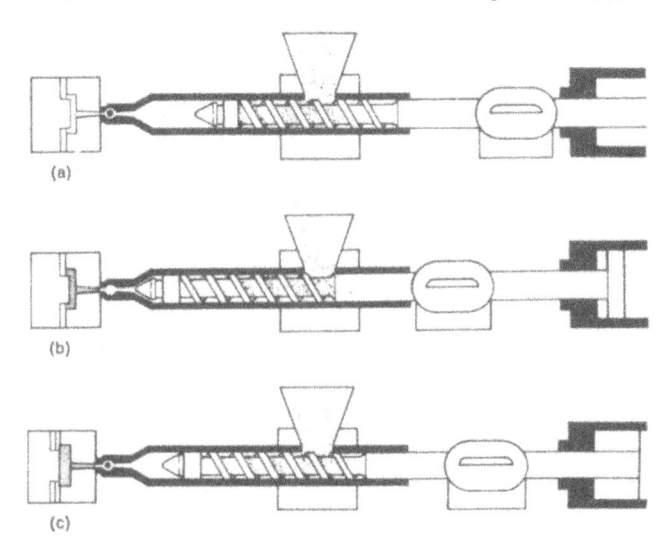

Fɪɢ. 12.13. Typical cycle sequence on a Farrel machine operating on the USM system of structural foam molding.

is opened, the ram advances forward, and the closed mold is filled under pressure. In Fig. 12.13(c), injection is complete, the nozzle shuts, the screw begins to refill at constant back-pressure, and foaming commences within the mold cavity and volume increases at a controlled rate.

When using the system of partial mold opening, molds must be designed as *positive*, or *vertical slash* types, to permit containment of the expanding material without leakage of the melt. Obviously, the final dimension of the part will be that produced by the partially open mold. Through holes cannot be produced using the usual butt-surface technique of closed mold molding. Instead, a through cored hole would require a mold section telescoping the opposite side of the mold by at least the amount of partial opening. The designer must also be certain the boundaries of the movable area do not detract from the appearance of the finished molded part. Three methods of using the USM system are illustrated and described in Fig. 12.14.

Cycle Time

Cycle times of all structural foam moldings are *necessarily longer* than for conventional, solid polymer injection moldings for the simple reason that molded foam has much lower thermal conductivity than a solid plastics part. For this reason, a molder of structural foam can use the older and slower machines because a few extra seconds of opening and closing speed is inconsequential in cycles of 2 to 4 minutes for a $1/4$ in. wall section.

To optimize production rates, particularly for relatively small parts, some molders use a *carrousel* type machine having multiple mold clamping stations, each fed in turn from a single stationary injection unit. A carrousel system is illustrated in Fig. 12.15. The makers of blow molding machinery* also produce carrousel equipment for use in structural foam molding. Four, six, or eight stations are the usual number.

When molding large parts, where moving or mounting a large mold becomes impractical, it is common to use a movable injection unit and a number of stationary clamp units arranged in a horizontal plane. The Schloemann-Siemag system for this arrangement is illustrated in Fig. 12.16. Another simple procedure also used by Siemag and by Mannesmann is to feed two mold clamp units sequentially from a stationary injection unit by means of a changeover valve system, as shown in Fig. 12.17.

The Molds

Designers of molds for structural foam products, such as those in Fig. 12.18, must be informed about which foaming system is to be used, the material to

---

*i.e., Wilmington Plastics Machinery Co., Greensboro, NC and Hettinga Equipment Co., Des Moines, IA

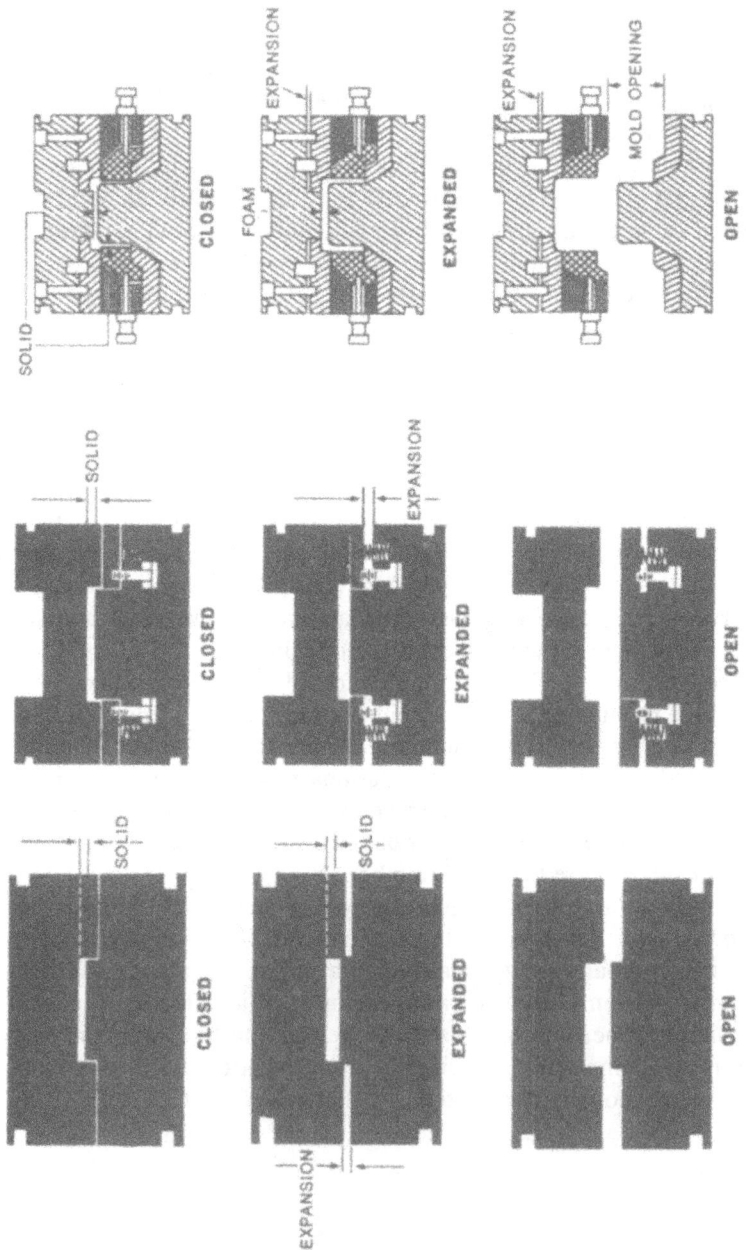

Fig. 12.14.   Three methods to increase the mold cavity volume when using the USM system.

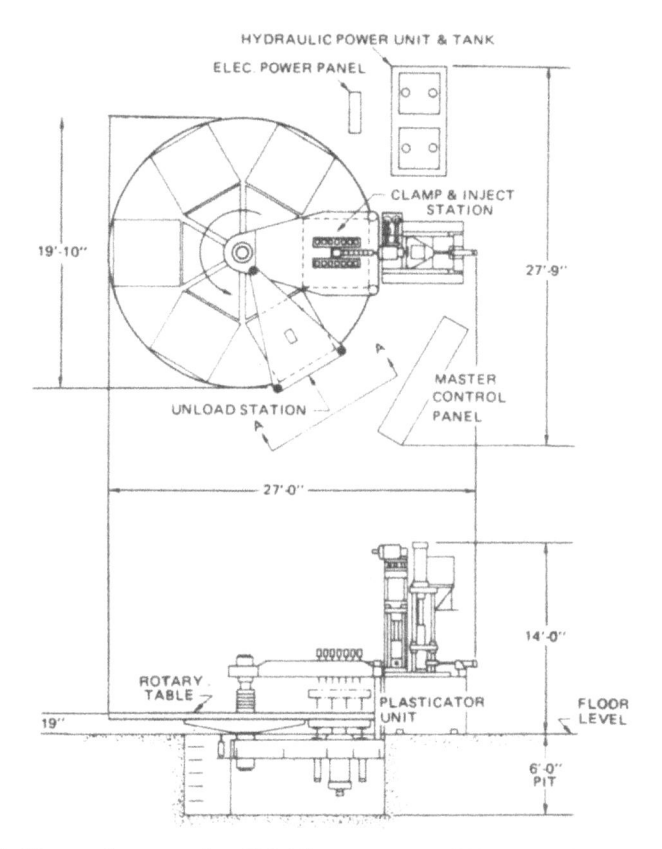

FIG. 12.15. Carrousel system of mold feeding rotates a mold into a molding and clamping position while other molds are cooling and ejecting. Note that this is a very large machine. (*Courtesy Wilmington Plastic Machinery Co., Greensboro, NC*)

be molded, and the characteristics of the machine available for the job. Ordinarily, the mold designer will not be called upon to choose the foaming system. However, the designer needs to know whether nitrogen or a chemical blowing agent (CBA) will be used. Chemical blowing agents can be in powder, concentrate in a liquid carrier, or solid bar stock form. Any of these are quite stable at room temperature, but convert to a *gas* at melt temperatures. If a CBA is chosen, the mold designer must be aware of which of the three forms is to be used; (1) liquid injection into the melt, or (2) pulverized powder which must be carefully tumbled and blended with base material, or (3) pelletized CBA mixed with similar pelletized base material. A sub-division of powder is to use a CBA bar (similar in size and shape to a large rectangular bar of soap), which is fed through a pulverizing unit mounted on the injection cylinder, where the pulverizer feeds the CBA at a uniform

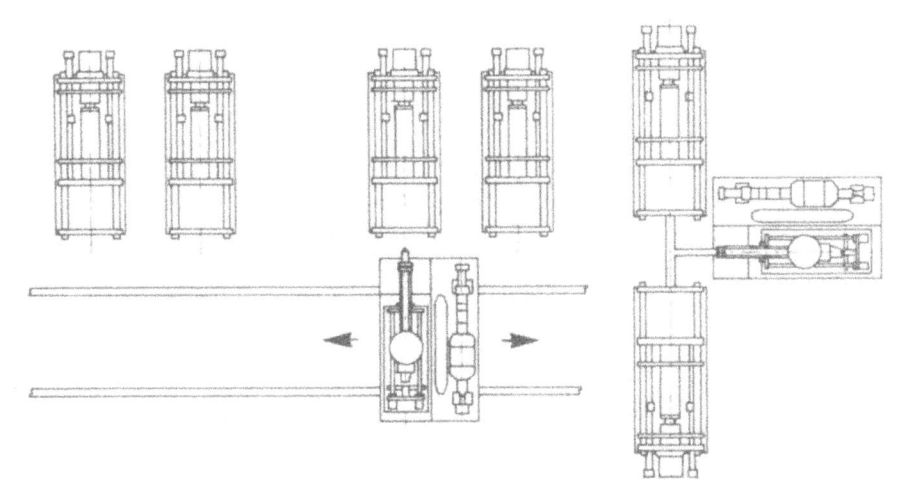

FIGS. 12.16 and 12.17. The Siemag system of multiple mold feeding for structural foam uses multiple station mold clamping and a movable injection unit as at left, or a valve. arrangement to feed alternate molds from a stationary injection unit, as at right. (*Courtesy Siemag*)

rate as the screw is plasticizing the next shot. In any case, the choosing of a specific CBA compatible with the molding material is a critical choice, but outside the scope of this text.

The material to be molded can be any one of a long list of thermoplastic materials, each with its own peculiarities of molding and each with its own compatible CBA. Incidentally, it is theoretically possible to cause any polymer material to foam, and the industry is coming close to providing this capability.

If the mold is to be a simple expanding type used to gain accurate control of skin thickness and core density, it will be relatively easy to design, because this process makes use of conventional injection mold design with a movable core that retracts after mold filling and to permit expansion of the inner core material (refer to Fig. 12.12). The movable core techniques are well covered in the chapter on Injection Molding.

It is important to note that patents cover some of the structural foam molding processes and, while we do not encourage unlicensed use of patented processes, we do encourage the designer to study the patent literature as a helpful approach to the design of molds for structural foam molding. The designer should be sufficiently alert to the possibility of patent infringement so that management can be warned to secure the necessary permissions before embarking on an expensive tooling project.

Uniform density is fundamental to quality structural foam molding. It is achieved by getting the material into the mold *very fast* and *very hot*. Slow

Figs. 12.18. Appliances or business machines are volume users of structural foam molded parts such as those illustrated. (*Courtesy General Electric Co., Pittsfield, MA*)

filling will permit foaming to start too soon and thus obstruct the flow. It is neither economical nor practical to convert standard injection molds to mold structural foam, because the thin sections commonly used will block the flow. Any *packing* during filling defeats foam uniformity. When ribs or other mold sections are too thin, the pressure needed to fill the cavity increases markedly and the clamp pressures may be insufficient to keep the mold from flashing. High-density or variable-density areas will result from restricted flow; such parts will warp in the paint ovens and under adverse environmental conditions. It is of extreme importance that the melt flow into all cavities be balanced to insure uniform and rapid filling of all areas of each cavity. In very large parts, multiple gating will become an important means of uniform filling of the cavity.

Venting and the use of *counterpressure* is related to quality molding. These items are discussed more fully in the following section on mold design features.

Mold Materials

Structural-foam molds are usually designed to withstand 1,000 psi, but the actual pressure will be in the range of 300 to 400 psi (a 2:1 to 3:1 safety factor). These low molding pressures permit the use of lower cost molds and "soft" metals. Any mold which has a wall thickness sufficient to contain cooling lines, guide pins, bushings, and assembly screws will probably be adequate for the stresses encountered in structural foam molding. Most experienced designers will draw a full size cross section of the mold just to see if it "looks right" for strength. Steel or machined aluminum is the usual material choice for cavity, force, and base components. A small proportion of molds are made of cast aluminum, and even smaller proportion are beryllium copper or kirksite castings. In fact, a long time molder of structural foam says he has never seen a beryllium copper tool. All mold makers presently building these molds are prepared to build molds from any of these materials, by either the casting or machining process. However, with present day machining capabilities with EDM and CNC equipment, the use of casting has been in constant decline.

The following listing shows the approximate changes in mold material usage currently as compared with 1978:

material	current	vs.	1978
Machined steel	50%		25%
Machined aluminum	43%		60%
Cast aluminum	5%		10%
Beryllium copper and kirksite	2%		5%

The use of steel has increased primarily because of its *abuse-resistance*. Nearly all tooling scheduled to make 5,000 parts or more, or to be operated for more than a few weeks, will be made of steel. Steel has a lower thermal conductivity than aluminum, thus the cooling portion of the cycle will lengthen by 10 to 15%. Steel molds represent *higher* initial cost because it takes longer to machine a given contour. Both of these disadvantages are offset by the mold retaining a fine finish during long production runs, and tolerating more abuse in handling, set-up, and operation. Of course, a large steel mold (e.g. for a 4-foot square pallet) will be considerably heavier and more difficult to handle than would an aluminum mold (heavier by the ratio of their respective specific gravities). Combining steel and aluminum can be a useful compromise where aluminum is used as the major component for contour areas, and steel inserts, such as pre-hard P-20, are placed in the high-abuse areas.

Machined aluminum (rolled plate). This is widely used because of its uniform texture, ready availability, and ease of machining. Its high thermal conductivity (as compared to steel), light weight, and dependable service also make it a good candidate for a structural foam mold. In those borderline applications where a choice between steel and aluminum is difficult to make, we recommend choosing the long wearing, abuse resistant steel. The molder should make this choice because the mold maker will naturally choose the easier machining and less costly tool. The most frequently used grades of aluminum are 6061 (soft) or the 7075 (hard). Rolled aluminum plate is available up to 10 and 20 in. thicknesses, but any requirement over 6 in. should be verified with a supplier. Thicker plate can be produced on special order with a corresponding lengthening of delivery time.

Advances in photographic processes of texturing now produce wood grain, brick, parquet, or overall patterns of any desired texturing on practically any base material. For multiple cavity molds of relatively small size, beryllium copper hobs may be used to impress texture detail into a machined aluminum or steel cavity. In any case, aluminum has poor wear resistance. It also mars and scratches easily, making it difficult to maintain highly polished areas. All the soft materials peen easily, meaning parting lines are easily damaged. Removal of the resulting flash on the molded part can quickly become a major expense.

Cast Aluminum. This can also be used for woodgrain surface textures or for highly decorative parts. Excellent detail can be achieved by the use of French sand or ceramic patterns. However, as mentioned above, modern photographic processes of texturing are so good that castings are now seldom used to achieve the grain patterns. Obviously, the extra patterns and preparation time for casting makes the photographic processes more financially attractive. Products such as furniture components may employ several decorative patterns or be restyled frequently. Cast aluminum can facilitate inexpensive design changes. Adversely, cast-aluminum molds may be poor quality castings with voids and defects that become problems in use. This is particularly true with deep cavities and narrow ribs. Matching force to cavity costs more with cast molds because accurate dimensions must be machined. Deep or narrow ribs are potential problems with cast aluminum; aluminum casters prefer to limit depth to six inches. A plus factor for cast aluminum is the potential for cast-in-place copper tubing for the cooling channels. On the other hand, aluminum plate is easily and quickly drilled; thus, the cost of cooling channels may be a stand-off provided straight holes are acceptable. Obviously, cast-in-place copper tubing can follow complicated part contours quite closely, with the result of the cooling media being much closer to the molding surface. Closer cooling means faster heat transfer.

Beryllium copper (BeCu). This alloy offers the highest thermal conductivity of the four materials mentioned, and it can be hardened like steel to provide abuse- and wear-resistance. When very fine texture detail is essential, BeCu is best because of its fine detail reproduction and its abuse-resistance. However, the cost per pound of BeCu is approximately eight times the cost of aluminum. In relatively small molds, this cost is inconsequential.

Kirksite. This is a castable alloy with a low pouring temperature that permits the use of cast-in copper tubing for cooling. Although it machines well and is inexpensive (in relation to aluminum or steel) its use has been greatly reduced over the past several years to the point where only 1 mold in 100 is made by kirksite casting.

Electroformed. Electrical deposition molds have been used for those "impossible" shapes that must be produced with the greatest accuracy and with detail that cannot be produced in other materials. However, the current extensive use of EDM machining techniques has practically replaced electroforming. Electroforming is covered in greater detail elsewhere in this text.

A hard surface coating of electroless nickel. This is often applied to where it is needed to withstand wear and abrasion. It must be remembered that hard surface coatings are no more resistant to distortion than are the base metal and that wear and abrasion resistance is their strong point. TFE or hybrid TFE coatings are occasionally used for improved mold release on the non-ejection side of the mold (normally the core side). The use of Armorclad® will produce a very hard surface on aluminum, but it is generally used on only one side of the mold when the mold is designed with a telescoping section of core and cavity. The principle of dissimilar metals applies here.

Finally, it is well to remember that the *cheapest* material may turn out to be the *most costly* in the long run. The best way to select material for the molds for structural foam molding is to consider (1) the production life of the tool, (2) the essential surface finish required, and (3) the overall quality of the part demanded by the customer. When these three times are determined, an intelligent choice of material is possible. Just remember that surface appearance and quality of structural foam moldings is directly related to the mold surface—its hardness, polish, texture, and endurance.

Mold Design Criteria

Cooling. Of all the cooling methods available, *bubblers* are preferred as the most efficient and they are commercially available in various sizes. However, many other cooling methods are available and used in their proper places. We enumerate: cast channels, drilled channels, cast-in copper tubing, cooling plates attached to the mold, and cast-in pockets or flood zones. The flood zone is the last choice as the least efficient and most difficult method to obtain a uniform cooling effect. In general, place the inlet so the coolant enters at the lowest level of the mold. The outlet is then placed at the highest level. This arrangement provides back pressure and eliminates trapped air, provided a directed flow is used to be sure that coolant uniformly traverses each channel. Trapped air within the system will result in hot spots. A turbulent flow of the coolant will remove the most heat in the shortest time. Standpipes in the flood pocket will help eliminate air pockets. However, heat cannot be removed from a structural foam molding unless and until the heat reaches the cold wall of the mold. Structural foam is an excellent insulator and, conversely, a poor transmitter of heat. Thus, only the heat which actually reaches the cold wall of the mold can be effectively dissipated. Poor thermal conductivity of structural foam accounts for the lengthy molding cycles where thick walls (1 to 4 in.) are encountered.

Gates. By definition, the point at which the melt enters the cavity of the mold is called a *gate*. The gate may be, and frequently is, the large end of the tapered sprue, and we call this a direct gate. Because of the nozzle design and the shut-off pin, the points of entry will be $5/8$-in. diameter or more. *Valve gates* with a *hot-runner* supply are commonly used, particularly where it may be necessary to sequentially feed material for a particular design, such as a long and narrow part where the length of flow exceeds practical limits. *Tab gating* is also in use. All of these gate types are well covered in the chapter on injection molds, but we mention them here as a reminder.

In any case, the gate(s) must be sufficiently large to permit the entering material to flow freely and fast. All the material suppliers will supply recommendations for gate sizes. We mention a few here for some commonly used materials: For polycarbonate foam, use 0.20 sq in./lb of material; for thermoplastic polyester and polyphenylene oxide, 0.15 sq in./lb of material. If at all possible, pass all of the melt through one gate. Multiple gates *always* create problems of flow and knit lines, thus multiple gates should be used only when good design requires them.

Components. Commercial sources are extensively used for mold bases, core pins, bushings, nozzles, ejector pins, and other standardized components for molds. Standard extrusion nozzles and nozzle bushings are readily available. A typical nozzle used for structural foam has a ⅝-inch-diameter opening ("0" dia.). Hot-runner manifolds with valve gates are also readily available either as standard items or custom made for a specific job. Knockout or ejector pins should be as large as possible, and certainly larger than in conventional solid polymer molds. A small diameter ejector pin (¼ in. or less) may pierce the foam instead of ejecting the part. In fact, compression of the foam is also undesirable, thus ejector pads or stripper plate construction should be considered. If it is absolutely necessary to place an ejector pin on an "appearance surface" be sure to disguise it with some decorative treatment of which the customer approves. *Sleeve ejectors* around core pins must have adequate area to prevent crushing of the foam. Do not depart from the material supplier's recommendations for wall thickness, boss diameters or rib contour unless at least one other knowledgeable designer agrees to the deviation. Even so, consider special treatment of these potentially troublesome deviations. Of course, *undercuts*, *backdraft*, and other special techniques have their application in structural foam molds.

Counterpressure and venting. We treat these two items in the same discussion because they are interrelated. *Counterpressure* in the mold is a relatively new use, and probably evolved from the older practice of evacuating (drawing a vacuum) on the mold activity. Users of counterpressure report a great improvement in surface finish as opposed to injecting a foam melt into a cavity at atmospheric pressure. Our first discussion is about venting and the numbers associated with it.

Venting. A *vent* is any opening which will allow the escape of air or gas while restricting the flow of the molding material. In general, a vent is needed in one or all of the following: (1) at the parting line and at the last point of fill (trapped air can explode and burn the part), or (2) on the far side of cored holes, or (3) at the top of a high boss, or (4) at the bottom of a deep recess. Of course, venting points will not become ascertainable until the gating point(s) are fixed. Thus, a review of the previous section on gates is in order. It is quite common to make a mold without vents, then cut the vents after the sample run. Vents are always located at or near the mold location which is the last to be filled with the molten polymer. Parting line vent locations can easily be ascertained during the sampling process, by simply placing 0.005 to 0.010 in. thick masking tape at 2 in. intervals around the mold. Outflow between the taped spots will help locate the point of final fill. Products with large openings and products with stepped designs are often difficult to vent while still obtaining a fast and uniform fill. A good designer will anticipate trouble in these areas and design a mold to avoid it. For example, quite frequently a movable pin (ejector pin) will be used for venting even though not needed for distortion free ejection. Finally, be sure any vent is not "dead end." In other words, a vent 0.005 to 0.010 in. deep and ¼ to ½ in. wide will be that size for ½ to ¾ in. from the parting line, then it tapers larger until it reaches a free space. The side walls of vents should be well drafted so material which frequently collects there can be quickly and easily dislodged. Digging or chiseling of solid material to remove it from the vents can (a) damage the mold, (b) become quite expensive when repeated several hundreds of times, and (c) quickly ruin a designer's reputation as a designer.

We now turn our attention to venting as it must be used if counterpressure in the mold cavity is part of the process. In order to use counterpressure, the mold *must* be sealed at every opening, and sufficient to hold 200 psi pressure. It should be obvious that counterpressure must have a regulated exhaust (venting, if you please) so that the 200 psi is essentially a constant pressure. Without venting, the pressure agent would build up to a very high pressure and prevent filling of the cavity with the polymer melt. In this circumstance, the vent opening is sealed off by a valve controlled by air or hydraulic pressure. The valve control is controlled by means of adjustable limit switches actuated by movement of the injection ram. As the ram injects the polymer, a vent in the yet unfilled area of the mold is opened just before the polymer reaches it. As soon as the polymer passes that particular vent, counterpressure rebuilds until the next vent is reached. Obviously, placement of vents and valves are critical to the success of this process.

Containment of the counterpressure is also critical to success. Every possible avenue of escape must be blocked. This includes ejector pins, side cores, slides, parting lines, and so forth. O-ring seals are used to accomplish the sealing except at the parting line. For example, the seal around ejector pins is accomplished by cutting a pocket on the outside of the mold and around the ejector pin hole. Then the O-ring pocket is cut around the pin. During assembly, it is important that the sharp edge of the ejector pin does not cut or nick the O-ring. The O-rings and their assembly and use is well documented by the makers of O-rings, and we need not repeat the information here. Location of the inlet for pressurizing medium (usually nitrogen) is critical, inasmuch as it must be at some point where the flow of the melt will not prematurely cover and seal the inlet. Control of the pressurizing medium is external to the mold, and is a mechanical process of using the proper limit switches, valves, and shut-offs. Such controls would not normally be the responsibility of the mold designer, but familiarity with the process and technique is essential to providing the "right connections."

Draft. A minimum of ½-degree draft is desirable, but do not forget the use of *negative draft* and *undercuts* for specific purposes. When *vertical textured surfaces* are involved, draft angles must be appropriate to release the texture. The usual formula is 1 degree of draft for each 0.001 in. depth of texture. Thus, depth of texture should either be specified by the designer, or the draft should match the texture applier's specifications. No-draft core pins can be used with sleeve ejectors. For some through holes, partial drilling may be required because inadequate flow and fill might otherwise result. The simplest rule is to use all of the draft permissible. If 1 degree is good, then 2 degrees should be better. However, excessive draft can also result in a larger than necessary cross section at the base of a rib with a result of cycle lengthening for cooling.

Construction. The design and construction of structural-foam molds follows the accumulated experience for compression, injection, transfer, and die-cast molds. We have attempted in this discussion of structural foam molds to specify those points at which some deviation must be considered. One of the principal differences is the greater use of machined plates for foam molds. In general, the construction illustrated by Fig. 12.19 is employed. Mold sizes are limited by the press platen area, the capacity of the extruder, and the cavity layout that will allow fast and uniform fill. Ribs and natural runners within the product should be utilized to the maximum by following the line of least resistance to flow. Cross flow always creates special problems. For obvious reasons of simplicity of machine design and operation, the melt manifold is located on the stationary side of the mold. The exception, of course, is the stationary mold and movable injection unit illustrated earlier.

Differential expansion can create horrendous problems, especially when

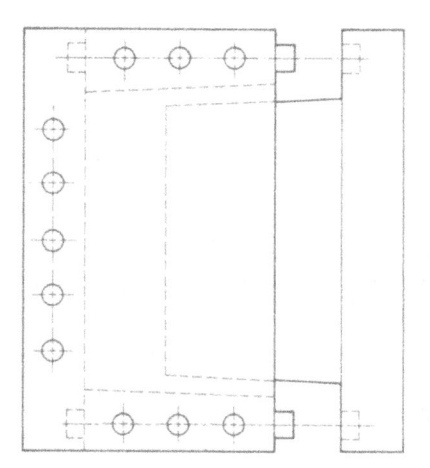

Fig. 12.19. Simplified sketch of a mold assembly of aluminum plates. Force side cooling channels not shown, but would use bubblers or internal drilling from top.

dissimilar materials, such as steel and aluminum, are used in the same mold. Aluminum expands farther and faster than steel. Thus, an aluminum core matched to a steel cavity would require more clearance when cold, or else the core will expand and be damaged by interference when hot. Calculations of expansions are essential to a good mold design. Rubbing surfaces, such as locating guides, cams, or dowels must rub against a material of different hardness to prevent (or at least minimize) galling. For example, 6061 aluminum rubbing against 6061 is sure to sieze and gall. Mild steel against mild steel can be just as likely to gall.

Mold sections are designed for 1,000 psi pressure, as stated earlier. When multiple grille openings are part of the product design, a runner across the grille area should be incorporated, as shown in Fig. 12.20. Gating into thin sections is also depicted in Fig. 12.20. Inasmuch as Fig. 12.20 notes 0.250 in. as a typical wall, we need to note that recent trends are away from the 0.250 in. wall which has been the standard for several years. Currently, much work is being done with 0.156 in. walls because of great improvement in the precision of controllers available for creating more uniform and thinner skins and still leave some wall for foam.

Since the success of structural-foam molding depends on an ''instant'' fill

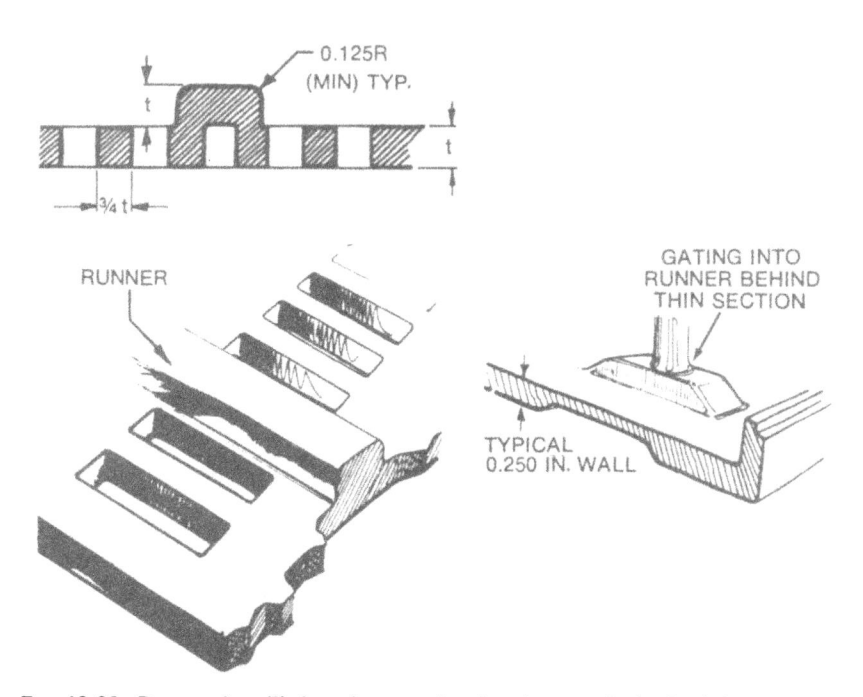

FIG. 12.20. Runners in grilled products can be placed across the back of the grille to enhance the part filling. (*Courtesy General Electric Co., Pittsfield, MA*)

of the cavity before expansion takes place; it becomes obvious that thinner walls will require better techniques of molding. *Packing*, *excess weight*, and *unfilled parts* are not acceptable in any wall thickness. For thinner walls, the incorporation of ribs or small runners in the direction of fill will be a great help. Increasing the wall thickness is probably the best answer when only a limited number of parts are required. But, when hundreds of thousands of parts are to be made in structural foam, any reduction of material, any reduction in cycle time, any improvement in quality, no matter how slight, is worth the extra effort needed to effect optimum conditions of production.

Runners. A discussion of runners would normally be expected before the gates, the mold, the wall thicknesses, and so forth. However, runners as sized and designed for what happens at the far end, so we must first consider part size, number of cavities, and press capacity, among other things. The high shot volume and the large mold-mounting platens are ideally suited to mounting and filling several mutually independent molds simultaneously. This task is solved by means of the runner system shown in Fig. 12.21, which consists of a runner, one or more injection units, appropriate connecting "pipes" and is built up on the unit structure principle.

The injection unit, also called a valve-gate, is hydraulically opened and closed; it can be controlled as desired and independently of all other valve-gates. This technical advance, called *sequential valve gating*, results in the maximum injection speed for each cavity or each gate, and assures that the melt delivered at the gate is identical with the melt delivered at other gates. Uniform and lowest density can thus be achieved. The use of multiple gates

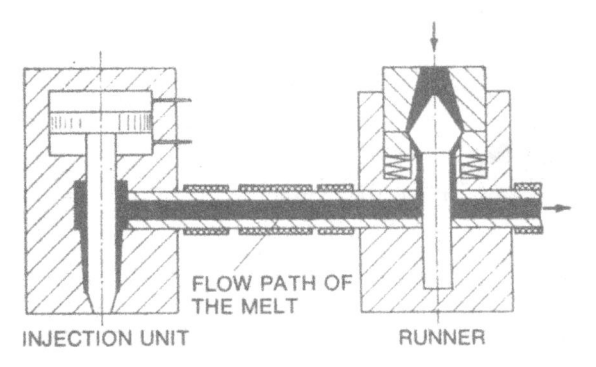

Fig. 12.21. Schematic of a valve gating system shows, left to right, the injection unit (valve gate), melt flow path, and the main sprue and runner. (*Courtesy Krauss-Maffei Corp., Grand Rapids, MI*)

reduces the flow path in the mold where the foaming is taking place. This technique is especially useful when injecting foam behind a surfacing or decorative film.

Sequential filling of cavities is particularly required when several parts of different wall thicknesses are molded in the same cycle. Resistance to filling is not the same at each cavity. With sequential filling, each individual mold cavity is completely filled before the next cavity starts to fill. Uniform and predictable densities are therefore obtained.

Figure 12.22 schematically shows the sequential molding of several parts in the same cycle. In order to fill the first mold, the injection units (valves) A and B are opened. The valves are then closed just before the mold is filled so that the remaining space can be filled by pressure of the foaming agent. Only after valves A and B have been closed is valve C opened, melt injected to desired quantity, and the valves closed to permit final foaming.

The technical advantage of the process just described is that the capacity of the machine is economically used to produce more parts in a given period of time. It seems impossible that anyone would overlook the obvious requirement that all parts must use the same plastic material. Thus, whole product groups, such as cabinets, doors, and drawers, can be manufactured in individual molds, simultaneously or sequentially, in one injection cycle. In these circumstances, the runner feed system can be part of the machine, thus reducing the investment in future molds which can use the same melt delivery system. For the more sophisticated systems, such as in high volume

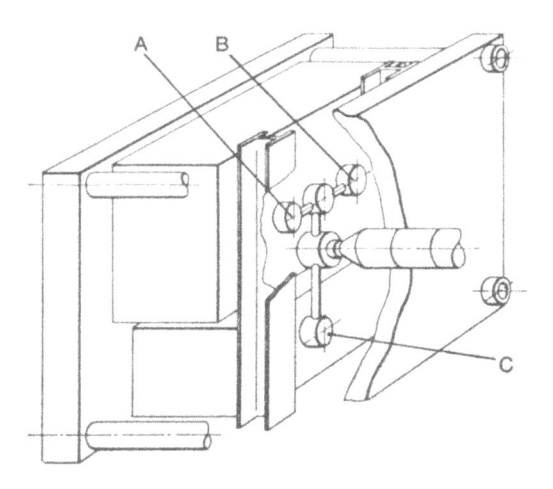

FIG. 12.22. Sequential mold filling of two or more molds is through valve gates A and B, followed by valve gate C. (*Courtesy Krauss-Maffei Corp., Grand Rapids, MI*)

appliances where quick change is a necessity, automatic mold changers have been incorporated as part of the machine investment. Such quick-change or automatic devices make it quite simple to access the runner system in the event of blockage or breakdown in that area.

There appears to be no limit to the size of structural foam molds. A 43 lb five foot square shipping pallet for automotive axle assemblies was recently demonstrated to this author. For some 20 years, this author has stated, ''there is no reason we cannot mold a room (or an entire house) if someone will install a press large enough. Then someone will design a mold for it and someone will build the mold.'' Notice that ability and willingness to finance a press is placed first. Figures 12.23 through 12.28 illustrate a variety of molds and their construction. Because of the low pressures used with these particular molds, inexpensive sections are assembled to form the total mold section. However, it should be noted that intricate molds with higher pressures and more complicated cycles are developments currently in progress. A study of the illustrations will provide guidance in the design of molds for structural foam products.

Fɪɢ. 12.23. Illustrated is the injection half of a structural foam mold showing the position of movable cores, water line connections, and other details. The two holes at top center are for the injection nozzle. (*Courtesy Johnson Controls*)

FIG. 12.24. Mold for a structural foam door-frame, mounted in a 300-ton machine. Note that molded parts are carried from machine by a conveyor. (*Courtesy Johnson Controls*)

MOLDS FOR EXPANDABLE STYRENE

Expanded Styrene

It is customary in talking about expandable polystyrene (EPS) to classify processing into three broad areas—Custom or shape. Block, Cup or thin-wall—on the basis of the major end markets for EPS materials.

Shape molding covers a broad range of applications such as packaging, consumer products, and novelties. Other relatively new areas of application include the use of EPS in disposable patterns for the production of metal castings* (similar to the lost wax process) and medium density EPS as a semi-structural material. A wide diversity of products fall into this category, and there is considerable variation in the methods, molds, and machines used to produce them. The molds used in shape molding are the subject of this section.

*See the applicable patents.

FIG. 12.25a. A 21 lb one-piece molded structural foam television cabinet.

FIG. 12.25b. Mold for cabinet in Fig. 12.25a is shown here. It uses aluminum and beryllium copper combined with hydraulically actuated side cores on all four sides of mold. Gating is at six points on the back open edge.

FIG. 12.26. An aluminum mold for a structural foam molding of a Kodak cabinet.

FIG. 12.27a. Polypropylene structural foam is used for the side frame of this Bassett chair. Mold is the fabricated aluminum mold shown in Fig. 12.27b.

FIG. 12.27b. Fabricated aluminum mold for chair side frame shown in Fig. 12.27a.

FIG. 12.28a. Injection half of machined aluminum mold for game table shown in Fig. 12.28b.

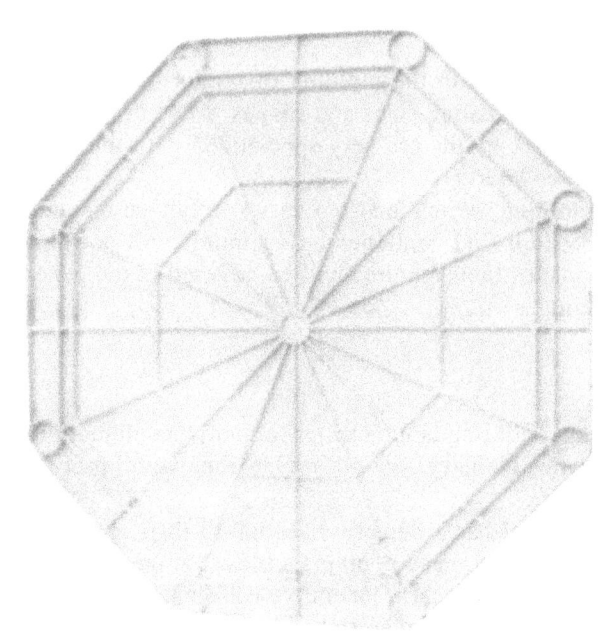

FIG. 12.28b. An early example of structural foam molding in an eight-sided game table. Original molder has passed from the scene, but mold survives and is still in use.

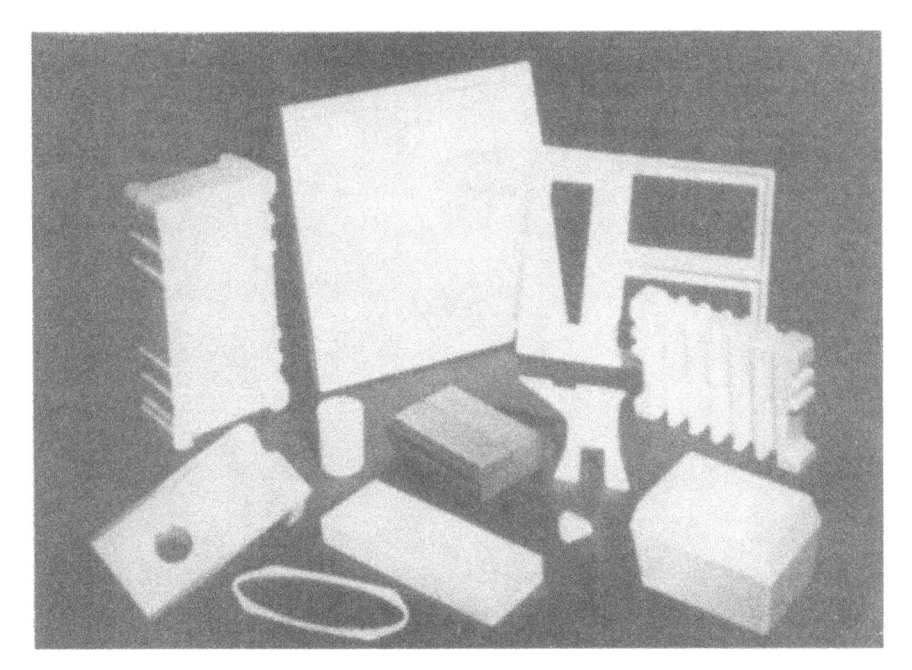

FIG. 12.29. Molded expandable polystyrene (EPS) parts show a broad range of applications. Primary markets are packaging, consumer products, and novelties. (*Courtesy Johnson Controls*)

Block molding refers to the molding of large blocks of foam which are later fabricated into desired sizes and shapes. Most of this material ends up as insulation. Block molds are very specialized with their own technology and will not be covered.

Thin-wall molding, which is also a very specialized technique, is used in producing EPS cups and containers. Cup molds will not be discussed explicitly, but most of their features will be covered in the general discussion on shape molding.

Processing

Since EPS shape molding is unfamiliar to many people both inside and outside of the plastics industry, we will review the overall process before looking at EPS molds.

Raw EPS material has a density of about 40 lb/ft^3 and resembles granulated white sugar. Each bead contains a blowing agent that causes it to expand when subjected to heat. Generally, EPS material is pre-expanded before molding by heating the material and allowing it to freely expand to the density desired in the finished product. Following pre-expansion, the ma-

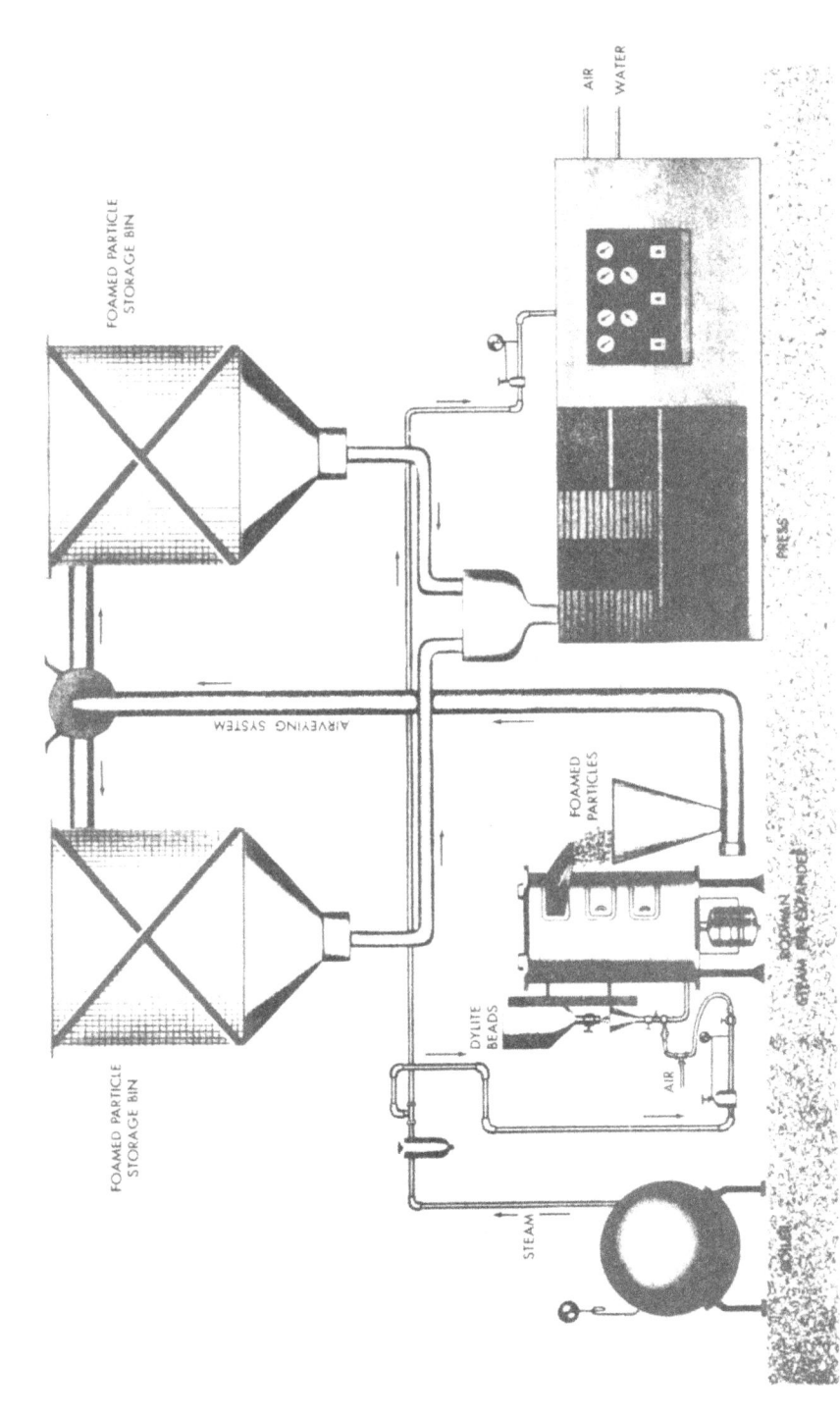

FIG. 12.30. Flow diagram for EPS material when processed by "shape molding" machinery. Steps in the process are: pre-expansion, screening, aging, and molding. (*Courtesy Arco/Polymers*)

FIG. 12.31. The most widely used EPS shape molding machines are horizontal acting, hy-draulically actuated, and capable of automatic operation. (*Courtesy Johnson Controls*)

terial is screened to eliminate any lumping. Then it is aged in large bags or bins for approximately one day before molding. Aging serves to reduce both the shrinkage and cycle times of molded parts.

The molding process itself consists of four steps: filling, fusing, cooling, and part removal. During fill, the mold cavity is completely filled with the pre-expanded material. Fusion takes place when the material is heated sufficiently to be softened and the blowing agent is activated. The individual beads then become welded together to form the desired part. Next, the part is cooled sufficiently to contain its internal pressure and to prevent post-expansion when it is removed from the mold. Finally, the finished part is ejected from the cavity.

EPS Shape Molding Machines

Nearly all machines being operated today are horizontal acting. Horizontal machines facilitate part ejection and take-away. They are also more easily guarded than vertical machines. Most presses are hydraulically actuated, but some pneumatic machines are also being produced. Machines usually are capable of automatic operation. There are, however, considerable differences from machine to machine in the sophistication of controls.

Molding Techniques

In addition to characterizing EPS molding methods by the major market segments for materials, we can also categorize molding into four different techniques: steam chest molding, probe molding, autoclave molding, and radio frequency molding. The latter three techniques account for a very small portion of the material market since shape, cup and block molding fall under the steam chest molding category.

In steam chest molding, Fig. 12.32, a chamber is formed behind the mold cavities and cores. This chamber, or in some cases chambers, serves the following functions:

1. It creates a passageway for steam used to heat the mold itself. Molds are preheated before or after fill or both by passing steam rapidly through the chamber.
2. It creates a pressure vessel in order to force steam directly into the cavity and through the EPS material. After pre-heating the mold, the drain allowing steam to leave the chamber is closed. This raises the pressure in the chamber, forcing steam through vents in the mold into cavities. Direct injection of some moisture into the material is required for adequate fusion.
3. It forms a passageway for water used in cooling the mold.
4. At times, it functions to create a pressure vessel for air which is used in releasing or ejecting the molded parts.

Probe molding is accomplished by distributing steam within EPS material through perforated metal tubes. Prior to fusion, the probes are withdrawn

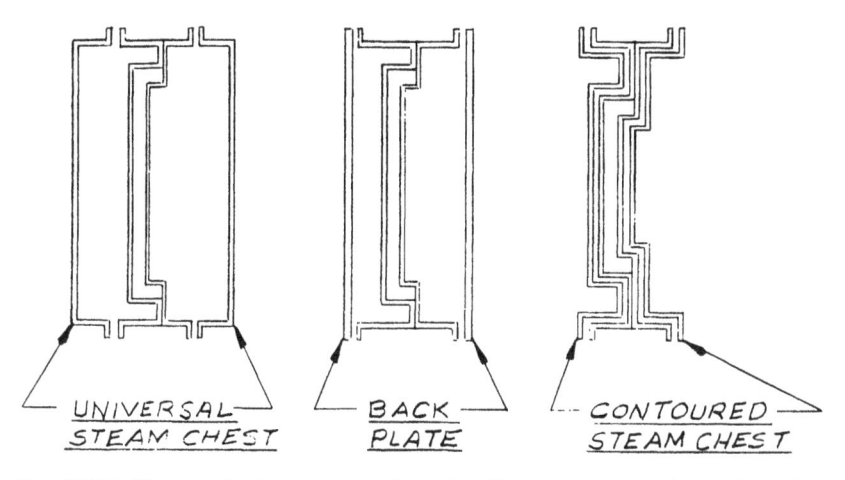

Fig. 12.32. Three methods are commonly used to form a steam chamber: universal steam chests, back plate, or contoured steam jacket.

and the expanding foam fills the temporary voids. Probe molding is used almost exclusively for molded-in-place applications, where uniformity of fusion and surface finish are not of critical importance.

Autoclave molding involves the use of a perforated mold cavity filled with material which is then placed in a steam autoclave. With this method, mold costs are kept to a minimum. Its chief drawback is its poor productivity, and the process is used only rarely, usually to obtain sample parts.

Radio frequency molding* is done by utilizing high frequency electrical energy to turn water into steam. Plastics or other low loss materials are used for the molds.

The last three molding methods have seen little commercial application to date and consequently are of a rather theoretical nature.

Overall Approach

A relatively unique approach is necessary in the design and production of EPS molds relative to those for conventional plastics molding processes. The single largest factor contributing to this uniqueness is the necessity for thermal cycling an EPS mold with each machine cycle. An additional factor is the uniqueness of the EPS material both in its pre-expanded and molded form.

In selecting the overall approach to be used in the building of the mold, it is necessary to consider a wide range of factors. These factors stem from a great diversity of molding techniques, EPS molding machinery, and applications. They are:

Cost of the mold
Design of the product
Universal steam chests
Machine design
Filling method

Cost of the Mold

In selecting an overall approach for the construction of an EPS mold, the cost of the mold and its relationship to the volume of parts to be produced are of primary importance. This relationship plays the dominant role in determining the degree of sophistication to be utilized in the processing method, machinery and tooling. It influences the number of cavities to be produced and thereby the size of the mold. It also affects the degree of automation and longevity for which the mold is designed, and even has a strong influence on the product design itself.

*See applicable patents.

Design of the Product

The function of a product determines the dimensional tolerances of the product. These tolerances influence the choice of method to be used in manufacturing the mold. Part function and design determine the finish required on the mold cavities. Functional but throwaway packaging does not require the same finish as an appearance item. Part design also influences the method and location of fill, fusion, and ejection of the part.

Universal Steam Chests

The majority of EPS molds made today are designed for operation with some type of universal or standard steam chest system. Therefore, the size, design, and features of the universal chests must be considered in the initial concept of the mold. They will influence the number and arrangement of cavities in the mold. The presence or absence of certain features in the chests will determine whether or not such features must be included in the mold. Locations and methods of fill and ejection are affected.

Machine Design

The machine on which the mold will be run must be considered, particularly the controls available to operate certain fill or ejection methods.

Filling

Because of the great variety of techniques available for filling and their strong interrelationship with some of the other considerations listed earlier, filling must be considered in the broad concept of the tool.

Design of the Product

We have reviewed the broad considerations that are necessary in developing an EPS mold concept. Now, before going into greater detail regarding the actual construction of a mold, lets look at one of these considerations, the design of the product, in greater detail.

A part design must satisfy the functional requirements of an application, and simplify and facilitate the molding of the part to the greatest degree possible.

1. Wall thicknesses should be as uniform as possible. Uniform walls allow for faster fusion and cooling. Uniform walls make mold cavities easier to fill. Eliminating thick sections provides a double savings by

reducing the amount of material in the part and by reducing the cycle time of the part. Thin wall sections can cause fill problems and should be avoided. Minimum wall thickness depends on the bead diameter of the material to be molded. Sections less than 5/16 in. thick must have ideal fill and venting situations if they are to be filled.

2. Since sharp corners are difficult to fill and easily damaged, they are best avoided. Radii should be as generous as possible.

3. In order to minimize part removal problems, as much draft as possible should be allowed. Two degree draft is normally considered minimal.

4. Undercuts are undesirable in an EPS foam part and should be avoided due to the difficulty of adequately heating, cooling, and sealing moving cores.

5. Molds with flat parting lines are easier and less expensive to tool.

6. If one side of the product is flat, a flat mold face can be used for the core side of the mold, at considerable savings. In some cases such "flat backs" already exist, and, therefore, a new mold is reduced to a new mold half for the contour side.

STEAM CHESTS

Now that we have covered the prerequisites for the design of an EPS mold, we will look at the mold itself, starting first with the steam chest.

The mold component which creates and seals off the steam chamber behind the mold cavity is generally one of three different types: (1) Universal steam chest, (2) backplate, (3) Contoured steam jacket.

Universal steam chests are essentially reusable, permanent mold halves that can be interchanged with a variety of mold faces. They are normally purchased as part of a molding machine and are used on the machine throughout its life. Universal steam chests are not always interchangeable within the EPS molding industry. Many different systems exist, most of which have been developed by various machinery manufacturers; it is hoped that standard universal chests will be achieved in the future.

The primary function of universal chests is the reduction of mold costs by incorporating many mold features into the chests themselves, thereby eliminating them from individual molds. A second important function is to reduce set-up times, by reducing utility connections and easing mold installation.

Universal steam chest systems fall into two general categories depending on whether or not slide runner fill methods are utilized. If they are used, only molds with a few cavities are filled with guns. The guns are then mounted to the molds sides, and do not pass through the chest. With this system,

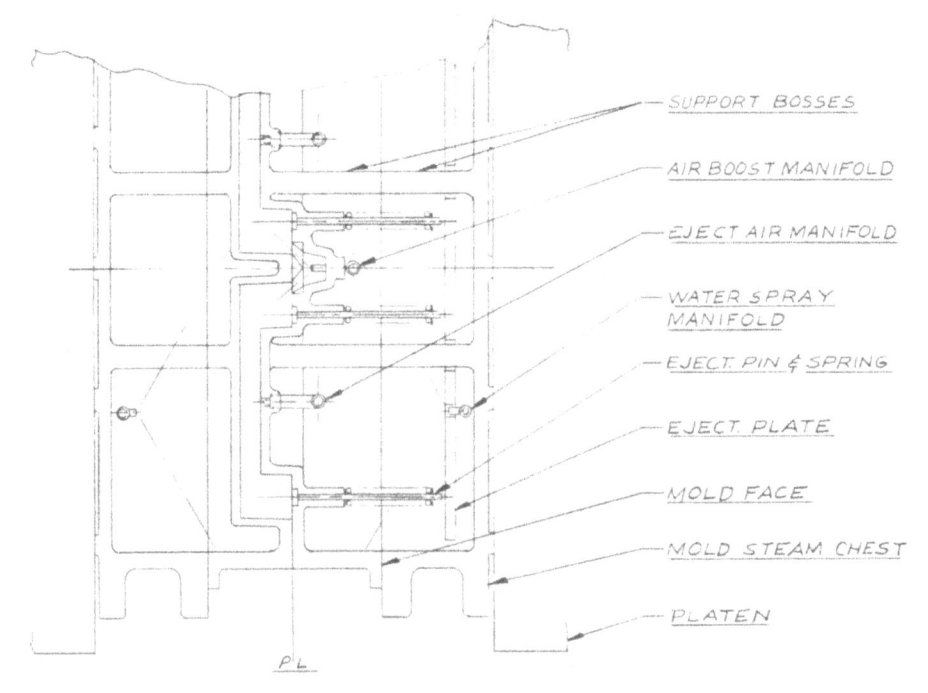

FIG. 12.33. Cross section of a typical EPS mold using a universal steam chest. The system shown relies on slide runner fill for multiple cavity molds.

ejection plates for the activation of spring-loaded ejection pins are built into the chest. Spray manifolds are also an integral part of the chest.

The other chest systems are found on machines utilizing only plug fill methods. In this case fill guns pass through the steam chests. Most chests are therefore of a modular construction with a removable back plate to accommodate various fill locations. In general, ejector pins also pass through the chest and are activated externally. Cooling must, in some cases, be incorporated into the mold. In certain systems, the side walls of chests with removable bottoms are also modular, and different "frame" combinations can be used to yield different chamber depths.

When back plates are used, a mold is referred to as self-contained, since it is a complete unit in itself. Such a mold can be mounted on any press with adequate platens and clearance for utility connections. With this type of mold virtually all of the features that will be discussed as part of a universal chest have to be designed into the mold itself.

For very high volume production, contoured steam jackets can be used. These closely follow the contour of the mold cavity, minimizing the volume of the steam chamber, and hence cycle times and steam consumption. However, since most of the energy consumed is used heating the chest, mold faces, and EPS, rather than just filling the chamber, reduction of steam consumption is limited. To date, applications of contoured chests have been limited, for the most part, to high volume, thin-walled parts such as coffee cups.

Due to the popularity of universal steam chests and slide runner fill in the United States, our general explanatory approach toward chest and mold design will incorporate these methods.

Mounting of Chests

Chests must be constructed so that they can be mounted to a specific machine. The mounting method must take into consideration their expansion and contraction, and still allow them to remain in a reasonably stable location. It is always desirable to insulate the chests from the machine platens, as this reduces heat loss from the mold during fusion. This can be accomplished through either air spaces or a layer of insulation material between the chests and plates.

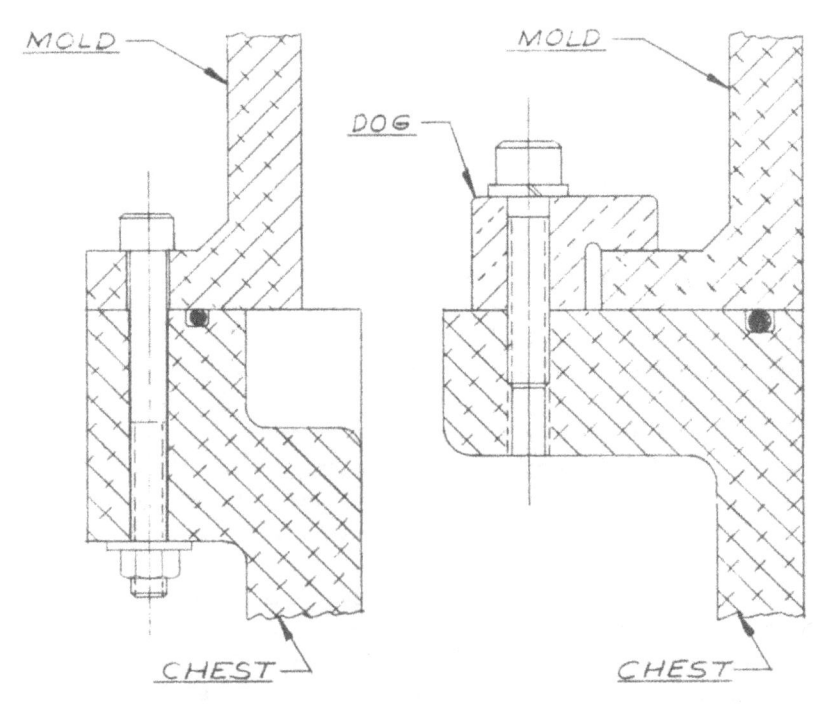

FIG. 12.34. EPS molds are mounted to steam chests by means of bolts or dogs. Note the O-ring seals of steam chamber.

Mold Mounting

Chests must provide a means of mounting mold faces. Generally, molds are either bolted to the chests or held down by means of "dogs," and sealed with steam resistant "O" rings.

Whichever system is used, the mounting means must allow for differences in the expansion and contraction of the chest and mold face. When bolts are used, generous clearance for expansion is provided in the mold face or the chest. Dogs naturally allow for movement. Despite this movement the mold faces must remain in a stable position. Line up or centering devices can be used to resolve this problem.

Steam chests must also allow for a steam proof joint between the mold face and the chest. This can be accomplished through the use of silicone rubber gasketing in a groove cut into the face of the chest.

Mounting molds to chests by means of "dogs" serves an additional purpose besides allowing for thermal expansion. "Dogs" are easily loosened, and mold faces changed, reducing set-up time for a mold.

Strength

In most shape molding of EPS, steam pressure in the chests does not exceed 30 psi. It is customary, however, to design steam chests for 50 psi internal pressure.

FIG. 12.35. Interior of a universal steam chest before mold face has been mounted. Note the eight support pillars. (*Courtesy Johnson Controls*)

Materials

Universal steam chests are most commonly made of cast aluminum. This is, no doubt, due to the fact that most face plates are made from aluminum, thus minimizing thermal expansion differences. Modular chest systems have utilized a number of materials, particularly in the frames. Bottom plates for modular chests have generally been aluminum, as have back plates for self-contained molds. Contoured jackets have usually been made from the same material as the mold cavities, and have therefore been machined bronze, beryllium copper, or stainless steel.

Steam Entry

The location and means of steam entry should allow steam to flow evenly through the chest, thereby uniformly heating the mold faces. "Dead spots" with little flow should be avoided. The location of the steam inlets, deflectors and drains will determine the efficiency of the system.

Drains

Multi-purpose drains are a necessary part of universal steam chests, but are often considered part of the machine rather than the mold. They are, however, an important design consideration. Drains, when in the open position, serve to help vent air from the cavities during mold fill. During presteam, they serve as a vent for the steam condensate as the mold increases in temperature.

Drains also serve to remove the cooling water during the cooling cycle. They should be sufficiently large to prevent water accumulation in the chest. Greater cooling of the bottom half of the mold can create problems in obtaining uniform fusion and ejection and cause water absorption by the part.

As a safety precaution, it is advisable to design drains so that they are normally in the open position, and that in the case of machine failure, they are safely neutralized.

Cooling

Today spray cooling is used almost universally, except in molds with contoured jackets which are flood cooled. Cooling manifolds are usually made from copper tubing which is soldered together into "spray manifolds." These spray manifolds can be mounted in the bottom of a chest or behind it. Clearance holes are provided in ejection plates for the spray nozzles, enabling them to extend through the retracted plate.

Spray nozzles are much more effective than tubing with drilled holes, as they distribute water more uniformly to the back of the mold. Threaded nozzles are preferred to other types since they can be more easily removed for cleaning.

The cooling system must allow for uniform cooling over the face of the mold, in order to make molding conditions uniform from cavity to cavity. Non-uniform temperatures can cause problems with part fusion and part ejection. Spray densities of 3 gal/min./sq ft of mold area provide good mold cooling.

Ejection

Universal chest systems can have an ejection plate built into one or both of the chests. The plate is actuated by the motion of the moving platen and by an air cylinder. The ejector plate is used in turn to activate spring-loaded ejection pins in the mold face plate. It is also possible to extend ejection pins through the rear of a chest and actuate these by bumper bars through the motion of the press. This method is popular with modular chest systems.

MOLD CONSTRUCTION

Now that we have covered the steam chest, we will look at the face plate part of the mold which contains the cavities.

The cavity and core portions of the mold are constructed as hollow shells. They must be heated and cooled with each molding cycle, and minimum

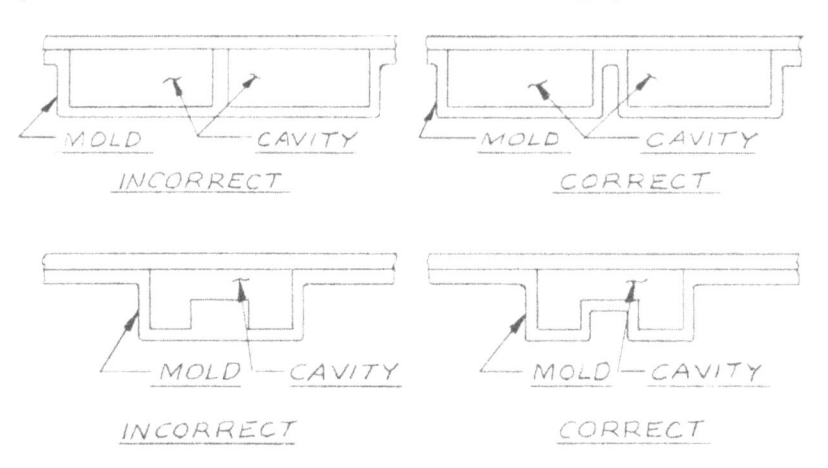

FIG. 12.36. Correct and incorrect mold features. Eliminate the incorrect common walls between cavities or any other situation limiting access of steam to parts of mold. Core thick mold sections so mold has a uniform wall.

Fɪɢ. 12.37. Photograph of the reverse side of an EPS mold face. Note the use of ribbing and bosses for adequate support of cavity against pressures generated by the expanding material. (*Courtesy Johnson Controls*)

wall thicknesses allow for fast heating and cooling. Areas with thick metal sections or areas which are difficult to heat and cool should be avoided.

Aluminum molds are usually constructed with ¼–⅜ in. metal thickness in the cavities. Additional strength is added to the mold by the use of ribbing and bosses. The mold bosses must, of course, be supported by bosses within the universal chest, or by a backplate in the event of a self-contained mold. Typical spacing for bosses is 6–8 in. for aluminum molds.

Molding pressures developed by expanding EPS material depend upon the density of the material. Pressures developed in the cavity will exceed steam pressures in the chest. Molds are commonly designed to withstand 50 psi pressure within the mold cavities.

Mold Material

The vast majority of EPS molds being produced are made of aluminum. Basic advantages offered by aluminun tooling are the following: low cost, easily cast, good heat transfer, light weight, and tolerance to water and steam environment. In addition, aluminum molds are easily altered and repaired.

For very high volume applications, bronze, beryllium copper, and stainless steel have also been utilized, but to date, such applications have been limited.

Not only must the cavity and chest material be impervious to the water and steam atmosphere, but any hardware used in the molds must also be impervious to it. Included would be feeder parts, ejection pins and springs, cooling manifold parts, and drain parts.

Methods of Manufacture

Since EPS cavities and mold faces are essentially uniform thickness shells, fabrication from stock is not usually practical. Fabrication would necessitate the machining of both the cavity and back side. Most EPS molds therefore are made from castings. Casting processes are an economical method of creating a uniform thickness mold, and multiple cavities, compared with machining.

A multi-cavity mold face can be produced in a single casting, or individual cavities can be cast and mounted to plates. The single casting offers several advantages, primarily in cost and overall quality of construction. However, core castings are frequently mounted, since closer tolerances can thus be obtained.

Castings can be produced by sand, plaster, or ceramic methods. Sand castings tend to be sound, but cannot be used where fine detail is required. They are suitable where cavities are to be machined. Plaster castings give good cavity finishes and can be used with a minimum of cavity cleaning and polishing. Plaster process castings usually contain porosity within the casting which may be objectionable if subsequently opened. Ceramic castings give excellent detail, and are fairly sound, but are substantially more expensive than other castings.

Shrinkage

Shrinkage of EPS parts is substantial. It also continues over a long period of time following the molding of the part. Factors affecting part shrinkage are: density of the material, age of beads at molding, and the water absorption by the part. The most commonly used shrinkage factor for EPS is 1/16 in./ft.

When producing patterns for cast EPS molds, metal shrinkage must also be incorporated into the pattern. For aluminum molds the total shrink factor EPS plus aluminum is then 3/16 in./ft.

Fill Methods

There are two primary methods of fill used in shape molding. Both methods rely on compressed air to carry the pre-expanded material into the cavities on an air stream.

Fill guns (Fig. 12.38) also called *plug feeders*, are the most widely used fill method. A fill gun has a two-positioned plug, which in the retracted position, allows material to enter the mold cavity. In the closed position, the plug forms part of the mold cavity. The plug is activated by an air cylinder that is controlled by the molding machine. The gun also has a venturi for conveying material to the mold.

Fill guns are often sold as machine options, or are available from various sources as off-the-shelf items. They are available in diameters ranging from ¼ to 1 in. as well as in rectangular shapes. Most guns are designed to be mounted to the steam chest and feed the mold cavities from the rear. Others are designed primarily for mold mounting and are used from the side.

Fill guns are usually used on a one gun per cavity basis. In some cases, however, sprues are used between cavities, and therefore a single gun is used to fill more than one cavity. This lessens the number of guns required, but creates a trimming operation and scrap allowances. Occasionally

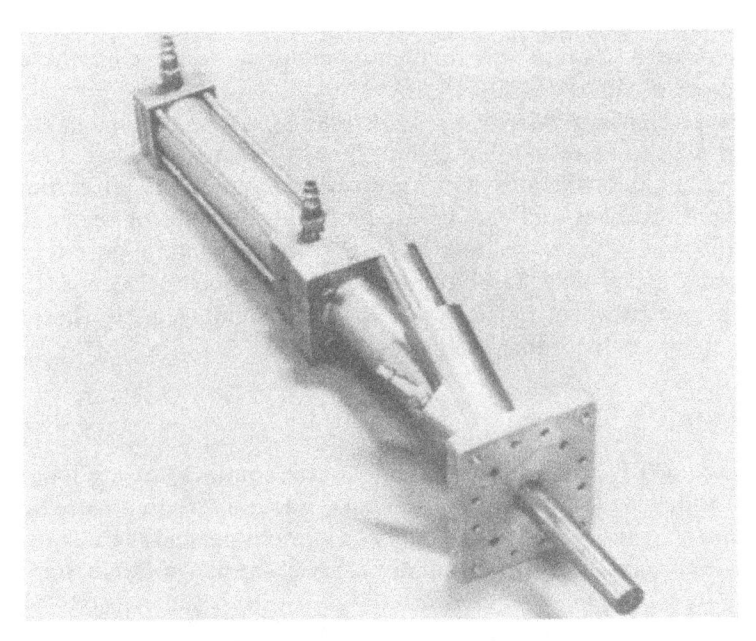

FIG. 12.38. A fill gun for filling a mold cavity with EPS beads. Many sizes and types are commercially available. (*Courtesy Kohler-General*)

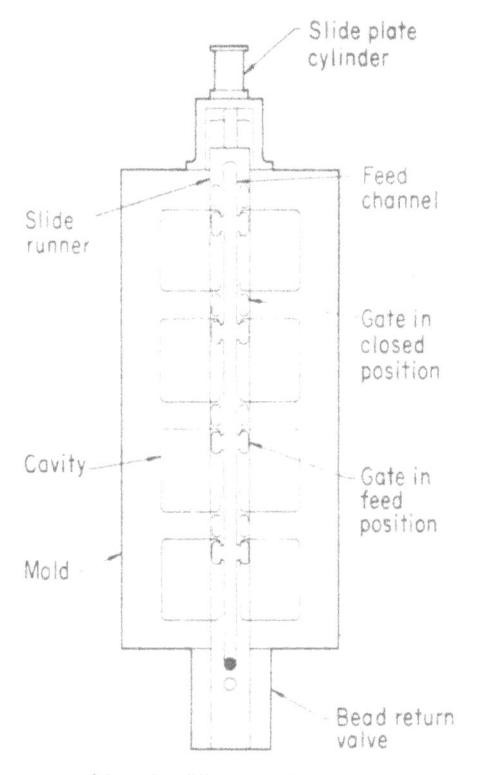

FIG. 12.39. Slide runner multi-cavity fill system frequently used in molds for EPS.

it is possible to produce parts by gating from one part directly into another, with the parts shipped as "breakways" or hot-wire cut into individual pieces.

There are several special types of guns available. One gun* has a vent built into its face, this helps to reduce back pressure in cavities where it is difficult to get sufficient venting.

Another type of gun has a plug capable of three positions, the third position being extended so that the plug operates as an ejection pin. Some guns feature keys to prevent the plugs from rotating. Plugs can then be contoured to follow the cavity shape.

The second prevalent fill method is the slide runner or slide fill system (Fig. 12.39). It is an effective method of filling multiple-cavity molds. The system consists of a channel across the face of the mold, and a sliding plate adjacent to the channel, either in the same mold half or the opposite mold half. In one position, gates in the plate align with the cavities, and

*See applicable patents.

material is thus allowed to flow into the cavities. In the closed position, the gates no longer line up with the cavities and a part of the runner block forms a part of the mold cavity.

The slide runner is activated by an air cylinder with about a 1 in. stroke. During fill, material is airveyed from the machine hopper to the channel and "boosted," using a special air manifold, into the cavities, through the open gates. The plate is shifted after the cavities have filled, and the remaining material in the channel is returned to the hopper.

Usually, the slide plate is made from aluminum plate and treated with some type of coating to prevent galling. Standard parts are available for slide runner systems, although the plate itself is usually custom made for each mold.

Venting

In many cases, EPS molds are not vented across the mold parting lines. Mechanical standoffs built into a molding machine can be used to create an artificial vent across the mold face. This method, however, is not as accurate as venting the mold face, particularly as a mold ages and distorts to some degree. Venting the mold face enables fill while the press is in high pressure clamp, insuring even and accurate venting.

Vents are usually cut around much of the mold cavity to a depth of .005–.010 in. Due to the large amount of air entering the cavity during fill, generous venting is necessary.

Vents can be direct into channels cut into the mold face, which further direct the air in a downward direction. This is desirable since steam escapes through the venting during fusion, and is thus directed toward the floor.

Fusion

In order to fuse an EPS part, it is necessary to allow steam to contact the EPS material. This is accomplished by two basic methods by allowing steam to pass through the cavity walls: core vents and steam holes. A core vent is a slotted plug which can be inserted in the face of an EPS mold.

This vent has slots about .005 in. wide, running across its face, which allow steam to pass through it and into the mold cavity.

Core vents are installed by drilling through the mold, countersinking a second hole to the height of the vent from the cavity side, and pressing the vent into place. Core vents are usually aluminum, and should be of the same material as the mold face to eliminate galvanic action.

Core vents are available in various diameters. They should be located approximately on 2-in. centers to allow sufficient steam for fusion. In

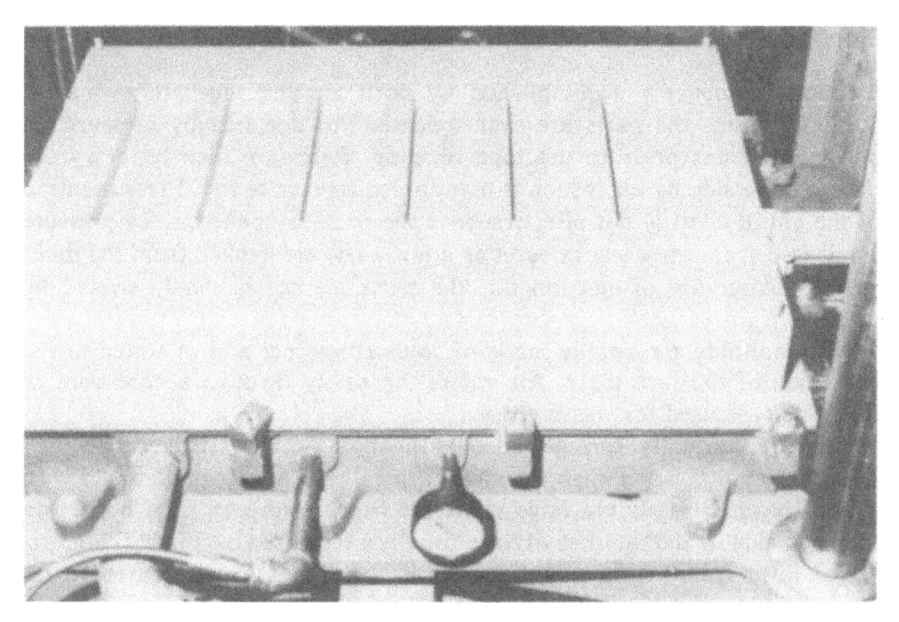

FIG. 12.40. Face of a multi-cavity EPS mold. Note clamp dogs, steam line connection, pressure gage, and mounting method. (*Courtesy Johnson Controls*)

general, vents on the cavity side should not be placed directly opposite those on the core side.

Drilled holes can also be used to allow steam from the chest to penetrate into the mold cavity. They should be kept as small as possible, to minimize problems caused by material entering and plugging the holes. Typical diameters for steam holes are 1/32 in. Steam holes do not pass as much steam as core vents, and must be spaced at closer intervals, typically 1 in.

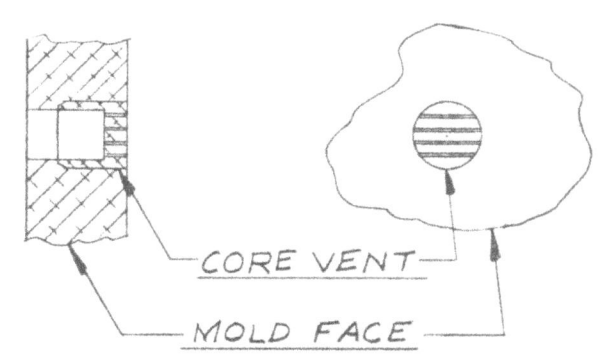

FIG. 12.41. Core vents are used in EPS mold to allow steam to pass through the mold face and into the cavities. Core vent small openings will appear on molded part as small ribs.

Ejection

Ejection is normally accomplished by both air and mechanical means. In many cases, the parts are first "released" to one side by pressurizing one steam chest prior to machine opening. Parts are then broken from the opposite side by air, which is manifolded to each cavity. Pressurization of the entire chest is not effective once the press is opened, since pressure is released from the chest as soon as a few parts are broken from the mold cavities. After use of ejection air, the parts are mechanically ejected by pins.

Air manifolds are usually made of soldered copper and mounted to the back side of the face plate. Air enters the cavity through a core vent of the same type used for steam entry.

Pins are normally spring-loaded with about a 2 in. stroke. The port side of the pin is larger than the shaft; typical pins are ¼ in. diameter with a 1 in. head. Pins should have sufficient bearing surface. Cast bosses on the back side of the mold cavity are usually necessary (see Fig. 12-33).

For lower production quantities, air ejection is sometimes used to reduce the cost of a mold. For some parts with generous draft, air ejection may even be sufficient for automatic operation.

Grippers (reverse taper pins) are of assistance when attempting to retain a part to one side prior to ejection.

Cocking of parts (non-uniform ejection) in mold cavities can be a serious problem, and, when possible, efforts should be made to eject the parts from their extreme corners.

To facilitate ejection, most aluminum molds are teflon coated. Teflon is easily damaged if a mold is mistreated and must be replaced periodically even when a mold is handled correctly. Life expectancy of the teflon coating on an EPS mold is about a few thousand hours running time. Generally, molds that hold a polish do not require a teflon coating.

Special Considerations

As pointed out earlier, many special techniques are used in shape molding, to solve special problems. Most of the standard methods have already been covered and we will now examine a few of the special techniques.

Dual Steaming

Part fusion can be improved by isolating the steam used in heating EPS material from the steam used in heating the mold. Improved fusion is the result of less steam condensate in the molded product. This is accomplished by withholding the cook steam entering the mold cavity until the mold heating steam has brough the mold to the molding temperature.

To date, this method has only been used in cup molding where fusion is critical to product acceptability. Associated with the techniques are certain difficulties not found in conventional shape molding, these are created by the elimination of the steam holes and core vents. Venting is reduced with all venting now being accomplished on the parting line. It is also difficult to incorporate ejection manifolds and pins into the mold without the leakage of steam into the mold cavity. The molds are more expensive and require special machine controls.

Floor Molds

When very large, slow cycling parts are encountered, it may be desirable to produce a mold with its own clamping method. These molds are set on the floor rather than in a press and may be operated manually or even semiautomatically, if provided with a control system. They are usually hinged on one side with a mechanical lock on the opposite side. This method is used in block molding and for other large items such as boats.

Special Fill Methods

"Crush fill" is a technique used to fill a part, considered too thin to be conventionally molded. The mold is constructed so that the core half telescopes into the cavity half. The cavity is filled prior to completely closing the mold. After fill, the press closes compressing the material in the cavity. This procedure has two drawbacks: Fusion is less complete since steam has difficulty in flowing through the material. Material is crushed into core vents and steam holes, further reducing part fusion and increasing maintenance costs.

A relatively new method of fill* is a modified slide runner system wherein the sliding plate is replaced by a series of individual plugs. The plugs are spring-loaded and activated by the chest mounted ejection plate. In their neutral position, material is allowed to enter the mold cavities. After fill, the ejection plate moves the plugs forward sealing off the gates into the mold cavities.

Stack Molding

When producing high volumes of shallow easy-to-mold parts of similar thickness, it is sometimes feasible to run two molds on a machine simul-

*See the applicable patents.

taneously, referred to as "stack" molding. This can be accomplished by using an additional platen on the machine, or by building the stack assembly entirely into the molds. Generally, the cavity halves of the molds are mounted back-to-back, as a floating assembly. After fusion, the parts are pressurized to make them adhere to the forces on the stationary and moving platens. Then the parts are ejected from the forces.

Conclusion

It has been impossible here to cover in great depth all the details of EPS mold design and construction. In many cases, detail has been sacrificed for completeness in an effort to provide an overview of the systems, methods and mold designs that are being used.

A final point, worth remembering, is that EPS tooling demands a high unity of construction wherein the various mold components not only fill their primary functions, but also work harmoniously with the rest of the mold during other segments of the molding cycle.

REFERENCES

Bender, R. J., *Handbook of Foamed Plastics*, Libertyville, IL: Lake Publishing Company, 1965.

DiPierro, Paul A., Building molds to accommodate large RIM parts, *Plastics Machinery and Equipment*, July 1982.

DuBois, J. Harry, Reaction injection molding, *Plastics Machinery and Equipment*, Dec. 1974.

DuBois, J. Harry, Designing molds for structural foam products, *Plastics Machinery and Equipment*, Oct. 1975.

Galli, Ed., RIM molds: Some tricks of the trade, *Plastics Machinery and Equipment*, Sept. 1984.

Krauss-Maffei, *Reaction Injection Molding Technique for the Automotive Industry*, Munich.

Liedtke, The evolution and productionizing of reaction injection molding, *SPE NATEC*, Dec. 1974.

Lettner, Horst H., *Advances in Injection Molding of Thermoplastics Containing Foaming Agents*, Munich: Krauss-Maffei.

Metzger, S. H., Jr., and D. J. Propelka, Reaction Injection Molding and Advances in Reaction Molding, Pittsburgh, PA: Mobay Chemical Co., 1975.

Sneller, Joseph A., *RIM takes the hurdles to productivity in stride: Fast demolding and in-mold finishings are just two ways*, *Modern Plastics*, p. 66, July 1986. Includes bibliography of vendors and suppliers.

Titlebaum, R. P., Liquid reaction molding, *SPE NATEC*, Dec. 1974.

Wood, Richard, Structural foam machines, An overview, *Plastics Machinery and Equipment*, Dec. 1974.

Chapter 13 / Care, Maintenance and Repair of Plastics Molds

Charles C. Davis, Jr.

Care is another word for preventive maintenance, and good care is essential to extend the useful life of plastics molds. Good care delays the day when the mold, after long service, must be refurbished.

CAUSES OF WEAR AND DAMAGE

Neglect causes damage and wear resulting in a greatly reduced mold life. Mold wear is seldom the direct action of the movement of the plastics material over the mold surface. An exception is in the case of compression molds using mineral or glass-filled materials where a scouring action takes place on the mold surfaces and causes abrasive wear. In transfer and injection molding of thermosetting materials, wear is often detected in the high flow areas such as in the sprues, runners, gates and portions of the cavities and cores that are directly opposite the gates.

Conversely, in injection and blow molds, plastics abrasive wear is discovered only in some very high production molds only after months of fast cycle operation. In such instances, wear is usually limited to the pinch off area of the blow mold and the gating area. In injection molds for thermoplastics, wear appears on the surface opposite the gate. In the latter case, wear is minimal except where an abrasive filler is compounded with the resin. Long-term studies show that the major cause of mold wear and damage to thermoplastics molds is not the result of abrasion from the plastics materials.

Most often damage results from continuing to run the mold after flashing occurs. Major damage results from closing the mold on the material itself, bits of flash, chips from sheared undercuts, cracked parts of runner or torn gates, stringers from nozzles or from sprues that are too hot or improperly controlled, and the inexcusable crime of allowing the mold to clamp on granules of raw material.

The sources of damage include:

(1) The result of closing the mold on foreign objects such as inserts, tools, screws, nuts, broken ejector pins, etc., caught between the parting line surfaces. In many cases the smallest bits of foreign objects cause the greatest damage, especially if they are caught at the cavity edge. These small bits are not detected by properly set low-pressure closing protective devices. Clamping on a tiny object, even if it is a plastics granule, concentrates the entire press clamp tonnage on this very small area—exceeding the elastic limit of any mold material regardless of quality and hardness.

(2) The result of the use of screw drivers, knives or cutters, etc., used to assist in the removal of sticking parts, flash, short shots, etc., from non-automated molds.

(3) The result of contact with water on unplated surfaces. Water forms in the molds from condensation, seepage through porous metals, leaky pipe fittings and "O"-rings. Careless handling of hoses and feed lines during hook-up leaves water on the mold surface. This is not harmful if detected immediately and carefully removed. Corrosion is progressive and even if the molds are stored after being sprayed with an antioxidant, a few drops of water or condensation can cause tremendous and costly damage.

(4) Attack from acids after exposure to corrosive materials which may form when some thermoplastics are decomposed by over-heating. Over-heating can occur in the plasticizing cylinder, the hot runner system or in the mold cavities, as the result of too small gates, inadequate venting or cooling systems.

(5) The result of accidents caused by a number of things. Mistakes in mold installation, continuing to produce in a malfunctioning machine, continuing to operate a mold which has started to squeak or squeal will result in damage. Damage will be done to molds that refuse to close or open normally or will not eject properly if their operation is allowed to continue without ending the problem. An alert and well-trained foreman can hear trouble starting as he passes through the molding room. The aforementioned types of mold damage are not limited to the pressroom. Tool-room technicians contribute their share by the improper assembly of molds, failure to tighten screws, leaving metal chips, grinding dust or polishing abrasive on or near sliding surfaces.

(6) Fatigue is a major cause of mold damage which leads to a breakdown of those mold components that are subjected to the maximum stresses while cycling correctly over long periods of time. This usually occurs, if it is going to occur at all, after 100,000 and before 300,000 molding cycles. In a multiple-cavity mold, components identical to those that break first may last infinitely longer. Fatigue can manifest itself in components subject to compressive loads as well as those in torque, tension, or bending.

Thermal shock has a questionable contribution to the deterioration of molds. So far as our analysis goes, unless rapid heating and cooling of the mold is part of the cycle, there is ordinarily only a small effect in comparison with hydraulically and mechanically induced stresses. Molds run in the higher temperature range—500 to 800° F deteriorate more rapidly; this is an historical problem with the die casting dies.

PREVENTIVE MAINTENANCE

Molds that are still on the drawing board offer innumerable opportunities. You can build a cheap mold that will be in trouble and not last very long or you can build in the well known factors that provide long life and minimum mold maintenance will be experienced. A mold that has become unusable because of wear, abuse, and improper construction has passed the opportunity for preventive maintenance. The options here after this happens are few unless the economics permit major replacement with corrected design.

Plating

It is important to specify chrome plating for thermoset molds before any wear occurs. New molds for this work should be plated immediately after the samples and mold operation are approved. Never run production before plating in molds for thermosetting materials. Periodic checks after the chrome-plated mold is in full production are desirable to find evidence of mold wear, which will show up first in corners and high flow areas. A simple check may be made by swabbing a copper sulfate solution in the mold areas; if the copper plates out on the surface, the chrome has gone and must be replaced. Instruments are available that will measure the thickness of chrome plating and predict potential failure. Mold components should be replated as soon as the chrome has gone. Mold parts must then be stripped, repolished, and replated. As specified in Chapter 4, the superficial hardness of chrome plating is the equivalent of 68 Rockwell C whereas the average compression mold steel is about 56 Rockwell C. The relative resistance to wear has been determined to be approximately equal to the ratio of the squares of the hardness numbers—thus emphasizing the importance of plating to minimize wear.

Replacement Parts

High volume transfer molds for thermosets often use replaceable inserts made from a high Rockwell steel or hardenable carbides. It is equally good practice to provide for the subsequent installation of an insert when it is expected that substantial wear may occur. Without this advance planning, it may be difficult or impossible to install an insert when and if it becomes necessary.

Control of Flash

The best quality blow molds are made with replaceable pinch-off inserts.

At the first appearance of flash in any mold, cleaning is essential. There probably is a build-up of some sort on the parting line, back of the stripper plate or between slides. A check will determine whether or not the clamps or other fastenings have loosened. The initial appearance of flash suggests a complete check up of the press. Stress rods have been known to break within one of the nuts hiding the failure.

The most successful program of preventive maintenance is based on scheduling each mold out of production regularly into the tool room. The last shot molded should be attached to the mold with its record tag. In the tool room, the mold is disassembled as needed for complete cleaning and careful inspection. After completion of essential repairs, replating, etc., moving parts are lubricated, mold is sprayed with an antioxidant and returned to storage or production. This is the best and least costly maintenance; nothing is more costly than unanticipated shutdowns at peak production periods. You pay for this service as a protection or you pay later as a loss; loss expense is greater than the prevention thereof.

Flashing

When a clean and otherwise properly operating mold is flashing, corrections can be made by the toolmaker at small cost. Often this is simply a matter of skimming off the parting line or the refitting of inserts, adjustment of wedges or the installation of oversize knockouts. Such corrections are facilitated by making advance provision in the mold design for such maintenance. Flashing is a self-aggravating situation. The longer a mold is run and flashing, the faster the flash area increases since it gradually adds to the projected area that the machine is capable of clamping adequately.

Hand Tool Damage

Very strict enforcement is recommended of well defined and publicized rules of acceptable shop practice concerning the presence of ferrous tools in the molding room. Sharpened soft brass or copper bars with facilities for re-sharpening are mandatory. When hard brass or bronze tools are provided it must be recognized that they will work-harden and must be annealed and reshaped daily by a responsible person.

The best prevention of tool damage is a fully automatic mold running in continuous production with minimal start ups.

Water Hazard

A number of preventive measures can be taken to avoid corrosion by water. The most popular is plating, usually chromium on thermoset molds and nickel or chrome on thermoplastics molds. It must be recognized that plating can be porous, especially if the steel substrate is pitted or has been subjected to corrosion previously. It is not safe to depend on plating for protection from water damage. Where chillers are used for mold temperature control, condensation of moisture on the mold surfaces can sometimes occur even while they are in full operation. The best solution for this problem is an air conditioned pressroom—or at least a humidity controlled atmosphere. When a chiller or very cold water is used for cooling, it is important to anticipate the shutdown time and allow the mold to warm up to above room temperature. Only then is it practical to clean and spray the mold before closing. A cold mold should never be closed; let it warm up above room temperature before closing. Condensation will occur in the mold if it is closed while cold. When cooling lines are being disconnected, the best practice is to disconnect the supply line hose first and then use the air nozzle to blow the remaining water from the mold into the drain line. This procedure prevents subsequent movement of water from the channels into the parting line or other openings.

Not all damage to molds is confined to molding surfaces. Rust and salt deposits form inside of the heating or cooling channels in spite of the use of well designed water treatment systems. Delicate molds with correspondingly thin steel wall sections separating inner and outer surfaces have been found to rust through from the inside. The best practice results from nickel plating these inner areas. Electroless nickel plating solutions can be pumped through the water lines and, since this process does not depend on the passage of electrical current, the plating develops evenly if they have been thoroughly cleaned before the plating. This must be done before any rust has formed.

Acids

To reduce or eliminate the attack of the mold surfaces by corrosive plastics or the products of their decomposition, plating can be most helpful. The plater must know exactly what the attacking medium is to suggest the proper plate. Gold has been used for some very corrosive substances. Chrome is attacked by some decomposition products, and stainless steel is not immune to corrosion. Nickel plate appears to be second only to gold plate in corrosion resistance. It must be recognized that the plating may not have completely covered the surface to be protected. This fact necessitates extreme care in the molding room to avoid the overheating of the compounds which trigger the attack. Besides mold temperature control and prevention of excess heating in the plasticizing cylinder, corrosion may also be minimized by a reduction in the speed of mold filling and the provision for adequate venting. Slowing the rate of mold filling may be accomplished by machine control and by increasing the runner and gate dimensions. Frictional heat from an excessively small gate is easily cured. Venting at the extremities of the cavity, plus added vents along the way, reduce the back pressure and permit a temperature reduction along the way. In some cases it helps to vent the runner system.

Accidents

Mold fastenings loosen from many causes: clamping force causes mold compression, the impact and bursting pressures caused by rapid filling of the mold, tensile stresses as the mold is pulled open—all these forces are augmented by inertia and acceleration during fast cycle operation. This loosening of fastenings can happen inside the mold as well as at the point of fastening in the press. In far too many cases, a mold is set into the machine with only four clamps to hold each half to its platen. These clamps may be adequate in a static position but become entirely inadequate if the mold on the moving platen is quite heavy or if there are slides or split cavities of large size. Other factors are:

- When deep part, or parts, in the case of multiple cavities, is combined with low draft angles;
- When sticky materials are being molded;
- Molding in fast cycles; or a combination of all of the foregoing.

The strength of mold clamps must be calculated from the pull-back strength of the press and this force compensated by an adequate number and size of screws that can be torqued to a total equal to the pull-back force multiplied by a safety factor of two. It is obviously useless to use stronger fastening devices to hold the mold in the press than are used to hold the mold together. Conversely, since the internal fastening devices are usually difficult to reach, it is extremely important to make sure that their combined force is well in excess of the pull-back force. This is another burden that the mold designer must bear. Poorly designed and cheap molds fall apart under the production forces. In the prevention of clamp loosening, it is necessary for someone to check all mold fastenings daily. In ultra-high speed operation, a check after each shift is recommended. This check should be combined with a cleaning and lubrication of all sliding components using a minimum of lubricant. A wipe or spray-on and wipe-off is the best procedure.

Whenever a mold resists opening, closing, ejecting, or other operation after it has been lubricated, it must be shutdown, taken out of the machine, and disassembled without any additional exercise of force to overcome the inoperative condition.

Fatigue

This is a subject that must be understood and appreciated by molders and mold makers for tools operated in continuous long-term production. Molds that have delicately proportioned core or cavity members are particularly subject to fatigue. All molds, even those with the best possible design will fatigue after long periods of production. The designer and mold maker must consider this factor and study the steel data carefully before any material selection is made. Some steels are more fatigue resistant than others. The mechanisms of steel failure are being studied by certain groups of metallurgists, and it is hoped that all of the basic data may soon be available. A rule of thumb that is reported to be reasonably reliable is to obtain data on the percent elongation at the elastic limit of the steel under consideration, at the hardness that is anticipated for the desired service. When this figure is below 3%, another steel is recommended. The alternative is to use a steel of lower hardness to gain increased elongation. There are obstacles and unknowns in this course of action. Very few steel companies are willing to release these specific data and prefer to evaluate their steels in relative terms only. Experience has shown that some steels having an elongation above the 3% value have poor performance records in fatigue failure.

One preventive measure is recognized; the metallurgists recommend periodic stress relieving. This can greatly extend the life of questionable mold sections that have not yet exhibited cracks. It consists of heating the parts to the same temperature or just below the temperature at which the mold sections were tempered originally. The plating must be stripped before this annealing operation since the stress relief temperature is usually above the plating stability point. Experience is the only available teacher at this time to suggest a desired time interval for this operation. Expensive mold sections merit this preventive care. Mold components with sharp or nearly sharp fillets and cores with a high ratio of length-to-cross-section will require the most frequent stress relief.

REPAIR

Welding

When preventive measures fail or have not been undertaken, the salvage of the mold parts is normally done by plating or welding. Before any plan is made for welding, it is extremely important to consult with an experienced mold welder. He will usually want to weld with the actual mold material and this emphasizes the need for branding the steel sections with the material used. Most mold makers stamp this information on the bottom of the mold part. When the mold has cracked, it is important to cut away all of the cracked area to make sure that the added material is bonded to a sound structure.

There are two very important things to remember when welding heat-treated molds. The mold must be preheated before and reheated after it has been welded. The following procedure has often worked without trouble:

1. Preheat the mold to about 100°F below its tempering temperature. Heating it to above the tempering temperature will draw the temper and soften the mold. The maximum preheating temperature should be 900°F.
2. Weld. Reheat the mold if the temperature drops 300°F below the preheating temperature before welding is completed.
3. After welding is completed, allow the mold to cool to hand warm or at least to 200°F.
4. Reheat the mold to the preheating temperature and hold one hour per inch of thickness of the mold, but a minimum of two hours.
5. Allow the mold to cool to room temperature.

The two heating steps are important. Heating the mold in preparation for welding prevents thermal shock and perhaps cracking of the mold during welding.

Reheating after welding serves a dual purpose. It tempers both the re-hardened layer of steel and the weld metal. It also relieves the tremendous stresses imparted to the welded area through the temperature gradients present during the operation. Figure. 13.1 shows a simplified version of a weldment.

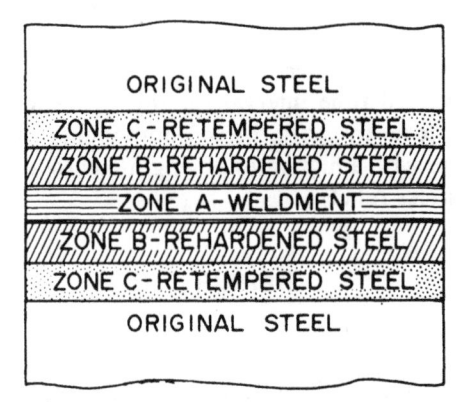

FIG. 13.1. Hardness variation zones around weldment. (*Courtesy Crucible Steel Co.*)

Reheating the mold after the welding is by far the most important. Some molders have experienced cracking of weldments after a short run. Cracking usually occurs in an area next to the weld, not exactly in the weld, pointing to the stresses set up in the surrounding area during welding.

Molders have also experienced cracking of welds which were heavily ground after they were reheated. Grinding is known to cause stresses, depending on the severity of the grinding operation. If not relieved, mold failures may occur at the ground surfaces.

It sometimes happens that the mold is too large to put into a furnace for preheating and post-heating. In such cases the area to be welded should be heated with torches. Care must be taken that the flame of the torch is not so hot that it causes the steel to heat to an excessively high temperature. Also, any impingement of sharp flames should be avoided.

There are times when it is unwise to weld a badly worn part since the proportions may be against success and the weld would warp or break the section or leave it vulnerable to breakage in production. It is unwise to weld when a small build-up is required over a large surface since this will induce warpage, particularly in the case of thin parts.

Electroforming

In cases where welding is questionable, it is preferable to electroform over the defective area (Figure 13.2). This process ordinarily plates on a

Fig. 13.2. A section of a cavity with electroformed nickel repair prior to machining. (*Courtesy Electromold Corp., Trenton, NJ*)

nickel alloy with a hardness of 45 Rockwell C. No heat treatment is required and stress is limited to a minimum. The bond strength of electroformed nickel can approach that of welding if the steel is suitable for the process. Stainless steels have been repaired successfully but other steels containing more than 0.5% of carbon have not been electroform repaired.

The electroform build-up is oversize the same as with welding. It may be machined to final dimension and chrome plated if a higher surface hardness is desired above that of the 45 Rockwell C nickel. In most cases the deposit of nickel can be feathered off to bare steel without the knit line being seen. Knit lines are hard to avoid when welding.

Chapter 14 / Thermodynamic Analysis of Molds*

Written by Wayne I. Pribble

INTRODUCTION

Inasmuch as cycle time for molding thermoplastics is largely controlled by the length of time required to remove residual heat and cool the plastic material so that it can be ejected; and cycle times for molding thermosets is largely controlled by the *cure* time, or the length of time required to add heat to the material in the mold to solidify the material so that it can be ejected, it becomes evident that *heat exchange* becomes a critical item in the rapid production of any particular molded part. Approximately 80% of any molding cycle involves heat exchange for either cooling or for heating.

If the heat exchange time can be reduced by 15% to 50%, it becomes obvious that much attention should be lavished on any means of determining an optimum set of conditions which will result in greatly reduced cycle times, or on the converse of greatly increased production rates. For about 30 years, most of the emphasis in mold design was just to create a configuration which would accurately shape the melt without a great deal of emphasis on just how *cheaply* that could be accomplished. However, with rapidly rising costs in the post-World War II period, competitive capitalism began to be keenly felt. The increase in the physical size of molded parts, the development of new plastic material formulae, the increase in equipment tonnage, and the increased cost of labor, materials, and overhead, made it a foregone con-

*The term MOLDCOOL™ is a trademark name of Application Engineering Corp., Elk Grove Village, IL. Another trademark name is MOLDFLOW™ by Moldflow Australia, Pty. Ltd., Bridgeport, CT. We recommend requesting brochures and cost information from all available vendors. The text in this chapter uses MOLDCOOL as an example for use in injection molds for thermoplastics. Programs are also available for *heat input* to molds for use in thermosetting plastics.

clusion that service companies would develop the *computer aided thermo-dynamic analysis of molds for plastics materials*. The mid-1970s saw the writing of analytical programs for such items as finite element analysis and three-dimensional modeling, as well as many other related programs. Finite element analysis is applied to mold stresses, molded part stresses, ultimate strengths, and rigidity. Such programs are usually segments of larger programs, but may also be individual programs themselves.

Fortunately for the plastics industry, several entrepreneurs saw the benefits and the challenge. Heat transfer formulae are well known, but all of them require ''mountains'' of calculations which are all but impossible when done by the ''handy-dandy'' calculator method (or as education in some university). Even though programs became increasingly available as computer power rapidly increased, very few took advantage of the available programs, either because of cost, or because confidential information would have to be released to third parties, or because lead time for mold designs did not allow time for something which had proven ''pretty good'' when performed by the ''seat-of-the pants'' method in the past. As a few ''competitors'' began to testify to the efficacy of the analysis by computer, and as the potential profitability began to be realized, and as marginally profitable jobs began running at a loss traceable to excessive cooling time or cure time, management realized the ''need to know.''

This chapter describes a computer-aided mold design program, called *MOLDCOOL*, which is readily available to mold designers and plastics processers on their premises, provided they have a computer terminal with a modem for telephone line transmission, and a printer. One obvious advantage of on-line analysis is the interaction while the mold design is in process. The use of *MOLDCOOL* as an example is not necessarily an endorsement of the program, but it does serve to illustrate the ease and simplicity of computer analysis. The writer is certain that potential users will want to investigate several of the fine systems which are available. *MOLDFLOW AUSTRALIA* has a powerful program also available, which has been licensed to the General Electric Co. in the U.S.A. To avoid confusion to the reader, we will confine ourselves to the *MOLDCOOL* version for our description.

MOLDCOOL™ SYSTEM

MOLDCOOL is available via a local telephone call (in most large cities) to the General Electric Mark III service network (the G.E. service name may change). Using English language, the user interacts with the computer service in a procedure which will optimize cooling, increase output of the injection molding machine, and improve the quality of finished products.

The user can choose from a menu of programs which:

1. matches the heat load with the cooling capacity of the mold so as to expedite the design of the mold cooling channels, or

2. discern areas in the mold in which cooling is deficient (or conversely, overcooled with respect to other parts of the mold) so that the sizes, the spacings, or the location of cooling channels can be modified during the design process, or

3. select optimum coolant *flow*, *temperature*, and *pressure* conditions to balance heat removal from the mold.

(Some programs allow functions other than cooling to be computer analyzed. Functions such as runner sizes, gate size and placement, flow of material in the mold, and/or the *effects* of changes in any one or all of the elements)

System Operations

The accompanying chart (Fig. 14.1) shows the sequence of programs in the computerized analysis of mold cooling. The following discussion refers to this chart.

In order to use any of the programs in the system, it becomes necessary to develop a mold file (computer data storage) and a cooling line file (also computer data storage). The creation of these two files by the user is an interactive program that inputs specific information about a specific mold (or part) so it can be processed or merged with general information about thermodynamics (already preprogrammed into the computer and stored for use at this time.)

Thus, when the user begins to design a mold for a particular project, the first step would be to access the *mold material data bank*. This data bank knows all the materials commonly used in mold construction, and it also knows the characteristics associated with heat transfer (thermodynamics). The user indicates his selection by selecting the appropriate code. The relevant information about each plate or section is stored in the current data for later use in the thermodynamic calculations by the program.

The second step is to access the *plastics material data bank*. When the user identifies the precise material to be molded, the data bank will supply all of the thermal data related to flow, heat transfer, shear rheology, etc. The preprogrammed data bank has been created with precise information about each material, and selection is made by manufacturer, grade, type, and number, with an identifying code for each.

The third step involves inputting data concerning the part geometry, the type mold (e.g., hot-runner, insulated runner, thermal tip, etc.), runner sizes

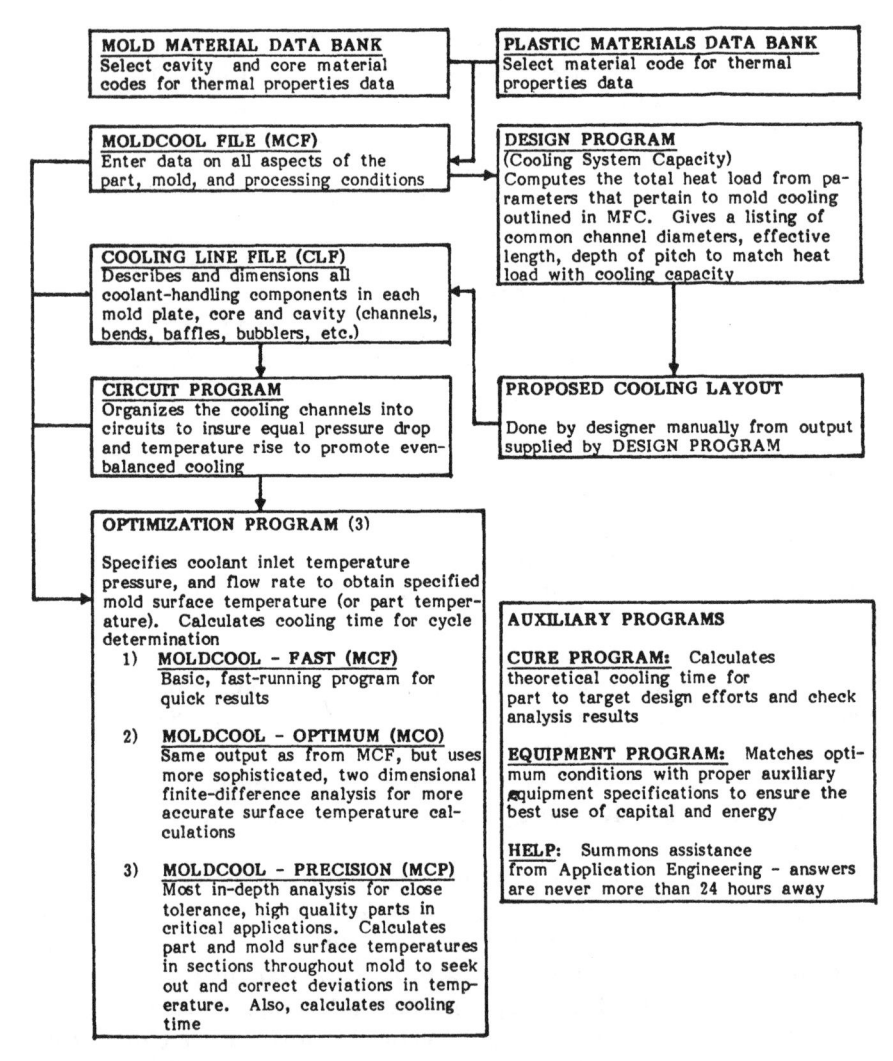

Fig. 14.1. *MOLDCOOL* computer-aided design program at a glance. (*Courtesy Application Engineering Corp., Elk Grove Village, IL*)

or capacity, surface temperatures desired, part weight, shot weight, and temperature of the injection melt.

With these three steps completed, the user and the computer are fully prepared to work together to design the cooling system for this particular mold. It should be obvious at this point in our description that it is a simple task to check several *materials for the mold*, several *molding materials*, differing runner sizes, different part geometry—the list is quite lengthy—

but all can be operated in the well-known "what if?" of spreadsheet programs on minicomputers.

Program Menu

This particular system offers the mold designer several program options. A usual starting point is the *design* program, which matches cooling capacity to the heat load that will be imposed on the mold. The program draws upon information in the *MOLDCOOL* file to compute the total heat load, and, utilizing mathematical procedures built into the design program, determines and makes a print-out of a list of the diameters, effective lengths, depth (distance from mold surface to center of passage), and pitch (spacing of consecutive passages centerline to centerline) to match the cooling capacity to the heat load.

From this printout of information, the mold designer can either confirm and/or modify a proposed coolant passage layout, or he can use the just printed information to make a proposed layout of the mold (hopefully, with his own CAD program operating on his minicomputer). Later, on another dial-up, the *proposed* cooling layout data is entered with appropriate control information such as bends, bubblers, baffles, heat pins, etc. to create a *cooling line* file containing the coolant handling components for each mold plate, the core, the cavity, backup plates (heater or cooling plates), insulating boards, sprue bushings, etc.

When a computer analysis is not available, an experienced designer can "eyeball" the various channels and determine the probable connections that will give approximately equal pressure drops between each inlet and its corresponding outlet. With computer analysis, the designer can run the *circuit* program which, in the case of *MOLDCOOL*, will handle up to 50 cooling channels. By combining the *MOLDCOOL* file and cooling line file, and analyzing the data, the computer will generate a printout giving the directions as to which channels can and should be in series, and which can and should be connected in parallel to achieve optimum quality in the molded part. By developing actual numbers from the data presented, the computer printout eliminates the estimates that would otherwise be made. Of course, the best computer program ever written is not very effective when the set-up man proceeds to hook up the mold to suit his own ideas. Even stamping "IN" and "OUT" is only a guide, and the designer should be aware of whether or not the connections are made as intended and as designed.

Optimization Program

The previous procedures focus on the layout of mold coolant passages. However, the *rate* of heat transfer from the mold also depends on the characteristics of the coolant flow through the passages.

The *optimization* segment of the program uses the information about flow, material dynamics, heat transfer, etc. to specify the *coolant temperature* at the inlet, the *pressure*, and the *flow rate* which will most adequately achieve the specified temperature at the mold surface, or the actual temperature of the plastic material in the mold. At this point, the designer needs to be reminded of the necessity for someone to specify the *ideal* temperature of the mold surface or the precise temperature at which the molded part is to be ejected. In those cases where the designer has made that decision or determination, do not keep it a secret. Make sure the criteria is recorded in the set-up and processing specifications sent to the shop.

The optimization segment is available in one of two formats (users choice). For most uses, the basic fast-running program (remember connection time to a time-sharing program is charged by the minute) will provide adequate control information on those parts where wider latitude is not detrimental. However, for those close-tolerance, high quality parts, where 100% is barely good enough, the precision in-depth analysis will perform a *finite element analysis*. In finite element analysis, the mold or the molded parts are broken into small segments called *finite elements* (triangular elements are the most frequently used, because triangles are more predictable—sizes can change, but as long as you have three sides, you have a triangle for which the mathematics is uniform from one to the other). Whichever analysis is chosen, the optimization segment calculates mold stresses, cooling time, heat removal capacities, and average temperature for part ejection.

Auxiliary Programs

For any combination of plastic material, processing temperature, part configuration, mold material, coolant characteristics, and coolant flow data, there can be and is a theoretical *minimum* cooling time. The *cure* segment of the program enables the designer to compare the cooling time projected from coolant passage layout and coolant characteristics with the theoretical minimum cooling time. If the difference is more than 10%, the designer would be well advised to close the gap, if possible. From the foregoing analysis in other segments of the program, it should be obvious that more channels can be added, spaced closer, a change in coolant flow or coolant characteristics can be made, modification of the part geometry (elimination of a heavy section, for example), and a whole host of other options are open to the designer. We would remind the designer of the 80/20 rule, which has its application in the selections of the options for closing the gap between theory and actuality. Eighty percent of the gap closing can be effected for 20% of the cost. The remaining 20% of striving for perfection should be evaluated by management people familiar with the economics of the particular job.

The *equipment* segment provides specifications for the auxiliary equipment needed to support optimum cooling conditions on the basis of the highest and best use of capital and energy. From this segment, the determination of the suitability of present equipment is made. Management can then decide on the economics of oversizing or undersizing, and on the cost of making that choice.

Help is always a welcome sight as a segment in any comprehensive program. Many programs for on-line communications provide immediate and on-line answers to typing HELP (here specify HELP with what other segment). In the case of *MOLDCOOL*, the owners of the program will be "fed" the information via their own computer link. Ideally, any real difficulty should be cleared up within one working day, and an answer or instructions supplied. Problems with infrequently used programs usually involve simply forgetting how to enter the correct commands or the correct data.

In choosing a *thermodynamic analysis program*, we recommend a careful analysis of the program itself. "User friendly" is a major requirement for the use of complicated or infrequently used programs. An on-screen, easily accessed menu should be readily available for each segment. On-screen instructions for data entry will save much connect-time when compared to "looking it up in the manual."

CONCLUSION

Many articles, several books, and a host of vendors are at your beck and call as you investigate the capabilities of *thermodynamic analysis*. At other points in this text, we have included the suggestion of Computer Aided Design (CAD). As this text goes to press, CAD programs are available in the $5,000 to $10,000 range, including computer, disk drives, video display, and printer or A/B size plotter. Inasmuch as a hard disk is now relatively economical for a microcomputer, we recommend 20MB or more. A plotter for C/D size, or E size, drawings will, of course, add $10,000 or more to the price. Interfacing these programs with CNC equipment, providing 3-D views, automatic generation of third views, automatic dimensioning, and all the other "goodies" available on the $75,000 (and up) systems will become reality. As a minimum, we recommend that the designer get as much experience as he can in the CAD field.

The same principles are discussed in the previous material of this chapter can and are being applied to *thermosets*, *blow molding*, *thermoforming*, *isothermal line charts*, and many other applications where the power of the computer can be utilized in a cost effective manner.

REFERENCE

Prost, Henry, Hyperthermal Runner Systems: New, Faster way to Mold Thermosets, *Plastics Technology*, May 1981.

Appendix

The mold designer should maintain an up-to-date file of manufacturers' catalogs to assist him in the proper selection of materials and parts. A complete list of all steel sizes available in his own shop or those which may be commercially available is an absolute necessity.

A manual containing tables of the functions of numbers, i.e., squares, square roots, logarithms, functions of angles and strengths of materials, thread standards, etc., will be found extremely useful.

A careful study of trade magazines and the articles and advertisements in them will reveal many of the fine points of design.

Some of the tables presented herein are designed to serve as a guide for the designer in collecting and organizing his own shop data. The stock steel sizes given here should not be followed too closely when financial considerations are involved, because they may not represent stocks obtainable at any given time.

The Appendix material should be supplemented by compilation of tables found in the trade magazines and in manufacturers' bulletins, catalogs, etc. These tables may be copied in a reference notebook or clipped from the magazines and bulletins and filed in the manner most convenient for ready reference.

METRICS

The metric units as used in this Appendix are those of long standing usage in international industry. The International System of Units (SI), which provides only one basic unit for each physical quantity, is becoming increasingly popular when complicated metric calculations are required. SI units and their definitions are readily available in any mechanical engineering handbook. We continue to use the older units because of the basic nature of this text, which is also used in countries not yet adjusted to SI units.

TABLE A.1. Decimal Equivalents of Fractions of an Inch.

		$\frac{1}{64}$	0.015625		$\frac{33}{64}$	0.515625
	$\frac{1}{32}$		0.03125		$\frac{17}{32}$	0.53125
		$\frac{3}{64}$	0.046875		$\frac{35}{64}$	0.546875
$\frac{1}{16}$			0.0625	$\frac{9}{16}$		0.5625
		$\frac{5}{64}$	0.078125		$\frac{37}{64}$	0.578125
	$\frac{3}{32}$		0.09375		$\frac{19}{32}$	0.59375
		$\frac{7}{64}$	0.109375		$\frac{39}{64}$	0.609375
$\frac{1}{8}$			0.125	$\frac{5}{8}$		0.625
		$\frac{9}{64}$	0.140625		$\frac{41}{64}$	0.640625
	$\frac{5}{32}$		0.15625		$\frac{21}{32}$	0.65625
		$\frac{11}{64}$	0.171875		$\frac{43}{64}$	0.671875
$\frac{3}{16}$			0.1875	$\frac{11}{16}$		0.6875
		$\frac{13}{64}$	0.203125		$\frac{45}{64}$	0.703125
	$\frac{7}{32}$		0.21875		$\frac{23}{32}$	0.71875
		$\frac{15}{64}$	0.234375		$\frac{47}{64}$	0.734375
$\frac{1}{4}$			0.250	$\frac{3}{4}$		0.750
		$\frac{17}{64}$	0.265625		$\frac{49}{64}$	0.765625
	$\frac{9}{32}$		0.28125		$\frac{25}{32}$	0.78125
		$\frac{19}{64}$	0.296875		$\frac{51}{64}$	0.796875
$\frac{5}{16}$			0.3125	$\frac{13}{16}$		0.8125
		$\frac{21}{64}$	0.328125		$\frac{53}{64}$	0.828125
	$\frac{11}{32}$		0.34375		$\frac{27}{32}$	0.84375
		$\frac{23}{64}$	0.359375		$\frac{55}{64}$	0.859375
$\frac{3}{8}$			0.375	$\frac{7}{8}$		0.875
		$\frac{25}{64}$	0.390625		$\frac{57}{64}$	0.890625
	$\frac{13}{32}$		0.40625		$\frac{29}{32}$	0.90625
		$\frac{27}{64}$	0.421875		$\frac{59}{64}$	0.921875
$\frac{7}{16}$			0.4375	$\frac{15}{16}$		0.9375
		$\frac{29}{64}$	0.453125		$\frac{61}{64}$	0.953125
	$\frac{15}{32}$		0.46875		$\frac{31}{32}$	0.96875
		$\frac{31}{64}$	0.484375		$\frac{63}{64}$	0.984375
$\frac{1}{2}$			0.500			

1 inch equals 25.40 millimeters.

TABLE A.2.   Millimeters into Inches.

Millimeters	Inches	Millimeters	Inches	Millimeters	Inches
1	.03937	34	1.35858	67	2.63779
2	.07874	35	1.37795	68	2.67716
3	.11811	36	1.41732	69	2.71653
4	.15748	37	1.45669	70	2.75590
5	.19685	38	1.49606	71	2.79527
6	.23622	39	1.53543	72	2.83464
7	.27559	40	1.57480	73	2.87401
8	.31496	41	1.61417	74	2.91338
9	.35433	42	1.65354	75	2.95275
10	.39370	43	1.69291	76	2.99212
11	.43307	44	1.73228	77	3.03149
12	.47244	45	1.77165	78	3.07086
13	.51181	46	1.81102	79	3.11023
14	.55118	47	1.85039	80	3.14960
15	.59055	48	1.88976	81	3.18897
16	.62992	49	1.92913	82	3.22834
17	.66929	50	1.96850	83	3.26771
18	.70866	51	2.00787	84	3.30708
19	.74803	52	2.04724	85	3.34645
20	.78740	53	2.08661	86	3.38582
21	.82677	54	2.12598	87	3.42519
22	.86614	55	2.16535	88	3.46456
23	.90551	56	2.20472	89	3.50393
24	.94488	57	2.24409	90	3.54330
25	.98425	58	2.28346	91	3.58267
26	1.02362	59	2.32283	92	3.62204
27	1.06299	60	2.36220	93	3.66141
28	1.10236	61	2.40157	94	3.70078
29	1.14173	62	2.44094	95	3.74015
30	1.18110	63	2.48031	96	3.77952
31	1.22047	64	2.51968	97	3.81889
32	1.25984	65	2.55905	98	3.85826
33	1.29921	66	2.59842	99	3.89763
				100	3.93700

Table of Cubic Equivalents
1 cc. = .061 cu. in.
1 cu. in. = 16.387 cc.

(*Courtesy Durez Resins & Molding Materials, North Tonawanda, NY*)

TABLE A.3.   Standard Socket-Head Cap Screws.

(Furnished in sizes indicated by ★)

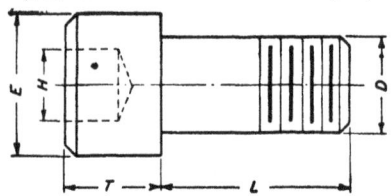

Dia. screw D ...	10	¼	⁵⁄₁₆	⅜	⁷⁄₁₆	½	⅝	¾
Thds. per in	32	20	18	16	14	13	11	10
Size tap drill ...	20 (.161)	¹³⁄₆₄ (.2031)	G (.261)	O (.316)	U (.368)	²⁷⁄₆₄ (.4219)	¹⁷⁄₃₂ (.5312)	²¹⁄₃₂ (.6562)
Dia. head E	⁵⁄₁₆	⅜	⁷⁄₁₆	⁹⁄₁₆	⅝	¾	⅞	1
Length head T ..	.190	¼	⁵⁄₁₆	⅜	⁷⁄₁₆	½	⅝	¾
Size hex. hole H	⁵⁄₃₂	³⁄₁₆	⁷⁄₃₂	⁵⁄₁₆	⁵⁄₁₆	⅜	½	⁹⁄₁₆

LENGTH UNDER HEAD L

L	10	¼	⁵⁄₁₆	⅜	⁷⁄₁₆	½	⅝	¾
⅜	★	★	★	..	..	..	..	..
½	★	★	★	★	★	★	..	..
⅝	★	★	★	★	★	★	..	..
¾	★	★	★	★	★	★	..	..
⅞	★	★	★	★	★	★	..	..
1	★	★	★	★	★	★	★	..
1¼	★	★	★	★	★	★	★	★
1½	★	★	★	★	★	★	★	★
1¾	★	★	★	★	★	★	★	★
2	★	★	★	★	★	★	★	★
2¼	..	★	★	★	★	★	★	★
2½	..	★	★	★	★	★	★	★
2¾	..	★	★	★	★	★	★	★
3	..	★	★	★	★	★	★	★
3¼	..	..	★	★	★	★	★	★
3½	..	..	★	★	★	★	★	★
4	..	..	..	★	★	★	★	★
4½	..	..	..	★	..	★	★	★
5	..	..	..	★	..	★	★	★
5½	..	..	..	..	..	★	★	★
6	..	..	..	..	..	★	★	★

(Courtesy of Danly Machine Corp., Chicago, IL)

TABLE A.4A. Spring Data Chart.

	13	19	25
Spring designation (wire size).	13	19	25
Wire size (sq.)	$\frac{5}{32}$	$\frac{3}{16}$	$\frac{1}{4}$
Carrying capacity (lb. per in. deflection)	375	550	1150
Space between coils	.195	.175	.147
A dia. over which spring works	$\frac{5}{8}$	$\frac{11}{16}$	$\frac{11}{16}$
B clearance hole	$\frac{11}{16}$	$\frac{3}{4}$	$\frac{3}{4}$
C dia. hole for spring	$1\frac{1}{16}$	$1\frac{3}{16}$	$1\frac{5}{16}$
D dia. stripper bolt head	$\frac{7}{8}$	1	1
E dia. counterbore for head of stripper bolt	$\frac{15}{16}$	$1\frac{1}{16}$	$1\frac{1}{16}$
J dia. head of stud	1	$1\frac{1}{8}$	$1\frac{1}{4}$

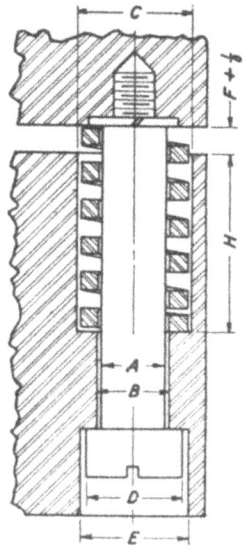

SPRING BOXING

TABLE A.4B. Spring Boxing.

Length of Spring Boxing F	WIRE SIZE 13		WIRE SIZE 19		WIRE SIZE 25	
	Free Lgth. of Spring	Depth of C'bore H	Free Lgth. of Spring	Depth of C'bore H	Free Lgth. of Spring	Depth of C'bore H
$\frac{1}{8}$	$1\frac{1}{8}$	$\frac{5}{8}$	$1\frac{1}{4}$	$\frac{3}{4}$	$1\frac{3}{4}$	$1\frac{1}{4}$
$\frac{1}{4}$	$1\frac{5}{8}$	$\frac{7}{8}$	$1\frac{7}{8}$	$1\frac{1}{8}$	$2\frac{1}{2}$	$1\frac{3}{4}$
$\frac{3}{8}$	$2\frac{1}{8}$	$1\frac{1}{8}$	$2\frac{1}{2}$	$1\frac{1}{2}$	$3\frac{3}{8}$	$2\frac{3}{8}$
$\frac{1}{2}$	$2\frac{5}{8}$	$1\frac{3}{8}$	$3\frac{1}{8}$	$1\frac{7}{8}$	$4\frac{1}{8}$	$2\frac{7}{8}$
$\frac{5}{8}$	$3\frac{1}{8}$	$1\frac{5}{8}$	$3\frac{3}{4}$	$2\frac{1}{4}$	5	$3\frac{1}{2}$
$\frac{3}{4}$	$3\frac{3}{4}$	2	$4\frac{3}{8}$	$2\frac{5}{8}$	$5\frac{3}{4}$	4
$\frac{7}{8}$	$4\frac{1}{4}$	$2\frac{1}{4}$	$4\frac{7}{8}$	$2\frac{7}{8}$	$6\frac{5}{8}$	$4\frac{5}{8}$
1	$4\frac{3}{4}$	$2\frac{1}{2}$	$5\frac{1}{2}$	$3\frac{1}{4}$	$7\frac{3}{8}$	$5\frac{1}{4}$
$1\frac{1}{2}$	$6\frac{3}{4}$	$3\frac{1}{2}$	8	$4\frac{3}{4}$	$10\frac{5}{8}$	$7\frac{1}{2}$

TABLE A.4C. Free Length of Side Springs.
(Allows $\frac{3}{8}$ In. Initial Compression)

Knockout Space G	Wire Size 13	Wire Size 19	Wire Size 25
$\frac{1}{4}$	$1\frac{3}{8}$	$1\frac{9}{16}$	$2\frac{1}{16}$
$\frac{1}{2}$	$1\frac{7}{8}$	$2\frac{3}{16}$	$2\frac{7}{8}$
$\frac{3}{4}$	$2\frac{1}{16}$	$2\frac{3}{4}$	$3\frac{11}{16}$
1	$2\frac{15}{16}$	$3\frac{3}{8}$	$4\frac{9}{16}$
$1\frac{1}{4}$	$3\frac{1}{2}$	4	$5\frac{3}{8}$
$1\frac{1}{2}$	$4\frac{1}{16}$	$4\frac{5}{8}$	$6\frac{3}{16}$
$1\frac{3}{4}$	$4\frac{9}{16}$	$5\frac{3}{16}$	7
2	$5\frac{1}{8}$	$5\frac{13}{16}$	$7\frac{13}{16}$
$2\frac{1}{2}$	$6\frac{1}{8}$	7	$9\frac{7}{16}$
3	$7\frac{1}{8}$	$8\frac{1}{4}$	$11\frac{1}{16}$
$3\frac{1}{2}$	$8\frac{1}{8}$	$9\frac{1}{2}$	$12\frac{3}{4}$
4	$9\frac{1}{8}$	$10\frac{5}{8}$	$14\frac{3}{8}$
5	$11\frac{1}{8}$	$13\frac{1}{8}$	16

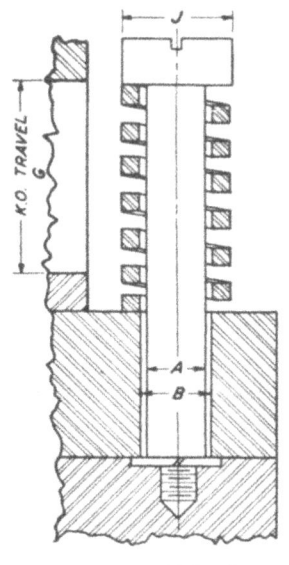

SIDE SPRING

(Courtesy of General Electric Co., Pittsfield, MA)

TABLE A.5.   Stock Sizes of Drill Rod—Diameters (Centerless Ground).

.039	.075	.106	.148	.182	.2344	.4219	.8125
.040	.077	.108	.151	.185	.250	.4375	.875
.041	.079	.110	.153	.188	.2656	.4531	1.000
.042	.081	.112	.155	.191	.2813.	.4688	
.045	.085	.115	.157	.194	.2969	.4844	
.050	.088	.120	.161	.197	.3125	.500	
.055	.092	.125	.164	.199	.3281	.5313	
.058	.095	.127	.168	.201	.3438	.5625	
.062	.097	.134	.172	.204	.3594	.5938	
.066	.099	.139	.175	.212	.375	.625	
.069	.101	.143	.178	.219	.3906	.6875	
.072	.103	.146	.180	.227	.4063	.750	

Note: Cold-drawn carbon steel drill rod is available in flat, square, round, hexagonal and octagonal shapes in sizes up to 1 inch.

TABLE A.6.   Mold Pin Sizes for Holding Inserts.

Thread Size	Dia. Screw	Tap Drill Size after Tapping	Per Cent of Full Thread	Dia. Pin to Hold Insert Down	Dia. Pin to Hold Insert Up (Taper)	Pin Size for Tap after Molding
0–80	.060	.0492	67	.047	.048 to .053	.0512
1–72	.073	.0610	67	.059	.059 to .064	.0630
1–64	.073	.0591	68	.057	.057 to .062	.0610
2–64	.086	.0730	64	.071	.071 to .076	.0760
2–56	.086	.0700	69	.068	.068 to .073	.0730
3–56	.099	.0827	70	.081	.081 to .086	.0860
3–48	.099	.0810	66	.079	.079 to .084	.0827
4–48	.112	.0937	68	.092	.092 to .097	.0960
4–40	.112	.0890	71	.087	.087 to .092	.0937
4–36	.112	.0860	72	.084	.084 to .089	.0906
5–44	.125	.1040	71	.102	.102 to .107	.1094
5–40	.125	.1024	70	.101	.101 to .106	.1065
5–36	.125	.0995	70	.098	.098 to .103	.1024
6–40	.138	.1160	68	.114	.114 to .119	.1200
6–36	.138	.1130	70	.111	.111 to .116	.1160
6–32	.138	.1094	70	.108	.108 to .113	.1130
8–36	.164	.1378	73	.136	.136 to .141	.1440
8–32	.164	.1360	69	.134	.134 to .139	.1406
8–30	.164	.1339	70	.132	.133 to .138	.1378
10–32	.190	.1610	71	.159	.159 to .164	.1660
10–30	.190	.1610	67	.159	.160 to .165	.1660
10–24	.190	.1520	70	.150	.150 to .155	.1562
12–28	.216	.1850	67	.183	.183 to .188	.1875
12–24	.216	.1800	67	.178	.178 to .183	.1800
14–24	.242	.2031	72	.201	.201 to .206	.2090
¼–20	.2500	.2031	72	.201	.201 to .206	.2031
⁵⁄₁₆–18	.3125	.2610	71	.259	.259 to .264	.2656
⅜–16	.3750	.3160	73	.314	.314 to .319	.3160
⁷⁄₁₆–14	.4375	.3750	67	.370	.373 to .378	.3750
½–13	.5000	.4219	78	.420	.420 to .425	.4219

TABLE A.7. Standard Draft Angles.

Depth	¼°	½°	1°	1½°	2°	2½°	3°	5°	7°	8°	10°	12°	15°	Depth
1/32	.0001	.0003	.0005	.0008	.0011	.0014	.0016	.0027	.0038	.0044	.0055	.0066	.0084	1/32
1/16	.0003	.0006	.0011	.0016	.0022	.0027	.0033	.0055	.0077	.0088	.0110	.0133	.0168	1/16
3/32	.0004	.0008	.0016	.0025	.0033	.0041	.0049	.0082	.0115	.0132	.0165	.0199	.0251	3/32
1/8	.0005	.0010	.0022	.0033	.0044	.0055	.0066	.0109	.0153	.0176	.0220	.0266	.0335	1/8
3/16	.0008	.0016	.0033	.0049	.0065	.0082	.0098	.0164	.0230	.0263	.0331	.0399	.0502	3/16
1/4	.0011	.0022	.0044	.0066	.0087	.0109	.0131	.0219	.0307	.0351	.0441	.0531	.0670	1/4
5/16	.0014	.0027	.0055	.0082	.0109	.0137	.0164	.0273	.0384	.0439	.0551	.0664	.0837	5/16
3/8	.0016	.0033	.0065	.0098	.0131	.0164	.0197	.0328	.0460	.0527	.0661	.0797	.1005	3/8
7/16	.0019	.0038	.0076	.0115	.0153	.0191	.0229	.0383	.0537	.0615	.0771	.0930	.1172	7/16
1/2	.0022	.0044	.0087	.0131	.0175	.0218	.0262	.0438	.0614	.0703	.0882	.1063	.1340	1/2
5/8	.0027	.0054	.0109	.0164	.0218	.0273	.0328	.0547	.0767	.0878	.1102	.1329	.1675	5/8
3/4	.0033	.0065	.0131	.0196	.0262	.0328	.0393	.0656	.0921	.1054	.1322	.1595	.2010	3/4
7/8	.0038	.0076	.0153	.0229	.0306	.0382	.0459	.0766	.1074	.1230	.1543	.1860	.2345	7/8
1	.0044	.0087	.0175	.0262	.0349	.0437	.0524	.0875	.1228	.1405	.1763	.2126	.2680	1
1¼	.0055	.0109	.0218	.0327	.0437	.0546	.0655	.1094	.1535	.1756	.2204	.2657	.3349	1¼
1½	.0064	.0131	.0262	.0393	.0524	.0655	.0786	.1312	.1842	.2108	.2645	.3188	.4019	1½
1¾	.0076	.0153	.0305	.0458	.0611	.0764	.0917	.1531	.2149	.2460	.3085	.3720	.4689	1¾
2	.0087	.0175	.0349	.0524	.0698	.0873	.1048	.1750	.2456	.2810	.3527	.4251	.5359	2

TABLE A.8. General Sizes and Considerations for Compression Molds (Min.).

Tons @ 2500 #/sq. in.	8	15	24	41	62	98	141	221	284	393
Size press (Ram dia.)	3	4	5	6½	8	10	12	15	17	20
Top and bottom plate for all molds	11/16	11/16	13/16	13/16	13/16	13/16	15/16	15/16	15/16	1 3/16
Number of hold-down slots in top and bottom plates	2	2	2	2	6	6	6	6 or 10	6 or 10	6 or 10
Size of socket-head screw to hold plates (minimum)	(4) 3/8–16	(6) 3/8–16	(6) 3/8–16	(6) 3/8–16	(6) ½–13	(6) ½–13	(6) ½–13	(6) 5/8–11	(10) 5/8–11	(10) 5/8–11
Number and diameter of hold-up bolts for transfer or loading shoe	(2) ½–13	(2) ½–13	(2) ½–13	(2) ½–13	(4) ½–13 or (2) 5/8–11	(4) ½–13 or (2) 5/8–11	(4) ½–13 or (2) 5/8–11	(4) 5/8–11	(4) 5/8–11	(4) 5/8–11
Size diameter guide pins	5/8 & ¾	5/8 & ¾	¾	7/8	1	1	1	1¼	1¼	1¼
Thickness of knockout bar	1 7/16	1 7/16	1 11/16	1 11/16	1 11/16	1 15/16	1 15/16	2 3/16	2 7/16	2 15/16
Length of knockout bar	14	14	18½	18½	25	25	25 or 29½	29½ or 33½	29½ or 33½	38½
Diameter of safety pins	½	½	½ or 5/8	½ or 5/8	5/8 or ¾	5/8 or ¾	¾	7/8	1	1
Minimum thickness of parallels	11/16	11/16	15/16	15/16	15/16	1 3/16	1 3/16	1 7/16	1 7/16	1 11/16
Length of loading shoe	14	14	14 or 18½	14 or 18½	18½ or 25	18½, 25 or 29½	25, 29½ or 33½	29½ or 33½	29½ or 33½	33½ or 38½
Thickness of retainer	15/16 or 1 3/16	15/16 or 1 3/16	1 3/16	1 3/16	1 7/16	1 7/16	1 7/16	1 11/16	1 11/16	1 15/16

(Courtesy Pribble Enterprises, Inc., New Haven, IN)

TABLE A.9. Selection of Taps for Class 2 and Class 3 Fits.

Size Inch	Threads per Inch			Class		Size Inch	Threads per Inch			Class	
	NC	NF	NS	2	3		NC	NF	NS	2	3
¼	20	..	..	CUT	CG	0	..	80	..	PG 1	PG 1
	..	28	..	CUT	PG 2	1	64	..	..	PG 1	PG 1
⁵⁄₁₆	18	..	..	CUT	CG		..	72	..	PG 1	PG 1
	..	24	..	CUT	PG 2		..	..	56	PG 1	PG 1
⅜	16	..	..	CUT	CG	2	56	..	..	PG 1	PG 1
	..	24	..	CUT	PG 2		..	64	..	PG 1	PG 1
⁷⁄₁₆	14	..	..	CUT	CG	3	48	..	..	CG	PG 1
	..	20	..	CUT	CG		..	56	..	CG	PG 1
½	13	..	..	CUT	CG	4	40	..	..	CG	PG 1
	..	20	..	CUT	CG		..	48	..	CG	PG 1
⁹⁄₁₆	12	..	..	CUT	CG		..	..	36	CG	PG 1
	..	18	..	CUT	CG	5	40	..	..	CG	PG 1
⅝	11	..	..	CUT	CG		..	44	..	CG	PG 1
	..	18	..	CUT	CG	6	32	..	..	CG	PG 1
11⁄₁₆	..	..	11	CUT	CG		..	40	..	CG	PG 1
	..	..	16	CUT	CG	8	32	..	..	CG	PG 1
¾	10	..	..	CUT	CG		..	36	..	CUT	PG 1
	..	16	..	CUT	CG	10	24	..	..	CUT	PG 1
⅞	9	..	..	CUT	CG		..	32	..	CUT	PG 1
	..	14	..	CG	CG	12	24	..	..	CUT	PG 1
	..	..	18	CG	CG		..	28	..	CUT	PG 1
1	8	..	..	CUT	CG	14	..	..	20	CUT	PG 1
	..	14	..	CG	CG		..	..	24	CUT	PG 1
1⅛	7	..	..	CUT	CG						
	..	12	..	CG	CG						
1¼	7	..	..	CUT	CG						
	..	12	..	CG	CG						
1⅜	6	..	..	CUT	CG						
	..	12	..	CG	CG						
1½	6	..	..	CUT	CG						
	..	12	..	CG	CG						

SYMBOLS

CUT —Cut thread taps, either in carbon or high-speed steel.
CG —Commercial-ground thread taps in high-speed steel.
PG —Precision-ground thread taps.
PG 01—Basic pitch diameter to basic minus .0005″.
PG 1—Basic pitch diameter to basic plus .0005″.
PG 2—Basic pitch diam. plus .0005″ to basic plus .001″.

NOTES

Taps specified in this table should normally produce the National Screw Thread Commission fits indicated in average material if used with reasonable care. The problem is one of selection. If specified tap does not give satisfactory gauge-fit in the work, a choice of some other grade of tap should be made. Grades may be considered as cut-thread, commercial-ground and precision-ground, respectively, the latter being divided into three optional groups according to pitch diameter limits.

Above table does not apply to bent shank tapper taps.

(Courtesy Greenfield Tap & Die Div., TRW, Inc., Greenfield, MA)

TABLE A.10.   Standard Countersinking Practice.

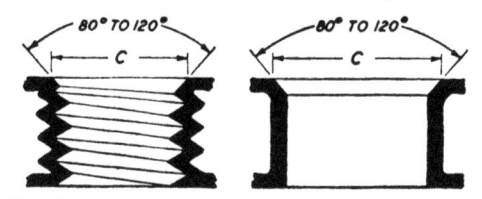

Bolt or Screw Size	Max. Dia. of Countersink—C	
	To Remove Burrs from Tapped Holes	To Break Corners of Other Holes
#0 (.060)	1/16 (.062)	
#1 (.073)	5/64 (.078)	
#2 (.086)	3/32 (.094)	
#3 (.099)	7/64 (.109)	
#4 (.112)	1/8 (.125)	
#5 (.125)	9/64 (.141)	
#6 (.138)	5/32 (.156)	
#8 (.164)	3/16 (.188)	
#10 (.190)	7/32 (.219)	
#12 (.216)	1/4 (.250)	
1/4	9/32 (.281)	5/16 (.312)
5/16	11/32 (.344)	3/8 (.375)
3/8	13/32 (.406)	7/16 (.438)
7/16	15/32 (.469)	1/2 (.500)
1/2	17/32 (.531)	9/16 (.562)
9/16	19/32 (.594)	11/16 (.688)
5/8	21/32 (.656)	3/4 (.750)
3/4	25/32 (.781)	7/8 (.875)
7/8	29/32 (.906)	1 (1.000)
1	1 1/32 (1.031)	1 1/8 (1.125)
1 1/8	1 5/32 (1.156)	1 3/8 (1.375)
1 1/4	1 9/32 (1.281)	1 1/2 (1.500)
1 3/8	1 13/32 (1.406)	1 5/8 (1.625)
1 1/2	1 17/32 (1.531)	1 3/4 (1.750)
1 3/4	1 25/32 (1.781)	2 (2.000)
2	2 1/32 (2.031)	2 1/4 (2.250)
2 1/4	2 9/32 (2.281)	2 5/8 (2.625)
2 1/2	2 17/32 (2.531)	2 7/8 (2.875)
2 3/4	2 25/32 (2.781)	3 1/8 (3.125)
3	3 1/32 (3.031)	3 3/8 (3.375)

TABLE A.11. Recommended Sizes of Holes for Production of Screw Threads with Normal Length of Engagement.

Bolt or Screw Size	Body Drill Size (Single holes only)	AMERICAN STANDARD SCREW THREADS					
		Coarse			Fine		
		Thd. per In.	Copper Aluminum Steel Die Casting	Brass Cast Iron Mal. Iron Hard Rubber Plastics	Thd. per In.	Copper Aluminum Steel Die Casting	Brass Cast Iron Mal. Iron Hard Rubber Plastics
#0	.0650				80	.0492	.0492
1	.0760	64	.0591	.0591	72	.0610	.0591
2	.0890	56	.0700	.0700	64	.0730	.0700
3	.1040	48	.0810	.0781	56	.0827	.0810
4	.1160	40	.0890	.0860	48	.0937	.0906
5	.1285	40	.0995	.0995	44	.1065	.1040
6	.1440	32	.1094	.1065	40	.1160	.1130
8	.1695	32	.1360	.1360	36	.1378	.1360
10	.1960	24	.1520	.1470	32	.1610	.1610
12	.2280	24	.1800	.1730	28	.1800	.1800
$\frac{1}{4}$	.2656	20	.2031	.1990	28	.2130	.2130
$\frac{5}{16}$	.3281	18	.2610	.2570	24	.2720	.2720
$\frac{3}{8}$	.4062	16	.3160	.3160	24	.3320	.3320
$\frac{7}{16}$	.4687	14	.3680	.3680	20	.3906	.3860
$\frac{1}{2}$	$\frac{17}{32}$	13	.4219	.4219	20	.4531	.4531
$\frac{9}{16}$	$\frac{19}{32}$	12	.4844	.4844	18	$\frac{33}{64}$	$\frac{1}{2}$
$\frac{5}{8}$	$\frac{11}{16}$	11	$\frac{17}{32}$	$\frac{17}{32}$	18	$\frac{37}{64}$	$\frac{9}{16}$
$\frac{3}{4}$	$\frac{13}{16}$	10	$\frac{21}{32}$	$\frac{21}{32}$	16	$\frac{11}{16}$	$\frac{11}{16}$
$\frac{7}{8}$	$\frac{15}{16}$	9	$\frac{49}{64}$	$\frac{49}{64}$	14	$\frac{13}{16}$	$\frac{51}{64}$
1	$1\frac{1}{16}$	8	$\frac{7}{8}$	$\frac{7}{8}$	14	$\frac{15}{16}$	$\frac{59}{64}$
$1\frac{1}{8}$	$1\frac{3}{16}$	7	$\frac{63}{64}$	$\frac{63}{64}$	12	$1\frac{3}{64}$	$1\frac{3}{64}$
$1\frac{1}{4}$	$1\frac{5}{16}$	7	$1\frac{7}{64}$	$1\frac{7}{64}$	12	$1\frac{11}{64}$	$1\frac{11}{64}$
$1\frac{3}{8}$	$1\frac{7}{16}$	6	$1\frac{13}{64}$	$1\frac{13}{64}$	12	$1\frac{19}{64}$	$1\frac{19}{64}$
$1\frac{1}{2}$	$1\frac{9}{16}$	6	$1\frac{21}{64}$	$1\frac{21}{64}$	12	$1\frac{27}{64}$	$1\frac{27}{64}$

TABLE A.12. Tons Pressure on Ram of Given Diameter and Given Line Pressure.

Ram Dia.	PRESSURE IN POUNDS PER SQUARE INCH																Gals. per In. Stroke	Gals. per Ft. Stroke	Area	Ram Dia.
	300	500	600	750	1000	1200	1500	1800	2000	2250	2500	2750	3000	3250	3500	4000				
2	.5	.8	.9	1.2	1.6	1.9	2.4	2.8	3.1	3.5	3.9	4.3	4.7	5.1	5.5	6.3	.014	.163	3.14	2
3	1.1	1.8	2.1	2.7	3.5	4.2	5.3	6.4	7.1	7.9	8.8	9.7	10.6	11.5	12.4	14.1	.031	.366	7.07	3
4	1.9	3.1	3.8	4.7	6.3	7.6	9.4	11.3	12.6	14.1	15.7	17.3	18.8	20.4	21.9	25.1	.054	.652	12.57	4
5	2.9	4.9	5.9	7.4	9.8	11.8	14.7	17.7	19.6	22.1	24.5	27	29.4	31.9	34.4	39.3	.085	1.02	19.63	5
6	4.2	7.1	8.5	10.6	14.1	17	21.2	25.4	28.3	31.8	35.3	38.9	42.4	45.9	49.5	56.5	.122	1.47	28.27	6
7	5.8	9.6	11.5	14.4	19	23	29	35	38	43	48	53	58	63	67	77	.167	2.00	38.49	7
8	7.5	12.6	15.1	18.8	25	30	38	45	50	57	63	69	75	82	88	101	.217	2.61	50.26	8
9	9.5	15.9	19.1	23.9	32	38	48	57	64	72	79	87	95	103	111	127	.275	3.30	63.62	9
10	11.8	19.6	23.6	29.5	39	47	59	71	79	88	98	108	118	128	137	157	.340	4.08	78.54	10
11	14.3	23.8	28.5	35.6	48	57	71	86	95	107	119	131	143	154	166	190	.411	4.93	95.03	11
12	17	28	34	42	57	68	85	102	113	127	141	156	170	184	198	226	.489	5.87	113	12
13	20	33	40	50	66	80	100	119	133	149	166	183	199	216	232	265	.576	6.90	133	13
14	23	38	46	58	77	92	115	139	154	173	192	212	231	250	269	308	.667	8.00	154	14
15	27	44	53	66	88	106	133	159	177	199	221	243	265	287	309	353	.766	9.20	177	15
16	30	50	60	75	101	121	151	181	201	226	251	276	301	327	352	402	.870	10.4	201	16
17	34	57	68	85	113	136	170	204	227	255	284	312	340	368	397	454	.98	11.8	227	17
18	38	64	76	95	127	153	191	229	254	286	318	350	382	414	445	509	1.10	13.2	254	18
19	43	71	85	106	142	170	213	255	284	319	354	390	425	461	496	567	1.23	14.7	284	19
20	47	79	94	118	157	188	236	283	314	353	393	432	471	511	550	628	1.36	16.3	314	20
21	52	87	104	130	173	208	260	312	346	390	433	476	519	563	607	693	1.50	18.0	346	21
22	57	95	114	143	190	228	285	342	380	428	475	523	570	618	665	760	1.65	19.7	380	22
23	62	104	125	156	208	249	312	374	415	467	519	571	623	675	727	831	1.80	21.6	415	23
24	68	113	136	170	226	271	340	407	452	509	565	622	678	735	792	905	1.96	23.5	452	24
25	74	123	147	184	245	295	368	442	491	552	614	675	736	798	859	982	2.12	25.4	491	25
26	80	133	159	199	265	319	398	478	531	597	664	730	796	863	929	1062	2.30	27.6	531	26
27	86	143	172	215	286	344	429	515	573	644	716	787	859	930	1002	1145	2.48	29.8	573	27
28	92	154	185	231	308	369	462	554	616	693	770	847	924	1001	1078	1232	2.67	32.0	616	28
29	99	165	198	248	330	396	495	594	661	743	826	908	991	1073	1156	1321	2.86	34.3	661	29
30	106	177	212	265	353	424	530	636	707	795	884	972	1060	1149	1237	1414	3.06	36.7	707	30
31	113	189	226	283	377	453	566	679	755	849	943	1038	1132	1227	1321	1510	3.27	39.2	755	31
32	121	201	241	302	402	483	603	724	804	905	1005	1106	1206	1307	1407	1608	3.48	41.8	804	32
33	128	214	257	321	428	513	641	770	855	962	1069	1176	1283	1390	1497	1711	3.70	44.4	855	33
34	136	227	272	340	454	545	681	817	908	1021	1135	1248	1362	1475	1589	1816	3.93	47.2	908	34

(Courtesy of The French Oil Mill Machinery Co., Piqua, OH)

TABLE A.13. Temperature of Saturated Steam.

Gauge Pressure Lb./Sq. In.	TEMPERATURE		Gauge Pressure Lb./Sq. In.	TEMPERATURE	
	Degree F.	Degree C.		Degree F.	Degree C.
0	212.0	100	90	331.0	166.1
5	227.1	108.4	100	337.8	169.9
10	239.4	115.2	110	344.0	173.3
15	249.7	121.0	120	349.9	176.6
20	258.7	126.0	130	355.5	179.4
25	266.7	130.4	140	360.8	182.7
30	273.9	134.4	150	365.8	185.5
40	286.6	141.5	160	370.6	188.1
50	297.5	147.5	170	375.2	190.6
60	307.2	152.9	180	379.6	193.1
70	315.9	157.7	190	383.8	195.4
80	323.8	162.1	200	387.9	197.7

TABLE A.14.   Weight of 1000 Pieces in Pounds Based on Weight of One Piece in Grams.

454 Grams = 1 Pound

Weight per Piece in Grams	Weight per 1000 Pieces in Pounds	Weight per Piece in Grams	Weight per 1000 Pieces in Pounds	Weight per Piece in Grams	Weight per 1000 Pieces in Pounds	Weight per Piece in Grams	Weight per 1000 Pieces in Pounds
1	2.2	26	57.2	51	112.3	76	167.4
2	4.4	27	59.4	52	114.5	77	169.6
3	6.6	28	61.6	53	116.7	78	171.8
4	8.8	29	63.8	54	118.9	79	174.0
5	11.0	30	66.0	55	121.1	80	176.2
6	13.2	31	68.2	56	123.3	81	178.4
7	15.4	32	70.4	57	125.5	82	180.6
8	17.6	33	72.6	58	127.7	83	182.8
9	19.8	34	74.8	59	129.9	84	185.0
10	22.0	35	77.0	60	132.1	85	187.2
11	24.2	36	79.2	61	134.3	86	189.4
12	26.4	37	81.4	62	136.5	87	191.6
13	28.6	38	83.7	63	138.7	88	193.8
14	30.8	39	85.9	64	140.9	89	196.0
15	33.0	40	88.1	65	143.1	90	198.2
16	35.2	41	90.3	66	145.3	91	200.4
17	37.4	42	92.5	67	147.5	92	202.6
18	39.6	43	94.7	68	149.7	93	204.8
19	41.8	44	96.9	69	151.9	94	207.0
20	44.0	45	99.1	70	154.1	95	209.2
21	46.2	46	101.3	71	156.3	96	211.4
22	48.4	47	103.5	72	158.5	97	213.6
23	50.6	48	105.7	73	160.7	98	215.8
24	52.8	49	107.9	74	162.9	99	218.0
25	55.0	50	110.1	75	165.1	100	220.2

Table of Weights (Avoirdupois)
1 gram = .0353 oz.
.0625 pounds = 1 ounce = 28.3 grams
454 grams = 1 pound

(Courtesy Durez Resins & Molding Materials, North Tonawanda, NY)

TABLE A.15. Specific Gravity into Ounces and Grams.

Specific Gravity	Ounces per Cu. Inch	Grams per Cu. Inch	Specific Gravity	Ounces per Cu. Inch	Grams per Cu. Inch
0.90	0.520	14.715	1.36	0.786	22.236
0.91	0.525	14.879	1.37	0.792	22.400
0.92	0.532	15.042	1.38	0.797	22.563
0.93	0.537	15.206	1.39	0.803	22.727
0.94	0.543	15.370	1.40	0.809	22.890
0.95	0.549	15.533	1.41	0.815	23.054
0.96	0.555	15.697	1.42	0.820	23.217
0.97	0.560	15.860	1.43	0.826	23.381
0.98	0.566	16.024	1.44	0.832	23.544
0.99	0.572	16.187	1.45	0.838	23.708
1.00	0.578	16.351	1.46	0.844	23.871
1.01	0.583	16.514	1.47	0.849	24.035
1.02	0.590	16.678	1.48	0.855	24.198
1.03	0.595	16.841	1.49	0.861	24.362
1.04	0.601	17.005	1.50	0.867	24.525
1.05	0.606	17.168	1.51	0.872	24.689
1.06	0.612	17.332	1.52	0.878	24.852
1.07	0.618	17.495	1.53	0.884	25.016
1.08	0.624	17.659	1.54	0.890	25.179
1.09	0.629	17.822	1.55	0.896	25.343
1.10	0.635	17.986	1.56	0.901	25.506
1.11	0.641	18.149	1.57	0.907	25.670
1.12	0.647	18.313	1.58	0.913	25.833
1.13	0.653	18.476	1.59	0.919	25.997
1.14	0.659	18.640	1.60	0.924	26.160
1.15	0.664	18.803	1.61	0.930	26.324
1.16	0.670	18.967	1.62	0.936	26.487
1.17	0.676	19.130	1.63	0.942	26.651
1.18	0.682	19.294	1.64	0.948	26.814
1.19	0.688	19.457	1.65	0.953	26.978
1.20	0.693	19.621	1.66	0.959	27.141
1.21	0.699	19.784	1.67	0.965	27.305
1.22	0.705	19.948	1.68	0.971	27.468
1.23	0.711	20.111	1.69	0.976	27.632
1.24	0.716	20.275	1.70	0.982	27.795
1.25	0.722	20.438	1.71	0.988	27.959
1.26	0.728	20.601	1.72	0.994	28.122
1.27	0.734	20.765	1.73	1.000	28.286
1.28	0.740	20.928	1.74	1.005	28.449
1.29	0.745	21.092	1.75	1.010	28.613
1.30	0.751	21.255	1.76	1.017	28.776
1.31	0.757	21.419	1.77	1.023	28.940
1.32	0.763	21.582	1.78	1.028	29.103
1.33	0.768	21.746	1.79	1.034	29.267
1.34	0.774	21.909	1.80	1.040	29.430
1.35	0.780	22.073	1.81	1.046	29.594

(Continued)

TABLE A.15. *(Continued)*

Specific Gravity	Ounces per Cu. Inch	Grams per Cu. Inch	Specific Gravity	Ounces per Cu. Inch	Grams per Cu. Inch
1.82	1.052	29.757	2.09	1.208	34.172
1.83	1.057	29.921	2.10	1.213	34.335
1.84	1.063	30.084	2.11	1.219	34.499
1.85	1.069	30.248	2.12	1.225	34.662
1.86	1.075	30.411	2.13	1.231	34.826
1.87	1.080	30.575	2.14	1.236	34.989
1.88	1.086	30.738	2.15	1.242	35.153
1.89	1.092	30.902	2.16	1.248	35.316
1.90	1.098	31.065	2.17	1.254	35.480
1.91	1.104	31.229	2.18	1.260	35.643
1.92	1.109	31.392	2.19	1.265	35.807
1.93	1.115	31.556	2.20	1.271	35.970
1.94	1.121	31.790	2.21	1.277	36.134
1.95	1.127	31.883	2.22	1.283	36.297
1.96	1.132	32.046	2.23	1.288	36.461
1.97	1.138	32.210	2.24	1.294	36.624
1.98	1.144	32.373	2.25	1.300	36.788
1.99	1.150	32.537	2.26	1.306	36.951
2.00	1.156	32.700	2.27	1.312	37.115
2.01	1.161	32.864	2.28	1.317	37.278
2.02	1.167	33.027	2.29	1.323	37.442
2.03	1.173	33.191	2.30	1.329	37.605
2.04	1.179	33.354	2.31	1.335	37.769
2.05	1.184	33.518	2.32	1.340	37.932
2.06	1.190	33.681	2.33	1.346	38.096
2.07	1.196	33.844	2.34	1.352	38.259
2.08	1.202	34.008	2.35	1.358	38.423

Factor used in converting to ounces per cubic inch = Specific gravity multiplied by 0.5778.
Factor used in converting to grams per cubic inch = Specific gravity multiplied by 16.35.
To compute:
Specific gravity—multiply the pounds per cubic foot by .01604.
Pounds per cubic foot—multiply specific gravity by 62.4.
Pounds per cubic inch—multiply specific gravity by .0361.

(Courtesy Pribble Enterprises, Inc., New Haven, IN)

TABLE A.16. Surface and Volume of Solids.

	Given	*Sought*

Prism

(RIGHT OR OBLIQUE, REGULAR OR IRREGULAR, PARALLELOPIPED)

Perimeter, P, perpendicular to sides; lateral length, L, — Lateral Surface $= PL$

Area of base, B; perpendicular height, h, — Volume $= Bh$

Area of section perpendicular to sides, A; lateral length, L, — Volume $= AL$

Cylinder

(RIGHT OR OBLIQUE, CIRCULAR OR ELLIPTIC)

Perimeter of base, P_b; perpendicular height, h, — Lateral Surface $= P_b h$

Perimeter, P, perpendicular to sides; lateral length, L, — Lateral Surface $= PL$

Area of base, B; perpendicular height, h, — Volume $= Bh$

Area of section perpendicular to sides, A; lateral length, L, — Volume $= AL$

Frustum of Any Prism or Cylinder

Area of base, B; perpendicular distance from base to center of gravity of opposite face, h, — Volume $= Bh$

Frustum of Cylinder

Area of section perpendicular to sides, A; maximum lateral length, L_1, and minimum, L_2, — Volume $= \frac{1}{2}A(L_1 + L_2)$

Pyramid or Cone

(RIGHT AND REGULAR)

Perimeter of base, P_b; slant height, L, — Lateral Surface $= \frac{1}{2}P_b L$

Area of base, B; perpendicular height, h, — Volume $= \frac{1}{3}Bh$

Pyramid or Cone

(RIGHT OR OBLIQUE, REGULAR OR IRREGULAR)

Area of base, B; perpendicular height, h, — Volume $= \frac{1}{3}Bh$
$= \frac{1}{3}$ the volume of prism or cylinder of same base and perpendicular height or $\frac{1}{2}$ the volume of hemisphere of same base and perpendicular height.

(Continued)

TABLE A.16.   (*Continued*).

	Given	*Sought*

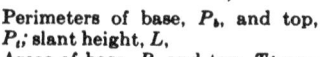

Frustum of Pyramid or Cone

(RIGHT AND REGULAR
PARALLEL ENDS)

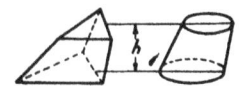

Perimeters of base, P_b, and top, P_t; slant height, L, Areas of base, B, and top, T; perpendicular height, h,

Lateral Surface
$=\tfrac{1}{2}L(P_b+P_t)$
Volume =
$\tfrac{1}{3}h(B+T+\sqrt{BT})$

Frustum of Any Pyramid or Cone

(PARALLEL ENDS)

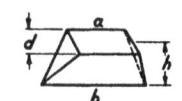

Areas of base, B, and top, T; perpendicular height, h,

Volume =
$\tfrac{1}{3}h(B+T+\sqrt{BT})$

Wedge

(PARALLELOGRAM FACE)

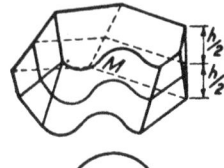

Length of edges, a and b; perpendicular height, h; perpendicular width, d,

Volume =
$\tfrac{1}{6}dh(2a+b)$

Prismatoid

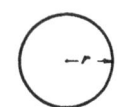

Areas of base, B, top, T, and of a section, M, parallel to and midway between base and top; perpendicular height, h,*

Volume =
$\tfrac{1}{6}h(B+T+4M)$

Sphere

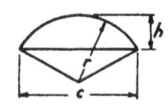

Radius, r,

Area $=4\pi r^2$
Volume $=\,{}^4/{}_3\pi r^3$

Spherical Sector

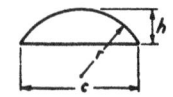

Radius, r; length of chord, c; height, h,

Area $=\dfrac{\pi r}{2}(4h+c)$
Volume $=\tfrac{2}{3}\pi r^2h$

Spherical Segment

Radius, r; length of chord, c; height, h,

Curved Surface =
$2\pi rh=\dfrac{\pi}{4}(4h^2+c^2)$

*This formula also applies to any of the foregoing solids with parallel bases, to pyramids, cones, spherical sections and to many solids with irregular surfaces.

Volume $=\dfrac{\pi}{3}h^2(3r-h)=\dfrac{\pi}{24}h(3c^2+4h^2)$

(*Continued*)

TABLE A.16. *(Continued)*

	Given	*Sought*

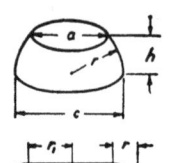

Spherical Zone

Radius, r; height, h; diameters, a and c,

Curved Surface $=2\pi rh$

Volume $=\dfrac{\pi h(3a^2+3c^2}{24}\quad +4h^2)$

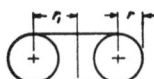

Ring of Circular Cross Section, Torus

Radius of ring, r, and of cross section, r_1,

Area $=4\pi^2 r\,r_1$

Volume $=2\pi^2 r^2 r_1$

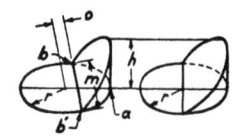

Ungula of Right, Regular Cylinder

Base = segment, bab' (smaller than semi-circle),

Radius of cylinder, r; height, h; length of chord, m; distance from chord to axis of cylinder, o,

Convex Surface $=$
$(rm-o\cdot\text{arc }bab')\dfrac{h}{r-o}$

Volume $=$
$(\dfrac{m^3}{12}-o\cdot\text{area }bab')\dfrac{h}{r-o}$

Base = semi-circle,

Radius of cylinder, r; height, h,

Convex Surface $=2rh$

Volume $=\frac{2}{3}r^2 h$

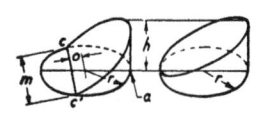

Base = segment, cac' (larger than semi-circle),

Radius of cylinder, r; height, h; length of chord, m; distance from chord to axis of cylinder, o,

Convex Surface $=$
$(rm+o\cdot\text{arc }cac')\dfrac{h}{r+o}$

Volume $=$
$(\dfrac{m^3}{12}+o\cdot\text{arc }cac')\dfrac{h}{r+o}$

Base = circle,

Radius of cylinder, r; height, h,

Convex Surface $=r\pi h$

Volume $=\frac{1}{2}r^2\pi h$

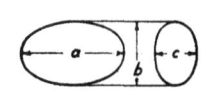

Ellipsoid

Axes, a, b and c,

Volume $=\frac{1}{6}\pi abc$

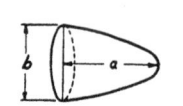

Paraboloid

Axes, a and b,

Volume $=\frac{1}{8}\pi ab^2$

The ratio of corresponding volumes of cone, paraboloid, sphere and cylinder of equal height is $\frac{1}{3}:\frac{1}{2}:\frac{2}{3}:1$

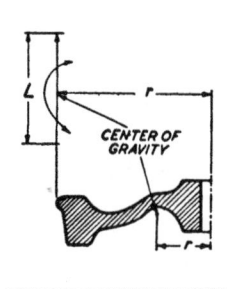

CENTER OF GRAVITY

Surface and Solids of Revolution

When a plane curve or straight line revolves around an axis of revolution in the same plane, a surface of revolution is produced. A solid of revolution results from the rotation of an area around an axis in its plane. Length of curve or straight line, L; normal distance from center of gravity to axis, r; angle of revolution, $\alpha°$; area of revolving plane, A,

Length of arc described by center of gravity
$=2\pi r\dfrac{\alpha}{360}$

Surface of revolution
$=2\pi rL\dfrac{\alpha}{360}$

Volume of solid
$=2\pi rA\dfrac{\alpha}{360}$

(From "New Departure Handbook," New Departure, Bristol, CT)

TABLE A.17. Injection Molding Cycle-Time Production Chart.

Cycle Time Sec	Min	Cycles per Hour	2	3	4	6	8	10	12	16	20	24	28	30	32
5		720	1,440	2,160	2,880	4,320	5,760	7,200	8,640	11,520	14,400	17,280	20,160	21,600	23,040
10		360	720	1,080	1,440	2,160	2,880	3,600	4,320	5,760	7,200	8,640	10,080	10,800	11,520
15	¼	240	480	720	960	1,440	1,920	2,400	2,880	3,840	4,800	5,760	6,720	7,200	7,680
20		180	360	540	720	1,080	1,440	1,800	2,160	2,880	3,600	4,320	5,040	5,400	5,760
25		144	288	432	576	864	1,152	1,440	1,728	2,304	2,880	3,456	4,030	4,320	4,608
30	½	120	240	360	480	720	960	1,200	1,440	1,920	2,400	2,880	3,360	3,600	3,840
35		102.8	205.6	308.4	411.2	616.8	822.4	1,028	1,232.6	1,644.8	2,056	2,467.2	2,878.4	3,084	3,289.6
40		90	180	270	360	540	720	900	1,080	1,440	1,800	2,160	2,520	2,700	2,880
45	¾	80	160	240	320	480	640	800	960	1,280	1,600	1,920	2,240	2,400	2,560
50		72	144	216	288	432	576	720	864	1,152	1,440	1,728	2,016	2,160	2,304
55		65.5	131	196.5	262	393	524	655	786	1,048	1,310	1,572	1,834	1,965	2,096
60	1	60	120	180	240	360	480	600	720	960	1,200	1,440	1,680	1,800	1,920
65		55.4	110.8	166.2	221.6	332.4	443.2	554	664.8	886.4	1,108	1,329.6	1,551.2	1,662	1,772.8
70		51.4	102.8	154.2	205.6	308.4	411.2	514	616.8	822.4	1,028	1,233.6	1,439.2	1,532	1,644.8
75	1¼	48	96	144	192	288	384	480	576	768	960	1,152	1,344	1,440	1,536
80		45	90	135	180	270	360	450	540	720	900	1,080	1,260	1,350	1,440
85		42.3	84.6	126.9	169.2	253.8	338.4	423	507.6	676.8	846	1,015.2	1,184.4	1,269	1,353.6
90	1½	40	80	120	160	240	320	400	480	640	800	960	1,120	1,200	1,280
95		37.9	75.8	113.7	151.6	227.4	303.2	379	454.8	606.4	758	909.6	1,061.2	1,137	1,212.8
100		36	72	108	144	216	288	360	432	576	720	864	1,008	1,080	1,152
105	1¾	34.3	68.6	102.9	137.2	205.8	274.4	343	411.6	548.8	686	823.2	960.4	1,029	1,097.6
110		32.7	65.4	98.1	130.8	196.2	261.6	327	392.4	523.2	654	784.8	915.6	981	1,046.4
115		31.3	62.6	93.9	125.2	187.8	250.4	313	375.6	500.8	626	751.2	876.4	939	1,001.6
120	2	30	60	90	120	180	240	300	360	480	600	720	840	900	960
125		28.8	57.6	86.4	115.2	172.8	230.4	288	343.6	460.8	576	691.2	806.4	864	921.6
130		27.7	55.4	83.1	110.8	166.2	221.6	277	332.4	443.2	554	664.8	775.6	831	886.4
135	2¼	26.7	53.4	80.1	106.8	160.2	213.6	267	320.4	427.2	534	640.8	747.6	801	854.4
140		25.7	51.4	77.1	102.8	154.2	205.6	257	308.4	411.2	514	616.8	719.6	771	822.4
150	2½	24	48	72	96	144	192	240	288	384	480	576	672	720	768
160		22.5	45	67.5	90	135	180	225	270	360	450	540	630	675	720
165	2¾	21.8	43.6	65.4	87.2	130.8	174.4	218	261.6	348.8	436	523.2	610.4	654	697.6
170		21.2	42.4	63.6	84.8	127.2	169.6	212	254.4	339.2	424	508.8	583.6	636	678.4
175		20.6	41.2	61.8	82.4	123.6	164.8	206	247.2	329.6	412	494.4	576.8	618	659.2
180	3	20	40	60	80	120	160	200	240	320	400	480	560	600	640
195	3¼	18.5	37	55.5	74	111	148	185	222	296	370	444	518	555	592
210	3½	17.1	34.2	51.3	68.4	102.6	136.8	171	205.2	273.6	342	410.4	478.8	513	547.2
240	4	15	30	45	60	90	120	150	180	240	300	360	420	450	480

Number of Cavities (column headings 2–32)

How to use the table: Here's a rapid method for calculating production rate, given cycle time and number of cavities; or establishing required cycle time and number of mold cavities, given a required production rate.

(Courtesy Marbon Chemical Div. Borg-Warner Corp.)

TABLE A.18.   Conversion Table for Conductivity Values.

Values are shown in the following conductivity units:
(a) Btu per hr per sq ft per Deg F per inch
(b) Cal per sec per sq cm per Deg C per cm

Where values are expressed in other units, conversion to the desired units may be calculated by use of the following table:

		Cal/sec/ sq cm/°C/ cm	Watt sq cm/°C/ cm	Btu/ sec/sq ft/ °F/in.	Btu/ hr/sq ft/ °F/in.	Btu/ hr/sq ft/ °F/ft
Cal/sec/sq cm/°C/cm	×	1	4.186	0.8064	2903	241.9
Watt/sq cm/°C/cm	×	0.2389	1	0.1926	693.5	57.79
Btu/sec/sq ft/°F/in.	×	1.24	5.191	1	3600	300
Btu/hr/sq ft/°F/in.	×	0.004134	0.001442	0.0002778	1	0.08333
Btu/hr/sq ft/°F/ft	×	0.0003445	0.01730	0.003333	12	1

Thermal conductivity expressed in any of the units in the left-hand column can be converted into any of the units in the headings of the columns by multiplying (×) by the number which is common to the row and column.
Thermal resistivity is the reciprocal of conductivity.

TABLE A.19.

Material	Thermal Conductivity		Material	Thermal Conductivity	
	Btu per hr per sq ft per °F per in.	Cal per sec per sq cm per °C per cm		Btu per hr per sq ft per °F per in.	Cal per sec per sq cm per °C per cm
Aluminum	1460.	0.5029	Phenolic Wood Flour	2.03	0.0007
Brass (70 Cu : 30 Zn)	770.	0.2653	Phenolic Cotton Fabric	2.03	0.0007
Cellulose Acetate	1.28	0.0004409	Phenolic Mica	3.19	0.0011
Copper	2665.	0.9181	Phenolic Cotton Flock	2.03	0.0007
Iron, cast, gray	350.	0.1206	Polymethyl Methacrylate	1.30	0.00045
Iron, wrought	420.	0.1447	Polystyrene	0.58	0.0002
Lead	241.	0.0830	Acetobutyrate	1.74	0.0006
Mica, pasted. segment	2.80	0.0009656	Phosphoric Acid—Cement	5.80	0.0020
Mica, sheet (clear)	3.85	0.0013263	Phenolic Asbestos Fabric	2.82	0.00097
Mycalex	3.774	0.0013	Melamine—Alpha		
Nickel	405.	0.1395	Cellulose	2.00	0.00069
Plastics—Laminated			Ethyl cellulose	1.45	0.0005
Phenolic—Cloth	2.3	0.00079	Saran	0.58	0.0002
Phenolic—Cloth	2.1	0.00072	Nylon	1.74	0.0006
Phenolic—Cloth	2.3	0.00079	Styraloy	1.25	0.00043
Phenolic—Paper	3.36	0.00116	Rock wool (Loose)	0.27	0.0000930
Phenolic—Glass	1.58	0.00054	Rubber, hard	1.1	0.000379
Phenolic—Glass	1.45	0.00050	Rubber, sponge	0.3 to 0.4	0.0001034 to 0.0001378
Melamine—Glass	2.43	0.00084	Steatite	17.4	0.006
Melamine—Glass	1.76	0.00061	Steel (1% C)	300.	0.1034
Phenolic—Glass	1.80	0.00062	Water	4.10	0.0014125
Silicone—Glass	1.01	0.00035	Wood, Pine (across grain)	0.762	0.000262
Plastics, Molded			Wood, Oak (across grain)	0.75	0.000258
Phenolic Asb. Fiber	3.77	0.0013	Wood, Maple (across grain)	2.97	0.00104
Cement—Asb.	6.97	0.0024	Zinc	770.	0.2653
Copolymer, Vinyl Chloride-Vinyl Acetate	1.16	0.0004			

TABLE A.20. Specific Heat of Various Materials.

Substance	Temp. °C	Specific Heat cals./g./°C	Substance	Temp. °C	Specific Heat cals./g./°C
Aluminum	−250	0.0039	Magnesium	−150	0.1762
	0	.2079		0	.231
	300	.248		300	.279
Aluminum, bronze (88.7 Cu, 11.3 Al)	20–100	.104	Manganese	−188 to −79	.082
Antimony	0	.0494		0	.1072
				325	.1783
Asbestos	20–98	.195	Marble	0–100	.21
Basalt	20–100	.20	Mercury	−263.3	.0055
Beryllium	0–300	.505		0	.0335
Borax (Boron)	0–100	.307		250	.0321
Brass (60 Cu, 40 Zn)	20–100	.0917	Mica (Mg)	20–98	.2061
Bronze (80 Cu, 20 Sn)	15–98	.086	Molybdenum	−257	.0004
Calcium	−185 to +20	.157		0	.0589
Calcspar	0–100	.20		475	.0750
Carboloy		.052			
Carbon, Charcoal	0–24	.165	MYCALEX, G-E 2821		.16
			MYCALEX, G-E 2803		.24
Diamond	−233	.0005	Nickel	−258	.0008
	0	.1044		0	.1032
	247	.303		500	.1270
Graphite	−243	.005			
	20	.17	Nitrogen, solid	−212	.39
			liquid	−200	.474
	138	.254	gas	15	.2477
Carborundum	3–44	.162	Oxygen, solid	−221.8	.336
Cellulose, dry		.37	liquid	−200	.394
Cement, powder	200–10	.20			
Chalk	20–99	.214	gas	15	.2178
			Paraffin	0–20	.6939
Chlorine	−113	.19	Phosphor Bronze (88 Cu, 12 Sn 0.94 P)		
Liquid	0–24	.226		20–100	.0874
Chromium	−150	.0599	Phosphorus, yellow	−136	.124
	0	.1044		9	.189
	100	.112			
			red	−136	.107
Clay, dry	20–100	.22		9	.190
Constantan	0	.098	Platinum	−255.6	.0012
Copper	−253	.0031		0	.0316
	0	.0910		500	.0349
	900	.1259			
			Porcelain	15–950	.26
Ebonite	20–100	.40	Potassium	−258.4	.032
Glass, normal thermometer	19–100	.1988		14	.18
Crown	10–50	.161	Quartz	12–100	.188
Flint	10–50	.117	Rock Salt	13–45	.219
Gold	−258.1	.0018			
			Silicon	−212	.029
	0	.0302		13.9	.168
	100	.0314		18–900.6	.210
Granite	12–100	.192	Silver	−238	.0146
Ice	−200	.168		0	.0557
	−160	.230			
				100	.0564
	−100	.325	Steel, ordinary (.004 C)	20	.107
	−20	.480		100	.117
	−10	.530	Sugar	20	.274
India Rubber (Para)	−100	.481	Sulphur	−188 to +18	.137
Invar (64 Fe, 36 Ni)	15–100	.120	Tin	−150	.0450
Iron, cast	20–100	.1189		0	.0536
wrought	15–100	.1152		100	.0577
pure	−256.2	.00067	Tungsten	−247.1	.0012
	0	.1043		−73.1	.0288
	500	.163		100	.0320
			Uranium	0–98	.0280
Lead	−270	.00001			
	0	.0297	Vulcanite	20–100	.3312
	300	.0356	Wood		.42
Lead, tin (solder)			Wood's metal	5–50	.0352
63.7 Pb, 36.3 Sn	12–99	.0407	Zinc	−252.4	.0071
46.7 Pb, 53.3 Sn	10–99	.0451		0	.0913
Leather, dry		.36		300	.1043

(*Courtesy General Electric Co., Pittsfield, MA*)

TABLE A.21.   Specific Heat of Water.

Temp. °C	Specific Heat* cals/gm/°C	Temp. °C	Specific Heat† cals/gm/°C
0	1.00874	110	1.0126
5	1.00477	120	1.0168
10	1.00184	130	1.0214
15	1.00000	140	1.0255
20	.99859	150	1.0310
25	.99765	160	1.0359
30	.99745	170	1.0422
35	.99743	180	1.0479
40	.99761	190	1.0550
45	.99790	200	1.0616
50	.99829	210	1.0695
55	.99873	220	1.0769
60	.99934	230	1.0857
65	1.00001	240	1.0939
70	1.00077	250	1.1035
75	1.00158	260	1.1126
80	1.00239	270	1.1230
85	1.00329	280	1.1329
90	1.00433	290	1.1442
95	1.00534	300	1.1549
100	1.00645		

* Mean: From observations of 7 scientists. † From observations of Dieterici.

(*Courtesy General Electric Co., Pittsfield, MA*)

TABLE A.22. Tap and Drill Data.

Size of Tap, No.	Size of Drill, No.	Size of Tap, No.	Size of Drill, No.	Size of Tap, No.	Size of Drill, No.	Size of Tap, No.	Size of Drill, No.
2 x 48	50	7 x 32	30	13 x 20	15	18 x 20	A
2 x 56	49	8 x 24	30	13 x 22	15	19 x 16	B
2 x 64	48	8 x 30	30	13 x 24	13	19 x 18	C
3 x 40	47	8 x 32	29	14 x 20	13	19 x 20	D
3 x 48	45	9 x 24	29	14 x 22	11	20 x 16	D
3 x 56	44	9 x 28	28	14 x 24	9	20 x 18	F
4 x 32	43	9 x 30	27	15 x 18	10	20 x 20	H
4 x 36	42	9 x 32	25	15 x 20	8	22 x 16	J
4 x 40	41	10 x 24	25	15 x 22	6	22 x 18	L
5 x 30	40	10 x 30	22	15 x 24	5	24 x 14	M
5 x 32	40	10 x 32	21	16 x 16	7	24 x 16	N
5 x 36	38	11 x 24	21	16 x 18	6	24 x 18	O
5 x 40	37	11 x 28	17	16 x 20	5	26 x 14	O
6 x 30	35	11 x 30	17	17 x 16	6	26 x 16	P
6 x 32	35	12 x 20	19	17 x 18	2	28 x 14	R
6 x 36	33	12 x 22	17	17 x 20	2	28 x 16	S
6 x 40	32	12 x 24	17	18 x 16	2	30 x 14	U
7 x 28	32	12 x 28	15	18 x 18	1	30 x 16	V
7 x 30	31						

Formula [(No. of screw) $\times$ 13] plus 60 = O.D. of screw.

TABLE A.23. Drills-Letter Sizes

Diameter, Inches		Decimals of 1 Inch	Diameter, Inches		Decimals of 1 Inch	Diameter, Inches		Decimals of 1 Inch
A	$\frac{15}{64}$	.234	J		.277	S		.348
B		.238	K	$\frac{9}{32}$	.281	T	$\frac{23}{64}$	.358
C		.242	L		.290	U		.368
D		.246	M	$\frac{19}{64}$	.295	V	$\frac{3}{8}$	.377
E	$\frac{1}{4}$	.250	N		.302	W	$\frac{25}{64}$	.386
F		.257	O	$\frac{5}{16}$	.316	X		.397
G		.261	P	$\frac{21}{64}$	.323	Y	$\frac{13}{32}$	.404
H	$\frac{17}{64}$	.266	Q		.332	Z		.413
I		.272	R	$\frac{11}{32}$	.339			

(*Courtesy General Electric Co., Pittsfield, MA*)

TABLE A.24. Drills-Number Sizes.

No.	Size of Drill in Decimals	No.	Size of Drill in Decimals	No.	Size of Drill in Decimals	No.	Size of Drill in Decimals
1	.2280	21	.1590	41	.0960	61	.0390
2	.2210	22	.1570	42	.0935	62	.0380
3	.2130	23	.1540	43	.0890	63	.0370
4	.2090	24	.1520	44	.0860	64	.0360
5	.2055	25	.1495	45	.0820	65	.0350
6	.2040	26	.1470	46	.0810	66	.0330
7	.2010	27	.1440	47	.0785	67	.0320
8	.1990	28	.1405	48	.0760	68	.0310
9	.1960	29	.1360	49	.0730	69	.0292
10	.1935	30	.1285	50	.0700	70	.0280
11	.1910	31	.1200	51	.0670	71	.0260
12	.1890	32	.1160	52	.0635	72	.0250
13	.1850	33	.1130	53	.0595	73	.0240
14	.1820	34	.1110	54	.0550	74	.0225
15	.1800	35	.1100	55	.0520	75	.0210
16	.1770	36	.1065	56	.0465	76	.0200
17	.1730	37	.1040	57	.0430	77	.0180
18	.1695	38	.1015	58	.0420	78	.0160
19	.1660	39	.0995	59	.0410	79	.0145
20	.1610	40	.0980	60	.0400	80	.0135

(Courtesy General Electric Co., Pittsfield, MA)

TABLE A.25. Coefficient of Thermal Expansion of Various Materials.

Material	Coef. of Expansion In/In./Deg C × 10⁻⁵	Temperature Deg C*	Material	Coef. of Expansion In/In./Deg C × 10⁻⁵	Temperature Deg C*
Aluminum 2S (99.2% Al)	2.394	30–60	Plastics—Laminated (Cont'd.)		
Alundum	0.87	25–900	Melamine—Glass cloth	{L = 1.0} {C = 1.1}	−30 to +30
			Silicone—Glass cloth		−30 to +30
Brass, 67 Cu, 33 Zn	1.85	30–60	Plastics—Molded		
Brass, cast	1.875	0 to 100	Phenolic Asb. fiber	3.0	−30 to +30
Brass, wire	1.930	0 to 100	Cement—Asbestos	0.8	−30 to +30
Bismuth	1.298	−180 to +15	Copolymer, Vinyl Chlor- ide-Vinyl Acetate	7.0	−30 to +30
Bronze (Phosphor) 95.5 Cu, 4 Sn, 1 Zn	1.89	30–60			
Bronze (Phosphor)	1.68	30–60	Urea-alpha Cellulose	1.5	−30 to +30
Bronze (Commercial) 90 Cu, 10 Zn	1.88	30–60	Phenolic-wood flour	3.0	−30 to +30
Brick, fire clay	0.81	25–100	Phenolic-cotton fabric	2.0	−30 to +30
			Phenolic-mica	2.0	−30 to +30
Cadmium	3.159	0–100	Phenolic-cotton flock	2.0	−30 to +30
Carborundum	{0.658} {0.474}	{25–100} {100–900}	Polymethyl Methacrylate	8.0	−30 to +30
			Polystyrene	6.0	−30 to +30
Copper (99.9+)	1.771	30–60	Acetobutyrate	10.0	−30 to +30
Fiber, Vulcanized	2.7	20–60			
			Phosphoric Acid-cement	0.9	−30 to +30
			Synthetic Hard Rubber		
Fluorspar	1.95	0–100	Mineral Filler	4.5	−30 to +30
			Phenolic Asb. fabric	2.0	−30 to +60
			Melamine-Alpha Cellulose	4.0	−30 to +30
German Silver (16 Cu, 15 Ni, 25 Zn)	1.836	0–100			
Glass, flint	0.788	50–60	Ethyl Cellulose	12.0	−30 to +30
Glass, plate	0.891	0–100	Saran	15.0	−30 to +30
Glass, Pyrex	{0.36} {1.51}	{21–471} {552–571}	Nylon	10.0	−30 to +30
			Styraloy	20.0	−30 to +30
Gold	{1.32} {1.43}	{−183 to +16} {16–100}	Resin Emulsion + Mineral and Flock Filler	3.0	−30 to +30
			Platinum	0.899	40
Granite	0.83		Porcelain	0.57	0–200
Graphite	0.786	40	Porcelain (Sillimanite for spark plugs)	0.409	25–600
Gun Metal	1.83				
Gutta Percha	19.83	20	Quartz (crystal)		
			Parallel to axis	0.797	0–80
			Perpendicular to axis	1.337	0–80
Ice	{5.1} {5.07}	{−20 to −1} {−10 to 0}			
Iridium	0.571	−183 to +19	Quartz (fused)	0.0256	−191 to +16
Iron	0.907	−190 to +17	Quartz (fused)	0.05	0 to 100
Iron, cast	0.850	−190 to +16	Quartz (fused)	0.0585	0 to 1200
Iron, wrought	1.14	−18 to +100			

(Continued)

Material	Coef. of Expansion In./In./Deg C × 10⁻⁵	Temperature Deg C*
Lava	{0.29–0.89 / 0.34–1.04}	{25–100 / 25–600}
Lead	{2.708 / 2.94}	{−183 to 14 / 18 to 100}
Limestone	0.9	25–100
Magnesium	{2.14 / 2.608}	{−183 to +15 / 18 to 100}
Magnesium Oxide	0.97–1.14	25 to 100
Marble	1.17	15 to 100
Mercury	3.0	−183 to −39
Mica (Muscovite)	0.9	25–625
Mica (Phlogopite)	{1.4 / 90.0}	{25–175 / 175–625}
Molybdenum	0.575	25–600
Monel 60 Ni, 12 Fe, 11 Cr, 2 Mn	1.4	30–60
Mycalex		
G-E No. 2802	1.09	20–350
G-E No. 2803	1.17	20–350
G-E No. 1364	0.96	20–350
G-E No. 2821	1.	20–350
Nickel	{1.012 / 1.397}	{−191 to +16 / 16 to 250}
Nickel Steel		
10% Ni	1.3	20
30	1.2	20
36 (Invar)	0.09	20
50	0.97	20
80	1.25	20
Paraffin	{10.66 / 13.03 / 47.71}	{0–16 / 16–38 / 38–99}
Phosphor Bronze	1.68	30–60
Plastics—Laminated		
Phenolic—Cotton cloth	{L = 1.6 / C = 1.6}	−30 to 30
Phenolic—Paper	{L = 2.3 / C = 3.1}	−30 to 30
Phenolic—Asb. cloth	{L = 1.8 / C = 1.8}	−30 to +30
Phenolic—Glass cloth	{L = 1.8 / C = 2.5}	−30 to +30

Material	Coef. of Expansion In./In./Deg C × 10⁻⁵	Temperature Deg C*
Rock Salt	4.04	40
Rubber (Hard)	6–8	20–60
Sandstone	0.7–1.2	20
Selenium	3.72	−180 to 0
Silicon	0.763	40
Silver	1.704	−191 to +16
Silver, German	1.8	30–60
Slate	0.6–1.0	20
Sodium	6.22	−188 to +17
Stainless Steel, 90–2 Cu +8 Cr ,0.4 Mn, 0.12 C	1.1	30–60
Steatite	{0.69–0.91 / 0.87–1.04}	{25–100 / 25–600}
Steel, 99 Fe, 1C	1.2	30–60
Stoneware	0.45	
Tellurium	1.675	40
Tin	{2.257 / 2.692}	{−183 to +16 / 18 to 100}
Tungsten	0.336	20 to 100
Wood—parallel to fiber		
ash	0.951	0–100
beech	0.257	2–34
chestnut	0.649	2–34
elm	0.565	2–34
mahogany	0.361	2–34
maple	0.638	2–34
oak	0.492	2–34
pine	0.541	2–34
walnut	0.658	2–34
Wood—across fiber		
beech	6.14	2–34
chestnut	3.25	2–34
elm	4.43	2–34
mahogany	4.04	2–34
maple	4.84	2–34
oak	5.44	2–34
pine	3.41	2–34
walnut	4.84	2–34
Zinc	{2.64 / 2.628}	{−180 to 0 / 10 to 100}

(*Courtesy General Electric Co., Pittsfield, MA*)

*Temperature given for each material is the range in which the previous column was determined.

Screw-Thread Fits.

Classification. Four classes of screw-thread fit are provided: Class 1, Class 2, Class 3, and Class 4. Of these Class 1 fit is the loosest and Class 4 the closest, with Classes 2 and 3 graduated between. These four fits are produced by the application of specific tolerances or tolerances together with allowances, the latter being applicable in the case of the screw only, to the basic pitch diameter of the thread which is the same for both internal and external threads of like size and pitch. The tolerances throughout are applied *plus* to the hole and *minus* to the screw.

Class 2 Fit. This class is a much higher quality of fit than Class 1 and is used extensively where interchangeability is desired. In some cases screws are made, that fit an internal thread very closely, or with a fit that has no play or looseness, such depending upon the accuracy of the pitch diameter of external and internal thread. This class of fit is recommended for general use.

Class 3 Fit. This class is one that is not used as extensively as Class 2, for it represents an exceptionally high grade of commercially threaded product and is recommended only in cases where the high cost of precision tools and continual checking of tools and product is warranted.

Class 4 Fit. This class is one that is not used on high production work and it is only used for selective assembly. This is a much higher grade of fit than Class 3, which is mostly used in automotive work. Class 4 Fit is used in airplane plants where a fine snug fit is required on various aircraft piece parts.

Index

Printed in the USA
CPSIA information can be obtained
at www.ICGtesting.com
CBHW061110080624
9768CB00004BA/10